猪的可持续营养

[美]Lee I. Chiba　编著

尹靖东　主译

中国农业出版社

Sustainable Swine Nutrition

By Lee I. Chiba

ISBN: 978-0-8138-0534-4

北京市版权局著作权合同登记号：图字 01-2014-2633 号

图书在版编目（CIP）数据

猪的可持续营养 /（美）李千叶（Chiba, L. I.）编著；尹靖东主译．—北京：中国农业出版社，2015.10

ISBN 978-7-109-20721-9

Ⅰ．①猪…　Ⅱ．①李…　②尹…　Ⅲ．①猪—家畜营养学　Ⅳ．①S828.5

中国版本图书馆 CIP 数据核字（2015）第 172940 号

中国农业出版社出版

（北京市朝阳区麦子店街18号楼）

（邮政编码100125）

策划编辑　邱利伟

责任编辑　周晓艳

北京通州皇家印刷厂印刷　　新华书店北京发行所发行

2015年10月第1版　　2015年10月北京第1次印刷

开本：787mm × 1092mm　1/16　印张：42.75

字数：528千字

定价：195.00元

（凡本版图书出现印刷、装订错误，请向出版社发行部调换）

参译人员名单

主　译：尹靖东（中国农业大学）

副主译：刘玉兰（武汉轻工大学）

孔祥峰（中国科学院亚热带农业生态研究所）

杨飞云（重庆市畜牧科学院）

参译人员：黄金秀（重庆市畜牧科学院）

易　丹（武汉轻工大学）

李振田（河南农业大学）

王金荣（河南工业大学）

晏向华（华中农业大学）

游金明（江西农业大学）

王自蕊（江西农业大学）

张桂杰（宁夏大学）

乔家运（天津市畜牧兽医研究所）

Lee I. Chiba博士简介

美国奥本大学动物科学系动物科学教授，先后于美国内布拉斯加大学获动物科学专业学士、硕士和博士学位。Chiba博士给本科生讲授动物营养、养猪生产课程，给研究生讲授单胃动物营养学和矿物质代谢课程。他的研究领域为生长猪瘦肉率、生产效率和猪肉感官品质的日粮调控，以及改善母猪繁殖性能的营养管理。Chiba博士曾是《动物科学》（*Journal of Animal Science*）杂志3个任期的编委、2个任期的副主编。现在他正在担任《动物科学》（*Journal of Animal Science*）第二任期栏目副主编和《家畜科学》（*Livestock Science*）的栏目副编辑。

序

我国有着悠久的养猪历史，很久以来一直是世界养猪规模最大的国家。养猪业在我国具有重要的经济地位，目前年产值已超过1.2万亿，占我国农业总产值的20%，是农业领域中单项产值最大的行业。养猪业为我国提供人均超过40kg的猪肉，为保障我国居民肉类供应发挥着重要作用；同时，养猪业及其相关产业吸纳了大量的劳动力就业。

随着畜牧业和现代生物技术的发展，养猪业也发生着巨大的变化，正逐渐成为一个技术密集和资金密集的行业。尽管我国养猪业取得了较大的发展，但也遇到了疫病流行、环境污染、人畜争粮、猪肉安全和猪肉品质下降等问题，这给养猪业的健康发展带来了很大隐患。因此，养好猪不只需要提高养猪生产效率，还要减少环境污染、控制疫病风险，并且最终实现提供优质猪肉的目标。国际上养猪业发达国家在提高生产效率和保障养猪业可持续性发展方面正发生着深刻的变化，深入了解这些国家在猪营养和饲养策略中取得的进展和努力方向，对于我国养猪业的健康发展有着重要的借鉴意义。

在本书翻译前期我参与了组织和安排，并阅读了2013年出版的Sustainable Swine Nutrition，感觉该书可为我们提供了解欧美养猪业可持续发展的思路和营养学策略。该书主编美国奥本大学的Lee I. Chiba博士，组织欧美猪营养学领域的35名科学家参与书稿撰写。全书分为“猪的营养学基础理论”和“可持续养猪的营养学”两部分，系统介绍了养猪业可持续发展的营养学策略的最新研究进展，以便使猪的饲养管理更符合其行为学和生理特点。

该书详细地介绍了利用饲料中的碳水化合物类型、添加剂及日粮配制技术，改善猪健康的营养策略。为了解决饲料资源紧缺问题和有效利用农副产品、生物柴油副产物以及非常规饲料原料，书中介绍了这些替代性原料在猪饲粮中的应用。满足各阶段猪营养需要是实现精确配制日粮、高效养猪的基础，为此书中阐述了如何准确估测饲料原料营养成分的有效利用率，并提出采用酶制剂、微生态制剂以及利用效率更高的微量元素添加剂，来提高猪对营养物质的利用效率；同时，作者也用很大的篇幅讨论环境对猪健康和营养物质利用效率的影响，介绍了欧美国家利用营养调控措施生产优质猪肉的进展。在书的最后，作者针对后备猪和种猪的营养和饲养策略，总结了当前最新的研究进展。

该书是在中国农业出版社组织下和我的托付下，尹靖东教授与刘玉兰教授结合研究方向，邀请了从事猪营养学研究的一线中青年专家，历时一年半翻译完成。尹靖东教授、刘玉兰教授、孔祥峰研究员和黄金秀研究员利用参加学术会议的机会，多次讨论译稿的修改工作。为保证译稿的准确和流畅，尹靖东教授、刘玉兰教授、孔祥峰研究员、杨飞云研究员和黄金秀研究员分头对照原书，对译稿进行了逐章逐句的校阅，并与译者反复研讨修改方式，尹靖东教授最后还对全书进行了审校和统稿。负责出版该书的中国农业出版社相关责任编辑为保证本书的质量和顺利出版，也付出了大量的努力和心血，经常在深夜还与译者讨论书稿的内容。

欧盟和美国业已或正在制定猪饲料中不添加抗生素添加剂的相关法律规定，预计随着食品安全意识的升级，饲料中禁用抗生素将成为养猪业的趋势，后抗生素时代如何饲养猪也已成为我国养猪业关注的焦点问题。该书的出版将有助于我国养猪业从业人员了解国际养猪科学前沿动态，在推动我国养猪业的健康发展方面发挥积极的促进作用。

虽然各位译者为该书的翻译付出很大努力，但由于书中内容涉及多个学科，并且是由多位不同背景的学者撰写，囿于译者的水平和精力所限，对原文的理解难免有不当之处，还需要读者指正，以期重印或再版时予以更正。

前言

猪的营养学是一门快速发展的科学。新的研究成果层出不穷，大大丰富了猪营养的基础理论体系。这些研究成果对养猪业的可持续发展具有极其重要的作用。由于这些成果纷繁复杂，因此为了更有效地利用这些信息，将目前的研究成果梳理并编辑成书就显得尤为重要。尽管我们有许多书籍覆盖了猪营养的诸多领域，但遗憾的是，很少有书籍专门针对如何实现养猪业的可持续发展这个问题。我希望这本书能够填补这方面的空白，为发展环境友好型的饲养措施，最终为养猪业的可持续发展奠定基础。

在商业化养猪生产中，采用饲粮配方和饲养措施的主要目的是获得最大的经济效益，而不一定是使动物获得最佳的生产性能。为了获得最大的经济效益，给猪提供的能量和必需营养素应尽可能满足而不是超过其需要量。这种“最优”饲养措施对能量和营养素的高效利用起着非常重要的作用。一方面，可以确保在未来的养猪业中，优质能量与营养素能持续供应；另一方面，可通过减少营养素的排泄，进而对建设环境友好型社会产生积极作用。采用这种饲养措施要考虑诸多因素，如猪遗传背景的不同，饲料营养成分的变异、利用率和稳定性，营养成分和非营养成分之间的互作，自由采食量，自然和社会环境等。系统全面地考虑这些因素是非常必要的。

随着世界人口的增长，以及新型工业化国家和经济发达程度较低的国家经济的发展，在未来，人和动物对优质能量和营养素资源的竞争将会更加激烈。因此，为养猪业寻找提供能量和营养素的非常规饲料原料显得非常重要。非常规饲料的饲养价值差异较大，其原因包括营养成分含量的差异及其他因素（如营养成分的利用率和稳定性、抗营养因子、营养成分之间及营养成分和非营养成分之间的互作、适口性等）。为了有效利用非常规饲料原料，我们需要整合所有营养学的基础理论和应用技术。此外，满足消费者对健康营养食品的需求，减少公众对环境问题的担忧，也是养猪业可持续发展不可缺少的一部分。因此，我们不仅要关注与获得最佳生产性能、最高能量和营养利用率相关的营养问题，还要关注猪胴体和肉品质、养猪生产对环境的影响等方面的问题。

作为一本猪营养方面的综合性书籍，《猪的可持续营养》包括了一些营养学的基本理论知识，如水、蛋白质或氨基酸、脂类、碳水化合物、能量代谢、维生素、矿物质及营养与免疫，本书的重点集中于上述领域目前的最新研究进展或相关问题。因此，《猪的可持续营养》会对一些基本理论知识进行简要回顾，重点是介绍最新的研究成果。此外，本书在各个相关研究领域，对如何实现养猪业的成功、可持续发展的问题，进行了系统的阐述和讨论。

Dr. David H. Baker是本书的编者之一，我们对他的去世表示深切的哀悼。Dr. Baker是伊利诺伊大学香槟分校营养科学和动物科学的名誉教授。他于2005年入选为美国国家科学院院士，而美国国家科学院院士对每一位科学家而言，是最高的、最有声望的学术荣誉。Dr. Baker从美国动物科学协会获得过六个重要的奖项，从美国家禽科学协会获得过五个重要奖项，从美国营养协会获得过两个重要奖项。此外，Dr. Baker还获得过美国农业部的杰出服务奖，以及农业科学和技术委员会的Charles A. Black奖。Dr. Baker发表了约600篇期刊论文，该记录在本领域至今没有被打破。Dr. Baker也是美国动物科学协会、家禽科学协会、美国营养协会的成员。他的事迹和成就必定激励着单胃动物营养研究领域及其他行业的同仁们的继续努力。

这本书的付梓与我同事的帮助是无法分开的，在此我对他们无私的帮助表示感谢。我真诚地感谢他们在这本书的编写过程中所付出的时间与汗水。我也要感谢我的研究生Sean D. Brotzge 和 Chhabi K. Adhikari，感谢他们在每一章的校正和参考文献格式的书写方面所做出的努力。

目录

第 *1* 章

猪营养中的水

1 导 语

水是猪日常饮食的一个关键组成，然而与此不相称的是，不管是科普书籍还是科学文献均很少关注水，只有出现问题时才会引起关注，因此水被称为“被遗忘的营养物质”。关于水对机体的重要性，Maynard（1979）给出了经典描述：“机体损失所有的脂肪和一半以上的蛋白质时仍能存活，但是失去10%的水分就会导致死亡。”

世界上大多数的猪肉主产区，水既充足又便宜，而且不会用于商业贸易，因此较少关注水（Fraser等，1990）。这也是相对于很多其他营养物质，人们对水的重要性缺乏了解的原因。此外，水还是一种特别难于研究的营养物质，对于著名的研究机构也是如此。因为那些用于能量、氨基酸、矿物质和维生素研究的传统方法，都很难应用在水上，甚至根本不适用。

而且，在实验室很难准确地测定水分含量。因为饲料、粪便、尿液或者猪肉样品中的水会与周围空气中的水不断交换，因此随着时间的推移，样品可能会增加或者丢失大量的水分。此外，样品中干物质含量的测定方法不仅去除了水分，也可能去除了氨气和短链脂肪酸等挥发性成分，从而产生另一种误差。干物质含量的测定虽然只需要最简单的试验设备，但要准确测定并不容易。尽管是这样的简单分子，水的研究却非常困难。

2 机体水含量

水分子是猪体内含量最为丰富的物质，一些组织液的水分含量可高达99%（Shields等，1983）。机体水分含量会随体重变化而不断改变，出生时约占体重的82.5%，到上市时降到体重的53%，这是由胴体瘦肉减少而脂肪增加引起的（Shields等，1983）。机体内的水主要分为三部分：细胞内的水（约占总水量的69%）、细胞间的水（约占总水量的22%）和其他存在于血管系统中的水（Mroz等，1995）。对于陆生生物，维持机体、组织和细胞内适宜的水平衡至关重要。水平衡与细胞、器官内以及细胞与器官之间的电解质平衡密切相关。电解质平衡是体内另外一种重要的稳态过程（Patience 等，1989）。

关于猪饮水的调控机制目前还不是很明确，可能与血容量减少和渗透压升高有关，也受调控采食量的信号分子的影响（Mroz等，1995）。此外，猪的行为刺激还会影响饮水量，如在无聊、饥饿和其他应激时，猪都会大量饮水（Fraseret等，1990）。

除胃外，各段的肠道既可以吸收水，也可以分泌水。水可以通过主动和被动吸收方式被吸收（Argenzio，1984）。当食糜逐步经过小肠和大肠时，渗透压梯度不断增加，使得大部分水被结肠末端吸收。当肠道内存在大量渗透性的活性离子时，会导致渗透压平衡紊乱，进而引起腹泻（Fraser等，1990）。

3 水的营养

因为水几乎直接或间接参与了体内所有的代谢过程，所以它无疑是所有生物体赖以生存的关键物质。水特殊的化学结构决定了其在机体内发挥着重要的生理、生化和营养功能。

水的比热高，使它成为热动态平衡中理想的热传导物质。例如，水的汽化热为2 258 J/g，是酒精等其他液体的2倍多，是正己烷和苯等有机溶剂的5倍（Lehninger，1982），是体内干物质的2.5倍。热应激时，水比其他液体或固体吸收的热量要高得多，从而体温变化更小。所以，水能有效维持机体体温的稳定。水的汽化热高还使它在机体肺蒸发散热过程中发挥着必不可少的作用。

水在维持酸碱平衡中也发挥着关键作用。水的pH是7，与大部分组织理想的生理pH非常接近。此外，水是碳酸氢盐缓冲体系的主要组成成分，通过生成CO_2和H_2O可以保持H^+和HCO_3^-的平衡。

$$H^++HCO_3^-=H_2CO_3=H_2O+CO_2$$

机体通过生成H_2O和CO_2，可排泄体内正常代谢产生的大量的酸，而水在其中发挥了极其重要的作用。血液中与血红蛋白有关的碳酸氢盐缓冲体系，也能清除CO_2等其他有毒分子，不会引起组织损伤和静脉pH变化。水在其中发挥着两个作用：一是化学作用，这已在前面进行了阐述；另一个作用是作为溶剂，将各种分子运送到全身。

作为一种溶剂，水是细胞间和器官间进行营养物质、化学能、代谢产物、废弃物交换的主要运输介质，同时还能将激素从产生或释放部位运送至靶细胞或靶器官。水特殊的化学结构——两极特性使它成为理想溶剂。例如，简单的盐易溶于水，但几乎不溶于苯和氯仿等其他液体（Lehninger，1982）。

水是机体内氧化和水解等化学反应的基础。氧化反应主要是降解饲粮中没有合成蛋白质的氨基酸及没有直接在体内沉积的碳水化合物和脂质。因为饲粮中约有2/3的蛋白质不能在体内沉积，且大部分碳水化合物会被氧化，这意味着它们在猪体内会产生大量的代谢水，据估计，生长猪体内产生的代谢水约占每日水平衡总量的12%（表1-1）。而饲粮中脂质被氧化的比例主要取决于猪不同生长阶段的生理状况及营养状态。因此，水不仅是动物体内的主要组成部分，还是许多生命活动所必不可少的物质。

表1-1　估算45 kg生长猪的水平衡

项目	摄入量（mL/d）	项目	排泄量（mL/d）
饮水量[1]	5 552	粪便中的水量[4]	672
代谢水量[2]	788	尿中的水量[4]	2 839
饲料中的水量[2]	252	消化过程损失的水量[5]	185
组织合成的水量[3]	74	其他损失的水量[6]	2 335
		排出的总水量	6 031
		体增重存留的水量[7]	6 35
摄入的总水量	6 666	排出水和存留水的总量	6 666

[1]猪的体重为45 kg，日采食量为2.1 kg，日增重为0.98 kg，日饮水量为5.55 kg（Shaw等，2006）。推测的蛋白沉积量为160 g/d，灰分的沉积量为35 g/d，脂肪的沉积量为150 g/d（均不是实测值）（Oresanya 等，2008）。

[2]该饲粮含有12%的水、5%的乙醚提取物（其中有85%的乙醚提取物或脂肪被消化，且可消化的乙醚提取物或脂肪转化成体脂的沉积效率为90%）、18%的粗蛋白（其中80%被消化，且80%是真蛋白质、20%是非蛋白氮；35%的可消化蛋白质在体内沉积，其余的均被分解代谢）。因而每天大概有9 g脂肪、157 g蛋白质和1 260 g碳水化合物被氧化代谢，每千克脂肪、蛋白质、碳水化合物产生的代谢水分别为1 190 mL、450 mL和560 mL（NRC，1981）。

[3]资料来源于Schiavon 和 Emmans，2000。

[4]假设饲粮消化率为82%，粪便含水量为64%。

[5]资料来源于Schiavon和 Emmans，2000。

[6]其他原因引起的水损失，其中大部分是由蒸发而损失的。

[7]组织沉积速度：每天的沉积总量为345 g，其中150 g脂肪、35 g灰分、160 g蛋白质；日增重980 g，日饮水量635 g。

4 水的平衡

◆ 4.1 水的摄入

饮水是猪获得水的最重要方式，但并不是唯一的来源。因为饲料中含有一定的自由水，所以猪在采食饲料后必然会摄入饲料中的自由水。氨基酸、碳水化合物和脂类氧化生成的代谢水也提供了猪每日需要水量的重要部分。弄清猪的饮水行为是非常困难的，因为猪的需水量受很多因素的影响（Fraser等，1990）。

那些影响猪的生理、生化和营养需要的因素都会改变饮水量，而这些因素本身又受到外界环境、健康状况、饲粮和饮水质量的影响。在自由饮水的情况下，猪饮用的水能满足各种活动需求。

Schiavon和Emmans（2000）曾提出一个简化模型来预测生长猪的饮水量。在这个模型中，水的摄入量会随着消化过程需要的水、粪和尿排出的水及生长沉积的水的增加而增加。反之亦然，水的摄入量会随着饲料水、氧化过程产生的代谢水、蛋白质和脂质合成过程释放的水的增加而减少。然而，他们指出，为了更准确地预测生长猪的饮水量，还需要进行更多的试验，如排泄体内多余的氮和电解质所需要的水量、通过尿液和粪便排泄矿物质所需要的水量、调节渗透压所需要的水量或其他诸多影响因子都需要进一步研究加以细化。

4.1.1 总的饮水量

猪每日摄入水量的最主要来源是饮水。出版文献指出，供水的唯一要求就是确保水供应充足且质量好。大家普遍认为，在水供应充足且质量好的情况下，猪能根据自身需求正常调节饮水量。但有些情况并非如此，有时猪也会大量饮水（Fraser等，1990），甚至超过其正常的生理需要（Vermeer 等，2009）。影响饮水量的主要因素包括体重、采食量和温度（Mroz等，1995）。

严格控制水平衡对机体来说至关重要，因为机体脱水或饮水过多都可以致命。大脑的下丘脑是渴感和饮水行为的调控枢纽（Koeppen和Stanton，2001），下丘脑的渗透压感受器能感受细胞外液的渗透压变化，血浆渗透压只要升高10 mOsm/kg 就能使机体产生渴感，进而导致饮水（Anderson和Houpt，1990）。血容量降低是引起渴感的另一个信号，因为血容量下降6%～7%也会产生渴感（Anderson和 Houpt，1990）。但根据饮水启动机制的不同，其他信号也会参与饮水调节。Mroz等（1995）发现，黏膜血流量、血管弹性及口腔干燥程度等信号都可能引起饮水。

关于猪饮水量的研究，多数都是在自由采食状态下进行估测的。在没有估算被浪费的水量的情况下，这些估测值指的就是与摄入水量不同的“耗水量”。饮水浪费会带来巨大的经济损失，主要是因为它会增加粪污量，从而增加每年的粪污处理费用。因此，应该重视饮水器的设计及其安装位置的选择，最大程度地减少水的浪费（Brumm，2010）。

4.1.2 影响饮水量的因素

体重、环境温度和采食量是影响猪饮水量的主要因素。与所有的营养物质一样，随着猪的生长，每日需水量也会不断增加。但是，目前还缺乏足够的证据来阐明体重与需水量之间的确切关系。Schiavon和Emmans（2000）报道，在严格控制的条件下，体重与饮水量的R^2只有0.45；同时他们还推测在商业化生产条件下，体重与饮水量之间的关系可能并不是很紧密。因为猪的自由饮水量还受其他许多因素的影响。

从直观上讲，环境温度升高会导致猪的饮水量增加。Schiavon和Emmans（2000）研究表明，气温每升高1℃，猪的饮水量就会增加0.12 L/d。然而，与此不一致的是，vandenheede和Nicks（1991）发现，温度从10℃升高到25℃时，育肥猪的饮水量由2.2 L/d增加到4.2 L/d。Mount 等（1971）研究表明，当温度从12～15℃升高到30～35℃时，体重为33.5 kg的猪饮水量提高了57%。而Straub等（1976）研究表明，在同样条件下，体重为90 kg的猪饮水量提高了63%。Yang等（1981）发现，猪体内的总水量是保持稳定的，但当温度从27℃升高到35℃时，水的周转速率就会提高。

值得注意的是，热应激期间猪会增加与饮水器咬玩的时间，从而增加了水的浪费，这可能导致对热应激期间需水量的估测偏大。

目前文献报道的水/饲料摄入比估算值变异非常大，低至1.5：1，高至5：1以上。造成以上差异的部分原因可能是环境条件、饲粮特性或行为影响的不同。同时，各试验方法的迥然不同也可能是造成以上差异的重要原因。当生长猪饲养在适宜环境中、饲喂典型的商业化饲粮且不考虑行为因素的影响时，其水/饲料摄入比为2.5：1（Shaw等，2008）；在同样的条件下，育肥猪的水/饲料摄入比要低些，约为2：1。

有时因为无法准确计算浪费的水量，会导致表观饮水量增加，所以应特别注意各种不同科学术语的差别。摄水量指的是猪实际摄入的水量，而耗水量表示的是供水系统减少的水量。例如，由壁挂式乳头饮水器浪费的水量一般为25%～50%，甚至更高（Li等，2005）。

大家普遍认为，饮水量会随饲粮蛋白质水平的增加而增加。该观点在诸多试验上得到证实（Suzuki等，1998；Pfeiffer 等，1995），且从生理学上也很好理解。因为饲粮添加过量的蛋白质，使肾脏需要排泄更多的尿素，进而会增加排出的水。然而，也有很多研究表明，饲粮蛋白质水平与饮水量并不是呈简单的线性关系（Albar和Granier，1996；Tachibana和Ubagai，1997；Shaw等，2006）。因此，综合现有资料得出，降低饲粮蛋白质水平并不是一种减少饮水量的有效途径。Mroz等（1995）也指出，许多有关摄水量与饲粮蛋白质水平的研究结果还受到饲粮中同时变化的矿物质水平的干扰。

众所周知，提高饲粮盐浓度会提高饮水量（Seynaeve 等，1996）。有趣的是，当矿物质本身的水含量较高时，提高矿物质的添加量也会使猪的饮水量增加（Maenz等，1994）。

正如许多其他动物一样，饥饿会诱导猪的饮水量增加。例如，Yang等（1984）采用限制饲喂的方法发现，随着猪摄入饲粮量的增加，水/饲料摄入比从5.1：1降到3.3：1。

猪会因为玩耍、饥饿或应激诱导的渴感消耗大量的水，所以采用简单测定的水消耗量来估算需水量，可能存在一定误差（Fraser等，1990；Vermeer等，2009）。

4.1.3 饲料水

猪可从饲料中获得一部分水。实际摄入的饲料水是饲料采食量乘以饲料含水量。饲料水在数值上占猪每日摄水量的比例不大，有时还不到总量的5%。

4.1.4 代谢水

每克脂肪、蛋白质或碳水化合物氧化产生的代谢水，分别平均为1.10 g、0.44 g和0.60 g。当然，代谢水产生的准确量受到特殊的脂肪酸、氨基酸或碳水化合物的化学结构的影响（Patience，1989）。

4.1.5 组织合成释放的水

体组成的合成过程中也会产生一部分水。体内沉积1 g蛋白质可释放0.16 g水，而沉积1 g脂肪只能释放0.07 g水（Schiavon和Emmans，2000）。

◆ 4.2 水的排泄

4.2.1 肾脏排泄

机体通过尿液排泄的水量受尿液中溶质含量和肾脏浓缩尿液能力的影响，其中猪肾脏浓缩尿液的能力约为1 mOsm/L（Brooks和Carpenter，1990）。尿液中最主要的溶质包括氮（主要是尿素氮，但不是全部）、钙、磷、钠、氯化物、镁和钾。这些不可挥发的阳离子和阴离子分别与代谢形成的阴离子和阳离子结合形成化合物（Patience，1989）。

肾小管的渗透性受脑垂体释放的抗利尿激素（antidiuretic hormone，ADH）的影响。当心房的受体感受到血容量下降时，脑垂体就会释放ADH；在ADH的作用下，肾小管重吸收的水增多，从而使血容量又恢复正常水平（Berdanier，1995）。除ADH外，肾素-血管紧张素系统可以刺激ADH和醛固酮的释放，增加钠离子和氯离子的重吸收，促进血管收缩，以维持体液平衡。醛固酮由肾上腺分泌，可以维持钠离子和氯离子的平衡（Berdanier，1995）。

4.2.2 粪便排泄

计算粪便排泄的水有多种方法，其中最简单、最不准确的方法是根据粪便的平均水含量来估测（表1-1）。而更复杂的方法是分析粪便的各组分并测定每种组分的含水量，但目前尚缺乏足够的数据来说明这种方法的准确程度（Schiavon和Emmans，2000）。

◆ 4.3 水的平衡

人们不能简单地给猪饲喂不同量的水，而应明确猪达到最佳生长性能所需要的水量。因为猪在细胞内、细胞间和细胞外拥有一个巨大的、动态的水平衡系统，所以任何的生长试验都需要对猪各种来源的水进行定量分析，包括饮水、饲料水、代谢水及通过粪便、尿液、呼吸和汗水排出的水。根据猪的生长状况不同，各种来源的水量必然会有所不同。要测定这么多来源的水非常困难，也从未进行过这种测定试验。但Schiavon 和Emmans（2000）试图通过计算猪消化、排粪、排尿、蒸发和生长所需要的水量来模拟摄水量。

根据Shaw 等（2006）的试验数据，表1-1试图阐述生长猪在常温环境下自由采食典型商品饲粮时体内的水平衡。从表中的数据来看，脂肪、蛋白质和碳水化合物代谢产生的水占每日摄水量的比例较高（约12%）；而饲料水所占的比例很低，仅为4%；尿液排泄的水（47%）比其他方式排泄的水（39%）稍多；其他方式排泄的水主要是由蒸发损失的水。很显然，外界环境温度对这部分水的影响非常大。粪便排出的水约占总排水量的11%，它受饲粮组成的影响较小，但受腹泻等胃肠疾病的影响非常大。

5 水的需要量

研究动物对水需要量的方法很多（Fraser等，1990）。传统的营养需要量评价方法是给猪饲喂不同营养水平的饲粮后，再根据生产性能的结果来评价营养需要量，但这种传统方法很难用于水的需要量研究。因为猪对水的需要量受到许多因素的影响，如环境温度、饲粮特性（如蛋白质、矿物质水平）及瘦肉或脂肪的增重量等都会影响猪的需水量。

第二种方法是让猪自由饮水，将猪达到最佳生长性能时的饮水量确定为猪的需水量。虽然文献中经常使用这种方法，但因为它不能保证猪的饮水量是建立在生理需求的基础上，所以该方法也存在很大问题。与其他物种一样，猪也会因为应激、饥饿等因素摄取大量的水（Patience等，1987）。一项研究发现，让体重为40 kg的猪自由采水，其日饮水量为1.70 ~ 16.8 L/d（Patience等，1987），说明用该方法确定需水量是不合适的。

第三种方法是确定预防猪脱水的饮水量。然而，猪会动用代价高的代谢方式来防止脱水，如通过排泄高渗尿来维持水平衡。有人认为，只有在血容量下降时才会产生高渗尿，高渗尿的产生机制极其精确，简单的需水量研究很少考虑。

有学者建议，猪的需水量可以用水/饲料摄入比来确定（Brumm等，2000）。但该方法忽视了体重、环境温度和饲粮组成这些影响需水量的关键因素的影响（Mroz等，1995）。如果能充分了解该方法的局限性，则可将水/饲料摄入比作为确定需水量的有效实用性方法。

常温环境下推荐的水/饲料摄入比如下：早期生长猪为2.5∶1，而末期育肥猪为2.0∶1。还有些研究表明，育肥猪的水/饲料摄入比可能低至1.5∶1。根据这些推荐值建议，在常温环境下25 kg猪的平均饮水量约为3.2 L，到130 kg出栏重时增至5.5 L。然而，必须再次强调的是，由于不同养殖场的饲料采食量和环境温度的差异很大（Mroz等，1995），不同群体猪行为的独特需求也不同（Fraser 等，1990），因此猪对水的实际需要量也存在很大差异。

6 猪的供水

是否给猪提供了适量的水会给养猪生产带来很多问题（Gonyou和Zhou，2000；Brumm等，2000）。供水不足会降低猪的生长性能，严重不足还会降低饲料消化率（Mroz等，1995）。而供水过量会造成水资源浪费，增加不必要的粪污量；当粪污用作肥料时，又会相应地增加粪污的运输费用。如果通过饮水给药，过多的水浪费还会增加药物费用。饮水器被粪便污染后，会降低猪的饮水量，进而降低生长性能。因此，养猪生产管理中选择合适的饮水器非常重要。

◆ 6.1 乳头式饮水器

给猪供水的方法有很多。传统的乳头式饮水器安装在猪栏的后墙或靠近后墙，方便猪自由饮水。采用乳头式饮水器，水的浪费问题很严重。采用普通的乳头式饮水器，猪要浪费约25%的水，大大增加那些必须从猪栏中清理掉的不必要的粪污量（Li等，2005）。Li等（2005）按照厂商推荐值设定乳头式饮水器的水流速度，安装高度根据猪的生长进行调整，发现在很多典型的商业化养殖生产中，乳头式饮水器安装的高度和水流速度经常会超过饮水器的规定值，从而使水的浪费量高达50%～60%。推荐乳头式饮水器的底端高出圈舍内最小猪肩部50 cm以上（Gill和Barber，1990），该数值是用公式150 × $BW^{0.33}$计算出来的（Petherick，1983）。水流速度过快也会增加水的浪费。

虽然应该避免乳头式饮水器的水流速过快，从而达到最少程度的浪费；但是，水的流速过慢也会带来一些严重的问题。例如，有研究报道，保育舍内的刚断奶仔猪由于不能获取充足的水，无法排出体内摄入的饲粮盐分，曾出现盐中毒现象。Neinable和Hahn（1984）发现，猪处于热应激状态时，即使低温下水流速适宜，也可能会造成供水不足。表1-2列出了乳头式饮水器的水流速一般推荐值，这在保证充足饮水量的同时避免大量浪费。

表1-2 乳头式饮水器的水流速率对于不同生理阶段猪的推荐值

猪的类型	水流速率的推荐值（mL/min）	
	最小值	最大值
妊娠母猪	500	1 000
泌乳母猪	1 000	2 000
断奶仔猪	750	1 000
生长育肥猪	750	1 000

与壁挂型乳头式饮水器不同，摇摆型乳头式饮水器悬挂在猪舍的天花板上，可以减少水的浪费，但具体的减少量目前还不清楚。Brumm等（2000）研究表明，与壁挂型饮水器相比，摇摆型饮水器能降低11%的用水量，但该研究并未测定水浪费的量。

◆ 6.2 盘式饮水器

如果能恰当调整好壁挂型盘式饮水器，浪费的水量就会极少。但饮水器的高度必须随着猪的生长而增加，否则会造成饮水器的污染，进而减少猪的饮水量。盘式饮水器不能安装在猪圈角落，以避免加大污染风险。Brumm 等（2000）研究表明，盘式饮水器的水消耗量比摇摆型乳头式饮水器的少25%。

◆ 6.3 干湿喂料器

另一种可供选择的供水方法是采用干湿喂料器，这种装置可让猪选择采食干料或湿料（装置的命名也是由此得来）。与壁挂型乳头式饮水器相比，采用干湿喂料器可使水的浪费量减少35%。在气候温度适宜的条件下，不需要提供额外的水源；但在炎热气候条件下，由于经常出现热应激，就需要提供其他类型的饮水器。饮水器类型的选择将决定水的浪费程度。表1-2列出了不同生理状态的猪使用乳头式饮水器时水流速率的推荐值（Patience等，1995）。

◆ 6.4 液态饲喂

液态饲喂相比于传统的固态饲喂具有很多优势，包括提高生长速度和改善饲料转化率（Hurst等，2008）。相比较玉米，当饲粮中含有小麦和大麦时，使用液态饲喂的优势会更加明显（De Lange等，2006）。人们越来越关注，发酵液态饲喂不仅可以改善仔猪健康水平，还可以提高很多植物性饲料中植酸磷的生物学利用率。

关于采用液体饲喂系统进行饲喂时最合适的水/饲料比，目前各研究报道不一。为了满足液态饲喂器的要求，保证饲料能从混合器中流出让猪采食，应有个最低水量的要求。但对于超出最低水量的部分，各研究报道的推荐值变异很大。例如，Barber等（1991a，1991b）发现，当水/饲料比提高到3：1或3.5：1时，干物质采食量增加；但是当这个比例进一步增至6：1时，干物质采食量下降。目前养猪商业生产实践中，水/饲料比值一般为2.5～3.5：1。

7 饮水的管理

◆ 7.1 妊娠母猪

妊娠母猪的饮水量主要受行为因素的影响，尤其是饥饿引起的渴感。与许多其他动物一样，妊娠母猪采食不足时会过度饮水，所以不同研究报道的妊娠母猪的饮水量数值差异很大，如饮水量为5.6 L/（d·kg）饲料（Lightfoot和Armsby，1984）、2.5 L/（d·kg）（Friend，1971）、14.9 L/d（Bauer，1982）、7.2 L/d（Madec等，1986）和25.8 L/d（Kuperus，1988）。而且，对于饮水量的变化范围，各研究报道差异也很大。Pollman等（1979）报道，每日饮水量的变化范围为3.4～46.2 L/d。要确定干奶母猪的具体饮水需要量非常困难，因为存在饥饿诱导的过量饮水，因此建议保证干奶母猪能全天自由饮用新鲜水。

◆ 7.2 泌乳母猪

饮水问题是泌乳母猪管理中碰到的常见问题，至少在某些情况下，水的摄入量不足会降低母猪泌乳性能。大部分对泌乳母猪饮水的研究只是简单地测定水的摄入量，并没有确定这个饮水量是否充足。Fraser等（1990）对12篇不同报道结果进行了总结，得出最合理的结论如下：无论是在同一个研究中，还是在不同研究之间，饮水量变异都很大，在所考察的研究中平均饮水量在8.1～25.1 L/d范围内变化。

Fraser和Phillips（1989）得到一组有趣的数据：泌乳母猪分娩后前5 d的饮水量低与仔猪生长速度的降低有关。饮水量少的母猪哺育的仔猪增重低，而饮水量多的母猪哺育的仔猪增重高。虽然无法阐明该研究结果的因果关系，但是可能需要关注的是，早期哺乳阶段母猪的饮水量对仔猪的健康生长至关重要。

需要强调的是，保证母猪分娩后前5 d的饮水量非常重要，这就意味着要让母猪尽可能地容易获取饮水器内的水，无论母猪是站立还是躺着，因为产后昏睡对分娩母猪来说也很重要。由于母乳中的水含量约占81%，因此要尽可能保证哺乳母猪摄入充足的水量，这个必要性不言而喻。

◆ 7.3 哺乳仔猪

关于哺乳仔猪需要补充的水量，目前各研究结果不一，特别是出生后前1～2周需要补充的水量。Fraser等（1990）指出，仔猪早期的饮水量为0～100 mL/d，甚至是100 mL/d以上。Deligeorgis等（2006）研究表明，仔猪平均在出生后16 h才开始接触饮水器；出生后48 h内接触饮水器的仔猪体重比不接触饮水器的仔猪要重得多，而且饮水器的安装位置也会影响水的摄入。

因为母猪产奶量、环境温度和教槽料的采食量都会影响仔猪的饮水量，所以通常建议从仔猪出生开始就让其自由饮水。由于北美的仔猪断奶日龄从不足21 d增加到25 d甚至更高，因此在母猪产床上不断供应新鲜的饮水以提高教槽料的采食量，将变得越来越重要。

◆ 7.4 断奶仔猪

在哺乳后期，仔猪将每天从母乳中摄入700～1000 mL甚至是1 L以上的水。但刚刚断奶后，仔猪的饮水量会出现异常地降低，从断奶第1天的1.0～1.5 L/d下降到第4天的0.4～1.0 L/d（McLeese等，1992；Maenz等，1993，1994；Torrey等，2008）。以上结果表明，在断奶后的前几天内，仔猪对水的摄入量：①不遵循采食量的变化模式；②可能是限制断奶后仔猪生长速度的重要因素。此外，目前还未发现仔猪饮水量与断奶后1周内的腹泻程度有关（Maenz等，1994）。Phillips和Phillips（1999）发现，不管选择哪种饮水系统，都无法改变这种饮水量的异常变化规律。

因为仔猪需要摄入充足的水来保证采食量达到最大，而且断奶后采食量低下的问题很严重，所以需要更加重视仔猪在这个重要转折期的饮水。要确保仔猪熟悉饮水器的位置并自由饮水，是早期断奶仔猪管理的关键部分（Dybklaer等，2006）。然而，仅让仔猪可以自由饮水，并不能保证猪摄入充足的水。从这个方面来讲，使用盘式饮水器具有很大优势，因为它们可让仔猪看见水，有利于充足饮水，但目前这个优势还未得到试验证实。

◆ 7.5 生长育肥猪

生长育肥猪的自由饮水量受很多因素影响，对此Brooks 和Carpenter（1990）曾进行过详细的综述。在常温环境下，水/饲料摄入比（校正过浪费的水）在早期生长阶段（25 ~ 50 kg体重）为2.5∶1，到育肥后期下降到2∶1（体重大于80 kg；Mroz等，1995；Li等，2005）。这些标准中存在的变异通常是由未测定的浪费或环境温度差异造成的。水的摄入量不足不仅会降低猪的生长速度和饲料消化率，而且其降低幅度需要通过大幅度限制采食量才能达到（Mroz等，1995）。

8 水的质量

◆ 8.1 常规质量标准

按照一定标准给猪提供充足的、高质量的饮水是所有养猪生产者追求的目标。然而，“质量”这个术语包涵很多方面。猪的饮水质量标准通常都采用人的饮水质量标准，包括水的气味、透明度、颜色和味道等指标，但是所有这些指标在猪上的要求与人的不同。

由于饮水标准的使用与测定都很重要，因此在考虑水质量时应当谨慎。虽然在理论上应该使用最优质的饮用水，但因为猪对饮水质量的适应范围比较宽，所以人的最低标准的饮用水，都非常适合用于猪。水的质量指标可以划分为三大类：感官指标、化学指标和微生物指标。

感官指标本身在养猪生产中的实际意义不大。因为猪的耐受力特别强，除了水中可察觉的异常颜色和气味外，猪都能接受。但需要特别注意的是，浑浊的且带有特殊颜色和气味的水可能是其他方面问题的征兆。

◆ 8.2 感观指标

8.2.1 浑浊度

浑浊度更多表现的是水的优劣感官特性，而不是质量特性。浑浊度高仅代表水中悬浮的胶状物质含量多，如泥沙或黏土。浑浊度低的水对猪的影响非常小。但水中悬浮的微生物可能也会造成浑浊度变高，对猪产生较大的影响。因此，猪饮用水的浑浊度可以定性测定，不需要定量测定。如果水的浑浊度小于5浊度单位（NTU），则猪可以饮用；如果大于5浊度单位（NTU），则需要检测水中化学物质和微生物的含量，以进一步确定水浑浊的原因。

8.2.2 颜色

水的颜色可以用真色度单位（TCU）来衡量。颜色不是猪饮用水重点考察的指标，除非饮用水受到有毒有害物质的污染。相对于颜色来说，检测溶解性总固体、硫酸盐、硬度及微生物等指标对猪的意义更重大。

8.2.3 气味

水的气味可以用嗅阈值（TON）来度量，猪饮用水一般不存在气味问题，因为新鲜的水几乎没有气味。但是，如果水中存在异常的气味，则需进一步分析这个气味产生的根源，最可能的原因是微生物污染或水中存在有机化合物。

◆ 8.3 化学指标

8.3.1 溶解性总固体

虽然溶解性总固体（total dissolved solids，TDS）的测定不是很准确，但仍可用作衡量水是否可以被猪饮用的指标。TDS主要包括水中存在的碳酸氢根，氯化根，硫酸根的钠盐、钙盐、镁盐等物质。随着TDS的增加，猪发生腹泻的概率会不断提高。一般而言，如果TDS较低（<1 000 mg/L），则矿物质污染基本上可以不考虑，也不需要进一步检测。如果TDS为1 000 ~ 3 000 mg/L，则可能会导致猪出现短暂性腹泻（特别是仔猪，当水中主要的阴离子是硫酸根时，更容易导致猪腹泻）；当水中的TDS高达3 000 ~ 5 000 mg/L时，猪也可能接受（NRC，1974），但需要仔细观察猪的反应；当TDS超过5 000 mg/L时，则必须在猪饮用前进行详细检查。简而言之，TDS是个粗略的检测指标，TDS越低表明水的矿物质污染越少，水质较好。虽然猪对不同水质的适应性强，但在可选择的情况下，最好选用TDS最低的水。

8.3.2 导电率

顾名思义，导电率是度量水导电能力的指标。导电率越高，说明水中溶解的矿物质离子浓度越高。因为导电率是非特异性的，所以提高导电率对水质的有效影响比较少。但是如果水的导电率很高，则需要进一步检测水样中含有的确切离子及含量。

导电率，有时又称为电导系数，以导电度单位（μS）/cm计量，乘以K值便可换算成TDS，公式如下：

$$\text{TDS（mg/L）= 电导率（}\mu\text{S/cm）} \times K$$

K值取决于水的组成，一般为0.55～0.75。例如，如果水中主要的污染物是钠离子和氯化物，则K值等于0.67；如果水中主要污染物是硫酸盐，K值会变高。

在猪的营养学中，首选的是直接检测TDS，而不是电导率。但不管是TDS还是电导率，都不能提供充分的、明确的信息用以准确鉴定水质问题。

8.3.3 水的pH

水的pH是衡量水酸碱性的指标，绝大部分水的pH都为6.5～8.5。如果水的pH升高，就可降低用氯处理的作用效果；如果水的pH降低，则水中某些药物会出现沉淀。因为水中添加pH调节剂会影响一些药物的作用，所以采用饮水给药时必须特别谨慎（Dorr等，2009）。

8.3.4 硬度

硬度可以衡量水中多价阳离子的含量，主要是钙离子和镁离子，它们以碳酸盐、碳酸氢盐、硫酸盐和氯化物的形式存在。目前还不清楚硬度是否会影响动物健康，但硬度太高不利于清洗污垢，而且需要使用更多的肥皂或清洁剂来清洗。另外硬度增高还会增加水供应设备、处理设备和加热设备中水垢积累。因此，水的硬度过高会使热水器、乳头式饮水器和过滤器出现故障问题。美国地质勘查局认为，以$CaCO_3$的浓度计量硬度时，浓度低于60 mg/L的是软水，而浓度高于180 mg/L的是超硬水（Chinn，2009）。

8.3.5 硫酸盐

硫酸盐是大部分地下水都存在的天然矿物质，但其浓度一般都较低，对猪不会产生负面影响。然而，在某些情况下，硫酸盐会超过猪饮用水的阈值为1 000 mg/L。虽然猪肠道内存在大量转运蛋白，可吸收硫酸盐，但硫酸盐经常又会被重分泌到大肠，进而导致渗透性腹泻（Maenz和Patience，1997）。这种渗透性腹泻的持续时间比较短，因为猪在数周之后就可以适应，这个适应的具体时间取决于水中硫酸盐的浓度。硫酸盐导致的腹泻问题在刚断奶仔猪上最为严重。因为刚断奶的仔猪从未接触过硫酸盐，生理上可能对硫酸盐的反应更加敏感。无论硫酸盐是以硫酸镁还是硫酸钠的形式存在，都能导致渗透性腹泻，且其程度取决于水中硫酸盐的浓度。

然而，现有诸多研究表明，不管硫酸盐是否引起腹泻，都不会影响猪的生长性能。猪看起来不太健康，主要是由脏而不是硫酸盐引起的，因为它们本身对水中高浓度的硫酸盐具有非常强的耐受力（表1-3）。McLeese 等（1992）研究表明，随着水中硫酸盐浓度的提高，仔猪的腹泻增多，表现为腹泻评分增加，但仔猪的生长性能不受任何影响；此外，还需要注意的是，该试验选用的是4周龄时断奶的仔猪。水中有些细菌能利用硫酸盐，产生氧气，同时产生H_2S 或 HS^- 。H_2S有臭鸡蛋味，有时会残留在水中。

表1-3　提高水中溶解性总固体和硫酸盐的浓度对断奶仔猪生长性能的影响[1]

项目	硫酸盐浓度（mg/L）		
	83	1 280	2 650
溶解性总固体（mg/L）	217	2 350	4 390
钙（mg/L）	24	184	288
镁（mg/L）	15	74	88
钠（mg/L）	24	446	947
硬度（mg/L）	124	767	1 080
pH	8.4	8.1	8.0
平均日增重（g）	430	430	440
平均日采食量（g）	550	560	570
增重 / 耗料比	0.782	0.786	0.772
平均日摄水量（L/d）	1.60	1.84	1.81
腹泻得分[2]	1.07	1.30	1.46

[1]资料来源于McLeese等，1992。

[2]1～3分；1=粪便形态正常，3=粪便呈水样。

当饮用水中的盐含量较高时，可通过降低饲料中的盐含量来降低猪腹泻的发生率及其程度。但是，使用这种方法的时候，必须特别谨慎。因为在降低饲料中盐浓度的同时，也会降低钠离子和氯离子的浓度；如果处理不当，就会导致钠离子或者氯离子缺乏，进而降低猪的采食量。

8.3.6 铁离子和锰离子

目前尚未发现饮用水中的铁和锰含量升高会直接导致猪出现健康问题，但是会引起猪生产管理上的问题。地表水中的铁和锰以还原型离子的形式存在，因而是可溶性的。但是，井水中的铁和锰与氧气接触后，发生氧化反应，变得不可溶。氧化铁呈典型的红褐色，而氧化锰的颜色更深，接近黑色。如果水中存在氧化铁或氧化锰，会使厕所和水槽发生永久性变色，甚至会沉积在热水器、乳头式饮水器、氯化器和其他设备内。水中的铁浓度不应超过0.3 mg/L，尽管水中的铁浓度低至0.1 mg/L就能沉淀，使仪器设备变色（Chinn，2009）。水中的锰含量应不超过0.05 mg/L（Chinn，2009）。

水中的铁还能促进需要铁的细菌生长。微生物会引起水的恶臭，并降低井水的输出量，这些现象都是由于水中或井水管道的菌液累积导致的。

8.3.7 硝酸盐和亚硝酸盐

人们特别关注饮用水中硝酸盐和亚硝酸盐的含量。因为硝酸盐和亚硝酸盐能与血红蛋白结合，形成高铁血红蛋白，从而降低血红蛋白的输氧能力，所以婴儿的饮用水中含硝酸盐和亚硝酸盐时，易患蓝婴症。牛比猪更容易受硝酸盐的影响，因为瘤胃细菌能将硝酸盐转化成亚硝酸盐，这样对机体的伤害更大。Garrison等（1966）研究表明，猪饮用水中含200 mg/L的硝酸盐就会降低猪的生长速度，破坏维生素A的代谢。Sorensen 等（1994）研究表明，从仔猪断奶到出栏，给猪饲喂高达2 000 mg/L的硝酸盐时，对猪的生长性能、血液中血红蛋白或高铁血红蛋白的水平都没有显著影响。上述两个试验的水中硝酸盐含量均高于大部分供水中的一般观测值，这些数据表明婴儿的硝酸盐浓度参考值可能不适用于断奶后的仔猪。

猪饮用水中的硝酸盐和亚硝酸盐总量的推荐值是不超过100 mg/L，且亚硝酸盐的含量不应超过10 mg/L（Chinn，2009）；但人饮用水中的推荐值不超过猪饮用水推荐值的10%。

8.3.8 钠离子

除了低钠饮食的人需要特别注意钠的摄入量外，人们一般都不会关注饮水中的钠离子含量。但是，由于像钠离子这样的阳离子在水中都要与阴离子结合，如果结合的阴离子是硫酸根，则容易导致腹泻（因为硫酸钠又称为芒硝，是一种强力泻药）；如果结合的阴离子是氯离子，则基本上没有负面影响；如果结合的阴离子是碳酸根或碳酸氢根，则水的pH会变高。需要注意的是，简单的水离子交换软化剂可将钙离子和镁离子替换成钠离子，从而提高水中的钠离子含量。

8.3.9 镁离子

人们也很少关注水中的镁离子含量。正如前文所述，镁离子会增加水的硬度；而且也要与阴离子结合，以保持电荷平衡。如果结合的阴离子是硫酸根，形成的硫酸镁又被称为泄盐，会使猪发生严重腹泻（见前文对硫酸盐的介绍）。

采用离子交换软化处理可以去除水中的镁离子，镁离子与钠离子的交换比例为1∶2。虽然钠离子与镁离子交换后可以降低水的硬度，但对猪的腹泻发生率没有任何影响，因为硫酸钠也是强力泻药。

8.3.10 氯离子

不管是地下水还是地表水，氯离子含量一般都不高。如果氯离子含量高于400 mg/L，水就会有金属味，但这是否会对猪产生不良影响，目前还不清楚。在饮水中的氯离子水平较高的时候，可以相应减少饲粮中的食盐含量。但只有在水中的钠离子含量也高或饲粮中含有非氯化钠源的钠的情况下，才能采取该方法。

◆ 8.4 微生物指标

水的微生物指标通常是影响水质量的主要指标。水中存在的病原微生物可导致猪群发病，使猪无法达到最佳生产性能。因为地表水被病原微生物的污染概率更高，所以使用地表水的风险非常高。但是，地下水也可能含有病原微生物，如可能含有沙门氏菌、志贺氏杆菌、霍乱弧菌和弧形杆菌等病原菌，或肠道病毒等病原病毒，或隐孢子虫和鞭毛虫等原生生物。此外，某些水藻类也能导致肠胃炎的发生。

9 水的处理

为了使饮用水达到最低质量标准，要对其进行处理，处理技术非常多。人的饮用水处理系统非常复杂，能够去除大量的有机和无机污染物。但是，由于这些处理的成本非常高，而且需要高素质人员的监管；因此在农业生产中，除非特殊情况，水基本上不进行这些复杂的处理。然而，随着畜禽饮水的要求不断提高，一些水的处理技术已经应用于养殖领域。

◆ 9.1 胶状物的去除

人们通常推荐使用活性炭来改善水的物理属性。活性炭能吸附许多影响水的味道、颜色和气味的成分，同时也能除去一些有机杂质。

◆ 9.2 水的软化

水可以被软化，且其最常见、最简单的处理方法是离子交换法，也就是将钙离子和镁离子替换成钠离子。一般情况下，没有必要对水进行软化；但当水的硬度过大、引起饮水器堵塞或影响到其他设备的时候，则需要软化处理。然而，用于洗衣服或洗澡的水可以进行软化，以减少肥皂的使用。

◆ 9.3 硫酸根离子的去除

反渗透法是减少或去除水中硫酸根离子的唯一可行的办法。但是由于使用这种方法的初期投资成本和后期运行成本都非常高，因此养猪户一般不会采用这种方法。

◆ 9.4 铁离子和锰离子的去除

使用特殊的过滤器，可以降低饮用水中的铁离子和锰离子含量。但是，先往水中加氯气，然后再经过沉淀池的沉淀，对铁离子的去除也有一定的效果（表1-4），而且成本非常低。为了最大程度地去除水中的铁离子，建议先加氯处理，再将水放入沉淀池。当水的pH大于7.5时，该方法的使用效果最佳（Vigneswaran和Visananthan，1995）。加氯处理也有利于防止沉淀池中水的微生物污染。沉淀池需要时常清洗，以清除池中沉积的铁和锰。建议清除铁离子的最后步骤是在沉淀之后使用过滤器进行过滤。

表1-4 加氯处理并沉淀7 d后对水成分的影响（mg/L）

项目	天数	
	0	7
酸碱度	7.92	8.06
溶解性总固体	2 388	2 378
硬度	761	760
硫酸盐	1 268	1 248
钠离子	446	432
钙离子	183	189
锰离子	75	72
氯离子	40	45
钾离子	10	11
硝酸盐	1	1
亚硝酸盐	0.3	0.3
铁离子	2.5	0.6

[1]资料来源于Tremblay等，1989；大草原猪中心的年度报告，被Patience等于995年引用。

一般不推荐采用加氯和过滤器的方法来清除饮水中的锰离子，因为只有在水的pH大于9.5时，该方法才能起作用，而这个pH远远高于大部分饮用水推荐的pH；而且采用上述的方法来去除锰离子，需要沉淀时间也会更长（Vigneswaran和Visananthan，1995）。因此，采用加氯和过滤器的方法来清除铁离子比清除锰离子更加有效。

◆ 9.5 水的消毒

任何水处理系统的关键都是对供水进行消毒。水的质量如何，应该随时被监测，当水中的微生物污染增加时，就需要进行消毒。幸运的是，在全球大部分猪肉产区，饮用水都没有被细菌、病毒和原生动物污染，因此不需要消毒。

最常见的消毒形式是加氯处理，可以在水中加入氯气、次氯酸钠（液体）或次氯酸钙（固体）进行处理。在水中加氯的目的是，产生可以用于消毒的化合物次氯酸（HOCl）和次氯酸离子（OCl^-），其化学平衡式为：

$$HOCl \longleftrightarrow OCl^- + H^+$$

次氯酸是一种最强的消毒剂，易在pH较低的水中形成。水中游离氯离子浓度以及氯离子与水相互作用的时间是影响加氯消毒效果的两个因素。水中“游离氯离子”的理想浓度为0.3～0.5 mg/L，在该浓度范围内，假设水的pH为7.5，则氯离子与水的相互作用时间一般为25～60 min。此外，水的pH、温度以及干扰化合物存在与否等其他因素也都会影响加氯消毒的效果。

加氯消毒时，如果水中存在有机物质，则需要添加更多的氯，使游离氯离子达到理想水平。消毒人员应注意在水中添加充足的氯，以达到消毒的效果；但不能过量添加，因为水中添加过量的氯可能会导致猪死亡，所以添加时要特别谨慎。加氯消毒对水中的隐孢子虫没有作用。

在养猪生产中，使用氯的替代品来处理饮用水越来越普遍。例如，在人的饮用水供应系统中常使用的臭氧、高锰酸钾、紫外线、氯胺和二氧化氯等（Vigneswaran和Visananthan，1995）。

水中存在需铁细菌时，可以采用冲击氯化的方法进行消毒，虽然每次冲击氯化的作用是暂时的，但重复多次之后就能达到最终的消毒效果。另一种消毒的方法是将普通的漂白剂与水按25∶900的比例混合后，倒入水井中并过夜。此时的水不能饮用，必须在水井被彻底冲洗、水恢复干净后，才能饮用。

10 饮水质量对猪生产性能的影响

关于饮水质量对猪生产性能的影响曾引起人们广泛的关注。总体来说，猪对饮水质量的变化具有非常强的适应能力。例如，饮水中硫酸根离子大量升高会导致猪出现渗透性腹泻，进而使猪变脏，但对猪的生长性能没有不良影响。而人们常常会错误地认为，猪的腹泻及其导致的变脏会引起生长性能下降。

关于饮水中硫酸根离子含量高对猪生长性能的影响，目前已有大量研究报道（Anderson和Stothers，1978；McLeese等，1992；Maenz等，1994；Patience等，2004）。其中，最有说服力的结果来自于在有1 200头母猪从产仔到上市的商品猪场中进行的试验，猪场井水的硫酸根离子浓度为1 650 mg/L、溶解性总固体浓度为3 078 mg/L（Patience等，2004）。该试验中，有一半的猪直接饮用井水，另一半的猪饮用经反渗透法处理的水（硫酸根离子浓度降至29 mg/L）。试验从仔猪断奶当天开始，此时的断奶体重约6 kg，试验为期35 d。结果发现，不同的水质量对所观测的仔猪性能指标没有任何影响，但不同处理组的腹泻指数差异非常大。以上结果证实了其他大量研究中报道的，猪对饮水中硫酸根离子具有非常高的耐受能力的结论。但目前尚不清楚渗透性腹泻是否会增加猪对胃肠道病原微生物的易感性。

作者：John F. Patience

译者：晏向华

参考文献

Albar, J., and R. Granier. 1996. Incidence du taux azoté de l'aliment sur la consommation d'eau, la production de liseir et les rejets azot és en engraissement. Journ ées Rech. Porcine en France 28:257–266.

Anderson, C. R., and T. R. Houpt. 1990. Hypertonic and hypovolemic stimulation of thirst in pigs. Am. J. Physiol. 258:R149–R154.

Anderson, D. M., and S. C. Stothers. 1978. Effects of saline water high in sulfates, chlorides and nitrates on the performance of young weanling pigs. J. Anim. Sci. 47:900–907.

Argenzio, R. A. 1984. Intestinal transport of electrolytes and water. In Duke's Physiology of Domestic Animals.M. J. Swenson, ed. Cornell University Press, Ithaca, NY.

Barber, J., P. H. Brooks, and J. L. Carpenter. 1991a. The effects of water to feed ratio on the digestibility, digestible energy and nitrogen retention of the grower ration. Anim. Prod. 52:601.

Barber, J., P. H. Brooks, and J. L. Carpenter. 1991b. The effect of four levels of food on the water intake and water to food ratio of growing pigs. Anim. Prod. 52:602.

Bauer, W. 1982. Consumption of drinking water by non-pregnant, highly pregnant and lactating gilts. Archiv. f`ur Experimentelle Veterin ä rmedizin 36:823–827 (Cited by Mroz et al., 1995).

Berdanier, C. D. 1995. Advanced Nutrition: Macronutrients. CRC Press, Inc. Boca Raton, FL.

Brooks, P. H., and J. L. Carpenter. 1990. The water requirement of growing–finishing pigs—theoretical and practical considerations.Pages 115–136 inRecent Advances in Animal Nutrition, 1990. W. Haresign and D. J. A. Cole, ed. Butterworths, London.

Brumm, M. C. 2010. Water recommendations and systems for swine. PIG 07-02-08 in National Swine Nutrition Guide.D. Meissinger, ed. U.S. National Pork Center of Excellence, Ames, IA.

Brumm, M. C. 2010.Water systems for swine. Pork Information Gateway, U.S. Pork Center of Excellence, Ames, IA.

Brumm, M. C., J. M. Dahlquist, and J. M. Heemstra. 2000. Impact of feeders and drinker devices on pig performance, water use and manure volume. Swine Health Prod. 8:51–57.

Chinn, T. D. 2009. Water supply. Pages 1–132 in Environmental Engineering: Water,

Wastewater, Soil and Grundwater Teatment and Remediation.N. L. Nemerow, F. J. Agardy, P. Sullivan, and J. A. Salvato, ed. John Wiley and Sons, Hoboken, NJ.

de Lange, C. F. M., C. H. Zhu, S. Niven, D. Columbus and D. Woods. 2006. Swine liquid feeding: Nutrition considerations. Proc.Western Nutr. Conf., Winnipeg, MB, Canada.

Deligeorgis, S. G., K. Karalis, and G. Kanzouros. 2006. The influence of drinker location and colour on drinking behavior and water intake of newborn pigs under hot environments. Appl. Anim. Behav. Sci. 96:233–244.

Dorr, P. M., D. Madson, S. Wayne, A. B. Scheidt, and G. W. Almond. 2009. Impact of pH modifiers and drug exposure on the solubility of pharmaceutical products commonly administered through water delivery systems. J. Swine Health Prod.17:217–222.

Dybklaer, L., A. P. Jacobsen, F. A. Togersen, and H. D. Poulsen. 2006. Eating and drinking activity of newly weaned piglets:Effects of individual characteristics, social mixing, and addition of extra zinc to the feed. J. Anim. Sci. 84:702–711.

Fraser, D., J. F. Patience, P. A. Phillips, and J. M. McLeese. 1990. Water for piglets and lactating sows: quantity, quality and quandaries. Pages 137–160 in In Recent Advances in Animal Nutrition . W. Haresign and D. J. A. Cole, ed. Buterworths,London.

Fraser, D., and P. Phillips. 1989. Lethargy and low water intake by sows during early lactation: A cause for low piglet weight gains and survival? Appl. Anim. Beahviour Sci. 24:13–22.

Friend, D. W. 1971. Self-selection of feeds and water by unbred gilts. J. Anim. Sci. 32:658–666.

Garrison, G. W., R. D. Wood, C. H. Chaney, and D. G. Waddill. 1966. Effects of nitrate and nitrites in drinking water on the utilization of carotene in swine. Kentucky Anim. Sci. Res. Rpt. 164:85.

Gill, B. P., and J. Barber. 1990. Water delivery systems for growing pigs. Farm Build. Progr. 102:19–22.

Gonyou, H. W., and Z. Zhou. 2000. Effects of eating space and availability of water in feeders on productivity and eating behavior of grower/finisher pigs. J. Anim. Sci. 78:865–870.

Hurst, D., L. Clarke, and I. J. Lean. 2009. Effect of liquid feeding by different water-to-feed ratios on the growth performance of growing-finishing pigs. Animal 2:9:1297–1302.

Koeppen, B. M., and B. A. Stanton. 2001. Renal Physiology. 3rd ed. Mosby Inc., St. Louis, MO.

Kuperus, W. 1988. Water intake by pregnant and lactating sows. Pig Experimental Station Raalte Report (Cited by Mroz et al.,1995).

Lehninger, A. L. 1982.Principles of Biochemistry . Worth Publ., New York.

Li, Y. Z., L. Chenard, S. P. Lemay, and H. W. Gonyou. 2005. Water intake and wastage at nipple drinkers by growing–finishing pigs. J. Anim. Sci. 83:1413–1422.

Maenz, D. D., and J. F. Patience. 1997. Presteady–state and steady–state function of the ileal brush border $SO4^{-2}$–OH^{-}exchanger.Biochem. Cell Biol. 75:229–236.

Maenz, D. D., J. F. Patience, and M. S. Wolynetz. 1993. Effect of water sweetener on the performance of newly weaned pigs offered medicated and unmedicated feed. Can. J. Anim. Sci. 73:669–672.

Maenz, D. D., J. F. Patience, and M. S. Wolynetz. 1994. The influence of the mineral level in drinking water and the thermal environment on the performance and intestinal fluid flux of newly−weaned pigs. J. Anim. Sci. 72:300–308.

Maynard, L. A., J. K. Loosli, H. F. Hintz, and R. G. Warner. 1979.Animal Nutrition. 7th ed. McGraw Hill Inc., New York.

McLeese, J. M., J. F. Patience, M. S. Wolynetz, and G. I. Christison. 1991. Evaluation of ground water supplies used on Saskatchewan swine farms. Can. J. Anim. Sci. 71:191–203.

McLeese, J. M., M. L. Tremblay, J. F. Patience, and G. I. Christison. 1992. Water intake patterns in the weanling pig: Effect of water quality, antibiotics and probiotics. Anim. Prod. 54:135–142.

Mount, L. E., C. W. Holmes, W. H. Close, S. R. Morrison, and I. B. Start. 1971. A note on the consumption of water by the growing pig at several environmental temperatures and feeding levels. Anim. Prod. 13:561–563.

Mroz, Z., A. W. Jongbloed, N. P. Lenis, and K. Vreman. 1995. Water in pig nutrition: Physiology, allowances and environmental implications. Nutr. Res. Rev. 8:137–164.

NRC. 1974. Nutrients and toxic substances in water for livestock and poultry. National Acadamies Press, Washington, DC.

NRC. 1981. Water−environment interactions. Pages 39–50 in effect of environment on nutrient requirements of domestic animals.National Acadamies Press, Washington, DC.

Nienaber, J. A., and G. LeRoy Hahn. 1984. Effects of water flow restriction and environmental factors on performance of nursery−age pigs. J. Anim. Sci. 59:1423–1429.

Nyachoti, C. M., J. F. Patience, and I. R. Seddon. 2005. Effect of water source (ground versus surface) and treatment on nursery pig performance. Can. J. Anim. Sci. 85:405–407.

Oresanya, T. F., A. D. Beaulieu, and J. F. Patience. 2008. Investigations of energy metabolism in weanling barrows: The interaction of dietary energy concentration and daily feed (energy) intake. J. Anim. Sci. 86:348–363.

Patience, J. F. 1989. The physiological basis of electrolytes in animal nutrition. Pages 211–228 inRecent Advances in Animal Nutrition . W. Haresign and D. J. A. Cole, ed. Butterworths, London.

Patience, J. F., A. D. Beaulieu, and D. A. Gillis. 2004. The impact of ground water high in sulphates on the growth performance,nutrient utilization, and tissue mineral levels of pigs housed under commercial conditions. J. Swine Health Prod. 12:228–236.

Patience, J. F., P. A. Thacker, and C. F. M. de Lange. 1995. Swine Nutrition Guide. 2nd ed. Prairie Swine Centre Inc., Saskatoon,SK.

Patience,J.F.,J.F. Umboh, R. K. Chaplin, and C. M. Nyachoti. 2005. Nutritional and physiological responses of growing pigs exposed to a diurnal pattern of heat stress. Livest. Prod. Sci. 96:205–214.

Patience, J. F., M. S. Wolynetz, D. W. Friend, and K. E. Hartin. 1987. A comparison of two urine collection methods for female swine. Can. J. Anim. Sci. 67:859–863.

Petherick, J. C. 1983. A note in the allometric relations in Large White × Landrace pigs. Anim. Prod. 36:497–500.

Pfeiffer, A., H. Henkel, M. W. A. Verstegen, and I. Philipczyk. 1995. The influence of protein intake on water balance, flow rate and apparent digestibility of nutrients at the distal ileum in growing pigs. Livest. Prod. Sci. 44:179–187.

Phillips, P. A., and M. H. Phillips. 1999. Effect of dispenser on water intake in pigs at weaning. Trans. ASAE 42:1471–1473.

Schiavon, S., and G. C. Emmans. 2000. A model to predict water intake of a pig growing in a known environment on a known diet.Br. J. Nutr. 84:873–883.

Seynaeve, M., R. de Wilde, G. Janssens, and B. de Smet. 1996. The influence of dietary salt level on water consumption, farrowing, and reproductive performance of lactating sows. J. Anim. Sci. 74:1047–1055.

Shaw, M. I., A. D. Beaulieu, and J. F. Patience. 2006. Effect of diet composition on water utilization in growing pigs. J. Anim. Sci.84:3123–3132.

Shields, R. G., Jr., D. C. Mahan, and P. L. Graham. 1983. Changes in swine body composition from birth to 145 kg. J. Anim. Sci.57:43–54.

Sorensen, B., B. Jensen, and H. D. Poulsen. 1994. Nitrate and pig manure in drinking water to early weaned piglets and growing pigs. Livest. Prod. Sci. 39:223–227.

Straub, G., J. H. Wenger, E. S. Tawfik, and D. Steinhauf. 1976. The effects of high environmental temperatures on fattening performance and growth of boars. Livest. Prod. Sci. 3:65–74.

Suzuki, K., X. C. Cheng, H. Kamo, T. Shimizu, and Y. Sato. 1998. Influence of low protein diets on water intake and urine and nitrogen excretion in growing pigs. Anim. Sci. Tech. 69:267–270.

Tachibana, F., and H. Ubagai. 1997. Effect of reducing crude protein and energy content in diets with amino acid supplementation on nitrogen balance, performance and carcass characteristics in pigs. Anim. Sci. Technol. 68:640–649.

Thacker, P. A. 2001. Water in swine nutrition. Pages 381–400 in Swine Nutrition. 2nd ed. A. J. Lewis and L. L. Southern, ed. CRC Press, New York.

Torrey, S., E. L. M. Toth Tamminga, and T. M. Widowski. 2008. Effect of drinker type on water intake and waste in newly weaned piglets. J. Anim. Sci. 86:1439–1445.

vandenheede, M., and B. Nicks. 1991. L'approvisionnement en eau des porcs: Un element a ne pas negliger. J. Ann. Med. Vet.135:123–128.

Vermeer, H. M., N. Kujiken, and H. A. M. Spoolder. 2009. Motivation for additional water use of growing–finishing pigs. Livest.Sci. 124:112–118.

Yang, T. S., M. A. Price, and W. V. McFarlane. 1981. The effect of level of feeding on water turnover in growing pigs. Appl. Anim.Ethol. 7:259–270.

第2章
猪的能量和能量代谢

1 导 语

养猪生产中，饲料成本占养猪生产总成本的60%以上，而其中能量成本所占比例最大。由于能量对饲料成本影响很大，且能量会显著影响到动物的生产性能。因此，用于描述饲料能值和动物能量需要量的能量体系取得了长足进步。不同动物生产部门对饲料原料的竞争，以及这些原料不管是用于生产生物能源还是作为人的食物来源，都要与现行的高效率、低环境影响的生产系统并行不悖，这就需要我们对饲料能值和动物能量需要量进行更明确定义，从而为动物生产可持续发展创造有效而又便利的条件。

本章主要目的是：①以有效能体系阐述能量在猪体内的利用过程（总能-消化能-代谢能-净能）来对饲料进行评价；②对猪生产中能量需要进行剖分，解析猪生长、繁殖过程中对能量的需要量；③探讨影响能量摄入的因素，以及饲料品质、动物特性和环境因素对能量摄入的调控。本章中，能量单位使用的是国际单位——焦耳（J）。

2 猪对能量的利用

◆ 2.1 原理和定义

猪摄入的所有饲料能量并不能都存留在体内，其中部分能量通过粪便、尿、发酵气体和体热损失掉。根据能量利用过程中的损失，可定义不同的能值和能量体系：消化能（digestible energy，DE），即动物摄入的总能（gross energy，GE）与粪能的差值；代谢能（metabolizable energy，ME），即消化能减去尿能及发酵产气的能量后剩余的能量；净能（net energy，NE），即代谢能与热增耗（heat increment，HI）的差值。

2.1.1 总能

燃烧热（或总能），既是最基本的饲料能量表述形式，也是饲料本身属性。饲料总能含量可用氧弹式测热计测定：将少量饲料原料完全氧化，测定所释放的热量。不同原料间总能含量变化范围很大，从蔗糖蜜的15 kJ/g到油或脂肪的39 kJ/g（以干物质为基础）（Sauvant等，2004）。化学组成不同导致了饲料原料间总能含量的差异。在有机物中，碳水化合物（如淀粉、糖和饲粮纤维）总能含量相对较低，而脂肪则具有较高的总能。除氧弹式测热计外，总能也可根据原料化学成分的组成，利用预测方程估测。INRA-AFZ表（Sauvant等，2004）给出了下列预测方程：

$$GE = 17.3 + 0.0617CP + 0.2193EE + 0.0387CF - 0.1867ash \quad (2\text{-}1)$$

式中，GE指MJ/kgDM；CP、EE、CF和ash分别指粗蛋白、乙醚浸出物（粗脂肪）、粗纤维和灰分（以干物质百分比表示）。

总能也可通过含能量的营养物质（g）直接估测，Noblet等（2004）提出了下列预测方程：

$$GE = 23.0CP + 38.9EE + 17.4starch + 16.5sugars + 18.8NDF + 17.7residue \quad (2\text{-}2)$$

式中，starch、sugars、NDF分别指淀粉、糖类和中性洗涤纤维；residue（残留物）是有机物与方程中其他营养成分间的差值。从公式2-2可得出，碳水化合物能值最低，蛋白质居中，脂类物质最高。公式2-2虽是经验方程，但却能很好地反映各种营养物质的能值。例如，淀粉和糖类的能值存在差异，其原因是由碳水化合物的聚合程度不同所致。葡萄糖能值为15.7 kJ/g（180 g/mol），在葡萄糖的长链聚合物中，每单位葡萄糖能量不变，但由于在聚合过程中释放掉水分使其重量变小（从180 g/mol减少

18 g/mol），因此，葡萄糖的长链聚合物中单位葡萄糖的理论能值应是15.7×180/(180−18)=17.4 kJ/g。总能含量通常受蛋白质中氨基酸组成的影响，而脂类物质中的脂肪酸组成对总能的影响相对较小。对氨基酸而言，总能变化范围较大，从天冬氨酸的14 kJ/g到亮氨酸、异亮氨酸、苯丙氨酸的31.6 kJ/g（van Milgen，2002）。

2.1.2 消化能

饲料中的消化能指总能减去消化道内消化后损失的能量，即总能减去粪能。虽然后肠发酵所产气体、热量与消化过程息息相关，但在计算消化能时，并没有考虑到这部分的能量损失。消化能与总能的比值相当于能量消化率（digestibility coefficient，DCe）。测定消化能时，通常将猪置于消化笼内进行；粪便量通过至少5 d的全收粪法获得，或通过在饲料中添加指示剂进行估测。对可单独饲喂的全价饲料或饲料原料（如谷物），可直接测定其消化能。但是，很多原料在饲粮中的用量有限，既要考虑猪对其的耐受性也要考虑其实际用量。对这些饲料原料，可采用套算法或回归法测定消化能：分别测定对照和试验饲粮消化能，对照组饲粮是试验饲粮的主要组成，试验组饲粮以对照组饲粮为基础，含有一定量的待测原料。这里有两个前提：首先，测定的两种饲粮消化能值的不同，仅仅只是因为待测原料；其次，饲粮中的矿物质和维生素（MV）不提供能量，即使饲粮中灰分会影响到能量消化率（后文将进一步讨论），因此对照和试验饲粮中MV成分保持一致非常重要。待测原料的消化率计算如下：

$$DCe(\%)=100[DEexp-DEcrtl\times\%crtl/(1-MV0)]/[GEexp-GEcrtl\times\%crtl/(1-MV0)] \quad (2-3)$$

GEexp和DEexp指试验饲粮总能和消化能（MJ/kg DM），GEcrtl和DEcrtl指对照组饲粮总能和消化能（MJ/kg DM），MV0为对照组饲粮中MV含量（%，干物质基础），%crtl是对照组饲粮（即对照组饲粮减去其MV含量，即MV0）占试验组饲粮的比例。然后，用测定的该待测原料总能值，乘以根据公式2-3估计的消化率（DCe），从而得出该待测原料的消化能值。一种对照饲粮，既可用作多种待测原料消化能值估测试验的对照，也可用作相同待测原料不同含量水平试验的对照。在没有测热计的情况下，粪能值可通过粪中近似组分计算。Noblet和Jaguelin给出下列方程（未发表数据）：

$$GE\ feces=18.73-0.192Ash+0.223EE+0.065CP \quad (2-4)$$

GE是干物质基础MJ/kg，化学组成以干物质百分比表示。

2.1.3 代谢能

饲料代谢能等于消化能减去尿能和发酵气体（主要是甲烷）的能量。通常将猪置于代谢笼内进行尿能测定，既费力又耗时，因此生长猪和成年母猪的尿能（饲料干物质基础，MJ/kg）预测方程已被提出（Le Goff和Noblet，2001；Noblet等，2004）：

$$尿能 = 0.19+0.031Nuri \quad (2-5)$$

$$尿能 = 0.22+0.031Nuri \quad (2-6)$$

Nuri是尿氮含量（g/kg干物质采食量）。尿氮含量取决于可消化氮与沉积氮间的差值，换言之，即取决于饲料中蛋白质含量与猪以蛋白质形式储存能量的能力。因此，尿能随猪不同生理阶段和饲料特征而变化。一种饲料或原料在实际应用中如果只使用一个代谢能值，那么就要计算标准尿能损失和标准代谢能值，即将尿氮损失看作可消化氮或总氮的固定比例。

甲烷产量需要在呼吸室内测定，仔猪和生长猪由于甲烷而损失的能量非常少，很多情况下可将其忽略。但对成年猪而言，后肠发酵非常发达（后面章节中将进一步讨论），甲烷释放量比生长猪高4～5倍，因此，在估测代谢能值时应予以考虑。

2.1.4 净能

净能是代谢能与热增耗的差值，热增耗与饲料利用（即采食、消化和代谢过程中的耗能）和“正常”生理活动下的能量消耗密切相关（图2-1）。净能与代谢能的比值（k）即代谢能转化为净能的效率，也可表示为1-（热增耗/代谢能），但对于某一特定饲料的热增耗与代谢能的比值，取决于代谢能的摄入量和一些生理因素。例如，当代谢能摄入量低于动物维持能量需要时，相比于高于维持能量需要，热增耗要小一些（Noblet等，1993，1994 a，1994 b；Birkett和de Lange，2001）；相对于蛋白质沉积，代谢能用于脂肪沉积时热增耗要少（Noblet等，1999）。随着代谢能摄入量的增加，代谢能用于脂肪沉积的比例明显高于蛋白质沉积。因此，至少从理论上来说，代谢能摄入量越高，热增耗与代谢能比值就越小。为了维持特定饲料或原料只有一个净能值的理念，有必要将测定条件标准化：①蛋白质和氨基酸能够满足动物的营养需要；②体增重部分的组成维持恒定；③在一个特定的生理状态下。对生长猪而言，净能摄入量是特定生产水平下的沉积能量与安静状态下的绝食产热（fasting heat production，FHP）之和（Noblet

等，1994 a）。净能值和相应 k 值等于满足动物维持和生长需要的能量之和。沉积能量可利用比较屠宰法测定，更多的是采用代谢能减去测热法估测的产热量（heat production，HP）来计算。FHP既可通过绝食动物直接测定，也可通过查阅文献获得。此外，也可通过测定不同代谢能水平下的产热量来推算FHP。当代谢能摄入量为零时，此时产热量应为FHP（图2-2；FHPr）。该方法虽在过去已广泛使用，但存在较大局限性。首先，采用推算法测定FHP时，是在采食量为自由采食量的60%～100%情况下测定产热量，然后将采食量外推为0来计算FHP。这种方法导致斜率和截距产生偏差。其次，FHP不是恒定的，且受禁食前饲喂水平的影响，尤其对生长猪而言（Koong等，1982；de Lange等，2006；Labussière等，2011）。很显然，动物可以根据饲料摄入量和生长强度来调整其基础能量消耗。有些学者研究发现，推算的FHPr明显低于测定的FHP，从而导致净能值和 k 值偏低，热增耗偏高（图2-2）。此外，他们也观测到不同饲喂水平下热增耗（产热量减去测定的绝食产热，以单位代谢能表示）是恒定的，此外，饲喂水平对FHP

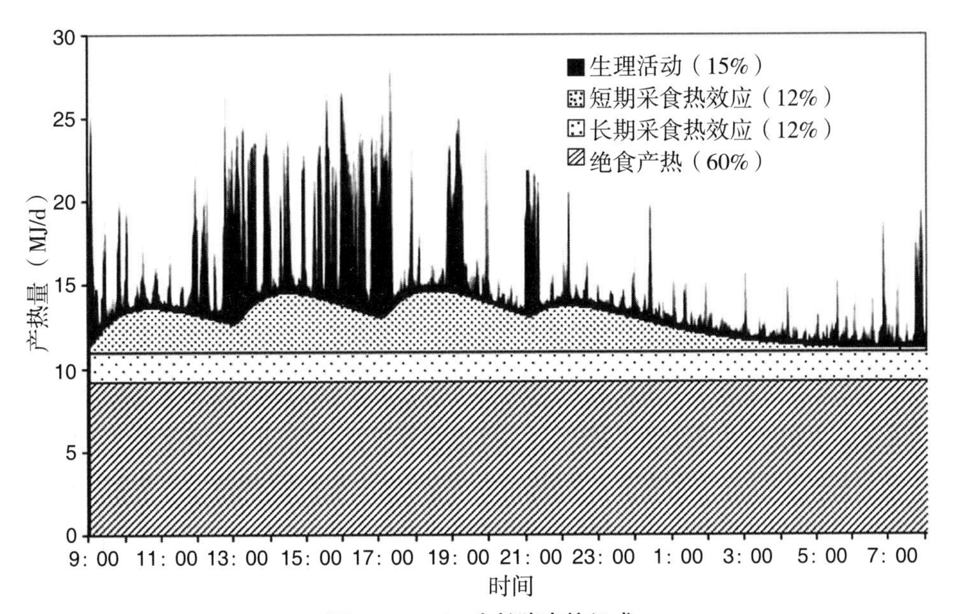

图2-1　60 kg生长猪产热组成

代谢能摄入量2.4 MJ/kg $BW^{0.60}$/d；日饲喂四次：09：00，13：00，17：00和21：00；TEF：采食热效应。（资料来源于van Milgen和Noblet，2000）

和HP的影响程度同时也受动物特征（如基因型）的影响（Renaudeau等，2007）。总而言之，利用FHPr计算净能值存在争议。Noblet等（2010）指出，动物采食后，利用间接测热法立即测定FHP是更可取的方法，若测定条件不满足，FHP可参考相关文献。产热量与气候条件密切相关，动物处在低于温度适中区的环境温度下，其产热量会增加，而沉积能量会降低。因此，为了减少估测净能值和 *k* 值的偏差，建议环境温度高于温度适中区。

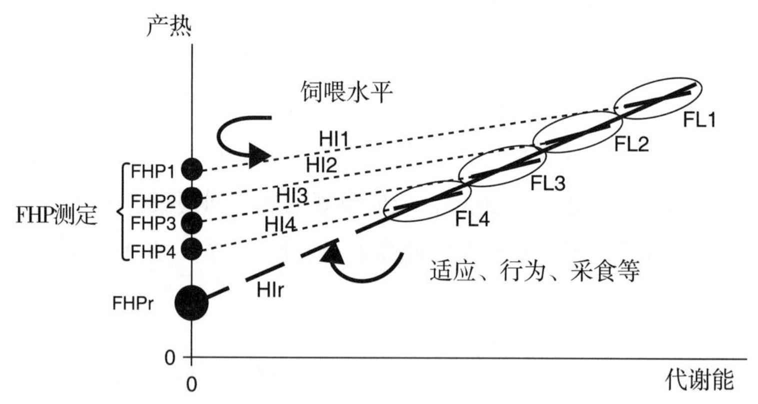

图2-2　饲喂水平（FLi）对单胃动物产热和绝食产热（FHP）的影响

每一个FHPi代表在不同饲喂水平下采食后立即测定的绝食产热量。FHPr（r为回归系数）通过对产热和代谢能的回归分析得到。斜率是热增耗的回归系数，每个饲喂水平下产热量（HPi）和绝食产热量（FHPi）的差值是热增耗。（资料来源于Koong等，1982；de Lange等，2007；Labussière等，2009b，2011）

从应用角度来看，为避免在计算饲料净能值时产生偏差，需要进行能量平衡试验：采用相似的动物（同性别、同品种及同体重范围），动物处于温度适中区、行为差异小，饲喂平衡饲粮且采食量相同，这样动物就能够表现出生长潜能。在这些情况下，错误估测FHP会影响到绝对净能值，但不会影响到不同原料净能值大小顺序（即净能值大的仍然大，净能值小的仍然小），也意味着当待测原料化学特征与全价平衡饲粮差距很大时，原料净能值不能被测定。

消化能和代谢能的测定相对容易，且花费相对合理，而净能测定就复杂得多，并且耗资巨大，因此，最好选择标准条件下建立的预测方程来估测净能值。

Noblet实验室提出了基于消化能或者代谢能含量，并结合化学组成来估测猪全价饲料或饲料原料的净能值的预测方程（Noblet等，1994 a），其他预测因素（即自变量）也可被采用，如化学组分、现有饲养标准以及消化率试验结果等。

产热量既可通过直接测热法测定，也可通过间接测热法的气体交换估测，或利用摄入的代谢能减去比较屠宰法获得的沉积能计算而得。后者可用于家禽等小动物，但是对大体型动物而言，实施起来非常困难且测定结果不准确。在猪研究方面，常用的方法是间接测热法，测定指标包括氧气消耗量、二氧化碳和甲烷生成量，然后将这些测定值与尿能结合起来，用于计算产热量（Brouwer，1965）。这种方法通过将同一动物在不同饲喂水平下（包括绝食）的测量结果结合起来（不需要适应期），从而缩短了测试时间。析因法将日产热量分成不同部分，可用来进一步解释能量平衡数据（van Milgen等，1997；图2-1）。

总之，饲料的净能值和对应的 *k* 值应该通过标准化、准确方法进行测定。净能值取决于FHP值、测定条件（如气候、活动等）以及能量沉积的组成形式。这表明文献中报道的猪净能值和对应的 *k* 值要慎重对待，并且相关数据不应该直接相互比较。该观点也同样适用于代谢能，代谢能值很大程度上取决于蛋白质的分解。此外，气体能量损失也影响代谢能的估测。

◆ 2.2 能量的消化利用

2.2.1 饲粮组成的影响

大多数猪饲粮的消化率为70%～90%，而饲料原料消化率的变化范围更大，通常为10%～100%（Sauvant等，2004）。饲粮消化率的变异与饲粮纤维含量密切相关，因为饲粮纤维相比其他营养成分难以消化（纤维的消化率<50%，脂肪或蛋白质的消化率为80%～90%，淀粉和糖的消化率为100%；表2-1和表2-2），并且可降低其他营养成分（如粗蛋白和脂肪）的表观消化率（Noblet和Perez，1993；Le Goff和Noblet，2001）。因此，饲粮消化率与纤维含量呈线性负相关（表2-1和表2-3）。饲粮消化率与纤维（表2-1和表2-3，公式2-7）的相关系数，本质是NDF或总饲粮纤维含量降低了饲粮营养浓度，至少对生长猪来说是这样的，即使部分纤维可被生长猪消化，但基本不能给动物提供消化能。

表2-1　不同纤维来源对生长猪纤维消化率的影响[1]

项目	纤维来源			
	甜菜渣	大豆皮	麦麸	麦秸
消化率（%）[2]				
中性洗涤纤维	60.1	67.9	40.4	15.0
酸性洗涤纤维	54.0	62.2	19.0	11.2
非淀粉多糖	69.5	79.1	45.8	16.3
能量消化率变化[3]	−0.80	−0.83	−1.25	−1.77

[1]资料来源于Chabeauti等（1991）。
[2]试验饲粮中纤维部分替代基础饲粮中淀粉。
[3]每增加1%非淀粉多糖，能量消化率下降的程度（DCe，%）。

表2-2　生长猪（G）和成年母猪（S）对高纤维原料中纤维和能量的消化率（%）[1]

消化率（%）	麦麸		玉米糠		甜菜渣	
	G	S	G	S	G	S
非淀粉多糖	46	54	38	82	89	92
非纤维素多糖	54	61	38	82	89	92
纤维素	25	32	38	82	87	91
非淀粉多糖+木质素	38	46	32	74	82	86
能量	55	62	53	77	70	76

[1]资料来源于Noblet和Bach Knudsen，1997。

纤维的消化利用随植物来源不同而变化（表2-1和表2-2），而且影响能量消化率。因此，表2-3中混合饲粮消化率的预测方程，不适用于特定原料消化率的估测（Noblet等，2003）。此外，公式2-8（表2-3）表明矿物质含量对饲粮消化率起负面影响，部分原因是某些饲料原料中矿物质与纤维结合在一起，更直接的原因是碳酸钙和磷酸盐类矿物质可能造成肠道磨损，从而造成消化率的下降（灰分提高1%，消化率下降0.5%，INRA，未发表数据）。

表2-3 饲粮组成对能量消化率（DCe，%）、消化能转化代谢能效率及混合饲粮中代谢能转化生长净能（k_g）或维持（km）净能效率的影响[1]

	方程	来源[2]
2-7	DCe = 98.3 − 0.90 NDF	1
2-8	DCe = 102.6 − 1.06 Ash − 0.79 NDF	1
2-9	DCe = 96.7 − 0.64 NDF	1
2-10	ME/DE = 100.3 − 0.21 CP	1
2-11	k_g = 74.7 + 0.36 EE + 0.09 ST − 0.23 CP − 0.26 ADF	2
2-12	km = 67.2 + 0.66 EE + 0.16 ST	3

[1]CF = 粗纤维，CP = 粗蛋白质，NDF = 中性洗涤纤维，EE = 粗脂肪，ST = 淀粉，ADF = 酸性洗涤纤维。

[2]资料来源：1 = Le Goff 和Noblet（2001；n = 77种饲粮，方程2-7和2-8由60 kg生长猪而得，方程2-9由成年母猪而得），2 = Noblet等（1994a；n = 61种饲粮；由45 kg生长猪而得），3 = Noblet等（1993；n = 14种饲粮；由成年母猪维持水平下而得）。

2.2.2 加工工艺的影响

饲料加工工艺可改变能量消化率。例如，制粒可使饲料能量消化率提高大约1%（Skiba等，2002；Le Gall等，2009）。对一些饲料，提高能量消化率显得尤为重要，能量消化率的提高与饲料化学和物理特性（颗粒大小）有关。如表2-4所示，能量消化率的改善，是由脂类（由玉米、全脂油菜籽或亚麻籽提供）消化率的提高所致。因此，这些原料能值很大程度上取决于加工处理。制粒可使高油玉米（油的含量为7.5%）的消化能增加0.45 MJ/kg（Noblet和Champion，2003）。将粗磨全脂油菜籽（干物质基础）粉碎或制粒，其消化能可分别达到10.0 MJ/kg和23.5 MJ /kg（Skiba等，2002）。到目前为止，有关定量研究制粒或其他加工工艺（膨化、酸化或添加酶制剂）改善猪饲粮中常规原料消化率的报道，还非常少。

表2-4 制粒对生长猪脂肪和能量消化率的影响（%）

项目	粉料	制粒
玉米－豆粕型饲粮[1]		
脂肪	61.0	77.0

（续）

项目	粉料	制粒
能量	88.4	90.3
小麦－豆粕－全脂油菜籽饲粮[2]		
脂肪	27.0	84.0
能量	73.1	87.4
小麦－玉米－大麦－豆粕饲粮		
能量[3]	75.8	77.3
能量		
玉米	87.0	90.0
全脂油菜籽	35.0	83.0
亚麻籽（膨化）[4]	51.0	84.0

[1]三种饲粮平均值，含有玉米81%、豆粕15.5%（Noblet和Champion，2003）。

[2]一种饲粮，含有小麦60%、豆粕15%、全脂油菜籽20%；油菜籽经过粗粉碎（Skiba等，2002）。

[3]四种饲粮平均值，含有不等量的高纤维原料（小麦麸和甜菜渣；Le Gall等，2009）。

[4]资料来源于Nobler等，2008。

另外，加工工艺对饲料在不同位置（如小肠或后肠）消化的影响，也需要进一步研究。除此之外，还要关注加工工艺的部分负面效应，如DDGS干燥过程中过度加热会导致美拉德反应，从而降低消化率（Cozannet等，2010）。

2.2.3 体重和生理阶段的影响

能量消化率除受饲粮成分本身影响外，还受其他因素的影响。对生长猪而言，消化率随体重（BW）的增加而提高（Noblet，2005；表2-5）。将采食量稍超过维持水平的成年母猪与接近自由采食的生长猪相比时，可发现体重对消化率的影响最大（Fernandez等，1986；Noblet和Shi，1993；Le Goff和Noblet，2001；表2-6）。高纤维含量的饲粮或原料对消化率的影响最大（表2-3中的公式2.7 和2.9；表2-5），而体重大的猪或成年母猪，纤维对饲粮消化率的负面影响逐渐减少。这是因为随着体重的增加，纤维可以为猪提供更多的能量。

表2-5 饲粮组成和体重对猪能量消化率的影响（%）[1,2]

体重（kg）	对照饲粮	+ 玉米淀粉	+ 饲粮纤维	+ 菜籽油
44	85.3	90.6	71.6	86.0
103	87.2	91.6	75.6	88.7
148	87.2	92.2	78.0	88.9

[1]资料来源于Noblet和Shi（1994）。

[2]对照饲粮由谷物和豆粕组成，其他饲粮由对照组饲粮+30%玉米淀粉，或8%菜籽油，或30%高纤维原料混合物（1/4小麦麸，1/2大豆皮，1/4甜菜渣）。

表2-6 猪体重和生理阶段对能量消化率的影响

项目	试验一[1]		试验二[2]	
	生长猪	干奶母猪	生长猪	哺乳母猪
体重（kg）	60	227	62	246
采食量（g DM/d）	2 044	2 119	2 062	4 850
能量消化率（%）	77.2[a]	80.5[b]	79.9[a]	84.9[b]

[1]三种饲粮平均值，以玉米、小麦、大麦、豌豆、豆粕、葵花粕、玉米蛋白粉、动物脂肪和不等量的小麦麸为基础（INRA数据）。

[2]三种饲粮平均值，以玉米、小麦、大麦、豌豆、豆粕、葵花粕、玉米蛋白粉、动物脂肪和不等量的小麦麸为基础（Etienne等，1997）。

[a,b]同一试验中上标不同字母表示差异显著（$P<0.05$）。

根据77种饲粮的测定结果，Le Goff和Noblet（2001）计算发现，对于60 kg生长猪和成年母猪，每克NDF能够分别提供3.4 kJ和6.8 kJ的能量。成年母猪和生长猪消化能之差与生长猪体内未消化有机物的量呈正比（平均4.2 kJ/g；Noblet等，2004；图2-3）。

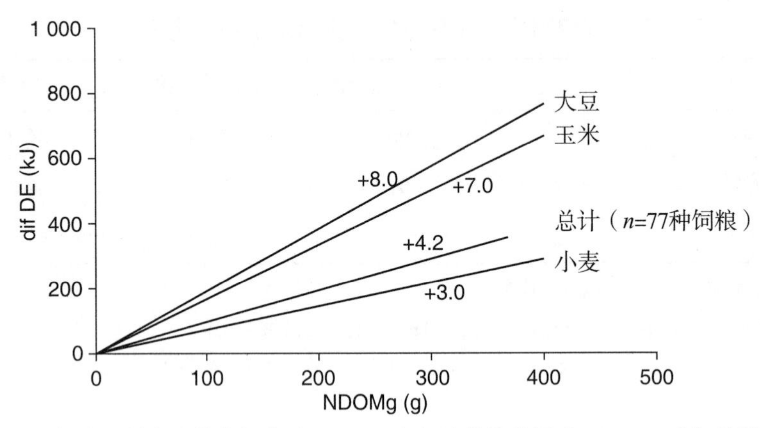

图2-3　生长猪原料未消化有机物（NDOMg）与消化能差异（dif DE，成年母猪消化能-生长猪消化能）的关系

（资料来源于Noblet等，2003a）

随着体重增加，猪对能量的消化率随之提高，这是由于猪对纤维消化率的提高所致（表2-2）。体重大的猪纤维消化率的提高，一方面是由于其后肠消化能力增强，更重要的是饲料在消化道内的流通速度降低（饲料在消化道的通过时间：成年母猪80 h，生长猪35 h；Le Goff等，2002）。与生长猪相比，饲粮纤维对成年猪蛋白质和脂肪消化率（即内源损失）的抑制作用降低，这是饲粮纤维对成年猪能量消化率的抑制作用降低的原因（Le Goff和Noblet，2001）。哺乳母猪采食能力强，每天采食量高达6 ~ 9 kg，其能量消化率也高于生长猪（表2-7）。成年母猪能量消化率受生理阶段和饲喂水平的影响甚微。这意味着在空怀期成年母猪（采用妊娠期间的饲养水平）上获得的能量消化率数据也同样适用于怀孕和哺乳母猪。这些数据高于生长猪的能量消化率（表2-7）。

表2-7　生长猪和成年猪对不同原料的消化能值（饲喂基础）[1]

原料	消化能（MJ/kg）		
	生长猪	成年猪	a[2]
小麦	13.85	14.10	3.0
大麦	12.85	13.18	2.5
玉米	14.18	14.77	7.0

（续）

原料	消化能（MJ/kg）		
	生长猪	成年猪	a[2]
豌豆	13.89	14.39	6.0
豆粕	14.73	15.61	8.0
油菜籽粕	11.55	12.43	3.5
葵花粕	8.95	10.25	3.5
麦麸	9.33	10.29	3.0
玉米蛋白饲料	10.80	12.59	7.0
大豆皮	8.37	11.46	8.0

[1]资料来源于Sauvant等，2004。

[2]消化能差异（成年猪消化能—生长猪消化能，kJ）/未消化有机物（生长猪，g）。

在特定饲粮纤维水平下，母猪和生长猪在能量消化率和消化能上的差异与纤维来源及其理化特性有关。表2-1和图2-3将来自于麦麸、玉米糠和甜菜渣的纤维对能量消化率的影响进行了比较。Noblet和Le Goff（2001）详细报道了不同植物源性的饲粮纤维对生长猪和成年母猪能量消化率的影响。结果表明，生长猪对饲粮纤维消化能力有限，且不同来源纤维间差异很小，而成年猪可有效地消化饲粮纤维，并受饲粮纤维化学特性的影响（如木质素含量）。从表2-7可知，不同生理阶段，禾本科谷物（如小麦、大麦及小麦麸）、十字花科类谷物（如油菜籽）或菊科谷物（如向日葵）源性纤维对消化能影响不大，而豆科类谷物（如豌豆、大豆及羽扇豆）源性纤维对消化能影响较大，尤其是豆科类谷物外壳纤维对消化能影响更大。因此，成年母猪和生长猪消化能的差异与生长猪未消化有机物成正比，且各科植物性原料对应特定回归系数（图2-3）。

有关比较仔猪和生长猪间消化率差异的研究很少，从实际生产考虑，仔猪通常饲喂易消化、低纤维含量饲粮。因此，在能量消化利用方面，仔猪可视为与生长猪相同。生长猪，特别是饲养到较大体重时（即育肥后期），其对能量的利用，理论上适用于各个生长阶段。根据实际生产需要，生长猪和仔猪采用相同能值，而不必考虑体重差异。也就是说，对于饲料被赋予两个消化能值：

一个是仔猪和生长猪阶段，一个是成年母猪阶段（Sauvant等，2004 a，2004 b；表2-7）是很合理的，尤其对消化饲料。鉴于在消化试验中体重对消化率影响比较小，因此大约60 kg体重生长猪的消化率，最能代表断奶-生长-育肥全期的能量消化率情况。

◆ 2.3 消化能转化成代谢能的效率

代谢能等于消化能减去尿能和发酵产生的可燃气体（甲烷和氢）能量。尿能（主要以尿氮形式）源于超过沉积所需的氨基酸脱氨基生成。在特定生产阶段，尿氮排泄量主要取决于饲粮中蛋白质含量。因此，代谢能与消化能之比与饲粮蛋白质含量线性相关（表2-3，公式2-10）。在大多数情况下，全价饲料代谢能与消化能之比接近0.96，但这个平均值不适用于单个的饲料原料。若饲粮蛋白质含量在典型饲粮的蛋白含量范围（10%～25%）之外时，公式2-10不能适用。因此最好的解决方法就是根据公式2-1和2-2通过尿氮（干物质基础，g/kg饲料摄入）预估尿能（干物质基础kJ/kg饲料摄入），公式中可消化氮的沉积效率为50%，总氮的沉积效率为40%（Sauvant等，2004）。

生长猪损失的甲烷能平均占消化能的0.4%（Noblet等，1994 a)。维持水平下的母猪，损失的甲烷能占消化能的比例会更大，达到1.5%（Noblet和Shi，1993）；而摄入高纤维饲粮时，损失的甲烷能甚至能占到3%（Noblet等Shi，1993；Jorgensen等，2001）。一般来说，甲烷生成量随体重和饲粮纤维含量的增加而升高（Noblet和Shi，1993；Jorgensen等，2001），根据Le Goff等（2002 a）综述的文献数据和Noblet实验室尚未发表的数据，生长猪和成年母猪发酵1 g饲粮纤维产生的甲烷能分别是0.67 kJ和1.33 kJ。

◆ 2.4 影响代谢能利用的因素

2.4.1 生理阶段的影响

代谢能利用效率除与热增耗（即在采食、消化和某些机体活动中的能量损失）密切相关外，还与营养物质的转化（如由葡萄糖合成脂肪）相关。代谢能利用效率（1-HI，或*k*）可直接测定，在大多数情况下，是通过回归方法计算得到。*k*值受诸多因素的影响。首先，*k*值取决于能量的最终利用方式，如沉积脂肪的*k*值（*kf*；约80%）高于沉积蛋白质的 *k*值（*kp*；约60%）。其次，若代谢能低于维持需要时，其利用效率与高于维持需要情况下的利用效率不同（Noblet等，1993，1994，1999；表2-8）。当代谢能仅用作维持生命活动，则情况将变得更加复杂。禁食期间，猪动员体内营养物质（蛋白质、脂肪和糖原）供能，用于维持生命活动，其中包括动员体内营养物质所消耗的能量。当饲粮代谢能水平满足动物维持能量需要（MEm）时，全部营养成分均由饲粮提供。在这种情况下，维持能量需要包含了动物采食、消化及吸收过程中的耗能。这意味着FHP与MEm线性关系的斜率（k_m）是相对效率值（即饲粮能量用于维持的效率与动用体内储存能量效率之比），而不是绝对值。因此，当饲粮能量用于维持效率高于动用体内储存能量效率时，k_m将大于1。

表2-8　猪对代谢能的利用效率[1]

阶段	生产目的	代谢能组成	效率（%）	来源
成年	维持	代谢能	77	1
生长 / 妊娠	蛋白质沉积	代谢能	60	2,4
	脂肪沉积	代谢能	80	2,4
	体增重	代谢能	74	3
妊娠	子宫增重	代谢能	50	4
泌乳	产奶	代谢能	72	5
生长	体增重 + 维持	脂肪	90	3
		淀粉	82	3
		蛋白质	58	3
		饲粮纤维	58	3

[1]1 = Nobelt等（1993c），2 = Noblet等（1999），3 = Noblet等（1994b），4 = Noblet和Etienne（1987b），5 = Noblet和Etienne（1987a）。

2.4.2 饲粮组成的影响

若动物特性和饲喂水平均处于标准情况下（见方法论章节），则*k*值因饲粮组成不同而异（表2-3，公式2-11），如*k*值随饲粮脂肪和淀粉含量增加而升高；

反之，随纤维和蛋白质含量增加而降低。对生长猪而言，*k*值变化是因为其对不同营养成分代谢能利用率不同所致，其中脂肪（约90%）和淀粉（约82%）的*k*值最高，而纤维（约60%）和粗蛋白的*k*值最低（Schiemann等，1972；Just等，1983；Noblet等，1994；van Milgen等，2001）。Noblet等（1994）通过测定61种饲粮得到平均*k*值为74%，标准的谷物-豆粕型饲粮的*k*值为75%。动物对不同营养成分能量利用效率存在差异，这意味着粗蛋白或纤维的热增耗高于淀粉或粗脂肪（表2-9）。

表2-9　不同能量体系中淀粉、蛋白质和脂肪的能值[1]

项目	淀粉	粗蛋白质[3]	粗脂肪[3]
消化能（kJ/g）[2]	17.5（100）	20.6（118）	35.5（202）
代谢能（kJ/g）[2]	17.5（100）	18.0（103）	35.3（202）
净能（kJ/g）[2]	14.4（100）	10.2（71）	31.5（219）
产热（kJ/g）	3.1	7.8	3.8

[1]资料来源于Noblet等，1994a；$n = 61$种饲粮。
[2]圆括号内数值表示相对于淀粉的百分比（%）。
[3]假设90%的粗蛋白质或粗脂肪可消化，而淀粉100%可消化。

Noblet等（1994b）用不同体重和不同体增重组成的猪进行了试验，发现在大多数实际情况下，代谢能的利用效率不受体增重组成的影响，不同营养素*k*值的排序基本相似。此外，对维持能量水平情况下的成年母猪，不同营养成分的*k*值排序与生长猪相似，但绝对值略高（Noblet等，1993，1994；表2-3，公式2-12）。van Milgen等（2001）比较了饲粮蛋白质用于蛋白质或脂肪沉积过程中的热增耗，发现两种途径中的热增耗相似，且能量利用效率相同。这表明，无论饲粮蛋白质最终以何种方式被利用，其净能值是恒定的。

综上所述，饲粮粗蛋白含量上升会增加产热量，脂肪则与之相反（Noblet等，2001），所以低粗蛋白或高脂肪饲粮被认定为低热增耗饲粮。而饲粮纤维对产热量的影响，尚无定论。一些研究结果表明，纤维的代谢能利用效率低，产热量随纤维含量增加而升高（Noblet等，1989；Ramonet等，2000；Solund

Olesen等，2001；Rijnen等，2003）。然而，也有研究表明，随饲粮纤维含量增加，产热量恒定不变或降低（Rijnen等，2001；Le Goff等，2002）。从生物化学角度来看，产热量应随纤维含量升高而增加，大多数研究也证实了这个结论。然而，饲粮纤维含量升高会改变动物行为（如生理活动减少）和新陈代谢过程，从而导致产热量降低（Schrama等，1998）。此外，饲粮纤维的效应也与纤维本身特性有关。

2.4.3 气候因素的互作

环境温度高于下限临界温度（lower critical temperature，LCT）时，猪能够通过调整采食来调节热量的释放，以维持体温恒定（见“温度调节的能量需要”部分）。当环境温度低于下限临界温度时，猪会减少向周围环境释放热量，同时提高产热量以维持体温恒定。在这种情况下，热增耗不是能量损失，而是满足维持体温的需要（Quiniou等，2001）。高蛋白质或高纤维饲粮的热增耗高。由于高蛋白饲粮成本高，且对环境有负面影响，故高蛋白质饲粮应用较少。因此，只有应用高纤维饲粮才是满足猪温度调节能量需要的潜在解决方案。表2-10在怀孕母猪上证实了这种可能性。怀孕母猪的特点是特别能适应高纤维饲粮（见“能量的消化利用”），采用限制饲喂，常处于低温环境（< 20℃，低于下限临界温度）。

表2-10　饲粮纤维水平和环境温度对怀孕母猪能量利用的影响[1,2,3]

项目	对照饲粮	+ 稻草	+ 苜蓿
代谢能摄入量（MJ/d）	29.6（100）	32.0（108）	34.0（115）
产热量（MJ/d）			
21.5℃	26.2（100）	27.1（103）	26.9（103）
10.5℃	34.9（100）	34.6（99）	34.5（99）

[1]资料来源于Noblet等，1989b。

[2]圆括号内数值表示占对照饲粮的比例（%）。

[3]所有怀孕母猪采食等量的对照饲粮，试验怀孕母猪每天补充600 g稻草或苜蓿，平均体重为205 kg。

◆ 2.5 能量评价体系

2.5.1 消化能和代谢能体系

饲料原料消化能值除用猪直接测定外，也可从文献（Stein等，2007）或饲料营养价值表中获得（NRC，1998；Sauvant等，2004）。饲料营养价值表数据仅适用于化学组成与实际应用原料相似的饲料原料。在使用消化率或消化能预测方程时，要考虑化学组成不同所带来的影响（Noblet等，2003；www.evapig.com）。如前所述，能量消化率受动物体重的影响，因此，使用消化能值更合适一些。从实用角度来看，建议使用两个消化能值："60 kg"生长猪消化能值，适用于仔猪和生长育肥猪；成年猪消化能值，适用于妊娠和哺乳母猪。饲料营养价值表的数值大多来自30～60 kg生长猪，并不适用于成年猪。根据生长猪和成年猪消化能的差值与生长猪未可消化有机物间的比例关系，Noblet等（2003）等提出了根据生长猪消化能值估测成年猪消化能值的方法（图2-3）。Sauvant等（2004）在其饲料能量价值表中应用了该方法（表2-7）。

配合饲料消化能值可由各种原料的消化能值相加而得，前提是原料间不存在互作。大多数情况下，这种假设是合理的（Noblet和Shi，1994）。若饲粮实际组成未知，消化能值可根据化学指标分析值，利用预测方程进行估测（Noblet和Perez，1993；Le Goff和Noblet，2001）：

$$DE\ (MJ/kg\ DM) = 17.69 + 0.146\ EE + 0.071\ CP - 0.132\ NDF - 0.341 Ash \qquad (2\text{-}13)$$

化学指标分析值以占干物质百分比（%）表示，方程适用于全价饲料而不能用于原料消化能的估测。

近红外技术和体外法也可用于饲料消化能值的估测（Boisen和Fernandez，1997；Noblet和Jaguelin-Peyraud，2007），若可消化营养成分含量和消化率（常量）已知，消化能值可利用下列方程估测：

$$DE\ (MJ/kg\ DM) = 0.232\ DCP + 0.383\ DEE + 0.174 Starch + 0.162\ Sugars + 0.178\ DRes \qquad (2\text{-}14)$$

DCP和DEE指可消化粗蛋白和可消化粗脂肪含量，DRes指可消化的残留物（可消化有机物含量减去方程中涉及的可消化营养成分含量），以占干物质百分比表示（Le Goff和Noblet，2001）。消化能也可根据饲料有机物的化学分析值，通过下列方程直接进行估测：

$$DE = 0.225\ CP + 0.317\ EE + 0.172\ Starch + 0.032\ NDF + 0.163\ Residue \quad (2\text{–}15)$$

“Residue”是有机物含量与方程中涉及的营养成分含量的差值，在所有预测方程中，饲粮纤维对估测值的准确性影响很大。方程2-14和2-15适用于原料或配合饲料，但存在误差，因为没有考虑到饲粮纤维性质和脂肪组成。

猪饲料代谢能值预测方法与消化能值类似。但是，代谢能直接测定不能采用常规方法进行，且代谢能值取决于蛋白质代谢，因此，建议根据消化能值和标准尿能值计算代谢能值（生长猪方程2-5）。

2.5.2 净能体系

目前猪净能体系，是在假设代谢能在维持和沉积效率相近的前提下，结合代谢能在维持、生长（Just等，1983；Noblet等，1994）或育肥（Schiemann等，1972）等方面的利用效率而发布的。荷兰采用的净能体系（CVB，1994），是根据Schiemann等（1972）提出的方程和文献数据改编而来。NRC（1998）采用的净能体系，结合了代表猪各个生产阶段的模型动物（仔猪）的净能直接测定值（Galloway和Ewan，1989）和原料净能测定值（非平衡饲粮）。Emmans（1994）基于对代谢能含量的校正，提出了一个通用模型。Boisen和Verstegen（1998）建议猪饲料净能值应称为“生理能量”，其依据是将体外消化法估测的可消化养分能值与养分潜在生成ATP的生化系数相结合。

Noblet等（1994）提出的净能体系，是根据大量的测定数据（61种饲粮）和回归分析而得。使用该体系计算净能值时，假设FHP是恒定的（750 kJ/kg $BW^{0.60}$/d），则净能值是沉积能量与FHP之和（相关预测方程见表2-11）。该预测方程在Noblet实验室得到更进一步的验证（Le Bellego等2001；Noblet等2001；van Milgen等，2001；Noblet等，2005）。该预测方程以传统饲料营养价值表数据为基础，且适用于猪生产各阶段的配合饲料和单个原料。必须指出的是，当计算生长育肥猪和成年母猪的净能值时，二者应采用不同的消化能值或可消化营养含量，故最终有两个净能值。因此，可靠的能量及其他营养成分的消化率数据，对准确预测猪饲料净能值非常必要。而实际上，能量及其他营养成分消化率数据的可靠性，是准确预测猪饲料净能值的最大限制因素。

表2-11 生长猪饲料消化能、代谢能和净能的估测方程

（61种饲粮；MJ/kg DM，%DM）[1]

编号	方程[2]
2-16	DE = 0.232 × DCP + 0.387 × DEE + 0.174 ST + 0.168 SU + 0.167 DRes
2-17	ME = 0.204 DCP + 0.393 DEE + 0.174 ST + 0.165 SU + 0.154 × DRes
2-18	NEg(2) = 0.121 DCP + 0.350 DEE + 0.143 ST + 0.119 SU + 0.086 DRes
2-19	NEg(4) = 0.703 DE − 0.041 CP + 0.066 EE − 0.041 CF + 0.020 ST
2-20	NEg(7) = 0.730 × ME − 0.028 × CP + 0.055 × EE − 0.041 × CF + 0.015 × ST

[1]资料来源于Noblet等，1994c。

[2]DCP = 可消化粗蛋白，EE = 粗脂肪，DEE = 可消化粗脂肪，ST = 淀粉，SU = 蔗糖，DRes = 可消化残留物（即总可消化有机物与方程中涉及的可消化营养成分之差）。

Noblet和van Milgen（2004）将不同净能体系的局限性和相互之间的可比性进行了综述。简言之，Noblet等（1994）提出的净能体系最适用于预测猪饲料净能值和生产性能。NRC（1998）的净能体系，与其他净能体系提出的净能值差异很大，尤其是对一些饲料原料。

2.5.3 不同能量体系的比较

不同营养成分的代谢能转化为净能的效率差异很大（表2-9和表2-11，方程2-16、2-17和2-18）。因此饲料在消化能或代谢能体系下和在净能体系下的优先顺序不同也是合乎常理的。净能最能代表饲料的真实能值和动物的能量需要。消化能和代谢能体系往往高估了蛋白质和纤维类饲料的能值，而低估了淀粉类和脂肪类原料的能值（表2-12）。

表2-12 生长猪对不同原料消化能、代谢能和净能的相对值[1,2]

项目	消化能	代谢能	净能	净能/代谢能(%)
动物脂肪	243	252	300	90
玉米	103	105	112	80
小麦	101	102	106	78
参考饲粮	*100*	*100*	*100*	*75*
豌豆	101	100	98	73
大豆（全脂）	116	113	108	72

（续）

项目	消化能	代谢能	净能	净能/代谢能(%)
小麦麸	68	67	63	71
豆粕	107	102	82	60

[1]资料来源于Sauvant等（2004）。

[2]各能量体系中，原料能值数据表示为占饲粮（包含小麦、豆粕、脂肪、麦麸、豌豆、矿物质和维生素）能值的百分数。

营养评价体系的优劣取决于其估测动物反应的能力，即根据获得每单位生产性能消耗的饲料量而不是根据饲粮组成来评定。表2-13中数据阐述了动物增重耗能与能量体系之间的关系。该表格数据证实，与消化能和代谢能相比，利用Noblet等（1994）预测方程计算的净能，能够更好地预测动物生产性能。换言之，净能可最准确估测饲料能值。

表2-13 根据不同能量体系和饲粮特征生长育肥猪生产性能的表现[1,2]

项目	消化能	代谢能	净能
脂肪添加（%；试验一）			
0	100	100	100
2	100	100	100
4	99	99	100
6	98	98	100
粗蛋白水平(30 ~ 100 kg; 试验二)			
正常	100	100	100
低	96	97	100
粗蛋白水平(90 ~ 120 kg; 试验三)			
正常	100	100	100
低	97	98	100

[1]资料来源于Noblet（2006）及未发表数据。

[2]相似的日增重或增重组成的能量需要量（或增重消耗的能量），数值表示为相比对照处理的能量需要（认为是100）。

2.5.4 结论

饲料能值首先取决于其化学性质。在消化能水平上，饲料能值主要由纤维（或多或少起到稀释作用）和脂肪（高能）含量决定；代谢能水平上，能值变化主要与饲粮粗蛋白含量有关，饲粮粗蛋白含量可影响尿氮和能量损失；在净能水平上，能值差异主要是因为粗蛋白质含量不同所致。营养成分对能值的影响，可通过表2-14中的方程系数说明。由该表可见，脂肪对净能的贡献最大，而粗蛋白最低；能值还取决于对饲料的加工方式，特别是对饲料脂肪成分的加工处理，对饲料消化率有重要影响；最后，能值还取决于动物类型，成年猪与生长猪相比，饲粮纤维能够提供更多的能量。

表2-14　生长猪饲粮中营养成分对能量的实际贡献情况（kJ/g）[1,2]

编号	项目	蛋白质	脂肪	淀粉	糖	残留物
2-21	总能	22.6	38.8	17.5	16.7	18.6
2-22	消化能	22.5	31.8	18.3	16.1	0.5
2-23	代谢能	19.7	32.2	18.2	15.9	0.5
2-24	净能	11.8	28.9	14.8	11.5	−0.9

[1]根据Noblet等（1994）数据重新计算而得。

[2]测定值经45 kg生长猪采食61种饲粮饲喂而采集，相关系数由多元线性回归方程（无截距）而得，残留物指总有机物与粗蛋白、脂肪、淀粉和糖之和的差值。

3 猪能量需要量

猪能量需要量以不同能量基础来表示。在自由采食情况下，能量需要量通常根据采食量（食欲）、生长潜能、气候因素及经济效益等方面考虑并进行调整。在这种情况下，动物会根据本身的能量需要而调整采食量，因此对其能量需要量精确定量是非常困难的。对限饲的生长猪或繁殖母猪，需要根据预期的生产性能或估计的需要量而设定一个饲喂水平。这些推荐量以平均值体现，不能反映基因型、生产水平、气候环境或动物行为的影响。在一些更复杂的分析方法（析因法或模型）中，能量需要量的各个组成部分（如维持、生理活动、温度调节、生长和产奶）被分别测定。本节将主要讨论这种方法。

在大多数试验和文献中，主要是以消化能（DE）或代谢能（ME）为基础来表示能量需要量。在测定能量需要量时，试验饲粮多采用传统饲粮（即谷物-豆粕基础饲粮）。对生长猪而言，代谢能、消化能转化为净能的效率分别约为75%和72%。因此，净能需要量（以饲粮能量浓度、日能量需要量、能量需要量各组分等来表示）可通过消化能或代谢能需要量乘以0.72或0.75而得。研究表明，对生长猪，代谢能转化为净能的绝对效率值因体重阶段或基因型的不同而略有差异（Noblet等，1994）。对成年母猪，当以维持水平饲喂时，其代谢能转化为净能的绝对效率值高于生长猪（Noblet等，1993）。绝对效率值差异与饲粮特性无关，并且所有营养物质在不同环境下差异大小都是一致的（Noblet等，2006）。这表明，不同生产阶段猪（包括妊娠、哺乳母猪和生产潜力不同的生长猪）净能需要量可通过相似方法计算而得。由于净能的估测方程大多来源于生长猪，且这些方程适用于猪的其他生长阶段，因此净能需要量是根据“生长猪”净能值来表示的（Noblet等，2006）。然而，生长猪净能值在不同体重或生理阶段有差异，因此建议使用两个净能值：生长猪净能值（包括仔猪）和成年母猪净能值（包括妊娠或哺乳母猪）（Sauvant等，2004）。

◆ 3.1 维持能量需要

一般认为维持能量需要（MEm，以代谢能表示）与代谢体重（BW^b）呈比例关系。生长猪 *b* 值最宜为0.60（Noblet等，1994a，1999），该指数（0.60）优于常用的代谢体重指数（0.75）。对处于环境温度为热中性区的生长猪，维持代谢能需要为1 MJ/kg $BW^{0.60}$/d（Noblet等，1999）。由于代谢能转化为净能的平均效率为75%，因而维持净能需要为0.750 MJ/kg$BW^{0.60}$/d。该值是很多学者根据相关试验测定的FHP的平均值（Le Bellego等，2001；van Milgen等，2001；Noblet等，2001；Le Goff等，2002；de Lange等，2006；Lovatto等，2006；Barea等，2010；Koong等，1982；Tess等，1984）。

以每千克$BW^{0.60}$表示维持能量需要时，猪不同生长阶段维持能量需要几乎是恒定的，品种或性别间差异很小。维持能量需要仅仅在极端品种间存在差异，如生长速度慢或脂肪型猪（如梅山猪）维持能量需要量低，生长速度快和瘦肉型猪（Noblet等，1999）及生长激素处理的猪（Noblet等，1992）维持能量需要量高。因此，大多数猪维持能量需要在标准条件下（即常规畜舍、热中性区环境和接近自由采食）可认为是相同的。维持能量需要包括0.200 MJ ME/kg $BW^{0.60}$/d的标准生理活动水平的耗能，其中大约一半用于站立（每天约4h），其余用于躺卧时的活动（van Milgen和Noblet，2003）。1 MJ ME/$BW^{0.60}$/d的维持能量需要是在呼吸测热室内（生理活动降低，热中性区环境）获得的。因此，较高的维持能量需要（1.05 MJ ME/kg $BW^{0.60}$/d）适用于饲养于常规畜舍、生理活动较多的猪（表2-15），但不适用于哺乳仔猪（Noblet和Etienne，1987）和早期断奶仔猪（Noblet和Le Dividich，1982）。这些特定阶段维持能量需要，应根据早期生长阶段体温调节和生理活动方面的数据，采用更适宜的方法重新计算。

表2-15　猪的能量需要量[1]

阶段	能量需要（MJ）	来源[2]
生长	维持能量需要 = 1.05 × kg $BW^{0.60}$	1
	体增重能量需要 = 23.0 × 蛋白质增重（kg）+ 39.9 × 脂肪增重（kg）	2
	瘦肉组织增重能量需要 = 8.5 ～ 10.5 MJ/kg	2
	脂肪组织增重能量需要 = 31 ～ 33 MJ/kg	2
	代谢能温度调节见图 2-5	3
妊娠	维持能量需要 = 0.440 × kg $BW^{0.75}$	4
	母体增重能量需要 = 9.7 × 体增重（kg）+ 54 × P2 增加（mm）	5
	子宫增重能量需要 = 4.8 × 胎重增加（kg）	6
	站立 100 min 代谢能需要 = 0.035 × kg $BW^{0.75}$	7
	温度调节能量需要（℃）=（0.010 ～ 0.020）× kg $BW^{0.75}$	8
泌乳	维持能量需要 = 0.460 × kg $BW^{0.75}$	9
	产奶能量需要 = 20.6 × 窝增重（kg）− 0.376 × 窝仔数	10

[1]能量转化效率见表2-8；BW = 体重。

[2]1 = Noblet等（1991，1999），2 = Noblet等（1999）及Karege（1991），3 = Quiniou等（2001），4 = Noblet和Etienne（1987b），5 = Dourmad等（1996，1997，1998），6 = Noblet等（1985b），7 = Noblet等（1993a），8 = Noblet等（1989b；环境温度低于20℃），9 = Noblet和Etienne（1987a），10 = Noblet和Etienne（1989）。

繁殖母猪维持能量需要与代谢体重成比例关系，代谢体重指数采用经典数值0.75。在热中性区和“标准”活动水平下，妊娠和哺乳母猪测定的维持能量需要见表2-15。哺乳母猪由于生产水平更高，故其维持能量需要高于妊娠母猪。妊娠母猪由于生理活动水平变异很大，故其维持能量需要也变异很大（Noblet等，1993；见“生理活动能量消耗”部分）。

◆ 3.2 生长能量需要量

从营养角度来看，生长体现为蛋白质、脂肪、矿物质和水分的沉积，并伴随蛋白质和脂肪沉积的代谢能需要。体蛋白质和脂肪沉积代谢能需要量（ME_p），可通过蛋白质和脂肪沉积量及代谢能用于沉积蛋白质和脂肪的效率（k_p和kf）来进行估测。对于传统谷物-豆粕型饲粮，Noblet等（1999）提出k_p和kf分别是60%和80%。体蛋白质和体脂肪能量含量大约是23.8 kJ/g和39.5 kJ/g，因此沉积1 g蛋白质或脂肪代谢能需要量为40 kJ或50 kJ。

从技术和经济角度来看，体组织的生长，如胴体瘦肉组织增加伴随脂肪组织降低，是非常重要的。对体增重部分瘦肉和脂肪组织化学组成的测定及相关饲料能量成本的计算表明，脂肪增重的饲料成本是蛋白质增重的3.5倍（表2-16）。生长猪在其能量含量和组织生长所需能量方面存在较大差异，结果导致体增重代谢能需要量直接取决于增重部分瘦肉与脂肪比率或脂类含量。在大多数实际猪生产中，增重部分的蛋白质含量是相对恒定的（为16%～17%）。

表2-16 生长猪组织和体增重的化学组成以及组织增重的能量需要[1]

组成	公猪			去势公猪		
	瘦肉[2]	脂肪	eBW	瘦肉	脂肪	eBW
水（%）	69.9	18.7	58.5	65.6	14.9	51.6
灰分（%）	1.0	0.2	3.1	1.0	0.2	3.0
蛋白质（%）	17.9	5.4	16.7	18.2	4.1	16.0
脂肪（%）	10.2	75.4	21.1	15.3	81.8	30.4
能量（kJ/g）	8.5	31.3	12.3	10.4	33.3	15.6

（续）

组成	公猪			去势公猪		
	瘦肉[2]	脂肪	eBW	瘦肉	脂肪	eBW
代谢能需要（kJ/g）[3]	12.2	39.8	17.2	14.9	42.6	21.6

[1]资料来源于Noblet等（1994）及未发表数据；公猪体重阶段在20～95 kg，基于比较屠宰法测定。

[2]瘦肉包括肌内脂肪；eBW为空腹体重。

[3]蛋白质和脂肪分别按40 MJ/g和50 MJ/g计算。

生长猪增重部分化学和组织组成取决于多种因素，本章不详细讨论。简言之，瘦肉型猪相比脂肪型猪、公猪相比母猪或阉公猪、低体重猪相比高体重生长育肥猪、能量限制猪相比自由采食猪，体增重部分能量含量要低（Campbell和Taverner，1988；Bikker等，1996；Noblet等，1994；Quiniou等，1999；表2-6）。体重为20～100 kg，空腹体增重部分的能量含量范围，从瘦肉型猪的10 MJ/kg到脂肪型猪的20～22 MJ/kg（Noblet等，1994）。总之，这会促使人们通过遗传选育或营养调控来降低生长猪体脂肪含量。

◆ 3.3 繁殖的能量需要

Noblet等（1989，1997）综述了怀孕母猪的能量需要量。妊娠阶段能量需要量等于维持、子宫生长和机体储备重建等方面能量需要量之和。特定条件下，如额外的生理活动或处于低温环境中，额外能量需要也必须予以考虑。表2-15列出了估测妊娠母猪能量需要量的依据。对单栏饲养的妊娠母猪，其下限临界温度是20～22℃；而对使用秸秆垫料或群饲的妊娠母猪，其下限临界温度会更低一些。实际生产中，大约2/3能量用于满足妊娠母猪的维持需要，而特殊的妊娠需要（如子宫组织增长）对能量需要量增加的影响可忽略不计。然而，如果要考虑因额外代谢体重而增加的维持能量需要，维持能量需要会更高，额外代谢体重是指子宫增长和乳腺发育。母体组织的能量需要由目标体增重及其组成所决定，由于断奶时母猪体况不同，经产母猪需要达到的目标体增重不尽相同。

总之，妊娠母猪能量需要量因体重、舍饲条件及配种时体况的不同而异

（Dourmad等，2008）。因此，在群饲情况下，如果妊娠母猪的饲喂量与其他猪相同，会导致生产性能发生较大变异，尤其是分娩时的体况。实际上，维持能量需要、子宫生长的能量需要、温度调节和生理活动的能量需要是要优先满足的。那么，母体组织中能量沉积直接取决于饲料能量供给量与优先满足能量需要量间的差值。对体况较差的母猪，生理活动增加或处于温度较低的环境中，均会导致能量沉积降低（Noblet等，1997）。因此，生理活动或行为变化能够显著影响妊娠母猪的能量平衡。通常认为子宫生长遵循一个指数曲线（Noblet等，1985），即妊娠的前2/3时期，子宫组织能量需要量低，而妊娠后1/3阶段能量需要量增加。根据妊娠阶段体重的增加，维持需要量将日益增加。因此，如果妊娠期日采食量保持不变，沉积在妊娠组织中的能量会大幅下降，甚至在妊娠最后2～3周出现能量负平衡（Dourmad等，1988；Young等，2004）。妊娠后期能量的额外需要表明在此阶段需要提高饲喂水平。妊娠后期增加饲喂量还能为分娩后采食量的快速增加做准备（表2-17）。

表2-17 妊娠阶段对母猪能量利用和生理活动的影响[1]

项目	妊娠阶段（周）		
	5～6	9～10	14～15
体重（kg）	182	207	224
能量平衡（MJ/d）			
代谢能摄入量	28.6	28.4	28.8
产热量	22.6	23.1	26.4
能量保留量			
总计	6.0	5.2	2.4
子宫	0.4	1.3	2.6
母体组织	5.7	4.2	−0.2
蛋白质形式	2.5	2.1	2.7
脂肪形式	3.5	3.2	−0.3
站立时间（min）	288	263	247
生理活动产热量（MJ/d）	5.7	6.2	6.9

[1]资料来源于Young等，2004；$n=12$头母猪。

影响哺乳母猪能量需要的主要因素是产奶量。母猪的产奶量通常很难测定，一般通过窝增重估测（Noblet和Etienne，1989；表2-15）。产奶量取决于母猪的基因潜能、窝仔猪数、哺乳天数（Etienne等，1998；Noblet等，1998）。哺乳期能量需要为维持需要和产奶需要之和。代谢能转化为奶能的效率平均为72%（表2-15）。图2-4阐述了窝增重对哺乳母猪能量需要量的影响。当生产水平非常高时（平均窝增重大于3 000 g/d），饲料需要量将超过8 kg/d。该计算方法也表明，额外的代谢能需要量与额外的窝增重成正比，每千克窝增重平均需要26 MJ代谢能，或者大约2 kg饲料。这种方法能简单方便地估算哺乳母猪的能量需要：维持能量需要1.9 ~ 2.2 kg饲料（200 ~ 250 kg母猪），再加上每千克窝增重需要2 kg饲料。在大多数实际情况下，哺乳母猪不能采食到足够饲料以满足其能量需要，因此哺乳期体重会下降。初产母猪能量不足和伴随的体重下降在生产中影响更大（这部分内容超出了本章范围）。因此，哺乳母猪自由采食非常重要。饲喂高能饲粮（通过降低饲粮纤维和增加脂肪含量）可以增加母猪能量摄入量。然而，很大一部分额外能量以乳脂的形式排出，对母猪本身直接效果小（Noblet等，1998）。

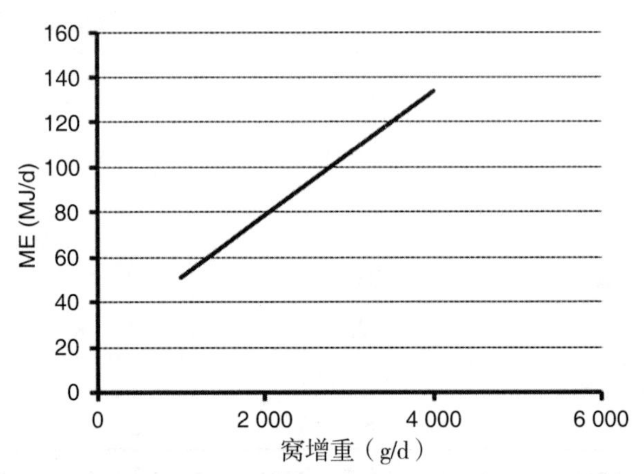

图2-4 窝增重对哺乳母猪能量需要量（MJ ME/d）的影响

母猪体重220 kg，哺育6（窝增重1000 g/d）至13头（窝增重>3000 g/d）仔猪，哺乳期21 d。

◆ 3.4 生理活动所需的能量

如上所述，生理活动所消耗的能量无法精确估测。从方法学角度来看，生理活动所耗能量是热增耗不可控的变异来源，可导致能量需要量的预测不准确，尤其是在显著影响动物行为的情况下。即使是在舍内（活动较少、站立时间较短），猪生理活动消耗的能量也占总能量消耗的比例很大，因为猪每单位生理活动量消耗的能量比其他畜禽高4～5倍（Noblet等，1993）。表2-18为Noblet实验室总结的相关数据。虽然少量的生理活动不可避免，并且包括在维持能量需要中，但是还需要考虑特定生理活动下的能量需要量，如妊娠母猪或户外饲养猪的常规行为。妊娠期是猪活动量最多，且活动量变异最大的阶段。Noblet实验室研究表明，每多站立1 min，热增耗会增加0.30 kJ/kg $BW^{0.75}$（Noblet等，1993；Ramonet等，2000；Le Goff等，2002；Young等，2004)。如表2-17所示，母猪站立时间从50 min到500 min时，由此导致的饲料需要量差异约为700 g/d。通常来说，妊娠母猪生理活动占总能量需要的比例较高（占代谢能摄入量的20%），并且变异较大（占代谢能摄入的10%～40%）。生理活动的变异性是导致母猪分娩时体况差异较大的重要原因。然而，在生长猪上，采食量接近自由采食水平时，热增耗变异小，占代谢能摄入的比例也小（8%～10%，表2-18）。

表2-18　猪生理活动的产热量

项目	生长阶段	仔猪	生长猪			怀孕母猪
	圈舍条件	群饲	群饲	群饲	单栏饲喂	单独饲喂
	饲喂方式	自由采食	自由采食	自由采食	受控饲喂	限饲
环境温度（℃）		23	19 ～ 22	12	24	24
体重（kg）		27	62	61	62	260
代谢能摄入量（MJ/d）		21.7	31.1	33.5	29.2	35.6
产热量（MJ/d）		11.2	17.9	19.7	16.9	29.5
生理活动产热量(MJ/d)		2.0	2.3	3.3	2.5	6.7
产热量（%）		17.9	12.8	16.9	14.7	22.6
代谢能摄入量(%)		9.2	7.4	10.0	8.5	18.7
来源[1]		1	2	2	3	4

[1]1 = Collin等（2001a），2 = Quiniou等（2001），3 = Le Bellego等（2001a），4 = Ramonet等（2000）。

◆ 3.5 体温调节能量需要

当环境温度低于下限临界温度时，猪产热量增加以维持其体温。Noblet等（2001）综述了猪生产中体温调节的概念和下限临界温度值。当环境温度低于下限临界温度时，产热量增加所需能量一般通过提高采食量来补偿，因此自由采食时猪体增重能够维持在正常水平（Quiniou等，2001；见“猪能量摄入调节”章节）。对仔猪而言，在新生期和断奶后头几天，下限临界温度一般较高，分别为32～34℃和26～28℃。因此，在这两个阶段，仔猪更易受到冷应激（Noblet和Le Dividich，1981，1982)。在其他生长阶段，下限临界温度较低（20～24℃），体温调节所需能量取决于饲养环境（如舍内或舍外、地板类型、群体规模）和采食量。生长育肥猪维持生产性能所需的额外采食量见图2-5。妊娠母猪由于下限临界温度较高（>22℃），且圈舍设施简陋，故经常处于低于下限临界温度的环境中。当妊娠母猪所处环境温度低于下限临界温度时，环境温度每降低1℃，产热量增加10～20 kJ/kg $BW^{0.75}$（Geuyen等，1984）。对单栏饲喂和/或较瘦的母猪，产热量相对更高（Noblet等，1997）。对于体重为200 kg母猪，环境温度每下降1℃产生的冷应激而增加的产热量，需要消耗大约70 g饲料来补偿。哺乳母猪由于采食量和生产水平高，产热量也高，故其下限临界温度较低（<15℃）。此外，供暖措施可为仔猪供暖并提高成活率，因此低温对于哺乳母猪来说并不是问题。目前，总体来说，很多国家通过改善圈舍保温条件和建筑质量，从而降低了冷应激影响。但是，在热带和亚热带国家或温带国家的夏季，热应激产生的影响却越来越大。热增耗在寒冷时期并不是能量的浪费，其有助于满足体温调节的能量需要（Quiniou等，2001；表2-10）。因此，从实际角度来讲，高热增耗饲粮（如高纤维饲粮）在低温环境下比适温和高温下更有优势（Noblet等，1985，1989，2001）。

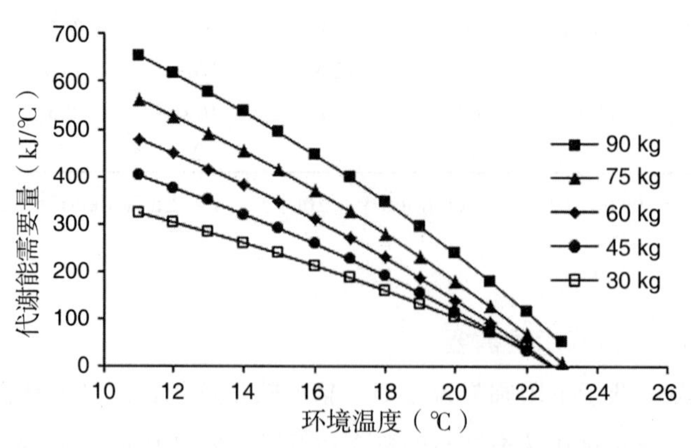

图2-5　不同体重（30～90 kg）生长猪温度调节的能量需要量

4 猪能量摄入反应

猪的生长取决于其自身因素（体重、性别、基因等）、营养供给及气候环境。在生长（Campbell和Taverner，1988；Quiniou等，1999；van Milgen等，2008）和妊娠阶段（Dourmad等，1996，2008），能量在蛋白和脂肪沉积之间分配。在哺乳阶段，能量优先用于产奶，甚至动用体内储备。母猪在能量负平衡情况下仍能维持产奶量，前提是体内储备消耗不能过多（Noblet等，1998）。

根据析因法的观点，能量会优先满足维持需要，其次是蛋白质沉积，最后是脂肪沉积。但这种优先次序并不非常恰当，因为蛋白质和脂肪沉积之间存在相互关系。能量摄入不足时，蛋白和脂肪的沉积同时受到限制，由此看来，脂肪沉积是必不可少的。研究表明，蛋白沉积和能量摄入之间存在线性-平台或者曲线-平台的关系，而脂肪增重与能量摄入之间接近于线性关系。蛋白沉积最大时的斜率难以测定（图2-6）。如前所述，蛋白质和脂肪的沉积速度与瘦肉和脂肪组织沉积速度及体增重密切相关（表2-19）。其重要性体现在：①单位代谢能（g/MJ ME）对脂类或脂肪组织的增重效果明显高于蛋白质或瘦肉组织，因此，随能量水平的增加，胴体脂肪组织含量上升（表2-20）；②在蛋白沉积的线性反应阶段，瘦肉型、公猪（与阉猪相比）以及小猪（与大猪相比）蛋白质沉积的斜率更大；③随能量摄入增加，蛋白沉积速度会达到平台期（PDmax），能量如果继续增加将会被用于脂肪沉积。Black等（1986）、Quiniou等（1999）、van Milgen和Noblet（2003）对以上内容进行了综述，其中部分内容见表2-19。妊娠母猪也表现出相似的结果（表2-19），其蛋白质沉积速度低于育肥猪。蛋白质沉积对能量摄入的反应（斜率及PDmax）同样也受环境温度影响。高温环境下，蛋白质沉积曲线的斜率降低，而脂肪沉积的斜率则有轻微增加（Le Bellego等，2002）。因此，在高温情况下，采食量降低，理论上来说更倾向于蛋白沉积。但是实际上，对自由采食的猪，在适温或高温条件下的胴体脂肪含量相似。

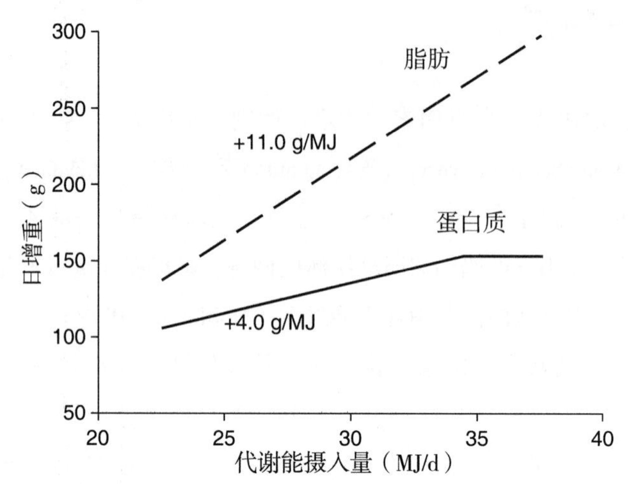

图2-6 45～100 kg阉公猪不同代谢能摄入量下蛋白质和脂肪的增重情况（综合两个阉公猪品种数据）

表2-19 猪对能量摄入的反应

项目	生长猪		怀孕母猪
	小公猪	阉猪	
体重范围（kg）	45 ～ 100	45 ～ 100	205[1]
增加代谢能带来的变化（g/MJ）			
蛋白质	6.1	4.7	2.3
脂肪	13.2	13.2	NA[2]
瘦肉	21.0	16.5	10.0
脂肪组织	9.7	9.7	12.0
体增重	36.0	28.5	24.0
来源	1	1	2

[1]配种时体重。

[2]NA表示为无变化。

[3]1＝Quiniou等（1996a，1996b；大白 × 皮特兰杂交后代），2＝Dourmad等（1996）。

5 生长猪饲料转化效率

从技术层面衡量，饲料转化效率是养猪生产效率的重要指标，由单位增重所需饲料或能量（料肉比，F：G）计算而得。饲料转化效率也可由以下公式2-21计算：

$$F:G = (ME\ intake / [ME]) / (Energy\ gain / [E]_{ADG})$$

$$F:G = (ME\ intake / [ME]) / ((ME\ intake - MEm) \times kg / [E]_{ADG})\qquad (2\text{-}21)$$

$$F:G = (1/[ME]) \times (1/k_g) \times [E]_{ADG} \times (FL/[FL-1])$$

式中，ME代表饲粮代谢能水平；k_g代表代谢能用于增重的效率（表2-8），$[E]_{ADG}$代表体增重部分的能量浓度，FL代表饲喂水平（为维持能量需要的倍数）。此公式表明，当饲料代谢能水平增加，或FL增加（意味着更低比例的能量用于维持需要）时，F：G降低。但是，如前所述，当能量摄入量增加时，$[E]_{ADG}$增加，从而导致F：G增加。因此，FL对$[E]_{ADG}$和FL/[FL-1]的影响是相反的。当FL在相当大的范围内变动时，F：G保持相对稳定（表2-20）。但是，当饲料采食量高时，尤其是超过蛋白质沉积速度达到平台期所需的能量水平时（图2-6），$[E]_{ADG}$增加非常快，FL/[FL-1]的影响则变小，因此F：G值变大（表2-20）。相反，当采食量很低时，FL/[FL-1]的影响很大，从而导致F：G值增加。实际上，F：G值最低时的代谢能摄入量通常低于自由采食水平，尤其是对蛋白质沉积较低或食欲较大的猪而言。因此，建议轻度控制能量的摄入，特别是在育肥阶段。这也表明，在特定FL水平下，体增重部分脂肪含量下降，导致$[E]_{ADG}$下降，F：G值也随之降低。总之，改善F：G的最佳手段就是通过调节体增重中脂肪-蛋白比例来降低胴体的脂肪含量。

表2-20 能量供应对生长猪生产性能和体组成的影响（45～100 kg BW）[1,2]

项目	能量供应（MJ ME/d）				
	22.6	26.7	29.4	32.2	37.6
体增重（g/d）	622	738	820	931	1013
饲料成本（MJ ME/kg 体增重）	36.4	36.2	35.8	34.6	37.1

（续）

项目	能量供应（MJ ME/d）				
	22.6	26.7	29.4	32.2	37.6
体蛋白质含量[3]	17.2	16.6	16.5	16.3	16.0
体瘦肉含量[3]	56.1	54.2	53.6	53.7	52.6
体脂含量[3]	18.6	21.0	22.0	22.4	22.8
脂肪组织含量[3]	12.0	14.2	15.1	15.2	15.7

[1] 资料来源于Quiniou等（1996）。

[2] BW＝体重。

[3] 屠宰时占空腹体重（即体重－肠道内容物）比例，相当于活体重95%。

F：G的大小取决于饲粮代谢能和体重范围。在实际生产过程中，25～100 kg猪的F：G为2.5～3.0，而且公猪低于母猪，母猪低于阉公猪（表2–21）。这意味着猪每采食1 kg饲粮可增重350～400 g，相当于1MJ代谢能可增重25～30 g（表2–22）。由于养猪的主要目的是生产瘦肉，因此提高单位饲料的瘦肉产量或能量在瘦肉中的沉积非常重要。表2–22列出公猪和阉公猪生长推荐值（15～17 g瘦肉增重/MJME）。这些数据表明，生长猪摄入的代谢能，约40%和13%分别转化为体增重和瘦肉增重的形式而储存。

表2–21 公猪、阉猪和后备母猪的生长性能比较[1]

项目	公猪	后备母猪	阉猪
采食量（kg/d）	2.41	2.45	2.70
体增重（g/d）	1 069	988	1 032
饲料转化效率（kg/kg）	2.26	2.48	2.62

[1] 资料来源于Quiniou等，2010；生长阶段为63～152日龄。

表2-22 去势对猪生长效率的影响（10～100 kg体重）[1]

项目	公猪	阉猪[2]	
代谢能摄入（MJ/d）	35.1	37.1	（106）
体增重（g/d）	1 096	1 014	（92）
蛋白质增加（g/d）	150	144	（96）
脂肪增加（g/d）	232	255	（110）
体增重（g/MJ 代谢能）	31.2	27.3	（87）
瘦肉增重（g/MJ代谢能）	16.9	15.0	（89）
体能量增加（MJ/MJ代谢能）	0.39	0.40	（101）
瘦肉能量增加（MJ/MJ代谢能）	0.14	0.13	（91）

[1]资料来源于Quiniou等，1995；Noblet，未发表数据。

[2]括号内数据表示为占公猪生长效率的比例（%）。

世界上大部分地区，为避免肉质变差而把公猪去势。但是，有些地区出于动物福利和经济效益等方面考虑，而不把公猪去势（如澳大利亚、英国）。公猪也可在屠宰前几周采取免疫去势（Dunshea等，2001）。母猪、阉公猪、公猪的能量利用效率数据表明，饲料采食量取决于性别（公猪=母猪<阉公猪），其中公猪的饲料成本最低（表2-21）。去势使猪沉积蛋白质的能力降低，脂肪沉积能力增加。此外，由于阉公猪食欲增加，导致脂肪沉积更加明显。公猪每单位饲料能量摄入的体增重、瘦肉增重、瘦肉能量增益较高。阉公猪每单位能量摄入量的体能量增益较高，其原因是阉公猪摄入的能量大部分以脂肪的形式储存（表2-22）。在阉公猪上的这些结果表明，饲料效率的改善并不意味着能量利用效率的提高；同样，基因的改良也不意味着胴体瘦肉率和生长速度的改进。

6 猪能量摄入的调节

自由采食情况下，评价猪是否具有采食足够的饲料或能量来满足需要或达到目标生长速度、蛋白质沉积及脂肪沉积的能力非常重要。本节不讨论猪采食量调节的所有方面，而只是简要地介绍影响能量摄入量的几个主要的与猪有关的因素，如体重、生理阶段、性别，以及主要的环境因素，如饲料能量浓度和环境温度。

表2-23描述了仔猪、生长猪及哺乳母猪在常规饲养环境下的采食模式。简单地说，随体重增加，猪每天的采食次数减少，而且猪在所有生长阶段均表现为昼行性（夜晚活动量占总活动量的比例不足1/3）。对体重大的猪或哺乳母猪，其昼行性表现得更为明显（图2-7），如这些猪在早上和傍晚有两个采食高峰期。这种采食行为也受环境条件的影响，如在白天高温、夜晚凉爽的情况下，猪在夜晚的采食行为可能更多，这在哺乳母猪上更明显（Quiniou等，2000；Renaudeau等，2003）。而妊娠母猪通常限饲，饲喂后很短时间内就将饲料采食殆尽，除非每天只饲喂一次纤维含量高、体积大的饲粮。

表2-23 猪的采食行为

项目	生长阶段	仔猪	生长猪				哺乳母猪
	品种	杂交猪	眉山猪	皮特兰猪	杂交猪		杂交猪
	圈舍条件	群饲	单独饲喂	单独饲喂	群饲	群饲	单独饲喂
体重范围（kg）		20 ~ 30	20 ~ 60	20 ~ 60	30 ~ 90	30 ~ 90	270
环境温度（℃）		23	24	24	19 ~ 22	29	22
采食量（g/d）		1 502	1 659	1 622	2 395	1 820	6 600
日采食次数		14.4	14.4	7.3	11.2	10.1	7.4
每次采食量（g）		114	125	250	248	205	972
白天采食量(%)		67	61	64	65	62	80
来源[1]		1	2	2	3	3	4

[1]1 = Collin等（2001），2 = Quiniou等（1999），3 = Quiniou等（2000），4 = Quiniou等（2000）。

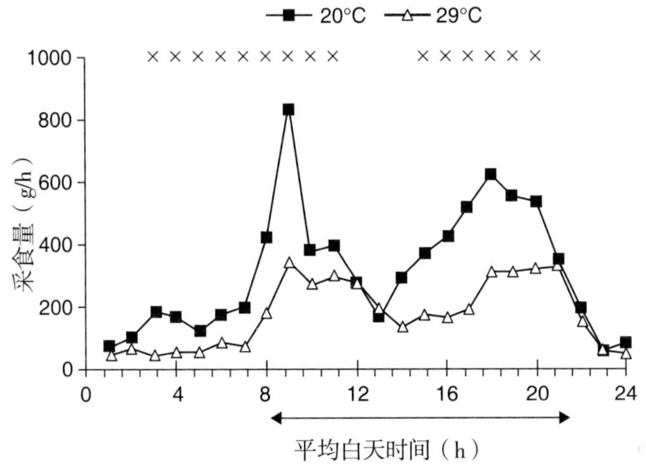

图2-7 温度对大白猪哺乳母猪每日采食动力学的影响

（资料来源于Renaudeau等，2002）

生长猪的自由采食量随体重的增加呈二次线型增加（NRC，1998；图2-8），增加速度受猪生长潜力的影响（如基因、性别；Quiniou等，1999a）。同样，采食量增加的速度也受环境温度的影响。高温时增加的速度变慢，这也意味着相比体重小的猪，体重大的猪的采食量更容易受热应激的影响（Nienaber等，1997；图2-8）。哺乳母猪自由采食量取决于体型或胎次，初产母猪采食量较低（O'Grady等，1985；Dourmad等，1994；Neil等，1996）。与生长猪类似，哺乳母猪对热应激特别敏感，高温会显著降低采食量（Schoenherr等，1989）。由于哺乳母猪采食量很高，以至于在适温区其采食量也会受到环境温度的影响（图2-9）。环境温度越高，气温每升高1℃，猪采食量下降的幅度越明显。在20～25℃时，环境温度每升高1℃，哺乳母猪采食量会降低200 g；在25～30℃时，采食量会降低500 g（Quiniou和Noblet，1999）。而对25 kg的仔猪，相应的值分别为10 g/℃和30 g/℃（Collin等，2001）；60 kg生长猪对应的值分别为40 g/℃和70 g/℃（Quiniou等，2000）。高温对采食量的负面影响在热带地区的高湿度环境下会更加严重（Renaudeau等，2003）。一般来说，高温条件下生长猪和哺乳母猪更难将所产热量散发出去，从而有中暑的风险。因此，适宜的措施包括降低能量摄入和减少饲料代谢利用过程中不可避免的产热。这些措施对哺乳母猪的效果更明显。

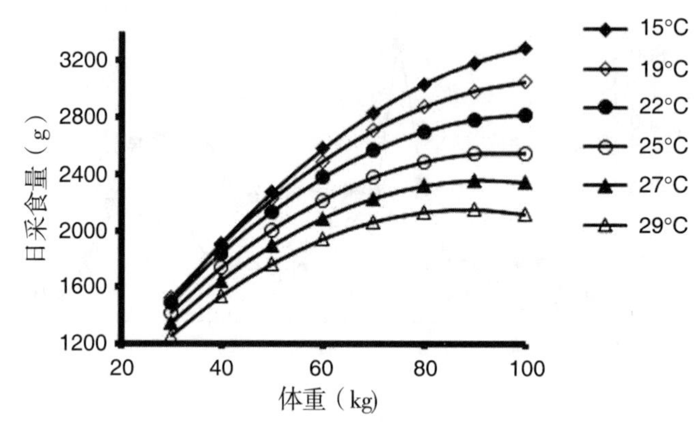

图2-8　体重（BW）和环境温度对生长猪自由采食量的影响

（数据来源于Quiniou等，2000a）

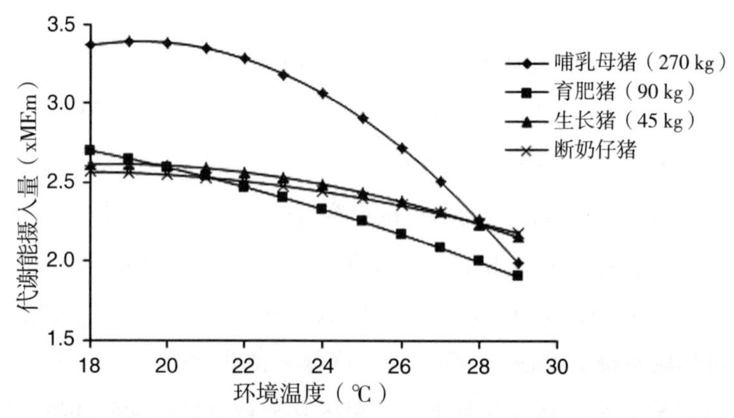

图2-9　环境温度对仔猪（Collin等，2001）、生长猪（Quiniou等，2000）和哺乳母猪（Quiniou和Noblet，1999）采食量的影响（采食量以维持代谢能需要量的倍数来表示）

饲粮能量浓度可通过添加纤维性饲料原料或脂肪含量高的原料来调节（表2-12）。在实际生产中，能量摄入量是限制猪生长的一个重要因素。因此，采食量、生长速度、饲料转化率、体组成和饲料能量水平之间的关系，一直以来受到极大关注。有综述文献总结了相关数据：在每个研究中，至少设置四个能量水平，并且蛋白质和能量的比例尽可能保持一致。在大部分研究中，猪在良好的环境条件下单独饲喂。能量浓度的增加通常会降低采食量，但采食量的降低通常没有能量浓度增加的影响大，因此能量摄入量总是增加的（图2-10）。但在大部分试验中，消化能的摄入量在饲粮能量浓度达到一定水平后会进入平台期，或者说试验中最低梯度的能量浓度都很高时，消化能摄入量随着饲粮能量浓度水平升高增加的幅度也就微乎其微。与此结果相吻合的是，体增重在同样高能量浓度时也达到平台期。饲粮能量浓度对能量摄入量的影响在哺乳母猪中同样也能观察到，但在母猪上较少受到关注，其原因是多余的能量会以乳脂形式分泌（Noblet等，1998）。饲粮能量浓度促进能量摄入量增加的正面效果，可以在热应激情况下生长猪和哺乳母猪的饲养管理中使用。高能量饲粮能缓解高温对猪生长性能的影响（Le Bellego等，2002；Renaudeau等，2002）。

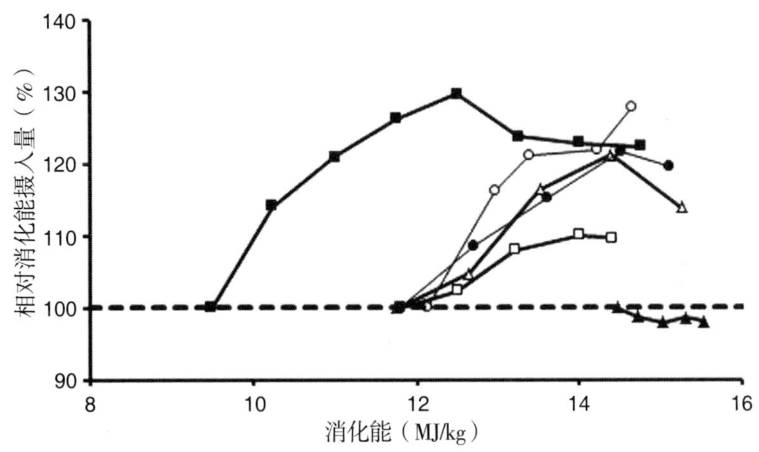

图2-10 消化能水平对生长猪自由采食时能量摄入量的影响（以占每个研究中最低消化能水平的比例来表示）

（资料来源于Chadd，1999；Stein等，1999；Smith等，1999；Campbell和Taverner，1989；Noblet，未发表数据）

7 结 语

能量体系是建立在饲料具有能值这一概念的基础上，因此，饲料能值与动物能量需要可在同等条件下进行比较。然而，本章的阐述表明，实际情况要复杂得多，如同一种饲料因猪的品种或饲料加工技术不同而导致能值存在差异。根据能量利用过程，不同的能量体系被提出。在饲料配制方面，净能是迄今为止最能反映饲料真实“能值”和动物真正“能量需要”的体系。在实际情况下，能量供应、环境和动物之间存在相互作用，解决复杂互作的唯一途径就是建立模型（Whittemore和Fawcett，1976；Black等，1986；Birkett和de Lange，2001；van Milgen等，2008）。尽管在建立模型方面已取得了很大进展，但是猪营养模型在不同情况下的变异非常大。为使模型能在世界范围内广泛应用，营养模型必须在精确性和稳健性之间找到平衡。鉴于此，颠覆本章所提及的能量营养的经典概念还为时尚早。

作者：Jean Noblet和Jaap Van Milgen

译者：张桂杰

参考文献

Barea, R., S. Dubois, H. Gilbert, P. Sellier, J. van Milgen, and J. Noblet. 2010. Energy utilization in pigs selected for high and low residual feed intake. J. Anim. Sci. 88:2062–2072.

Brouwer, E. 1965. Report of subcommittee on constants and factors. Page 441 in Proc. 3rd Symp. Energy Metabolism. K. L.Blaxter, ed. Academic Press, London.

Bikker, P., M. W. A. Verstegen, and R. G. Campbell. 1996. Performance and body composition of finishing gilts (45–85 kg) as affected by energy intake and nutrition in earlier life. II. Protein and lipid accretion in body components. J. Anim. Sci. 74:817–826.

Birkett, S., and K. de Lange. 2001. Limitations of conventional models and a conceptual framework for a nutrient flow representative of energy utilization by animals. Br. J. Nutr. 86:647–659.

Black, J. L., R. G. Campbell, I. H. Williams, K. J. James, and G. T. Davies. 1986. Simulation of energy and amino acid utilization in the pig. Res. Dev. Agric. 3:121–145.

Boisen, S., and M. W. A. Verstegen. 1998. Evaluation of feedstuffs and pig diets. Energy or nutrient–based evaluation systems? Ⅱ. Proposal for a new nutrient–based evaluation system. Acta Agric. Scandinavica, Sec. A Anim. Sci. 40:86–94.

Campbell, R. G., and M. R. Taverner. 1988. Genotype and sex effects on the relationship between energy intake and protein deposition in growing pigs. J. Anim. Sci. 66:676–686.

Chabeauti, E., J. Noblet, and B. Carré. 1991. Digestion of plant cell walls from four different sources in growing pigs. Anim. Feed Sci. Techn. 32:207–213.

Collin, A., J., van Milgen, and J. Le Dividich. 2001b. Modelling the effect of high, constant temperature on feed intake in young growing pigs. Anim Sci. 72:519–527.

CVB. 1994. Veevoedertabel. Centraal Veevoederbureau in Nederland, Lelystad, The Netherlands.

Cozannet, P., Y. Primot, C. Gady, J. P. Métayer, M. Lessire, F. Skiba, and J. Noblet. 2010. Energy value of wheat distillers grains with solubles for growing pigs and adult sows. J. Anim. Sci. 88:2382–2392.

de Lange, K., J. van Milgen, J. Noblet, S. Dubois, and S. Birkett. 2006. Previous feeding level influences plateau heat production following a 24 h fast in growing pigs. Br. J. Nutr. 95:1082–1087.

Dourmad, J. Y. 1991. Effect of feeding level in the gilt during pregnancy on voluntary feed intake during lactation and changes in body composition during gestation and lactation. Livest. Prod. Sci. 27:309-319.

Dourmad, J. Y., M. Etienne, A. Prunier, and J. Noblet. 1994. The effect of energy and protein intake of sows on their longevity. Livest. Prod. Sci. 40:87-97.

Dourmad, J. Y., M. Etienne, and J. Noblet. 1996. Reconstitution of body reserves in multiparous sows during pregnancy: Effect of energy intake during pregnancy and mobilization during the previous lactation. J. Anim. Sci. 74:2211-2219.

Dourmad, J. Y., J. Noblet, M. C. Père, and M. Etienne. 1998. Mating, pregnancy and pre-natal growth. Pages 129-153 in A *Quantitative Biology of the Pig*. I. Kyriazakis, ed. CAB International, Wallingford, UK.

Dourmad, J. Y., M. Etienne, A. Valancogne, S. Dubois, J. van Milgen, and J. Noblet. 2008. INRAPorc: A model and decision support tool for the nutrition of sows. Anim Feed Sci. Technol. 143:372-386.

Dunshea, F. R., C. Colantoni, K. Howard, I. McCauley, P. Jackson, K. A. Long, S. Lopaticki, E. A. Nugent, J. A. Simons, J.Walker, and D. P. Hennessy. 2001. Vaccination of boars with a GnRH vaccine (Improvac) eliminates boar taint and increases growth performance. J. Anim. Sci. 79:2524-2535.

Emmans, G. 1994. Effective energy: A concept of energy utilization applied across species. Br. J. Nutr. 71:801-821.

Etienne, M., J. Y. Dourmad, and J. Noblet. 1998. The influence of some sow and piglet characteristics and of environmental conditions on milk production. Pages 285-299 in *The Lactating Sow.* M.W. A. Verstegen, P. J. Moughan, and J.W. Schrama, eds. Wageningen Pers, Wageningen, The Netherlands.

Fernandez, J. A., H. Jorgensen, and A. Just. 1986. Comparative digestibility experiments with growing pigs and adult sows. Anim. Prod. 43:127-132.

Galloway, S. T., and R. C. Ewan. 1989. Energy evaluation of tallow and oat groats for young swine. J. Anim. Sci. 67:1744-1750.

Geuyen, T. P. A., J. M. F. Verhagen, and M.W. A. Verstegen. 1984. Effect of housing and temperature on metabolic rate of pregnant sows. Anim. Prod. 38:477-485.

Jorgensen, H., K. E. Bach Knudsen, and P. K. Theil. 2001. Effect of dietary fibre on energy metabolism of growing pigs and pregnant sows. Pages 105-108 in Energy Metabolism in Animals. A. Chwalibog and K. Jakobsen, eds. EAAP Publ. No. 103. Wageningen Pers, Wageningen, The Netherlands.

Just, A., H. Jorgensen, and J. A. Fernandez. 1983. Maintenance requirement and the net energy value of different diets for growth in pigs. Livest. Prod. Sci. 10:487-506.

Karege, C. 1991. Influence de l'âge et du sexe sur l'utilisation de l'énergie et la composition corporelle du porc en croissance. PhD. Diss. Université de Montpellier, France.

Koong, L. J., J. A. Nienaber, J. C. Pekas, and J. T. Yen. 1982. Effects of plane of nutrition on organ size and fasting heat production in pigs. J. Nutr. 112:1638–1642.

Labussière, E, G. Maxin, S. Dubois, J. van Milgen, G. Bertrand, and J. Noblet. 2009. Effects of feed intake on protein and fat deposition in milk fed veal calves. Animal 3:557–567.

Labussière, E., J. van Milgen, C. de Lange, and J. Noblet. 2011. Maintenance energy requirements of growing pigs and calves are influenced by feeding level. J. Nutr. 141:1855–1861.

Le Bellego, L., J. van Milgen, and J. Noblet. 2001. Energy utilization of low protein diets in growing pigs. J. Anim. Sci. 79:1259–1271.

Le Bellego, L., J. van Milgen, and J. Noblet. 2002. Effect of high temperature and low protein diets on performance of growing–finishing pigs. J. Anim. Sci. 80:691–701.

Le Bellego, L., J. van Milgen, J. Noblet. 2002. Effects of high temperature on protein and lipid deposition and energy utilization in growing pigs. Anim. Sci. 75:85–96.

Le Dividich, J., and J. Noblet. 1986. Effect of dietary energy level on the performance of individually housed early–weaned piglets in relation to environmental temperature. Livest. Prod. Sci. 14:255–263.

Le Goff, G., and J. Noblet. 2001. Comparative digestibility of dietary energy and nutrients in growing pigs and adult sows. J. Anim. Sci. 79:2418–2427.

Le Goff, G, L. Le Groumellec, J. van Milgen, and J. Noblet. 2002. Digestive and metabolic utilization of dietary energy in adult sows: influence of level and origin of dietary fibre. Br. J. Nutr. 87:325–335.

Lovatto, P., D. Sauvant, J. Noblet, S. Dubois, and J. van Milgen. 2006. Effects of feed restriction and subsequent re–feeding on energy utilization in growing pigs. J. Anim. Sci. 84: 3329–3336.

Neil, M., B. Ogle, and K. Anner. 1996. A two diet system and ad libitum lactation feeding of the sow. 1. Sow performance. Anim. Sci. 62:337–347.

Nienaber, J. A., G. L. Hahn, T. P. McDonald, and R. L.Korthals. 1996. Feeding patterns and swine performance in hot environments. Trans. ASAE 39:195–202.

Noblet, J., and J. Le Dividich. 1981. Energy metabolism of the newborn pig during the first 24 hrs of life. Biol. Neonate 40:175–182.

Noblet, J., and J. Le Dividich. 1982. Effect of environmental temperature and feeding level on energy balance traits in early weaned piglets. Livest. Prod. Sci. 9:619–632.

Noblet, J., J. Le Dividich, and T. Bikawa. 1985a. Interaction between energy level in the diet and environmental temperature on the utilization of energy in growing pigs. J. Anim. Sci.

61:452–459.

Noblet, J., W.H. Close, R.P. Heavens, and D. Brown. 1985b. Studies on the energy metabolism of the pregnant sow. Uterus and mammary tissue development. Br. J. Nutr. 53:251–265.

Noblet, J., and M. Etienne. 1987a. Metabolic utilization of energy and maintenance requirements in lactating sows. J. Anim. Sci. 64:774–781.

Noblet, J., and M. Etienne. 1987b. Metabolic utilization of energy and maintenance requirements in pregnant sows. Livest. Prod. Sci. 16:243–257.

Noblet, J., Y. Henry, and S. Dubois. 1987. Effect of protein and lysine levels in the diet on body gain composition and energy utilization in growing pigs. J. Anim. Sci. 65:717–726.

Noblet, J., J.Y. Dourmad, J. Le Dividich, and S. Dubois. 1989. Effect of ambient temperature and addition of straw or alfafa in the diet on energy metabolism in pregnant sows. Livest. Prod. Sci. 21:309–324.

Noblet, J., and M. Etienne. 1989. Estimation of sow milk nutrient output. J. Anim. Sci. 67:3352–3359.

Noblet, J., J. Y. Dourmad, and M. Etienne. 1990. Energy utilization in pregnant and lactating sows: Modelling of energy requirements. J. Anim. Sci. 68:562–572.

Noblet, J., P. Herpin, and S. Dubois. 1992. Effect of recombinant porcine somatotropin on energy and protein utilization in growing pigs: Interaction with capacity for lean tissue growth. J. Anim. Sci. 70:2471–2484.

Noblet, J., and J. M. Perez. 1993. Prediction of digestibility of nutrients and energy values of pig diets from chemical analysis. J. Anim. Sci. 71:3389–3398.

Noblet, J., and X. S. Shi. 1993. Comparative digestibility of energy and nutrients in growing pigs fed ad libitum and adult sows fed at maintenance. Livest. Prod. Sci. 34:137–152.

Noblet, J., X. S. Shi, and S. Dubois. 1993a. Energy cost of standing activity in sows. Livest. Prod. Sci. 34:127–136.

Noblet, J., X. S. Shi, and S. Dubois. 1993c. Metabolic utilization of dietary energy and nutrients for maintenance energy requirements in pigs: Basis for a net energy system. Br. J. Nutr. 70:407–419.

Noblet, J., C. Karege, and S. Dubois. 1994a. Prise en compte de la variabilité de la composition corporelle pour la prevision du besoin énergétique et de l'efficacité alimentaire chez le porc en croissance. Journées Rech. Porcine en France 26:267–276.

Noblet, J., H. Fortune, X. S. Shi, and S. Dubois. 1994b. Prediction of net energy value of feeds for growing pigs. J. Anim. Sci. 72:344–354.

Noblet, J., X. S. Shi, and S. Dubois. 1994c. Effect of body weight on net energy value of feeds for growing pigs. J. Anim. Sci. 72:648–657.

Noblet, J., and K. E. Bach Knudsen. 1997a. Comparative digestibility of wheat, maize and sugar beet pulp non-starch polysaccharides in adult sows and growing pigs. Pages 571–574 in Digestive Physiology in Pigs. J. P. Laplace, C. Février, and A. Barbeau, eds. INRA, Paris.

Noblet, J., J. Y. Dourmad, M. Etienne, and J. Le Dividich. 1997b. Energy metabolism in pregnant sows and newborn pigs. J. Anim. Sci. 75:2708–2714.

Noblet, J., M. Etienne, and J. Y. Dourmad. 1998. Energetic efficiency of milk production. Pages 113–130 in The Lactating Sow. M. W. A. Verstegen, P. J. Moughan, and J. W. Schrama, eds. Wageningen Pers, Wageningen, The Netherlands.

Noblet, J., C. Karege, S. Dubois, and J. van Milgen. 1999. Metabolic utilization of energy and maintenance requirements in growing pigs: Effect of sex and genotype. J. Anim. Sci. 77:1208–1216.

Noblet, J., and G. Le Goff. 2001. Effect of dietary fibre on the energy value of feeds for pigs. Anim. Feed Sci. Technol. 90:35–52.

Noblet, J., J. Le Dividich, and J. van Milgen. 2001. Thermal environment and swine nutrition. Pages 519–544 in Swine Nutrition. 2nd ed. A. J. Lewis and L. L. Southern, eds. CRC Press, Boca Raton, FL.

Noblet, J., L. Le Bellego, J. van Milgen, and S. Dubois. 2001. Effects of reduced dietary protein level and fat addition on heat production and nitrogen and energy balance in growing pigs. Anim. Res. 50:227–238.

Noblet, J., B. Sève, and C. Jondreville. 2004. Nutritional values for pigs. Pages 25–35 in Tables of Composition and Nutritional Value of Feed Materials: Pigs, Poultry, Cattle, Sheep, Goats, Rabbits, Horses, Fish. D. Sauvant, J. M. Perez, and G. Tran, eds. Wageningen Academic Publishers, Wageningen and INRA ed., Versailles, The Netherlands.

Noblet, J., and J. van Milgen. 2004. Energy value of pig feeds: Effect of pig body weight and energy evaluation system. J. Anim. Sci. 82(E. Suppl.):E229–E238.

Noblet, J. 2006. Recent advances in energy evaluation of feeds for pigs. Pages 1–26 in Recent Advances in Animal Nutrition 2005. P. C. Garnsworthy, and J. Wiseman, eds. Nottingham Univ. Press, Nottingham, UK.

Noblet, J. 2007. Recent developments in energy and amino acid nutrition of pigs. Pages 21–48 in Gaining the Edge in Pork and Poultry Production: Enhancing Efficiency, Quality and Safety. J. A. Taylor-Pickard and P. Spring, eds.Wageningen Academic Publishers, Wageningen, The Netherlands.

Noblet, J., and Y. Jaguelin-Peyraud. 2007. Prediction of digestibility of organic matter and energy in the growing pig from an *in vitro* method. Anim. Feed Sci. Technol. 134:211–222.

NRC. 1998. Nutrient Requirements of Swine. 10th rev. ed. Natl. Acad. Press, Washington, DC.

O'Grady, J. F., P. B. Lynch, and P. A. Kearney. 1985. Voluntary feed intake by lactating sows. Livest. Prod. Sci. 12:355-365.

Quiniou, N., J. Noblet, J. van Milgen, and J. Y. Dourmad. 1995. Effect of energy intake on performance, nutrient and tissue gain and protein and energy utilisation in growing boars. Anim. Sci. 61:133-143.

Quiniou, N., J. Y. Dourmad, and J. Noblet. 1996a. Effect of energy intake on the performance of different types of pig from 45 to 100 kg body weight. 1. Protein and lipid deposition. Anim. Sci. 63:277-288.

Quiniou, N., J. Y. Dourmad, and J. Noblet. 1996b. Effect of energy intake on the performance of different types of pig from 45 to 100 kg body weight. 2. Tissue gain. Anim. Sci. 63:289-296.

Quiniou, N., and J. Noblet. 1999. Influence of high ambient temperatures on performance of multiparous lactating sows. J. Anim. Sci. 77:2124-2134.

Quiniou, N., S. Dubois,Y. Le Cozler, J. F. Bernier, and J. Noblet. 1999a. Effect of growth potential (body weight and breed/castration combination) on the feeding behaviour of individually-kept growing pigs. Livest. Prod. Sci. 61:13-22.

Quiniou, N., J. Noblet, J. Y. Dourmad, and J. van Milgen. 1999b. Influence of energy supply on growth characteristics in pigs and consequences for growth modelling. Livest. Prod. Sci. 60:317-328.

Quiniou, N., J. Noblet, and S. Dubois. 2000a. Voluntary feed intake and feeding behaviour of group-housed growing pigs are affected by ambient temperature and body weight. Livest. Prod. Sci. 63:245-253.

Quiniou, N., D. Renaudeau, S. Dubois, and J. Noblet. 2000b. Influence of high ambient temperatures on food intake and feeding behaviour of multiparous lactating sows. Anim. Sci. 70:471-479.

Quiniou, N., J. Noblet, J. vanMilgen, and S. Dubois. 2001. Modelling heat production and energy balance in group-housed growing pigs exposed to cold or hot ambient temperatures. Br. J. Nutr. 85:97-106.

Quiniou, N., V. Courboulay, Y. Salaün, and P. Chevillon. 2010. Conséquences de la non castration des porcs mâles sur les performances de croissance et le comportement: comparaison avec les mâles castrés et les femelles. Journées Recherche Porcine 42:113-118.

Ramonet, Y., J. van Milgen, J. Y. Dourmad, S. Dubois, M. C. Meunier-Salaün, and J. Noblet. 2000. The effect of dietary fibre on energy utilisation and partitioning of heat production over pregnancy in sows. Br. J. Nutr. 84:85-94.

Renaudeau, D., N. Quiniou, and J. Noblet. 2001. Effect of high ambient temperature and dietary protein level on performance of multiparous lactating sows. J. Anim. Sci. 79:1240-1249.

Renaudeau, D., and J. Noblet. 2001. Effect of exposure to high ambient temperature and dietary protein level on sows milk production and performance of piglets. J. Anim. Sci. 79:1540–1548.

Renaudeau, D., J. L. Weisbecker, J. Noblet. 2003. Effects of season and dietary fibre on feeding behaviour of lactating sows in a tropical climate. Anim. Sci. 77:429–437.

Renaudeau, D., B. Bocage, and J. Noblet. 2006. Influence of energy intake on protein and lipid deposition in Creole and Large White growing pigs. Anim. Sci. 82:937–945.

Sauvant, D, J. M. Perez, and G. Tran. 2004. Tables of Composition and Nutritional Value of Feed Materials: Pigs, Poultry, Cattle, Sheep, Goats, Rabbits, Horses, Fish. Wageningen Academic Publishers, Wageningen, The Netherlands.

Schoenherr, W. D., T. S. Stahly, and G. L Cromwell. 1989. The effects of dietary fat or fiber addition on yield and composition of milk from sows housed in warm or hot environment. J. Anim. Sci. 67:482–495.

Schiemann, R., K. Nehring, L. Hoffmann, W. Jentsch, and A. Chudy. 1972. Energetische Futterbevertung und Energienormen. VEB Deutscher Landwirtschatsverlag, Berlin, Germany.

Skiba, F., J. Noblet, P. Callu, J. Evrard, and J. P. Melcion. 2002. Influence du type de broyage et de la granulation sur la valeur énergétique de la graine de colza chez le porc en croissance. Journées Rech. Porcine en France 34:67–74.

vanMilgen, J., J. Noblet, S. Dubois, and J. F. Bernier. 1997. Dynamic aspects of oxygen consumption and carbon dioxide production in swine. Br. J. Nutr. 78:397–410.

van Milgen, J., J. F. Bernier, Y. Le Cozler, S. Dubois, and J. Noblet. 1998b. Major determinants of fasting heat production and energetic cost of activity in growing pigs of different body weight and breed/castration combination. Br. J. Nutr. 79:509–517.

van Milgen, J., and J. Noblet. 2000. Modelling energy expenditure in pigs. Pages 103–114 in Modelling Nutrient Utilization in Farm Animals. J. P. McNamara, J. France, and D.E. Beever, eds. CAB International, Oxon, UK.

van Milgen, J., J. Noblet, and S. Dubois. 2001. Energetic efficiency of starch, protein, and lipid utilization in growing pigs. J. Nutr. 131:1309–1318.

van Milgen, J., and J. Noblet. 2003. Partitioning of energy intake to heat, protein and fat in growing pigs. J. Anim. Sci. 81(E. Suppl. 2):E86–E93.

van Milgen, J., A. Valancogne, S. Dubois, J. Y. Dourmad, B. Sève, and J. Noblet. 2008. InraPorc: A model and decision support tool for the nutrition of growing pigs. Anim Feed Sci. Technol. 143:387–405.

Whittemore, C. T., and R. H. Fawcett. 1976. Theoretical aspects of a flexible model to simulate protein and lipid growth in pigs. Anim. Prod. 22:87–96.

第3章

脂类及猪对脂类的利用

1 导 语

猪饲粮脂类种类繁多，化学性质各异，但有一个共同的特点，即不溶于水，溶于乙醚、氯仿、苯等非极性溶剂。脂类可分为两大类，即简单脂类和复合脂类。简单脂类包括脂肪酸（fatty acid，FA）、甘油三酯、类固醇、前列腺素和蜡质等；复合脂类包括磷脂、糖脂和脂溶性维生素等。

脂类具有重要的营养生理功能。例如，脂类（如脂肪酸、甘油三酯）是动物体内能量储存和转运的主要形式；皮下脂肪具有防止体热散失和防止机械损伤的作用；脂类是细胞膜和生物活性成分（如维生素和激素）的重要组成成分，参与细胞代谢、细胞识别和细胞免疫等生理活动。

饲粮脂类，通常也称为饲粮脂肪，主要包括甘油三酯和脂肪酸。其他一些脂类虽然具有重要的生物学功能，但在饲粮中含量很低。猪饲粮中常见脂肪酸的碳原子数一般为14～22，其中以16和18个碳原子数为主。饲粮中大部分不饱和脂肪酸含有顺式双键。猪和其他哺乳动物可以通过Δ-9去饱和酶把饱和脂肪酸变成不饱和脂肪酸，但是这些动物缺乏催化ω-3或ω-6位去饱和的酶。因此，亚油酸（C18：2ω-6）和α-亚麻酸（ALA；C18：3ω-3）都是植物来源的多不饱和脂肪酸（PUFA），为动物的必需脂肪酸。饲粮脂肪被猪摄入后，会进一步通过去饱和及延长碳链来产生长链多不饱和脂肪酸（LCPUFA），如花生四烯酸（ARA；C20：4ω-6）、二十碳五烯酸（EPA；C20：5ω-3）和二十二碳六烯酸（DHA；C22：6ω-3）。这些长链多不饱和脂肪酸可进一步合成其他具有重要生物活性的脂类。

脂类在猪营养中的基本作用和对猪生长性能的影响，前人已经进行了系统的介绍（Azain，2001），因此此处不再赘述。本章将重点介绍脂类在猪营养和代谢中特殊的作用，包括对猪肉品质的影响和在生物医学应用方面的最新研究进展。此外，需要说明的是，有关脂类对免疫功能和猪肉品质的影响，在本书的其他章节（第8章和17章）会进行系统阐述。

2 猪饲粮中脂肪的主要来源

饲粮脂肪通常指的是饲粮中的甘油三酯。这些酯类可以以植物油（如豆油）、动物油（如牛油或猪油）或者以动、植物油组合的形式添加到饲粮中。脂类也是饲料原料的组成成分之一（表3-1）。在美国，玉米通常占饲粮组分的65%~80%，而玉米含有3.5%~4.0%乙醚浸出物（即粗脂肪）。谷物类（大麦、玉米、燕麦、小麦）通常含有2%~4%乙醚浸出物（风干基础）。油粕类（菜籽粕、棉籽粕和豆粕）一般是采用溶剂浸提法生产的，其乙醚浸出物含量低于2%。植物性饲料原料，不管是谷物、油粕还是植物油，所含的脂肪酸通常为十八碳脂肪酸，含有1~3个双键（表3-2）。用油菜籽、橄榄油和花生生产加工而成的产品富含油酸（18：1ω-9）。用玉米、棉籽、燕麦、红花、大豆和向日葵生产加工而成的产品富含亚油酸。用亚麻籽加工而成的产品富含α-亚麻酸。

表3-1 一些常用饲料原料的脂肪含量

原料	脂肪含量（%）
面包店副产物	11.3
大麦	1.9
菜籽粕（溶剂浸提法）	3.5
玉米	3.9
棉籽粕（溶剂浸提法）	1.5
玉米酒精糟液（DDGS）	9.0
鱼粉（鲱鱼）	9.4
燕麦	4.0
家禽副产物粉	12.6
豆粕（溶剂浸提法）	1.0
小麦	2.0
干乳清粉	0.9

资料来源于NRC，1998。

表3-2 常用饲料原料的主要脂肪酸（占总脂肪酸的百分比，%）组成和碘价（风干基础）

来源	16：0	18：0	18：1ω-9	18：2ω-6	18：3ω-3	碘值
植物来源						
玉米	12.3	3.5	20.2	60.4	1.9	128
DDGS	13.8	4.2	23.3	54.5	2.5	122
油菜籽	3.9	1.9	64.1	18.7	9.2	190
大豆	11.0	4.0	23.4	53.2	7.8	130
亚麻籽	4.8	4.7	19.9	15.9	52.7	180
动物来源						
黄脂膏	16.2	10.5	47.5	17.5	1.9	75
牛油	24.9	18.9	36.0	3.1	0.6	44
精选白脂膏	21.5	14.9	41.1	11.6	0.4	60
禽油	21.6	6.0	37.3	19.5	1.0	78

DDGS指玉米酒精糟液。

资料来源于NRC，1998；White，2000。

猪饲料中动物来源的饲料原料包括血液制品、肉骨粉、家禽副产物粉、乳清粉和鱼粉。这些饲料原料一般作为蛋白源，但是家禽副产品和鱼粉中脂肪含量很高。其他动物来源的原料，如牛油、精选白脂膏和禽油，主要作为脂肪来源，这些原料中其他营养素含量很低或不含其他营养素。

不同植物来源的饲料原料，其脂肪含量差异较大。谷物类原料（大麦、玉米、燕麦和小麦）中总脂肪含量为2%～4%（风干基础），主要含有不饱和脂肪酸，其中亚油酸含量最为丰富。植物性蛋白源（如豆粕和菜籽粕）中的脂肪已被浸提，因此其脂肪很低。

其他饲料原料，如面包店副产物或DDGS中的脂肪含量相对较高。

碘价常被用于测定脂肪酸的不饱和度。碘价指的是每100 g脂肪所能结合的碘的克数。碘价是影响猪肉脂肪品质的一个重要的因素（详见第17章）。

3 猪在脂类利用方面的生物学特性

众所周知，脂类参与了机体的能量平衡和细胞膜的组成，促进了机体组织的生长发育。然而，一些特殊的脂肪酸，尤其是长链多不饱和脂肪酸及其代谢产物的生物学功能尚未研究清楚。目前研究已证实，在猪的生长发育过程中，这些特殊的脂肪酸对能量代谢、肠道发育、免疫功能和基因表达调控等均发挥着重要作用。有关这方面的研究可拓宽我们对生物活性脂类功能的认识，从而促进养猪业的进步，并为生物医学的发展奠定基础。

◆ 3.1 猪的脂类代谢特点

与其他动物相比，猪在脂类代谢方面具有独特的特点。虽然乳脂是哺乳仔猪氧化代谢的主要底物（占饲粮能量的60%）（Girard等，1992），但是新生仔猪氧化脂肪酸的能力很低。研究表明，新生仔猪肝细胞摄取的90%的油酸会被重新酯化，而不是通过β-氧化途径氧化供能。因此尽管乳脂含量是提高的，但是哺乳仔猪仍然会出现低酮血症（Pégorier等，1981；Adams等，1997）。Pégorier等（1981）发现，哺乳仔猪血液中的酮体浓度低于0.25 mmol/L。血液中酮体浓度低，是由于酮体合成量减少，而不是酮体利用率增加所致。通常，生理浓度的

β-羟基丁酸提供的能量仔猪能量需要量的比例不足5%（Tetrick等，1995）。

新生仔猪利用脂肪酸的能力较差，其原因可能有两个方面：一是脂肪酸氧化途径的关键酶的基因表达量低，二是这些关键酶具有独特的蛋白结构（Nicot等，2001）。肝中肉碱棕榈酰基转移酶Ⅰ（carnitine palmitoyltransferase Ⅰ，CPTⅠ）为脂肪酸氧化的限速酶，该酶负责把脂肪酸从胞浆转运至线粒体。与其他哺乳动物相比，新生仔猪的CPTⅠ有一个非典型结构。猪的CPTⅠ蛋白为哺乳动物肝型-CPTⅠ（L-CPTⅠ）和肌肉型CPTⅠ（M-CPTⅠ）的天然杂合体，其L-CPTⅠ部分有酰基辅酶A结合位点，M-CPTⅠ部分有肉碱和丙二酰辅酶A的结合位点（丙二酰辅酶A为CPT Ⅰ的生理性抑制剂）（Nicot等，2002）。除了CPTⅠ的结构独特外，酮体生成的关键酶——线粒体3-羟基-3-甲基戊二酰辅酶A合成酶（mHMGCS）的活性极低，从而导致仔猪酮体生成量低。mHMGCS的酶活性低，并不是因为mRNA表达量低，而是由于翻译率低所致（Barrero等，2001）。

仔猪有相当高比例的脂肪酸氧化在过氧化物酶体中进行（Yu等，1997）。在仔猪肝细胞中，40%以上的β-氧化是在过氧化物酶体中进行的。这表明，猪对脂肪酸氧化代谢的调控与其他哺乳动物存在相当大的差异。过氧化物酶体增殖物激活受体α（PPARα）在促进脂肪酸氧化方面发挥着重要作用。PPARα控制着L-CPTⅠ、mHMGCS和酰基辅酶A氧化酶（ACO）等脂肪酸氧化关键基因的表达。激活PPARα可促进这些关键基因的转录，进而促进过氧化物酶体和线粒体的脂肪酸氧化。在猪上，PPARα内源性配体和药物clofibrate可以通过激活PPARα，从而促进这些关键基因的表达（Yu等，1997；Peffer等，2005），但是却不导致过氧化物酶体的增殖（Cheon等，2005）。

此外，猪肝脏中脂肪酸的合成量相当少，脂肪的从头合成主要在脂肪组织中进行。事实上，脂肪组织中脂肪合成酶的mRNA表达量比肝脏中高7倍（Ding等，2000）。这表明，猪肝脏中脂肪酸合成及脂肪转运方式与其他品种动物存在差异。弄清这种差异对全面了解动物的脂肪代谢，以及调节激素和转录因子在脂肪代谢中的作用至关重要（Bergen和 Mersmann，2005）。与其他动物相比，新生仔猪脂肪合成能力较低。随着日龄的增长，猪的脂肪合成能力逐步提高。因此，出栏猪绝大部分的脂肪是在育成阶段沉积的（Azain等，2001）。温度可影响猪的脂肪沉积。在高环境温度中，猪肝脏和脂肪组织中脂类代谢会增强，从而促进血浆甘油三酸酯的吸收和在脂肪组织中储存，进而导致猪沉积更多脂肪（Kouba等，2001）。

◆ 3.2 长链多不饱和脂肪酸对猪发育的影响

尽管长链多不饱和脂肪酸和必需脂肪酸占总脂肪的比例很小，但是对新生仔猪的发育却是至关重要的。这些脂肪酸对大脑、视网膜（Carlson等，1999；Novak等，2008）和其他机体组织的生长发育发挥着重要作用。研究发现，人的发育和功能上的一些缺陷与组织中长链多不饱和脂肪酸的含量低存在一定的相关性。然而，目前有关满足猪生长和发育的长链多不饱和脂肪酸最佳需要量尚不清楚。NRC（1998）推荐了猪饲粮中亚油酸（ω-6 PUFA的前体）和亚麻酸（ω-3 PUFA的前体）的需要量，但长链多不饱和脂肪酸的需要量尚不清楚。近年来的研究发现，在动物出生之后，循环系统和各个器官中的脂肪酸组成发生了很大变化，表现为亚油酸和ARA含量增加，而棕榈油酸和油酸含量下降。这种发育性变化受母猪饲粮脂肪酸组成的显著影响。

在妊娠和哺乳母猪饲粮中添加ω-3长链多不饱和脂肪酸，可显著提高血浆、乳汁和生殖器官（子宫内膜、孕体）ω-3长链多不饱和脂肪酸的含量（Brazle等，2009；Farmer等，2009；Bazinet等，2003），从而改变新生仔猪和断奶仔猪大脑和肠道的脂肪酸组成，进而改变仔猪大脑和肠道的结构和生理功能（Rooke等，2001a；Gabler等，2009；Farmer等，2009；Boudry等，2009；Bazinet等，2003）。在妊娠母猪饲粮添加ω-3长链多不饱和脂肪酸，可促进孕体发育，提高仔猪出生前后的存活率（Rooke等，2001b；Spencer等，2004）。此外，在怀孕母猪饲粮中长期添加ω-3长链多不饱和脂肪酸，能有效提高断奶仔猪对葡萄糖的吸收，其机制可能与仔猪肠道脂肪酸组成的改变有关。因此，肠道脂肪酸组成的改变，可能有利于新生仔猪迅速适应出生后饲粮的变化，有利于营养物质的吸收（Gabler等，2009）。

在仔猪配方奶粉和饲粮中添加长链多不饱和脂肪酸，可改变仔猪组织的脂肪酸组成（Dullemeijer等，2008；Hess等，2008）。ω-3脂肪酸含量的变化会影响ω-6脂肪酸（尤其是ARA）的含量（Rooke等，2001a），进而使组织中ω6：ω3的比例降低（Farmer和Petit，2009）。ω6：ω3比例降低所造成的影响目前尚不清楚，但据报道，ω6：ω3比例可影响仔猪肠道刷状缘膜脂质组成、流动性和酶活性（Daveloose等，1993）。从分娩前第8天开始，给怀孕母猪饲喂低ω6：ω3饲粮，可有效改善母猪分娩后第1天的采食量，这与低ω6：ω3饲粮缓解围产期炎症反应有关（Papadopoulos，2009a，2009b）。然而，给哺乳仔猪饲喂含高水平ω-6 PUFA的饲粮，会降低DHA在仔猪大脑中的沉积，可能会导致神经发育受损（Novak等，2008）。

Brazle 等（2009）研究了饲粮中添加长链多不饱和脂肪酸对组织（孕体和子宫内膜）脂肪酸组成和生物学功能的时间依赖性的影响。他们发现，在配种前30 d添加长链多不饱和脂肪酸，可能是母猪饲粮脂肪酸组成影响孕体发育和成活的一个关键时期（Brazle等，2009）。组织长链多不饱和脂肪酸的富集也受长链多不饱和脂肪酸添加剂量的影响。然而，组织长链多不饱和脂肪酸的最大富集量所需的时间，因组织和长链多不饱和脂肪酸的不同而不同（Hess等，2008）。哺乳仔猪添加EPA后第8天，虽然肠上皮细胞中EPA的富集量达到最大（接近20倍），然而EPA的最大富集量只是ARA最大富集量的50%。

◆ 3.3 长链多不饱和脂肪酸对猪免疫系统的影响

有关长链多不饱和脂肪酸在缓解猪炎症反应、改善免疫应答方面的生物学功能，Leskanich和Noble（1999）已进行了详细的阐述。在妊娠后期和泌乳母猪饲粮中添加ω-3 长链多不饱和脂肪酸，可使ω-3 长链多不饱和脂肪酸在仔猪的免疫组织和母猪的乳汁中富集（Fritsche，1993a，1993b；Mitre等，2005），使仔猪肺泡巨噬细胞释放前列腺素E（PGE）、血栓素B和白细胞三烯B的量降低。ω-3 PUFA在组织中富集，还可增强胰岛素的敏感性，改善糖尿病患者的高血糖症，其原因可能与ω-3 PUFA的免疫调节作用直接相关（Nettleton和Katz，2005；Huang等，2009）。母体将ω-6或ω-3 PUFA转移到胎儿，以及通过母乳将PUFA转移到仔猪，可增加仔猪血液中白细胞、免疫球蛋白G（IgG）和伪狂犬病抗体水平，从而提高仔猪被动和主动免疫水平（Mitre等，2005）。然而，这种改善可能受仔猪日龄的影响。其原因是仔猪刚出生时，单核细胞系相对不成熟（Binter等，2008）。给仔猪（Kastel 等，1999）或无菌仔猪（Revajov′等，2001）口服ω-3 长链多不饱和脂肪酸，可大幅提高血液中CD4和CD8淋巴细胞、B淋巴细胞和单核细胞的总数。在断奶仔猪饲粮中添加ω-3 长链多不饱和脂肪酸，可降低大肠杆菌脂多糖（LPS）刺激后血液中白细胞介素1β（IL-1β）、PGE_2和皮质醇的含量，提高血液中类胰岛素生长因子（IGF）-Ⅰ的含量（Liu等，2003）。

研究还发现，共轭亚油酸（CLA，亚油酸的位置和几何异构体）对猪的免疫功能有广泛的影响。然而，在不同的试验模型中，CLA对免疫反应的影响存在一定的差异。Lai等（2005）研究发现，在断奶仔猪饲粮中添加CLA，可提高伴刀豆球蛋白（ConA）诱导的淋巴细胞增殖率，提高卵清蛋白抗体水平，这可能与PGE_2的产生有关。Bontempo 等（2004）和Corino等（2009）研究表明，在母猪饲粮中添加CLA，可提高初乳中IgG、IgA和IgM滴度，提高仔猪刚断奶时的血清IgG滴度，且该滴度会维持到断奶后2周。此外，Lai等（2005）发现，在仔猪饲粮中添加CLA，可抑制LPS诱导的炎性细胞因子的表达；另外，在仔猪外周血单核细胞（peripheral blood mononuclear cells，PBMC）中添加CLA，也可抑制IL-1β、IL-6、TNF-α的产生和表达。CLA对IL-1β、IL-6和TNF-α等炎性细胞因子的抑制作用，很大程度上是决定于t10c12-CLA。CLA的抗炎特性可能是通过PPARγ依赖机制介导（Lai等，2005）。t10c12-CLA可提高猪中性粒细胞的吞噬能力，这是通过PPARγ依赖途径，使PBMC产生TNF-α介导的。PPARγ可能通过调节单核细胞和巨噬细胞的活性，在调节猪的炎症和免疫反应方面发挥着重要作用（Kang等，2007）。鉴于CLA的免疫调节作用，在养猪生产中，CLA具有作为抗生素替代品的潜力，亦可提高动物对疫苗的反应，提高动物对疾病的抵抗力（Azain，2003）。

◆ 3.4 长链多不饱和脂肪酸对猪基因表达的影响

脂肪酸可改变猪脂肪组织、肝脏和肌肉的基因表达（Ding等，2003）。长链多不饱和脂肪酸及其衍生物（类二十烷酸）是PPAR的内源性配体。长链多不饱和脂肪酸对 PPARα及其靶基因的影响目前尚不清楚。然而，在仔猪饲粮中添加DHA，可降低脂肪细胞决定和分化因子1（adipocyte determination and differentiation-dependent factor 1，ADD1）在肝脏中的mRNA表达量，提高乙酰辅酶A氧化酶（ACO）在肝脏和肌肉组织中的mRNA表达量（Hsu 等，2004）。这表明，长链多不饱和脂肪酸通过激活PPARα，改变了与脂肪酸代谢相关基因的表达。在猪饲粮中添加亚麻酸或DHA，可通过激活PPARγ提高成肌细胞中脂肪细胞型脂肪酸结合蛋白1（aP2）和脂联素的表达（Yu等，2008；Luo等，2009），也可通过PPARγ信号机制，降低炎性细胞因子基因的表达（Zhan等，2009）。在其他组织（如肌肉和脾脏），长链多不饱和脂肪酸也提高了PPARγ的mRNA表达量。

4 新生仔猪脂类代谢

◆ 4.1 新生仔猪脂类代谢的特点

4.1.1 脂类的供能作用

新生仔猪在获得足够的母乳之前，会面临机体能量代谢负平衡。新生仔猪的临界温度为34℃，而环境温度可能比这个温度低10℃，因此充足的能量供应对维持体温恒定是至关重要的（Le Dividich和Noblet，1983）。新生仔猪刚出生时，体脂含量≤2%，且绝大部分脂类为组成细胞膜的结构性脂类，仅有极少一部分脂类可为新陈代谢供给燃料。虽然新生仔猪的能量利用率非常高，但母乳的摄入量很有限（Le Dividich等，1994）。由于母乳摄入量有限，因此新生仔猪产热量较少，从而导致体温下降。除了能量摄入不足外，新生仔猪分解代谢饲粮脂肪酸的能力也很低。氧化能力低可能导致通过β-氧化产生的能量低。低能量可能会降低氮沉积，并导致仔猪生长迟缓（Straarup等，2006）。目前，有许多研究通过提高母猪饲粮能量水平，或给新生仔猪供给充足能量，以提高新生仔猪的能量摄入量，增加脂肪的吸收和氧化。在妊娠早期的母猪饲粮中添加不同的植物油，能改变仔猪出生体重的分布，改善低出生体重仔猪的能量状况。饲粮脂肪的组成也能影响胎儿的发育。在妊娠早期饲粮中添加单不饱和脂肪酸，能降低低出生体重仔猪的比例，而PUFA的作用则恰好相反（Laws等，2009）。给新生仔猪口服中链甘油三酯或添加中链甘油三酯到配方奶中，也能提高出生初期的能量利用效率。

4.1.2 生物活性脂类

除了作为能源物质之外，饲粮脂肪中长链多不饱和脂肪酸还对新生仔猪生长和发育发挥着重要作用。虽然目前对长链多不饱和脂肪酸的准确需要量尚不清楚，但是有报道称，长链多不饱和脂肪酸缺乏可能导致新生仔猪肠黏膜的物理性质（Daveloose等，1993）及行为反应（Ng和Innis，2003）发生改变。这种改变可能与组织（如肠黏膜和额叶皮质）中磷脂的长链多不饱和脂肪酸改变有关。此外，营养不良可导致仔猪肠道结构和功能损伤，而通过使长链多不饱和脂肪酸在肠道磷脂中富集，有利于这种损伤的修复（López-Pedrosa等，1999）。

◆ 4.2 应用于新生仔猪的特殊脂类

4.2.1 短链脂肪酸

短链脂肪酸（SCFA），尤其是丁酸，主要由结肠细菌发酵产生。作为结肠的主要能源物质，SCFA对结肠健康，尤其是在维持肠屏障功能方面发挥着重要作用。SCFA水平取决于结肠菌群的数量及其多样性、食糜在肠道中停留的时间和碳源的类型。断奶仔猪和生长育肥猪从后肠吸收和代谢SCFA的能力都很强。Jorgensen等（1997）通过灌注试验发现，灌注的SCFA（乙酸、丙酸和丁酸等的混合物）从粪便中排出的比例不到1%。Yen等（1991）估算，猪从肠道吸收的SCFA提供的能量占机体总产热量的24%。此外，灌注SCFA不影响营养物质和能量的消化率。SCFA能通过促进肠上皮细胞增殖、抑制肠细胞凋亡，使肠道适应性增强，肠道吸收面积增加（Bartholome等，2004；Kien等，2007）。在配方奶粉中添加丁酸钠，能显著提高新生仔猪空肠末端和回肠的隐窝深度、绒毛高度以及黏膜厚度。SCFA可刺激新生仔猪胰腺的分泌，并提高血浆胰多肽和胆囊收缩素的含量（Kotunia等，2004）。回肠灌注不同剂量的SCFA，对猪消化间期胰腺分泌具有短期的调节作用（Sileikiene等，2008）。SCFA还对生长育肥猪有齿食道口线虫感染有影响（Petkevicius等，2004）。一些研究也发现，丁酸钠能提高新生仔猪和断奶仔猪的日增重（Kotunia等，2004；Gálfi和Bokori，1990）。然而，短链脂肪酸在新生仔猪结肠的产生和代谢尚不清楚。

4.2.2 中链甘油三酯

由6～12个碳原子脂肪酸组成的甘油三酯，称为中链甘油三酯。热带油（如椰子油和棕榈仁油等）中富含中链甘油三酯。此外，樟脑树果实中也含有中链甘油三酯。与典型的长链甘油三酯不同，中链甘油三酯分子量低，在体液中的溶解性高，可直接通过肝门静脉途径被快速吸收，而不需要肉碱通过线粒体转运和代谢，并具有低的酯化亲和力（Odle，1997）。新生仔猪能有效地消化和吸收中链甘油三酯。猪脂类消化的第一步在胃中进行，由胃脂肪酶催化。胃脂肪酶的产物可促进中链甘油三酯在肠道的消化。灌注研究已证实，即使在胰脂肪酶缺乏的情况下（Guilot等，1994），猪肠道仍然可以更迅速地水解并吸收中链甘油三酯（Guillot等，1993）。这表明，胃脂肪酶的水解作用在猪消化吸收中链甘油三酯中起着重要作用。与长链甘油三酯相比，虽然中链甘油三酯的消化速度更快，吸收更完全，但是不同链长的中链脂肪酸的利用存在较大差异。中链甘油三酯链长的变化，会大大影响其在新生仔猪体内消化、吸收和代谢的速度和程度（Odle，1997）。当中链甘油三酯的碳链长度减少时，其相对吸收率显著增加。研究一致表明，C6：0的消化吸收率最高（Wieland等，1993a，1993b；Odle等，1994）。其原因可能是，随着链长度的增加，中链甘油三酯和中链脂肪酸的溶解度下降，以及含有C6：0的中链甘油三酯水解的速率较快（Dicklin等，2006）。此外，乳化可使中链甘油三酯利用率大大提高。研究表明，乳化可使所有不同链长的中链甘油三酯的消化率和吸收率都有类似程度的提高（Wieland等，1993a，1993b；Dicklin等，2006）。

鉴于新生仔猪对中链甘油三酯有较高的消化率和吸收率，目前一些学者对含有中链甘油三酯的油脂在哺乳仔猪上的应用进行了研究。Cera等（1990）报道，与含有大豆油或烤大豆的饲粮相比，含有中链甘油三酯或椰子油的饲粮的脂肪表观消化率提高了11%～17%。与饲喂含有其他脂肪源饲粮的猪相比，饲喂含有椰子油饲粮的猪血清甘油三酯含量提高，血清尿素含量降低。虽然与碳链较长的中链甘油三酯相比，碳链较短的中链甘油三酯消化和吸收快得多，但是动物试验却表明，从C7至C10的中链甘油三酯的氧化效率没有差异（Odle等，1992）。与此相似，Murry等（1999）研究发现，中链甘油三酯对仔猪的生长性能和养分表观消化率没有影响，但可降低体脂肪含量，提高血浆中中链甘油三酯含量。基于此，Odle（1997）建议，添加中链甘油三酯的剂量不要超过6.5 mmol/kg代谢体重（$BW^{0.75}$），采用乳化三己酸甘油酯（按照体积比30%），对从出生至48h以内的仔猪效果最好。采用L-肉碱也有一定的益处。Casellas等（2005）研究报道，与对照组相比，口服甘油三酯使低出生体重仔猪死亡率降低了1.9倍。然而，但另一些研究表明，口服甘油三酯对新生仔猪没有任何效果（Lee和Chiang，

1994）。

4.2.3 共轭亚油酸

目前有关新生仔猪对共轭亚油酸（CLA）利用的研究很少。Corl等（2008）研究表明，在低脂（3%）或高脂（25%）配方奶粉中添加1%CLA，对新生仔猪增重、肝脏和肌肉脂肪酸氧化无影响，但却降低了体脂沉积。体脂沉积的降低可能与CLA抑制脂肪组织中脂肪酸吸收和脂肪酸合成有关。Zhou等（2007）通过血管基质细胞（从新生仔猪皮下脂肪分离）培养试验证明，CLA（混合异构体或t10c12 CLA）减少了脂肪前体细胞数和总细胞数，降低了脂类沉积，并抑制了脂肪细胞特异性基因（如ADD-1、PPARγ、aP2、LPL和IR等）的表达。有关CLA对免疫系统的影响也在新生仔猪上进行了研究。Bassaganya-Riera等（2001）研究表明，CLA增加了新生仔猪外周血CD8+淋巴细胞亚群数量，并刺激了淋巴细胞增殖。Bassaganya-Riera等（2006）也发现，CLA缓解了右旋糖醋硫酸酯钠（DSS）引起的新生仔猪生长迟缓，缓解了DSS导致的肠道损伤和临床症状。

4.2.4 长链多不饱和脂肪酸

在母猪饲粮中添加长链多不饱和脂肪酸（如DHA、ARA和EPA）对新生仔猪发育的影响，在前文已经阐述。在哺乳期和断奶期间，给仔猪补充DHA，能使DHA在大脑、视网膜、肝脏、肺脏、脂肪组织、血浆、红细胞以及肠黏膜磷脂中富集（Huang和Craig-Schmidt，1996；Craig-Schmidt等，1996；Mathews等，2002；Blanaru等，2004；Huang等，2007；Dullemeijer等，2008；Hess等，2008）。补充ARA，能使ARA在血浆、肝脏、脂肪组织、红细胞和肠黏膜磷脂中富集（Mathews等，2002；Blanaru等，2004；Huang等，2007；Hess等，2008），但未在大脑和视网膜中富集（Huang等，2007）。长链多不饱和脂肪酸在组织中富集与饲粮长链多不饱和脂肪酸存在剂量依赖关系，这一现象可在肠黏膜磷脂中观察到（Hess等，2008）。

Merritt等（2003）和Hess等（2008）研究表明，添加长链多不饱和脂肪酸对仔猪生长、采食、临床化学、血液学、器官重量和病理组织学均无影响。然而，Blanaru等（2004）发现，给新生仔猪添加长链多不饱和脂肪酸，可改善神经系统发育，提高免疫反应，增加骨量。骨量的增加与饲粮中ARA/DHA的比例有关。当ARA和DHA的比值为0.5：0.1/100 g脂肪时，骨量增加，但超过这个值则没有效果（Mollard等，2005）。Bassaganya-Riera等（2007）发现，在配方奶粉中添加ARA和DHA，可调节新生仔猪抗原特异性T细胞反应。

仔猪对添加长链多不饱和脂肪酸的反应具有组织特异性。DHA在血浆和红细胞中含量的增加，与在神经（脑和视网膜）和内脏（肝脏和脂肪）组织中含量的增加呈正相关关系。然而，ARA在血浆和红细胞中的增加仅与在内脏组织的增加呈正相关，而与在神经组织中的增加无相关性（Huang等，2007）。由于DHA和ARA在渗入磷脂中时，二者并不存在明显的竞争，因此，ARA对添加了DHA和ARA的饲粮并没有影响（Craig-Schmidt等，1996）。此外，饲粮中高水平的亚油酸可影响DHA在大脑中的富集，从而抑制次级神经突的生长（Novak等，2008）。

5 生长猪对脂类的利用

◆ 5.1 生长猪脂类需求量

5.1.1 断奶对脂类代谢的影响

断奶仔猪饲粮中含有3%～10%的脂肪。猪对脂类的利用率因脂肪来源（动物油脂或植物油脂）和猪生长阶段的不同而异。研究表明，饲粮中脂肪的消化率受断奶日龄、断奶后时间、脂肪来源和结构的影响（Cera等，1988；Straarup等，2006）。豆油和椰子油比牛油和猪油更易消化。添加乳化剂可提高牛油和猪油的消化率，但对生长性能影响非常小（Jones等，1992）。虽然在哺乳期间，仔猪的胰腺重量、脂肪酶活性（以每单位组织湿重计算）和总脂肪酶活性增加，但是断奶会导致胰脂肪酶活性降低，尤其是在断奶后的第1周。断奶后7 d，无论饲粮中是否补充油脂，胰脂肪酶和肠脂肪酶活性都会有提高（Cera等，1990）。鱼油可促进胰液分泌，增加脂肪酶的产量，但是脂肪酶产量的增加对脂肪消化率并没有影响（Hedemann等，2000）。Hedemann等（2004）指出，这种低的消化能力与其他因素相互作用，增加了仔猪断奶后腹泻的风险。

5.1.2 脂肪来源对脂类代谢的影响

饲粮脂肪来源也影响粗脂肪和脂肪酸的表观消化率。事实上，脂肪消化率随着不饱和脂肪酸与饱和脂肪酸比值的增加而升高，随着脂肪中游离脂肪酸含量的增加而降低（Powles等，1995）。然而，脂肪来源对猪的生长无影响（Mitchaothai等，2008；Apple等，2009a），也未改变胴体分割肉的产量和胴体组成，却显著影响了脂肪组织的脂肪酸组成。例如，饲喂牛油（富含饱和脂肪酸），会产生富含饱和脂肪酸的猪肉（Mitchaothai等，2008；Apple等，2009a，2009b，2009c）。与之相似，饲喂豆油、向日葵油和禽油，会降低猪肉中单不饱和脂肪酸含量，增加PUFA含量。然而，猪肉中PUFA含量增高，可导致猪肉变软，从而导致经济效益下降（Apple等，2009c）。饲粮脂肪对胴体脂肪酸组成和沉积影响的程度，因脂肪酸链长（Smith等，1996）、猪的性别和基因型（Smith等，1999；Kloareg等，2007）的不同而异。

5.1.3 乳化对脂类消化的影响

早期研究表明，脂肪和饲粮的物理特性可能影响仔猪对脂肪的利用率，奶脂的高消化率归因于其高度乳化状态（Frobish等，1967）。乳化程度影响脂肪酶的活性和微生物脂肪酶对脂肪的水解作用。Lairon等（1980）发现，脂肪酶活性与磷脂/甘油三酯的比值有关，而最佳比例取决于乳化程度。Jones等（1992）在饲料中添加溶血卵磷脂或卵磷脂作为不同脂肪源（大豆油、牛油、猪油、椰子油）的乳化剂，结果表明，无论是加入卵磷脂还是溶血卵磷脂，牛油的消化率均提高，相比之下卵磷脂比溶血卵磷脂效果更好。Xing等（2004）发现，在含有猪油的饲粮中添加溶血卵磷脂能提高断奶仔猪生长性能。因此，乳化剂对消化率的影响似乎与脂肪的来源有关。然而，也有研究表明，乳化剂没有提高断奶仔猪和生长育肥猪对大豆油、椰子油和猪油的利用率（Jones等，1992；Overland等，1993a，1993b）。Overland等（1994）和Averette Gatlin等（2005）也发现，添加乳化剂对含有8%氢化脂肪饲粮的表观脂肪消化率没有影响。

5.1.4 脂类沉积

猪脂肪沉积速率因日龄和遗传背景的不同而异。营养重分配剂可进一步影响脂肪的沉积速率。一般来说，仔猪脂肪沉积率低，而即将上市的猪的脂肪沉积则急剧增加。育肥猪1/3～1/2的日增重可能是脂肪沉积。猪饲粮中添加脂肪不一定改变脂类沉积速率，但影响所沉积的脂肪的组成（Madsen等，1992）。饲粮中添加脂肪，可抑制脂肪的从头合成（Smith等，1996）。与在啮齿类动物上的研究报道不同，在猪上，饲粮脂肪对脂肪酸从头合成的抑制作用似乎不受饲粮脂肪饱和程度的影响。事实上，饱和脂肪比不饱和脂肪可能具有更强的抑制作用（Allee等，1972；Camara等，1996；Smith等，1996）。饲粮脂肪来源对脂类代谢和体组成的影响与猪基因型密切相关（Freire等，1998）。目前，饲喂玉米酒糟（DDGS）对猪胴体品质（尤其是对胴体碘值）的影响，是养猪业中普遍关注的一个重要问题。在屠宰前饲喂15%（或以上）的DDGS（含9%～10%脂肪）4周或更长时间，能显著提高胴体碘值，而饲喂含有玉米（3.5%脂肪）的饲粮则没有这种影响。

5.1.5 脂类对肉品质的影响

近10年来，随着消费者对健康、营养丰富食品的追求，猪肉品质受到越来越多的关注。影响猪肉品质的关键因素是脂肪的含量、分布以及组成。改善猪肉品质的首要目标是在降低背膘厚度的同时增加肌内脂肪含量、改变胴体脂肪酸组成。虽然饲粮脂肪来源不影响生长性能和胴体组成，但却会影响皮下脂肪和肌内脂肪的脂肪酸组成（Corino等，2008）。提高饲粮PUFA水平可使PUFA在猪肉中富集。研究表明，饲喂玉米DDGS含量高达30%的饲粮，可使生长育肥猪腹脂、背膘和肌内脂肪PUFA显著增加（Xu等，2009）。PUFA含量的增加可导致胴体脂肪碘价增高（Stein和Shurson，2009）。当背膘碘值超过80，猪肉就会变肥、变软，从而造成一定的经济损失（Apple等，2009a，2009b）。氢化脂肪是植物油经过化学氢化以降低PUFA含量的产品，已用在猪饲料中来增加猪肉硬度。研究表明，育肥猪饲粮中添加氢化脂肪，可减少脂肪碘值，增加猪肉硬度。其中，氢化或部分氢化的效果要比不饱和脂肪与饱和脂肪混合的效果好。饲粮中添加部分氢化的油菜籽油（富含Δ6至Δ11 C18：1反式脂肪酸异构体），可增加背膘和肌内脂肪中反式单不饱和脂肪酸的含量，增加背膘硬度（Gläser等，2002）。

此外，早期断奶仔猪饲粮中添加CLA，可降低脂肪组织中单不饱和脂肪酸含量，提高总饱和脂肪酸含量（Demaree等，2002）。同样，在育肥猪饲粮中添加0.6%～1%的CLA，也可提高猪肉饱和脂肪含量及猪肉硬度（Weber等，2006；White等，2009；Gatlin等，2002b；Averette Gatlin等，2006）。猪肉饱和脂肪的增加是因为硬脂酰辅酶A去饱和酶受到抑制的缘故。因此，CLA可增加猪肉硬脂酸的含量，减少油酸的含量（Demaree等，2002；Smith等，2002；Tischendorf等，2002；Morel等，2008；White等，2009）。饲粮CLA也可降低背膘脂肪酸合成酶及脂蛋白脂酶的活性，提高眼肌脂肪细胞型脂肪酸结合蛋白mRNA的表达水平（Jiang等，2010）。长期添加CLA可增大眼肌面积，增加肌内脂肪含量及CLA在猪肉中的沉积（Dugan等，2004；Sun等，2004）。

在育肥猪饲粮添加ω-3PUFA，可使猪肉富含ω-3PUFA，改善营养物质的吸收。因此，人食用富含ω-3PUFA的猪肉，可有效降低心血管疾病的发生（Coates等，2009）。有关如何使用CLA和ω-3PUFA，将在随后的章节进行详述。

◆ 5.2 特殊脂类在生长育肥猪上的应用

5.2.1 共轭亚油酸

尽管饲粮中添加CLA可以提高饲料转化效率（Ostrowska等，1999），但对生产性能和采食量却没有影响（Ostrowska等，1999；Lai等，2005；Weber等，2006；Morel等，2008）。研究表明，在饲粮中添加CLA，可以降低脂肪含量，增加脂肪硬度和CLA的含量，从而提高猪肉品质。Ostrowska等（1999）和Larsen等（2009）研究发现，在育肥猪饲粮中，随着CLA含量的升高，脂肪沉积量呈线性或二次曲线降低。与此类似，Thiel-Cooper等（2001）也发现，饲喂含CLA饲粮，可降低猪第10根肋骨处的背膘厚。Larsen等（2009）发现，在猪饲粮中添加CLA，使胴体瘦肉沉积呈二次曲线增长，提高腹脂硬度和熏肉的硬度，但是不能改善熏肉的可切割性。此外，CLA也可以提高脂肪的饱和度，进而延长熏肉的保质期，但对熏肉的水分、蛋白与脂肪含量没有影响，对熏肉和腰肉的风味与外观几乎没有影响（Janz等，2008）。感官测试分析结果表明，消费者对于饲粮添加CLA所生产的猪肉产品与普通商品猪肉的喜好程度几乎没有差异。

饲粮CLA也能沉积于组织，改变血浆和组织中脂肪酸组成。研究发现，CLA可以提高背阔肌与皮下脂肪硬脂酸和软脂酸的百分比（占总脂肪酸），降低油酸、γ-亚麻酸、亚麻酸和ARA的百分比（Morel等，2008）。在背膘和网膜等组织中沉积的CLA异构体种类，与饲粮中CLA异构体的种类相似。然而，在肝脏和心脏等器官中沉积的CLA异构体种类，却与饲粮中CLA异构体的种类存在很大的差异（Kramer等，1998）。CLA在猪肉或猪肉产品中沉积，对消费者具有重要的意义（Dugan等，2004）。

5.2.2 长链多不饱和脂肪酸

猪肉脂肪酸组成极易因饲粮脂肪酸组成的改变而改变，尽管这种改变可能受脂肪酸从头合成及猪育肥阶段的影响。DHA在组织中的沉积与饲粮亚麻酸的含量具有显著的相关性（Blank等，2002）。饲粮添加DHA可以提高火腿肌、腰肌、腹肌、肩肌中EPA、DHA、ω-3和W-6二十二碳五烯酸（DPA）的含量（Marriott 等，2002；Meadus等，2009）。Haak 等（2008）也表明，给育肥猪饲喂鱼油，可以显著提高EPA和DHA的比例（与总脂肪酸的百分比）。但是，添加鱼油和亚麻籽油对猪肉最终pH、滴水损失、感官特点、脂质氧化或肉色都没有影响（Haak等，2008）。猪采食ω-3 PUFA的持续时间可影响猪肉的脂肪酸组成和氧化稳定性，但是猪在育肥早期摄入ω-3 PUFA停下后，猪肉中ω-3 PUFA的浓度可以很大程度地恢复，特别是在ω-3 PUFA持续供应的情况下（Jaturasitha等，2009）。因此，饲粮中PUFA的供应时间长短是比较灵活的。脂肪酸组成的恢复主要与脂肪酸有关。一般而言，DHA比EPA更易恢复，然而对DPA却没有影响。这些结果表明，在不影响猪肉质的情况下，在猪饲粮中添加ω-3 PUFA是提高猪肉营养价值的一种非常有效的方法。

猪肉中ω-3 PUFA含量提高，对提高消费者的健康水平具有重要意义。近来研究表明，人类摄入富含ω-3 PUFA的猪肉，可显著降低心血管疾病的发生率（Coates等，2009）。然而，在育肥猪的饲粮中使用较高水平的DHA（1%）和PorcOmega（15%）时，会导致猪肉感官品质下降（Meadus等，2009；Howe等，2002）。有关猪肉感官品质下降是否对消费者对猪肉的喜好产生影响，尚需要进一步评估。

6 种猪对脂类的利用

有关PUFA在公猪和母猪繁殖中的作用已经有相当多的报道（Wathes 等，2007），这些作用与必需脂肪酸维持细胞膜的功能、必需脂肪酸是类二十烷酸的前体以及ω-6/ω-3比例对发育有重要影响有关。繁殖公猪饲粮中添加ω-3脂肪酸，可提高精液量、精子活力和形态。母猪饲粮中添加各种类型的脂肪，可提高新生仔猪的成活率与断奶体重。有关ω-3脂肪酸对母猪及其后代（仔猪）的影响，虽然有一些报道，但是尚未进行系统的研究。

◆ 6.1 饲粮脂肪对母猪的影响

6.1.1 饲粮脂肪对乳脂和新生仔猪成活率的影响

20世纪70年代到80年代早期的研究结果表明，在妊娠晚期，母猪饲粮中添加脂肪可提高新生仔猪的成活率（Seerley等，1974；Pettigrew，1981）。断奶前仔猪的死亡，大多数在出生之后头3 d。新生仔猪出生时，体内能量贮备较少（体脂含量低于3%）。母猪在妊娠后期饲喂添加了脂肪的饲粮，可提高新生仔猪的成活率，从而提高每窝仔猪的成活率。新生仔猪成活率的提高是因为更多的胎盘营养转运给了胎儿。给妊娠母猪饲粮中添加脂肪，可提高新生仔猪体脂、肝糖原含量和空腹血糖水平。在选育的母猪上，脂肪的添加效果没有上述的明显（Shurson和Irvin，1992）。因此，随着基因的改良，脂肪对母猪的影响可能不如过去明显。

饲粮脂肪对母猪繁殖的影响，主要体现在对乳脂含量的影响上（Averette等，1999；Lauridsen和Danielsen，2004）。在妊娠后期和哺乳期母猪饲粮中添加脂肪，可提高乳脂含量、仔猪平均日增重和平均断奶体重（Gatlin等，2002a）。由于仔猪可获得更多可利用的能量，因此可使仔猪的断奶体重提高。这些早期的研究表明，仔猪成活率的提高与乳脂含量的增加主要是受脂肪添加与否的影响，而不受脂肪类型的影响。最近的研究表明，饲喂含有ω-3脂肪的饲粮，对于母猪（Mateo等，2009）和公猪的繁殖性能均有益处。

6.1.2 脂肪对妊娠周期和窝仔数的影响

饲粮脂肪除了可提高新生仔猪存活率和体重外，也可影响妊娠周期。ω-3脂肪酸对妊娠周期具有特殊的影响，这种影响目前在人（Allen和Harris，2001）、鼠（Olsen等，1990）和羊（Capper等，2006）上均已得到证实。延长妊娠周期可以增加初生仔猪的能量贮备，从而提高成活率。Rooke 等（2001a）研究表明，与植物油相比，在母猪（196头）妊娠期间添加鲑鱼油，可很大程度上延长妊娠周期。虽然所产仔猪的出生重降低，但是仔猪断奶前的死亡率也降低。然而，其他的研究没有证实这个结果。

由于窝仔数本身的变异很大，因此研究饲粮脂肪对窝仔数的影响相当困难。初步研究表明，在配种之前和妊娠早期，饲喂母猪含ω-3脂肪酸的饲粮，可增加窝仔数（Spencer等，2004）。然而，该报道却未被正式出版的文献所证实（Mateo等，2009）。

在母猪妊娠和哺乳期间饲喂ω-3脂肪酸，ω-3脂肪酸可通过胎盘传递给胎儿（Rooke等，2001a，2001b；Spencer等，2004），或通过乳汁传递给新生仔猪（Fritsche等，1993b；Mateo等，2009）。这种脂肪酸的传递对人类健康具有重要意义，如可提高断奶前成活率、免疫功能和智力。给母猪饲喂含鲑鱼油的饲粮，可大大降低仔猪的压死率，从而提高仔猪断奶前的存活率（Rooke等，2001a）。ω-3脂肪酸对仔猪存活率的影响，与在动物模型和人上的研究结果类似。许多研究表明，哺乳动物在子宫里或新生期间接触到ω-3 脂肪酸，其认知能力和智力都会有不同程度的提高（Innis，2007）。与没有饲喂DHA的猪相比，饲喂含DHA饲粮的猪在迷宫测试中的表现更好（Ng和Innis，2003）。与玉米油相比，鱼油可缓解断奶仔猪脂多糖诱导的炎症反应（Liu等，2003），表明鱼油可发挥免疫调节作用。最近的结果表明，在妊娠期和哺乳期，给母猪饲喂含鱼油的饲粮，可提高仔猪肌糖原含量和营养物质吸收率，从而降低仔猪断奶后体重损失（Gabler等，2007）。

◆ 6.2 饲粮脂肪对公猪的影响

有关饲粮脂肪对公猪繁殖力影响的研究相对较少，然而，已有的研究结果表明，饲喂公猪含ω-3脂肪酸的饲粮，可提高精子活力，降低精子畸形率。早期研究表明，公猪连续饲喂含金枪鱼油（30 g/d）的饲粮6周后，其精子活力显著提高（Rooke等，2001c）。与此类似，公猪连续饲喂含鲨鱼肝油（40 g/d）的饲粮4周后，精液产量也有相应的提高（Mitre等，2004）。Estienne 等（2008）研究发现，公猪连续饲喂深海鱼油（约50 g/d）16周后，精液活力并未得到改善，然而精子数量却显著增加。

作者：Xi Lin, Mike Azain
和Jack Odle

译者：刘玉兰

参考文献

Adams, S. H., X. Lin, X. X. Yu, J. Odle, and J. K. Drackley. 1997. Hepatic fatty acid metabolism in pigs and rats: major differencesin endproducts, O_2 uptake, and beta-oxidation. Am. J. Physiol. 272:R1641–R1646.

Allee, G. L., D. R. Romsos, G. A. Leveille, and D. H. Baker. 1972. Lipogenesis and enzymatic activity in pig adipose tissue asinfluenced by source of dietary fat. J. Anim. Sci. 35:41–47.

Allen, K. G., and M. A. Harris. 2001. The role of n-3 fatty acids in gestation and parturition. Exp. Biol. Med. 226:498–506.

Apple, J. K., C. V. Maxwell, B. R. Kutz, L. K. Rakes, J. T. Sawyer, Z. B. Johnson, T. A. Armstrong, S. N. Carr, and P. D. Matzat.2008. Interactive effect of ractopamine and dietary fat source on pork quality characteristics of fresh pork chops duringsimulated retail display. J. Anim. Sci. 86:2711–2722.

Apple, J. K., C. V. Maxwell, D. L. Galloway, S. Hutchison, and C. R. Hamilton. 2009a. Interactive effects of dietary fat sourceand slaughter weight in growing–finishing swine: I. Growth performance and longissimus muscle fatty acid composition.J. Anim. Sci. 87:1407–1422.

Apple, J. K., C. V. Maxwell, D. L. Galloway, C. R. Hamilton, and J.W. Yancey. 2009b. Interactive effects of dietary fat source andslaughter weight in growing–finishing swine: II. Fatty acid composition of subcutaneous fat. J. Anim. Sci. 87:1423–1440.

Apple, J. K., C. V. Maxwell, D. L. Galloway, C. R. Hamilton, and J. W. Yancey. 2009. Interactive effects of dietary fat source andslaughter weight in growing–finishing swine: III. Carcass and fatty acid compositions. J. Anim. Sci. 87:1441–1454.

Averette, L. A., J. Odle, M. H. Monaco, and S. M. Donovan. 1999. Dietary fat during pregnancy and lactation increases milk fatand insulin-like growth factor I concentrations and improves neonatal growth rates in swine. J. Nutr. 129:2123–2129.

Averette Gatlin, L., M. T. See, J. A. Hansen, and J. Odle. 2003. Hydrogenated dietary fat improves pork quality of pigs from twolean genotypes. J. Anim. Sci. 81:1989–1997.

Averette Gatlin, L., M. T. See, and J. Odle. 2005. Effects of chemical hydrogenation of supplemental fat on relative apparent lipiddigestibility in finishing swine. J. Anim. Sci. 83:1890–1898.

Averette Gatlin, L., M. T. See, D. K. Larick, and J. Odle. 2006. Descriptive flavor analysis of bacon and pork loin from lean-genotypegilts fed conjugated linoleic acid and supplemental fat. J. Anim. Sci. 84:3381–3386.

Azain, M. J. 2001. Fat in swine nutrition.Pages 95–105 in swine nutrition. A. J. Lewis and L. L. Lee Sothern, eds. CRC Press,Boca Raton, FL.

Azain, M. J. 2003. Conjugated linoleic acid and its effects on animal products and health in single-stomached animals. Proc. Nutr.Soc. 62:319–328.

Barrero, M. J., C. S. Alho, J. A. Ortiz, F. G. Hegardt, D. Haro, and P. F. Marrero. 2001. Low activity of mitochondrial HMG-CoAsynthase in liver of starved piglets is due to low levels of protein despite high mRNA levels. Arch. Biochem. Biophys.385:364–371.

Bartholome, A. L., D. M. Albin, D. H. Baker, J. J. Holst, and K. A. Tappenden. 2004. Supplementation of total parenteral nutritionwith butyrate acutely increases structural aspects of intestinal adaptation after an 80% jejunoileal resection in neonatal piglets.J. Parenter. Enteral., Nutr. 28:210–22.

Bassaganya-Riera, J., R. Hontecillas-Magarzo, K. Bregendahl, M. J. Wannemuehler, and D. R. Zimmerman. 2001. Effects ofdietary conjugated linoleic acid in nursery pigs of dirty and clean environments on growth, empty body composition, andimmune competence. J. Anim. Sci. 79:714–721.

Bassaganya-Riera, J., and R. Hontecillas. 2006. CLA and n-3 PUFA differentially modulate clinical activity and colonic PPARresponsivegene expression in a pig model of experimental IBD. Clin.Nutr. 25:454–65.

Bassaganya-Riera, J., A. J. Guri, A. M. Noble, K. A. Reynolds, J. King, C. M.Wood, M. Ashby, D. Rai, and R. Hontecillas. 2007.Arachidonic acid– and docosahexaenoic acid–enriched formulas modulate antigen-specific T cell responses to influenza virusin neonatal piglets. Am. J. Clin. Nutr. 85:824–836.

Bazinet, R. P., E. G. McMillan, and S. C. Cunnane. 2003. Dietary alpha-linolenic acid increases the n-3 PUFA content of sow'smilk and the tissues of the suckling piglet. Lipids 38:1045–1049.

Bergen, W. G., and H. J. Mersmann. 2005. Comparative aspects of lipid metabolism: impact on contemporary research and use ofanimal models. J. Nutr. 135:2499–2502.

Binter, C., A. Khol-Parisini, P. Hellweg, W. Gerner, K. Schäfer, H. W. Hulan, A. Saalm¨uller, and J. Zentek. 2008. Phenotypic andfunctional aspects of the neonatal immune system as related to the maternal dietary fatty acid supply of sows. Arch. Anim.Nutr. 62:439–453.

Blanaru, J. L., J. R. Kohut, S. C. Fitzpatrick-Wong, and H. A. Weiler. 2004. Dose response of bone mass to dietary arachidonicacid in piglets fed cow milk-based formula. Am. J. Clin.

Nutr. 79:139–147.

Blank, C., M. A. Neumann, M. Makrides, and R. A. Gibson. 2002. Optimizing DHA levels in piglets by lowering the linoleic acidto alpha-linolenic acid ratio. J. Lipid. Res. 43:1537–1543.

Bontempo, V., D. Sciannimanico, G. Pastorelli, R. Rossi, F. Rosi, and C. Corino. 2004. Dietary conjugated linoleic acid positivelyaffects immunologic variables in lactating sows and piglets. J. Nutr. 134:817–824.

Boudry, G., V. Douard, J. Mourot, J. P. Lallès, and I. Le Huärou-Luron.2009. Metabolic activity of the enteric microbiota influencesthe fatty acid composition of murine and porcine liver and adipose tissues. J. Nutr. 139:1110–1117.

Brazle, A. E., B. J. Johnson, S. K. Webel, T. J. Rathbun, and D. L. Davis. 2009. Omega-3 fatty acids in the gravid pig uterus asaffected by maternal supplementation with omega-3 fatty acids. J. Anim. Sci. 87:994–1002.

Camara, M., J. Mourot, and C. Eevrier. 1996. Influence of two dairy fats on lipid synthesis in the pig: Comparative study of liver,muscle and the two back fat layers. Annals Nutr. Metabolism 40:287–295.

Capper, J. L., R. G. Wilkinson, A. M. Mackenzie, and L. A. Sinclair. 2006. Polyunsaturated fatty acid supplementation duringpregnancy alters neonatal behavior in sheep. J. Nutr. 136:397–403.

Carlson, S. E. 1999. Long-chain polyunsaturated fatty acids and development of human infants. Acta Paediatr. Suppl. 88:72–77.

Casellas, J., X. Casas, J. Piedrafita, and X. Manteca. 2005. Effect of medium- and long-chain triglyceride supplementation on smallnewborn-pig survival. Prev. Vet. Med. 67:213–221.

Cera, K. R., D. C. Mahan, and G. A. Reinhart. 1988. Weekly digestibilities of diets supplemented with corn oil, lard or tallow byweanling swine. J. Anim. Sci. 66:1430–1437.

Cera, K. R., D. C. Mahan, and G. A. Reinhart. 1990. Evaluation of various extracted vegetable oils, roasted soybeans, medium-chaintriglyceride and an animal-vegetable fat blend for postweaning swine. J. Anim. Sci. 68:2756–2765.

Changhua, L., Y. Jindong, L. Defa, Z. Lidan, Q. Shiyan, and X. Jianjun. 2005. Conjugated linoleic acid attenuates the productionand gene expression of proinflammatory cytokines in weaned pigs challenged with lipopolysaccharide. J. Nutr. 135:239–244.

Cheon, Y., T. Y. Nara, M. R. Band, J. E. Beever, M. A. Wallig, and M. T. Nakamura. 2005. Induction of overlapping genes byfasting and a peroxisome proliferator in pigs: Evidence of functional PPARalpha in nonproliferating species. Am. J. Physiol.Regul. Integr. Comp. Physiol. 288:R1525–535.

Coates, A. M., S. Sioutis, J. D. Buckley, and P. R. Howe. 2009. Regular consumption of n-3 fatty acid-enriched pork modifiescardiovascular risk factors. Br. J. Nutr. 101:592–597.

Corino, C., M. Musella, and J. Mourot. 2008. Influence of extruded linseed on growth, carcass composition, and meat quality ofslaughtered pigs at one hundred ten and one hundred sixty kilograms of liveweight. J. Anim. Sci. 86:1850–1860.

Corino, C., G. Pastorelli, F. Rosi, V. Bontempo, and R. Rossi. 2009. Effect of dietary conjugated linoleic acid supplementation insows on performance and immunoglobulin concentration in piglets. J. Anim. Sci. 87:2299–3005.

Corl, B. A., S. A. Mathews Oliver, X. Lin, W. T. Oliver, Y. Ma, R.J. Harrell, and J. Odle. 2008. Conjugated linoleic acid reducesbody fat accretion and lipogenic gene expression in neonatal pigs fed low– or high–fat formulas. J. Nutr. 138:449–454.

Craig–Schmidt, M. C., K. E. Stieh, and E. L. Lien. 1996. Retinal fatty acids of piglets fed docosahexaenoic and arachidonic acidsfrom microbial sources. Lipids 31:53–59.

Daveloose, D., A. Linard, T. Arfi, J. Viret, and R. Christon. 1993. Simultaneous changes in lipid composition, fluidity and enzymeactivity in piglet intestinal brush border membrane as affected by dietary polyunsaturated fatty acid deficiency. Biochim.Biophys.Acta. 1166:229–237.

Demaree, S. R., C. D. Gilbert, H. J. Mersmann, and S. B. Smith. 2002. Conjugated linoleic acid differentially modifies fattyacid composition in subcellular fractions of muscle and adipose tissue but not adiposity of postweaning pigs. J. Nutr.132:3272–3279.

Dicklin, M. E., J. L. Robinson, X. Lin, and J. Odle. 2006. Ontogeny and chain–length specificity of gastrointestinal lipases affectmedium–chain triacylglycerol utilization by newborn pigs. J. Anim. Sci. 84:818–825.

Ding, S. T., A. P. Schinckel, T. E. Weber, and H. J. Mersmann. 2000. Expression of porcine transcription factors and genes relatedto fatty acid metabolism in different tissues and genetic populations. J. Anim. Sci. 78:2127–2134.

Ding, S. T., A. Lapillonne, W. C. Heird, and H. J. Mersmann. 2003. Dietary fat has minimal effects on fatty acid metabolismtranscript concentrations in pigs. J. Anim Sci. 81:423–431.

Dugan, M. E., J. L. Aalhus, and J. K. Kramer. 2004. Conjugated linoleic acid pork research. Am. J. Clin. Nutr. 79:1212S–1216S.

Dullemeijer, C., P. L. Zock, R. Coronel, H. M. Den Ruijter, M. B. Katan, R. J. Brummer, F. J. Kok, J. Bee*km*an, and I. A. Brouwer.2008. Differences in fatty acid composition between cerebral brain lobes in juvenile pigs after fish oil feeding. Br. J. Nutr.100:794–800.

Estienne, M. J., A. F. Harper, and R. J. Crawford. 2008. Dietary supplementation with a source of omega–3 fatty acids increasessperm number and the duration of ejaculation in boars. Theriogenology 70:70–76.

Farmer, C., and H. V. Petit. 2009. Effects of dietary supplementation with different forms of flax in late–gestation and lactation onfatty acid profiles in sows and their piglets. J. Anim. Sci. 87:2600–2613.

Freire, J. P., J. Mourot, L. F. Cunha, J. A. Almeida, and A. Aumaitre. 1998. Effect of the source of dietary fat on postweaninglipogenesis in lean and fat pigs. Ann Nutr. Metab. 42:90–95.

Fritsche, K. L., D. W. Alexander, N. A. Cassity, and S. C. Huang. 1993a. Maternally-supplied fish oil alters piglet immune cellfatty acid profile and eicosanoid production. Lipids 28:677–682.

Fritsche, K. L., S. C. Huang, and N. A. Cassity.1993b. Enrichment of omega-3 fatty acids in suckling pigs by maternal dietaryfish oil supplementation. J. Anim. Sci. 71:1841–1847.

Frobish, L. T., V. M. Hays, V. C. Speer, and R. C. Ewan. 1967. Digestion of sow milk fat and effect of diet form on fat utilization.J. Anim. Sci. 26(Suppl. 1):1478. (Abstr.)

Gabler, N. K., J. D. Spencer, D. M. Webel, and M. F. Spurlock. 2007. In utero and postnatal exposure to long chain (n-3) PUFAenhances intestinal glucose absorption and energy stores in weanling pigs. J. Nutr. 137:2351–2358.

Gabler, N. K., J. S. Radcliffe, J. D. Spencer, D. M.Webel, and M. E. Spurlock. 2009. Feeding long-chain n-3 polyunsaturated fattyacids during gestation increases intestinal glucose absorption potentially via the acute activation of AMPK. J. Nutr. Biochem.20:17–25.

Gálfi, P., and J. Bokori. 1990. Feeding trial in pigs with a diet containing sodium n-butyrate. Acta Vet. Hung. 38:3–17.

Gatlin, L. A., J. Odle, J. Soede, and J. A. Hansent. 2002a. Dietary medium- or long-chain triglycerides improve body condition oflean-genotype sows and increase suckling pig growth. J. Anim. Sci. 80:38–44.

Gatlin, L. A., M. T. See, D. K. Larick, X. Lin, and J. Odle. 2002b. Conjugated linoleic acid in combination with supplementaldietary fat alters pork fat quality. J. Nutr. 32:3105–3112.

Girard. J., P. Ferré, J. P. Pégorier, and P. H. Duée. 1992. Adaptations of glucose and fatty acid metabolism during perinatal periodand suckling-weaning transition. Physiol. Rev. 72:507–562.

Gläser, K. R., C.Wenk, and M. R. Scheeder. 2002. Effects of feeding pigs increasing levels of C 18:1 trans fatty acids on fatty acidcomposition of bac*kf*at and intramuscular fat as well as bac*kf*at firmness. Arch Tierernahr. 56:117–130.

Guillot, E., P. Vaugelade, P. Lemarchal, and A. Rérat. 1993. Intestinal absorption and liver uptake of medium-chain fatty acids innon-anaesthetized pigs. Br. J. Nutr. 69:431–442.

Guillot, E., P. Lemarchal, T. Dhorne, and A. Rerat. 1994. Intestinal absorption of medium chain fatty acids: *in vivo* studies in pigsdevoid of exocrine pancreatic secretion. Br. J. Nutr. 72:545–553.

Haak, L., S. De Smet, D. Fremaut, K. van Walleghem, and K. Raes. 2008. Fatty acid profile and oxidative stability of pork asinfluenced by duration and time of dietary linseed or fish oil supplementation. J. Anim. Sci. 86:1418–1425.

Hedemann, M. S., and B. B. Jensen. 2004. Variations in enzyme activity in stomach and pancreatic tissue and digesta in pigletsaround weaning. Arch Anim. Nutr. 58:47–59.

Hedemann, M. S, A. R. Pedersen, and R. M. Engberg. 2000. Exocrine pancreatic secretion is stimulated in piglets fed fish oilcompared with those fed coconut oil or lard. J. Nutr. 131:3222–3226.

Hess, H. A., B. A. Corl, X. Lin, S. K. Jacobi, R. J. Harrell, A. T. Blikslager, and J. Odle. 2008. Enrichment of intestinal mucosalphospholipids with arachidonic and eicosapentaenoic acids fed to suckling piglets is dose and time dependent. J. Nutr.138:2164–2171.

Howe, P. R., J. A. Downing, B. F. Grenyer, E. M. Grigonis-Deane, and W. L. Bryden. 2002. Tuna fishmeal as a source of DHA forn-3 PUFA enrichment of pork, chicken, and eggs. Lipids 37:1067–1076.

Hsu, J. M., P. H. Wang, B. H. Liu, and S. T. Ding. 2004. The effect of dietary docosahexaenoic acid on the expression of porcinelipid metabolism-related genes. J. Anim. Sci. 82:683–689.

Huang, M. C., J. T. Brenna, A. C. Chao, C. Tschanz, D. A. Diersen-Schade, and H. C. Hung. 2007. Differential tissue doseresponses of (n-3) and (n-6) PUFA in neonatal piglets fed docosahexaenoate and arachidonoate. J. Nutr. 137:2049–2055.

Huang, M. C., and M. C. Craig-Schmidt. 1996. Arachidonate and docosahexaenoate added to infant formula influence fatty acidcomposition and subsequent eicosanoid production in neonatal pigs. J. Nutr. 126:2199–208.

Huang, T.,M. L.Wahlqvist, T. Xu, A. Xu, A. Zhang, and D. Li. 2010. Increased plasma n-3 polyunsaturated fatty acid is associatedwith improved insulin sensitivity in type 2 diabetes in China. Mol. Nutr. Food Res. 54:S112–S119.

Innis, S. M. 2007. Dietary (n-3) fatty acids and brain development. J. Nutr. 137:855–859.

Janz, J. A., P. C. Morel, R.W. Purchas, V. K. Corrigan, S. Cumarasamy, B. H.Wilkinson, andW. H. Hendriks. 2008. The influenceof diets supplemented with conjugated linoleic acid, selenium, and vitamin E, with or without animal protein, on the qualityof pork from female pigs. J. Anim. Sci. 86:1402–1409.

Jaturasitha, S., R. Khiaosa-ard, P. Pongpiachan, and M. Kreuzer. 2009. Early deposition of n-3 fatty acids from tuna oil in lean andadipose tissue of fattening pigs is mainly permanent. J. Anim. Sci. 87:693–703.

Jiang, Z. Y., W. J. Zhong, C. T. Zheng, Y. C. Lin, L. Yang, and S. Q. Jiang. 2010. Conjugated linoleic acid differentially regulatesfat deposition in bac*k*fat and longissimus muscle of finishing pigs. J. Anim. Sci. 88:1694–1705.

Jones, D. B., J. D. Hancock, D. L. Harmon, and C. E. Walker. 1992. Effects of exogenous emulsifiers and fat sources on nutrientdigestibility, serum lipids, and growth performance in weanling pigs. J. Anim. Sci. 70:3473–3482.

Jørgensen, H., T. Larsen, X. Q. Zhao, and B. O. Eggum. 1997. The energy value of short-chain fatty acids infused into the caecumof pigs. Br. J. Nutr. 77:745–756.

Kang, J. H., S. S. Lee, E. B. Jeung, and M. P. Yang. 2007. Trans-10, cis-12-conjugated linoleic acid increases phagocytosis ofporcine peripheral blood polymorphonuclear cells *in vitro*. Br. J. Nutr. 97:117–125.

Kastel, R., V. Revajová, D. Magic, J. Pistl, M. Levkut, L. Bindas, J. Sajbidor, and M. Horváth. 1999. Effect of oil containing n-3polyunsaturated fatty acids (PUFA) on the immune response and growth factors in piglets. Acta Vet. Hung. 47:325–334.

Kien, C. L., R. Blauwiekel, J. Y. Bunn, T. L. Jetton, W. L. Frankel, and J. J. Holst. 2007. Cecal infusion of butyrate increasesintestinal cell proliferation in piglets. J. Nutr. 137:916–922.

Kloareg, M., J. Noblet, and J. van Milgen. 2007. Deposition of dietary fatty acids, de novo synthesis and anatomical partitioningof fatty acids in finishing pigs. Br. J. Nutr. 97:35–44.

Kotunia, A., J.Woliński, D. Laubitz, M. Jurkowska, V. Romé, P. Guilloteau, and R. Zabielski. 2004. Effect of sodium butyrate on thesmall intestine development in neonatal piglets fed [correction of feed] by artificial sow. J. Physiol. Pharmacol. 55(Suppl. 2):59–68.

Kouba, M., D. Hermier, and J. Le Dividich. 2001. Influence of a high ambient temperature on lipid metabolism in the growing pig.J. Anim. Sci. 79:81–87.

Kramer, J. K., N. Schat, M. E. Dugan, M. M. Mossoba, M. P. Yurawecz, J. A. Roach, K. Eulitz, J. L. Aalhus, A. L. Schaefer, andY. Ku, Y. 1998. Distributions of conjugated linoleic acid (CLA) isomers in tissue lipid classes of pigs fed a commercial CLAmixture determined by gas chromatography and silver ion-high-performance liquid chromatography. Lipids 33:549–58.

Lai, C., J. Yin, D. Li, L. Zhao, and X. Chen. 2005. Effects of dietary conjugated linoleic acid supplementation on performance andimmune function of weaned pigs. Arch. Anim. Nutr. 59:41–51.

Lairon, D., G. Nalbone, H. Lafont, J. Leonardi, J. L. Vigne, C. Chabert, J. C. Hauton, and R. Verger. 1980. Effect of bile lipids onthe adsorption and activity of pancreatic lipase on triacylglycerol emulsions. Biochim.Biophys.Acta. 618:119–128.

Larsen, S. T., B. R. Wiegand, F. C. Parrish, Jr., J. E. Swan, and J. C. Sparks. 2009. Dietary conjugated linoleic acid changes bellyand bacon quality from pigs fed varied lipid sources. J. Anim. Sci. 87:285–95.

Lauridsen, C., and V. Danielson. 2004. Lactational dietary fat levels and sources influence milk composition and performance ofsows and their progeny. Livest. Prod. Sci. 91:95–105.

Laws, J., J. C. Litten, A. Laws, I. J. Lean, P. F. Dodds, and L. Clarke. 2009. Effect of type and timing of oil supplements to sowsduring pregnancy on the growth performance and endocrine profile of low and normal birth weight offspring. Br. J. Nutr.101:240–249.

Le Dividich, J., and J. Noblet. 1983. Thermoregulation and energy metabolism in the neonatal pig. Ann Rech Vet. 14:375–381.

Le Dividich, J., P. Herpin, and R. M. Rosario-Ludovino. 1994. Utilization of colostral energy by the newborn pig. J. Anim. Sci.72:2082–2089.

Lee, H. F., and S. H. Chiang. 1994. Energy value of medium-chain triglycerides and their efficacy in improving survival of neonatalpigs. J. Anim. Sci. 72:133–138.

Leskanich, C. O., and R. C. Noble. 1999. The comparative roles of polyunsaturated fatty acids in pig neonatal development. Br. J.Nutr. 81:87–106.

Liu, Y. L., D. F. Li, L. M. Gong, A. M. Gaines, and J. A. Carroll. 2003. Effects of fish oil supplementation on the performanceand immunological, adrenal and somatrophic responses of weaned pigs after *Escherichia coli* lipopolysaccharide challenge.J. Anim. Sci. 81:2758–2765.

López-Pedrosa, J. M., M. Ramírez, M. I. Torres, and A. Gil. 1999. Dietary phospholipids rich in long-chain polyunsaturated fattyacids improve the repair of small intestine in previously malnourished piglets. J. Nutr. 129:1149–55.

Luo, H. F., H. K.Wei, F. R. Huang, Z. Zhou, S.W. Jiang, and J. Peng. 2009. The effect of linseed on intramuscular fat content andadipogenesis related genes in skeletal muscle of pigs. Lipids 44:999–1010.

Madsen, A., K. Jakobsen, and H. P. Mortensen. 1992. Influence of dietary fat on carcass quality in pigs. A review.Acta Agric.Scand. A 42:220–225.

Marriott, N. G., J. E. Garrett, M. D. Sims, and J. R. Abril. 2002. Composition of pigs fed a diet with docosahexaenoic acid.J. Muscle Foods. 13:265–277.

Mateo, R. D., J. A. Carroll, Y. Hyun, S. Smith, and S. W. Kim. 2009. Effect of dietary supplementation of n-3 fatty acids andelevated concentrations of dietary protein on the performance of sows. J. Anim. Sci. 87:948–959.

Mathews, S. A., W. T. Oliver, O. T. Phillips, J. Odle, D. A. Diersen-Schade, and R. J. Harrell. 2002. Comparison of triglyceridesand phospholipids as supplemental sources of dietary long-chain polyunsaturated fatty acids in piglets. J. Nutr. 132:3081–3089.

Meadus,W. J., P. Duff, B. Uttaro, J. L. Aalhus, D. C. Rolland, L. L. Gibson, and M. E. Dugan. 2010. Production of docosahexaenoicacid (DHA) enriched bacon. J. Agric. Food Chem. 58:465–472.

Merritt, R. J., N. Auestad, C. Kruger, and S. Buchanan. 2003. Safety evaluation of sources of docosahexaenoic acid and arachidonicacid for use in infant formulas in newborn piglets. Food Chem. Toxicol. 41:897–904.

Mitchaothai, J., H. Everts, C. Yuangklang, S. Wittayakun, K. Vasupen, S. Wongsuthavas, P. Srenanul, R. Hovenier, and A. C.Beynen. 2008. Digestion and deposition of individual fatty acids in growing–finishing pigs fed diets containing either beeftallow or sunflower oil. J. Anim.

Physiol. Anim. Nutr. (Berl). 92:502–510.

Mitre, R., C. Cheminade, P. Allaume, P. Legrand, and A. B. Legrand. 2004. Oral intake of shark liver oil modifies lipid compositionand improves motility and velocity of boar sperm. Theriogenology. 62:1557–1566.

Mitre, R., M. Etienne, S. Martinais, H. Salmon, P. Allaume, P. Legrand, and A. B. Legrand. 2005. Humoral defence improvementand haematopoiesis stimulation in sows and offspring by oral supply of shark-liver oil to mothers during gestation andlactation. Br. J. Nutr. 94:753–762.

Mollard, R. C., H. R. Kovacs, S. C. Fitzpatrick-Wong, and H. A. Weiler. 2005. Low levels of dietary arachidonic and docosahexaenoicacids improve bone mass in neonatal piglets, but higher levels provide no benefit. J. Nutr. 135:505–512.

Morel, P. C., J. A. Janz, M. Zou, R. W. Purchas, W. H. Hendriks, and B. H. Wilkinson. 2008. The influence of diets supplementedwith conjugated linoleic acid, selenium, and vitamin E, with or without animal protein, on the composition of pork fromfemale pigs. J. Anim. Sci. 86:1145–1155.

Murry, A. C. Jr., S. Gelaye, J. M. Casey, T. L. Foutz, B. Kouakou, and D. Arora. 1999. Type of milk consumed can influenceplasma concentrations of fatty acids and minerals and body composition in infant and weanling pigs. J. Nutr. 129:132–138.

Ng, K.-F., and S. M. Innis. 2003. Behavioral responses are altered in piglets with decreased frontal cortex docosahexaenoic acid.J. Nutr. 133:3222–3227.

Nicot, C., F. G. Hegardt, G. Woldegiorgis, D. Haro, and P. F. Marrero. 2001. Pig liver carnitine palmitoyltransferase I, with low*Km* for carnitine and high sensitivity to malonyl-CoA inhibition, is a natural chimera of rat liver and muscle enzymes.Biochemistry 40:2260–2266.

Nicot, C., J. Relat, G. Woldegiorgis, D. Haro, and P. F. Marrero. 2002. Pig liver carnitine palmitoyltransferase. Chimera studiesshow that both the N- and C-terminal regions of the enzyme are important for the unusual high malonyl-CoA sensitivity.J. Biol. Chem. 277:10044–10049.

Nettleton, J. A., and R. Katz. 2005. n-3 long-chain polyunsaturated fatty acids in type 2 diabetes: A review. J. Am. Diet Assoc.105:428–440.

Novak, E. M., R. A. Dyer, and S. M. Innis. 2008. High dietary omega-6 fatty acids contribute to reduced docosahexaenoic acid inthe developing brain and inhibit secondary neurite growth. Brain Res. 1237:136–145.

NRC. 1998. Nutrient requirements of swine. 10th rev. ed. Natl. Acad. Press, Washington, DC.

Odle, J., N. J. Benevenga, and T. D. Crenshaw. 1992. Evaluation of [1-14C]-medium-chain fatty acid oxidation by neonatal pigletsusing continuous-infusion radiotracer kinetic methodology. J. Nutr. 122:2183–2189.

Odle, J., X. Lin, T. M. Wieland, T. A. van Kempen. 1994. Emulsification and fatty acid chain length affect the kinetics of[14C]-medium-chain triacylglycerol utilization by neonatal piglets. J. Nutr. 124:84–93.

Odle, J. 1997. New insights into the utilization of medium-chain triglycerides by the neonate: Observations from a piglet model.J. Nutr. 127:1061–1067.

Olsen, S. F., H. S. Hansen, and B. Jensen. 1990. Fish oil versus arachis oil food supplementation in relation to pregnancy durationin rats. Prostaglandins Leukot.Essent. Fatty Acids 40:255–260.

Ostrowska, E., M. Muralitharan, R. F. Cross, D. E. Bauman, and F. R. Dunshea. 1999. Dietary conjugated linoleic acids increaselean tissue and decrease fat deposition in growing pigs. J. Nutr. 129:2037–2042.

Overland, M., M. D. Tokach, S. C. Cornelius, J. E. Pettigrew, and J. W. Rust. 1993a. Lecithin in swine diets: I. Weanling pigs.J. Anim. Sci. 71:1187–1193.

Overland, M., M. D. Tokach, S. G. Cornelius, J. E. Pettigrew, and M. E. Wilson. 1993b. Lecithin in swine diets: II. Growing–finishing pigs. J. Anim. Sci. 71:1194–1197.

Overland, M., Z. Mroz, and F. Sundstøl. 1994. Effect of lecithin on the apparent ileal and overall digestibility of crude fat and fattyacids in pigs. J. Anim. Sci. 72:2022–2028.

Papadopoulos, G. A., D. G. Maes, S. van Weyenberg, T. A. van Kempen, J. Buyse, and G. P. Janssens. 2009a. Peripartal feedingstrategy with different n-6:n-3 ratios in sows: Effects on sows' performance, inflammatory and periparturient metabolicparameters. Br. J. Nutr. 101:348–357.

Papadopoulos, G. A., T. Erkens, D. G. Maes, L. J. Peelman, T. A. van Kempen, J. Buyse, and G. P. Janssens. 2009b. Peripartalfeeding strategy with different n-6:n-3 ratios in sows: Effect on gene expression in bac*kf*at white adipose tissue postpartum.Br. J. Nutr. 101:197–205.

Peffer, P. L., X. Lin, and J. Odle. 2005. Hepatic beta-oxidation and carnitine palmitoyltransferase I in neonatal pigs after dietarytreatments of clofibric acid, isoproterenol, and medium-chain triglycerides. Am. J. Physiol. Regul. Integr. Comp. Physiol.288:R1518–R1524.

Pégorier, J. P., P. H. Duée, R. Assan, J. Peret, and J. Girard. 1981. Changes in circulating fuels, pancreatic hormones and liverglycogen concentration in fasting or suckling newborn pigs. J. Dev. Physiol. 3:203–217.

Petkevicius, S., K. D. Murrell, K. E. Bach Knudsen, H. Jørgensen, A. Roepstorff, A. Laue, and H. Wachmann. 2004. Effects ofshort-chain fatty acids and lactic acids on survival of Oesophagostomum dentatum in pigs. Vet. Parasitol. 122:293–301.

Pettigrew, J. E. 1981. Supplemental fat for peripartal sows: A review. J. Anim. Sci. 53:107–117.

Pettigrew, J. E., and R. L. Moser. 1991. Fat in Swine Nutriton. Pages 133–145 in Swine

Nutriton. E. R. Miller, D. E. Ullrey, andA. J. Lewis, eds. Butterworth–Heineman, Stoneham, MA.Powles, J., J. Wiseman, D. J. A. Cole, and S. Jagger. 1995. Prediction of the apparent digestible energy value of fats given to pigs.Anim. Sci. 61:149–154.

Revajová, V., J. Pistl, R. Kastel, L. Bindas, D. Magic, Sr., M. Levkut, A. Bomba, and J. Sajbidor. 2001. Influencing the immuneparameters in germ–free piglets by administration of seal oil with increased content of omega–3 PUFA. Arch. Tierernahr.54:315–327.

Rooke, J. A., A. G. Sinclair, and M. Ewen. 2001a. Changes in piglet tissue composition at birth in response to increasing maternalintake of long–chain n–3 polyunsaturated fatty acids are non–linear. Br. J. Nutr. 86:461–470.

Rooke, J. A., A. G. Sinclair, and S. A. Edwards. 2001b. Feeding tuna oil to the sow at different times during pregnancy hasdifferent effects on piglet long–chain polyunsaturated fatty acid composition at birth and subsequent growth. Br. J. Nutr. 86:21–30.

Rooke, J. A., C. C. Shao, and B. K. Speake. 2001c. Effects of feedingtuna oil on the lipid composition of pig spermatozoa and invitro characteristics of semen. Reproduction 121:315–322.

Seerley, R. W., T. A. Pace, C. W. Foley, and R. D. Scarth. 1974. Effect of energy intake prior to parturition on milk lipids andsurvival rate, thermostability and carcass composition of piglets. J. Anim. Sci. 38:64–69.

Shurson, G. C., and K. M. Irvin. 1992. Effects of genetic line and supplemental dietary fat on lactation performance of Duroc and Landrace sows. J. Anim. Sci. 70:2942–2949.

Sileikiene, V., R. Mosenthin, E. Bauer, H. P. Piepho, M. Tafaj, D. Kruszewska, B. Weström, C. Erlanson–Albertsson, and S. G.Pierzynowski. 2008. Effect of ileal infusion of short–chain fatty acids on pancreatic prandial secretion and gastrointestinalhormones in pigs. Pancreas 37:196–202.

Smith, D. R., D. A. Knabe, and S. B. Smith. 1996. Depression of lipogenesis in swine adipose tissue by specific dietary fatty acids.J. Anim. Sci. 74:975–983.

Smith, S. B., H. J. Mersmann, E. O. Smith, and K. G. Britain. 1999. Stearoyl–coenzyme A desaturase gene expression duringgrowth in adipose tissue from obese and crossbred pigs. J. Anim. Sci. 77:1710–1716.

Smith, S. B., T. S. Hively, G. M. Cortese, J. J. Han, K. Y. Chung, P. Castenada, C. D. Gilbert, V. L. Adams, and H. J. Mersmann.2002. Conjugated linoleic acid depresses the delta9 desaturase index and stearoyl coenzyme A desaturase enzyme activity inporcine subcutaneous adipose tissue. J. Anim. Sci. 80:2110–2115.

Spencer, J. D., L.Wilson, S. K.Webel, R. C. Moser, and D. M.Webel. 2004. Effect of feeding protected n–3 polyunsaturated fattyacids (*Fertilium*) on litter size in gilts. J. Anim. Sci. 82(Suppl. 2):81. (Abstr.)Stein, H. H., and G. C. Shurson. 2009. Board–invited review: the use and application of distillers dried grains with solubles inswine diets. J. Anim. Sci. 87:1292–1303.

Straarup, E. M., V. Danielsen, C. E. Høy, and K. Jakobsen. 2006. Dietary structured lipids for post-weaning piglets: Fat digestibility,nitrogen retention and fatty acid profiles of tissues. J. Anim. Physiol. Anim. Nutr. (Berl.) 90:124–135.

Sun, D., X. Zhu, S. Qiao, S. Fan, and D. Li. 2004.Effects of conjugated linoleic acid levels and feeding intervals on performance,carcass traits and fatty acid composition of finishing barrows. Arch. Anim. Nutr. 58:277–286.

Tetrick, M. A., S. H. Adams, J. Odle, and N. J. Benevenga. 1995. Contribution of D-(-)-3-hydroxybutyrate to the energy expenditureof neonatal pigs. J Nutr.125264–272.

Thiel-Cooper, R. L., F. C. Parrish, Jr., J. C. Sparks, B. R.Wiegand, and R. C. Ewan. 2001. Conjugated linoleic acid changes swineperformance and carcass composition. J. Anim. Sci. 79:1821–1828.

Tischendorf, F., P. Möckel, F. Schöne, M. Plonné, and G. Jahreis. 2002. Effect of dietary conjugated linoleic acids on thedistribution of fattyacids in serum lipoprotein fractions and different tissues of growing pigs.J. Anim. Physiol. Anim. Nutr.(Berl). 86:313–325.

Wathes, D. C., D. R. E. Abayasekara, and R. J. Aitken. 2007. Polyunsaturated fatty acids in male and female reproduction. Biol.Reprod. 77:190–201.

Weber, T. E., R. T. Richert, M. A. Belury, Y. Gu, K. Enright, and A. P. Schinckel. 2006. Evaluation of the effects of dietary fat,conjugated linoleic acid, and ractopamine on growth performance, pork quality, and fatty acid profiles in genetically leangilts. J. Anim. Sci. 84:720–732.

White, H. M., B. T. Richert, J. S. Radcliffe, A. P. Schinckel, J. R. Burgess, S. L. Koser, S. S. Donkin, and M. A. Latour. 2009. Feeding conjugated linoleic acid partially recovers carcass quality in pigs fed dried corn distillers grains with solubles. J.Anim. Sci. 87:157–166.

White, P. J. 2000. Fatty acids in oilseeds (Vegetable oils).Pages 209–238 in fatty acids in foods and their health implications.C. K. Chow, eds. Marcel Dekker, Inc, New York.

Wieland, T. M., X. Lin, and J. Odle. 1993a. Emulsification and fatty-acid chain length affect the utilization of medium-chaintriglycerides by neonatal pigs. J. Anim. Sci. 71:1869–1874.

Wieland, T. M., X. Lin, and J. Odle. 1993b. Utilization of medium-chain triglycerides by neonatal pigs: Effects of emulsificationand dose delivered. J. Anim. Sci. 71:1863–1868.

Xing, J. J., E. van Heugten, D. F. Li, K. J. Touchette, J. A. Coalson, R. L. Odgaard, and J. Odle. 2004. Effects of emulsification,fat encapsulation, and pelleting on weanling pig performance and nutrient digestibility. J. Anim. Sci. 82:2601–2609.

Xu, G., S. K. Baidoo, L. J. Johnston, D. Bibus, J. E. Cannon, and G. C. Shurson. 2010. Effects of feeding diets containingincreasing levels of corn distillers dried grains with solubles

(DDGS) to grower–finisher pigs on growth performance, carcasscomposition, and pork fat quality. J. Anim. Sci. 88:1398–1410.

Yen, J. T., J. A. Nienaber, D. A. Hill, andW. G. Pond. 1991. Potential contribution of absorbed volatile fatty acids to whole–animalenergy requirement in conscious swine. J. Anim. Sci. 69:2001–2012.

Yu, Y. H., E. C. Lin, S. C. Wu, W. T. Cheng, H. J. Mersmann, P. H. Wang, and S. T. Ding. 2008. Docosahexaenoic acid regulatesadipogenic genes in myoblasts via porcine peroxisome proliferator–activated receptor gamma. J. Anim. Sci. 86:3385–9332.

Yu, X. X., J. K. Drackley, and J. Odle. 1997. Rates of mitochondrial and peroxisomal beta–oxidation of palmitate change duringpostnatal development and food deprivation in liver, kidney and heart of pigs. J. Nutr. 127:1814–1821.

Zhan, Z. P., F. R. Huang, J. Luo, J. J. Dai, X. H. Yan, and J. Peng. 2009. Duration of feeding linseed diet influencesexpression of inflammation–related genes and growth performance of growing–finishing barrows. J. Anim. Sci. 87:603–611.

Zhou, X., D. Li, J. Yin, J. Ni, B. Dong, J. Zhang, and M. Du. 2007. CLA differently regulates adipogenesis in stromal vascularcells from porcine subcutaneous adipose and skeletal muscle. J. Lipid Res. 48:1701–1709.

第4章

氨基酸及猪对氨基酸的利用

1 导 语

近年来，为了提高蛋白质的利用效率、减少环境中氮的排放，学者们已开展了大量研究。这些努力促进了具有最佳蛋白质水平和氨基酸比例的饲粮的研发，旨在最大程度地提高动物的生长性能与泌乳性能。然而，很少有研究关注影响饲粮理想氨基酸模式进入组织蛋白质沉积的过程，以及这些过程最终调节动物如何利用饲粮蛋白质。

对上述过程的进一步了解，更加说明了将饲粮氨基酸模式等同于动物组织氨基酸组成这一传统方法的局限性。例如，通过氨基酸和组织池是远不能了解氨基酸有效利用系数的，当仅仅根据体组织氨基酸组成时，会导致对饲粮氨基酸需要量和模式的析因估计不准确。本章主要综述和探讨对调节或影响不同生长阶段猪氨基酸利用起重要作用的关键过程。

饲粮氨基酸可被小肠上皮细胞和肠道微生物利用，首过代谢可引起氨基酸模式的改变。近期对新型氨基酸转运载体系统及其分子结构调节作用的发现，推动了生理学上对细胞膜表面氨基酸相互作用及其影响细胞氨基酸利用的前期研究。对胎儿氨基酸代谢的最新了解，让人们发展了给妊娠母猪饲喂蛋白质和氨基酸的新策略，目的是最大程度地降低营养素的浪费。氨基酸有效性和仔猪哺乳需求不同时，乳腺的氨基酸代谢会发生改变，这促使研究者不断完善泌乳母猪的氨基酸需要量。衰老过程通过体组织对激素刺激的反应能力，最终影响肌肉对氨基酸的利用。

2 细胞对氨基酸的转运：氨基酸利用的前提

饲粮氨基酸在细胞内的有效性受位于细胞膜上的氨基酸转运载体蛋白的协同活性调控，氨基酸转运载体蛋白负责氨基酸的跨膜运输（Broër，2008；Palacin等，1998；Shennan等，2000）。氨基酸转运的调控是非常复杂的，因为很多转运载体不仅可以转运多种氨基酸，而且可以协同转运不同氨基酸同时进出细胞（Shennan，2000）。本章主要综述目前已知的与猪营养相关的转运系统和转运蛋白（有些是由其他动物的研究进展推断得来），并没有探讨真核细胞氨基酸转运的机制。

氨基酸通过跨细胞途径或细胞旁路途径穿过上皮细胞。慢速转运的亲水性化合物，通过细胞旁路途径经细胞间紧密连接被摄取（Urakami等，2003）。由于不同组织中紧密连接的屏障特性不同，即屏障强度（一般以电阻表示）和对电荷的选择性不同，因此大多数上皮细胞之间的连接是“紧密的”，这样才有助于氨基酸通过跨细胞途径转运（Colegio等，2002）。氨基酸转运的跨细胞途径包括非载体介导的自由扩散途经和载体介导的转运途经，后者需要大量的载体（或转运蛋白），其中需要消耗能量者为主动转运，不需要消耗能量者为被动转运。载体蛋白介导的被动转运过程包括简单扩散和易化扩散。氨基酸通过简单扩散途经穿过细胞膜脂质双分子层的速度，取决于特定氨基酸在脂质双分子层疏水端的可溶性（也被称为渗透系数）和细胞膜两侧氨基酸的浓度梯度。不带电荷的小分子氨基酸能自由通过细胞膜，但是赖氨酸、精氨酸、组氨酸、鸟氨酸、天冬氨酸和谷氨酰胺酸等带电荷的氨基酸通过细胞间隙或膜孔转运。通过简单扩散过程的氨基酸转运的重要程度，可能随着组织和器官的不同而异，这取决于该组织和器官中载体蛋白介导的转运能力。另一种形式的被动转运是易化扩散途经，其特点是需要载体或者转运蛋白介导氨基酸的摄取。主动转运也需要载体或者转运蛋白，与促进扩散不同的是，主动转运需要消耗能量。因此，根据其转运机制（是否依赖能量）和底物特异性的不同，易化扩散和主动转运途径中的氨基酸转运载体或转运蛋白可被分为不同的转运系统（Hyde等，2003；Broër等，2004；Hundal和Taylor，2009）。主动转运过程中的能量依赖型转运载体作为二级主动转运载体起作用，通过偶合氨基酸的转运和Na^+向胞内的运动过程，促进特定氨基酸的浓集性摄入和胞内累积。因此，Na^+依赖性氨基酸转运载体是共转运系统，因为其同向转运两种溶质（如氨基酸和Na^+）。非能量依赖性的氨基酸转运载体是不依赖于Na^+的，为反向转运系统或单向转运系统。

反向转运系统包括三级主动转运载体，可以反向转运两种氨基酸，有助于胞外氨基酸的摄取，同时通过二级主动转运载体转出胞内积累的胞质氨基酸（Hundal和Taylor，2009）。例如，在图4-1中，$b^{0,+}$载体可促进胞外赖氨酸的摄取，同时转出通过$B^{0,+}$载体转入胞内累积的亮氨酸。$B^{0,+}$是Na^+依赖性的共转运系统。与主动转运过程相似，氨基酸的易化扩散过程是空间特异性的，但不同的是易化扩散过程不需要代谢能。因此，氨基酸转运是在单向转运载体介导下顺着浓度梯度或电化学梯度被动进行的，也可以是双向进行的。

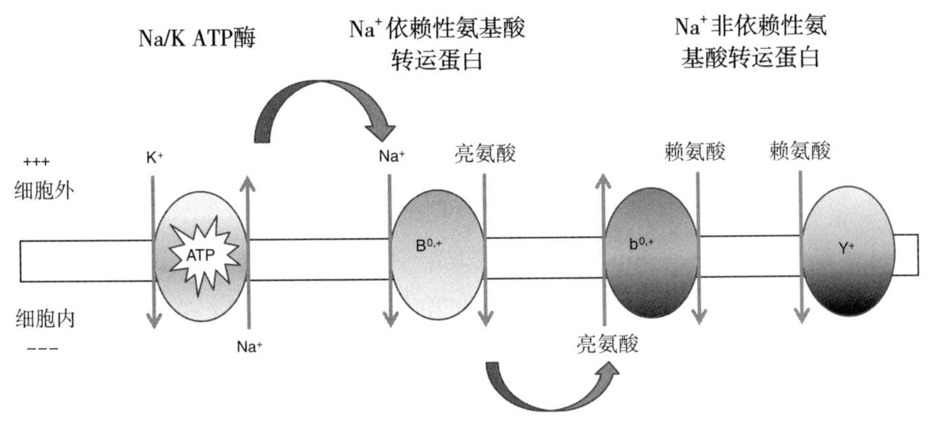

图4-1 细胞膜上的初级（Na/K ATP酶）、二级（Na^+依赖性氨基酸转运蛋白）和三级（Na^+非依赖性氨基酸转运蛋白）主动转运机制

二级主动转运蛋白（如$B^{0,+}$系统）将氨基酸从胞外（细胞外侧）池转运到胞内（细胞内侧）池，而三级主动转运蛋白（$b^{0,+}$和y^+系统）使个别氨基酸进行重分配，但是不改变总氨基酸池的大小。+++和---分别表示细胞外区室净带正电荷、细胞内区室净带负电荷。根据Hundal和Taylor（2009）修改和重新做图，经美国生理协会同意后再版。

氨基酸进入器官细胞的能力既取决于特定氨基酸与转运蛋白结构域的亲和力（即*Km*），也取决于细胞壁上功能性转运载体的数量（即Vmax；Souba和Pachitti，1992）。氨基酸转运系统的特点因其转运底物的不同而异（Souba和Pacitti，1992）。有些转运系统在机体内普遍表达，还有些转运系统只在特定组织中表达（Souba和Pacitti，1992）。

◆ 2.1 阳离子（碱性）氨基酸转运

碱性氨基酸转运载体（CAT）蛋白是与碱性氨基酸亲和、异位密切相关的转运蛋白。机体中存在不同的碱性氨基酸转运系统，CAT蛋白是碱性转运载体中y^+系统的特有成员。CAT转运蛋白是典型的pH非依赖性的，其转运活性由细胞膜的超极化来激活。这些转运蛋白似乎是由细胞膜转运侧存在的氨基酸来激活（Deves和Boyd，1998；Closs，2002）。Na^+非依赖性系统y^+在机体内普遍表达，并且只能特异性地转运碱性氨基酸（即精氨酸、组氨酸、赖氨酸和鸟氨酸）。

◆ 2.2 碱性氨基酸和中性氨基酸共有的转运

赖氨酸可以通过其他载体蛋白进行转运，这些载体蛋白不具有与碱性氨基酸的特异性亲和力，如CAT蛋白。无论是否存在Na^+，赖氨酸均可以被转运（Vilella等，1990；Wilson和Webb，1990），Na^+依赖性转运蛋白可以利用Na^+来增加转运蛋白的亲和力（Souba和Pacitti，1992；Soriano-Garcia等，1999；Vilella等，1990）。Na^+依赖性赖氨酸摄取是通过$B^{0,+}$系统实现的（Souba和Pacitti，1992）。$ATB^{0,+}$转运载体通过Na^+依赖性和Cl^-依赖性机制转运疏水性（双极性）的碱性和中性氨基酸（Broër，2008），其与中性氨基酸的亲和力强于碱性氨基酸（Sloan和Mager，1999）。细胞的氨基酸饥饿可上调体外培养的爪蟾卵细胞中$ATB^{0,+}$的表达，补充氨基酸可下调细胞中$ATB^{0,+}$的表达（Taylor等，1996）。

除$B^{0,+}$系统外，在上皮细胞中还存在另外两种广泛表达的系统。这两种系统为赖氨酸和中性氨基酸共有的。其中$b^{0,+}$系统的分子形式是杂聚肽碱性氨基酸转运蛋白rBAT/$b^{0,+}$AT，y^+L系统的分子形式是y^+LAT1/4F2hc和y^+LAT2/4F2hc。y^+LAT1/4F2hc和y^+LAT2/4F2hc均是Na^+非依赖性中性和碱性氨基酸转运蛋白，由一个起催化作用的轻链（y^+LAT）和一个由二硫键连接的重链（4F2hc）组成（Torrents等，1998）。y^+LAT1/4F2hc主要在肾脏和肠上皮细胞中表达，而y^+LAT2/4F2hc在脑、心脏、睾丸、肾脏、小肠和腮腺等多个组织中广泛分布（Broër 等，2000a）。这两种转运蛋白均可参与细胞内碱性氨基酸转出细胞，以及细胞外中性氨基酸转入细胞（Broër，2008）。rBAT/$b^{0,+}$AT是具有广谱特异性的Na^+非依赖性中性和碱性氨基酸转运蛋白，由一个具有催化作用的轻链$b^{0,+}$AT1和一个共价结合的Ⅱ型糖蛋白重链rBAT组成，其间由一个二硫键连接（Dave等，2004；Broër，2008）。$b^{0,+}$系统由一条重链rBAT和一条轻链$b^{0,+}$AT组成。在生理条件下，rBAT/$b^{0,+}$AT作为三级主动转运蛋白，将赖氨酸和精氨酸转入细胞的同时，将中性氨基酸转出细胞（Bauch等，2003）。在肾脏和肠细胞中，rBAT/$b^{0,+}$AT和y^+LAT1/4F2hc作为功能单位，负责碱性氨基酸的重新摄取，同时将中性氨基酸分泌到肾盂或肠腔中（Bauch等，2003；Chillaron等，1996；Sperandeo等，2008）。

为了使rBAT/$b^{0,+}$AT能够反向转运赖氨酸/中性氨基酸，通过B^0AT1转运蛋白（B^0系统的分子实体）集中（Na^+协同转运）顶端的中性氨基酸转运活性，以保证细胞内存在充足的中性氨基酸。转运蛋白B^0AT1是Na^+依赖性氨基酸转运蛋白，能主动转运亮氨酸和缬氨酸等大分子支链中性氨基酸，但不能转运酸性氨基酸或者碱性氨基酸。B^0AT1转运蛋白位于肾脏和肠上皮细胞的基顶膜（Broër等，2004），可将肾盂或肠腔中的中性氨基酸重新摄取到细胞中。

3 肠道内氨基酸的利用

◆ 3.1 肠道内氨基酸吸收和转运的机制

饲粮蛋白质的利用涉及一系列的步骤，包括胃和小肠中蛋白质的消化以及小肠对小肽和游离氨基酸的吸收。蛋白质消化后，氨基酸利用的第一步是发生在肠腔–基顶膜表面（或称为上皮细胞刷状缘），在这里三肽和二肽直接被吸收，或者在跨基顶膜转运之前分别被水解为其组分即二肽和单一氨基酸（图4–2）。通过基顶膜转入细胞内的氨基酸进行如下代谢途径：①代谢；②原位合成蛋白质（如肽酶、载脂蛋白和黏蛋白）；③转出到肠腔，同时将肠腔中的氨基酸转入细胞内；④通过基底膜进入门脉血。另外，细胞内氨基酸的存在和利用率，取决于肠系膜动脉通过基底膜供给的氨基酸量。正如之前讨论过的一样，氨基酸通过跨细胞途径或细胞旁路途径穿过上皮细胞。然而，在小肠的“渗漏性”上皮细胞中，细胞旁路途径是营养素转运的主要途径（Colegio等，2002），因此可以推测，尤其是当细胞外氨基酸浓度较高时会发生氨基酸的转运。下面将主要讨论多肽和氨基酸转运的跨细胞机制。但是了解细胞旁路途径也很重要，虽然目前对细胞旁路途径的研究和了解较少，但是千万不要忘记，细胞旁路途径在饲粮氨基酸利用和低分子量蛋白质直接吸收过程中可能发挥着重要作用。

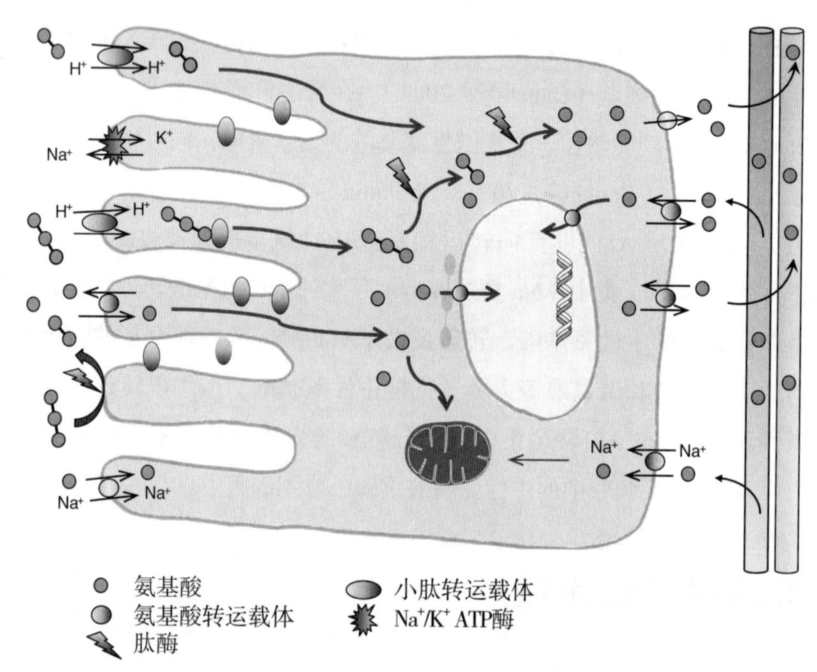

图4-2 肽和氨基酸跨膜转运及其在小肠柱状上皮细胞摄取的机制

3.1.1 多肽

早期研究表明，肠腔内的肽酶活性只能分解出小部分的游离氨基酸（Adibi，1971；Silk等，1976）。后来，在人肠腔的刷状缘和可溶物之间发现了两类分布几乎相同的三肽酶和二肽酶活性，得出的结论是存在两类完全不同的黏膜肽酶：一类位于细胞质中，另一类位于细胞的刷状缘膜上（Silk等，1985）。总体来说，很大比例的三肽和二肽在刷状缘上被水解后，分别以其组分二肽和氨基酸的形式，通过小肽和游离氨基酸转运机制被摄取。

生理学和分子生物学研究表明，肠道小肽转运载体Pept-1是肠黏膜刷状缘膜上唯一的小肽转运载体（Abidi，2003），其不能转运游离氨基酸（Adibi，1971，2003），这一点在猪体内可能相同。这就是关于肽转运的二元假说。与刷状缘肽酶亲和力高的小肽主要被刷状缘膜表面的酶水解为游离氨基酸后被吸收，而与刷状缘肽酶亲和力低的小肽直接被吸收，然后被细胞质中的肽酶水解（Silk等，1985；图4-2）。利用Northern blot技术检测猪的所有组织（如半腱肌、背最长肌、肾脏、肝脏、胃、盲肠、结肠和小肠），发现小肽转运载体PepT1只在小肠中表达，而且空肠中的丰度最高，其次是十二指肠和回肠。因为出现在门静脉血中的大部分氨基酸是游离氨基酸形式，所以基顶膜摄取和胞内代谢后，基底膜氨基酸的转运很可能是单一氨基酸转运机制参与的。因此，PepT1蛋白可能位于基顶膜上。另外，PepT1是pH依赖性的，因此受肠腔刷状缘膜表面H^+的激活。因为目前很难从基底膜上分离囊泡开展摄取方面的研究，所以对肠上皮细胞跨基底膜转运的了解远没有对基顶膜转运的了解多。尽管如此，通过上皮柱状细胞基底膜的小肽转运还是有可能的，但是目前对小肽参与的肠道吸收的数量和营学方面的重要性尚不清楚。

3.1.2 单一氨基酸

目前，已经在多种动物模型上研究了肠道氨基酸吸收的机制和解剖学部位。但是，以猪为模型的研究较少。

3.1.2.1 基顶膜转运

除了$b^{0,+}$系统外，所有的基顶膜转运载体均是离子依赖性的，且能够浓集转运。因为人肾脏中$b^{0,+}$系统的缺失会导致遗传性高氨基酸尿-高胱氨酸尿症的发生（Feliubadaló等，1999），所以在过去的几年里对$b^{0,+}$系统的研究很多。杂聚肽转运载体rBAT/$b^{0,+}$AT是肾脏和肠道中碱性氨基酸和胱氨酸的主要转运载体。近年来，有学者研究了猪肠道组织中rBAT/$b^{0,+}$AT转运载体在不同解剖部位的发育性表达规律（Xiao等，2004；Feng等，2008；Wang等，2009）。藏猪的$b^{0,+}$AT在小肠和肾脏中高表达，而在心脏、脑、肺脏和背最长肌中低表达（Wang等，2009）。在猪的十二指肠、空肠、回肠和结肠中，也检测到了$b^{0,+}$AT的表达（Xiao等，2004；Feng等，2008）。如图4-3所示，$b^{0,+}$氨基酸转运系统的活性是Na^+非依赖性的，可介导基顶膜对精氨酸、组氨酸、赖氨酸（AA^+）和胱氨酸（CSSC）等碱性氨基酸的摄取，同时转出细胞内的丝氨酸和苏氨酸等中性氨基酸（图4-3中用AA^0表示；Broër，2008）。图4-4列出的是第7和21日龄哺乳藏猪的肠道、肾脏和肌肉中$b^{0,+}$AT基因的解剖部位分布和发育性表达情况。

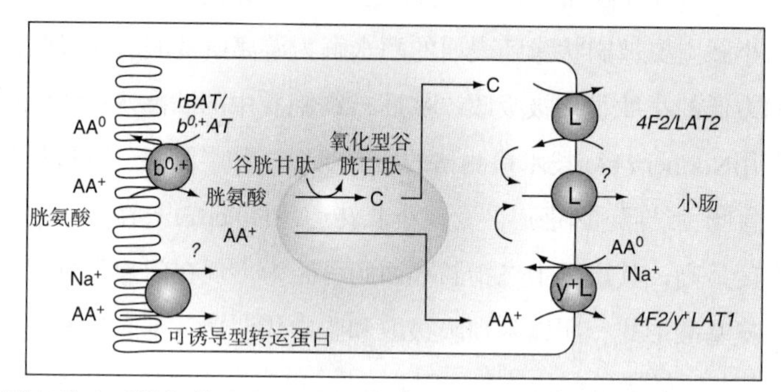

图4-3 肠道细胞中碱性氨基酸（AA^+）和中性氨基酸（AA^0）的基顶膜转运和基底膜转运

rBAT/$b^{0,+}$AT是杂聚肽转运蛋白；$b^{0,+}$是Na^+非依赖性的氨基酸转运系统；L为L型氨基酸转运载体；y^+L为y^+L型氨基酸转运载体；4F2/LAT2为L系统的分子实体；4F2/y^+LAT1为y^+L系统的分子实体。（资料来源于Broër，2008；经美国生理协会批准后再版）

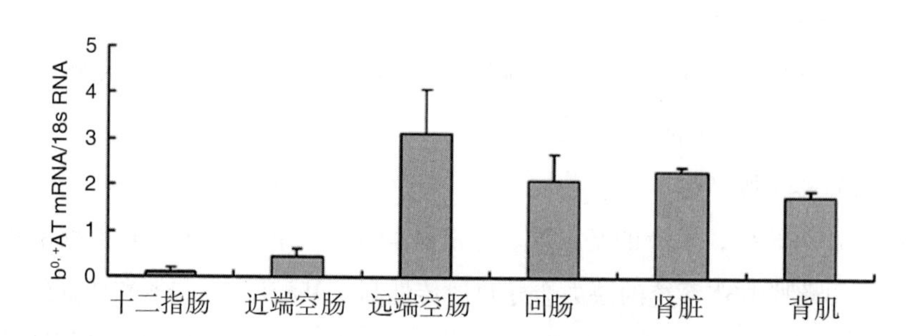

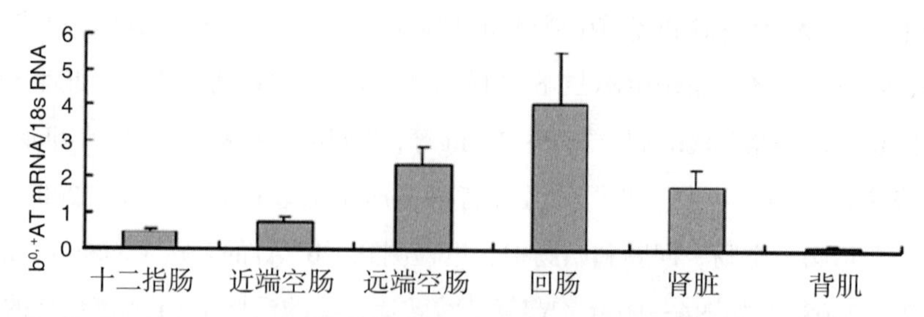

图4-4 7日龄（上图）和21日龄（下图）哺乳藏猪不同组织中$b^{0,+}$AT的mRNA相对丰度

利用Northern blot技术检测丰度，用光密度法定量各条带。（资料来源于Wang等，2009；经Elsevier批准后再版）

与7日龄相比，21日龄仔猪的空肠$b^{0,+}$AT表达水平更低，回肠表达水平更高。Xiao等（2004）报道了类似的表达规律，即回肠和空肠中$b^{0,+}$ATmRNA的丰度比十二指肠高，在35日龄时十二指肠和空肠中$b^{0,+}$ATmRNA的表达趋于稳定。此外，Feng等（2008）报道，从1日龄到150日龄，猪回肠中$b^{0,+}$ATmRNA的丰度呈线性增加。另外$b^{0,+}$系统还可以促进鸡空肠上皮细胞对赖氨酸的吸收（Soriano-Garcia等，1999；Angelo等，2002）。除了$b^{0,+}$AT外，在Na^+存在时，还存在一种可诱导的转运蛋白用于赖氨酸的转运（Broër，2008）。在牛的小肠中，赖氨酸是通过Na^+依赖性和Na^+非依赖性过程转运的，在小肠远端，如回肠中赖氨酸的摄取能力更强，但是在空肠中赖氨酸的亲和力更高（Wilson和Webb，1990）。因为大多数中性氨基酸的基顶膜转运是逆浓度梯度进行的，所以需要依赖Na^+。目前已发现两种主要的基顶膜转运系统，即B^0和ASC系统，已鉴定出的其分子实体分别是B^0AT和ASCT1。

3.1.2.2 基底膜转运

尽管Na^+非依赖性系统y^+的转运载体蛋白可能只存在于基底膜并在此发挥作用（图4-3），但是其在鸡（Angelo等，2002）和牛（Wilson和Webb，1990）肠道组织吸收赖氨酸过程中均发挥着重要作用。碱性氨基酸通过单向转运蛋白CAT-1转出细胞，同时通过y^+LAT1和y^+LAT2反向转入中性氨基酸。在马的小肠黏膜中已检测到了碱性氨基酸转运载体CAT-1的mRNA（Woodward等，2010）。鉴于卵母细胞和体外培养脑细胞内CAT-1与精氨酸结合的偏好性（Broër等，2000）和血液中高浓度的谷氨酰胺，推测小肠基底膜上y^+LAT2的主要功能是将肠上皮细胞内的精氨酸转运到血液，同时转入谷氨酰胺，这在某种程度上解释了小肠上皮细胞对谷氨酰胺的高摄取率。

大多数中性氨基酸是顺浓度梯度被离子非依赖性氨基酸转运蛋白（包括Asc-1、LAT2、y^+LAT1和y^+LAT2）转运的（Krehbiel和Matthews，2003；图4-5）。LAT2是L系统的分子实体之一，据报道其存在于小鼠和人的肠上皮细胞基底膜（Rossier等，1999）以及马的肠黏膜上（Woodward等，2010）。芳香族氨基酸通过Na^+非依赖性单向转运蛋白TAT1，选择性地顺浓度梯度转运。最后，同向浓集转运以及将中性氨基酸从肠系膜血液中转运到细胞中，是通过Na^+依赖性ATA2转运蛋白和SNAT2进行的。牛（Wilson和Webb，1990）和鳗鱼

（Vilella等，1990）饲粮中的氨基酸存在竞争性抑制作用，表明整个小肠可能都需要利用这些赖氨酸转运系统。

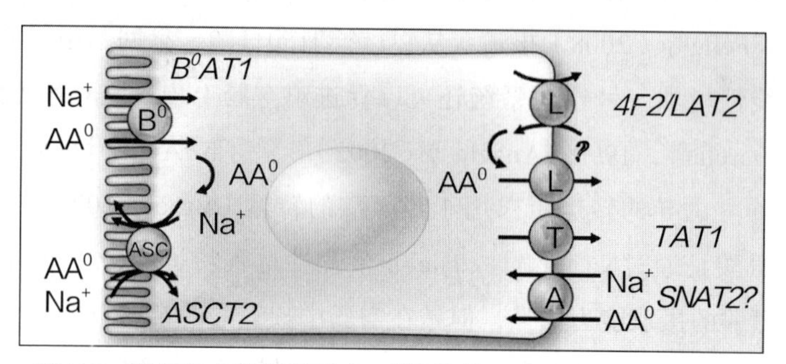

图4-5 肠细胞中性氨基酸（AA⁰）的基顶膜转运和基底膜转运

AA⁰为中性氨基酸；B⁰为ASC的基底膜转运系统；B⁰AT1为B⁰系统的分子实体；ASCT2为ASC系统的分子实体；L、T、A分别为L、T、A型氨基酸转运载体；4F2/LAT2为L系统的分子实体；TAT1为T系统的分子实体；SNAT2为Na^+依赖性转运蛋白。（资料来源于Broër，2008；经美国生理协会批准后再版）

在小肠的不同肠段中，Na^+非依赖性转运系统和Na^+依赖性转运系统与氨基酸的亲和力和功能均不同，其转运载体也不同（Wilson和Webb，1990）。因此，推测竞争性抑制导致的赖氨酸吸收效率低下可以在不同肠段相互补偿。例如，空肠中Na^+依赖性和Na^+非依赖性赖氨酸转运系统的亲和力均较高，对赖氨酸的转运能力较低；但是回肠中相同转运系统的亲和力较低，对赖氨酸的转运能力较高（Wilson和Webb，1990）。这些系统对饲粮中氨基酸的平衡是否存在协同调节作用，以确保最多的赖氨酸吸收，目前尚不明确。

3.1.3 肠腔与动脉氨基酸的利用

毫无疑问，采食过程中肠道营养主要是通过肠腔侧供给。氨基酸在基底膜远肠腔侧（浆膜侧）被逆向交换，或者被转出细胞后进入静脉毛细血管系统，这清晰地说明肠系膜动脉供应的氨基酸是供肠细胞利用的。目前，对决定肠腔和动脉氨基酸直接用于满足肠道氨基酸需要的相对比例的机制尚不明确。肠腔和动脉氨基酸对肠道营养供给的相对贡献，进食期间和进食后均在不断发生变化，每种氨基酸之间也不同，同时也受营养状况的影响（Bos等，2003），因此明确肠道利用饲粮氨基酸和动脉氨基酸的分配比例是很困难的。

Stoll等（2000）报道，在空肠近端的肠腔中存在大量的亮氨酸，此处肠黏膜也摄取大量的亮氨酸，但其摄取量沿着小肠从前到后逐渐降低。另外，在小肠后段中，即使肠腔可利用氨基酸减少，也会有大量的蛋白质存留。因此，Stoll等（1998）指出，后端肠道是从循环中而不是饲粮中获得大部分的氨基酸用于蛋白质的合成，这在某种程度上解释了远端肠道比近端肠道蛋白质周转速度降低的原因（Stoll等，2000）。但是，近端肠道蛋白质周转速度快可能还与蛋白酶活性高有关。另外，近端肠道主要是依靠动脉血供应氨基酸，而不是依靠肠腔供应氨基酸。

与游离脂肪酸不同，氨基酸的分子量较小，反射系数低，很容易通过毛细管孔隙从细胞外间隙扩散到静脉毛细血管。因此，门脉血中出现的氨基酸即代表净吸收的氨基酸。

3.1.4 大肠氨基酸的利用

小肠在氨基酸利用和吸收中发挥的作用毋庸置疑（Metges，2000），但是已经有足够的研究表明，猪的大肠也具有吸收氨基酸的能力。在空肠后端灌注^{15}N标记的细菌后，在静脉血中可检测到标记的赖氨酸（Niiyama等，1979），提示大肠可吸收微生物合成的赖氨酸。从消化道中手术分离盲肠，在分离到的盲肠囊上安装瘘管，可以检测到天冬酰胺、丝氨酸、苏氨酸、酪氨酸、精氨酸、组氨酸、赖氨酸和天冬氨酸的原位吸收（Olszewski和Buraczewski，1978）。在猪的近端结肠（位于盲肠末端），通过任何转运系统尤其是B^{0+}系统，疏水性中性氨基酸比亲水性中性或碱性氨基酸更容易被吸收（Sepúlveda和Smith，1979）。在体外直接处理小鼠的结肠肠腔表面，可检测到放射性同位素标记甘氨酸的吸收，证明小鼠结肠中存在$B^{0,+}$系统（Ugawa等，2001）。然而，在小肠的四个肠段、盲肠和结肠中均有mRNA的表达（Hatanaka等，2001），结肠和盲肠中mRNA的相对丰度高于小肠远端。与此相反，小肠的四个肠段中均检测到了H^+偶联小肽转运蛋白Pept-1的mRNA表达，但在盲肠和结肠中未检测到。以上数据表明，$ATB^{0,+}$mRNA只在小鼠的肠道后端表达。然而，从数量上估计，微生物合成的氨基酸只有大约10%被大肠吸收，说明猪的大肠，至少在饲喂常规谷物基础饲粮条件下，其营养学意义不大（Torrallardona等，2003a，2003b）。

◆ 3.2 肠上皮细胞的氨基酸代谢和利用的个体发育：从新生期到断奶早期

3.2.1 谷氨酰胺和谷氨酸

猪乳中含有高浓度的游离的和肽结合的谷氨酰胺和谷氨酸，这对仔猪的生长、发育和小肠功能至关重要（Wang等，2008；Kim和Wu，2009）。这两种氨基酸在哺乳仔猪肠道中可被完全代谢（Reeds等，1996），从而为肠细胞提供能量，也可为合成其他氨基酸（脯氨酸、精氨酸、鸟氨酸和瓜氨酸）、DNA和蛋白质提供氮源（Bertolo和Burrin，2008）。在被肠细胞摄取后，谷氨酰胺酶（GSE）催化谷氨酰胺脱氨生成谷氨酸，而谷草转氨酶（GOT）、谷丙转氨酶（GPT）和谷氨酸脱氢酶（GDH）催化谷氨酸转氨基生成α-酮戊二酸。然后，生成的α-酮戊二酸进入三羧酸循环，被完全代谢成CO_2和ATP（Burrin和Stoll，2009）。肠细胞摄取的部分谷氨酰胺和谷氨酸也能用来合成瓜氨酸（精氨酸的中间代谢物）。吡咯啉-5-羧酸合成酶（P5CS）催化的第一步反应，是将谷氨酰胺和谷氨酸转化为Δ^1-L-吡咯啉-5-羧酸（P5C）。然后，P5C被转运到细胞质中，在吡咯啉-5-羧酸还原酶（P5CR）作用下转化为脯氨酸，或者在线粒体中被鸟氨酸δ-转氨酶（OAT）催化进行转氨基生成鸟氨酸。但是，由于仔猪肠细胞中P5CS的活性较低，因此哺乳期的大部分时间均缺乏这两种代谢途径（Wu，1997；Davis和Wu，1998）。谷氨酸最终可在N-乙酰谷氨酸合成酶（NAGS）催化下合成N-乙酰谷胺酸（NAG）。NAG是鸟氨酸氨甲酰转移酶（OTC）的变构激活剂，因此在精氨酸和脯氨酸的代谢中起着调控作用（图4-6）。

3.2.2 精氨酸

精氨酸是仔猪实现最大化生长的必需氨基酸（Southern等，1983）。事实上，一些研究表明，母乳中精氨酸的供给可能是限制哺乳仔猪实现体增重最大化的因素（Kim和Wu，2004；Wu等，2004）。母乳只能为1周龄仔猪提供不到40%的精氨酸需要量（Wu等，2004）。因此，精氨酸的内源性合成对维持哺乳仔猪的精氨酸平衡发挥着关键作用（Flynn和Wu，1996）。哺乳仔猪的精氨酸是在小肠中利用饲粮脯氨酸合成的，断奶仔猪的精氨酸是在肾脏中利用内源性瓜氨酸合成的（Wu和Morris，1998；Bertolo等和Burrin，2008）。精氨酸的代谢特性随着年龄的增长发生改变，是因为猪肠上皮细胞中参与精氨酸代谢的酶（如精氨酸

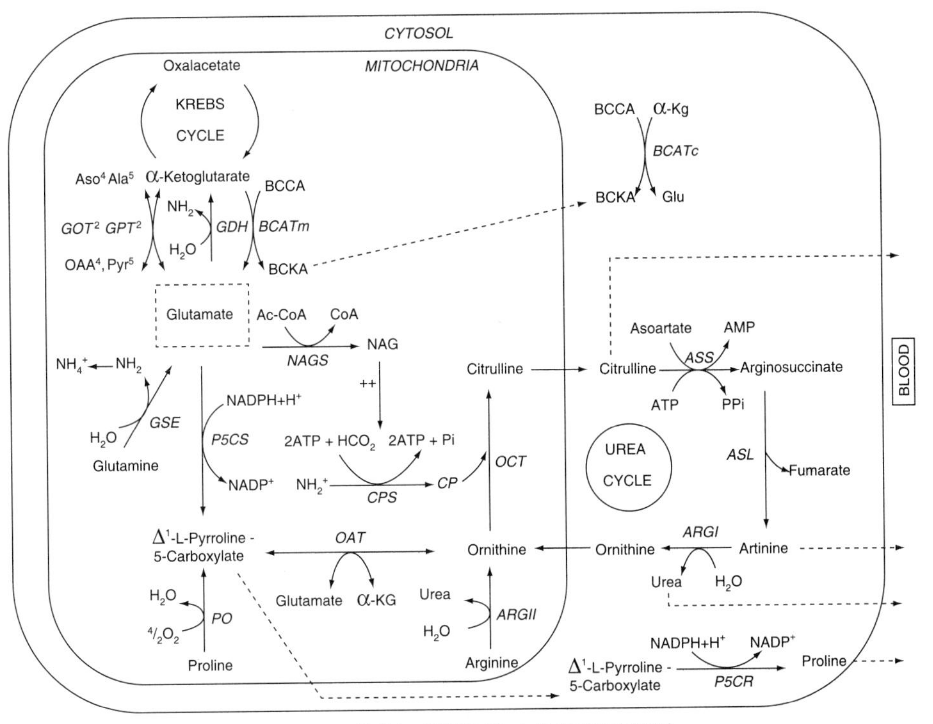

图4-6 哺乳仔猪肠细胞中的氨基酸代谢

CYTOSOL为细胞质；MITOCHONDRIA为线粒体；BLOOD为血液；Oxalacetate为草乙酸盐；KREBSCYCLE为三羧酸循环；α-Ketoglutarate为α-酮戊二酸；GOT为谷草转氨酶；GPT为谷丙转氨酶；OAA为草酰乙酸；Pyr为吡咯烷酮；GDH为谷氨酸脱氢酶；BCAA为支链氨基酸；BCATm和BCATc分别为线粒体和细胞质中的支链氨基转移酶（BCAT）；BCKA为α-酮酸；Glutamate为谷氨酸；Ac-CoA为乙酰辅酶A；CoA为辅酶A；NAGS为N-乙酰谷氨酸合成酶；NAG为N-乙酰谷氨酸；GSE为谷氨酰胺酶；Glutamine为谷氨酰胺；NADPH为还原性辅酶Ⅱ；P5CS为吡咯啉-5-羧酸合成酶；Δ^1-L-pyrroline-5-carboxylate为Δ^1-L-吡咯啉-5-羧酸；Proline为脯氨酸；PO为脯氨酸氧化酶；Citrulline为瓜氨酸；OCT为氨甲酰转移酶；Ornithine为鸟氨酸；Urea为尿素；Arginine为精氨酸；ARGⅠ和Ⅱ为精氨酸酶Ⅰ和Ⅱ；Asoartate为天冬氨酸；ASS为精氨酸琥珀酸盐合成酶；Argininosuccinate为琥珀酸；UREA CYCLE为尿素循环；ASL为精氨酸琥珀酸盐酶；Fumarate为延胡索酸。

酶Ⅰ和Ⅱ以及P5CR）的活性和表达量产生了发育性变化（Wu等，1996，2004；Wu，1997）。在哺乳期动物中，饲粮脯氨酸首先在脯氨酸氧化酶（PO）催化下转化为Δ^1-L-吡咯啉-5-羧酸（P5C）。然后，鸟氨酸δ-转氨酶（OAT）催化

P5C发生转氨基作用，与谷氨酸作用生成鸟氨酸，随后鸟氨酸在氨甲酰基转移酶（OCT）作用下生成瓜氨酸。瓜氨酸从线粒体转运到细胞质中，在精氨酸琥珀酸合成酶（ASS）作用下生成精氨酸琥珀酸盐。精氨酸琥珀酸盐再在精氨酸琥珀酸酶（ASL）作用下分解为游离精氨酸和延胡索酸。前者进入门静脉，后者进入线粒体作为三羧酸循环的中间代谢物。门静脉精氨酸经过肝脏和肾脏后，最终供整个机体代谢利用（Wu和Morris，1998；Bertolo和Burrin，2008）。

与哺乳仔猪不同，断奶仔猪的肠细胞中催化瓜氨酸转化为精氨酸的ASS和ASL活性很低，而降解精氨酸的精氨酸酶（ARG Ⅰ 和 Ⅱ）的活性很高（Wu，1997）。因此，断奶仔猪的小肠不能直接合成精氨酸。饲粮中的精氨酸、谷氨酰胺、谷氨酸和脯氨酸先在肠细胞中被转化为瓜氨酸，然后进入门静脉血（Bertolo和Burrin，2008；Wu，1997），最后瓜氨酸被肾脏细胞转化为精氨酸，说明肾脏细胞中ASS和ASL的活性很高（Brosnan，2003）。小肠净合成的瓜氨酸，为肝脏精氨酸酶作用下的饲粮精氨酸旁路代谢提供了一种有效机制，防止精氨酸被肝脏完全分解（Bertolo和Burrin，2008）。饲粮精氨酸在细胞质和线粒体中精氨酸酶（ARG Ⅰ 和 Ⅱ）作用下转化为瓜氨酸。利用谷氨酰胺合成瓜氨酸需要线粒体中谷氨酰胺酶（GSE）和吡咯啉-5-羧酸合成酶（P5CS）的参与，GSE将谷氨酰胺转化为谷氨酸，P5CS将谷氨酸转化为P5C。断奶仔猪的肠细胞最终也可利用饲粮中的氨基酸，在鸟氨酸δ-转氨酶（OAT）和细胞质中的Δ1-L-吡咯啉-5-羧酸还原酶作用下合成脯氨酸。抑制猪肠细胞中的OAT活性，可导致精氨酸合成的脯氨酸减少80%~85%（Wu等，1996）。但是，当精氨酸不足时，OAT只作用于鸟氨酸的合成过程（Nelson等，2008）。

3.2.3 其他必需氨基酸

肠上皮细胞是新生仔猪肠道降解大部分支链氨基酸的重要部位，但不降解其他必需氨基酸（Chen等，2009）。在哺乳仔猪中，经过肠道的首过代谢，有40%的亮氨酸、30%的异亮氨酸和40%的缬氨酸进入到门脉组织（Stoll等，1998）。一旦被肠细胞摄取，支链氨基酸会在细胞质和线粒体中的支链氨基转移酶（BCAT）作用下，转化为相应的α-酮酸（BCKA）。发生转氨基作用后，线粒体支链α-酮酸脱氢酶复合体（BCKAD）催化所有的三种α-酮酸，发生氧化脱羧反应，生成酰基辅酶A衍生物。但是，仔猪肠上皮细胞中BCKAD的活性很低，所以大部分的α-酮酸被释放到细胞外间隙（Chen等，2009）。因此，支链氨基酸不太可能是仔猪小肠的重要能量底物。前面已经提到，黏膜中支链氨基酸的分解代谢可为丙氨酸和谷氨酸的合成提供氮源（图4-6），也可以生成α-酮酸（BCKA）。尽管BCKA对肠细胞的作用尚不明确，但BCKA可以减少肠细胞中蛋白质的降解，这与鸡骨骼肌中的报道一致（Nakashima等，2007）。

Stoll等（1998）报道，0～21日龄仔猪的肠上皮细胞，可大量分解代谢组氨酸、赖氨酸、蛋氨酸、苯丙氨酸、苏氨酸和色氨酸。然而，近期的研究（Chen等，2007，2009）表明，猪的肠上皮细胞不会大量氧化这些必需氨基酸，因为缺乏降解这些必需氨基酸的关键酶，包括苏氨酸脱氢酶、苏氨酸脱水酶、酵母氨酸脱氢酶和苯丙氨酸羟化酶。因此，小肠中组氨酸、赖氨酸、蛋氨酸、苯丙氨酸、苏氨酸和色氨酸的代谢可能是在肠黏膜上的肠道微生物作用下完成的，而不是Chen等（2009）提出的被仔猪肠上皮细胞吸收。Chen等（2009）提出的微生物可以改变和利用必需氨基酸，可为饲粮添加抗生素增加仔猪骨骼肌中蛋白质的沉积提供一种机制解释（Bergen和Wu，2009）。只有当鼠和鸡缺乏支链氨基酸和其他氨基酸时，BCAT和谷氨酰胺转氨酶L和K才能催化蛋氨酸和苯丙氨酸发生少量的降解（Wu和Thompson，1989b；Wu等，1991）。例如，在培养基中分别添加2 mmol/L的亮氨酸、异亮氨酸和缬氨酸，可完全抑制断奶前（Chen等，2009）和断奶后（Chen等，2007）仔猪肠上皮细胞中蛋氨酸和苯丙氨酸的转氨基作用。

4 生长期氨基酸的利用

◆ 4.1 氨基酸参与的胰岛素信号通路

新生期动物进食后，所有组织中的蛋白质合成水平都很高，其特点是生长速度快（Suryawan等，2006）。骨骼肌中蛋白质的沉积量显著增加，这是由进食后胰岛素和氨基酸的含量增加调控的；但在肝脏和心脏等其他组织中，蛋白质的沉积只受氨基酸含量的影响。骨骼肌对胰岛素和氨基酸刺激均能产生反应，使得骨骼肌比其他器官的氨基酸利用率高，蛋白质合成速度快（Davis等，2002）。人们才刚刚开始了解胰岛素和氨基酸调节蛋白质合成的分子机制。大量研究表明，激素和营养素刺激共同通过哺乳动物雷帕霉素靶蛋白（mTOR）通路，诱导新生动物的基因表达和蛋白质合成（Kimball等，2006；Avruch等，2008；Wang和Proud，2009；Wu，2009）。

胰岛素通过PI-3-激酶-Akt途径激活mTOR（图4-7）。胰岛素与胰岛素受体结合后激活该受体，然后反过来激活磷脂酰肌醇3-激酶（PI3K）。PI3K催化膜结合的PIP_2[磷脂酰肌醇（4,5）-二磷酸]转化为PIP_3[磷脂酰肌醇（3,4,5）-三磷酸]。PIP_3与蛋白激酶B（PKB）结合，并激活该酶。PKB通过蛋白质磷酸化直接激活

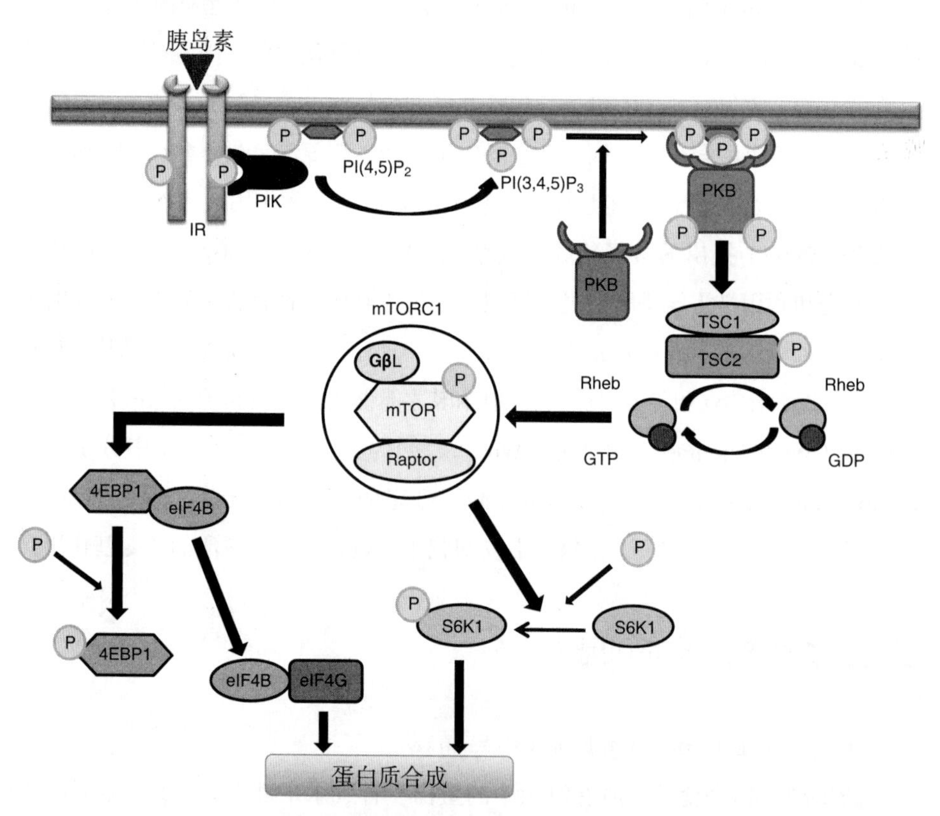

图4-7　胰岛素信号通路介导的蛋白质合成机制

P为磷酸；IR为胰岛素受体；PIK为磷脂酰肌醇激酶；PI（4,5）P_2为（磷脂酰肌醇（4,5）-二磷酸）；PI（3,4,5）P_3为磷脂酰肌醇(3,4,5)-三磷酸；PKB为蛋白激酶B；mTOR为哺乳动物雷帕霉素靶蛋白；mTORC1为mTOR复合体-1；GβL为G蛋白β亚基样蛋白；raptor为mTOR调控相关蛋白；Rheb为mTOR信号通路中上游调控蛋白；GTP为三磷酸鸟苷；GDP为二磷酸鸟苷；TSC1/TSC2为结节状硬化复合体；4EBP1为4E-结合蛋白-1；eIF4B为真核细胞翻译起始因子-4B；eIF4G为真核细胞翻译起始因子4G；S6K1为核糖体蛋白S6激酶。

mTOR，或通过抑制TSC1/TSC2（结节状硬化复合体）的作用间接激活mTOR。TSC1和TSC2形成一个功能性复合体，在激活状态下能够使mTOR的激活剂Rheb（脑中富含的Ras同系物）失活。mTOR激活后，可形成mTOR复合体-1，包括mTOR、raptor（mTOR调节相关蛋白）和G-BetaL（G蛋白β亚基样蛋白）。mTOR复合体-1能调节eIF4EBP1（真核细胞翻译起始因子-4E-结合蛋白-1）和核糖体蛋白S6K1（S6激酶）的磷酸化。非磷酸化的4EBP1与eIF4E（真核细胞翻译起始因子-4E）结合后，可抑制蛋白质合成的起始过程。4EBP1被mTOR磷酸化后，与eIF4E的亲和力降低，4EBP1与eIF4E分离。EIF4E与eIF4F的其他组分结合成具有活性的复合体，启动蛋白质的翻译过程。另一方面，S6K1的激活能使40S核糖体S6蛋白磷酸化，促进40S核糖体亚基与活化的翻译多聚核糖体结合，增加蛋白质的合成。

氨基酸或者亮氨酸调节体内mTOR通路活性的分子机制尚不明确。由于饲粮氨基酸水平不影响PKB和TSC2的磷酸化过程（Suryawan等，2008），因此氨基酸不依赖于PKB和TSC1/2而激活mTOR复合体-1。氨基酸能通过增加S6K1和4EBP1的活性（Suryawan等，2008），或者通过调节raptor与mTOR的相互作用（Hara等，2002；Corradeti和Guan，2006），从而激活mTORC1，诱导蛋白质的合成。

◆ 4.2 生长猪蛋白质合成的发育性调控

在新生动物所有组织中，骨骼肌的蛋白质合成速率最大，而且在出生后的一段时间内其下降也最迅速（Davis等，2002）。肌肉蛋白质合成的这一发育性下降现象，在快速颤动和糖酵解的肌肉中更为显著，并伴随着细胞中核糖体数量的减少（Davis等，2002），以及进食后胰岛素和氨基酸升高引起的核糖体翻译mRNA效率的下降（Suryawan等，2006）。除此之外，与骨骼肌蛋白质合成调节相关的多种氨基酸和胰岛素信号分子的激活，也是随着发育过程而被调控。肌肉中蛋白质合成的正向调节分子（mTOR、S6K1和4EBP1）的活性，随着年龄的增加而下降；但是蛋白质合成的负向调节分子（如TSC2）的活性，在生长猪中较高（Suryawan等，2006）。7日龄仔猪的raptor丰度及其与mTOR结合力比26日龄仔猪高，说明肌肉蛋白质合成相关的蛋白质活性和基因表达均与仔猪的年龄有关（Suryawan等，2006）。

5 妊娠期氨基酸的分配

目前，对妊娠母猪氨基酸利用的了解还有限。但是，仍有一些研究描述了妊娠期胎儿组织中蛋白质和氨基酸沉积的时态变化。这些试验结果对设计妊娠母猪高效利用饲粮氨基酸的营养策略具有潜在应用价值。妊娠期间，氨基酸被用于满足母体需要（包括维持需要和乳腺组织生长需要）、非乳腺组织沉积和孕体需要（包括胎儿和胎盘生长）。

◆ 5.1 胎儿生长

早期研究表明，给妊娠早期母猪饲喂低至0.5%蛋白质的饲粮，不会影响仔猪的初生重（Antinmo等，1976）。近期研究（Wu等，1998）表明，给妊娠母猪饲喂含0.5%蛋白质（可以忽略不计）的饲粮，能够维持血浆氨基酸浓度。但是母体蛋白质降解增加，氨基酸氧化速率下降，从而为胎儿利用节约氨基酸。在这一研究中，羊水和胎儿血浆氨基酸浓度降低，说明胎盘对氨基酸的转运减少，很可能是因为胎盘基顶膜氨基酸转运基因（包括CAT-1）的表达下调，这与在蛋白质营养不良的妊娠大鼠中的报道一致（Malandro等，1996）。妊娠60 d时蛋白质不足，会导致羊水中精氨酸和鸟氨酸的浓度分别降低37%和48%，胎儿血浆氨浓度会相应升高，说明精氨酸和鸟氨酸在胎儿营养和脱氨毒中发挥着作用。妊娠期后半段时间，胎儿的生长显著加快（Wu等，1999；McPherson等，2004；Ji等，2005；图4-8）。

妊娠第70天之前每个胎儿的蛋白质日沉积量为0.25 g，妊娠第70天之后平均每天增加至4.63 g。妊娠第40 ~ 110天，蛋白质沉积量占51% ~ 64%，而脂肪沉积量只占12.4% ~ 16%。其中肠道生长增加明显，在妊娠第40天和第110天时分别占体重的2.5%和6.2%，是生长最快的器官（McPherson等，2004）。如之前所述，胎猪小肠是利用谷氨酰胺和脯氨酸从头合成精氨酸的重要部位。据报道，在整个妊娠期间，精氨酸是胎猪中含量最高的氨基酸类的氮载体（Wu等，1999）。妊娠期间，胎儿的氨基酸组成是不断变化的，甘氨酸和羟脯氨酸含量大量增加，脯氨酸和精氨酸含量也显著增加（表4-1），其他氨基酸占总氨基酸的比例下降。表4-1中未列出谷氨酰胺和谷氨酸，这两种氨基酸共占胎儿总氨基酸的13.5%，且在妊娠期间未发生改变。

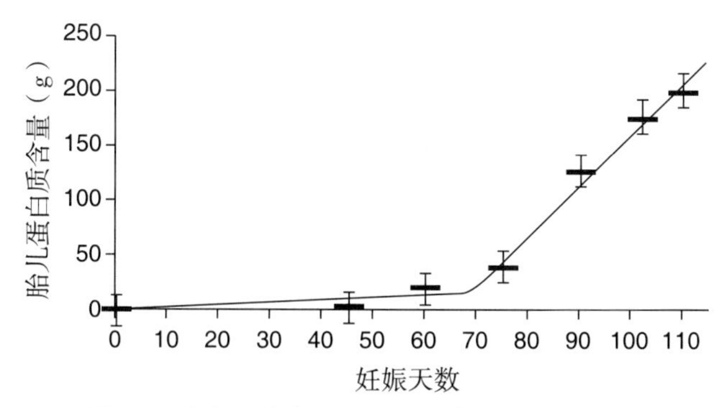

图4-8 妊娠天数与胎儿蛋白质含量（g）的关系

拐点为妊娠第68.5天。在妊娠第68.5天之前上述关系描述为$y = 0.249 \times (x-68.5) + 17.078$，在妊娠第68.5天之后上述关系描述为$y= 4.629 \times (x - 68.5)+ 17.078$。（资料来源于Ji等，2005）

表4-1 胎猪的氨基酸组成（g氨基酸/100 g总氨基酸）

氨基酸	孕龄（d）				
	40	60	90	110	114
丙氨酸	6.06	6.13	6.45	6.83	6.76
精氨酸	6.30	6.60	6.93	6.80	6.78
甘氨酸	6.16	7.34	9.88	10.7	11.30
组氨酸	2.57	2.14	2.15	2.15	2.17
羟脯氨酸	0.84	1.85	3.15	3.50	3.64
异亮氨酸	3.64	3.48	3.15	3.04	3.03
亮氨酸	8.42	7.81	7.29	7.13	7.07
赖氨酸	8.60	6.97	6.26	6.04	6.04
蛋氨酸	2.32	2.14	2.03	1.99	1.95
苯丙氨酸	4.64	4.14	3.81	3.65	3.60
脯氨酸	6.15	8.29	8.20	8.21	8.19
苏氨酸	4.21	3.87	3.66	3.47	3.39

（续）

氨基酸	孕龄（d）				
	40	60	90	110	114
色氨酸	1.20	1.24	1.24	1.20	1.16
缬氨酸	5.68	4.91	4.74	4.51	4.41

每个妊娠日龄的动物数：第40天、第60天和第90天均为6头，第110天为5头，第114天为4头。

资料来源于Wu等，1999。

根据羟脯氨酸数据估计，妊娠第40天时胶原蛋白占胎儿总蛋白含量的7%，在妊娠第110～114天增加至29%。妊娠后期子宫摄取的精氨酸、脯氨酸和羟脯氨酸只能最低限度地满足胎儿生长的需要。在妊娠第110天时，子宫静脉吸收的脯氨酸仅次于所有必需氨基酸中的最高者，而摄取的精氨酸几乎占脯氨酸的一半（图4-9）。

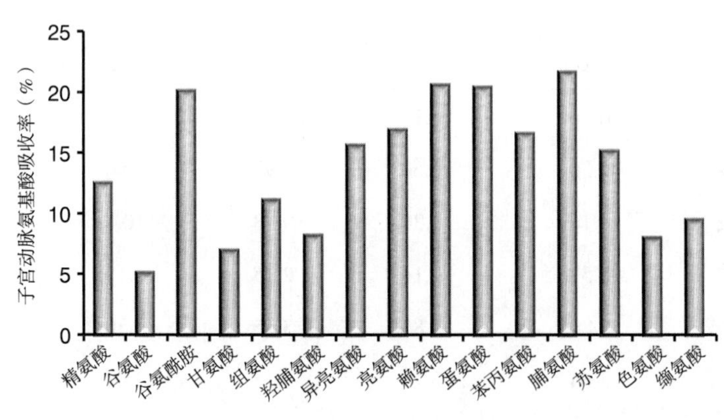

图4-9　子宫动脉的氨基酸吸收

数据以百分比表示，根据Wu等（1999）报道中妊娠第110～114天的数据计算得来，子宫动脉氨基酸吸收率＝子宫动静脉氨基酸的浓度差×100/动脉氨基酸浓度。

另外，尽管子宫摄取的瓜氨酸和鸟氨酸的比例较低（在图4-9中未列出），但子宫对它们的净吸收量分别是其在胎儿组织中沉积量的55倍和15倍，说明胎儿可利用瓜氨酸和鸟氨酸合成精氨酸（图4-10）。因此，瓜氨酸和鸟氨酸可以节约精氨酸。在所有必需氨基酸中，赖氨酸和蛋氨酸的吸收率最高，其次是亮氨酸和苯丙氨酸。就子宫动静脉浓度差而言，亮氨酸最高，其次是赖氨酸、精氨酸和苏氨酸（Wu等，1999）。因此，亮氨酸、赖氨酸、苏氨酸、蛋氨酸和苯丙氨酸好像是用于胎儿发育的重要必需氨基酸。目前尚不明确氨基酸吸收率及其组成是否随着妊娠期发生改变，这种改变可反映妊娠期胎儿生长对氨基酸的相对利用率。子宫摄取的精氨酸、脯氨酸和羟脯氨酸仅能满足胎儿对精氨酸的需要量，说明胎猪可自身合成精氨酸，这与之前所述一致。在所有必需氨基酸中，子宫摄取的赖氨酸在胎儿体内接近100%用于胎儿沉积，其次是苯丙氨酸和蛋氨酸（图4-10）。

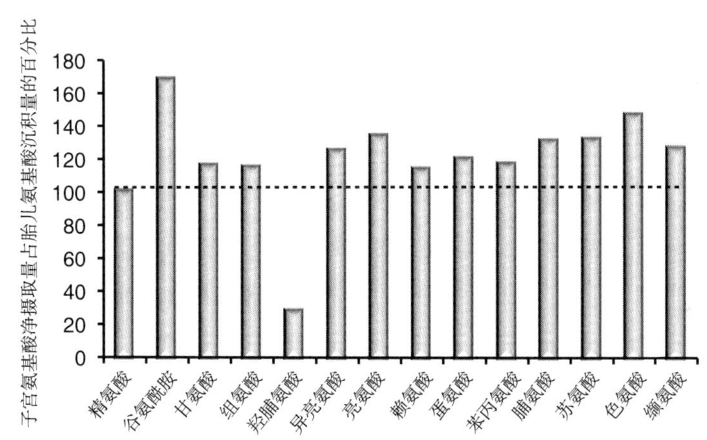

图4-10 妊娠第110～114天子宫氨基酸净摄取量占胎儿氨基酸沉积量的比例

根据Wu等（1999）的数据计算而来，虚线表示子宫氨基酸净摄取量占胎儿氨基酸沉积量的100%。瓜氨酸和鸟氨酸分别占5.537%和1.502%，此图未列出。

◆ 5.2 乳腺生长

初产母猪和经产母猪的乳腺发育主要是在妊娠期，初产母猪的乳腺在泌乳期也会发育。了解妊娠期乳腺发育，对最大程度地提高乳腺实质（产奶的乳腺上皮细胞）的发育至关重要；事实上，仔猪窝生长速度与乳腺大小呈正相关（Nielsen等，2001）。妊娠期乳腺生长的特征是，妊娠后1/3期乳腺实质质量和乳腺组织DNA浓度显著增加（Kensinger等，1982；Sørensen等，2002），这与胎儿蛋白质沉积速率的增长一致。虽然有人推测妊娠期乳腺增长与胎儿蛋白质沉积可能受某些因素协同调控，如刺激乳房发育的激素（Ji等，2006），但是乳腺组织发育也可能在某种程度上受胎儿信号的调控（Kensinger等，1986）。乳腺中含有大量脂肪，粗脂肪浓度从妊娠第45天的94%线性降低到妊娠第112天的58%；相反，粗蛋白质浓度从5%增加到39%。与胸部乳腺和腹股沟部乳腺相比，腹部乳腺中的蛋白质含量更高、脂肪含量更低（Ji等，2006）。因此，单个乳腺的平均蛋白质日沉积速量，在妊娠期缓慢发育阶段（第1～75天）为0.08 g，在妊娠期迅速发育阶段（第75～112天）则平均达到1.05 g。假设总共有12头仔猪、14个乳腺，将饲粮氨基酸在胎儿和乳腺池之间进行分配，以满足组织蛋白质沉积的需要量，在妊娠期缓慢发育阶段，乳腺和胎儿蛋白质合成利用的氨基酸分别占27%和73%；在妊娠期迅速发育阶段，为了满足胎儿对蛋白质沉积的需要，这种分配发生了改变，乳腺沉积只占21%，而胎儿沉积占79%。虽然变化较小，但是这种改变可使妊娠前期和中期的饲粮氨基酸比妊娠后期更多地用于乳腺组织蛋白质沉积。这说明了最大程度地降低乳腺脂肪沉积、提高氨基酸用于乳腺蛋白质沉积的重要性，尤其是在妊娠期缓慢发育阶段，这种最大化依赖于适当的氨基酸平衡。例如，缓慢发育阶段和迅速发育阶段的过渡与组织学结构的变化一致，妊娠早期脂肪小叶胶原组织形成丰富的网状结构，妊娠中期输乳管的延伸，以及妊娠后期上皮结构的发育（Hovey等，1999）。因此，乳腺蛋白质组成可能随着基质细胞蛋白质的改变而改变，与上皮细胞蛋白质相比，基质细胞蛋白质包含更多的羟脯氨酸。虽然妊娠早期和妊娠后期乳腺组织氨基酸的组成和比例尚不明确，但是这可能为析因法计算妊娠期饲粮氨基酸比例提供帮助。另外，对决定妊娠期乳腺氨基酸摄取的因素也缺乏研究。考虑到妊娠后期乳腺发育明显加快，所以乳腺氨基酸的利用率和氨基酸缺乏很可能会增加。

最后，妊娠期的能量摄入会降低妊娠期迅速发育阶段（75～105 d）乳腺实质组织的蛋白质沉积，因此很可能会影响氨基酸的利用率。初产母猪摄入过多能量，会降低DNA合成、乳腺实质组织的RNA和总蛋白。虽然提高饲粮蛋白质水平不会增加乳腺实质组织中的总蛋白，但是会降低乳腺实质外基质的重量。因此，妊娠期75～105 d提高饲粮蛋白质水平，不利于乳腺的发育，但是高能量饲粮对乳腺分泌组织的发育却是有害的（Weldon等，1991）。

6 泌乳期氨基酸的分配

只有为泌乳母猪提供足量的、适当比例的氨基酸，才能使饲粮蛋白质的利用率达到最高。泌乳期间，大概有1/3血液循环氨基酸被乳腺利用（Guan等，2004），并直接用于合成乳蛋白，或者在乳腺组织中沉积和代谢。饲粮氨基酸过量或不足导致的氨基酸不平衡，均可降低动物对饲粮氮的利用率，限制乳蛋白的合成（Pérez Laspiur 等，2009），增加环境中氮的排放（Otto 等，2003）。乳蛋白是在乳腺上皮细胞中合成的，饲粮氨基酸可能成为乳蛋白合成的限制性因素（Guan等，2002）。过去几年，人们对泌乳母猪氨基酸营养的研究都集中在使仔猪生长最大化。然而，近年来人们意识到，研究母猪氨基酸营养必须考虑乳腺的氨基酸需要量、功能和代谢。

◆ 6.1 乳腺的氨基酸利用

6.1.1 乳腺氨基酸转运的机制及其调控

猪乳腺组织氨基酸的转运过程在体内（Trottier，1997；Guan等，2002；Nielsen等，2002；Guan等，2004）和体外（Hurley等，2000；Jackson等，2000）试验中都有过研究。泌乳期日益生长的仔猪对母乳的需要量增加（Hartmann等，1997)，通过增加动静脉氨基酸浓度差而不是增加血液氨基酸来增加猪乳腺净吸收氨基酸的量，说明有些氨基酸吸收的调控是在自身转运水平上进行的（Nielsen等，2002)。从泌乳期第4天到第18天，氨基酸转运蛋白ASCT1和$B^{0,+}$的mRNA丰度分别增加了2倍和1.3倍多（Pérez Laspiur等，2004）。

赖氨酸是乳蛋白合成的第一限制性氨基酸，特别是当饲粮中以玉米和豆粕作为主要蛋白质来源时（Tokach等，1993；Richert等，1997)，而且泌乳期猪乳腺对赖氨酸的摄取率是所有必需氨基酸中最高的（Trottier等，1997）。因此，上调或者下调编码赖氨酸转运蛋白的基因能影响乳腺细胞摄取赖氨酸的量，从而调节赖氨酸的利用率，也可能影响泌乳期饲粮中总的氮利用率。猪乳腺组织中赖氨酸转运系统的种类尚未明确。通过移植培养猪乳腺发现，赖氨酸是通过不同于经典y^+系统的Na^+非依赖性系统来转运的（Shennan和McNeillie，1994；Hurley等，2000）。这个想法是依据两个试验观察结果得出的：①猪乳腺组织是通过一个*Km*接近1.4 mmol/L的Na^+非依赖性转运机制来吸收赖氨酸的，是其他组织中y^+系统*Km*的3～10倍（Deves和Boyd，1998）；②猪乳腺组织吸收赖氨酸是非特异性的，因为赖氨酸的吸收被超过生理浓度50%的L-亮氨酸、L-丙氨酸和L-蛋氨酸所抑制（Calvert和Shennan，1996；Hurley等，2000）。在小鼠乳腺中，精氨酸通过两种系统转运：一个是特异性转运碱性氨基酸的系统（即经典的y^+系统），另一个是可以同时转运碱性氨基酸和中性氨基酸的系统（Sharma和Kansal 2000）。事实上，有研究表明，小鼠乳腺组织中通过y^+系统摄取精氨酸的*Km*是0.76 mmol/L（Sharma和Kansal，2000），几乎是其他组织吸收精氨酸的*Km*的10倍（Dev es和Boyd，1998）。猪乳腺组织通过y^+系统摄取精氨酸的*Km*也可能比其他组织中吸收精氨酸的*Km*大很多。超过生理浓度的中性氨基酸只能部分抑制赖氨酸的摄取，而生理浓度的精氨酸能强烈抑制赖氨酸的摄取，所以说泌乳母猪乳腺中y^+系统具有重要的生理学和营养学意义。Perez Laspiur等（2004）研究表明，猪的在体乳腺细胞中表达编码系统y^+氨基酸转运蛋白CAT-1和CAT-2b的基因，CAT-2b通过自适应调节模式应对可利用氨基酸（Perez Laspiur等，2009）。

大量研究表明，碱性氨基酸和支链氨基酸转运之间存在相互作用。例如，泌乳母猪乳腺组织对缬氨酸的摄取（Jackson等，2000）和泌乳大鼠乳腺组织对赖氨酸的摄取（Shennan等，1994；Calvert和Shennan，1996）会被生理浓度的亮氨酸强烈抑制，猪乳腺组织对缬氨酸的摄取也会被赖氨酸强烈抑制（67%被抑制）（Hurley等，2000）。Calvert和Shennan（1996）指出，乳腺中碱性氨基酸和中性氨基酸之间的这种相互作用可能有重要的生理学意义。例如，泌乳母猪饲粮中普遍添加了纯化的氨基酸。母猪饲粮中过量添加赖氨酸会导致缬氨酸的缺乏（Richert等，1996，1997），但其具体机制尚不明确。与此相反，过量添加缬氨酸会降低泌乳母猪乳腺中赖氨酸的跨膜转运（Guan等，2002）。在泌乳母猪中，经过乳腺的动静脉氨基酸浓度差随着饲粮蛋白质水平的增加而增加，但是当蛋白质水平超过蛋白质需要量时（除了亮氨酸和异亮氨酸），经过乳腺的动静脉氨基酸浓度差会降低（Guan等，2004）。因此，给母猪饲喂高水平蛋白质饲粮后，乳腺对碱性氨基酸和其他中性氨基酸的转运减少，但是亮氨酸和异亮氨酸的转运不会减少。虽然本章的前面讨论过氨基酸有共同的转运蛋白，但是泌乳母猪中碱性氨基酸和支链氨基酸相互作用的机制尚不明确。Perez Laspiur等（2004）的研究表明，猪乳腺组织中存在转运蛋白ATB$^{0,+}$（B$^{0,+}$系统），但是ATB$^{0,+}$的mRNA丰度不随饲粮蛋白质摄入量和泌乳时间的改变而改变（Perez Laspiur等，2009）。提示缺乏高水平的中性氨基酸时通过ATB$^{0,+}$摄取的赖氨酸抑制是竞争性抑制，而不是非竞争性抑制。

前面描述的杂聚肽碱性氨基酸转运蛋白rBAT/b$^{0,+}$AT（系统b$^{0,+}$）、y^{+}LAT1/4F2hc和 y^{+}LAT2/4F2hc（系统y^{+}L），解释了乳腺基底膜分界面上赖氨酸、精氨酸和大分子中性支链氨基酸相互作用的实质。研究发现，这些转运蛋白在泌乳母猪乳腺组织中均有表达（Manjarın 等，2010）。蛋白质过量时，y^{+}LAT1/4F2hc和 y^{+}LAT2/4F2hc很可能上调，乳腺细胞对中性氨基酸的净摄取量增加，对赖氨酸的摄取量降低（图4-11）。这一假设是基于饲粮中蛋白质过量会选择性地增加猪乳腺中亮氨酸和异亮氨酸的摄取，但是减少赖氨酸的转运（Guan等，2004）。此外，亮氨酸（Calvert和Shennan，1996）和缬氨酸（Richert等，1997）能选择性地抑制母猪乳腺组织对赖氨酸的摄取。母猪蛋白质摄入量不足或者仔猪对母乳的需要量较高会导致氨基酸的缺乏，rBAT/b$^{0,+}$AT可能会上调以确保乳腺细胞能摄取足够的赖氨酸，用于乳蛋白的合成。赖氨酸/中性氨基酸逆浓度转运要求细胞内中性氨基酸浓度的较高，细胞内的中性氨基酸是通过本章前面描述的B^{0}系统（Na^{+}协同转运）的ATB$^{0,+}$或者B^{0}AT1浓集转运而来的。B^{0}AT1是Na^{+}依赖性氨基酸转运蛋白，能转运亮氨酸和缬氨酸等大分子的支链中性氨基酸，而不能转运酸性或者碱性氨基酸。

Perez Laspiur等（2004）的研究表明，ASCT1转运蛋白是Na^{+}依赖性氨基酸转运蛋白，能转运小分子中性氨基酸，如丙氨酸、丝氨酸和半胱氨酸（Utsunomiya-Tate等，1996）。但是ASCT1不能通过基顶膜净转运中性氨基酸，因为ASCT1通过基顶膜转运中性氨基酸的同时必须有反向基质氨基酸的交换。虽然如此，Perez Laspiur等（2009）发现，泌乳期ASCT1的mRNA丰度增加，说明ASCT1在猪乳腺细胞氨基酸利用中发挥着重要的调控作用。

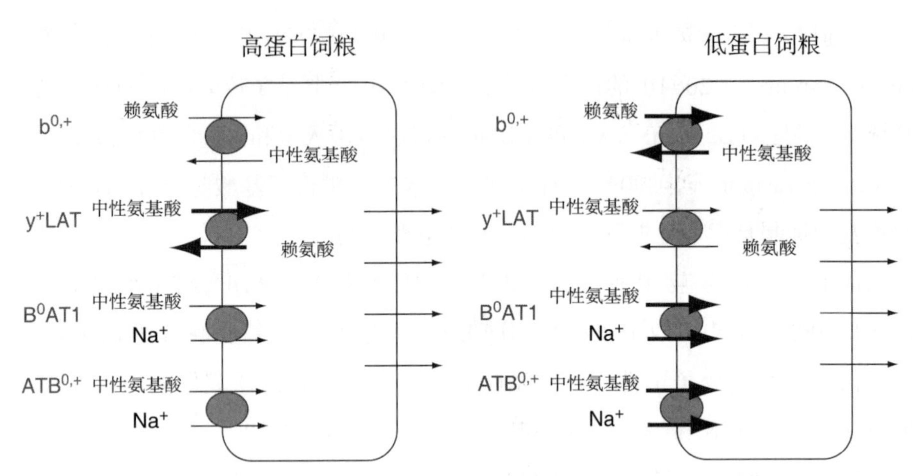

图4-11　饲粮蛋白质水平过高时特定的赖氨酸转运蛋白调控的假想模型（左）和蛋白质缺乏或需乳量高时特定的赖氨酸转运蛋白调控的假想模型（右）

$b^{0,+}$为Na^+非依赖性氨基酸转运蛋白；y^+LAT为系统y^+L转运蛋白；B^0AT1为Na^+依赖性氨基酸转运蛋白；$ATB^{0,+}$为系统$B^{0,+}$转运蛋白。

6.1.2 氨基酸利用的激素调节

迄今为止，已经对妊娠期和泌乳期啮齿类动物的乳腺组织（Verma和Kansal，1993；Sharma 和Kansal，1999，2000）、哺乳和未哺乳动物的乳腺组织（Shennan和McNeillie，1994；Trottier 等，1997）进行了比较。妊娠期小鼠的乳腺组织与泌乳期的乳腺组织相比，氨基酸转运系统的V_{max}更小（Verma和Kansal，1993）。说明泌乳会改变氨基酸转运能力，编码氨基酸转运蛋白的基因表达量也会增加。人们普遍认可吮吸母乳（Mephan，1983）和按摩乳房（Auldist 等，1995；King等，1997）会刺激母体产乳。另外，转运系统y^+（Sharma和Kansal，2000）和L（Sharma和Kansal 1999）在未哺乳的乳腺组织中被抑制，泌乳初期在促乳激素（胰岛素和促乳素）的作用下上调（Sharma和Kansal，1999，2000），说明胰岛素和促乳素在调节乳腺细胞碱性氨基酸转运中发挥着一定作用。在仔猪需要母乳时，促乳素促进营养素进入乳腺，而不是进入脂肪组织，然后在乳腺中合成乳蛋白，包括β-酪蛋白、乳清酸蛋白和α-乳白蛋白（Ben-Jonathan等，2006）。在泌乳母猪中，促乳素是关键的催乳激素，是诱导泌乳发生必不可少的激素，与奶牛不同，促乳素还是泌乳母猪维持泌乳必不可少的激素（Farmer，2001）。

泌乳期的啮齿类动物中，乳汁积累几个小时会降低乳腺动静脉氨基酸浓度差（Vina 等，1981)。Shennan和McNeillie（1994）通过乳腺移植培养，发现24h未哺乳的乳腺中氨基异丁酸（AIB）的摄取比哺乳的乳腺少，AIB是氨基酸类似物，是L转运系统的专一性底物。他们还发现，在促乳素的刺激下，大鼠乳腺通过L系统摄取的AIB增加。最近Theil等（2005）研究发现，72h未哺乳的母猪乳腺中，促乳素受体基因表达量下降。Shennan和McNeillie（1994）的早期研究发现，乳汁积累会通过降低乳腺上皮细胞基底膜上促乳素受体的数量来降低氨基酸的摄入量，因此泌乳期乳腺组织中促乳素与受体结合能调节氨基酸的摄取量和利用率。

6.1.3 **乳腺的增长**

初产母猪从分娩到泌乳期第21天乳腺组织大幅度增长，总的乳腺DNA增加2倍（Kim等，1999a）。Kim等（1999a）研究发现，初产母猪的乳腺会经历肥大到增生的过程。经产母猪从分娩到泌乳期第21天乳腺组织的DNA浓度未增加，说明经产母猪乳腺组织几乎没有净生长（Manjarın和Trottier，未发表），但是从分娩到泌乳期第21天乳腺组织的总RNA浓度呈线性增加，说明细胞内蛋白质合成增加（Manjarın和Trottier，未发表)。氨基酸进入乳腺合成乳蛋白和基本蛋白质。虽然现在还无法对结构蛋白进行定量评估，但已有研究指出，泌乳期乳腺摄取必需氨基酸的总量比分泌到乳汁中必需氨基酸的总量多（Trottier等，1997；Trottier和Guan，2000；Guan等，2004），泌乳期约26%的必需氨基酸被乳腺留用。Kim等（1999b）指出，哺乳仔猪体内每天净沉积0.7 g必需氨基酸，约占乳腺留用的必需氨基酸的14%。因此，乳腺代谢通路利用了大量的必需氨基酸，包括必需氨基酸及其他化合物的氧化和合成。母猪乳腺每天吸收的188.5 g必需氨基酸中，49 g被乳腺留用，约占总摄取量的25%（Trottier 等，1997）。乳腺细胞留用的必需氨基酸可能被用来合成非必需氨基酸，为乳糖和脂肪酸合成提供能量，或者用于结构蛋白合成和乳腺重组（Spires等，1975）。因此，精氨酸、亮氨酸、异亮氨酸和缬氨酸等必需氨基酸在乳腺中的沉积最多，母猪乳汁中脯氨酸、天冬氨酸和天冬酰胺等非必需氨基酸比摄入的多（Trottier等，1997）。

◆ 6.2 泌乳期猪乳腺的氨基酸代谢

如图4-12所示，泌乳期大部分被乳腺截留的氨基酸用于代谢途径，这些代谢途径是氨基酸的必需损失还是乳汁合成过程中的一部分尚不明确。然而，整个氨基酸代谢过程图解有利于我们针对感兴趣的基因进行研究，增加哺乳仔猪中氨基酸转化为其他营养物质的利用率。

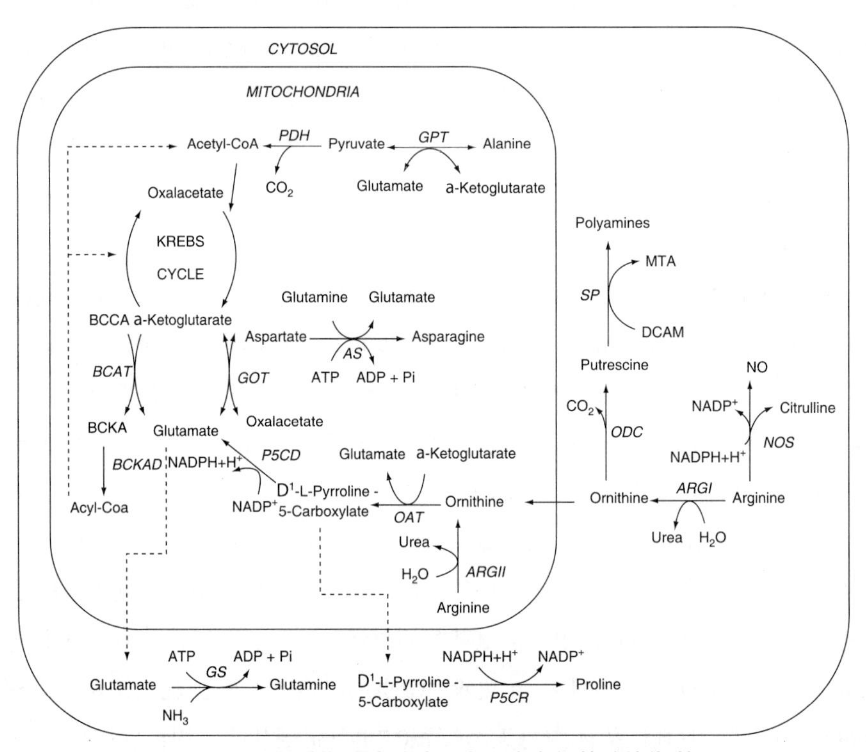

图4-12 泌乳期母猪乳腺上皮细胞中氨基酸的代谢

CYTOSOL为细胞质；MITOCHONDRIA为线粒体；Acetyl-CoA为乙酰辅酶A；Pyruvate为丙酮酸盐；PDH为丙酮酸脱氢酶；GPT为谷丙转氨酶；Alanine为丙氨酸；Glutamate为谷氨酸；α-Ketoglutarate为α-酮戊二酸；Oxalacetate为草乙酸盐；KREBSCYCLE为三羧酸循环；BCAA为支链氨基酸；BCAT为支链氨基转移酶；BCKA为α-酮酸；Aspartate为天冬氨酸；Glutamine为谷氨酰胺；Asparagine为天冬酰胺；GOT为谷草转氨酶；BCKD为支链α-酮酸脱氢酶复合体；NADPH为还原性辅酶Ⅱ；P5CD为Δ^1-L-吡咯啉羧酸脱氢酶；Δ^1-L-pyrroline-5-carboxylate为Δ^1-L-5-吡咯啉羧酸；OAT为鸟氨酸转氨酶；Ornithine为鸟氨酸；Urea为尿素；ARGI和Ⅱ为精氨酸酶；Arginine为精氨酸；GS为谷氨酰胺合成酶；P5CR为吡咯啉-5-羧酸还原酶；Proline为脯氨酸；Polyamines为多胺；SP为亚精胺合成酶；Putrescine为腐胺；ODC为鸟氨酸脱羧酶；NOS为NO合成酶。

6.2.1 **精氨酸**

一些研究表明，泌乳母猪乳腺组织既可以通过精氨酸酶途径将精氨酸分解成脯氨酸、鸟氨酸和尿素，也可以通过精氨酸酶、NO合成酶途径将精氨酸分解成少量多胺和NO（O'Quinn等，2002）。泌乳母猪乳腺组织中有两种不同的精氨酸酶，即精氨酸酶Ⅰ（胞质酶）和精氨酸酶Ⅱ（线粒体酶），都能将精氨酸分解成尿素和鸟氨酸。胞质中产生的鸟氨酸被鸟氨酸脱羧酶（ODC）和亚精胺合成酶（SP）用于合成多胺（Wu和Morris，1998），或者被转入线粒体内，在鸟氨酸转氨酶（OAT）作用下转化为Δ^1-L-吡咯啉羧酸。Δ^1-L-吡咯啉羧酸被Δ^1-L-吡咯啉-5-羧酸脱氢酶（P5CD）转化为谷氨酸，或者转运到细胞质中，在Δ^1-L-吡咯啉-5-羧酸还原酶（P5CR）作用下转化为脯氨酸。泌乳母猪乳腺组织中P5CR的活性是P5CD的56倍，因此大部分来源于精氨酸的P5C被转化为脯氨酸，小部分被转化为谷氨酸和谷氨酰胺。除此之外，因为猪乳腺缺乏Δ^1-L-吡咯啉-5-羧酸合成酶，所以乳腺组织不能利用谷氨酸和谷氨酰胺合成脯氨酸（O' Quinn等，2002）。

泌乳期乳腺中合成NO是精氨酸降解的次要途径（O'Quinn等，2002）。精氨酸和O_2在NOS的作用下生成NO，NO合成后迅速扩散到组织中调节血流量和乳腺细胞摄取血浆营养物质的量（Meininger和Wu，2002；Kim和Wu，2009）。

6.2.2 **支链氨基酸（BCAA）**

母猪乳腺中亮氨酸、缬氨酸和异亮氨酸等支链氨基酸的摄取量（泌乳期第13～20天76 g/d）比乳蛋白中这些支链氨基酸的合成量（46 g/d）大很多（Trottier等，1997），因此泌乳母猪乳腺每天可以分解约30 g的支链氨基酸。一些研究表明，乳腺细胞中支链氨基酸的分解代谢与其他器官中相似，包括两个酶催化步骤（Li等，2009）。第一步是亮氨酸、异亮氨酸和缬氨酸在支链氨基转移酶（BCAT）作用下发生转氨基作用。哺乳动物体内有两种BCAT：一种是线粒体中的BCAT（BCATm），能在哺乳动物全身表达，另一种是细胞质中的BCAT（BCATc）。有些研究表明，BCATc几乎只存在于神经组织中（Sweatt等，2004；Hutson等，2005），但是Li等（2009）在乳腺组织中发现了线粒体的BCAT亚型和细胞质的BCAT亚型。因此，母猪乳腺中的支链氨基酸转氨基作用既可以发生在线粒体中，也可以发生在乳腺细胞的细胞质中。该转氨基反应中，亮氨酸、异亮氨酸和缬氨酸的α-氨基转移到α-酮戊二酸上，生成谷氨酸和相应的α-酮酸（支链酮酸；分别是α-酮异己酸、α-酮基-β-戊酸甲酯和α-酮异戊酸），然后支链α-酮酸脱氢酶复合体（BCKD）催化这三种α-酮酸氧化脱羧，生成酰基辅酶A衍生物。BCKD是一个多酶复合体，位于线粒体膜的内表面（Harper等，1984）。如果支链氨基酸转氨基作用发生在细胞质中（在胞质BCAT亚型作用下），α-酮酸（BCKA）必须转移到线粒体中才能完成氧化反应。支链氨基酸氧化的第二步是乙酰辅酶A的氧化，是在两种不同脱氢酶作用下完成的。这一步后，支链氨基酸代谢途径被分为两类，一类生成乙酰辅酶A（亮氨酸和异亮氨酸），另一类生成琥珀酰辅酶A（缬氨酸和异亮氨酸），最终都进入三羧酸循环（Nelson和Cox，2008）。

6.2.3 谷氨酸和谷氨酰胺、天冬氨酸和天冬酰胺

谷氨酸/谷氨酰胺和天冬氨酸/天冬酰胺具有重要的营养作用，因为它们是母猪泌乳高峰期乳汁中丰度最高的游离氨基酸和与蛋白质结合的氨基酸（Wu和Knabe，1994）。泌乳期乳腺中谷氨酸和谷氨酰胺的摄取率最高，但是天冬氨酸/天冬酰胺的吸收量低于其在乳汁中的含量，说明乳腺细胞可以合成天冬氨酸/天冬酰胺（Trottier等，1997）。Li等（2009）研究发现，乳汁中大部分天冬氨酸是谷氨酸在谷草转氨酶（GOT）催化作用下转氨基得来的，谷氨酸也可以被胞质中的谷氨酰胺合成酶（GS）转化为谷氨酰胺。有趣的是，尽管GOT的活性比GS的活性高，但是母猪乳腺组织中谷氨酰胺的合成量比天冬氨酸合成量高（Li等，2009）。最后，谷氨酸和丙酮酸在谷丙转氨酶（GPT）的作用下发生转氨基作用，生成丙氨酸和α-酮戊二酸。尽管如此，谷氨酸的合成比丙氨酸的合成多，说明泌乳母猪体内的转氨基作用偏向于生成谷氨酸（Li等，2009）。

作者：Nathalie L. Trottier
和Rodrigo Manjarfn
译者：晏向华

参考文献

Adibi, S. A. 1971. Intestinal transport of dipeptides in man: Relative importance of hydrolysis and intact absorption. J. Clin. Invest.50: 2266–2275.

Adibi, S. A. 2003. Regulation of expression of the intestinal oligopeptide transporter (Pept-1) in health and disease. Am. J. Physiol.285: G779–G788.

Angelo, S., A. M. Rojas, H. Ramirez, and R. Deves. 2002. Epithelial cells isolated from chicken jejunum: An experimental modelfor the study of the functional properties of amino acid transport system $b^{0,+}$. Comp. Biochem. Physiol. A 132: 637–644.

Atinmo, T., C. Baldijão, W. G. Pond, and R. H. Barnes. 1976. Decreased dietary protein or energy intake and plasma growth hormone levels of the pregnant pig, its fetuses and developing progeny. J. Nutr. 106:940–946.

Auldist, D. E., D. Carlson, L. Morrish, C. Wakeford, and R. H. King. 1995. The effect of increased suckling frequency on mammary development and milk yield of sows. Page 137 in Manipulating Pig Production V. D. P. Hennessy and P. D. Cranwell, eds.Proceedings of the Australian Pig Science Association, Werribee, Australia.

Avruch, J., X. Long, S. Ortiz-Vega, J. Rapley, A. Papageorgiou, and N. Dai. 2008. Amino acid regulation of TOR complex 1. Am.J. Physiol. Endocrinol. Metab. 296:592–602.

Bauch, C., N. Forster, D. Loffing-Cueni, V. Summa, and F. Verrey. 2003. Functional cooperation of epithelial heteromeric aminoacid transporters expressed in madin-darby canine kidney cells. Biochem. J. 278:1316–1322.

Ben-Jonathan, N., E. R. Hugo, T. D. Brandebourg, and C. R. LaPensee. 2006. Focus on prolactin as a metabolic hormone. TrendsEndocrinol. Metab. 17:110–116.

Bergen, W. G., and G. Wu. 2009. Intestinal nitrogen recycling and utilization in health and disease. J. Nutr. 139:821–825.

Bertolo, R.F., and D. G. Burrin. 2008. Comparative aspects of tissue glutamine and proline metabolism. J. Nutr. 138:2032–2039.

Bos, C., B. Stoll, H. Fouillet, C. Gaudichon, X. Guan, M. A. Grusak, P. J. Reeds, D. Tomé, and D. G. Burrin. 2003. Intestinal lysinemetabolism is driven by the enteral availability of dietary lysine in piglets fed a bolus meal. Am. J. Physiol. Endocrinol.Metab. 285:E1246–E1257.

Broër, A., C. A. Wagner, F. Lang, and S. Broër. 2000a. The heterodimeric amino acid transporter 4F2hc/y^{+}LAT2 mediates arginineefflux in exchange with glutamine. Biochem. J. 349:787–795.

Broër, A., C. A. Wagner, F. Lang, and S. Broër. 2000b. Neutral amino acid transporter ASCT2 displays substrate-induced Na^{+}exchange and a substrate-gated anion conductance. Biochem. J. 346:705–710.

Broër, A., K. Klingel, S. Kowalczuk, J. E. J. Rasko, J. Cavanaugh, and S. Broër. 2004. Molecular cloning of mouse amino acidtransport system B^{0}, a neutral amino acid transporter related to Hartnup disorder. Biochem. J. 279:24467–24476.

Broër, S. 2008. Amino acid transport across mammalian intestinal and renal epithelia. Physiol. Rev. 88:249–286.

Brosnan, J. T. 2003. Interorgan amino acid transport and its regulation. J. Nutr. 133:S2068–2072.

Burrin, D. G., and B. Stoll. 2009. Metabolic fate and function of dietary glutamate in the gut. Am. J. Clin. Nutr. 90(Suppl):850S–856S.

Calvert, D. T., and D. B. Shennan. 1996. Evidence for an interaction between cationic and neutral amino acids at the blood-facingaspect of lactating rat mammary epithelium. J. Dairy Res. 63:25–33.

Chen, L. X., Y. L. Yin, W. S. Jobgen, S. C. Jobgen, D. A. Knabe, W. X. Hu, and G. Wu. 2007. *In vitro* oxidation of essential aminoacids by intestinal mucosal cells of growing pigs. Livest. Prod. Sci. 109:19–23.

Chen, L. X., P. Li, J. Wang, X. L. Li, H. Gao, Y. Yin, Y. Hou, and G. Wu. 2009. Catabolism of nutritionally essential amino acidsin developing porcine enterocytes. Amino Acids 37:143–152.

Chillaron, J., R. Estevez, C. Mora, C. A. Wagner, H. Suessbrich, F. Lang, J. L. Gelpi, et al. 1996. Obligatory amino acid exchangevia systems b$^{o,+}$-like and y^{+}L-like. A tertiary active transport mechanism for renal reabsorption of cystine and dibasic aminoacids. Biochem. J.271:17761–17770.

Closs, E. I. 2002. Expression, regulation and function of carrier proteins for cationic amino acids. Curr. Opin. Nephrol. Hypertens.11:99–107.

Colegio, O. R., C. M. van Itallie, H. J. McCrea, C. Rahner, and J. M. Anderson. 2002. Claudins create charge-selective channelsin the paracellular pathway between epithelial cells. Am. J. Physiol. Cell Physiol. 283:C142–C147.

Corradetti, M. N., and K. L. Gua. 2006. Upstream of the mammalian target of rapamycin: Do all roads pass through mTOR?Oncogene. 25:6347–60.

Dave, M. H., N. Schulz, M. Zecevic, C. A. Wagner, and F. Verrey. 2004. Expression of heteromeric amino acid transporters alongthe murine intestine. Am. J. Physiol. 258:597–610.

Davis, P. K., and G. Wu. 1998. Compartmentation and kinetics of urea cycle enzymes in porcine enterocytes. Comp. Biochem.Physiol. B Biochem. Mol. Biol. 119:527–37.

Davis, T., D. Burrin, M. Fiorotto, and H. Nguyen. 1996. Protein synthesis in skeletal muscle and jejunum is more responsive tofeeding in 7-than in 26-day old pigs. Am. J. Physiol. 270:E802–E809.

Davis, T. A., M. L. Fiorotto, D. G. Burrin, P. J. Reeds, H. V. Nguyen, P. R. Beckett, R. C. vann, and P. M. O'Connor. 2002.Stimulation of protein synthesis by both insulin and amino acids is unique to skeletal muscle in neonatal pigs. Am. J. Physiol.Endocrinol. Metab. 282:E880–E890.

Dev'es, R., and C. A. R. Boyd. 1998. Transporters for cationic amino acids in animal cells: Discovery, structure, and function.Physiol. Rev. 78:487–545.

Farmer, C. 2001. The role of prolactin for mammogenesis and galactopoiesis in swine. Livest. Prod. Sci. 70:105–113.

Feliubadal'o, L., M. Font, J. Purroy, F. Rousaud, X. Estivill, V. Nunes, E. Golomb, et al. 1999. Non-type I cystinuria caused bymutations in SLC7A9, encoding a subunit ($b^{0,+}$AT) of rBAT. Nat. Genet. 23:52–57.

Feng, D. Y., X. Y. Zhou, J. J. Zuo, C. M. Zhang, Y. L. Yin, X. Q. Wang, and T. Wang. 2008. Segmental distribution and expressionof two heterodimeric amino acid transporter mRNAs in the intestine of pigs during different ages. J. Sci. Food Agric.88:1012–1018.

Flynn, N. E., and Wu, G. 1996. An important role for endogenous synthesis of arginine in maintaining arginine homeostasis inneonatal pigs. Am. J. Physiol. Regul. Integr. Comp. Physiol. 271:R1149–R1155.

Guan, X., B. J. Bequette, G. Calder, P. K. Ku, K. N. Ames, and N. L. Trottier. 2002. Amino acid availability affects amino acidtransport and protein metabolism in the porcine mammary gland. J. Nutr. 132:1224–34.

Guan, X., J. E. Pettigrew, P. K. Ku, K. N. Ames, B. J. Bequette, and N. L. Trottier. 2004. Dietary protein concentration affectsplasma arteriovenous difference of amino acids across the porcine mammary gland. J. Anim. Sci. 82:2953–63.

Hara, K., Y. Maruki, X. Long, K. Yoshino, N. Oshiro, S. Hidayat, C. Tokunaga, J. Avruch, and K. Yonezawa. 2002. Raptor, abinding partner of target of rapamycin (TOR), mediates TOR action. Cell. 110:177–189.

Harper, A. E., R. H. Miller, K. P. Block. 1984. Branched-chain amino acid metabolism. Annu. Rev. Nutr. 4:409–454.

Hartmann, P. E., N. A. Smith, M. J. Thompson, C. M. Wakeford, and P. G. Arthur. 1997. The lactation cycle in the sow: Physiologicaland management contradictions. Livest. Prod. Sci. 50:75–87.

Hatanaka, T., T. Nakanishi, W. Huang, F. H. Leibach, P. D. Prasad, V. Ganapathy, and M. E. Ganapathy. 2001. Na^+-andCl^--coupled active transport of nitric oxide synthase inhibitors via amino acid transport system B0, +. J. Clin. Invest. 107:1035–1043.

Hovey, R. C., T. B. McFadden, and R. M. Akers. 1999. Regulation of mammary gland growth and morphogenesis by the mammaryfat pad: A species comparison. J. Mammary Gland Biol. Neoplasia. 4:53–68.

Hundal, H. S., and P. M. Taylor. 2009. Amino acid transceptors: Gate keepers of nutrient exchange and regulators of nutrientsignaling. Am. J. Physiol. Endocrinol. Metab. 296:E603–E613.

Hurley, W. L., H. Wang, J. M. Bryson, and D. B. Shennan. 2000. Lysine uptake by mammary gland tissue of the lactating sow. J. Anim. Sci. 78:391–395.

Hutson, S. M., A. J. Sweatt, and K. F. Lanoue. 2005. Branched-chain amino acid metabolism: Implications for establishing safeintakes. J. Nutr. 135:1557S–1564S.

Hyde, R., P. M. Taylor, and H. S. Hundal. 2003. Amino acid transporters: Roles in amino acid sensing and signalling in animalcells. Biochem. J. 373:1–18.

Jackson, S. C., J. M. Bryson, H. Wang, and W. L. Hurley. 2000. Cellular uptake of valine by lactating porcine mammary tissue. J. Anim. Sci. 78:2927–2932.

Ji, F., G. Wu, J. R. Blanton Jr., and S. W. Kim. 2005. Changes in weight and composition in various tissues of pregnant gilts andtheir nutritional implications. J. Anim. Sci. 83:366–375.

Ji, F., W. L. Hurley, and S. W. Kim. 2006. Characterization of mammary gland development in pregnant gilts. J. Anim. Sci.84:579–587.

Kensinger, R. S., R. J. Collier, F. W. Bazer, C. A. Ducsay, and H. N. Becker. 1982. Nucleic acid, metabolic and histological changesin gilt mammary tissue during pregnancy and lactogenesis. J. Anim. Sci. 54:1297–1308.

Kensinger, R. S., R. J. Collier, F. W. Bazer, and R. R. Kraeling. 1986. Effect of number of conceptuses on maternal hormoneconcentrations in the pig. J. Anim. Sci. 62:1666–1674.

Kim, S. W., W. L. Hurley, I. K. Han, and R. A. Easter. 1999a. Changes in tissue composition associated with mammary glandgrowth during lactation in sows. J. Anim. Sci. 77:2510–2516.

Kim, S. W., W. L. Hurley, I. K. Han, H. H. Stein and R. A. Easter. 1999b. Effect of nutrient intake on mammary gland growth inlactating sows. J. Anim. Sci. 77:3304–3315.

Kim, S. W., and G. Wu. 2004. Dietary arginine supplementation enhances the growth of milk fed young pigs. J. Nutr. 134:625–630.

Kim, S. W., and G. Wu. 2009. Regulatory role for amino acids in mammary gland growth and milk synthesis. Amino Acids.37:89–95.

Kimball, S. R., and L. S. Jefferson. 2006. New functions for amino acids: Effects on gene transcription and translation. Am. J.Clin. Nutr. 83:500S–507S.

King, R. H., B. P. Mullan, F. R. Dunshea, and H. Dove. 1997. The influence of piglet body weight on milk production of sows.Livest. Prod. Sci. 47:169–174.

Krehbiel, C. R. and J. C. Matthews. 2003. Absorption of amino acids and peptides. Page 60 inAmino Acids in Animal Nutrition2nd ed. J. P. F. D'Mello, ed. CAB International, Wallingford, UK.

Li, P., D. A. Knabe, S. W. Kim, C. J. Lynch, S. M. Hutson, and G. Wu. 2009. Lactating porcine mammary tissue catabolizesbranched-chain amino acids for glutamine and aspartate synthesis. J. Nutr. 139:1502–1509.

Malandro, M. S., M. J. Beveridge, M. S. Kilberg, and D. A. Novak. 1996. Effect of low-protein diet-induced intrauterine growthretardation on rat placental amino acid transport. Am. J. Physiol. Cell Physiol. 271:C295–C303.

McPherson, R. L., F. Ji, G. Wu, J. R. Blanton, Jr., and S. W. Kim. 2004. Growth and compositional changes of fetal tissues in pigs.J. Anim. Sci. 82:2534–2540.

Meininger, C. J., and G. Wu. 2002. Regulation of endothelial cell proliferation by nitric oxide. Methods Enzymol. 352:280–295.

Mepham, T. B. 1983. Physiological aspects of lactation. Pages 4–28 inBiochemistry of Lactation. T. B. Mepham, ed. ElsevierScience Publishers, Amsterdam, The Netherlands.Metges, C. C. 2000. Contribution of microbial amino acids to amino acid homeostasis of the host. J. Nutr. 130:1857S–1864S.

Nakashima, K., Y. Yakabe, A. Ishida, M. Yamazaki and H. Abe. 2007. Suppression of myofibrillar proteolysis in chick skeletalmuscles byα -ketoisocaproate. Amino Acids. 33: 499–503.

Nelson, D. L., and M. M. Cox. 2008. Lehninger, Principles of Biochemistry. 5th ed. W. H. Freeman and Co., New York, NY.

Nielsen, O. L., A. R. Pedersen, and M. T. Sørensen. 2001. Relationships between piglet growth rate and mammary gland size ofthe sow. Livest. Prod. Sci. 67:273–279.

Nielsen, T. T., N. L. Trottier, H. H. Stein, C. Bellaver, and R. A. Easter. 2002. The effect of litter size and day of lactation on aminoacid uptake by the porcine mammary gland. J. Anim. Sci. 80:2402–2411.

Niiyama, M., E. Deguchi, K. Kagota, and S. Namioka. 1979. Appearance of15N labeled intestinal microbial amino acids in thevenous blood of the pig colon. Am. J. Vet. Res. 40: 716–718.

NRC. 1998. Nutrient Requirements of Swine . 10th rev. ed. The National Academies Press, Washington, DC.

Olszewski, A., and S. Buraczewski. 1978. Absorption of amino acids in isolated pig caecum in situ. Effect of concentration ofenzymatic casein hydrolysate on absorption of amino acids. Acta Physiol. Pol. 29:67–77.

O'Quinn, P. R., D. A. Knabe, and G. Wu. 2002. Arginine catabolism in lactating porcine mammary tissue. J. Anim. Sci. 80:467–474.

Otto, E. R., M. Yokoyama, P. K. Ku, N. K. Ames, and N. L. Trottier. 2003. Nitrogen balance and ileal amino acid digestibility ingrowing pigs fed diets reduced in protein concentration. J. Anim. Sci. 81:1743–1753.

Palacin, M., R. Estevez, J. Bertran, and A. Zorzano. 1998. Molecular biology of mammalian amino acid transporters. Physiol. Rev.78:969–1054.

P'erez Laspiur, J., J. L. Burton, P. S. D. Weber, R. N. Kirkwood, N. L. Trottier. 2004. Short communication: Amino acid transportersin porcine mammary gland during lactation. J. Dairy Sci. 87:3235–3237.

P'erez Laspiur, J., J. L. Burton, P. S. D. Weber, J. Moore, R. N. Kirkwood, and N. L. Trottier. 2009. Dietary protein intake andstage of lactation differentially modulate amino acid transporter mRNA abundance in porcine mammary tissue. J. Nutr.139:1677–1684.

Reeds, P. J., L. J. Wykes, J. E. Henry, M. E. Frazer, D. G. Burrin, and F. Jahoor. 1996. Enteral glutamate is almost completelymetabolized in first pass by the gastrointestinal tract of infant pigs. Am. J. Physiol. 270:E413–E418.

Richert, B.T., M. D. Tokash, R. D. Goodband, J.L. Nelssen, J. E. Pettigrew, R.W. Walker, L.J. Johnston. 1996. Valine requirementof the high-producing sow. J. Anim. Sci. 74:1307–1313.

Richert, B. T., M. D. Tokach, R. D. Goodband, J. L. Nelssen, R. G. Campbell, and S. Kershaw. 1997. The effect of dietary lysineand valine fed during lactation on sow and litter performance. J. Anim. Sci. 75:1853–1860.

Rossier, G., C. Meier, C. Bauch, V. Summa, B. Sordat, F. Verrey, and L. C. Kuhn. 1999. LAT2, a new basolateral 4F2hc/CD98-associated amino acid transporter of kidney and intestine. Biochem. J. 274:34948–34954.

Sep ′ulveda, F. V., and M. W. Smith. 1979. Different mechanisms for neutral amino acid uptake by newborn pig colon. J. Physiol.286: 479–490.

Sharma, R., and V. K. Kansal. 1999. Characteristics of transport systems for L-alanine in mouse mammary gland and theirregulation by lactogenic hormones: Evidence for two broad spectrum systems. J. Dairy Research. 66: 385–398.

Sharma, R., and V. K. Kansal. 2000. Heterogeneity of cationic amino acid transport systems in mouse mammary gland and theirregulation by lactogenic hormones. J. Dairy Research. 67:21–30.

Shennan, D. B., S. A. McNeillie, E. A. Jamison, and D. T. Calvert. 1994. Lysine transport in lactating rat mammary tissue: Evidence for an interaction between cationic and neutral amino acids. Acta Physiol. Scand. 151:461–466.

Shennan, D. B., I. D. Millar, and D. T. Calvert. 1997. Mammary-tissue amino acid transport systems. Proc. Nutr. Soc. 56:177–191.

Shennan, D. B. 1998. Mammary gland membrane transport systems. J. Mammary Gland Biol. Neoplasia. 3:247–258.

Shennan, D. B., and M. Peaker. 2000. Transport of milk constituents by the mammary gland. Physiol. Rev. 80:925–951.

Silk, D. B., A. Nicholson, and Y. S. Kim. 1976. Hydrolysis of peptides within lumen of small intestine. Am. J. Physiol. 231:1322–1329.

Silk, D. B. A., G. K. Grimble and R. G. Rees. 1985. Protein digestion and amino acid and peptide absorption. Proc. of the Nutr.Soc. 44:63–72.

Sloan, J. L., and S. Mager. 1999. Cloning and functional expression of a human $Na^+$and Cl^- dependentneutral and cationicamino acid transporter $B^{0,+}$. J. Biol. Chem. 274:23740–23745.

Sørensen, M. T., K. Sejrsen, and S. Purup. 2002. Mammary gland development in gilts. Livest. Prod. Sci. 75:143–148.

Soriano-Garcia, J. F., M. Torras-LlortMoreto, and R. Ferrer. 1999. Regulation of L-methionine and L-lysine uptake in chickenjejunal brush border by dietary methionine. Am. J. Physiol. 2777:R1654–R1661.

Souba, W. W., and A. J. Pacitti. 1992. How amino acids get into cells: mechanisms, models, menus and mediators. J. Parenter.Enteral. Nutr. 16:569–578.

Southern, L. L., and D. H. Baker. 1983. Arginine requirement of the young pig. J. Anim. Sci. 57:402–412.

Sperandeo, M. P., G. Andria, and G. Sebastio. 2008. Lysinuric protein intolerance: Update and extended mutation analysis of the SLC7A7 gene. H. Mutation. 29:14–21.

Spires, H. R., J. H. Clark, R. G. Derrig, and C. L. Davis. 1975. Milk production and nitrogen utilization in response to post-ruminalinfusion of sodium caseinate in lactating cows. J. Nutr. 105:1111–1121.

Stoll, B., J. Henry, P. J. Reeds, H. Yu, F. Jahoor, and D. G. Burrin. 1998. Catabolism dominates the first-pass intestinal metabolismof dietary essential amino acids in milk protein-fed piglets. J. Nutr. 128:606–614.

Stoll, B., X. Chang, M. Z. Fan, P. J. Reeds, and D. G. Burrin. 2000. Enteral nutrient intake determines the rate of intestinal proteinsynthesis and accretion in neonatal pigs. Am. J. Physiol. 279:G288–G294.

Suryawan, A., J. Escobar, J. W. Frank, H. V. Nguyen, and T. A. Davis. 2006. Developmental regulation of the activation ofsignaling components leading to translation initiation in skeletal muscle of neonatal pigs. Am. J. Physiol. Endocrinol. Metab. 291:E849–E859.

Sweatt, A., M. Wood, A. Suryawan, R. Wallin, M. C. Willingham, and S. M. Hutson. 2004. Branched-chain amino acid catabolism:Unique segregation of pathway enzymes in organ systems and peripheral nerves. Am. J. Physiol. 286:E64–E76.

Taylor, P. M., S. Kaur, B. Mackenzie, and G. J. Peter. 1996. Amino-acid dependent modulation of amino acid transport in *Xenopuslaevis* oocytes. J. Exp. Biol. 199:923–931.

Theil, P. K., R. Labouriau, K. Sejrsen, B. Thomsen, and M. T. Sorensen. 2005. Expression of genes involved in regulation of cellturnover during milk stasis and lactation rescue in sow mammary glands. J. Anim. Sci. 83:2349–2356.

Tokach, M. D., R. D. Goodband, J. L. Nelssen, and L. J. Kats. 1993. Valine: A deficient amino acid in high lysine diets for thelactating sow. J. Anim. Sci. 71(Suppl. 1):68. (Abstr.)

Torrallardona, D., C. I. Harris, and M. F. Fuller. 2003a. Pigs' gastrointestinal microflora provide them with essential amino acids. J. Nutr. 133:1127–1131.

Torrallardona, D., C. I. Harris, and M. F. Fuller. 2003b. Lysine synthesized by the gastrointestinal microflora of pigs is absorbed,mostly in the small intestine. Am. J. Physiol. Endocrinol. Metab. 284:E1177–E1180.

Torrents, D., R. Estevez, M. Pineda, E. Fernandez, J. Lloberas, Y. B. Shi, A. Zorzano, and M. Palacin. 1998. Identification andcharacterization of a membrane protein (y^+L amino acid transporter-1) that associates with 4F2hc to encode the amino acidtransport activity y^+L: A candidate gene for lysinuric protein intolerance. Biochem. J. 273:32437–32445.

Trottier, N. L. 1997. Nutritional control of amino acid supply to the mammary gland during lactation in the pig. Proc. Nutr. Soc.56:581–591.

Trottier, N. L., C. F. Shipley, and R. A. Easter. 1997. Plasma amino acid uptake by the mammary gland of the lactating sow. J. Anim. Sci. 75:1266–1278.

Trottier, N. L., and X. Guan. 2000. Research paradigms behind amino acid requirements of the lactating sow: Theory and futureapplication. J. Anim. Sci. 78 (Suppl. 3):48–58.

Ugawa, S., Y. Sunouchi, T. Ueda, E. Takahashi, Y. Saishin, and S. Shimada. 2001. Characterization of a mouse colonic system $B^{0,+}$amino acid transporter related to amino acid absorption in colon. Am. J. Physiol. Gastrointest. Liver Physiol. 281:G365–G370.

Urakami, M., R. Ano, Y. Kimura, M. Shima, R. Matsuno, T. Ueno, and M. Akamatsu. 2003. Relationship between structure and permeability of tryptophan derivatives across human intestinal epithelial (Caco-2) cells. Z. Naturforsch. 58c:135–142.

Utsunomiya-Tate, N., H. Endou, and Y. Kanai. 1996. Cloning and functional characterization of a system ASC-like Na^+-dependentneutral amino acid transporter. Biochem. J. 271:14883–14890.

Verma, N., and V. K. Kansal. 1993. Characterization of the routes of methionine transport in mouse mammary glands. IndianJ. Med. Res. [B] 98:297–304.

Vilella, S., G. A. Ahearn, G. Cassano, M, Maffia, and C. Storelli. 1990. Lysine transport by brush-border membrane vesicles ofeel intestine: Interaction with neutral amino acids. Am. J. Physiol. 259:R1181–R1188.

Viña, J. R., I. R. Puertes, and J. Vi ña. 1981. Effect of premature weaning on amino acid uptake by the mammary gland of lactatingrats. Biochem. J. 200:705–709.

Wang, W., W. Gu, X. Tang, M. Geng, M. Fan, T. Li, W. Chu, et al. 2009. Molecular cloning, tissue distribution and ontogeneticexpression of the amino acid transporter $b^{0,+}$cDNA in the small intestine of Tibetan suckling piglets. Comp. Biochem.Physiol. B154:157–164.

Wang, J., L. X. Chen, P. Li, X. L. Li, H. J. Zhou, F. L. Wang, D. F. Li, et al. 2008. Gene expression is altered in piglet small intestineby weaning and dietary glutamine supplementation. J. Nutr. 138:1025–1032.

Wang, X., and C. G. Proud. 2009. Nutrient control of TORC1, a cell-cycle regulator. Trends Cell. Biol. 19:260–267.

Weldon, W. C., A. J. Thulin, O. A. MacDougald, L. J. Johnston, E. R. Miller, and H. A. Tucker. 1991. Effects of increased dietaryenergy and protein during late gestation on mammary development in gilts. J. Anim. Sci. 69:194–200.

Wilson, J. W., and K. E. Webb, Jr. 1990. Lysine and methionine transport by bovine jejunal and ileal brush border membranevesicles. J. Anim. Sci. 68:504–514.

Woodward, A. D., S. J. Holcombe, J. P. Steibel, W. B. Staniar, C. Colvin, and N. L. Trottier. 2010. Cationic and neutral amino acidtransporter transcript abundances are differentially expressed in the equine intestinal tract. J. Anim. Sci. 88:1028–1033.

Wu, G., and J. R. Thompson. 1989a. Is methionine transaminated in skeletal muscle? Biochem. J. 257:281–284.

Wu, G., and J. R. Thompson. 1989b. Methionine transamination and glutamine transaminases in skeletal muscle. Biochem.J. 262:690–691.

Wu, G., J. R. Thompson, and V. E. Baracos. 1991. Glutamine metabolism in skeletal muscle from the broiler chick (*Gallusdomesticus*) and the laboratory rat (*Rattus norvegicus*). Biochem. J. 274:769–774.

Wu G., and D. A. Knabe. 1994. Free and protein-bound amino acids in sow's colostrum and milk. J. Nutr. 124:415–424.

Wu, G., D. A. Knabe, N. E. Flynn, W. Yan, and S. P. Flynn. 1996. Arginine degradation in developing porcine enterocytes. Am. J.Physiol. Gastrointest. Liver Physiol. 271:G913–G919.

Wu, G. 1997. Synthesis of citrulline and arginine from proline in enterocytes of postnatal pigs. Am. J. Physiol. Gastrointest. LiverPhysiol. 272:G1382–G1390.

Wu, G., and S. M. Morris, Jr. 1998. Arginine metabolism: Nitric oxide and beyond. Biochem. J. 336:1–17.

Wu, G., W. G. Pond, T. Ott, and F. W. Bazer. 1998. Maternal dietary protein deficiency decreases amino acid concentrations infetal plasma and allantoic fluid of pigs. J. Nutr. 128: 894–902.

Wu, G., T. L. Ott, D. A. Knabe, and F. W. Bazer. 1999. Amino acid composition of the fetal pig. J. Nutr. 129:1031–1038.

Wu, G., D. A. Knabe, and S. W. Kim. 2004. Arginine nutrition in neonatal pigs. J. Nutr. 134:S2783–S2790.

Wu, G. 2009. Amino acids: Metabolism, functions, and nutrition. Amino Acids 37:1–17.

Xiao, X. J., E. A. Wong, and K. E. Webb. 2004. Developmental regulation of fructose and amino acid transporter gene expressionin the small intestine of pigs. FASEB J. 18:269. (Abstr.)

第5章

碳水化合物及其在猪上的应用

1 导 语

碳水化合物是自然界中存在的由碳、氢和氧按照$C_n:H_{2n}:O_n$比例组成的化合物。碳水化合物是猪饲粮中最丰富的单一的饲料能量，占总能量摄入量的60%～70%。根据其聚合度，饲料中的碳水化合物一般分为糖（Sugars）、寡聚糖和多聚糖，而多聚糖又包括了从完全可消化淀粉到不可消化的非淀粉多糖（NSP）的抗性不同的一大类淀粉（Cummings和Stephen，2007；Englyst 等，2007）。目前已经清楚，饲粮中的碳水化合物类物质在肠道及机体其他部位有着不同的代谢归宿和生理特性（Bach Knudsen和Jørgensen，2001）。淀粉和糖在小肠中由消化酶水解而被消化为单糖，并被吸收和代谢；而非淀粉多糖、抗性淀粉（RS）和寡聚糖主要是在大肠（盲肠和结肠）中被微生物发酵。碳水化合物被微生物发酵，最终生成的主要是短链脂肪酸（SCFA）和乳酸（LA），这些终产物可被结肠上皮细胞吸收，并在结肠上皮细胞、肝脏、脂肪和肌肉细胞中代谢（Bergman，1990）。

猪一生中所摄食的饲粮中的碳水化合物的组成变异很大。母猪乳汁中的碳水化合物主要是乳糖。而给生长猪和成年母猪饲喂的饲粮中的碳水化合物，就其化学结构及组织方式而言，则远复杂得多。猪通过改变肠道形态结构、消化酶活性和微生物水解活性来应对饲粮的这一变化。本章的主要目的是概述猪饲粮中存在的饲料碳水化合物及其消化、吸收和利用情况。

2 碳水化合物的命名与分类

碳水化合物是一类多样性非常丰富的分子，根据其分子大小（聚合度，DP），按照化学分类方法可分为糖（DP 1~2）、寡聚糖（DP 3~9）和多聚糖（DP ≥10），后者又包括淀粉和非淀粉多糖（Cummings和Stephen，2007；Englyst 等，2007）。糖单体分子的组成和糖残基之间连接的类型影响碳水化合物在机体内的代谢途径。基于化学分类，可将碳水化合物成分分为两种营养类型：一是可消化碳水化合物，即可被宿主酶消化并可在小肠中被吸收的碳水化合物（包括单糖、双糖和淀粉）；二是非消化性碳水化合物（NDC），即可供栖息在大肠中的微生物发酵的碳水化合物（包括寡聚糖、RS和NSP）。饲料中的碳水化合物及其消化终产物举例见表5-1。

表5-1 饲料中的碳水化合物分类及其在猪肠道中的可能消化终产物

分类	聚合度	举例	内源酶	被吸收的分子
单糖	1	葡萄糖		葡萄糖
	1	果糖		果糖
双糖	2	蔗糖	+	葡萄糖 + 果糖
	2	乳糖	+	葡萄糖 + 半乳糖
寡聚糖	3	棉子糖	−	SCFA
	4	水苏糖	−	SCFA
	3 ~ 9	果寡糖	−	SCFA
多聚糖	≥ 10	淀粉	+	葡萄糖
	≥ 10	非淀粉多糖	−	SCFA

◆ 2.1 糖

糖（DP 1～2）是由单糖和双糖组成的水溶性成分（图5-1）。蔗糖是植物产品中含量最丰富的一种糖（Bach Knudsen，1997），而乳糖是牛奶中碳水化合物的主要成分。单糖和麦芽糖在未发芽的植物体中的含量一般都较低。

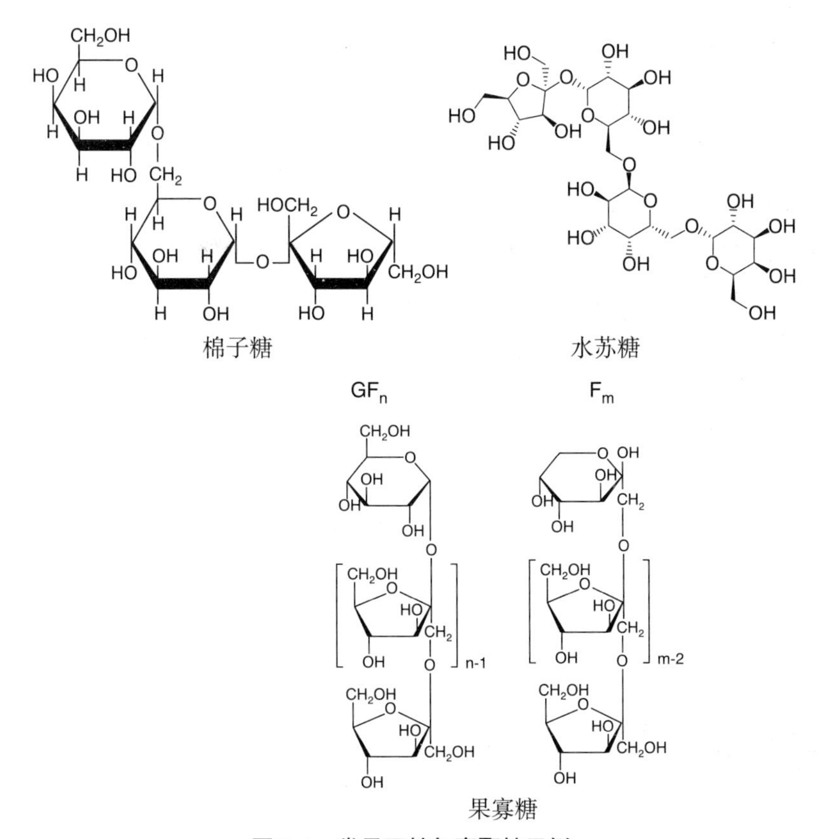

图5-1　常见双糖与寡聚糖示例

G=葡萄糖残基；F=果糖残基；n和m=果糖残基的重复数。

◆ 2.2 寡聚糖

寡聚糖（DP 3 ~ 9）是由3 ~ 9个单体通过α或β糖苷键连接在一起组成的水溶性化合物（图5-1）。寡聚糖存在于根、块茎和种子以及多种豆类、锦葵、菊科和芥菜类植物的副产品中（Bach Knudsen和 Li，1991），它们可作为饲料原料被混合使用（Flickinger 等，2003）。

◆ 2.3 淀粉

天然淀粉是不溶于水的半结晶物质，在多种植物组织中以颗粒形式存在（Gallant 等，1992；图5-2）。纯淀粉主要由α-葡聚糖组成（约占干物质的99%），以直链淀粉和支链淀粉的形式存在。直链淀粉大致为线性α（1 ~ 4）分子（约占99%），其分子量为$1 \times 10^5 \sim 1 \times 10^6$；而支链淀粉的分子更大（分子量为$1 \times 10^7 \sim 1 \times 10^9$），具有更大的支链，约含有95%的α（1 ~ 4）连接和5%的α

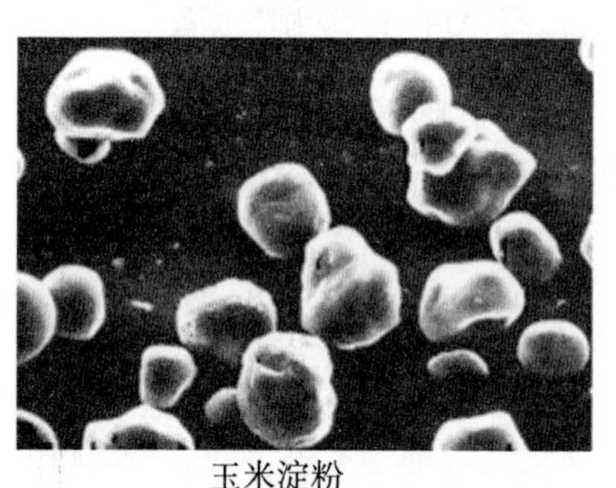

玉米淀粉

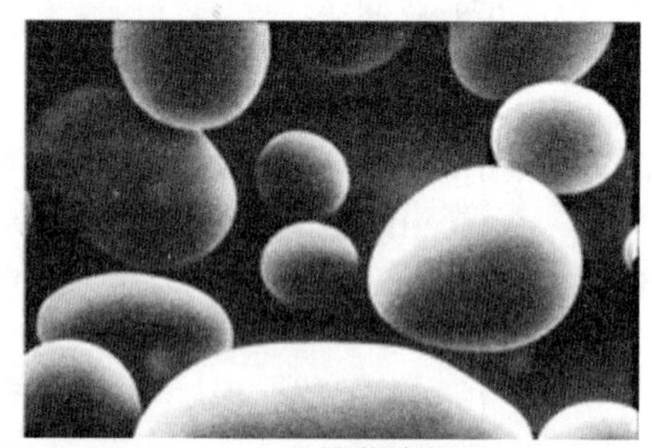

马铃薯淀粉

直链淀粉

支链淀粉

图5-2　来源于玉米和马铃薯的淀粉颗粒以及构成淀粉颗粒的两种主要的聚合体（直链淀粉和支链淀粉）

（1～6）连接（Biliaderis，1991）。在淀粉颗粒中，两类α-葡聚糖以不同的比例存在。当直链淀粉与支链淀粉的比例较低时（约<15%），淀粉就呈蜡状；中等直链淀粉的含量占16%～35%；当直链淀粉含量超过约36%时，为高直链淀粉（或淀粉）（Ring 等，1988）。通过X-射线的衍射研究，可将淀粉分为A、B和C三种类型（Biliaderis，1991）。A型淀粉存在于谷类中，在一般情况下具有一个开放的结构；而B型淀粉则更紧凑，存在于马铃薯等块茎中；C型淀粉是A型淀粉和B型淀粉的组合，存在于豆类中（Würsch等，1986）。所有的淀粉均可以潜在地被α-淀粉酶消化，但是有一些淀粉则可以抵抗小肠的消化。抗性淀粉的主要形式是物理包被的淀粉，如在完整的细胞壁结构内的淀粉（称为RS_1）、生淀粉颗粒（RS_2）和老化直链淀粉（RS_3）（Englyst 等，1992）。

◆ 2.4 非淀粉多糖

非淀粉多糖是由一系列主要存在于初级植物细胞壁或次级植物细胞壁中的可溶性多糖和不溶性多糖组成（Selvendran，1984；Carpita 和 Gibeaut，1993；McDougall 等，1996a；图5-3）。单子叶谷类和双子叶豆类植物对猪的营养是最重要的（Bach Knudsen，1997；Theander等，1989）。细胞壁多糖的组分包括戊糖（阿拉伯糖和木糖）、己糖（葡萄糖、半乳糖和甘露糖）、6-脱氧已糖（鼠李糖和岩藻糖）以及糖醛酸（葡萄糖醛酸和半乳糖醛酸，或其4-O-甲基醚）。尽管细胞壁多糖仅由10个常见的单糖构成，但每个单糖可以两个环（吡喃糖和呋喃糖）的形式存在，并且这些残基可以通过糖苷键在三个、四个或五个有效羟基中的任何一个之间以两种方向（α或β）连接。其结果是，细胞壁多糖可形成数量巨大的三维结构，从而提供一个更大的功能性表面。非淀粉多糖也可以与木质素和木栓质连接，提供疏水性表面。此外，多糖上的荷电基团，即糖醛酸上的酸性基团，可以影响离子特性，并发生不同程度的酯化。

谷物及其副产品中主要的细胞壁多糖是纤维素、阿拉伯木聚糖以及混合连接的β（1->3）（1->4）-D-葡聚糖（β-葡聚糖），但在不同谷物中其相对比例、结构以及与其他成分相互连接等方面存在着显著差异（Theander等，1989；Bach Knudsen，1997）。富含蛋白质的饲料原料的细胞壁成分比谷物细胞壁中的情况更为复杂。豌豆和大豆子叶的初级细胞壁中碳水化合物的组成主要是果胶多糖（鼠

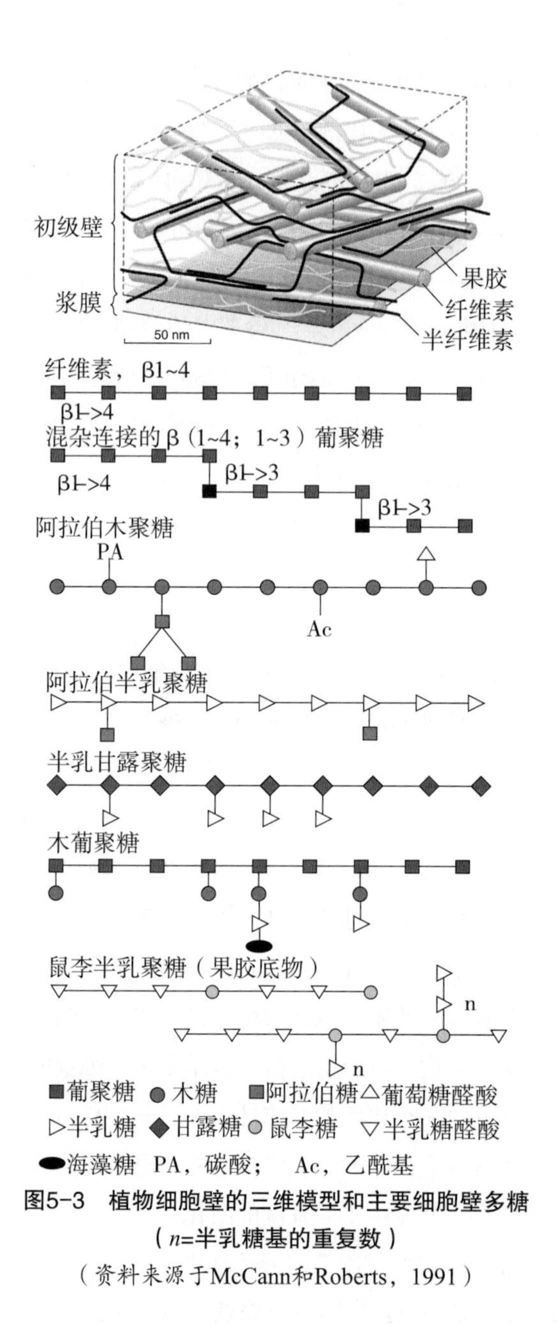

图5-3 植物细胞壁的三维模型和主要细胞壁多糖

（*n*=半乳糖基的重复数）

（资料来源于McCann和Roberts，1991）

李半乳糖醛酸）、纤维素、木葡聚糖和糖蛋白。豌豆中的阿拉伯聚糖和大豆中的阿拉伯半乳聚糖，既可以游离存在，又可以与鼠李半乳糖醛酸连接（Theander等，1989；Bach Knudsen，1997）。种子外壳的细胞壁中除了含有纤维素和酸性木聚糖外，也含有大量果胶多糖，而其中的木质素含量较低；与谷物外壳相比，其中的木质素含量则更低。与无胚乳种子相反，所有胚乳豆科种子中含有半乳甘露聚糖，在种子发育期间其沉积于胚乳细胞壁上，在以后的种子发芽过程中常与淀粉一起被利用。

◆ 2.5 木质素

木质素虽然不是碳水化合物，但在这里仍被当做是一种碳水化合物，因为它与细胞壁多糖紧密联系，而且许多旧的和目前仍然常用的纤维测定的分析方法也包括木质素的测定。因此，在本文中如果不包括木质素，则难于讨论碳水化合物的理化性质及其在胃肠道中的降解。木质素是由松柏醇、对香豆和芥子醇聚合形成的（Davin等，2008）。这些苯基丙烷单元通过醚和碳-碳键的不规则三维模式相连接，其中的任何一个碳原子均可以作为芳香环的一部分。木质素既可以直接通过糖残基也可以间接通过阿魏酸酯化多糖，共价连接到多糖上（Liyama等，1994；Davin 等，2008；图5-4）。木质素常将聚合物固定到一定位置，以便黏合和锚定纤维素微纤维和其他基质多糖。通过这种方式使得细胞壁变得非常坚硬，难以被大肠中的微生物降解。

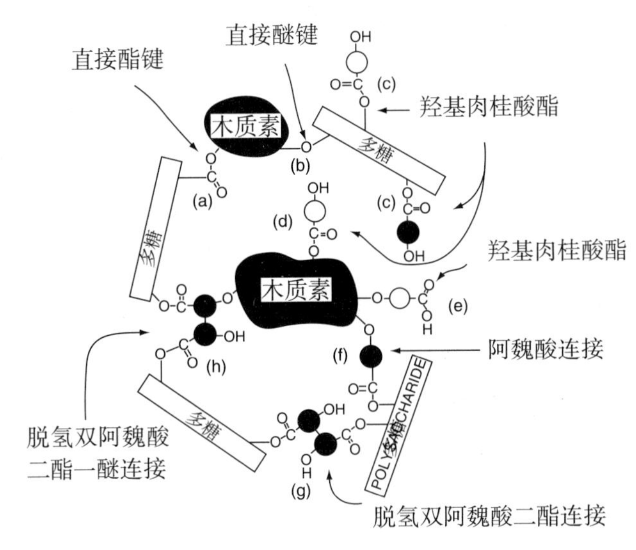

图5-4 多糖间可能的相互连接和细胞壁中的木质素

（资料来源于Liyama 等，1994）

3 纤维的理化特性

纤维的理化特性、水合性质和黏度与构成细胞壁聚合物的类型及其分子间的关联有关（McDougall 等，1996b）。水合性质包含泡胀度、溶解性、持水量和水结合容量（WBC）。持水量和水结合容量在文献中可以互换使用，因为二者均反映了纤维源在其基质内固定水的能力。聚合物增容过程的第一步是泡胀，进入的水分在纤维的大分子中扩散，直到其完全分散（如细胞壁，图5-3详述的三维空间；Thibault 等，1992）。具有有序、规则结构的多糖（如纤维素或直链阿拉伯木聚糖）不可能存在增溶溶解作用，因为线性结构增加了非共价键的强度，稳定了有序的构象。在这些条件下，只有溶胀才可以发生（Thibault 等，1992）。当大多数多糖溶解于水时，呈黏稠的溶液状（Morris，1992）。黏度取决于初级结构、聚合物的分子量和浓度。大分子可增加稀释溶液的黏度，其能力主要取决于大分子占据的体积。虽然一系列的多糖从分析定义来讲是可溶的，但其在体内的溶解性可能被饲料基质限制，从而限制了其黏度增加的特性。

4 饲粮中碳水化合物与木质素的测定方法

截至目前，传统的Weende概略养分分析系统已使用了150多年（Henneberg和 Stohmann，1859）。该系统包括分析干物质、灰分、脂肪、粗蛋白（N×6.25）和粗纤维，粗纤维的测定是用1.25%硫酸和1.25%氢氧化钠回流脂肪萃取残余物后的残渣，并以不可溶残留物中的灰分进行校正。无氮浸出物（NFE）是干物质减去灰分、蛋白质（N×6.25）、脂肪和粗纤维的总和得到的（图5-5）。以基于特异性酶法、比色法和色谱分析法等现代分析技术（图5-6），碳水化合物组分可以根据其化学和营养性质进行分类（表5-1和图5-1~5-3）。常用的方法包括：测定糖和寡聚糖的酶法或色谱法、测定淀粉和抗性淀粉的酶法、测定整体饲粮纤维或分类测定可溶性多糖、不溶性多糖和木质素的比重法或酶化学法（Bach Knudsen和Li，1991；Englyst 等，1994；Theander 等，1994）。由于测定纤维的分析方法不同，分析原理也各不相同，因此文献中报道的测定值同样也存在变化。如图5-7所示，酶化学法的测定值

（Bach Knudsen，1997）均大于van Soest及其同事建立的洗涤法的测定值（van Soest，1963；van Soest和Wine，1967），更大于粗纤维法的测定值。

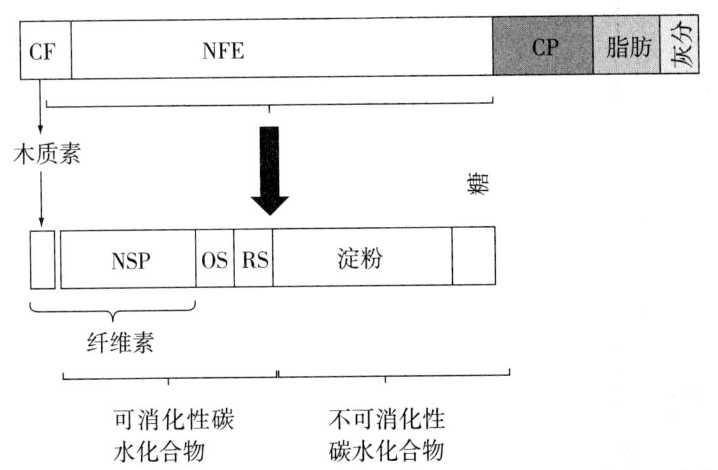

图5-5　碳水化合物和木质素组成了概略分析（Weende）中的粗纤维（CF）和无氮浸出物（NFE）

CP为粗蛋白；NFE为无氮浸出物；NSP为非淀粉多糖；OS为寡聚糖；RS为抗性淀粉。

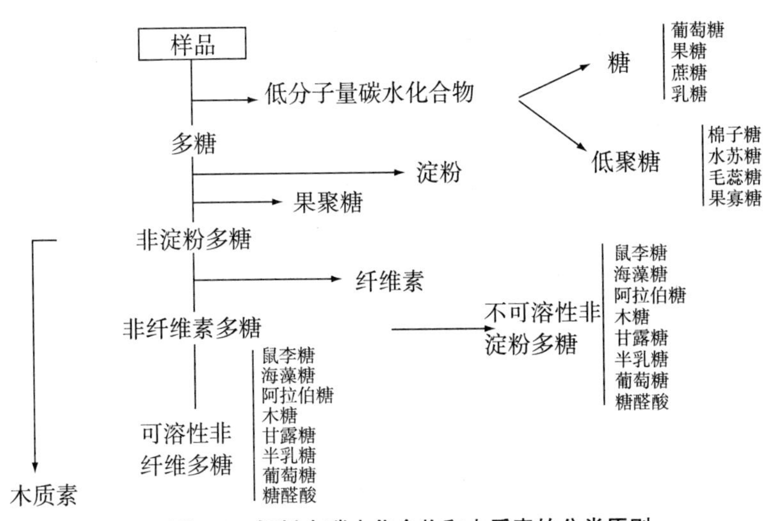

图5-6　饲料中碳水化合物和木质素的分类原则

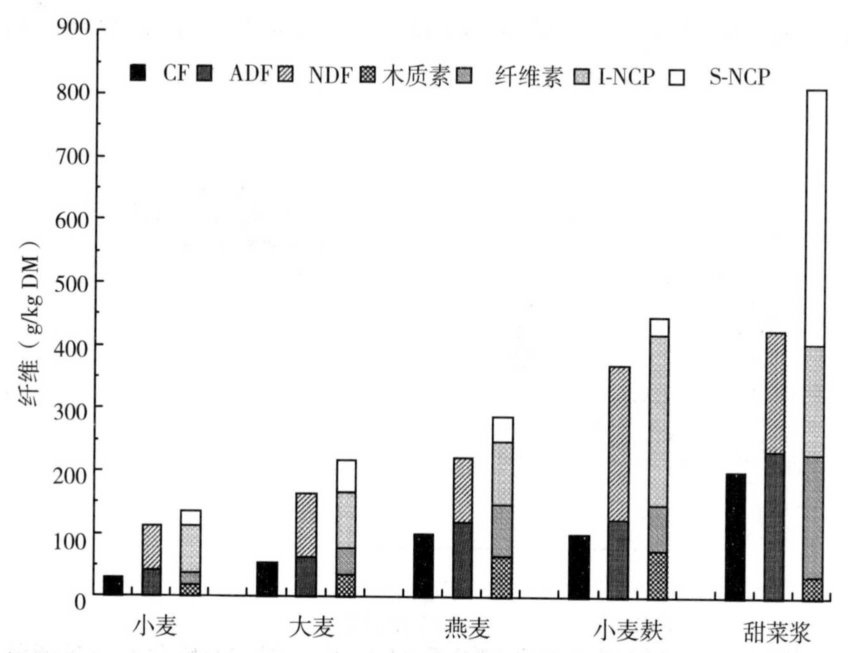

图5-7 以粗纤维（CF）法、van Soest酸性洗涤纤维（ADF）和中性洗涤纤维（NDF）法测定的不同饲料原料中的纤维含量分析值，以及利用酶-化学-比重Uppsala和Englyst法测定的克拉松木质素、纤维素、非可溶性非纤维素多糖（I-NCP）和可溶性非纤维素多糖（S-NCP）的含量分析值

5 饲料原料中的碳水化合物与木质素

饲料中的碳水化合物不是以纯的化学实体形式存在的，而是以糖、寡聚糖和多糖的混合物形式存在，多糖则主要与蛋白质和木质素等其他生物聚合物结合（图5-6）。虽然现代养猪业依赖于相对较少的饲料原料，但大多数是来源于谷类（玉米、小麦、大麦、燕麦、黑麦和大米）、谷物副产品（不同的加工部分，生物燃料和酒精行业等的残留物）、谷物替代品（如甘薯粉）、豆类（如豌豆、蚕豆和羽扇豆）、蛋白浓缩物（如大豆、油菜、向日葵和棉花籽的粉或饼）以及糖和淀粉工业的副产品。饲料原料中的各种碳水化合物的各种各样的组成（表5-2，图5-7），使得生产具有不同组分的复合饲料成了可能。例如，利用大米作为碳水化合物的主要原料可生产低膳食纤维饲粮（Hopwood等，2004）；而与此相反，植物性食品和农业企业的副产品可用于生产高膳食纤维饲粮（Serena等，2008b）。

表5-2 饲料原料中碳水化合物和木质素的通常含量（g/kg DM）[1]

饲料原料	可消化 CHO		不可消化 CHO						KL[4]	纤维
	糖	淀粉	OS	果聚糖[2]	RS	S-NCP[3]	I-NCP[4]	纤维素[4]		
大米	2	837	2	＜1	3	9	1	3	8	22
玉米	17	680	3	6	10	9	66	22	11	108
小麦	13	647	6	15	4	25	74	20	19	138
大麦	16	585	6	4	2	56	88	43	35	221
燕麦	13	466	5	3	2	40	110	82	66	298
小麦麸	37	220	16	20	2	29	273	72	75	449
大麦皮	21	172	12	7	2	20	267	192	115	594
玉米DGGS		35	ND	ND	ND	25	183	68	47	323
小麦DGGS		92	ND	ND	ND	55	135	61	86	337
豌豆	39	432	49	ND	22	52	76	53	12	192
蚕豆	32	375	54	ND	32	50	59	81	20	210
大豆粕	77	27	60	ND	ND	63	92	62	16	233
油菜饼	72	15	16	ND	ND	43	103	59	90	295
棉籽饼	12	18	54	ND	ND	61	103	92	83	340
豌豆壳	15	88	ND	5	ND	121	148	452	9	677
马铃薯浆	＜1	122	ND	ND	127	280	95	202	35	612
甜菜渣	38	5	ND	0	ND	290	27	203	37	737
菊苣根	156	ND	ND	470	ND	76	24	48	11	158

[1]CHO，碳水化合物；OS，寡聚糖；RS，抗性淀粉；S-NCP，可溶性非纤维素多糖；I-NCP，不可溶性非纤维素多糖；KL，克拉松木质素；DDGS，谷物干酒糟及可溶物；ND，未测出。

[2]果聚糖是寡聚糖（DP3～9）和多聚糖（DP>10）的混合物。

[3]S-NCP与可溶性纤维是同义词。

[4]不可溶性非纤维素多糖、纤维素和克拉松木质素的总和，是不可溶性纤维。

资料来源于Bach Knudsen，1997；Serena和Bach Knudsen，2007，以及未发表数据。

6 饲料原料与普通饲料的加工

饲料原料中的淀粉常存在于谷物和豆类植物中，与蛋白质结合，其中多数是相对疏水的，该蛋白质-淀粉网状结构被细胞壁包围。因此，在消化过程中，淀粉常被保持在被摄入的食物颗粒内部，从而不被水溶解。块茎和豆类中的淀粉在肠腔液的极端环境中可被很好地保护，即使是在谷物中，只有谷物的物理结构发生改变，淀粉才能接触到α-淀粉酶。谷物或豆类的物理加工（研磨、裂化和滚筒制粉）和热加工（制粒、膨化和挤压蒸煮）是提高淀粉利用率（水渗透和随后的α-淀粉酶消化）的主要加工方法。研磨是一个物理过程，即通过降低颗粒的大小增加表面积，从而使消化液可以更充分地与底物接触。例如，边长为1 cm的立方体，其具有6 cm^2的表面积；如果将其分为边长为0.1 cm的立方体，其表面积将提高到60 cm^2，或者说增加了10倍。另一种物理加工方式是水热处理，可将淀粉的物理形态从晶体改变为凝胶结构（图5-8）。这可扩大淀粉颗粒的表面积，促使其有效进入极性溶液与α-淀粉酶相互作用（Biliaderis，1991）。蒸煮后的冷却过程可重新改变多糖的物理状态（老化），这可能会降低其消化率（图5-7）。与蜡质类型相比，含有高比例直链淀粉（高直链淀粉）的淀粉，加热处理后其消化率通常更容易降低（Brown 等，2001）。因为淀粉酶凝沉是一个不可逆的过程，可形成复杂的结构（类似于纤维素）。

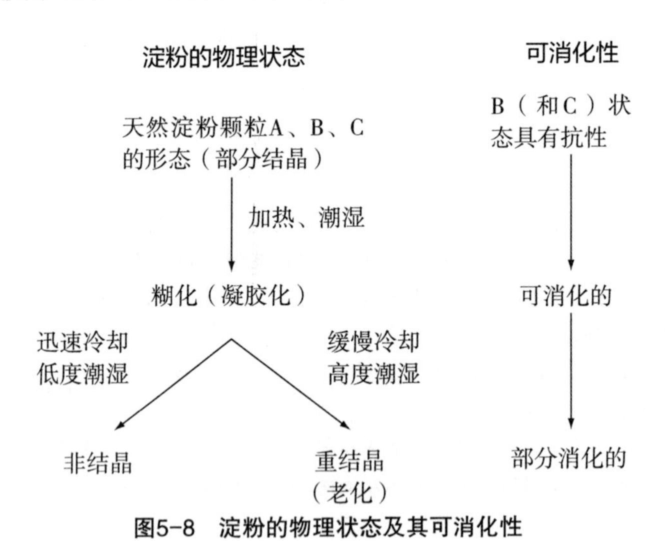

图5-8 淀粉的物理状态及其可消化性

（资料来源于Cummings，1997；基于Biliaderis，1991；Cummings和Englyst，1995的数据）

在用于动物饲料前，植物性饲料和农业企业的副产品所代表的是饲料原料加工的另一种形式。例如，工业生产油、生物燃料、糖、淀粉、啤酒和果胶过程中产生的副产品均经过了这种或那种方式的物理处理和化学处理（Serena和Bach Knudsen，2007）。一般来讲，这些类型的饲料原料，代表着一类特别异质的植物残留物，其来源于不同植物种属和原植物（如谷物、块茎、根、果实、茎、壳以及外皮）。在加工过程中，其暴露于各种各样的不同的理化处理，用于提取有经济价值的重要成分。形成的残余物会因此而具有不同的基质，富含蛋白质的物质（如大豆粕、菜籽粕等）或植物细胞壁（如非淀粉多糖和木质素）。

7 碳水化合物在小肠中的消化

任何比单糖更大的分子不能被小肠上皮细胞转运，因此饲料中的大多数碳水化合物需要被降解为低分子量的化合物后才能被吸收。对于双糖和淀粉来说，这一过程涉及唾液和胰液中分泌的α-淀粉酶、存在于小肠刷状缘上的各种寡聚糖酶和蔗糖酶以及乳糖酶（Kidder和Manners，1980；Gray，1992）。然而，对碳水化合物水解作用的贡献，来源于永久定植在胃肠道中的菌群。Jensen和Jørgensen（1994）报道，总厌氧菌从胃中的10^7 ~ 10^9活菌数逐渐增加到小肠远端中的10^9活菌计数。曾有报道指出，在胃的食糜中存在相当水平的乳酸和SCFA，在小肠远端食糜中的含量更高（Argenzio和Southworth，1974；Bach Knudsen等，1991）。

◆ 7.1 糖

蔗糖易于被降解，并从猪的肠腔中被吸收，即使蔗糖的水平非常高对其消化率也没有限制（Ly，1992，1996；表5-3）。葡萄糖也是如此，可直接被小肠吸收，但果糖的吸收率较低，也不完全。乳源性二糖乳糖也是这种情况。仔猪的乳糖酶活性很高，但断奶后降低到一定水平，不足以降解饲粮中高含量的乳糖。利用插管猪的研究表明，当采食的乳糖水平加倍时，其吸收系数降至几乎一半（Rérat等，1984b）。

表5-3 猪胃肠道中糖和寡聚糖消化率的经典值

来源	糖			寡聚糖				
	葡萄糖	果糖	蔗糖	棉子糖	水苏糖	毛蕊糖	果聚糖	总计
葡萄糖	98.3	—	—	—	—	—	—	—
果糖	—	86.6	—	—	—	—	—	—
蔗糖	—	—	98.3	—	—	—	—	—
豌豆								
干燥的	—	—	96	68	74	58	—	66
烤干的	—	—	94	65	53	23	—	42
豆粕	—	—	87	38	49	—	—	49
羽扇豆								
蓝色	—	—	93	84	93	77	—	—
黄色	—	—	92	64	88	74	—	—
菊粉							40[1]	40[1]
菊苣根[2]	—	—	—	—	—	—	33[3]	33[3]

[1]三种饲粮的平均值，为53～208 g/kg DM。
[2]菊苣根中的果聚糖是寡聚糖和多糖的混合物。
[3]三种饲粮的平均值，为79～156 g/kg DM。
资料来源于Canibe 和 Bach Knudsen，1997b；Gdala等，1994；Hedemann和Bach Knudsen，2009；Ly，1992，以及未发表数据。

◆ 7.2 寡聚糖

虽然猪的机体内缺乏能够降解大多数寡糖键的酶，但对大豆、羽扇豆和豌豆中的棉子糖以及果聚糖（寡聚糖和菊粉的混合物）、低聚果糖和低聚反式半乳糖的研究表明，在小肠中存在高度变异的、相对较高的消化系数（Gdala等，1994；Canibe和Bach Knudsen，1997b；Gdala和Buraczewska，1997；Gdala等，1997a；Houdijk等，1999，2002；表5-3）。消化系数与剂量可能紧密相关，因为低含量水平比更高的含量水平（Hedemann和Bach Knudsen，2009）表现出更大的消化系数（Houdijk等，1999）。

◆ 7.3 淀粉

内源性消化液中唯一的糖酶是唾液和胰腺的α-淀粉酶，其可消化淀粉中的α-1,4-葡 糖 苷 键（Kidder和Manners，1980；Moran，1985；Gray，1992）。从数量上讲，唾液中的α-淀粉酶对淀粉降解的贡献较少，因为它对酸不稳定，在胃中（pH通常为2～4）被快速降解（Argenzio和Southworth，1974；Duke，1986）。因此，绝大多数的淀粉是被肠腔内的胰腺α-淀粉酶降解。胰液中分泌的碳酸氢盐可将十二指肠中的pH提高至5～6的水平，pH沿小肠长度的延伸进一步增加，在回肠中达到中性（Argenzio和Southworth，1974；Duke，1986）。较高的内源性分泌物可将饲粮残余物进一步稀释至约10%的干物质，这有助于该极性溶液渗透进饲料颗粒中，从而保证了淀粉的有效裂解。由于α-淀粉酶不能破坏支链淀粉中存在的α-1,6键，破坏靠近分枝点的α-1,4连接的能力又存在空间位阻，因此α-淀粉酶消化的终产物是麦芽糖、麦芽三糖和α-限定糊精。之后，寡聚糖进一步被存在于肠道刷状缘膜表面的大分子糖蛋白成分寡糖酶降解为葡萄糖（Lentze，1995）。淀粉糖化酶（淀粉葡萄糖苷酶、麦芽糖酶-葡萄糖淀粉酶）能够从α-限定糊精的非还原端依次打开单一的α-1,4-连接的葡萄糖残基，但当α-1,4-连接的葡萄糖位于该糖的末端时就被阻止。当经过其他酶的作用被暴露后，α-糊精酶是唯一能够裂解非还原末端α-1,6连接的糖酶（Moran，1985；Gray，1992）。较短的α-1,4连接的寡聚糖，如麦芽糖和麦芽三糖，然后通过麦芽糖酶与葡萄糖分开，这是一种高效的α-1,4葡萄糖苷酶。随后，葡萄糖终产物通过特定的葡萄糖载体进行转运。该载体是仅在小肠表达的完整的刷状缘糖蛋白，与单糖具有强亲和性。葡萄糖运输到肠上皮细胞的实际驱动力是由钠-钾ATP酶提供的，即将细胞内的Na^+泵至基顶膜（Lentze，1995）。葡萄糖扩散可能是从基底外侧（浆膜）表面进入绒毛轴的毛细血管，并进一步进入门静脉。

利用回肠瘘管猪研究测定的淀粉消化率的典型值如表5-4所示。经过精细研磨的谷物，淀粉的开放结构容易被α-淀粉酶水解。大多数研究表明，当食糜到达小肠末端时，大量的淀粉已经被吸收。制粒、挤压蒸煮等水热处理，能够改变淀粉的天然结构，使谷物淀粉更容易被消化。例如，在面包和挤压蒸煮方面的研究表明，淀粉消化率均在98%以上（Glitsø等，1998；Bach Knudsen 和Canibe，2000；Bach Knudsen等，2005；Sun等，2006；Le Gall等，2009）。然而，饲粮的颗粒大小可降低小肠中淀粉的消化率，粗糙颗粒包被了细胞内的营养物质，从而阻止了内容物不能被内源性酶消化。例如，粗加工大麦的淀粉消化率较精细研磨大麦的淀粉消化率约低4个绝对单位（表5-4）。

表5-4　饲喂不同类型生淀粉以及加工淀粉和纤维时猪的摄入量和小肠中淀粉的消化率[1]

项目	淀粉				非淀粉多糖	
	类型	形态	摄入量（g/d）	消化率（%）	摄入量（g/d）	消化率（%）
谷物混合物	A	R	556	96.0	182	20.0
+小麦麸	A	R	310	95.7	351	11.0
+甜菜浆	A	R	193	95.3	633	37.0
小麦粉	A	R	978	99.4	45	30.0
+小麦麸	A	R	1 003	98.7	86	10.0
+燕麦麸	A	R	1 086	98.6	77	36.0
燕麦片	A	R	885	97.0	123	21.0
燕麦粉	A	R	878	98.6	81	25.0
燕麦麸	A	R	814	98.9	202	15.0
小麦—细粉（2.9%>2 mm）	A	R	874	96.3	212	8.0
小麦—粗粉（12.0%>2 mm）	A	R	832	96.3	215	6.0
大麦—细粉（0.7%>2 mm）	A	R	819	96.5	286	18.0
大麦—粗粉（23.3%>2 mm）	A	R	832	92.2	303	－7.0
黑麦粉面包	A	G	838	98.9	116	19.0

（续）

项目	淀粉				非淀粉多糖	
	类型	形态	摄入量（g/d）	消化率（%）	摄入量（g/d）	消化率（%）
全谷物黑麦面包	A	G	777	98.0	202	7.0
高纤维小麦面包	A	G	841	98.8	350	21.8
高纤维黑麦面包	A	G	721	98.3	321	22.7
豌豆—干燥	C	R	457	88.9	225	40.0
豌豆—烤干	C	G	443	85.7	197	24.0
蚕豆 1	C	R	—	81.5	—	—
蚕豆 2	C	R	—	86.4	—	—
马铃薯	B	R	712	39.8	—	—
马铃薯	B	G	751	98.3	—	—

NSP，非淀粉多糖；A，A型淀粉；B，B型淀粉；C，C型淀粉；R，未加工的；G，胶状的。

资料来源于Bach Knudsen和Hansen，1991；Bach Knudsen等，1993； Bach Knudsen等，2005； Canibe和Bach Knudsen，1997b； Gdala和Buraczewska，1997；Glits等，1998；Graham等，1986；Sun等，2006，以及未发表数据。

淀粉的类型是显著影响其消化率的另外一个因素，豆类淀粉（C型）比谷类淀粉更难消化。在对豆科植物的研究中发现，豌豆淀粉的消化率大约为88%（Canibe和Bach Knudsen，1997b；Gdala和Buraczewska，1997；Sun 等，2006），蚕豆淀粉的消化率大约为84%（Gdala和Buraczewska，1997）。烘烤并不能增加豌豆淀粉的可消化性（Canibe和Bach Knudsen，1997b），而挤压蒸煮可将其消化率提高到生豌豆与谷物消化率之间的水平（Sun 等，2006）。马铃薯淀粉（B型）的结晶特性使其成了最不容易被猪消化的淀粉，其消化率为30%～50%（Sun 等，2006；van der Meulen 等，1997a）。生的马铃薯淀粉仅可偶尔低剂量用于养猪生产实际中。

◆ 7.4 非淀粉多糖

上述研究结果还表明，20%～25%的非淀粉多糖在通过小肠时被降解。这项研究进一步表明，不同纤维多糖之间存在巨大的差异。例如，线性β-葡聚糖和相对可溶性β-葡聚糖的回肠消化率总是大于不可溶性纤维素和不溶性阿拉伯木聚糖复合物（Bach Knudsen和Hansen，1991；Bach Knudsen 等，1993；Glitsø 等，1998；Bach Knudsen 等，2005；Le Gall 等，2009）。果胶多糖也存在很高的变异性，如一些研究显示其降解率较高（Canibe和Bach Knudsen，1997b）；而也有研究表明，果胶多糖几乎完全不能降解（Jørgensen 等，1996）。

◆ 7.5 物理效应

纤维成分有可能与前肠的消化过程有关，因为其代表了未被胃和小肠内源酶降解的饲料部分（Bach Knudsen和Jørgensen，2001）。表5-5明确显示，饲粮到达小肠末端过程中，随着食糜中可消化营养素（淀粉和糖）被耗尽，非淀粉多糖的浓度大幅增加。表5-5也清楚显示，非淀粉多糖浓度和食糜流速的增加，与饲粮中的纤维水平呈强相关。随着食糜在胃和小肠中蠕动，非淀粉多糖中的某些成分可被溶解，这将增加液体相的黏度（图5-9）。然而，不同纤维来源增加黏度的能力很大程度上取决于其化学和结构组成。因此，燕麦（也可能包括大麦）中的β-葡聚糖在很大程度上是解聚的（Johansen 等，1997），因此对黏度的影响较小；相反，可溶性阿拉伯木聚糖更难于降解（Le Gall 等，2009），因此导致较高的肠腔黏度（Bach Knudsen 等，2005；Lærke 等，2008；Le Gall 等，2009）。然而，利用回肠瘘管猪对78种饲料的研究结果表明（Bach Knudsen 等，未发表数据），可溶性纤维和不可溶性纤维对淀粉消化率均无任何大的影响，除非某些很特别的情况。此外，相对较少的利用回肠瘘管母猪进行的研究表明，母猪对淀粉的消化率与生长猪的水平相当，而且即使添加很高水平（429～455 g/kg DM；Serena 等，2008b）的可溶性纤维或不可溶性纤维也不影响淀粉的消化率。

表5-5 饲粮与回肠食糜中的食糜流速、指示剂参数和碳水化合物浓度[1]

项目	食糜流速（g/d）	指示剂参数	Dig CHO		NDC	
			蔗糖	淀粉	果聚糖	NSP
生长猪						
低饲粮纤维						
饲粮		100	6	517	—	56
回肠	2 126	652	7	17	—	366
中饲粮纤维						
饲粮		100	7	454	—	97
回肠	2 584	472	8	12	—	372
高饲粮纤维						
饲粮		100	29	492	14	211
回肠	3 785	345	8	28	20	514
成年母猪						
低饲粮纤维						
饲粮		100	21	501	9	140
回肠	5 560	347	10	59	3	267
高饲粮纤维						
饲粮		100	23	210	6	363
回肠	9 816	187	3	33	1	507

[1]Dig，可消化的；CHO，碳水化合物；NDC，不可消化性碳水化合物；NSP，非淀粉多糖。

资料来源于Bach Knudsen和Canibe，2000； Bach Knudsen 等，2005；Serena 等，2008b。

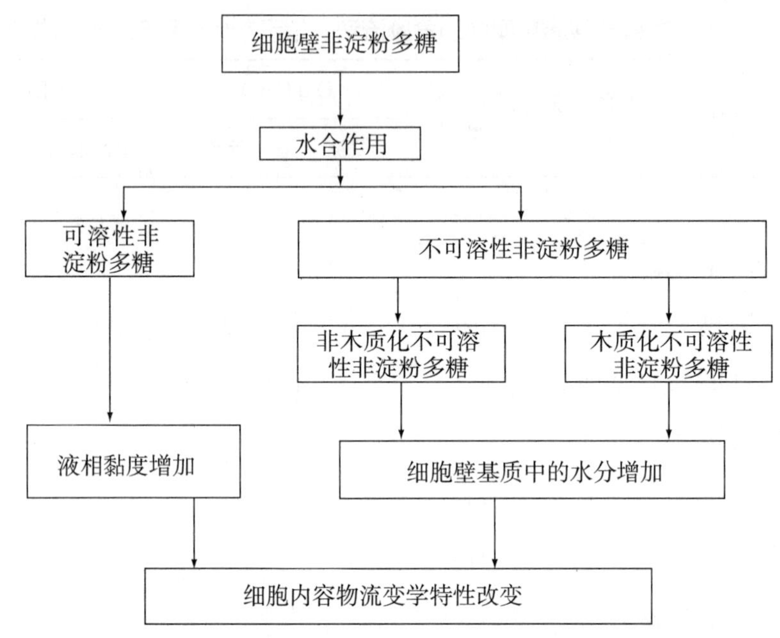

图5-9　可溶性和不可溶性非淀粉多糖对食糜流变学特性的影响

8 大肠中碳水化合物的消化

大肠是一个具有低氧浓度、低流速和高水分含量等特征的厌氧发酵室，所有这些条件都有利于细菌的生长，每克新鲜内容物中的活菌计数可能会达到10^{11} ~ 10^{12}（图5-10和图5-11）。

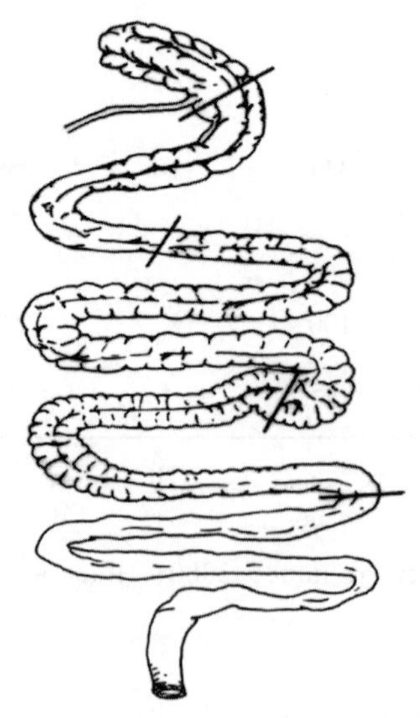

图5-10　大肠不同肠段中的发酵及其参数

微生物生态系统，包含数百种厌氧菌，每种厌氧菌占据一个特定小生境，它们之间存在无数的相互联系（Louis 等，2007；Flint 等，2008）。大肠中饲粮残渣发酵的产物包括：①SCFA，其通过被动扩散方式被吸收到门静脉，作为肠细胞生长和更新的底物，或随粪便排出体外（Bergman，1990）；②气体，其通过排气和呼吸进行排泄（Jensen和Jørgensen，1994）；③微生物菌体。细菌只能接触到小肠中未被消化的食物残渣。因此，沿着大肠的长度存在营养素梯度（图5-10）。在盲肠和近端结肠中碳水化合物的浓度最大，在远端结肠和直肠中的浓度最低。盲肠和近端结肠中较高的养分浓度导致微生物生长快、SCFA产生多，因此SCFA的浓度较高、pH较低（Bach Knudsen 等，1993；Jensen和Jørgensen，1994；Glitsø 等，1998）。随着食糜向后移动，大多数容易被利用的碳水化合物被分解，细菌的生长被抑制，从而减少了SCFA的生成，SCFA的浓度降低，pH接近中性（Bach Knudsen 等，1991；Bach Knudsen 等，1993）。在某些情况下，SCFA的组成也可能会发生改变，通常是从盲肠/近端结肠到远端结肠乙酸浓度增加、丙酸浓度降低（Bach Knudsen 等，1991）。

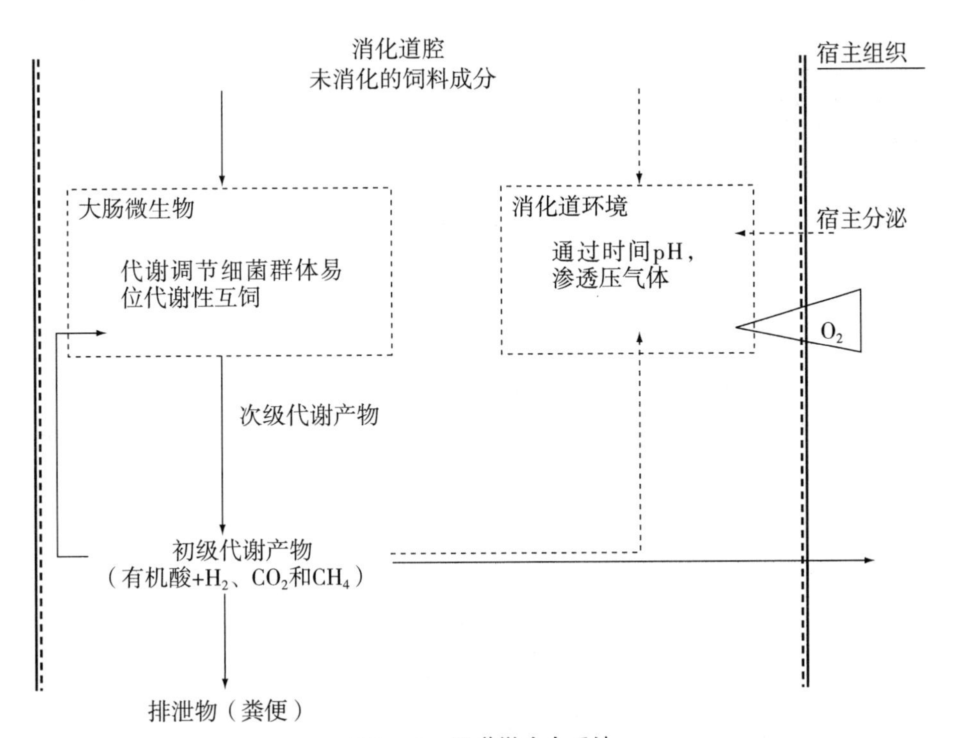

图5-11 肠道微生态系统

代谢流以实线箭头表示，其他影响以虚线箭头表示。（资料来源于Louis 等，2007）

◆ 8.1 糖与寡聚糖

除乳糖以及某些罕见情况下的果糖外，所有的糖几乎完全在小肠被消化（Ly，1992，1996；Bach Knudsen和Canibe，2000）。进入到大肠的糖的数量是非常有限的，并且在盲肠和近端结肠被迅速降解。在某些情况下，到达大肠的寡聚糖的量很大，但是这类碳水化合物在大肠中也会被快速降解。高剂量的可快速发酵碳水化合物可以是不能吸收的糖、淀粉和菊粉，但是可引起乳酸的快速产生，从而导致pH的大幅下降（Bach Knudsen 等，2003；Petkevicius 等，2003）。

◆ 8.2 淀粉

来源于细粉碎饲粮中的残留淀粉（RS_1）到达大肠后，在盲肠和近端结肠中可被迅速降解（Bach Knudsen 等，1993）；豌豆中的生淀粉颗粒（RS_2）也是如此。然而，当饲喂高粱-橡树子为基础的饲粮时，淀粉降解有些缓慢（Morales 等，2002）。有时可以观察到淀粉的不完全消化道消化，完整的籽粒在排便中出现。

◆ 8.3 非淀粉多糖

非淀粉多糖到达大肠后可以以各种不同的形态存在，它们具有不同的溶解度、链长，且与其他分子存在不同的连接方式。细胞壁多糖，如β-葡聚糖、阿拉伯木聚糖和果胶，从细胞壁结构中释放出来，变得可溶解。以β-葡聚糖为例，多糖甚至可以大量解聚（Johansen 等，1997）。这些寡聚糖和多聚糖在大肠中的降解速率和整体降解程度受其化学性质、溶解度和木质化程度的影响。β-葡聚糖、可溶性阿拉伯木聚糖和果胶（Bach Knudsen 等，1993；Canibe和Bach Knudsen，1997a；Glitsø 等，1998；Le Gall 等，2009）在盲肠和结肠近端中被降解，而更加不可溶的非淀粉多糖，如纤维素和不溶性阿拉伯木聚糖，则在结肠更远端的位置被更慢地降解（Bach Knudsen 等，1993；Canibe和Bach Knudsen，1997a；Glitsø 等，1998；Le Gall 等，2009）。图5-12中的结果可以表明，饲喂含有谷物和豆粕（仅有的植物成分）饲粮的猪，非淀粉多糖残留物从回肠到粪便消化率逐渐增加的进程（Gdala 等，1997a）。根据植物原料的组成，半乳糖和糖醛酸可被视为主要来源于豆粕的果胶多糖的标记物，木糖可被视为谷物中阿拉伯木聚糖的标记物，葡萄糖可被视为β-葡聚糖和纤维素的标记物。图5-13和表5-6显示了饲粮纤维组分的化学组成对碳水化合物在大肠中消化的影响，总结了饲喂各种植物原料时的研究结果，进一步说明了非木质化原料（如小麦粉、磨碎的燕麦、黑麦粉、燕麦麸和甜菜浆）中纤维素和阿拉伯木聚糖的消化率比木质化原料（黑麦和小麦的果皮/种皮以及小麦麸）中的高得多（Graham 等，1986；Bach Knudsen和Hansen，1991；Bach Knudsen 等，1993；Longland 等，1993；Glitsø 等，1998）。另外，因为多糖和木质素紧密连结，所以整个多糖-木质素复合物变得非常不可溶，主要的细胞壁多糖被降解为几乎相同的程度。这与非木质化原料存在较大差异，在非木质化原料中，纤维素的消化性比半纤维素更差（Bach Knudsen和Hansen，1991；Glitsø 等，1998）。从表5-6也可以很清楚地看出，较高的纤维添加水平导致solka-floc中结晶纤维素的消化率降低（Longland 等，1993）。然而，有一项研究未表明天然原料中纤维水平与纤维消化率之间的一致关系（Stanogias和Pearce，1985）。

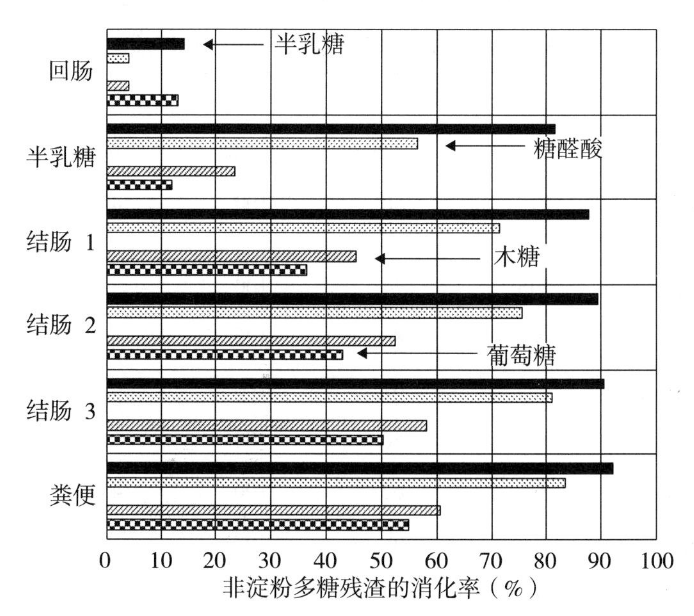

图5-12 饲喂含有谷物和豆粕（饲粮中唯一的植物成分）饲粮的猪从回肠到粪便非淀粉多糖残留物的消化率（结肠1、2和3分别为结肠的前、中和后三分之一）

（资料来源于Gdala 等，1997b）

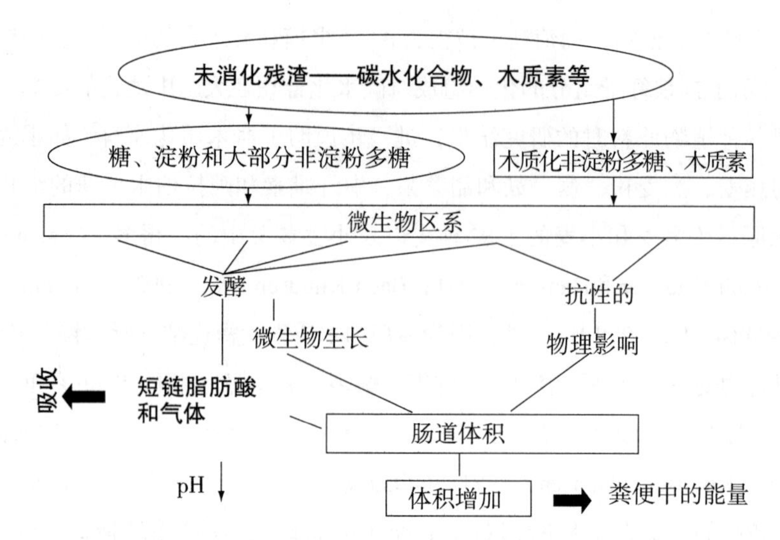

图5-13　大肠中碳水化合物降解的示意图以及对结肠和粪便重量、体积和能量的影响

表5-6　利用含有不同纤维水平的植物原料开展的试验测定的总非淀粉多糖、β-葡聚糖、纤维素和阿拉伯木聚糖的全消化道消化率[1]

植物来源	纤维（g/kg DM）	消化率（%）			
		总 NSP	β-葡聚糖	纤维素	AX
谷物混合物-豆粕	210	58	100	45	44
大麦-豆粕	148	74	100	56	66
小麦粉	35	83	100	60	85
+小麦糊粉	55	67	100	47	68
+小麦皮/壳	62	50	100	24	50
+小麦麸	62	62	100	44	62
磨碎燕麦	93	90	100	78	82
+燕麦麸	109	92	100	83	84
全谷黑麦	156	67	ND[1]	28	65
黑麦粉	94	87	ND	84	83

（续）

植物来源	纤维（g/kg DM）	消化率（%）			
		总 NSP	β－葡聚糖	纤维素	AX
黑麦糊粉	180	73	ND	35	73
黑麦皮/壳	177	14	ND	10	-1
小麦－细粉（2.9%>2 mm）	154	68	100	—	71
小麦－粗粉（12.0%>2 mm）	148	64	100	—	68
大麦－细粉（0.7%>2 mm）	185	61	100	—	62
大麦－粗粉（23.3%>2 mm）	148	57	100	—	53
半纯化＋纤维素粉—低	128	50	ND	51	ND
半纯化＋纤维素粉—高	229	12	ND	9	ND
半纯化＋甜菜浆—低	123	97	ND	93	ND
半纯化＋甜菜浆—高	211	96	ND	89	ND

[1]NSP＝非淀粉多糖；AX＝阿拉伯木聚糖；ND＝未测定。

资料来源于Bach Knudsen和Hansen，1991；Bach Knudsen 等，1993；Bach Knudsen 等，2005；Canibe和Bach Knudsen，1997b；Glitsø 等，1998；Graham 等，1986；Longland 等，1993。

与仔猪和生长猪相比，母猪或成年动物可以更好地对纤维成分进行大量地降解。成年动物通常具有较低的单位体重采食量、较慢的食糜流动、更大的肠道体积和较高的纤维素分解活性（Varel和Pond，1985；Glitsø 等，1998；Serena 等，2008a）。以前和最近的研究表明，母猪一般比生长猪具有更高的纤维消化率和代谢能值（Fernández 等，1986；Shi和Noblet，1993；Jørgensen 等，2007）。试验中可观察到两组动物对一些谷类副产品和粗饲料的利用存在很大的区别

（Fernández 等，1986；Jørgensen 等，2007）。另外，母猪比生长猪更容易降解具有复杂组分的纤维多糖，如玉米糠中的阿拉伯木聚糖（Noblet和Bach Knudsen，1997）。

长期饲喂高纤维饲粮可能会产生效果。Longland 等（1993）的结论是，饲喂高纤维饲粮的猪1周后可适应饲粮中的氮能平衡，但是需要3～5周时间才能适应抗性非淀粉多糖残留物的消化。

◆ 8.4 物理效应

饲粮纤维摄入量的增加必然会影响排便行为，不仅是因为pH降低、微生物生长加快和SCFA产生，而且也是由于纤维的机械刺激作用和持水性能（图5-13），其结果是结肠和粪便的体积增加，内容物通过时间减少（Glitsø 等，1998；Serena 等，2008a）。然而，各种纤维源对粪便重量和能量排泄的影响，与进入大肠的聚合物的类型紧密相关。因此，可溶性纤维，如可溶性阿拉伯木聚糖、β-葡聚糖和果胶在大肠中可被广泛降解（表5-6），对粪便湿重仅产生轻微的影响；相反，不可溶性纤维组分，如纤维素、不可溶性半纤维素和木质素可完全抵抗微生物的降解（Glitsø 等，1998；Serena 等，2008b；Le Gall 等，2009），因此对粪便干重产生很大的影响。从图5-14也可清楚看出，与生长猪相比，母猪肠道处理纤维的能力更强。

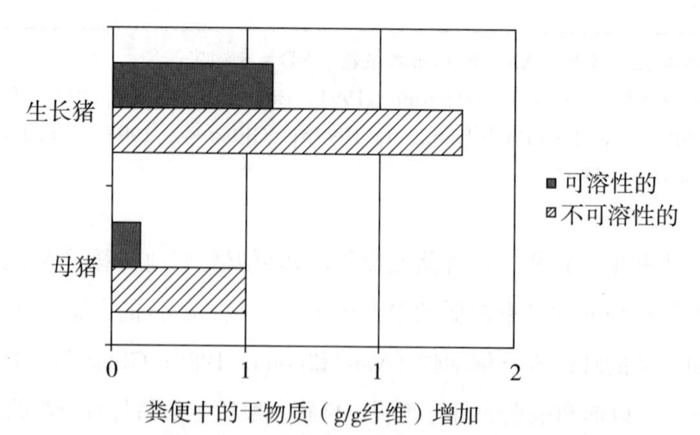

图5-14 可溶性纤维和不可溶性纤维对生长猪和母猪粪便增量的影响

（资料来源于Bach Knudsen和Hansen，1991；Serena 等，2008b）

9 小肠与大肠中营养素的定量消化

表5-7中的数据显示，饲粮纤维浓度对回肠和整个消化道中营养素的定量消化具有重要作用。大量的糖（接近100%）、淀粉（97%）、蛋白质（75%）和脂肪（72%），在通过小肠过程中消失。对于从小肠到大肠运输的有机物（碳水化合物）的量来说，纤维浓度是迄今为止最重要的因素，因为非淀粉多糖和淀粉的量约占未消化残留物量的50%。到达大肠的有机物（OM）在通过大肠时约有一半被发酵，但不同营养素之间存在很大的差异。脂肪不会消失，而37%的粗蛋白、59%的非淀粉多糖、71%的不明残留物和90%的淀粉会消失。表5-7还显示，随着纤维浓度的升高，大肠中有机残余物的降解量增加。例如，当纤维水平为150 g/kg DM时，降解量为170 g OM/d；当纤维水平为200 g/kg DM时，降解量为286 g OM/d。当给母猪饲喂含有429～455 g/kg DM的纤维时，有机物的降解可达到355～503 g/d的水平（Serena 等，2008b）。

表5-7 采食量及回肠和粪便中营养素的回收率（g/d），纤维对回肠和粪便中营养素回收率的影响[1]

项目	采食量（g/d）	回肠回收率（g/d）	纤维的影响			粪便回收率（g/d）	纤维的影响		
			截距	斜率	R^2		截距	斜率	R^2
干物质	2 000	536	113	3.1	0.75	273	−25	2.2	0.79
有机物	1 903	475	88	2.8	0.78	231	−38	2.0	0.80
蛋白质（N×6.25）	351	88	39	0.4	0.29	56	10	0.34	0.65
脂肪	130	36	25	0.1	0.06	35	21	0.1	0.15
碳水化合物									
糖	99	ND[3]	—	—	—	ND[3]	—	—	—
淀粉	984	31	13	0.11	0.08	3	−1	<0.1	0.15
NSP	244	191	5	1.3	0.76	79	−49	0.9	0.69

（续）

项目	采食量（g/d）	回肠回收率（g/d）	纤维的影响			粪便回收率（g/d）	纤维的影响		
			截距	斜率	R^2		截距	斜率	R^2
木质素[2]	36	36[3]	−2	0.3	0.54	36[3]	−2	0.27	0.34
残留物	59	100	6	0.7	0.31	29	−16	0.3	0.21

[1]表格中的数据综合了22个研究中的78种饲粮。采食量是基于2000 g干物质计算的，将报道的化学组成转换为大量营养素。回肠和粪便的回收率是基于文献（Bach Knudsen等，未发表数据）中报道的消化系数计算的。ND，未测定；NSP，非淀粉多糖。

[2]假设木质素在经过消化道时不被降解。

[3]回肠和粪便中的糖残留是残留物的一部分。

10 碳水化合物同化作用产物的吸收

一些研究表明，碳水化合物衍生的营养物质对机体的供给发生在两个阶段：第一个阶段是快速和营养素高流入阶段（吸收阶段），采食后持续4～5 h；第二个阶段是营养素低流入阶段（吸收后阶段），持续到下一次采食（Giusi-Peerier等，1989；R’erat，1996；Bach Knudsen 等，2000；Bach Knudsen 等，2005；图5-15）。在吸收阶段，还原糖明显是碳水化合物同化作用的主要产物，其水平是SCFA的4～8倍。然而，在吸收后阶段，SCFA变得越来越重要，SCFA的量与下一餐1 h前还原糖的量相当。

饲粮碳水化合物的组成影响碳水化合物同化作用产物的产生速度和类型。因此，葡萄糖和蔗糖似乎比淀粉和乳糖吸收更快（Rérat 等，1984a，1984b）。在另外一系列的试验中，麦芽糖似乎比淀粉能够更迅速地被吸收到门静脉中（Rérat等，1993）。淀粉的类型也可影响葡萄糖的吸收率，而当作为纤维分离物添加到饲粮时，纤维水平似乎只影响葡萄糖的吸收率。当将黏稠的瓜尔胶添加到一个半纯合饲粮时，可降低采食后门静脉的葡萄糖含量（Ellis 等，1995）。与此相反，不可溶性纤维源（小麦麸）和可溶性纤维源（甜菜纤维、燕麦麸和黑麦）似乎均不能改变营养素的吸收（Michel和Rerat，1998；Bach Knudsen 等，2000）。

在吸收的早期阶段LA出现在门静脉中，与SCFA相比，这与淀粉的吸收规律能够很好地同步（Bach Knudsen等，2000；Serena 等，2007）。在一些研究中，LA的吸收估计接近SCFA的吸收，但不能排除插管猪试验中LA的吸收被高估，因为大量的LA可以从肠道的葡萄糖氧化产生（Vaugelade等，1994）。

SCFA的吸收速度低于葡萄糖（图5-15）。据报道，只有当饲喂高含量的易于发酵的碳水化合物（乳糖或糖醇）时（Giusi-Peerier 等，1989；Rérat等，1993），或长期禁食后（Rérat 等，1987；Bach Knudsen等，2005），门静脉中SCFA的浓度才会发生昼夜变化。但是，以还原糖或短链脂肪酸吸收的能量的相对比例，受饲粮碳水化合物组成的强烈影响（表5-8）。因此，当通过纤维、RS、难吸收糖或糖醇提高大肠中可发酵碳水化合物的水平时，可观察到动物能量供给的大量增加（相对比例和绝对比例）（Giusi-Peerier 等，1989；Rérat 等，1993；Bach Knudsen等，2000；Bach Knudsen 等，2005）。例如，当大肠中有更多

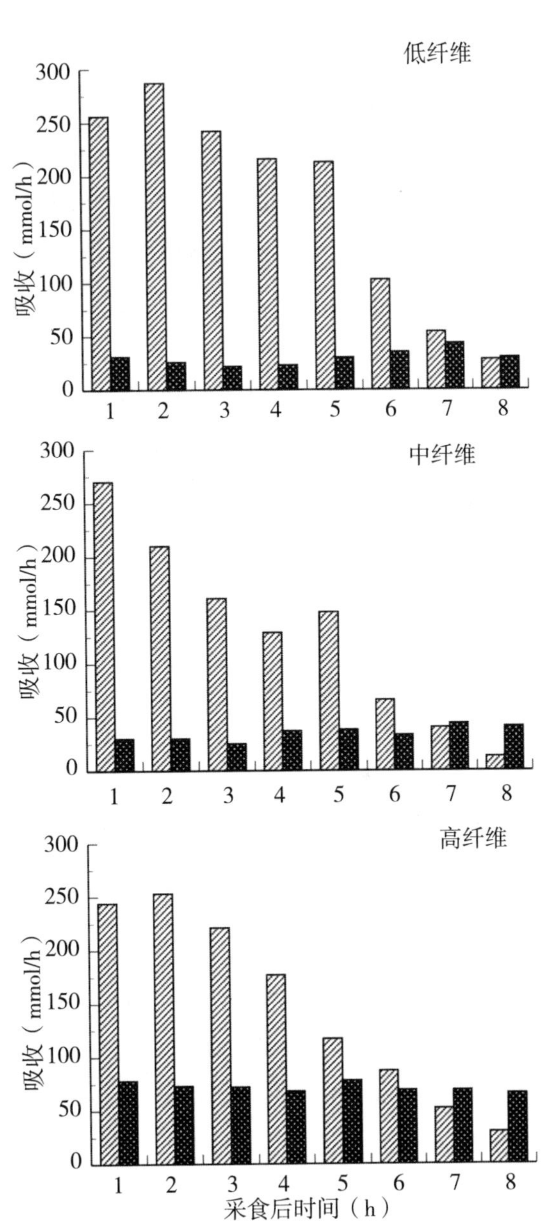

图5-15 采食低膳食纤维、中膳食纤维和高膳食纤维饲粮后葡萄糖（斜线图）和短链脂肪酸的吸收（mmol/h）

猪每天被饲喂3次，每8h饲喂1次。（资料来源于Bach Knudsen 等，2000；Bach Knudsen 等；2005）

碳水化合物发酵时，门静脉中的SCFA流量增加（$r = 0.90$），而葡萄糖流量降低（$r = -0.70$）。这也影响以葡萄糖或SCFA形式吸收能量的比例；当饲喂低NDC玉米淀粉饲粮时，只有约4%的吸收能量来源于SCFA酸；当饲喂高NDC马铃薯饲粮时，却有44%的吸收能量来源于SCFA。在饲喂高纤维饲粮（4.29 g/kg DM纤维）的母猪上，已观察到更多的能量供给（占52%）来源于SCFA，而饲喂低纤维饲粮（含177 g/kg DM纤维）的母猪，其能量供给仅有12%来源于SCFA（Serena等，2009）。

表5-8 饲粮颗粒、可消化淀粉和不可消化淀粉摄入量对葡萄糖和短链脂肪酸门静脉浓度和流速及以葡萄糖和短链脂肪酸形式吸收的能量比例的影响[1]

饲粮	采食量（g）				葡萄糖		SCFA		吸收的能量（%）	
	饲粮颗粒	可消化淀粉	NDC							
			RS	NSP	mmol/L	mmol/h	μmol/L	mmol/h	Glu	SCFA
LF 小麦面包	1 300	746	4	77	8.10	175	775	30	93.0	7.0
HF 小麦麸	1 300	663	3	140	7.69	127	854	30.8	90.5	9.5
HF 燕麦麸	1 300	605	3	140	7.66	132	908	37.1	89.1	10.9
HF 黑麦面包	1 250	676	13	254	6.60	157	1 140	76.9	82.4	17.6
HF 小麦面包	1 250	610	7	275	6.43	117	1 001	66.5	80.2	19.8
玉米淀粉	860	536	9	39	8.85	146	459	13.9	96.0	4.0
豌豆淀粉	860	535	15	36	6.90	105	454	17.8	93.1	6.9
玉米淀粉	1 250	762	20	66	8.14	185	480	19.1	95.7	4.3
玉米：马铃薯淀粉(1 ：1)	1 250	609	189	66	6.94	109	1 240	60.3	80.6	19.4
马铃薯淀粉	1 250	361	458	66	5.97	49	1 620	88.9	55.9	44.1

Dig，可消化的；NDC，非消化性碳水化合物；SCFA，短链脂肪酸；RS，抗性淀粉；NSP，非淀粉多糖；Glu，葡萄糖；LF，低纤维；HF，高纤维。

资料来源于Bach Knudsen 等，2000；Bach Knudsen 等，2005；van der Meulen 等，1997a；van der Meulen 等，1997b。

到达大肠的碳水化合物的类型可以改变SCFA的摩尔组成。用甜菜纤维替代麦麸可增加乙酸含量、减少丙酸含量（Michel和Rerat，1998），RS可增加丁酸含量而减少醋酸含量（van der Meulen 等，1997a），燕麦麸和黑麦可增加门静脉中丁酸的比例，而减少醋酸的比例（Bach Knudsen 等，2000；Bach Knudsen 等，2005）。

11 碳水化合物同化作用吸收产物的利用

碳水化合物衍生产物的吸收部位会影响能量的利用。从碳水化合物转化为SCFA的发酵方程可以算出，大约有25%的能量随H_2和CH_4的排出而被损失。与生长猪相比（消化能的0.5%～1.2%；Jørgensen，2007），母猪由H_2损失的能量是比较低的，由CH_4损失的能量较多、比例较高（消化能的1.5%～3.5%）。大肠吸收的SCFA的利用率比小肠吸收的葡萄糖的利用率较低，估计只有69%（表5-9），这是因为在代谢过程中SCFA的利用效率较低。然而，成年母猪SCFA的利用效率略有增加（90%），可能是因为能量主要用于维护而不是用于生长。向大肠中灌注SCFA的利用率是82%（Jørgensen等，1997），而高纤维饲粮发酵的SCFA的利用率计算值仅有73%的效率（Jørgensen等，1996）。其他试验表明，给母猪或生长猪无论是口服还是盲肠灌注SCFA，其利用效率均较低（表5-10）。

表5-9 大肠中发酵能量的利用

饲粮组成	体重范围（kg）	大肠中的发酵能量（DE 的 %）	与小肠有关的效率（RE/ME，%）	作者
马铃薯淀粉，纤维素	60 ～ 90	18 ～ 33	51	Just 等，1983
马铃薯，甜菜，青草，苜蓿粉	90 ～ 180	9 ～ 40	66	Hoffmann 等，1990
玉米淀粉，纤维素，大豆荚	30 ～ 105	13 ～ 27	63	Bakker 等，1994

（续）

饲粮组成	体重范围（kg）	大肠中的发酵能量（DE的%）	与小肠有关的效率（RE/ME，%）	作者
甜菜浆，玉米蒸馏谷物，向日葵粕等	38 ~ 47	3 ~ 27	82	Noblet 等，1994
豌豆纤维，果胶	40 ~ 125	7 ~ 29	73	Jørgensen 等，1996
大麦秆，大麦壳，小麦麸，马铃薯纤维，大豆纤维	50 ~ 70	4 ~ 29	76	Jørgensen（未发表数据）
小麦麸，甜菜浆，种子渣，啤酒渣，	46 ~ 125	8 ~ 35	69	Jørgensen（未发表数据）
豌豆壳，马铃薯浆，果胶渣，甜菜浆	160 ~ 243	5 ~ 40	90	Jørgensen（未发表数据）

RE，保留的能量；ME，可代谢的能量。

表5-10 大肠灌注短链脂肪酸的利用

饲粮类型	体重范围（kg）	灌注物	利用（ME的%）	作者
大麦/鱼粉	140 ~ 180	酒精	酒精：72	Jentsch 等，1968
		乳酸	乳酸：75	
		乙酸	乙酸：60	
大麦/豆粕	160 ~ 200	乙酸	乙酸：79	Roth 等，1988
		丙酸	丙酸：75	
谷物/燕麦粉副产品/小麦麸/鱼粉	55 ~ 120	乙酸	乙酸：65	Gädeken 等，1989
		丙酸	丙酸：71	
		丁酸	丁酸：67	
大麦/大豆粉	179 ± 17	乙酸和丙酸的混合物	混合物：70	Müller 等，1991
大麦/小麦淀/粉鱼粉/酪蛋白	60 ~ 120	乙酸、丙酸和丁酸的混合物	混合物：82	Jørgensen 等，1997

从可消化的碳水化合物到不可消化碳水化合物，碳水化合物组成变化的影响是双重的。首先，因为并不是所有的存在于纤维多糖中的能量都可以被降解，所以能量的消化率以及伴随的代谢能会有所降低；其次，由于大部分的可吸收能量来源于SCFA，其利用率低于葡萄糖，因此能量的利用率，即净能值作为代谢能的一部分就会降低。

12 意 义

生长猪和成年母猪肠道中具有高的酶活和微生物活性，可以应对不同饲料原料中存在的简单和复杂的碳水化合物。这保证了大部分的糖和淀粉被小肠内的酶降解。然而，高水平的果糖和乳糖可能导致这些糖的吸收不良。虽然在小肠中没有内源性酶来打开多数寡聚糖和非淀粉多糖中的连接键，但仍有40%左右的寡聚糖和20%的NSP在通过胃和小肠时消失。纤维多糖降解的主要部位是大肠，产生的SCFA可作为能量被重新利用。但是，与小肠中以葡萄糖形式吸收能量的利用率相比，大肠中以SCFA形式吸收能量的利用率较低，主要是由于发酵气体的能量损失，以及中间代谢过程中SCFA的利用率较低。对饲料原料和加工工艺的选择是提高饲料有效能值的途径。

作者：Knud Erik
Bach Knudsen,
HelleNygaardLærke
和Henry Jorgensen
译者：孔祥峰

参考文献

Argenzio, R., and M. Southworth. 1974. Site of organic acid production and absorption in gastrointestinal tract of pigs. Am. J. Physiol. 228:454–460.

Bach Knudsen, K. E. 1997. Carbohydrate and lignin contents of plant materials used in animal feeding. Anim. Feed Sci. Technol. 67:319–338.

Bach Knudsen, K. E., and N. Canibe. 2000. Breakdown of plant carbohydrates in the digestive tract of pigs fed on wheat or oat based rolls. J. Sci. Food and Agri. 80:1253–1261.

Bach Knudsen, K. E., N. Canibe, and H. Jørgensen. 2000. Quantification of the absorption of nutrients deriving from carbohydrate assimilation: Model experiment with catheterised pigs fed on wheat and oat based rolls. Br. J. Nutr. 84:449–458.

Bach Knudsen, K. E., and I. Hansen. 1991. Gastrointestinal implications in pigs of wheat and oat fractions 1. Digestibility and bulking properties of polysaccharides and other major constituents. Br. J. Nutr. 65:217–232.

Bach Knudsen, K. E., B. B. Jensen, J. O. Andersen, and I. Hansen. 1991. Gastrointestinal implications in pigs of wheat and oat fractions 2. Microbial activity in the gastrointestinal tract. Br. J. Nutr. 65:233–248.

Bach Knudsen, K. E., B. B. Jensen, and I. Hansen. 1993. Digestion of polysaccharides and other major components in the small and large intestine of pigs fed diets consisting of oat fractions rich in ß−d−glucan. Br. J. Nutr. 70:537–556.

Bach Knudsen, K. E., and H. Jørgensen. 2001. Intestinal degradation of dietary carbohydrates—from birth to maturity. Pages 109–120 in Digestive Physiology in Pigs. J. E. Lindberg and B. Ogle, eds. CABI Publishing, Wallingford, UK.

BachKnudsen, K. E., and B.W. Li. 1991. Determination of oligosaccharides in protein−rich feedstuffs by gas−liquid chromatography and high−performance liquid chromatography. J. Agri. Food. Chem. 39:689–694.

Bach Knudsen, K. E., S. Petkevicius, H. Jørgensen, and K. D. Murrell. 2003. A high load of rapidly fermentable carbohydrates reduces worm burden in infected pigs. Page 169 in Manipulating Pig Production. J. E. Paterson, ed. Australasian Pig Science Association, Werribee, VIC, Australia.

Bach Knudsen, K. E., A. Serena, A. B. K. Kjær, H. Jørgensen, and R. Engberg. 2005. Rye bread enhances the production and plasma concentration of butyrate but not in the plasma concentrations of glucose and insulin in pigs. J. Nutr. 135:1696–1704.

Bergman, E. N. 1990. Energy contributions of volatile fatty acids from the gastrointestinal tract in various species. Phys. Rev. 70:567–590.

Biliaderis, C. G. 1991. The structure and interactions of starch with food constituents. Can. J. Physiol. 69:60–78.

Brown, I. L., K. J. McNaught, D. Andrews, and T. Morita. 2001. Resistant starch: Plant breeding, applications development and commercial use. Pages 401–412 in Advanced Dietary Fibre Technology. B. V. McCleary and L. Prosky, eds. Blackwell Science Ltd, Oxford, UK.

Canibe, N., and K. E. Bach Knudsen. 1997a. Apparent digestibility of non-starch polysaccharides and short-chain fatty acids production in the large intestine of pigs fed dried or toasted peas. Acta Agric. Scand. 47:106–116.

Canibe, N., and K. E. Bach Knudsen. 1997b. Digestibility of dried and toasted peas in pigs. 1. Ileal and faecal digestibility of carbohydrates. Anim. Feed Sci. Technol. 64:293–310.

Carpita, N. C., and D. M. Gibeaut. 1993. Structural models of primary cell walls in flowering plants: Consistency of molecular structure with the physical properties of the walls during growth. Plant J. 3:1–30.

Cummings, J. H. 1997. The large intestine in nutrition and disease. Institute Danone, Brussels, Belgium.

Cummings, J. H., and H. N. Englyst. 1995. Gastrointestinal effects of food carbohydrate. Am. J. Clin. Nutr. 61:938S–945S.

Cummings, J. H., and A. M. Stephen. 2007. Carbohydrate terminology and classification. Eur. J. Clin. Nutr. 61(Suppl. 1): S5–18.

Davin, L. B. et al. 2008. Dissection of lignin macromolecular configuration and assembly: Comparison to related biochemical processes in allyl/propenyl phenol and lignan biosynthesis. Natl. Prod. Rep. 25:1015–1090.

Duke, G. E. 1986. Alimentary canal:Secretion and digestion, special digestion functions and absorption. Pages 289–302 in Avian Physiology. P. D. Sturkie, ed. Springer-Verlag, New York.

Ellis, P. R., F. G. Roberts, A. G. Low, and L. M. Morgan. 1995. The effect of high-molecular-weight guar gum on net apparent glucose absorption and net apparent insulin and gastric inhibitory polypeptide production in the growing pig: Relationship to rheological changes in jejunal digesta. Br. J. Nutr. 74:539–556.

Englyst, H. N., S. M. Kingman, and J. H. Cummings. 1992. Classification and measurement of nutritionally important starch fractions. Eur. J. Clin. Nutr. 46:S33–50.

Englyst, H. N., M. E. Quigley, and G. J. Hudson. 1994. Determination of dietary fibre as non-starch polysaccherides with gas-liquid chromatography, high-performance liquid chromatography or spectrophotometric measurements of constituent sugars. Analyst 119:1497–1509.

Englyst, K. N., S. Liu, and H. N. Englyst. 2007. Nutritional characterization and measurement of dietary carbohydrates. Eur. J. Clin. Nutr. 61(Suppl. 1):S19–39.

Fernández, J. A., H. Jørgensen, and A. Just. 1986. Comparative digestibility experiments with growing pigs and adult sows. Anim. Prod. 43:127–132.

Flickinger, E. A., J. van Loo, and G. C. Fahey, Jr. 2003. Nutritional responses to the presence of inulin and oligofructose in the diets of domesticated animals: A review. Crit. Rev. Food. Sci. Nutr. 43:19–60.

Flint, H. J., E. A. Bayer, M. T. Rincon, R. Lamed, and B. A. White. 2008. Polysaccharide utilization by gut bacteria: Potential for new insights from genomic analysis. Nat Rev Microbiol. 6:121–131.

Gallant, D. J., B. Bouchet, A. Buléon, and S. Pérez. 1992. Physical characteristics of starch granules and susceptibility to enzymatic degradation. Eur. J. Clin. Nutr. 46:S3–S16.

Gdala, J., and L. Buraczewska. 1997. Ileal digestibility of pea and faba bean carbohydrates in growing pigs. J. Anim. Feed Sci. 6:235–245.

Gdala, J., L. Buraczewska, A. M. J. Jansman, J. Wasilewko, and P. Leeuwen. 1994. Ileal digestibility of amino acids and carbohydrates in lupins for young pigs. Pages 93–96 in VIth Int. Symp. Dig. Physiol. in Pigs. No. 89. EAAP Publication No 89. W.-B. Souffrant and H. Hagemesiters, eds. Bad Doberan, Germany.

Gdala, J.,H.N. Johansen,K. E.BachKnudsen, I.H.Knap, P.Wagner, and O.B. Jergensen. 1997a. The digestibility of carbohydrates, protein and fat in the small and large intestine of piglets fed non-supplemented and enzyme supplemented diets. Anim. Feed Sci. Technol. 65:15–33.

Giusi-Peerier, A., M. Fiszlewicz, and A. Rérat. 1989. Influence of diet composition on intestinal volatile fatty acid and nutrient absorption in unanesthetized pigs. J. Anim. Sci. 67:386–402.

Glitsø, L. V., G. Brunsgaard, S. Højsgaard, B. Sandström, and K. E. Bach Knudsen. 1998. Intestinal degradation in pigs of rye dietary fibre with different structural characteristics. Br. J. Nutr. 80:457–468.

Graham, H., K. Hesselman, and P. Åman. 1986. The influence of wheat bran and sugar-beet pulp on the digestibility of dietary components in a cereal-based pig diet. J. Nutr. 116:242–251.

Gray, G. M. 1992. Starch digestion and absorption in nonruminants. J. Nutr. 122:172–177.

Gädeken, D., G. Breves, and H. J. Oslage. 1989. Efficiency of energy utilization of intracaecally infused volatile fatty acids in pigs. Pages 115–118 in Proc. 11th Symp. Energy

Metabol. Farm Animals. Lunteren, Netherlands.

Hedemann, M. S., and K. E. Bach Knudsen. 2010. Dried chicory root have minor effects on the digestibility of nutrients and the composition of the microflora at the terminal ileum and in faeces of growing pigs. Livest. Sci. 134:53–55.

Henneberg, W., and F. Stohmann. 1859. Über das erhaltungsfutter volljährigen rindviehs. J. Landwirtsch. 3:485–551.

Hopwood, D. E., D. W. Pethick, J. R. Pluske, and D. J. Hampson. 2004. Addition of pearl barley to a rice-based diet for newly weaned piglets increases the viscosity of the intestinal contents, reduces starch digestibility and exacerbates post-weaning colibacillosis. Br. J. Nutr. 92:419–427.

Houdijk, J. G., R. Hartemink, M. W. Verstegen, and M. W. Bosch. 2002. Effects of dietary non-digestible oligosaccharides on microbial characteristics of ileal chyme and faeces in weaner pigs. Arch. Tierernahr. 56:297–307.

Houdijk, J. G. M., et al. 1999. Apparent ileal and total-tract nutrient digestion by pigs as affected by dietary nondigestible oligosaccharides. J. Anim. Sci. 77:148–158.

Jensen, B. B., and H. Jørgensen. 1994. Effect of dietary fiber on microbial activity and microbial gas production in various regions of the gastrointestinal tract of pigs. Appl. Environ. Microbiol. 60:1897–1904.

Johansen, H. N., K. E. Bach Knudsen, P. J. Wood, and R. G. Fulcher. 1997. Physico-chemical properties and the digestibility of polysaccharides from oats in the gastrointestinal tract of pigs. J. Sci. Food Agric. 73:81–92.

Jørgensen, H. 2007. Methane emission by growing pigs and adult sows as influenced by fermentation. Livest. Sci. 109:216–219.

Jørgensen, H., T. Larsen, X.-Q. Zhao, and B. O. Eggum. 1997. The energy value of short-chain fatty acids infused into the carcum of pigs. Br. J. Nutr. 77:745–756.

Jørgensen, H., A. Serena, M. S. Hedemann, and K. E. Bach Knudsen. 2007. The fermentative capacity of growing pigs and adult sows fed diets with contrasting type and level of dietary fibre. Livest. Sci. 109:111–114.

Jørgensen, H., X.-Q. Zhao, and B.O. Eggum. 1996. The influence of dietary fibre and environmental temperature on the development of the gastrointestinal tract, digestibility, degree of fermentation in the hind-gut and energy metabolism in pigs. Br. J. Nutr. 75:365–378.

Kidder, D. E., and M. J. Manners. 1980. The level and distribution of carbohydrases in the small intestine mucosa of pigs from three weeks of age to maturity. Br. J. Nutr. 43:141–153.

Lærke, H. N. et al. 2008. Rye bread reduces plasma cholesterol levels in hypercholesterolaemic pigs when compared to wheat at similar dietary fibre level. J. Sci. Food Agri. 88:1385–1393.

Le Gall, M., A. Serena, H. Jorgensen, P. K. Theil, and K. E. Bach Knudsen. 2009. The role of whole-wheat grain and wheat and rye ingredients on the digestion and fermentation processes in the gut—a model experiment with pigs. Br. J. Nutr. 102:1590–1600.

Le Gall, M., K. Eybye, and K. E. Bach Knudsen. 2009. Molecular weight changes of arabinoxylans incurred by the digestion processes in the upper gastrointestinal tract of pigs. Livest. Sci. 134:72–75.

Lentze, M. J. 1995. Molecular and cellular aspects of hydrolysis and absorption. Am. J. Clin. Nutr. 61:946S–951S.

Liyama, K., T. B.-T. Lam, and B. A. Stone. 1994. Covalent cross-linkes in the cell wall. Plant Physiol. 104:315–320.

Longland, A. C., A. G. Low, D. B. Quelch, and S. P. Bray. 1993. Adaptation to the digestion of non-starch polysaccharide in growing pigs fed on cereal or semi-purified basal diets. Br. J. Nutr. 70:557–566.

Louis, P., K. P. Scott, S. H. Duncan, and H. J. Flint. 2007. Understanding the effects of diet on bacterial metabolism in the large intestine. J. Appl. Microbiol. 102:1197–1208.

Ly, J. 1992. Studies of the digestibility of pigs fed dietary sucrose, fructose or glucose. Arch. Anim. Nutr. 42:1–9.

Ly, J. 1996. The pattern of digestion and metabolism in high sugar feeds for pigs. Cuban J. Agric. Sci. 30:117–129.

McCann, M. C., and K. Roberts. 1991. Architecture of the primary cell wall. Pages 109–129 in Cytoskeletal Basis of Plant Growth and Form. C. W. Lloyd, ed. Academies Press, London.

McDougall, G. J., I.M.Morrison, D. Stewart, and J. R.Hillman. 1996a. Plant cell walls as dietary fibre: Range, structure, processing and function. J. Sci. Food Agric. 70:133–150.

McDougall, G. J., I. M. Morrison, D. Stewart, and J. R. Hillman. 1996b. Plant cell walls as dietary fibre: Range, structure, processing and function. J. Sci. Food Agric. 70:133–150.

Michel, P., and A. Rerat. 1998. Effect of adding sugar beet fibre and wheat bran to a starch diet on the absorption kinetics of glucose, amino nitrogen and volatile fatty acids in the pig. Reprod. Nutr. Dev. 38:49–68.

Morales, J., J. F. Perez, S. M. Martin-Orue, M. Fondevila, and J. Gasa. 2002. Large bowel fermentation of maize or sorghum-acorn diets fed as a different source of carbohydrates to landrace and iberian pigs. Br. J. Nutr. 88:489–498.

Moran, E. T., Jr. 1985. Digestion and absorption of carbohydrates in fowl and events through perinatal development. J. Nutr. 115:665–674.

Morris, E. R. 1992. Physico-chemical properties of food polysaccharides. Pages 41–55 in dietary fibre: A component of food: nutritional function in health and disease. T. F. Schweizer and C. A. Edwards, eds. Springer-Verlag, London.

Noblet, J., and K. E. Bach Knudsen. 1997. Comparative digestibility ofwheat, maize and sugar beet pulp non-starch polysaccharides in adult sows and growing pigs. Pages 571–574 in Digestive Physiology in Pigs. J. P. Laplace, C. Fevrier, and A. Barbeau, eds. INRA, Saint Malo, France.

Noblet, J., H. Fortune, X. S. Shi, and S. Dubois. 1994. Prediction of net energy value of feeds for growing pigs. J Anim. Sci. 71:3389–3398.

Petkevicius, S., K. E. Bach Knudsen, K. D. Murrell, and H. Wachmann. 2003. The effect of inulin and sugar beet fibre on oesophagostomum dentatum infection in pigs. Parasitol. 127:61–68.

Rérat, A. 1996. Influence of the nature of carbohydrate intake on the absorption chronology of reducing sugars and volatile fatty acids in the pigs. Reprod. Nutr. Dev. 1996:3–19.

Rérat, A., M. Fiszlewicz, A. Giusi, and P. Vaugelade. 1987. Influence of meal frequency on postprandial variations in the digestive tract of conscious pigs. J. Anim. Sci. 64:448–456.

Rérat, A., A. Giusi-Périer, and P. Vaissade. 1993. Absorption balances and kinetics of nutrients and bacterial metabolites in conscious pigs after intake of maltose- or maltitol-rich diets. J. Anim. Sci. 71:2473–2488.

Rérat, A. A., P. Vaissade, and P. Vaugelade. 1984a. Absorption kinetics of some carbohydrates in conscious pigs. 1. Qualitative aspects. Br. J. Nutr. 51:505–515.

Rérat, A. A., P. Vaissade, and P. Vaugelade. 1984b. Absorption kinetics of some carbohydrates in conscious pigs. 2. Quantitative aspects. Br. J. Nutr. 51:517–529.

Ring, S. G., J. M. Gee, M.Whittam, P. Orford, and I. T. Johnson. 1988. Resistant starch: Its chemical form in foodstuffs and effect on digestibility *in vitro*. Food Chem. 28:97–109.

Selvendran, R. R. 1984. The plant cell wall as a source of dietary fibre: Chemistry and structure. Am. J. Clin. Nutr. 39:320–337.

Serena, A., and K. E. Bach Knudsen. 2007. Chemical and physicochemical characterisation of co-products from the vegetable food and agro industries. Anim. Feed Sci. Technol. 139:109–124.

Serena, A., M. S. Hedemann, and K. E. Bach Knudsen. 2008a. Influence of dietary fiber on luminal environment and morphology in the small and large intestine of sows. J. Anim. Sci. 86:2217–2227.

Serena, A., H. Jorgensen, and K. E. Bach Knudsen. 2008b. Digestion of carbohydrates and utilization of energy in sows fed diets with contrasting levels and physicochemical properties of dietary fiber. J. Anim. Sci. 86:2208–2216.

Serena, A., H. Jorgensen, and K. E. Bach Knudsen. 2009. Absorption of carbohydrate-derived nutrients in sows as influenced by types and contents of dietary fiber. J. Anim. Sci. 87:136–147.

Serena, A., H. Jørgensen, and K. E. Bach Knudsen. 2007. The absorption of lactic acid is more synchronized with the absorption of glucose than with the absorption of short-chain fatty acids—a study with sows fed diets varying in dietary fibre. Livest. Sci. 109:118–121.

Shi, X. S., and J. Noblet. 1993. Digestible and metabolizable energy values of ten feed ingredients in growing pigs fed ad libitum and sows fed at maintenance level; comparative contribution of the hindgut. Anim. Feed Sci. Technol. 42:223–236.

Stanogias, G., and G. R. Pearce. 1985. The digestion of fibre by pigs 1. The effects of amount and type of fibre on apparent digestibility, nitrogen balance and rate of passage. Br. J. Nutr. 53:513–530.

Sun, T., H. N. Lærke, H. Jørgensen, and K. E. Bach Knudsen. 2006. The effect of heat processing of different starch sources on the *in vitro* and *in vivo* digestibility in growing pigs. Anim. Feed Sci. Technol. 131:66–85.

Theander, O., E. Westerlund, P. Åman, and H. Graham. 1989. Plant cell walls and monogastric diets. Anim. Feed Sci. Technol. 23:205–225.

Theander, O., P. Åman, E.Westerlund, and H. Graham. 1994. Enzymatic/chemical analysis of dietary fiber. J. AOAC 77:703–709.

Thibault, J.-F.,M. Lahaye, and F. Guillon. 1992. Physico-chemical properties of food plant cell walls. Pages 21–39 in dietary fibre: A component of food. Nutritional function in health and disease. T. F. Schweizer and C. A. Edwards, eds. Springer-Verlag, London.

van der Meulen, J., et al. 1997a. Effect of resistant starch on net portal-drained viscera flux of glucose, volatile fatty acids, urea, and ammonia in growing pigs. J. Anim. Sci. 75:2697–2704.

van der Meulen, J., J. G. M. Bakker, B. Smits, and H. d. Visser. 1997b. Effect of source of starch on net portal flux of glucose, lactate, volatile fatty acids and amino acids in the pig. Br. J. Nutr. 78:533–544.

van Soest, P. J. 1963. Use of detergents in the analysis of fibrous feeds. Ii. A rapid method for the determination of fiber and lignin. J. AOAC 46:829–835.

van Soest, P. J., and R. H. Wine. 1967. Use of detergents in the analysis of fibrous feeds. IV. Determination of plant cell-wall constituents. J. AOAC 50:50–55.

Varel, V. H., and W. G. Pond. 1985. Enumeration and activity of cellulolytic bacteria from gestating swine fed various levels of dietary fibre. Appl. Environ. Microbiol. 49:858–862.

Vaugelade, P., et al. 1994. Intestinal oxygen uptake and glucose metabolism during nutrient absorption in the pig. Proc. Soc. Exp. Biol. Med. 207:309–316.

Würsch, P., S. Del Vedovo, and B. Koellreutter. 1986. Cell structure and starch nature as key determinants of the digestion rate of starch in legume. Am. J. Clin. Nutr. 43:23–29.

第6章

维生素及猪对维生素的利用

1 导 语

维生素的发现主要集中在20世纪的前50年，即从1915年首次用维生素这个名词来描述大米皮层的提取物（可治疗脚气病）到1948年发现维生素B_{12}这段时间。在此期间，在人和动物营养研究领域，主要通过维生素的缺乏症和相关疾病来研究维生素的作用，这种研究方法一直沿用至今。特别是在动物营养研究领域，通常是以预防动物出现维生素缺乏症，而不是以动物发挥最佳生产性能的添加水平作为维生素的需要量。此外，目前动物生产中维生素（尤其是B族维生素）的推荐摄入量是依据20世纪50年代到60年代间的研究结果来制定的，而在此期间，研究人员认为动物可以从粪中获取大量的维生素（ARC，1981；McDowell，2000），其原因是有许多B族维生素可以由肠道微生物合成并随粪便排出体外。但是在现代畜禽养殖条件下，动物很难接触到粪便，因此通过粪便摄入的B族维生素很少（Greer和Lewis，1978；de Passille等，1989；Bilodeau等，1989）。除此之外，条缝地板养猪技术的广泛应用更加降低了粪便作为B族维生素来源的可能性。

目前，猪对某些维生素的需要量尚不清楚（NRC，1998；ARC，1981；INRA，1984；McDowell，2000；BSAS，2003），主要是因为缺乏最新的研究资料。随着现代集约化养殖业的发展，动物生产性能显著提高，需要给动物提供更高的营养水平来维持其最佳的生长性能、妊娠性能和泌乳性能。因此猪营养需要量可能要作出相应的调整，以满足动物生产性能提高的需要。然而目前关于维生素营养需要量的研究资料并没有得到更新，许多研究资料（大于2/3）是在1980年以前发表的，而1998年后的研究文献主要集中在个别的维生素，如维生素E、泛酸和叶酸（表6-1）。

表6-1 ARC（1981）和NRC（1998）引用的猪维生素需要量的文章数和1998年发表的猪维生素研究文章[1]

维生素	ARC（1981）	NRC（1998）	1998 年以来的发表文章的数目
脂溶性维生素			
维生素 A[2]	107	24	6
维生素 D	32	14	6
维生素 E	90	43	31
维生素 K	5	6	—
水溶性维生素			
硫胺素	13	11	4
生物素	22	35	4
核黄素	20	15	4
泛酸	23	19	13
吡哆醇	23	14	7
尼克酸	34	22	6
胆碱	22	20	9

（续）

维生素	ARC（1981）	NRC（1998）	1998 年以来的发表文章的数目
叶酸	12	12	13
维生素 B_{12}	86	15	5
维生素 C	42	13	8

[1] 包括4篇甜菜碱的文章。
[2] 包括β-胡萝卜素。

通常把13种维生素分为两大类：脂溶性维生素和水溶性维生素（表6-2）。表6-2列出了猪维生素营养的研究历史、代谢和饲粮特征。当饲粮中的维生素超过动物需要量时，脂溶性维生素（即维生素A、维生素D、维生素E和维生素K）可以在体内储存，而水溶性维生素（即维生素C和B族维生素）并不能在体内大量储存（除维生素B_{12}外）。因此，水溶性维生素需要每日提供，以保证其最大的代谢活性。了解饲粮维生素的这些特点对于确定动物特殊生理阶段（代谢需求变化快，如妊娠和断奶期间）维生素的需要量尤其重要。举例来说，动物可以依靠体内储备获取一部分脂溶性的维生素。鉴于此，在育肥阶段，可通过减少饲粮中脂溶性维生素的添加量，进而降低饲料成本。尽管生产实践中这种做法并不影响生产性能（McGlone，2000；Shaw等，2002），然而体内储备的维生素会被耗竭，从而导致肌肉中维生素（特别是水溶性维生素）的含量降低，最终影响肌肉的营养价值（Leonhardt等，1996；Lombardi-Boccia等，2005）。因此，很显然，饲料原料本身所含有的某些水溶性的维生素并不能满足动物的代谢需求。

表6-2 维生素的发现历史、活性、添加水平及代谢特性[1]

维生素	发现/分离/合成	生物活性形式	预混料/饲料中每月损失率(%)	NRC(1998)推荐量[2]					目前添加水平(mg/kg)[2,3]				
				W[4]	G[4]	F[4]	P[4]	L[4]	W	G	F	P	L
A	1909/1931/1947	视黄醇、视黄酸、视黄醛、胡萝卜素	1/6～7	2.2	1.5	1.3	4.0	2.0	12.4	9.4	8.4	11.9	12.3
E[5]	1922/1936/1938	生育酚、生育三烯酚	0.3/～	16	11	11	44	44	60	37	33	53	55
D	1918/1932/1959	钙化醇、胆钙化醇	10/10	220	175	150	200	200	1 750	1 393	1 268	1 650	1 678
K	1929/1936/1939	甲萘醌、叶绿醌、维生素 K_2	34～38/50	0.5	0.5	0.5	0.5	0.5	3.1	2.2	2.0	2.9	3.0
C[6]	1912/1928/1933	抗坏血酸	1～2/2～5	—	—	—	—	—	—	—	—	—	—
硫胺素	1890/1910/1936	硫胺素	17/5～20	1.3	1.0	1.0	1.0	1.0	2.0	1.0	1.0	1.8	1.8
核黄素	1920/1933/1935	FMN、FAD	5～40/2～10	3.8	2.8	2.0	3.8	3.8	8.3	5.6	5.0	7.2	7.4
尼克酸	1935/1935/1949	NAD、NADP	2～4/1～2	19	13	5	10	10	40	28	25	38	39
泛酸	1931/1938/1940	泛酸	1～8/0～5	11	8.5	7	12	12	22	16.5	15	20	21
吡哆醇	1934/1938/1939	吡哆醇、吡哆醛、吡哆胺	20/2～5	1.8	1.3	1.0	1.0	1.0	3.2	1.7	1.5	2.8	2.9
生物素	1931/1935/1943	生物素	5/1～2	70	50	50	200	200	240	110	100	300	300
胆碱	1932	乙酰胆碱、磷脂酰胆碱	1～2/0.5	550	350	300	1 250	1 000	550	380	300	1 000	1 000
叶酸	1941/1941/1946	甲基四氢叶酸、四氢叶酸	10～40/10～50	0.3	0.3	0.3	1.3	1.3	1.6	0.7	0.6	2.8	2.8
维生素 B_{12}	1926/1948/1972	甲基钴胺素、腺苷钴胺素	5～25/1～5	19	13	5	15	15	31	22	22	28	28

[1]资料来源于Le Grusse和Watier，1993；McGinnis，1994；Charlton和Ewing，2007；Riaz等，2009。

[2]除了维生素A（MIU）、维生素D（IU）、生物素（μg/kg）和维生素B_{12}外，其余维生素添加量的单位均为mg/kg饲粮（90%的干物质）。

[3]依据BASF（2001），胆碱研究数据来自Charlton和Ewing（2007）。

[4]断奶仔猪（W）、生长猪（G）、育肥猪（F）分别指体重为3～10 kg、10～50 kg、50～80 kg的猪。妊娠猪为P，泌乳猪为L。公猪的推荐量等同于妊娠母猪。

[5]目前泌乳母猪的维生素E添加量可能更高（150IU/kg）。为维持猪肉品质，生长猪的维生素E添加量也需更高。

[6]由于应激条件下动物合成有限，可能需要添加维生素C。

本章通过综述近年来猪维生素营养的研究进展，为各种饲养标准体系（ARC，1981；NRC，1998；McDowell，2000；Lewis和Southern，2001；BSAS，2003）补充相关的数据。

2 维生素与猪繁殖性能

◆ 2.1 公猪繁殖性能

传统上，对公猪繁殖力的研究通常从基因、行为、养殖环境等方面着手。有关公猪营养的研究非常少，可能与过去的几十年间，主要采用公猪和母猪自然交配的配种方式有关。事实上，在自然交配的情况下，公猪的营养对母猪繁殖性能或受精率的影响确实较小。这是因为一头公猪射出精子的数量是所有卵子（每只母猪排卵15～20个）受精所需精子数量的5～20倍。然而，目前，加拿大和欧洲等国家在动物生产中广泛（占80%以上）使用了人工授精技术，这使公猪营养的研究显得非常重要，包括采用充足的营养和特殊的营养使公猪每次的射精量达到最大。目前，微量营养素对公猪繁殖性能影响的研究很少（Audet等，2004）。

研究发现，在高强度采精情况下，给公猪饲喂NRC（1998）推荐的维生素摄入量时，并不能使公猪的精子产生量达到最高（Audet等，2004）。当维生素摄入量超过3～10倍的NRC推荐量时，并不影响公猪激素的分泌，但是改变了血液和精液中的维生素含量（Audet等，2009a）。除维生素B_6外，大多数维生素以不同的转移效率由血液转移到精液中（Audet等，2009b）。然而，在高强度采精或人工授精的情况下，饲粮中添加维生素对公猪的精液量和精液品质并没有显著影响（Audet等，2009b）。

◆ 2.2 母猪繁殖性能

经过数十年的停滞后，猪的生产力在过去的10～20年得到了持续提高。通过对西方品系种猪进行大量的基因筛选，培育了多产系种猪。该系品种的母猪可以多排出4～5枚卵母细胞，但妊娠初期胚胎的成活率有所降低。尽管如此，母猪每窝仍能多产1～2只仔猪（Driancourt等，1998）。虽然，我们很早就知道提高猪的排卵率会降低胚胎的成活率，但相关机制仍不清楚。研究发现，额外成熟的卵泡排出的卵母细胞的质量和受精率均较差，且胚胎细胞的发育与分化也比较差，这是影响母猪繁殖力进一步提高的主要问题（Driancourt等，1998）。

母猪窝产仔数多也会导致仔猪的初生重差异较大（Tribout等，2003；Le Cozler等，2004），而初生重差异大会影响仔猪后期的生长性能和胴体品质（Gondret等，2005）。人们曾尝试采用寄养和分批断奶等方式来降低仔猪的初生重差异，但效果甚微（Matte等，1991）。因此，应考虑在母猪妊娠阶段采取合理的方式来解决这一问题，这就需要了解妊娠阶段子宫的哪些生理变化导致了仔猪初生重的差异。研究表明，某些脂溶性维生素和水溶性维生素与妊娠期子宫的生理变化密切相关，尤其是在妊娠早期（Mahan和Vallet，1997；Matte等，2006）。在妊娠早期，胚胎死亡率占受精卵的15%～40%，母体-胎儿间的相互作用决定了胚胎的发育情况（Cox，1997）。子宫的分泌物（也称“子宫乳”）（Solymosi和Horn，1994）为胚胎提供平衡的营养素、激素和生长因子。成活的胚胎也会分泌激素和细胞因子，进而影响子宫的分泌。研究发现，在妊娠晚期和泌乳期，胎儿和仔猪完全依赖母体（通过子宫、初乳和常乳）供应的维生素。事实上，从时间上看，妊娠期和泌乳期一共有135 ～ 140 d，与断奶到屠宰（大约130 d）的时间基本一样。这表明后代从母体获得足够的维生素是非常重要的。通过检测母体分娩前血液及初乳中维生素的含量，可以估测母体转移维生素到仔猪的效率（表6-3）。结果表明，尽管不同维生素从母体转移到胎儿和仔猪的效率不同，但是在出生后早期阶段，新生仔猪对初乳中各种维生素的利用至关重要。

表6-3 母体以子宫-胎儿和初乳-乳汁方式转移不同维生素的效率

维生素[1]	产前（母体）/吃初乳前（仔猪）[2]	子宫转移维生素的重要性	吃初乳前（仔猪）/吃初乳后（仔猪）[3]	初乳－常乳转移维生素的重要性
视黄醇	3.9	—	2.7	++
维生素 E	2.3	—	4.7	++++
维生素 D	>5	—	>1.2[4]	+
维生素 C	0.3	+++	4.1	++++

（续）

维生素[1]	产前（母体）/吃初乳前（仔猪）[2]	子宫转移维生素的重要性	吃初乳前（仔猪）/吃初乳后（仔猪）[3]	初乳－常乳转移维生素的重要性
叶酸	1.9	—	3.1	+++
维生素 B_{12}	0.2	++++	2.0	++

[1]视黄醇（Håkansson等，2001）、维生素E（Håkansson等，2001；Loudenslager等，1986；Pinelli-Saavedra和Scaife，2005；Pehrson等，2001）、维生素D（Goff等，1984）、维生素C（Pinelli-Saavedra和Scaife，2005；Yen和Pond，1983）、叶酸（Barkow等，2001；Matte和Girard，1989）、维生素B_{12}（Simard等，2007）。

[2]产前母体血液维生素的含量与吃初乳前仔猪血液维生素的含量之比。

[3]吃初乳前与吃初乳后仔猪血液维生素的含量比值，吃初乳后仔猪指10日龄仔猪。

◆ 2.3 脂溶性维生素和维生素C

2.3.1 维生素A

尽管目前维生素A在繁殖中的作用仍不清楚，但可以确定维生素A对动物的繁殖是必需的。研究发现，卵泡的成熟、黄体和子宫上皮细胞的正常功能、子宫环境的维持、胚胎发育等生理过程都需要维生素A。此外，维生素A还可以刺激三级卵泡合成雌激素和黄体合成孕酮（Nune等，1995）。维生素A缺乏不仅会影响机体健康，而且会导致睾丸萎缩和卵泡直径降低（Palludan，1963）。由于母猪可利用体内储备的维生素A，从而导致了母猪对维生素A的需要量难以确定（NRC，1998）。Braude等（1941）研究报道，成年母猪饲粮中不添加维生素A并不影响前三次妊娠过程，但在第四次妊娠时会出现维生素A缺乏症。

20世纪60年代中期，东欧国家研究表明，注射维生素A可以提高母猪窝产仔数（ARC，1982）。与此相似，Coffey和Britt（1993）研究也发现，在配种前和妊娠早期的母猪饲粮中添加维生素A和/或β－胡萝卜素可提高窝产仔数和断奶仔猪数。然而，Pusateri等（1999）却发现，从断奶到下次分娩期间的任何时间，单次注射1×10^6 IU的维生素A对窝产仔数、窝重、仔猪重、弱仔数和木乃伊胎的数量均无影响。

近年来研究发现，在断奶和配种时，给青年母猪（1和2胎次）肌内注射高剂量的维生素A（250 000 IU或500 000 IU）可线性提高窝产仔数和断奶仔猪数，然而，对3～6胎次母猪的窝产仔数却没有影响。这表明母猪的年龄不同，满足其最大繁殖性能的维生素A的需要量不同。

维生素A可通过影响卵巢孕酮的产生，从而影响卵巢甾醇类物质的合成和子宫内的环境（Chew，1993）。孕酮能刺激母猪子宫合成大量不同功能的蛋白质（Roberts和Bazer，1988），而后者对孕体的营养状况非常重要（Buhi等，1979）。对猪而言，由于滋养层细胞并不能进入子宫上皮，而是附着于子宫表面，因而这些蛋白质显得尤为重要。研究已证实，维生素A转运蛋白（即视黄醇结合蛋白，RBP）存在于黄体期母猪的子宫分泌物和孕体中，而这种转运蛋白可以把维生素A从子宫内膜转运到胎儿中（Chew等，1993）。Adams等（1981）发现，孕酮处理的母猪子宫分泌物中维生素A含量升高，表明由RBP将类维生素A物质转运到孕体中的量增加。Antipatis等（2008）报道，在受精和妊娠早期降低维生素A摄入量，可以提高仔猪初生重的均匀度，这可能与孕酮含量的改变有关。也有研究表明，妊娠初期适当降低维生素A的摄入量并不影响受胎率、窝产仔数、孕酮分泌量、胚胎和新生猪器官与体重之间的异速生长关系（Antipatis等，2008）。

传统认为，β-胡萝卜素（或其他类胡萝卜素类物质）在动物体内唯一的功能是提供维生素A，因此，β-胡萝卜素在体内的其他功能很少被关注。因此，与维生素A相比，人们对β-胡萝卜素在繁殖和免疫功能方面的作用认识非常有限。研究发现，从配种到断奶，给母猪每周注射228 mg β-胡萝卜素可以显著降低胚胎死亡率，提高窝产仔数、出生和断奶时的窝重（Brief和Chew，1985）。与此相似，Coffey和Britt（1993）研究表明，在断奶时给经产母猪单次注射β-胡萝卜素，随着β-胡萝卜素注射水平（0 mg、50 mg、100 mg、200 mg）的增加，窝产仔数也呈线性增加趋势。但是不清楚产仔数增加是否与排卵数增加或胚胎死亡率降低相关（Chew，1993）。尽管许多研究发现β-胡萝卜素可以提高母猪的繁殖性能，但是仍不清楚β-胡萝卜素是直接调控繁殖过程，还是仅仅作为维生素A的来源（Chew，1993）。目前，有关β-胡萝卜素对公猪繁殖性能影响的研究非常少。

2.3.2 维生素D

有关维生素D对猪繁殖性能的研究非常少。事实上，有关妊娠和泌乳母猪的维生素D推荐量并非来自科学试验的结果。目前维生素D的推荐量从200 IU/kg（妊娠和泌乳母猪；NRC，1998）到800 IU/kg（妊娠和哺乳母猪；LUS，2002），或800 IU/kg（妊娠母猪）和1 000 IU/kg维生素D_3（泌乳母猪）（BSAS，2003）。BSAS（2003）指出，维生素D的摄入水平应是上述推荐量的2倍。由此可见，维生素D及其代谢物在母猪繁殖中的作用尚需进一步研究。

研究表明，维生素D在母体-孕体的相互作用中具有重要作用（Vigiano等，2003）。除影响钙磷吸收之外，1,25（OH）$_2D_3$的免疫调节和抗增殖作用也为我们所熟知（Hewison等，2000）。相比之下，尽管有报道表明维生素D与雌性和雄性动物的繁殖性能有关，但是1,25（OH）$_2D_3$对繁殖的影响知之甚少，比如，在维生素D缺乏的小鼠模型中，雌鼠的繁殖力显著下降（Vigiano等，2003）。维生素D对动物繁殖力的影响与钙依赖性的作用机制无关（Kwiecinksi等，1989）。最近研究发现，1α-OH酶在胎盘中有表达（Zehnder等，2002）。根据该发现，以及维生素D缺乏可以降低雌性动物的繁殖力，可以推测1,25(OH)$_2D_3$的局部合成对胚胎的着床和/或胎盘形成具有重要作用（Halloran和Deluca，1983；Hickie等，1983；Kwiecinksi等，1989），这可能与1,25（OH）$_2D_3$的免疫调节作用或调控胚胎着床相关的基因表达相关（Zehnder等，2001）。

Lauridsen等（2010a）研究了不同剂量的两种维生素D[维生素D_3和25（OH）D_3对母猪早期繁殖过程的影响。在第一个试验中，将160头后备母猪随机分为8组，从初情期到妊娠第28天，分别饲喂不同水平的两种维生素D[200 IU/kg、800 IU/kg、1 400 IU/kg、2 000 IU/kg饲粮维生素D_3；5μg/kg、20μg/kg、35μg/kg、50μg/kg饲粮25（OH）D_3]。第二个试验也是采用这8个饲粮处理组，而试验是在经产母猪上进行，试验期从配种第1天到断奶。结果发现，除了高剂量的维生素D组可以降低死胎数外（1 400 IU/kg和2 000 IU/kg维生素D组每窝死胎数分别为1.17和1.13；而200 IU/kg和800 IU/kg维生素D组每窝死胎数分别为1.98和1.99），饲粮维生素D对母猪的繁殖性能无影响。进一步研究发现，不论饲粮维生素D的来源和添加水平如何，维生素D几乎不能够转移到后代中，其依据是在哺乳仔猪中未能检测到25（OH）D_3。即使能够检测到25（OH）D_3，其在仔猪血液的含量也低于5nmol/L（Lauridsen等，2010a）。

研究发现，在所测试的家畜品种中，仔猪血浆中的25(OH)D_3含量最低，因而易出现维生素D缺乏症（Horst和Littledike，1982）。事实上，猪与其他家畜在血浆25（OH）D_3含量上的差异可能源于养殖方式的不同。比如，能接触阳光的猪，其血浆维生素D及其代谢物的含量是圈养猪（不能接触阳光）的2.2 ~ 20.3倍（Engstrom和Littledike，1986）。Goff等（1984）给分娩前20 d的母猪（5头）肌内注射5 000 000IU的维生素D_3，研究维生素D_3对母猪和仔猪血液维生素D_3及其代谢物含量的影响。尽管试验母猪的头数很少，但是他们发现，初生仔猪血浆中25（OH）D_3、24,25-（OH）$_2D_3$和25,26-（OH）$_2D_3$含量与分娩母猪血液中的含量具有高度相关性。然而，新生仔猪血浆的1,25-(OH)$_2D_3$含量非常低，与母猪血液中的含量不具有相关性。该研究认为，给母猪肌内注射维生素D_3是一种给仔猪提供维生素D_3（通过母乳）的有效方式（Goff等，1984）。然而，通过在母猪饲粮中添加200 ~ 2 000IU/kg的维生素D_3并不是给哺乳仔猪补充维生素D的有效营养手段（Lauridsen等，2010a）。

2.3.3 维生素E

繁殖动物严重缺乏维生素E会导致死胎和胎儿重吸收（Nielsen等，1979）。许多研究表明，谷物饲粮中添加维生素E可以提高窝产仔数（Adamstone等，1949）。Mahan（1994）研究了饲粮中不同水平的dl-α-生育酚乙酸酯（22IU/kg、44IU/kg、66IU/kg）对母猪繁殖性能和血液、初乳和常乳中α-生育酚含量的影响，该研究持续研究了母猪的5个胎次。结果表明，随着饲粮维生素E水平的提高，产仔数增加，而乳腺炎、子宫炎和乳缺乏症发病率则降低。此外，初乳和常乳中的α-生育酚含量也随维生素E水平的升高而升高（Mahan，1994）。如前所述，因为胎盘只能转移少量的维生素E给胎儿（表6-3），所以仔猪出生时维生素E比较缺乏（Mahan，1991）。但是出生后的很短时间（4 d），哺乳仔猪血浆维生素E的含量就能提高（Lauridsen等，2002b）。研究发现，提高泌乳母猪饲粮中维生素E的水平，可提高断奶时仔猪血液和组织中维生素E的含量（Mahan，1991；Lauridsen和Jensen，2005）。

近年来，研究人员报道了不同维生素E的来源（RRR-α-生育酚乙酸酯和all-rac-α-生育酚乙酸酯）对母猪繁殖性能、母猪和仔猪维生素E含量的影响。Mahan等（2000）研究表明，不同维生素E的来源对母猪的繁殖性能（如产仔数，乳腺炎和子宫炎及乳缺乏症的发病率、阴道溢液等）没有影响。而Mahan等

（2000）则发现，与合成维生素E（all-rac-α-生育酚乙酸酯）相比，天然的维生素E（RRR- α -生育酚乙酸酯）可以提高母猪血清、初乳和常乳及21日龄哺乳仔猪血清和肝脏中维生素E的水平。定量比较这两种不同维生素E来源的效果比较困难，因为血液中的维生素E易被新吸收的维生素E替代（Traber等，1998），所以难以估测进入血液和乳中总的维生素E水平。通过采用稳定性同位素标记的维生素E可以克服这一困难。相对于未标记的维生素E，稳定性同位素标记的维生素E在测定维生素E的相对活性方面具有很多优势。同时饲喂同位素标记的天然的或合成的维生素E，可以比较母猪对不同异构体的利用率。Lauridsen等（2002a,2002b）通过给母猪饲喂标记的维生素E（d_3-RRR-α-和d_6-all-rac-α-生育酚乙酸酯），发现母猪在两种维生素E源的利用上存在差异，其中RRR-α-生育酚乙酸酯较易利用。因此，权威文献上报道RRR-α-与all-rac-α-生育酚乙酸酯的生物效价比为1.36：1，该数据低估了二者真实的生物效价比。母猪血浆和乳汁中d_3-α-生育酚：d_6-α-生育酚为2：1，导致哺乳仔猪血液和组织中这两种生育酚的比值也为2：1（Lauridsen等，2002a，2002b）。在随后的母猪试验中，通过HPLC方法将8种维生素E的立体异构体进行分离鉴定，形成了5个峰：峰1，为4个2S-生育酚；峰2，为2RSS-生育酚；峰3，为2RRS-生育酚；峰4，为2RRR-生育酚（天然形式）；峰5，为2RSR-生育酚。同时，该试验也检测了母猪乳汁和血液及仔猪血液中α-生育酚及其立体异构体的含量。Lauridsen和Jensen（2005）在母猪分娩前一周和泌乳阶段（28 d）饲喂不同水平维生素E（70IU/kg、150IU/kg、250IU/kg饲粮的all-rac-α-生育酚乙酸酯，风干基础），分别于试验第2、16、28天收集母猪乳汁及哺乳仔猪血浆样品。母猪乳汁和仔猪血浆中α-生育酚各种异构体的含量见表6-4。其中RRR-型含量最高，而2S-型含量很低，尽管饲粮中RRR-型和2S-型占总all-rac-α-生育酚乙酸酯的比例分别为12.5%和50%。与以前的研究结果一样，提高饲粮all-rac-α-生育酚乙酸酯水平可以使母猪乳汁中α-生育酚的含量增加，但是对仔猪血浆的影响尚不清楚。然而，对样品进行脂类含量校正后，各种饲粮处理对样品α-生育酚含量的影响就有显著差异（Lauridsen和Jensen，2005）。总体来说，表6-4反映了母猪对天然的和合成的维生素E利用率是不同的，对天然的RRR-α-生育酚利用率较高。

表6-4 猪乳（mg/kg，泌乳第2天）和仔猪血液（mg/L，4日龄）中α-生育酚的所有立体异构体形式及含量[1]

	总α-生育酚[2]	2S	RSS	RRS	RSR	RRR
70IU						
猪乳	6.39	0.25	1.08	1.50	1.26	2.28
仔猪血浆	7.24	0.13	1.24	1.63	1.33	2.21
150IU						
猪乳	10.2	0.69	1.73	2.37	2.07	3.40
仔猪血浆	6.53	0.27	1.42	1.97	1.56	2.79
250IU						
猪乳	16.5	1.41	3.22	3.72	3.35	4.79
仔猪血浆	7.27	0.23	1.25	1.81	1.38	2.38

[1]从妊娠第108天到仔猪断奶（分娩后28d）给母猪饲喂不同水平的all-rac-α-生育酚乙酸酯（70 IU/kg、150 IU/kg、250IU/kg）。

[2]资料来源于Lauridsen和Jensen，2005。

尽管睾丸退化被认为是雄性动物维生素E缺乏的一个证明，但是长期以来关于公猪维持繁殖力所需维生素E的研究比较少。精液富含多不饱和脂肪酸，容易氧化并损伤精子，从而改变精子的活力。Brzezinska-Slebodzinska等（1995）认为，饲粮维生素E可以作为公猪精液的抗氧化剂，但是在精液中直接添加维生素E并不能保护精子免受氧化损伤（Jones和Mann，1977）。Brzezinska-Slebodzinska（1995）研究发现，给公猪饲喂1 000IU/d的维生素E达7周后，精液的过氧化程度降低。维生素E似乎需要整合到细胞膜的磷脂脂肪酰甲酯上才能发挥抗氧化功能，因此在饲粮中添加维生素E可能会解决这个问题（Marin-Guzman等，1997，2000）。

2.3.4 维生素C

由于目前尚不清楚维生素C对猪的繁殖性能是否有益，因此饲养标准并没有给出维生素C的推荐摄入量（ARC，1982；NRC，1998）。然而，在维生素C合成缺陷的母猪上的研究表明，组织中充足的维生素C对胎儿的发育非常重要。Wegger和Palludan（1994）研究发现，给猪按每千克体重饲喂50 mg维生素C时，卵母细胞的成熟、受精、胚胎和胎儿的发育等均能正常进行，且母猪和胎儿血浆中维生素C的含量分别为（0.58 ± 0.04）mg/dL和（1.27 ± 0.09）mg/dL，而未添加维生素C组母猪和胎儿血浆中维生素C的含量分别为（17 ± 8）mg/dL和（18 ± 6）μg/dL。在妊娠期的任何阶段，若停止饲喂维生素C达24 ~ 38 d后，胎儿出现水肿，皮下和骨膜下出血，且骨骼钙化显著降低。因此维生素C对维持卵巢的正常功能十分重要，尤其是对于三级卵泡的成熟和黄体功能。对于雄性动物，生殖器官的发育、精子成熟、间质细胞合成睾酮等均需要维生素C的参与。然而，有关饲粮中添加维生素C对公猪繁殖性能的影响的研究结果存在争议（Audet等，2004）。

和维生素E一样，维生素C在保护机体免受氧化损伤方面具有重要作用。维生素C可以保护精子细胞，清除颗粒细胞和黄体细胞内的活性氧自由基。Pinelli-Saavedra等（2008）从母猪怀孕开始一直到仔猪断奶（21 d断奶），在母猪饲粮中添加维生素E和维生素C，研究维生素E和维生素C在初乳、常乳、仔猪血液和组织中的沉积及对免疫反应的影响。结果发现，单独添加维生素C（10 g/d），提高了母猪淋巴细胞对ConA和PHA刺激的反应，但对仔猪淋巴细胞的活力没有影响。饲粮中添加维生素E（500 mg/kg）影响了维生素E在体内的沉积，而每千克饲粮中同时添加维生素E和维生素C（500 mgVE和10 gVC）提高了仔猪血液IgA和IgG的含量。Pinelli-Saavedra和Scaife（2005）发现，饲粮中添加维生素C（高达10 g/d）对母猪的繁殖性能和仔猪生长性能均没有影响。由于胎盘转移维生素C的效率随着母体血清维生素C含量的升高而降低，因此仔猪从母体获得维生素C的主要途径是乳汁而不是胎盘。研究证实，仔猪血浆维生素C与母乳维生素C的含量呈显著正相关（Hidiroglou和Batra，1995）。

◆ 2.4 水溶性维生素

2.4.1 硫胺素和泛酸

自从ARC（1981）和NRC（1998）发布以来，有关硫胺素和泛酸对母猪繁殖性能影响的研究未见任何报道。在1947年仅有一篇关于硫胺素的报道之后，ARC（1981）认为“动物繁殖对硫胺素的需要量与维持动物最佳生长性能所需的硫胺素量并无差别”。同样对于泛酸，ARC（1981）和NRC（1998）并没有最新的研究报道，最近的一篇关于泛酸对母猪繁殖性能的研究要追溯到1971年Teague等发表的文章。

2.4.2 生物素

一些研究表明生物素可以提高动物的繁殖性能，然而，也有一些研究表明生物素对动物繁殖性能没有影响（Brooks，1986；Kornegay，1986）。总体来说，这些研究报道可能是由于实验动物数量有限，以至于不能达到数据上的显著差异，也可能是由于实验动物自身繁殖水平比较低下所致。Lewis等（1991）以不同胎次的300窝仔猪为试验对象，发现母猪饲粮中添加330μg/kg的生物素并不影响窝产仔数，但是增加了21日龄断奶仔猪的数量。据我们所知，自从NRC（1998）发表以来，仅有一篇关于生物素对母猪繁殖性能影响的研究报道（Garcia-Castillo等，2006）。该研究表明，与对照组（0.7ppm*的生物素）相比，高剂量的生物素（10ppm和28ppm）对后备母猪的繁殖性能无影响，该试验是从母猪发情前期（70 kg）开始持续到初次泌乳结束。然而，由于该试验的每个处理组只有7头后备母猪，因此这篇最新的研究报道提供的信息尚需进一步验证。

2.4.3 核黄素

1988年以前的研究表明，核黄素在猪繁殖中具有重要作用。在20世纪80年代早期，研究发现，配种1周后的母猪子宫可以分泌大量的核黄素（Moffat等，1980）。Bazer和Zavy（1988）研究发现，妊娠第4天到第10天，给猪饲喂100 mg/d的核黄素，提高了母猪的窝产仔数，这可能与胚胎成活率的提高有关。然而，随后的其他试验并没有得出同样的结果（Luce等，1990；Tilton等，1991；Wiseman等，1991；Pettigrew等；1996）。Bazer和Zavy（1988）的研究发现，生物素可能对繁殖力低下的母猪（每窝少于10只）有效果。在确定母猪核黄素需要

*1ppm=1mg/L。

量之前，需了解造成这些试验结果矛盾的因素。通常，除了繁殖性能指标外，代谢指标也可以用于确定饲粮核黄素的最佳添加水平，包括血液总核黄素（FAD、FMN、核黄素）水平和红细胞谷胱甘肽还原酶活力（EGRAC）。EGRAC最初用于人的营养，现在也广泛用于猪的营养。该指标通常用系数来表示，1.0 ~ 1.2表示摄入的核黄素能够满足机体需要，1.2 ~ 1.3表示摄食的核黄素临界缺乏，大于1.3表示摄入的核黄素不足（Le Grusse和Watier，1993）。表6-5显示了在两个不同的试验中，饲粮核黄素水平对不同妊娠阶段母猪的EGRAC的影响（Frank等，1984；Pettigrew等，1996）。这两个试验的结果具有可比性，因为这两个试验有一个相同的核黄素水平，而且EGRAC的检测结果非常接近。由表6-5可知，每天摄入10 mg的核黄素似乎足以使EGRAC稳定且达到最低。尽管EGRAC值看起来是一个衡量核黄素缺乏的可靠指标，但是近来有些学者开始怀疑EGRAC衡量核黄素需要量的可靠性。事实上，EGRAC值与猪血液和肝脏中总核黄素含量并无相关性（Giguère等，2002）。鉴于此，尚需其他指标来衡量核黄素的添加量（Hoey等，2009）。

表6-5 不同饲粮维生素B_2水平对不同妊娠阶段母猪红细胞谷胱甘肽还原酶活性（EGRAC）的影响[1]

饲粮核黄素添加水平（mg/d）	EGRAC（3 周）	EGRAC（7 周）	EGRAC（14 周）
1.5[1]	1.45	1.91	2.82
5.5[1]	1.37	1.42	1.64
9.5[1]	1.16	1.21	1.20
10[2]	1.23	1.22	1.17
60[2]	1.20	1.22	1.18
110[2]	1.20	1.21	1.19
160[2]	1.20	1.23	1.18

[1]Frank等（1984）。

[2]Pettigrew等（1996）。

2.4.4 吡哆醇和尼克酸

有关吡哆醇对窝产仔数影响的研究报道并不多，而且这些研究资料大都发表于20世纪60年代以前。NRC（1998）和ARC（1991）并没有正式给出吡哆醇的推荐量，仅给出了一些建议。在80年代早期，Easter等（1983）和Russell等（1985）的研究表明，饲粮中添加2.0 ~ 3.0 mg/kg的吡哆醇可以满足组织需要并提高动物的繁殖性能。Knights等（1998）认为，饲粮中添加16 mg/kg吡哆醇可以提高断奶后母猪吡哆醇的水平和妊娠期母猪表观氮沉积率，缩短母猪的发情间期，但是并不影响母猪的其他繁殖性能。

90年代，有些研究报道了尼克酸对母猪的影响。然而，令人疑惑的是NRC（1998）并没有提到尼克酸。事实上，Ivers等（1993）发现饲粮中不添加尼克酸，或者添加少量的色氨酸（尼克酸的前体，0.12%）就足以满足妊娠期和泌乳期母猪对尼克酸的需要。

最近一项研究表明，尽管泌乳一周的母猪采食量较高（6.4 kg/d），且所采食的饲粮中含有尼克酸（45 mg/kg）、色氨酸（0.22%）、维生素B_6（3 mg/kg），但是分娩后母猪中尼克酸和吡哆醇的含量仍会短暂降低。研究人员认为，泌乳早期，母猪中尼克酸和吡哆醇可能会短时间缺乏。今后的研究需要确定色氨酸用于合成尼克酸的比例，从而可以准确测定泌乳母猪对尼克酸的需要量。

2.4.5 胆碱

自NRC（1998）发布以来，目前还没有关于胆碱对母猪繁殖性能影响的新的报道。Donovan等（1997）报道了母猪分娩前12h到分娩后28 d间初乳和常乳中不同结合形式的胆碱的含量。他们发现，初乳和常乳中大部分胆碱（75% ~ 85%）与磷脂结合在一起，仔猪似乎对这种形式的胆碱利用率较高。事实上，乳汁中的胆碱既可被仔猪代谢并氧化生成甜菜碱（再次甲基化），也可以被仔猪直接利用（参与膜的完整性和神经递质的功能）。后面的功能是通过节省内源性合成的胆碱和减少甲基化来体现的。仔猪对乳汁胆碱的这种利用方式可能与调控体内高同型半胱氨酸的合成相关（Balance等，2005；Simard等，2007）。

2.4.6 叶酸

由于可以预防心血管疾病和先天性畸形，叶酸在人类营养研究中引起了研究人员的极大兴趣。叶酸的代谢与维生素B_{12}密切相关。叶酸的主要功能是参与体内一碳单位的转移，而一碳单位参与嘌呤和嘧啶的合成，同时也参与再次甲基化过程（蛋氨酸循环）。因此，叶酸对于蛋白质沉积和组织的形成至关重要。在同型半胱氨酸再次甲基化生成蛋氨酸的反应过程中，叶酸可以提供甲基，而维生素B_{12}则是酶的辅助因子（Le Grusse和Watier，1993）。因此，叶酸的有益作用与参与调控机体同型半胱氨酸水平有关。同型半胱氨酸是正常蛋氨酸代谢的中间产物，是一种强氧化剂，其含量过高不利于血管的完整性和正常的胚胎发育（Piertzik和Brönstrup，1997），因此必须尽量降低同型半胱氨酸的产生。叶酸和（或）维生素B_{12}缺乏会导致机体或组织同型半胱氨酸升高，从而提高心血管疾病、流产和先天畸形的发生率（Piertzik和Brönstrup，1997）。

研究表明，给妊娠母猪饲喂叶酸可以提高10%的产仔率（Matte等，1984；Kovcin等，1988；Lindemann和Kornegay，1989；Thaler等，1989；Friendship和Wison，1991；Lindemann，1993）。同样，给泌乳母猪饲喂叶酸可促进哺乳仔猪的生长（Matte等，1992），这可能是因为仔猪通过初乳和常乳获得了大量的叶酸（Barkow等，2001）。叶酸提高产仔率可能与其在妊娠早期降低胚胎死亡率有关（Tremblay等，1989）。叶酸通过两种途径来发挥这种作用：直接影响胚胎发育（DNA、蛋白质和雌激素分泌；Matte等，1996；Guay等，2002a）；间接刺激子宫分泌促生长因子（如细胞转化生长因子β_2，TGFβ_2）和胚胎附植相关的因子（如前列腺素E_2，PEG_2；Matte等，1996；Giguère等，2000；Guay等，2004a；Guay等，2004b）。但是，叶酸对胚胎和子宫的作用效果也与母猪的胎次有关（Matte等，2006）。事实上，与初产母猪相比，在经产母猪上，叶酸提高产仔率（Lindemann和Kornegay，1989）、促进子宫分泌PEG_2（Duquette等，1997）和TGFβ_2（Guay等，2004b）的效果更显著一些（表6-6）。关于叶酸在不同胎次的母猪上的效果存在差异，有不同原因分析。目前的研究认为，叶酸对初产母猪效果较小的原因可能与维生素B_{12}（可以影响同型半胱氨酸的平衡）的代谢活性降低有关。Matte等（2006）指出，叶酸对母猪生理和繁殖性能的作用受母猪胎次的影响，这种影响似乎与同型半胱氨酸的平衡有关。有些研究支持了这一观点：同型半胱氨酸改变滋养层细胞的完整性（DiSimone等，2004；Kim等，2009）；促进胞外花生四烯酸（C20：4n-6，PEG_2的前体物质）的产生，进而降低胞内花生四烯酸的水平（Signorello等，2002）。Matte和Girard（1999）等研究了不同叶酸水平（0 ～ 20 mg/kg）对妊娠母猪的影响，发现对窝产活仔数为12 ~ 13头的妊娠母猪，饲粮中叶酸的最佳添加水平约为10 mg/kg。

表6-6 母猪胎次和叶酸对妊娠30d尿囊液成分、胚胎死亡率和每窝产活仔数的影响[1]

项目	初产母猪		经产母猪	
	−	+	−	+
每窝产活仔数	9.1	9.3	11.5	13.5
胚胎死亡率（%）	14.4	12.8	39.2	32.6
尿囊液总 PGE_2（ng）	1 574	1 750	1 890	2 318
尿囊液总 $TGF\beta_2$（ng）	81.1	66.0	66.7	138.1

[1]资料来源于Matte等，2006。
“+”添加叶酸，“−”不添加叶酸。

2.4.7 维生素B_{12}

目前维生素B_{12}在繁殖中的作用尚不清楚。研究维生素B_{12}需要量的方法与其他水溶性维生素不同，因为与其他水溶性维生素相比，诱导出维生素B_{12}缺乏症（如即将分娩的母猪流产）需要相当长的时间（至少1年）（Ensminger等，1951；Cunha等，1944；Frederick和Brisson，1961）。如前所述，叶酸可能通过影响同型半胱氨酸的含量来影响动物的繁殖功能，这同样也引起了人们对维生素B_{12}的关注。而在过去的30 ~ 40年里，维生素B_{12}在猪营养中的作用一直被忽视。NRC（1998）也指出，自从1940年以来，关于维生素B_{12}对猪繁殖性能的影响只有很少的几篇报道（Anderson和Hogan，1950；Frederick和Brisson，1961；Teague和Grifo，1966）。Guay等（2002a）研究表明，初产母猪血液维生素B_{12}的含量比经产母猪血液维生素B_{12}的含量低2 ~ 3倍，其原因可能与竞争利用维生素B_{12}有关。和其他维生素一样，初产母猪的繁殖可能与生长和维持竞争维生素B_{12}。当母猪达到体成熟后，更多的维生素B_{12}可供给动物的繁殖功能需求。此外，由于妊娠早期大量的维生素B_{12}转移至子宫，因此妊娠早期似乎需要更多的维生素B_{12}（Guay等，

2002b）。事实上，妊娠第15天，子宫角总维生素B_{12}的含量是血液总维生素B_{12}含量的180%～300%，而血液大约占体重的4%（Matte和Girard，1996）。尽管在子宫组织中发现了维生素B_{12}的一个转运载体（转钴胺蛋白Ⅰ），但是如此多的维生素B_{12}由母猪血液转移到子宫的生理机制仍不清楚（Pearson等，1998）。Guay等（2002b）通过检测妊娠早期（第0～15天）母猪血液维生素B_{12}含量的变化，发现饲粮添加0.16μg的维生素B_{12}可以有效提高母猪维生素B_{12}的水平。由于目前的维生素B_{12}的推荐量（0.15μg；NRC，1998）是由很少量的研究资料得出的，因此对维生素B_{12}的最佳需要量应进行重新测定。在这方面，ARC（1981）认为"现阶段发布的维生素B_{12}的需要量15μg/kg是暂时性的，根据试验结果，动物对维生素B_{12}的需要量可能增加10倍……"。近期，Simard等（2007）估测了初产母猪对维生素B_{12}的需要量。他们发现，妊娠期母猪饲粮中添加164μg/kg和93μg/kg的维生素B_{12}可分别使血浆维生素B_{12}含量达到最高和血浆同型半胱氨酸含量达到最低。妊娠期母猪饲粮中添加维生素B_{12}同样也影响维生素B_{12}由母体到仔猪（子宫或初乳）的转移，而且也会降低哺乳仔猪体内同型半胱氨酸的聚积（Simard等，2007）。

3 维生素与猪生长

如前所述，在现代养猪生产中，猪从断奶到上市屠宰的生存时间大约占了猪一生时间的一半（从受精到屠宰）。因此，从逻辑上讲，首先要确定断奶早期仔猪维生素的需要量（以每天或每千克体重为基础），同时要比较哺乳阶段乳汁和断奶后仔猪饲粮的维生素水平。表6-7列出了初乳、常乳和断奶仔猪饲粮中部分维生素的含量，以及为仔猪每天提供的维生素的量（每千克体重）。初乳维生素的含量比常乳高，如维生素B_{12}和维生素E的含量分别是常乳中的1.4倍和4倍。这种差异并不能全部归因于总固体物质的差异（常乳总固体物质的含量比初乳低20%～25%）（Pond和Houpt，1977；Yang等，2009）。以每天为仔猪提供的维生素来看（每千克体重），初乳的摄入非常重要，因为初乳为仔猪提供的维生素B_{12}和维生素E的含量分别是常乳的3.6倍和18.3倍。如前所述，初乳是仔猪从母体获得维生素非常重要的一种方式，因为分娩前部分维生素从母体转移到胎儿的效率非常低。因此，有人认为初乳提供的维生素的量可以作为分娩后仔猪维生素的需要量。

表6-7 初乳、乳汁及断奶饲粮中维生素的水平

维生素	初乳[1]		常乳[1]		NRC（1998）推荐断奶饲粮		BASF（2001）推荐断奶饲粮	
	含量（U/L）	提供给仔猪（U/d/kg BW）[2]	含量（U/L）	提供给仔猪（U/d/kg BW）[2]	添加量（U/kg）	提供给仔猪（U/d/kg BW）[2]	添加量（U/kg）	提供给仔猪（U/d/kg BW）[2]
视黄醇（mg）	2.5	0.77	0.7	0.08	0.67	0.03	3.76	0.19
维生素 E（mg）	10	5.5	2.5	0.3	16	0.8	60	3
维生素 C（mg）	24.7	7.7	11.1	1.4	N/A[3]	N/A	N/A	N/A
叶酸（μg）	44.6	13.7	13.4	1.6	300	15	1 600	80
维生素 B_{12}（μg）	6.1	1.87	4.3	0.53	17.5	0.88	31	1.55

[1]视黄醇（Håkansson等，2001）、维生素E（Håkansson等，2001；Loudenslager等，1986；Pinelli-Saavedra和Scaife，2005；Pehrson等，2001）、维生素C（Pinelli-Saavedra和Scaife，2005；Yen和Pond，1983）、叶酸（Barkow等，2001；Matte和Girard，1989）、维生素B_{12}（Simard等，2007）。

[2]依据：每天摄食初乳0.43L，0～1日龄仔猪体重为1.4 kg；摄食母乳0.8L，14～21日龄体重为6.5 kg；采食0.5 kg，21～35日龄体重为10 kg。

[3]N/A：无研究数据。但是饲粮中添加1 g/kg的维生素相当于50mg/d/kg BW。

NRC（1998）推荐的断奶后仔猪维生素的需要量（每天为每千克体重提供的维生素）要低于初乳维生素的含量（除了叶酸），但是高于常乳维生素的含量（除了维生素A）。加拿大BASF（2001）推荐仔猪的维生素需要量（每天每千克体重）均高于常乳和初乳维生素的含量（除了维生素A和维生素E，这两种维生素仍低于初乳的最高水平）。

一些研究报道了饲粮中添加“多维”对猪生长性能和其他指标的影响。在9 ~ 28 kg阶段，以达到最佳生长性能为衡量指标，高瘦肉率品系猪对多维（核黄素、尼克酸、泛酸、维生素B_{12}和叶酸）的需要量（>470%；NRC，1998）要高于中等瘦肉品系猪（Stahly等，2007）。然而，在生长育肥阶段（33 ~ 110 kg），尽管多维（核黄素、吡哆醇、泛酸和生物素）添加量达到GfE（1987）或NRC（1998）推荐的400%和800%时也没有改善猪的生长性能，但是使猪肉中维生素B_6、泛酸、生物素含量分别提高了37%、58%和129%（Böhmer和Roth-Maier，2007）。

Branner和Roth-Maier（2006）认为，益生素可能会影响肠道B族维生素的代谢。益生素被认为是促生长剂和抗菌药物的替代物，因为益生素可增加肠道屏障功能（Madsen等，2001）、刺激免疫和增强肠道防疫系统（Dugas等，1999）。事实上，有研究表明，益生素（屎肠球菌）并不影响肠道对硫胺素、核黄素、泛酸和生物素的利用。这些研究结果在近期的一项研究（Lessard等，2009）中得到了证实，该研究发现益生菌（乳酸片球菌和酿酒酵母菌）单独添加和复合添加均未影响断奶仔猪体内叶酸或维生素B_{12}的平衡。

◆ 3.1 脂溶性维生素和维生素C

尽管生产中维生素的缺乏不容易发生，但是应该注意仔猪断奶阶段维生素的缺乏，特别是脂溶性维生素的缺乏。因为在断奶阶段，仔猪对脂类的吸收能力相对较差。哺乳仔猪脂肪表观消化率非常高（96%；Cranwell和Moughan，1989），但是在断奶阶段会降到65% ~ 80%（Cera等，1988）。而且超早期断奶的推广及欧盟国家禁止使用抗生素等这些因素增加了仔猪的断奶应激。此外，断奶阶段仔猪的生长特别是肌肉组织生长相对较快，提高了仔猪对维生素的需要量。值得注意的是，维生素E和维生素C具有抗氧化活性，它们可以与其他抗氧化剂（硒）和促氧化剂（铁、铜、多不饱和脂肪酸）相互作用，这些情况在配制断奶仔猪和生长猪饲粮时需加以考虑。上述提到的问题会在本书其他章节中进行阐述。

3.1.1 维生素A

猪的肝脏可以储存维生素A，以备维生素A摄入不足时所需。维生素A需要量可通过脑脊液压、肝脏和血液维生素A含量等来衡量，这些指标均比体增重准确。猪在出生后头8周，饲粮中需要添加75～605μg/kg视黄醇乙酯；在生长育肥阶段，以日增重为衡量指标时，需添加35～130μg /kg视黄醇乙酯，而以肝脏和脑脊液压为检测指标时，则需添加344～930μg /kg视黄醇乙酯（NRC，1998）。然而，在商品教槽料中，维生素A的添加量超过了NRC（1998）标准。Ching等（2002）研究发现，饲粮中添加不同水平的维生素A（2 200IU/kg、13 200IU/kg、26 400IU/kg）并不影响仔猪断奶后35 d的生长性能。此外，Hoppe等（1992）也发现，与对照组（添加5 000IU维生素A）相比，饲粮中添加高水平的维生素A（10 000IU/kg、20 000IU/kg或40 000IU/kg）并没有提高生长猪的生长性能。同时，当饲粮维生素A的水平过高时，会影响断奶仔猪体内维生素E的含量，并对机体的抗氧化能力产生不利影响（Ching等，2002）。猪采食10倍于正常剂量的维生素A（195 000IU/kg）饲粮3～4 d后会出现维生素A过量症（突然跛足）(Reigner等，2004)。

如前所述，猪对β-胡萝卜素的吸收率非常低。然而，由于β-胡萝卜素在机体免疫调节中发挥着重要作用，因此β-胡萝卜素在断奶仔猪（免疫系统易受到抑制）中的作用受到特别关注。研究表明，给母猪饲喂β-胡萝卜素后，尽管母猪初乳中免疫球蛋白的含量没有差异，但是仔猪血液中免疫球蛋白会提高（Brief和Chew，1985）。β-胡萝卜素可提高猪细胞的有丝分裂（Hoskinson等，1992）。给猪（50～55 kg）单次注射β-胡萝卜素（0 mg、20 mg或40 mg），发现β-胡萝卜素在淋巴细胞内的分布为：细胞核含量最多，线粒体和微粒体含量居中，胞浆含量最低（Chew等，1991a）。但是注射β-胡萝卜素并没有改变血浆中视黄醇或α-生育酚的含量（Chew等，1991b）。饲粮中添加β-胡萝卜素似乎并不影响母猪和仔猪血清IgG的水平（Kostoglou等，2000）。有关β-胡萝卜素在猪免疫功能中的作用尚需进一步研究。

3.1.2 维生素D

初生仔猪血液25-OHD_3含量过低会导致佝偻病。这种维生素D缺乏症表现为骨骼发育受阻和肌病。维生素D的缺乏会降低钙、磷、镁等在体内的沉积（Miller等，1965）。因此，成年母猪轻度缺乏维生素D会导致骨矿物元素含量降低（软骨病），但是严重缺乏维生素D不仅会导致钙镁缺乏，还会出现抽搐。Bethke等（1946）认为生长猪每千克饲粮中维生素D的最低需要量为200IU/kg，而英国推荐（BSAS，2003）母猪的维生素D的需要量要高一些，从800IU（每千克体重<60 kg）到600IU（60 ~ 90 kg体重）。丹麦推荐量为800IU（仔猪，6 ~ 9 kg）到500IU（仔猪，9 ~ 30 kg）和400IU（25 ~ 100 kg体重）。

近来研究发现，给育肥猪饲喂含高剂量维生素D_3的饲粮可提高肌肉钙含量，进而提高猪肉的嫩度。Wiegand等（2002）在育肥猪屠宰前3 d，给其每天饲喂含250 000 IU或500 000 IU维生素D_3的饲粮，虽然血浆钙含量显著提高，但是猪肉的嫩度却没有得到改善。然而，Wilborn等（2004）却发现，饲喂40 000 IU/kg或80 000 IU/kg（风干基础）维生素D_3 44 d后可改善肉色并提高肌肉pH，但不影响肌肉钙的含量。与此相似，Lahucky等（2007）发现，给生长育肥猪饲喂高剂量的维生素D_3（500 000 IU/d，5 d）可显著提高血浆钙含量，对肌肉红度也有一定的影响。维生素D除影响肉品质外，也影响肌肉的营养价值。Jakobsen等（2007）发现肌肉和肝脏总维生素D的含量（如维生素D_3和其代谢物25-OHD_3）与饲粮总维生素D的水平存在剂量-效应关系。

与维生素A类似，过量的维生素D也容易使人畜中毒。摄入过量的维生素D对机体产生不良的影响，这些不良的影响均与血液钙含量异常升高有关。从饲喂的时间来看，若少于60 d，猪饲粮中维生素D_3的最高含量应限制在33 000IU/kg；若大于60 d，则应限制在2 200IU/kg（NRC，1998；McDowell，2000）。

3.1.3 维生素E

维生素E对断奶后仔猪的生长和健康十分重要。在小肠，维生素E是以游离醇的形式单独或与脂肪一起被吸收，因此商品all-rac-α-生育酚乙酸酯需水解后才能被肠道吸收，但是断奶仔猪肠道水解all-rac-α-生育酚乙酸酯的能力较差（Chung等，1992；Hedemann和Jensen，1999；Lauridsen等，2001）。尽管哺乳母猪采食充足的维生素E可以提高断奶时仔猪体内维生素E的储备（Mahan，1991，

1994；Lauridsen和Jensen，2005），但是断奶后仔猪组织（肝脏和肌肉）维生素E的含量会迅速降低（Lauridsen和Jensen，2005）。而对35～49日龄的仔猪，肝脏中维生素E的含量却没有发生改变。研究表明，猪的肝脏可以短期储存大量的α-生育酚（Jensen等，1990；Lauridsen等，2002b）。在其他组织中，维生素E含量下降，可能意味着在断奶仔猪饲粮中添加70IU/kg的all-rac-α-生育酚乙酸酯（或者all-rac-α-生育酚乙酸酯的生物学效价）不足以促使肝脏α-生育酚转移到其他组织（Lauriden和Jensen，2005）。随后的试验研究了饲粮中不同维生素E添加水平（85 mg/kg、150 mg/kg、300 mg/kg all-rac-α-生育酚乙酸酯）对血浆维生素E含量的影响，试验饲粮中分别含5%的动物脂肪、葵花油或鱼油。结果发现，在断奶后第1周，血浆维生素E含量显著降低，而饲粮中添加维生素E并不影响42日龄以前仔猪的血浆维生素E的含量（Lauridsen，2010b）。当给42日龄后的仔猪饲喂含150 mg/kg和300 mg/kg的all-rac-α-生育酚乙酸酯的饲粮时，血清维生素E的含量与饲粮维生素E的水平呈正相关，并维持在1.5～2.0 mg/L。因此，血浆或血清维生素E的含量（1.5～2.0 mg/L）可以作为检测猪维生素E适宜需要量的指标（Wilburn等，2008）。

一些研究表明，在饲粮中添加维生素E可提高猪肉品质（Jensen等，1998）。脂类氧化是肉质变坏的主要原因。肉品质的变化可通过味道、颜色、纹理、营养价值等指标来反映，有时也可通过所产生的有毒产物来反映。在饲粮中添加高于需要量的维生素E可以有效降低猪肉和肉产品的脂类氧化程度。肉中维生素E的蓄积量取决于肉的类型、饲粮维生素E的添加水平和添加时间（Jensen等，1998）。目前还没有关于维生素E的毒性报道，如在生长猪饲粮中添加高达500 mg/kg（Bonette等，1990）和700 mg/kg（Jensen等，1997）的维生素E也未发现有任何毒性。事实上，饲粮中添加700 mg/kg的α-生育酚乙酸酯并未使肌肉组织中的α-生育酚达到饱和状态（Jensen等，1997）。肌肉组织储存维生素E可使肌肉组织免受氧化攻击，从而提高猪肉储藏期间的稳定性。尽管饲粮维生素E在控制猪肉褪色效果方面存在争议，但是肉品质的其他指标（如肉色和滴水损失）会受到饲粮维生素E的影响（Jensen等，1998）。在预冻肉（Asghar等，1991；Monahan等，1994）和生鲜肉（Cheah等，1995）中，维生素E和滴水损失的降低存在一定的相关性，但具体机制还需进一步研究（Jensen等，1998）。同样也有研究表明，与对照组相比，育肥猪饲粮中添加高于营养水平的维生素E可进一步降低肌糖原含量，并有降低猪肉系水力的趋势（Rosenvold等，2002）。近来有研究显示，当饲粮维生素和微量元素的水平提高到推荐量的150%～200%时，猪肉的滴水损失显著降低（Apple，2007）。

3.1.4 维生素C

由于7日龄的猪就可以合成维生素C（Braude等，1950），因此通常情况下，猪饲粮中不需添加维生素C。然而，在应激条件下（如断奶），合成维生素C的关键酶1-古洛糖-γ-内酯氧化酶（GLO）活性可能会降低（Ching等，2001）。另外，研究发现，断奶后仔猪血液维生素C显著降低（Yen和Pond，1981；Mahan和Saif，1983），这表明维生素C合成不足或者出现断奶应激。Mahan等（1994）发现，在断奶后头2周，给断奶仔猪饲喂维生素C可改善其生长性能。DeRodas等（1998）研究了饲粮中添加维生素C对早期断奶仔猪（14日龄）生长性能和铁含量的影响。结果发现，75ppm的L-抗坏血酸基-2-多磷酸盐可以满足早期断奶仔猪对维生素C的需要量。相反，Fernandez-Duenas等（2008）研究表明，饲粮添加150 mg/kg维生素C或350 mg/kg β-胡萝卜素对仔猪的生长性能和抗氧化能力没有影响。

维生素E和维生素C之间具有协同作用。研究发现，维生素C可以节省体内维生素E的用量（Burton等，1990）。Lauridsen和Jensen（2005）研究发现，与对照组（不添加维生素C）相比，饲粮添加500 mg/kg维生素C可以显著提高断奶仔猪免疫细胞（表6-1）、肝脏和肌肉中α-生育酚的含量。此外，与对照组相比，饲粮添加维生素C提高了免疫细胞RRR-α-生育酚的含量，降低了RRS-生育酚的含量，进而提高了断奶期仔猪血浆IgM的含量。Zhao等（2002）发现，随着饲粮维生素C水平升高（<300 mg/kg），断奶仔猪血浆IgG的含量也呈线性增加。维生素C在猪免疫功能中的作用仍需进一步的研究。

维生素C和维生素E之间的协同作用也对肉品质有重要作用。Eichenberger等（2004）研究表明，与单独添加维生素C组相比，在阉猪（25～106 kg）饲粮中联合添加300ppm维生素C和200ppm的dl-维生素E乙酯显著提高了所有检测组织（除了猪蹄）中维生素E的含量。饲粮添加维生素C并不影响机体的抗氧化能力（通过TBARS值来衡量），而维生素E则提高了机体的抗氧化能力。而Pion等（2004）

则研究表明，在屠宰前48h在饮水中分别添加0 mg/kg、1000 mg/kg、2000 mg/L的维生素C并不影响猪肉品质。

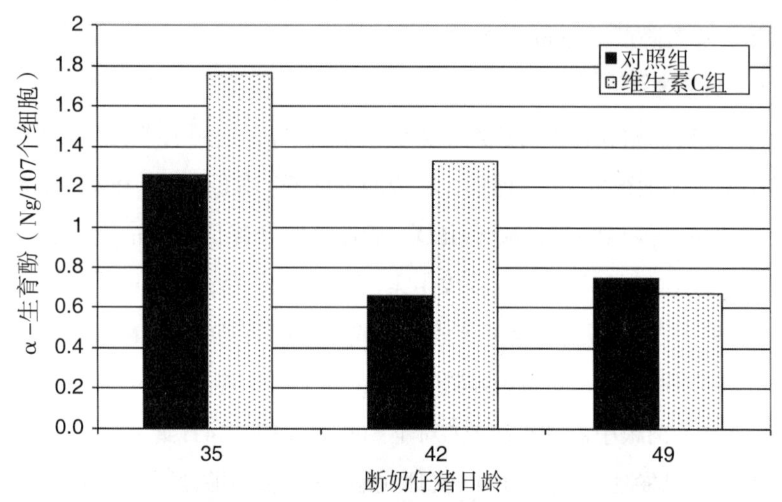

图6-1　维生素C对断奶仔猪免疫细胞α-生育酚含量的影响

（资料来源于Lauridsen和Jensen，2005）

维生素C对脯氨酸和赖氨酸的羟基化也是必需的，后两者是胶原蛋白不可缺少的组分，而胶原蛋白是软骨和骨骼生长的必需成分。基于此，维生素C可促进骨基质和牙本质的形成。Nakano等（1983）和NRC（1998）探讨了维生素C在预防或减轻猪骨软骨病（这种病与赖氨酸羟基化减少导致的胶原交联不足有关）中的作用。结果表明，饲粮中添加维生素C对预防骨软骨病没有效果。与此类似，Pointillart等（1997）研究表明，给47日龄的猪饲喂含1 500 mg/kg或1 000 mg/kg的维生素C达到4个月后，骨骼形成的标示物（除了骨钙蛋白）和血液及尿液中骨代谢指标均不受影响。通过这项长期试验，研究人员认为摄入高水平的维生素C对猪骨代谢或骨骼特征并无积极的影响。此外，Armocida等（2001）在猪高发的前腿损伤上的研究也表明，猪前腿患病率与维生素C含量的降低并无相关性。

3.1.5 维生素K

断奶仔猪和生长猪的维生素K需要量约为0.5 mg/kg。天然形式的维生素K有K_1（叶绿醌）和K_2（四烯甲萘醌）两种形式，价格比较贵。因此通常生产中常用合成的维生素K_3（甲萘醌）及其衍生物甲萘醌硫酸氢钠（MSB）作为饲粮维生素K的来源。Marchetti等（2000）研究发现，合成维生素K-甲萘醌亚硫酸氢烟酰胺（MNB）是猪饲粮中很好的维生素K来源。维生素K与血液凝固有关，它为四种凝血因子（Ⅱ、Ⅶ、Ⅸ和Ⅹ）由前体转化为活性形式所必需。因此，维生素K缺乏，导致凝血因子不能转化为活体形式，从而导致凝血时间延长。生产上可以通过检测凝血酶原时间来确定维生素K的缺乏与否。即使维生素K缺乏的时间很短，肝脏储存的维生素K也会很快被消耗掉（Kindberg和Suttie，1989）。NRC（1998）认为，动物可以耐受1 000倍需要量的维生素K。维生素K缺乏导致凝血酶原升高和凝血时间延长，并有可能导致内出血和死亡（NRC，1998）。一些影响血液凝固的因素，如钙过量、霉菌毒素、饲料发霉等可能会增加猪对维生素K的需要量（Hoppe，1988；NRC，1998）。目前还不清楚钙过量是否会降低肠道微生物维生素K的合成，或降低肠道对维生素K的吸收，或破坏了维生素K的活性。除了发霉玉米，抗生素和脂溶性维生素的不平衡也会影响维生素K的作用效果。维生素E和维生素K的交互作用似乎存在组织特异性（Tovar等，2006），因为与不添加维生素E组相比，饲粮添加维生素E降低了大鼠血液中叶绿醌的含量。此外，通过检测凝血酶原和部分凝血活酶时间的变化，发现饲粮中四种天然的维生素E和β-胡萝卜素分别有增强和降低出血的趋势（Takahashi等，1995）。

◆ 3.2 水溶性维生素

3.2.1 硫胺素和生物素

自ARC（1981）和NRC（1998）发布以来，只有一篇关于硫胺素与猪生长的研究报道。Woodworth等（2000）发现，与未添加硫胺素组相比，饲粮（玉米-豆粕-乳清粉为基础）中添加5.5 mg/kg的硫胺素并不影响断奶仔猪（5～25 kg BW）的生长性能。该试验添加的硫胺素的水平分别是ARC（1981）和NRC（1998）推荐量的2.5倍和5倍。

对于生物素，其在维持猪蹄完整性和提高猪蹄抗损伤能力方面的作用已经为人们所熟知。NRC（1998）和INRA（1984）认为饲粮中生物素的添加量为50～100μg/kg。尽管NRC（1998）提出“猪体内生物素的来源有很大一部分可能是来自肠道微生物合成”，但是事实上微生物合成的生物素并不能被猪利用，因为生物素主要在小肠被吸收（Mosenthin等，1990）。因此，与其他维生素（如叶酸和维生素B_{12}）相比，由饲粮提供的生物素对猪生长尤为重要，因为肝肠循环对于机体生物素的平衡几乎没有影响，且胆汁排泄的生物素低于饲粮生物素摄入量的2.2%（Zempleni等，1997）。

Kopinsky和Liebholz（1989）关于生长育肥猪生物素需要量的研究发现，50～100μg/kg的生物素可以有效预防蹄损伤。此外，添加100μg/kg的生物素可使机体各个组织的生物素含量达到最高（图6-2）。以上这些检查指标可以用来衡量猪对生物素的需要量。

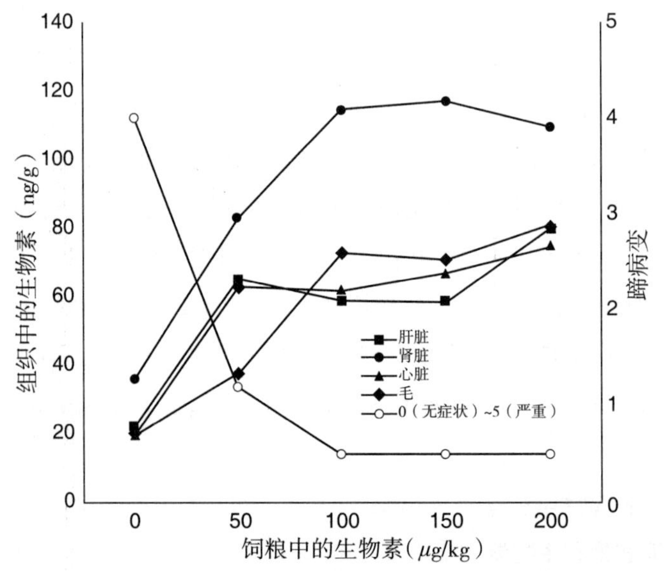

图6-2 不同组织中生物素的含量、蹄病变程度与饲粮生物素的关系

（资料来源于Kopinski和Liebholz，1989）

Partridge和McDonald（1990）研究表明，在15～88 kg猪饲粮中添加500μg/kg的生物素有改善饲料利用率的趋势，其原因可能是生物素提高了多不饱和脂肪酸的代谢效率。脂肪酸碳链延长需要二碳单位（如乙酸和丙二酸），其中丙二酸在脂肪酸碳链延长过程中的效率比乙酸高出20～30倍（Roland和Edwards，1971），而丙二酸的合成需要生物素的参与（Watkins，1989；Watkins和Kratzer，1987）。研究发现，饲粮中添加生物素改变了家禽（Watkins，1989；Watkins和Kratzer，

1987；Roland和Edwards，1971）、大鼠（Liu等，1994）和猪（Martelli等，2005）脂肪酸的类型。因此，胴体脂肪酸的含量和类型可能与动物体内生物素的含量有关。

3.2.2 维生素B_6、核黄素和尼克酸

仔猪断奶时血浆中吡哆醇的含量非常低（Matte等，1997，2001），其原因可能是母猪乳汁中的维生素B_6含量很低，大约只有0.40μg/mL（Benedikt等，1996）。该含量大约只有仔猪生长所需维生素B_6的一半（Coburn，1994）。断奶后，仔猪维生素B_6缺乏进一步加剧。其原因是与母乳相比，仔猪断奶后的饲粮蛋白质含量较高且相对不平衡，体内氨基酸相互转化和氧化速率增加所致。吡哆醛-5-磷酸（吡哆醇的一种活性形式）是蛋白质代谢相关酶的重要辅酶因子（Le Grusse和Watier，1993），吡哆醛-5-磷酸的利用率取决于仔猪的生长速率（Matte等，1997，2001）。

色氨酸在体内的代谢依赖于维生素B_6和色氨酸氧化。色氨酸代谢产生丙氨酸（一种生糖氨基酸）、乙酰辅酶A（完全氧化）或烟酰胺核苷酸（Le Grusse和Watier，1993）。若维生素B_6缺乏，色氨酸氧化产生的中间代谢物，如犬尿氨酸和黄尿素可能会在体内聚积，通过竞争转运色胺素穿过血脑屏障进而抑制大脑色氨酸的吸收，从而减少5-羟色胺的合成（Bender，1987）。5-羟色胺由色氨酸通过维生素B_6依赖的途径（脱羧）生成，因此饲粮提供维生素B_6可以刺激大鼠5-羟色胺的合成（Hartvig等，1995）。然而，饲粮中同时添加维生素B_6和色氨酸促进5-羟色胺合成的效果会更好（Lee等，1988）。仔猪饲喂含维生素B_6的饲粮并灌注色氨酸后，血液色氨酸消失速度增加且3-羟犬尿氨酸会短暂升高（Matte等，2001），表明维生素B_6可刺激色氨酸的氧化代谢。但关于饲粮维生素B_6与5-羟色胺的研究鲜见报道。

有研究报道了维生素B_6、色氨酸和其他营养素（如核黄素）间的相互作用对早期断奶仔猪体内葡萄糖平衡的影响（Matte等，1997，2001，2005，2008），包括色氨酸在维生素B_6依赖性的酶作用下脱羧生成5-羟色胺，或维生素B_6参与色氨酸氧化并生成尼克酸的过程。事实上，维生素B_6的缺乏会抑制上述反应的进行并生成大量的中间代谢产物（如黄尿酸），而黄尿酸则能显著降低胰岛素水平（Kotake等，1968）。

维生素B_6也可作为辅酶参与体内的转硫反应（如同型半胱氨酸生成半胱氨酸）（Le Grusse和Watier，1993）。Zhang等（2009）研究表明，断奶仔猪缺乏维生素B_6导致生长性能下降和吡哆醛-磷酸快速降低（3周内），并产生严重的高同型半胱氨酸血症。此外，他们也研究了不同水平的维生素B_6对生长性能和代谢指标的影响。结果发现，当摄入维生素B_6超过8 mg/kg时并不能改善断奶仔猪的生长性能，这与Matte等（2005）和Woodworth等（2000）的研究结果相似。该维生素B_6的添加水平是NRC（1998）推荐量的5倍，但与北美（BASF，2001）实际生产中的平均水平一致。以生长性能和（或）氮沉积率为衡量指标，仔猪断奶后（5～30 kg体重）维生素B_6的需要量为2～5ppm（Adams等，1967；Kösters和Kirchgessner，1976；ARC，1981；Bretzinger，1991；NRC，1998）。

如前所述，体内维生素B_6的代谢与其他B族维生素，特别是核黄素（维生素B_2）密切相关。维生素B_2的活性形式为FMN和FAD，二者参与维生素B_6及其活性形式（吡哆醛磷酸）和排出形式（4-吡哆酸）之间的相互转化（Le Grusse和Watier，1993）。同时，核黄素也是谷胱甘肽还原酶的辅助因子（NADPH的一个成分），该酶可以催化氧化型的谷胱甘肽重新生成还原型的谷胱甘肽。还原型的谷胱甘肽可还原过氧化物，在防止组织过氧化中发挥关键作用。饲喂维生素B_2缺乏的饲粮显著降低了肝脏GSH-Px的活力和硒含量（Brady等，1979）。与不摄入维生素B_2相比，每天摄入2 mg维生素B_2提高了许多组织GSH-Px的活力和硒含量（特别是以亚硒酸盐为硒源）（Parsons等，1985）。

前面已提到，动物（包括猪）体内的色氨酸与尼克酸是相互关联的（Le Grusse和Watier，1993）。色氨酸转变成尼克酸取决于饲粮、激素（Fukuwatari和Shibata，2007）和动物品种（对于人、鸡、猪和鸭来讲，1 mg尼克酸分别等价于60 mg、45 mg、50 mg和170 mg的色氨酸；Firth和Johnson，1956；Fukuwatari和Shibata，2007；Chen等，1996），但是与尼克酸的添加水平没有关系（Fukuwatari和Shibata，2007；Matte等，2008）。色氨酸用于体内代谢（合成尼克酸）和蛋白质合成（生长）的比例尚需进一步研究。研究发现，饲粮中添加0.05%的色氨酸提高了早期断奶仔猪血浆尼克酰胺的水平（>45%；Matte等，2004；Matte等，2007）。Ivers等（1993）给生长育肥猪饲喂含不同水平色氨酸（0～81 mg/kg）的饲粮，并没有观测到生长性能有差异。然而，Real等（2002）发现，饲粮添加55

mg/kg色氨酸提高了猪的饲料转化率和肉品质（滴水损失、pH和肉色）。Matte等（2008）发现，饲粮中添加15 mg/kg或60 mg/kg的尼克酸，有提高4～6周龄断奶仔猪采食量和6～10周龄仔猪日增重的趋势。

3.2.3 **泛酸**

近年来，泛酸对猪生长性能的影响引起了人们的关注。Grinstead等（1998）发现，在饲粮中添加0～120 mg/kg泛酸可线性提高早期断奶仔猪平均日增重（ADG）和平均日采食量（ADFI）。事实上，泛酸（8.9μg/g乳汁或47 mg/kg母乳干物质）对哺乳仔猪也非常重要。对于生长育肥猪，当饲粮中泛酸的添加量高于维持猪最大日增重的需要量（<120 mg/kg）时，能量由维持脂肪沉积转向维持蛋白质合成，从而导致猪的体成分发生改变（Stahly和Lutz，2001；Autrey等，2002；Santoro等，2006）。然而，有关泛酸对猪肉组成影响的研究结果并不一致。有些试验发现，30～90 mg/kg泛酸并未改变猪肉组成（Radcliffe等，2003；Yang等，2004；Saddoris等，2005；Groesbeck等，2007）。

3.2.4 **胆碱和甜菜碱**

尽管饲粮中胆碱的添加量比其他维生素高很多，但是人们仍然把胆碱视为B族维生素。胆碱并不符合维生素的经典定义，因为在猪体内可由甲基化途径合成，其甲基来自：S-腺苷甲硫氨酸（SAM）转化生成S-腺苷同型半胱氨酸（SAH）；合成磷脂酰胆碱需要三个分子的SAM（三个甲基）（Stead等，2006）。胆碱主要以磷脂形式参与构建哺乳动物的细胞膜（Zeisel等，2006）。胆碱的另一种形式为乙酰胆碱，具有神经递质功能。磷脂酰胆碱（又称卵磷脂）通过氧化生成甜菜碱为机体提供甲基。甜菜碱对动物的营养生理功能详见Eklund等（2005）。简单地说，甜菜碱是甘氨酸的三甲基衍生物，在体内由胆碱氧化生产或由饲粮提供。它是一种强效渗透压保护剂，在甲基转移反应中是一种甲基供体。后面这个代谢途径可以调节体内同型半胱氨酸的水平，也被称为甜菜碱-同型半胱氨酸转甲基途径（BHMT；Finkelstein，1990）。然而BHMT似乎并不能降低生长育肥猪体内同型半胱氨酸的水平（Skomial等，2004），这可能与BHMT在组织的分布有关（仅在猪肝脏和肾脏中；Delgado-Reyes等，2001），也可能是因为肝脏和肾脏BHMT相关的酶对饲粮含硫氨基酸、胆碱或甜菜碱不敏感造成的。

近来，研究人员报道了几种动物（包括猪）体内胆碱和甜菜碱的功能及这两种甲基供体与蛋氨酸的相互作用（Simon，1999）。自从NRC（1998）发布后，有研究用植物卵磷脂替代饲粮中的氯化胆碱，因为氯化胆碱可对其他营养素造成破坏。事实上，人类对卵磷脂的吸收要慢于对氯化胆碱的吸收。然而与人类不同，猪对这两种来源的胆碱的吸收并不存在差异（Jakob等，1998）。研究发现，饲粮中添加氯化胆碱（0 和0.2%）或卵磷脂（0和1.0%、1.5%、2.0%、2.5%）并不影响猪的生长性能和肉品质（Kuhn等，1998）。

3.2.5 叶酸和维生素B_{12}

研究表明，肠外途径获得叶酸或口服叶酸均可提高仔猪、生长猪和育肥猪的生长性能（Matte等，1990；Lindemann和Kornegay，1986；Matte等，1993），但是也有研究并未得出类似结果（Letendre等，1991；Gannon和Liebholz，1990）。最近，有研究表明，叶酸、维生素B_{12}和蛋氨酸的相互作用对生长育肥猪的生长性能和肉品质有影响（Giguère等，2008）。尽管添加叶酸对生长性能、肉品质、猪肉抗氧化能力没有影响，但却提高了肌肉叶酸的含量（+22%），并降低了同型半胱氨酸的水平。叶酸降低同型半胱氨酸水平的作用机制值得进一步研究，因为这对于人类营养和健康非常重要（Giguère等，2008）。

生长育肥猪维生素B_{12}的需要量是依据1966年前开展的试验研究制定的，是以增加动物蛋白质沉积潜力和饲粮蛋白质水平为指标确定维生素B_{12}的需要量的。摄入的氨基酸（如蛋氨酸和赖氨酸）比例不平衡时可通过添加维生素B_{12}得以缓解（ARC，1981）。给26日龄断奶仔猪肌内注射20μg/kg维生素B_{12}[相当于NRC（1988）推荐量的100倍]，4周后仔猪的体增重和采食量分别提高了14%和12%（Wilson等，1991）。House和Fletcher（2003）通过检测生长性能和代谢指标（主要是同型半胱氨酸），发现5 ~ 10 kg断奶仔猪需要35μg/kg的晶体维生素B_{12}。前已述及，饲粮中维生素B_{12}的含量超过25μg/kg时，生长育肥猪的生长性能、肉品质或代谢指标（特别是同型半管氨酸）均不受影响（Giguère等，2008）。然而，饲粮中添加0.150μg维生素B_{12}可使猪肉维生素B_{12}的含量提高55%。

4 结 语

维生素作为一种营养素，与氨基酸和脂肪酸一样为动物的生长和健康所必需。维生素代谢的重要性因猪生理阶段的不同（仔猪、生长猪或种猪）而有所变化。

现代集约化养殖技术提高了动物的生产力，因此需增强体内的合成代谢以维持生长、妊娠和泌乳处于最佳水平。可见，动物的营养需要量可能需要作出调整以适应动物生产力提高的需要，然而相关的研究资料非常有限且亟待更新。政府机构和个人采用的动物营养需要量或推荐量差别非常大。同样，维生素需要量的相关研究资料的缺乏，导致了以经验方式确定维生素的添加水平。尽管在生产实践中并不存在维生素缺乏的风险，但确定猪维生素的需要量仍将是一项具有挑战性的研究。与繁殖性能和生长性能指标相比，饲料成本是影响维生素添加量的重要因素。除此之外，其他相关问题也应该加以考虑。比如，妊娠期和哺乳期母体维生素转移到胎儿和仔猪的效率也是非常关键的问题，因为胎儿和哺乳仔猪的维生素完全是从母体获得的。另外，维生素也可能影响猪的健康状况，因为已发现有些维生素会影响动物的免疫反应。此外，维生素也影响肉品质，包括提高猪肉的稳定性（如维生素E）和生产富含维生素的猪肉（如有些B族维生素），后者是猪肉促销的一种市场手段。

作者：J. Jacques Matte和
Charlotte Lauridsen
译者：易丹

参考文献

Adams, C. R., C. E. Richardson, and T. J. Cunha. 1967. Supplemental biotin and vitamin B_6 for swine. J. Anim. Sci. 26:903. (Abstr.)

Adams, K. L., F.W. Bazer, and R. M. Roberts. 1981. Progesterone-induced secretion of a retinol-binding protein in the pig uterus. J. Reprod. Fertil. 62:39-47.

Adamstone, F. B., J. L. Krider, and M. F. James. 1949. Response of swine to vitamin E-deficient rations. Ann. N.Y. Acad. Sci. 52:260-268.

Anderson, G. C., and A. C. Hogan, 1950. Adequacy of synthetic diets for reproduction of swine. Proc. Soc. Exp. Biol. Med. 75:288-290.

Antipatis, C., A. M. Finch, and C. J. Ashworth. Effect of controlled alterations in maternal dietary retinol on foetal and neonatal retinol status on pregnancy outcome in pigs. Livest. Sci. 118:247-254.

Apple, J. K. 2007. Effects of nutritional modifications on the water-holding capacity of fresh pork: a review. J. Anim. Breed. Gen. 124:43-58.

ARC. 1981. The Nutrient Requirements of Pigs. Agricultural Research Council. Commonwealth Agricultural Bureaux, Slough, UK.

Armocida, A., P. Beskow, P., Amcoff, A. Kallner, and S. E*km*an. 2001. Vitamin C plasma concentrations and leg weakness in the forelegs of growing pigs. J. Vet. Med. Ser. A-Phys Path. Clin. Med. 48:165-178.

Asghar, A., J. I. Gray, A. M. Booren, E. A. Gomaa, and M.M. Abouzied, E. R. Miller, and D. J. Buckley. 1991. Effects of surpranutritional dietary vitamin E levels on subcellular deposition of α-tocopherol in the muscle and on pork quality. J. Agric. Sci. 57:31-41.

Audet, I., J. P. Laforest, G. P. Martineau, and J. J. Matte. 2004. Effect of vitamin supplements on some aspects of performance, vitamin status, and semen quality in boars. J. Anim Sci. 82:626-633.

Audet, I., N. Bérubé; J. L. Bailey, J. -P. Laforest, H. Quesnel, and J. J. Matte. 2009a. Effects of dietary vitamin supplementation and semen collection frequency on hormonal profile during ejaculation in the boar. Theriogenology 71:334-341.

Audet, I., N. Bérubé; J. L. Bailey, J.-P. Laforest, and J. J. Matte. 2009b. Effects of dietary vitamin supplementation and semen collection frequency on reproductive performance and semen quality in boars. J. Anim. Sci. 87:1960–1970.

Autrey, B. A., T. S. Stahly, and T. R. Lutz. 2002. Efficacy of pantothenic acid as a modifier of body composition in pigs. J. Anim. Sci. 80(Suppl. 1):163. (Abstr.)

Ballance, D. M., and J. D. House. 2005. Development of the enzymes of homocysteine metabolism from birth through weaning in the pig. J. Anim. Sci. 83(Suppl. 1):160. (Abstr.)

Barkow, B., J. J. Matte, H. Böhme, and G. Flachowsky. 2001b. Influence of folic acid supplements on the carry–over of folates from the sow to the piglet. Br. J. Nutr. 85:179–184.

Bässler, K. H. 1997. Enzymatic effects of folic acid and vitamin B_{12}. Int. J. Vit. Nutr. Res. 67:385–388.

BASF. 2001. Fortification en vitamines des aliments porcs et volailles au Canada. Page 17 in Séminaire Technique. M. Duval, ed. St–Pie de Bagot, QC, Canada.

Bazer, F. W. and M. T. Zavy. 1988. Supplemental riboflavin and reproductive performance of gilts. J. Anim. Sci. 66(Suppl. 1):324. (Abstr.)

Bender, D. A. 1987. Oestrogens and vitamin B_6–Actions and interactions. World Rev. Nutr. Diet. 51:140–188.

Benedikt, J., D. A. Roth–Maier and M. Kirchgessner. 1996. Influence of dietary vitamin B_6 supply during gravidity and lactation on total vitamin B_6 concentration (pyridoxine, pyridoxal and pyridoxamine) in blood and milk. Int. J. Vit. Nutr. Res. 66:146–150.

Bethke, R. M., W. Burroughs, O. H. M. Wilder, B. H. Edgington, and W. L. Robison. 1946. The comparative efficacy of vitamin D from irradiated yeast and cod liver oil for growing pigs, with observations on their vitamin D requirements. Ohio Agricultural Experiment Station Bulletin 667:1–29. Wooster, OH.

Bilodeau, R., J. J. Matte, A. M. B. de Passillé, C. L. Girard, and G. J. Brisson. 1989. Effects of floor type on serum folates, serum vitamin B_{12}, plasma biotin and on growth performances of pigs. Can. J. Anim. Sci. 69:779–788.

Böhmer, B. M., and D. A. Roth–Maier. 2007. Effects of high–level dietary B–vitamins on performance, body composition and tissue vitamin contents of growing finishing pigs. J. Anim. Physiol. Anim. Nutr. 91:6–10.

Bonette, E. D., E. T. Kornegay, M. D. Lindemann, and C. Hammerberg. 1990. Humoral and cell–mediated immune response and performance of weaned pigs fed four supplemental vitamin E levels and housed at two nursery temperatures. J. Anim. Sci. 58:1337–1345.

Brady, P. S., L. J. Brady, M. J. Parsons, D. E. Ullrey, and E. R. Miller. 1979. Effects of riboflavin deficiency on growth and glutathione peroxidase system enzymes in the baby pig. J. Nutr. 109:1615–1622.

Branner, G. R. and D. A. Roth–Maier. 2006. Influence of pre–, pro–, and synbiotics on the intestinal availability of different B–vitamins. Arch. Anim. Nutr. 60:191–204.

Braude, R., A. S. Foot, K. M. Henry, S. K. Kon, S. Y. Thompson, and T. H. Mead. 1941. Vitamin A studies with rats and pigs. Biochem. J. 35:693–707.

Braude, R., S. K. Kon, and J. W. G. Porter. 1950. Studies in the vitamin C metabolism of the pig. Br. J. Nutr. 4:186–197.

Bretzinger, J. 1991. Pyridoxine supply of early weaned piglets. Doctorate Thesis in Veterinary Medicine. Ludwig–Maximillians University of Munich, Munich, Germany.

Brzezinska–Slebodzinska, E., A. B. Slebodzinski, B. Pietras, and G. Wieczorek. 1995. Antioxidant effect of vitamin E and glutathione on lipid peroxidation in boar semen plasma. Biol. Trace Elem. Res. 47:69–74.

Brief, S., and B. P. Chew. 1985. Effects of vitamin A and β–carotene on reproductive performance in gilts. J. Anim. Sci. 60:998–1004.

Brooks, P. H. 1986. The role of biotin in intensive systems of pig production. In Proc. 6th Intl. Conf. Production Disease in Farm Animals. Belfast, Northern Ireland.

BSAS. 2003. Nutrient requirement standards for pigs. Whittemore C. T., M. J. Hazzledine, and W. H. Close, auth. BSAS, Penicuik, Midlothian, UK.

Buhi, W., F. W. Bazer, C. Ducsay, P. W. Chun, and R. M. Roberts. 1979. Iron content, molecular weight and possible function of the progesterone–induced purple glycoprotein. Fed. Proc. 38:733–733.

Burton, G. W., U. Wronska, L. Stone, D. O. Foster, and K. U. Ingold. 1990. Biokinetics of dietary RRR–α–tocopherol in the male guinea pig at three dietary levels of vitamin C and two levels of vitamin E. Evidence that vitamin C does not "spare" vitamin E *in vivo*. Lipids. 25:199–209.

Cera, K. R., D. C. Mahan, and G. A. Reinhart. 1988. Weekly digestibilities of diets supplemented with corn oil, lard, or tallow by weanling swine. J. Anim. Sci. 66:1430–1437.

Charlton S. J., and W. N. Ewing. 2007. The vitamins directory. Page 150 in *Context*. Products Ltd. Leicestershire, UK.

Cheah, K. S., A. M. Cheah, and D. I. Krausgrill. 1995. Effect of dietary supplementation of vitamin E on pig meat quality. Meat Sci. 39:255–264.

Chen, B. J., T. F. Shen, and R. E. Austic. 1996. Efficiency of tryptophan–niacin conversion in chickens and ducks. Nutr. Res. 16:91–104.

Chew, B. P., T. S. Wong, J. J. Michal, F. R. Standaert, and L. R. Heriman. 1991a. Kinetic characteristics of β–carotene uptake after an injection of β–carotene in pigs. J. Anim. Sci. 69:4883–4891.

Chew, B. P., T. S. Wong, J. J. Michal, F. R. Standaert, and L. R. Heriman. 1991b. Subcellular distribution of beta–carotene, retionol, and alpha–tocopherol in porcine lymphocytes after a single injection of beta–carotene. J. Anim. Sci. 69:4892–4897.

Chew, B. P. 1993. Effects of supplemental β-carotene and vitamin A on reproduction in swine. J. Anim. Sci. 71:247–252.

Ching, S., D. C. Mahan, T. G.Wisemann, and N. D. Fastinger. 2002. Evaluating the antioxidant status of weanling pigs fed dietary vitamin A and E. J. Anim. Sci. 80:2396–2401.

Chung, Y. K., D. C. Mahan, and A. J. Lepine. 1992. Efficacy of dietary D-α-tocopherol and Dl-α-tocopheryl acetate in for weanling pigs. J. Anim. Sci. 70:2485–2492.

Coburn, S. P. 1994. Advances in Research and Applications. Pages 259–300 in Vitamins and Hormones. vol 48. G. Litwack, ed. Academic Press, New York.

Coffey, M. T., and J. H. Britt. 1989. Effect of β-carotene injection on reproductive performance of sows. J. Anim. Sci. 67(Suppl.1):251. (Abstr.)

Coffey, M. T., and J. H. Britt. 1993. Enhancement of sow reproductive performance by β-carotene or vitamin A. J. Anim. Sci. 71:1198–1202.

Cox, N. M. 1997. Control of follicular development and ovulation rate in pigs. J. Reprod. Fertil. 52(Suppl.):31–46.

Cranwell, P. D., and P. J. Moughan. 1989. Biological limitations imposed by the digestive system to the growth performance of weaned pigs. Page 149 in Manipulating Pig Production, vol. II. J. L. Barnett and D. P. Hennessy, eds. Australian Pig Science Association, Victoria, Australia.

Cunha, T. J., O. B. Ross, P. H. Phillips, and G. Bohstedt. 1944. Further observations on the dietary insufficiency of a corn-soybean ration for reproduction of swine. J. Anim. Sci. 3:415–421.

De Rodas, B. Z., C. V. Maxwell, M. E. Davis, S. Mandali, E. Broe*km*an, and B. J. Stoeker. 1998. L-ascorbyl-2-polyphosphate as a vitamin C source for segrated and conventionally weaned pigs. J. Anim. Sci. 76:1636–1643.

Delgado-Reyes, C. V., M. A. Wallig, and T. A. Garrow. 2001. Immunohistochemical detection of betaine-homocysteine S-methyltransferase in human, pig, and rat liver and kidney. Arch. Biochem. Biophys. 393:184–186.

de Passillé, A. M. B., G. Pelletier, J. Menard, and J. Morisset. 1989. Relationships of weight gain and behavior to digestive organ weight and enzyme activities in piglets. J. Anim. Sci. 67:2921–2929.

DiSimone, N., P. Riccardi, N. Maggiano, A. Piacentani, M. D'Asta, A. Capelli, and A. Caruso, 2004. Effect of folic acid on homocysteine-induced trophoblast apoptosis. Mol. Hum. Reprod. 10:665–669.

Donovan, S. M., M. H. Mar, and S. H. Zeisel. 1997. Choline and choline ester concentrations in porcine milk throughout lactation. Nutr. Biochem. 8:603–607.

Driancourt, M. A., F. Martinat-Botté, and M. Terqui. 1998. Contrôle du taux d'ovulation chez la truie: l'apport des modèles hyperprolifiques. INRA Prod. Anim. 11:221–226.

Dugas, B., A. Mercenier, I. Lenoir-Wijnkoop, C. Arnaud, N. Dugas, and E. Postaire. 1999. Immunity and probiotics. Immunol. Today 20:387-390.

Duquette, J., J. J. Matte, C. Farmer, C. L. Girard and J. P. Laforest. 1997. Pre- and post-mating dietary supplements of folic acid and uterine secretory activity in gilts. Can. J. Anim. Sci. 77:415-420.

Easter, R. A., P. A. Anderson, E. J. Michel, and J. R. Corley. 1983. Response of gestating gilts and starter, grower and finisher swine to biotin, pyridoxine, folacin and thiamine additions to corn-soybean meal diets. Nutr. Rep. Internat. 28:945-950.

Eichenberger, B., H. P. Pfirter, C. Wenk, and S. Gebert. 2004. Influence of dietary vitamin E and C supplementation on vitamin E and C content and thiobarbituric acid reactive substances (TBARS) in different tissues of growing pigs. Arch. Anim. Nutr. 58:195-208.

Eklund M., E. Bauer, J. Wanatu, and R. Mosenthin. 2005. Potential nutritional and physiological functions of betaine in livestock. Nutr. Res. Rev. 18:31-48.

Engstrom, G. W., and T. Littledike. 1986. Vitamin D metabolism in the pig. Swine in Biomed. Res. 1091-1112.

Ensminger, M. E., J. P. Bowland, and T. J. Cunha. 1947. Observations on the thiamine, riboflavin, and choline needs of sows for reproduction. J. Anim. Sci. 6:409-423.

Ensminger, M. E., R. W. Colby, and T. J. Cunha. 1951. Effect of certain B-complex vitamins on gestation and lactation in swine. Washington Agric. Exp. Sta., Sta. Circ. 134:1-35.

Fernandez-Duenas, D. M., G. Mariscal, E. Ramirez, and Cuaron. 2008. Vitamin C and β-carotene in diets for pigs at weaning. Anim. Feed. Sci. Tech. 146:313-326.

Finkelstein, J. D. 1990. Methionine metabolism in mammals. J. Nutr. Biochem. 1:228-237.

Firth, J., and C. Johnson. 1956. Quantitative relationships of tryptophan and nicotinic acid in the baby pig. J. Nutr. 59:223-234.

Frank, G. R., R. A. Easte, and J. M. Bahr. 1984. Riboflavin requirement of gestating swine. J. Anim. Sci. 59:1567-1572.

Frederick, G. L., and G. J. Brisson. 1961. Some observations on the relationship between vitamin B_{12} and reproduction in swine. Can. J. Anim. Sci. 41:212-219.

Friendship, R. M., and M. R.Wilson. 1991. Effects of intramuscular injections of folic acid in sows on subsequent litter size. Can. Vet. J. 32:565-566.

Fukuwatari, T., and K. Shibata. 2007. Effect of nicotinamide administration on the tryptophan-nicotinamide pathway in humans. Intern. J. Vit. Nutr. Res. 77:255-262.

Gannon, N. J., and J. Leibholz. 1990. Manipulating Pig Production II. Page 136 in Proc. of the Biennal Conf. of the Australasian Pig Sci. Assoc., November 27-29, 1989. Sydney, Australia.

Garcia-Castillo, R. F., J. L. Jasso-Pitol, R. Morones-Reza, J. R. Kawas-Garza, and J. Salinas-Chavira. 2006. Adicion de altos niveles de biotina en dietas para cerdas puberes y gestantes. Agronomia Mesoamericana. 17:1–5.

Giguère, A., C. L. Girard, R. Lambert, J. P. Laforest, and J. J. Matte. 2000. Reproductive performance and uterine prostaglandin secretion in gilts conditioned with dead semen and receiving dietary supplements of folic acid. Can. J. Anim. Sci. 80:467–472.

Giguère, A., C. L. Girard, and J. J. Matte. 2002. Erythrocyte glutathione reductase activity and riboflavin nutritional status in early–weaned piglets. Internat. J. Vit. Nutr. Res. 72:383–387.

Giguère, A., C. L. Girard, and J. J. Matte. 2008. Methionine, folic acid and vitamin B–12 in growing–finishing pigs: Impact on growth performance and meat quality. Arch. Anim. Nutr. 62:193–206.

GfE (Gesellschaft für Ernärungsphysiologie). 1987. Energie–und Nährstoffbedarf landwirtschaftlicher Nutztiere, Schweine. DLGVerlag, Frankfurt/Main, Germany.

Goff, J. P., R. L. Horst, and E. T. Littledike. 1984. Effect of sow vitamin D status at parturition on the vitamin D status of neonatal piglets. J. Nutr. 114:163–169.

Gondret, F., L. Lefaucheur, I. Louveau, B. Lebret, X. Pichodo, and Y. Le Cozler. 2005. Influence of piglet birth weight on postnatal growth performance, tissue lipogenic capacity, and muscle histological traits at market weight. Livest. Prod. Sci. 93:137–146.

Greer, E. B., and C. E. Lewis. 1978. Mineral and vitamin supplementation of diets for growing pigs. I. Wheat–based diets and the effect of preventing cross–coprophagy on the response to supplementation. Aust. J. Exp. Agric. Anim. Husb. 18:688–697.

Greer, E. B., J. M. Leibholz, D. I. Pickering, R. E. Macoun, and W. L. Bryden. 1991. Effect of supplementary biotin on the reproductive performance, body condition and foot health of sows on three farms. Aust. J. Agric. Res. 42:1013–1021.

Grinstead, G. S., R. D. Goodband, J. L. Nelssen, M. D. Tokach, S. S. Dritz, and R. Stott. 1998. Effects of increasing pantothenic acid on growth performance of segregated early–weaned pigs. Pages 87–89 in Kansas State Swine Day Rep. Kansas State Univ. Agric. Exp. Stn. Coop. Ext. Serv., Manhattan, KS.

Groesbeck, C. N., R. D. Goodband, M. D. Tokach, S. S. Dritz, J. L. Nelssen, and J. M. Derouchey. 2007. Effects of pantothenic acid on growth performance and carcass characteristics of growing–finishing pigs fed diets with or without ractopamine hydrochloride. J. Anim. Sci. 85:2492–2497.

Guay, F., J. J. Matte, C. L. Girard, M. F. Palin, A. Giguère, and J. P. Laforest. 2002a. Effect of folic acid and glycine supplementation on embryo development and folate metabolism during early pregnancy in pigs. J. Anim. Sci. 80:2134–2143.

Guay, F., J. J. Matte, C.L. Girard, M.F. Palin, A. Giguère, and J. P. Laforest. 2002b. Effect of folic acid and vitamin B_{12} supplements on folate and homocysteine metabolism in pigs during early pregnancy in pigs. Br. J. Nutr. 88:253–263.

Guay, F., J. J. Matte, C. L. Girard, M. F. Palin, A. Giguère, and J. P. Laforest. 2004a. Effects of folic acid supplement on uterine prostaglandin metabolism and interleukin-2 expression on day 15 of gestation in white breed and crossbred Meishan sows. Can. J. Anim. Sci. 84:63-72.

Guay, F., J. J. Matte, C. L. Girard, M. F. Palin, A. Giguère, and J. P. Laforest. 2004b. Effect of folic acid plus glycine supplement on uterine prostaglandin and endometrial granulocyte-macrophage colony-stimulating factor expression during early pregnancy in pigs. Theriogenology. 61:485-498.

Håkansson, J., J. Hakkarainen, and N. Lundeheim. 2001. Variation in vitamin E, glutathione peroxidase and retinol concentrations in blood plasma of primiparous sows and their piglets, and in vitamin E, selenium and retinol contents in sows' milk. Acta Agric. Scand. A. Anim. Sci. 51:224-234.

Hall, D. D., G. L. Cromwell, and T. S. Stahly. 1991. Effects of dietary calcium, phosphorus, calcium:phosphorus ratio and vitamin K on performance, bone strength and blood clotting status of pigs. J. Anim. Sci. 69:646-655.

Halloran, B. P., and H. F. DeLuca. 1983. Effect of vitamin D deficiency on fertility and reproductive capacity in female in the rat. J. Nutr. 110:1573-1580.

Hartvig, P., K. G. Lindner, P. Bjurling, B. Langstrom, and J. Tedroff. 1995. Pyridoxine effect on synthesis rate of serotonin in the monkey brain measured with positron emission tomography. J. Neural. Transm. [Gen. Sect.] 102:91-97.

Hedemann, M. S., and S. K. Jensen. 1999. Vitamin E status in newly weaned piglets is correlated to the activity of carboxyl ester hydrolase. Page 181 in Proc. Aust. Pig Sci. Assoc. Adelaide, Australia.

Hewison, M., M. A. Gacad, J. Lemire, and J. S. Adams. 2001. Vitamin D as a cytokine and a hematopoietic factor. Rev. Endocrinol. Metab. Disorders. 2:217-227.

Hickie, J. P., D.M. Lavigne, andW. D.Woodward. 1983. Reduced fecundity of vitamin D deficient rats. Comp. Biochem. Physiol. A. 74:923-925.

Hidiroglou, M., and T. R. Batra. 1995. Concentration of vitamin C in milk of sows and in plasma of piglets. Can. J. Anim. Sci. 75:275-277.

Hoey, L., H. Mcnulty, and J. J. Strain. 2009. Studies of biomarker responses to intervention with riboflavin: A systematic review. Am. J. Clin. Nutr. 89:S1960-S1980.

Hoppe, P. P. 1988. Lack of vitamin K can kill weaners. Pig International, June, 16-18.

Hoppe, P. P., F. J. Schoner, and M. Frigg. 1992. Effects of dietary retinol on hepatic retinol storage and on plasma and tissue α-tocopherol in pigs. Internat. J. Vit. Nutr. Res. 62:121-129.

Horst, R. L., and E. T. Littledike. 1982. Comparison of plasma concentrations of vitamin D and its metabolites in young and aged animals. Comp. Biochem. Physiol. 73B:485-489.

Hoskinson, C. D., B. P. Chew, and T. S. Wong. 1992. Effects of injectable beta-carotene and vitamin A on mitogen induced lymphocyte proliferation in the pig *in vivo*. Biol. Neon. 62:325-336.

House, J. D., and C. M. T. Fletcher. 2003. Response of early weaned piglets to graded levels of dietary cobalamin. Can. J. Anim. Sci. 83:247-255.

INRA. 1984. Alimentation des animaux monogastriques: Porcs, lapins, volailles, Institut National de la Recherche Agronomique.

Ivers D. J., S. L. Rodhouse, M. R., Ellersieck, and T. L. Veum. 1993. Effect of supplemental niacin on sow reproduction and sow and litter performance. J. Anim. Sci. 71:651-655.

Jakob, S., R. Mosenthin, G. Huesgen, J. Kinkeldei, and K. J. Poweleit. 1998. Diurnal pattern of choline concentrations in serum of pigs as influenced by dietary choline or lecithin intake. Z. Ernähr. 37:353–357.

Jakobsen, J., H. Maribo, A. Bysted, H. M. Sommer, and O. Hels. 2007. 25-Hydroxyvitamin D_3 affects vitamin D status similar to vitamin D_3 in pigs-but the meat produced has a lower content of vitamin D. Br. J. Nutr. 98:908-913.

Jensen, M., A. Lindholm, and R. V. J. Hakkareinen. 1990. The vitamin E distribution in serum, liver, adipose and muscle tissues in the pig during depletion and repletion. Acta Vet. Scand. 31:129-136.

Jensen, C., J. Guidera, I. M. Skovgaard, H. Staun, L. H. Skibsted, S. K. Jensen, A. J. Moller, J. Bucklery, and G. Bertelsen. 1997. Effects of dietary alpha-tocopheryl acetate supplementation on alpha-tocopherol deposition in porcine m psoas major and m longissimus dorsi and on drip loss, colour stability and oxidative stability of pork meat. Meat Sci. 55:491-500.

Jensen, C., C. Lauridsen, and G. Bertelsen. 1998. Dietary vitamin E: quality and storage stability of pork and poultry. Trends in Food Sci. Technol. 9:62-72.

Kim, E. S., J. S. Seo, J. H. Eum, J. E. Lim, E. L., D. H. Kim, T. K. Youn, and D. R. Lee. 2009. The effect of folic acid on *in vitro* maturation and subsequent embryonic development of porcine immature oocytes. Mol. Reprod. Dev. 76:120-121.

Kindberg, C. G., and J. W. Suttie. 1989. Effect of various intakes of phylloquinone on signs of vitamin K deficiency and serum liver phylloquinone concentrations in the rat. J. Nutr. 119:175-180.

Knights, T. E. N., R. R. Grandhi, and S. K. Baidoo. 1998. Interactive effects of selection for lower bac*kf*at and dietary pyridoxine levels on reproduction, and nutrient metabolism during the gestation period in Yorkshire and Hampshire sows. Can. J. Anim. Sci. 78:167-173.

Kopinski, J. S. and J. Leibholz. 1989. Biotin studies in pigs. II. The biotin requirement of the growing pig. Br. J. Nutr. 62:761-772.

Kornegay, E. T. 1986. Biotin in swine production: A review. Livest. Prod. Sci. 14:65-89.

Kösters, W. W., and M. Kirchgessner. 1976. Change in feed intake of early-weaned piglets in response to different vitamin B_6 supply. Z. TierphysiologieTiernährung und Futtermittelkde. 37:247-254.

Kostoglou, P., S. C. Kyriakis, A. Papasteriadis, N. Roumpies, C. Alexopoulos, and K. Saoulidis. 2000. Effect of β-carotene on health status and performance of sows and their litters. J. Anim. Physiol. A. Anim. Nutr. 83:150-157.

Kotake,Y., T. Sotokawa,A.Hisatake, M. Abeand, and Y. Ikeda. 1968. Studies on xanthurenic acid insulin complex. II. Physiological activities. J. Biochem. (Tokyo) 63:578-581.

Kovčin, S., S. Fivkovič, M. Beukovič and M. Lalič. 1988. Uticaj folne kiseline na reprodukciju krmaˇca. Zbornik radova br. 17-18. Institut za stočarstvo, Novi Sad, Serbia.

Kuhn, M., B. Frohmann, A. Petersen, K. Rubesam, and C. Jatsch. 1998. Utilization of crude soybean lecithin as a native choline source in feed rations of fattening pigs. Fett/Lipid 100:78-84.

Kwiecinski, G. C., G. I. Petri, and H. F. DeLuca. 1989. 1, 25-dihydroxyvitamin D_3 restores fertility of vitamin D-deficient female rats. Am. J. Physiol. 256:E483-E487.

Lahucky, R., I. Bahelka, U. Kuechenmeister, K. Vasickova, K. Nuernberg, K. Ender, and G. Nuernberg. 2007. Effects of dietary supplementation of vitamin D_3 and E on quality characteristics of pigs and longissimus muscle antioxidative capacity. Meat Sci. 77:264-268.

Lauridsen, C., M. S. Hedemann, and S. K. Jensen. 2001. Hydrolysis of tocopherol and retinyl esters by porcine carboxyl ester hydrolase is affected by their carboxylate moiety and bile acids. J. Nutr. Biochem. 12:219-224.

Lauridsen, C., H. Engel, A. M. Craig, and M. G. Traber. 2002a. Relative bioactivity of dietary RRR- and all-rac-α-tocopheryl acetates in swine assessed with deuterium-labeled vitamin E. J. Anim. Sci. 80:702-707.

Lauridsen, C., H. Engel, S. K. Jensen, A. M. Craig, and M. G. Traber. 2002b. Lactating sows and suckling piglets preferentially incorporate RRR- over all-rac-α-tocophol into milk, plasma and tissues. J. Nutr. 132:1258-1264.

Lauridsen, C., and S. K. Jensen. 2005. Influence of supplementation of all-rac-α-tocopheryl acetate preweaning and vitamin C postweaning on α-tocopherol and immune responses of piglets. J. Anim. Sci. 83:1274-1286.

Lauridsen, C., T. Larsen, U. Halekoh, and S. K. Jensen. 2010a. Reproductive performance and bone status markers of gilts and lactating sows supplemented with two different forms of vitamin D. J. Anim. Sci. 88:202-213.

Lauridsen, C. 2010b. Evaluation of the effect of increasing dietary vitamin E in combination with different fat sources on performance, humoral immune responses and antioxidant status of weaned pigs. Anim. Feed Sci. Technol. 158:85-94.

Le Cozler, Y., X. Pichodo, H. Roy, C. Guyomarc'h, H. Pellois, N. Quiniou, I. Louveau, B. Lebret, L. Lefaucheur and F. Gondret. 2004. Influence du poids individuel et de la taille de la portée à la naissance sur la survie du porcelet, ses performances de croissance et d'abattage et la qualité de la viande. Journées Rech. Porcine en France. 36:443–450.

Lee, N. S., G. Muhs, G. C. Wagner, R. D. Reynolds, and H. Fisher. 1988. Dietary pyridoxine interaction with tryptophan or histidine on brain serotonin and histamine metabolism. Pharmacol. Biochem. Behav. 29:559–564.

Le Grusse, J., and B. Watier. 1993. Les vitamines. Données biochimiques, nutritionnelles et cliniques. Centre d' Études et d'Information sur les Vitamines. Produits Roche, Neuilly sur Seine, France.

Leonhardt, M., S. Gerbert, and C. Wenk. 1996. Stability of α–tocopherol, thiamine, riboflavin and retinol in pork muscle and liver during heating as affected by dietary supplementation. J. Food Sci. 61:1048–1052.

Lessard, M., M. Dupuis, N. Gagnon, E. Nadeau, J. J. Matte, J. Goulet, and J. M. Fairbrother. 2009. Administration of Pediococcus Acidilactici or Saccharomyces Cerevisiae Boulardii modulates development of porcine mucosal immunity and reduces intestinal bacterial translocation after Escherichia Coli challenge. J. Anim. Sci. 87:922–934.

Letendre, M., C. L. Girard, J. J. Matte and J.–F. Bernier. 1991. Effects of intramuscular injections of folic acid on folates status and growth performance of weanling pigs. Can. J. Anim. Sci. 71:1223–1231.

Lewis, A. J., and L. L. Southern. 2001. Page 1009 in Swine Nutrition. CRC Press. Boca Raton, Florida.

Lewis A. J., G. L. Cromwell, and J. E. Pettigrew. 1991. Effects of supplemental biotin during gestation and lactation on reproductive performance of sows: a cooperative study. J. Anim. Sci. 69:207–214.

Lindemann, M. D. 1993. Supplemental folic acid: A requirement for optimizing swine reproduction. J. Anim. Sci. 71:239–246.

Lindemann, M. D., and E. T. Kornegay. 1986. Effect of folic acid additions to weanling pig diets. Va. Tech. Livestock Res. Rep. 20–22.

Lindemann, M. D., and E. T. Kornegay. 1989. Folic acid supplementation to diets of gestating and lactating swine over multiple parities. J. Anim. Sci. 67:459–464.

Lindemann, M. D., J. H. Brendemuhl, L. I. Chiba, C. S. Darroch, C. R. Dove, M. J. Estienne, and A. F. Harper. 2008. A regional evaluation of injections of high levels of vitamin A on reproductive performance of sows. J. Anim. Sci. 86:333–338.

Liu, Y. Y., Y. Shigematsu, I. Bykov, A. Nakai, Y. Kikawa, T. Fukui, and M. Sudo. 1994. Abnormal fatty acid composition of lymphocytes of biotin–deficient rats. J. Nutr. Sci. Vitaminol.

40:283–288.

Lombardi–Boccia, G., S. Lanzi, and A. Aguzzi. 2005. Aspects of meat quality: Trace elements and B vitamins in raw and cooked meats. J. Food Comp. Anal. 18:39–46.

Loudenslager, M. J., P. K. Ku, P. A. Whetter, D. E. Ullrey, C. K. Whitehair, H. D. Stowe, E. R. Miller. 1986. Importance of diet of dam and colostrum to the biological antioxidant status and parenteral iron tolerance of the pig. J. Anim. Sci. 63:1905–1914.

Luce, W. G., R. D. Geisert, M. T. Zavy, A. C. Clutter, F. W. Bazer, C. W. Maxwell, M. D. Wiltmann, R. M. Blair, M. Fairchild, and J. Wiford. 1990. Effect of riboflavin supplementation on reproductive performance of bred sows. Page 269 in *Animal Science Report*. Oklahoma Agric. Exp. Stn., Stillwater, OK.

LUS. 2002. Normer for Naringsstoffer (in Danish), 10. udgave, Landsudvalget for Svin, Kbhv., Denmark.

Madsen, K., A. Cornish, P. Soper, C. McKaigney, H. Jijon, C., Yachimec, J. Doyle, L. Jewell, and C. DeSimone. 2001. Probiotic bacteria enhance murine and human intestinal epithelial barrier function. Gastroenterology 121:580–591.

Mahan D. C., and J. L. Vallet. 1997. Vitamin and mineral transfer during fetal development and the early postnatal period in pigs. J. Anim. Sci. 75:2731–2738.

Mahan, D. C. 1991. Assessment of the influence of dietary vitamin E on sows and offspring in three parities: Reproductive performance, tissue tocopherol, and effects on progeny. J. Anim. Sci. 69:2904–2917.

Mahan, D. C. 1994. Effects of dietary vitamin E on sow reproductive performance over a five–parity period. J. Anim. Sci. 72:2870–2879.

Mahan, D. C., Y. Y. Kim, and R. L. Stuart. 2000. Effect of vitamin E sources (RRR– or all–rac–α–tocopheryl acetate) and levels on sow reproductive performance, serum, tissue, and milk α–tocopherol contents over a five–parity period, and the effects on the progeny. J. Anim. Sci. 78:110–119.

Mahan, D. C., and L. J. Saif. 1983. Efficacy of vitamin C–supplementation for weanling swine. J. Anim. Sci. 56:631–639.

Marchetti, M., M. Tassinari, and S. Marchetti. 2000. Menadione nicotinamide bisulphite as a source of vitamin K and niacin activities for the growing pig. Anim. Sci. 71:111–117.

Marin–Guzman, J., D. C. Mahan, Y. K. Chung, J. L. Pate, and W. F. Pope. 1997. Effects of dietary selenium and vitamin E on boar performance and tissue responses, semen quality and subsequent fertilization rates in mature gilts. J. Anim. Sci. 75:2994–3003.

Marin–Guzman, J., D. C. Mahan, and J. L. Pate. 2000. Effect of dietary selenium and vitamin E on spermatogenic development in boars. J. Anim. Sci. 78:1537–1543.

Markant, A., M. Kuhn, O. P. Walz, and J. Pallauf. 1993. The intermediate relationship of nicotinamid and tryptophan in piglets. J. Anim. Physiol. Anim. Nutr. 70:225–235.

Martelli, G., L. Sardi, P. Parishini, A. Badiani, P. Parazza and A. Mordenti. 2005. The effects of a dietary supplement of biotin on Italian heavy pigs' (160 kg) growth slaughtering parameters, meat quality and the sensory properties of cured hams. Livest. Prod. Sci. 93:117–124.

Matte J. J., and C. L. Girard. 1989. Effects of intramuscular injections of folic acid during lactation on folates in serum and milk and performance of sows and piglets. J. Anim. Sci. 67:426–431.

Matte, J. J., and C. L. Girard. 1996. Changes of serum and blood volumes during gestation and lactation in multiparous sows. Can. J. Anim. Sci. 76:263–266.

Matte, J. J., and C. L. Girard. 1999. An estimation of the requirement for folic acid in gestating sows: The metabolic utilization of folates as a criterion of measurement. J. Anim. Sci. 77:159–165.

Matte, J. J., W. H. Close, and C. Pomar. 1991. The effect of interrupted suckling and split-weaning on reproductive performance of sows: A review. Livest. Prod. Sci. 30:195–212.

Matte, J. J., A. Giguère, and C. Girard. 2005. Some aspects of the pyridoxine (vitamin B_6) requirement in weanling piglets. Br. J. Nutr. 93:723–730.

Matte, J. J., C. L. Girard, and G. J. Brisson. 1984. Folic acid and reproductive performances of sows. J. Anim. Sci. 59:1020–1025.

Matte, J. J., C. L. Girard, and G. J. Brisson. 1992. The role of folic acid in the nutrition of gestating and lactating primiparous sows. Livest. Prod. Sci. 32:131–148.

Matte, J. J., C. L. Girard, and B. Sève. 2001. Effects of long term parenteral administration of vitamin B_6 on B_6 status and some aspects of the glucose and protein metabolism of early-weaned piglets. Br. J. Nutr. 85:11–21.

Matte, J. J., C. L. Girard, and G. F. Tremblay. 1993. Effect of long-term addition of folic acid on folate status, growth performance, puberty attainment, and reproductive capacity of gilts. J. Anim. Sci. 71:151–157.

Matte, J. J., F. Guay, and C. L. Girard. 2006. Folic acid and vitamin B_{12} in reproducing sows: New concepts. Can J. Anim. Sci. 86:197–205.

Matte, J. J., A. A. Ponter, and B. Sève. 1997. Effects of chronic parenteral pyridoxine and acute enteric tryptophan on pyridoxine status, glycemia and insulinemia stimulated by enteric glucose in weanling piglets. Can. J. Anim. Sci. 77:663–668.

Matte, J. J., C. Farmer, C. L. Girard, and J.-P. Laforest. 1996. Dietary folic acid, uterine function and early embryonic development in sows. Can. J. Anim. Sci. 76:427–433.

Matte, J. J., C. L. Girard, R. Bilodeau, and S. Robert. 1990. Effects of intramuscular injections of folic acid on serum folates, haematological status and growth performance of growing–finishing pigs. Reprod. Nutr. Develop. 30:103−114.

Matte, J. J., N. LeFloch, C. Relandeau, L. Le Bellego, A. Giguère, and M. Lessard. 2004. Is vitamin B_6 a modulator of the effect of supplementary tryptophan on tryptophan metabolism and growth responses in weanling pigs? J. Anim. Sci 82(Suppl.1):19−20.

Matte, J. J., N. Le Floc'h, A. Giguère, L. Le Bellego, and M. Lessard. 2007. La vitamine B_6 (pyridoxine) et son effet régulateur sur les réponses métabolique et zootechnique à un supplément de tryptophane aux porcelets en sevrage hâtif. Journées Rech. Porcine en France 39:119−124.

Matte, J. J., A. Giguère, D. Melchior, and N. LeFloc'h. 2008. Is niacin (vitamin B_3) a modulator of the effect of supplementary tryptophan on tryptophan metabolism and growth responses in early−weaned pigs? J. Anim. Sci. 86(E−Suppl. 2):177. (Abstr.)

McDowell, L. R. 2000. Vitamins in Animal Nutrition. Academic Press Inc. San Diego, CA.

McGinnis, C. H. Jr. 1994. Maintaining vitamin stability during extrusion. Feed Mix 2:10−13.

McGlone, J. J. 2000. Deletion of supplemental minerals and vitamins during the late finishing period does not affect pig weight gain and feed intake. J. Anim. Sci. 78:2797−2800.

Miller, E. R., D. A. Schmidt, J. A. Hoefer, and R. W. Luecke. 1965. Comparisons of casein and soy protein upon mineral balance and vitamin D_2 requirement of the baby pig. J. Nutr. 85:347−353.

Monahan, F. J., A. Asghar, J. I. Gray, D. J. Buckley, and P. A. Morrissey. 1994. Effect of oxidized dietary−lipid and vitamin E on the colour stability of pork chops. Meat Sci. 37:205−215.

Mosenthin, R., W. C. Sauer, L. Völker, and M. Frigg. 1990. Synthesis and absorption of biotin in the large intestine of pigs. Livest. Prod.Sci. 25:95−103.

Mosnier, E., J. J. Matte, M. Etienne, P. Ramaekers, B. Söve, and N. Le Floc'h. 2009. Tryptophan metabolism and related B vitamins in the multiparous sow fed ad libitum after farrowing. Arch. Anim. Nutr. 63:467−478.

Nakano, T., F. X. Alherne, and J. R. Thompson. 1983. Effect of dietary supplementation of vitamin C on pig performance and the incidence of osteochondrosis in elbow and stiftle joints in young growing swine. Can. J. Anim. Sci. 63:421−428.

Nielsen, H. E., V. Danielsen, M. G. Simesen, G. Gissel−Nielsen, W. Hjarde, T. Leth, and A. Basse. 1979. Selenium and vitamin E deficiency in pigs. I. Influence on growth and reproduction. Acta Vet. Scand. 20:276−288.

NRC. 1998. Nutrient Requirements of Swine, 10th rev. ed. Natl. Acad. Press. Washington, DC.

Nunetz, S. B., J. A. Medin, H. Keller, K. Wang, K. Ozato, W. Wahli, and J. Segars. 1995.

Retinoid X receptor b and peroxisome proliferator activated receptor activate an estrogen responsive element. Rec. Progr. Horm. Res. 50:465–469.

Palludan, B. 1963. Vitamin A deficiency and its effect on the sexual organs of the boar. Acta Vet. Scand. 4:166–155.

Parsons, M. J., P. K. Ku, D. E. Ullrey, H. D. Stowe, P. A. Whetter, and E. R. Miller. 1985. Effects of riboflavin supplementation and selenium source on selenium metabolism in the young pig. J. Anim. Sci. 60(Suppl. 2):451–461.

Partridge, I. G. and M. S. McDonald. 1990. A note on the response of growing pigs to supplemental biotin. Anim. Prod. 50:195–197.

Pearson, P. L., H. G. Klemcke, R. K. Christenson, and J. L. Vallet. 1998. Uterine environment and breed effects on erythropoiesis and liver protein secretion in late embryonic and early fetal swine. Biol. Reprod. 58:911–918.

Pehrson, B., H. Holmgren, and U. Trafikowska. 2001. The influence of parenterally administered α–tocopherol acetate to sows on the vitamin E status of the sows and suckling piglets and piglets after weaning. J. Vet. Med. A. 48:569–575.

Pettigrew, J. E., S. M. El–Kandelgy, L. J. Johnston, and G. C. Shurson. 1996. Riboflavin nutrition of sows. J. Anim. Sci. 74:2226–2230.

Pietrzik, K. and A. Brönstrup. 1997. Folate in preventive medicine: A new role in cardiovascular disease, neural tube defects and cancer. Ann. Nutr. Metabol. 41:331–343.

Pinelli–Saavedra, A., and J. R. Scaife. 2005. Pre– and postnatal transfer of vitamins E and C to piglets in sows supplemented with vitamin E and vitamin C. Livest. Prod. Sci. 97:231–240.

Pinelli–Saavedra, A., A. M. C. de la Barca, J. Hernandez, R. Valenzuela, and J. R. Scaife. 2008. Effect of supplementing sows' feed with alpha–tocopherol acetate and vitamin C on transfer of alpha–tocopherol to piglet tissues, colostrums, and milk: Aspects of immune status of piglets. Res. Vet. Sci. 85:92–100.

Pion, S. J., E. van Heugten, M. T. See, D. K. Larik, and S. Pardue. 2004. Effects of vitamin C supplementation on plasma ascorbic acid and oxalate concentrations and meat quality in swine. J. Anim. Sci. 84:2004–2012.

Pointillart, A., I. Denis, C. Colin, and H. Lacroix. 1997. Vitamin C supplementation does not modify bone mineral content or mineral absorption in growing pigs. J. Nutr. 127:1514–1518.

Pond, W. G., and K. A. Houpt. 1978. The Biology of the Pig. Cornell University Press, Ithaca, NY.

Pusateri, A. E., M. A. Die*km*an, and W. L. Singleton. 1999. Failure of vitamin A to increase litter size in sows receiving injections at various stages of gestation. J. Anim. Sci. 77:1532–1535.

Radcliffe, J. S., B. T. Richert, L. Peddireddi, and S. A. Trapp. 2003. Effects of supplemental pantothenic acid during all or part of the growing–finishing period on growth performance and carcass composition. J. Anim. Sci. 81(Suppl. 1):255. (Abstr.)

Real, D. E., J. L. Nelssen, J. A. Unruh, M. D. Tokach, R. D. Goodband, S. S. Dritz, J. M. DeRouchey, and E. Alonso. 2002. Effects of increasing dietary niacin on growth performance and meat quality in finishing pigs reared in two different environments. J. Anim. Sci. 80:3203–3210.

Reiner, G., B. Hertrampf, K. Kohler. 2004. Vitamin A–intoxification in the pig. Tierarztliche Praxis Ausgabe Grosstiere Nuttztiere. 32:218–224.

Riaz, M. N., M. Asif, and R. Ali. 2009. Stability of vitamins during extrusion. Crit. Rev. Food Sci. Nutr. 49:361–368.

Roberts, R. M., and F. W. Bazer. 1988. The functions of uterine secretions. J. Reprod. Fertil. 82:875–892.

Roland, D. A., Sr., and H. M. Edwards, Jr. 1971. Effect of essential fatty acid deficiency and type of dietary fat supplementation on biotin–deficient chicks. J. Nutr. 101:811–818.

Rosenvold, K., H. N. Larke, S. K. Jensen, A. Karlsson, K. Lundström, and H. J. Andersen. 2002. Manipulation of critical quality indicators and attributes in pork trough vitamin E supplementation level, muscle glycogen, reducing finishing feeding and preslaughter stress. Meat Sci. 62:485–496.

Russell, L. E., R. A. Easter, and P. J. Bechtel. 1985. Evaluation of the erythrocyte aspartate aminotransferase activity coefficient as an indicator of the vitamin B–6 status of postpubertal gilts. J. Nutr. 115:1117–1123.

Saddoris, K. L., L. Peddireddi, S. A. Trapp, B. T. Richert, B. Harmon, and J. S. Radcliffe. 2005. The Effects of Supplemental Pantothenic Acid in Grow–Finish Pig Diets on Growth Performance and Carcass Composition. Prof. Anim. Sci. 21:443–448.

Santoro, P., P. Macchioni, L. Franchi, F. Tassone, M. C. Ielo, and D. P. Lo Fiego. 2006. Effect of dietary pantothenic acid supplementation on meat and carcass traits in the heavy pig. Vet. Res. Comm. 30(Suppl. 1):383–385.

Simard, F., F. Guay, C. L. Girard, A. Giguere, J. P. Laforest, and J. J. Matte. 2007. Effects of concentrations of cyanocobalamin in the gestation diet on some criteria of vitamin B–12 metabolism in first–parity sows. J. Anim. Sci. 85:3294–3302.

Simon, J. 1999. Choline, betaine and methionine interactions in chickens, pigs and fish (Including crustaceans). World Poult. Sci. J. 55:353–374.

Skomial, J., M. Gagucki, and E. Sawosz. 2004. Urea and homocysteine in the blood serum of pigs fed diets supplemented with betaine and an enhanced level of B group vitamins. J. Anim. Feed Sci. 13(Suppl. 2):53–56.

Solymosi, F., and P. Horn. 1994. Protein content and amino acid composition of the uterine milk in swine and cattle. ActaVeterinaria Hungarica 42:487–494.

Stahly, T. S., and T. R. Lutz. 2001. Role of pantothenic acid as a modifier of body composition in pigs. J. Anim. Sci. 79(Suppl.1):68. (Abstr.)

Stahly, T. S., N. H. Williams, T. R. Lutz, R. C. Ewan, and S. G. Swenson. 2007. Dietary B vitamin needs of strains of pigs with high and moderate lean growth. J. Anim. Sci. 85:188–195.

Stead, L. M., J. T. Brosnan, M. E. Brosnan, D. E. vance, and R. L. Jacobs. 2006. Is it time to reevaluate methyl balance in humans? Am. J. Clin. Nutr. 83:5–10.

Takahashi, O. 1995. Hemorrhagic toxicity of a large dose of alpha-tocopherol, beta-tocopherol, gamma-tocopherol, and deltatocopherol, ubiquinone, beta-carotene, retinol actate and l-ascorbic acid in the rat. Food Chem. Tox. 33:121–128.

Teague, H. S., and A. P. Grifo. 1966. Vitamin B_{12} supplementation of sow rations. J. Anim. Sci. 25:895. (Abstr.)

Teague, H. S., A. P. Grifo, Jr., and W. M. Palmers. 1971. Pantothenic acid deficiency in the sow. J. Anim. Sci. 33(Suppl. 1):239. (Abstr.)

Thaler, R. C., J. R. Nelssen, R. D. Goodband, and G. L. Allee. 1989. Effect of dietary folic acid supplementation on sow performance through two parities. J. Anim. Sci. 67:3360–3369.

Tilton, S. L., R. O. Bates, and R. J. Moffatt. 1991. Effect of riboflavin supplementation during gestation on reproductive performance of sows. J. Anim. Sci. 69(Suppl. 1):482. (Abstr.)

Tovar, A., C. K. Ameho, J. B. Blumberg, J. W. Peterson, D. Smith, and S. L. Booth. 2006. Extrahepatic tissue concentrations of vitamin K are lower in rats fed a high vitamin E diet. Nutr. Met. 3:29.

Traber, M. G., A. Elsner, and R. Brigelium-Flohe. 1998. Synthetic as compared with natural vitamin E is preferentially excreted as α-CEHC in human urine: Studies using deuterated α-tocopheryl acetates. Am. J. Clin. Nutr. 60:397–402.

Tremblay, G. F., J. J. Matte, J. J. Dufour, and G. J. Brisson. 1989. Survival rate and development of foetuses during the first 30 days of gestation after folic acid addition to a swine diet. J. Anim. Sci. 67:724–732.

Tribout, T., J. C. Caritez, J. Gogue, J. Gruand, Y. Billon, M. Bouffaud, H.Lagant, J. LeDividich, F. Thomas, H. Quesnel, R. Guéblez, and J. P. Bidanel. 2003. Estimation, par utilisation de semence congelée, du progrès génétique réalisé en France entre 1977 et 1998 dans la race porcine Large White: résultats pour quelques caractères de reproduction femelle. Journées Rech. Porcines en France. 35:285–292.

Vigiano, P., S. Mangioni, F. Pompei, and I. Chiodo. 2003. Maternal-conceptus cross talk—a review. Placenta. 24:S56–S61.

Watkins, B. A. 1989. Influence of biotin deficiency and dietary trans-fatty acids on tissue lipids in chickens. Br. J. Nutr. 61:99–111.

Watkins, B. A., and F. H. Kratzer. 1987. Dietary biotin effects on polyunsaturated fatty acids in chick tissue lipids and prostaglandin E_2 levels in freeze-clamped hearts. Poult. Sci. 66:1818-1828.

Wegger, I., and B. Palludan. 1994. Vitamin C deficiency causes hematological and skeletal abnormalities during fetal development. J. Nutr. 124:241-248.

Wiegand, B. R., J. C. Sparks, D. C. Beitz, F. C. Parrish, Jr., R. L. Horst, A. H. Trenckle, and R. C. Ewan. 2002. Short-term feeding of vitamin D_3 improves color but does not change tenderness pf pork-loin chops. J. Anim. Sci. 2116-2121.

Wilborn, B. S., C. R. Kerth, W. F. Owsley, W. R. Jones, and L. T. Frobish. 2004. Improving pork quality by feeding supranutritional concentrations of vitamin D_3. J. Anim. Sci. 82:218-224.

Wilburn, E. E., D. C. Mahan, D. A. Hill, T. E. Shipp, H. Yang. 2008. An evaluation of natural (RRR-α-tocopheryl acetate) and synthetic (all-rac-α-tocopheryl acetate) vitamin E fortification in the diet or drinking water of weanling pigs. J. Anim. Sci. 86:584-591.

Wilson, M. E., J. E. Pettigrew, and R. D. Walker. 1991. Provision of additional vitamin B_{12} improved growth rate of weanling pigs. J. Anim. Sci. 69(Suppl. 1):359.

Wiseman, S. L., J. R.Wenninghoff, R. D. Sauer, and D. M. Danielson. 1991. The effect of supplementary riboflavin fed during the breeding and implantation period on reproductive performance of gilts. J. Anim. Sci. 69(Suppl. 1):359. (Abstr.)

Woodworth, J. C., R. D. Goodband, J. L. Nelssen, M. D. Tokach, and R. E. Musser. 2000. Added dietary pyridoxine, but not thiamin, improves weanling pig growth performance. J. Anim. Sci. 78:88-93.

Yang, H., J. Lopez, T. Radle, M. Cecava, D. Holzgraefe, and J. Less. 2004. Effects of adding pantothenic acid into reduced protein diets on performance and carcass traits of growing-finishing pigs. J. Anim. Sci. 82(Suppl. 2):39. (Abstr.)

Yang, Y. X., S. Heo, Z. Jin, J. H. Yun, J. Y. Choi, S. Y. Yoon, M. S. Park, et al. 2009. Effects of lysine intake during late gestation and lactation on blood metabolites, hormones, milk composition and reproductive performance in primiparous and multiparous sows. Anim. Reprod. Sci. 112:199-214.

Yen, J. T., and W. G. Pond. 1981. Efect of dietary vitamin C addition on performance, plasma vitamin C and hematic iron status in weanling pigs. J. Anim. Sci. 5:1292-1296.

Yen, J. T., and W. G. Pond. 1983. Response of swine to periparturient vitamin supplementation. J. Anim. Sci. 56:621-624.

Zehnder, D., R. Bland, M. C.Williams, R.W. McNinch, A. J. Howie, P. M. Stewart, and M. Hewison. 2001. Extra-renal expression of 25-hydroxyvitamin D3-1α-hydrolase. J. Clin. Endocrinol. Metab. 86:888-894.

Zehnder, D., N. K. Evans, M. D. Kilby, N. J. Bulmer, B. A. Innes, P. M. Stewart, and M. Hewison. 2002. The ontogeny of 25-dihydroxyvitamin D_3, 1α-hydroxylase expression in human placenta and deciduas. Am. J. Pathol. 161:105–114.

Zeisel, S. H. 2006. Choline: Critical role during fetal development and dietary requirements in adults. Ann. Rev. Nutr. 26:229–250.

Zempleni, J., G. M. Green, A. W. Spannagel, and D. M. Mock. 1997. Biliary excretion of biotin and biotin metabolites is quantitatively minor in rats and pigs. J. Nutr. 127:1496–1500.

Zhang, Z., E.Kebreab, M. Jing, J. C. Rodriguez–Lecompte, R.Kuehn, M. Flintoft, and J.D. House. 2009. Impairments in pyridoxine–dependent sulphur amino acid metabolism are highly sensitive to the degree of vitamin B_6 deficiency and repletion in the pig. Anim. 3:826_837.

Zhao, J. M., D. F. Li, X. S. Piao, W. J. Yang, and F. L. Wang. 2002. Effects of vitamin C supplementation on performance, iron status and immune function of weaned pigs. Arch. Anim. Nutr. 56:33–40.

第7章

矿物元素及其在猪饲粮中的应用

1 导 语

数百年前，人们就知道需要在猪饲粮中补充矿物元素。然而，营养学家往往会忽略动物对矿物元素的需要，特别是微量元素，因为他们认为饲粮成分将为动物提供必需矿物元素。而实际生产上的情况与之相反，通常会在动物饲粮中过量添加矿物元素，以避免因添加量不足引起的动物生产力低下。最近，普渡大学的研究人员证实，不同遗传背景的猪从出生到上市体重的生长规律存在差异（Hinson等，2009），说明至少在这个生长周期，它们的营养需要量可能也不尽相同。NRC（1998）提出，以瘦肉沉积率为评价指标，不同遗传背景的猪对能量和氨基酸的需要量不同，但对矿物元素的需要量却没有差异。实际上，在过去50年时间里，虽然猪的饲养模式、管理水平、饲料组成和遗传背景都发生了很大改变，但NRC（1998）推荐的矿物元素需要量基本没有变化。

根据机体需要的多与少，矿物元素可分为三大类：常量元素、微量元素和非必需矿物元素。常量元素是指机体需要量比较大（g或kg/d）的矿物元素，包括钙（Ca）、磷（P）、镁（Mg）、硫（S）、氯（Cl）、钾（K）和钠（Na）。尽管动物对铁（Fe）的需要量要比常量元素低得多，但要比绝大多数微量元素的高些。一般认为，Fe是一种微量元素。机体需要的其他微量元素（μg/d）还包括锌（Zn）、铜（Cu）、锰（Mn）、硒（Se）、钼（Mo）、碘（I）、钴（Co）等。必需矿物元素的功能已被证实，主要作为催化剂或调节剂参与调节机体的生理功能（O’Dell和Sunde，1997）。还有很多种矿物元素在体内具有一定生理功能，但并不是机体所必需的，且没有典型的缺乏症，如氟（F）、铬（Cr）、钒（V）、硅（Si）、镍（Ni）、砷（As）、锂（Li）、铅（Pb）、硼（B）等。铜和锌除具有正常的生理功能外，还具有药理功能，这使得它们不再是简单满足动物需要的必需微量元素，而且还可以作为一种药剂。

2 硫

目前尚不清楚硫在动物体内是否能被单独利用，但硫是硫胺素、生物素、胱氨酸、半胱氨酸、牛磺酸和蛋氨酸的组成成分。蛋氨酸是一种必需氨基酸，由植物和微生物合成，因此硫是植物和微生物合成蛋氨酸所必需的养分。皮肤、头发和指甲的含硫量非常高，因为构成这些组织的蛋白质是由许多含硫氨基酸组成的。硫可以与硒、铜、铁等微量元素相互作用。目前，由于一些玉米干酒精及可溶物（DDGS）在生产过程中采用硫酸进行净化，因此它们的含硫量非常多。最近，人们开始关注饲料中的硫含量。但硫是一种阴离子，现有的测定方法要么重复性差，要么准确性低；所以要准确分析饲料中的硫含量，目前仍然相当困难。最近，Kerr等（2008）采用热燃烧或电感耦合等离子体方法来测定硫含量。

3 钙

钙是一种众所周知的矿物元素，因为它在地球上的含量非常丰富，且在硬水中也含有一定量的钙（与镁共存）。随着生物的进化，需要合理地添加钙，以不阻碍磷的利用，并保证细胞内的钙含量正常（Bronner，1997）。成年动物大约有99%的钙以羟磷灰石、白磷钙石、碳酸盐或磷酸盐的形式存在于骨骼中。肠腔的钙含量是血液的20倍之多，但这些钙并不能被肠道全部吸收。

◆ 3.1 钙的吸收

哺乳仔猪摄入的乳钙是液体形式，但饲粮添加的钙则来源于固体形式的磷酸氢钙和磷酸二氢钙。这些固体形式的钙必须通过胃酸、肠道酶、肠道蠕动的作用，先进行消化和溶解，然后再被机体利用。这些过程常被错误地认为是“生物学利用率”，这样定义的“生物学利用率”指的是营养物质穿过肠道上皮的过程或营养物质吸收的过程，并不包括生物功能。在酸性食糜条件下，钙以离子形式存在与吸收；然而，随着肠道pH的变化，越来越多的钙会沉淀下来。因此，通过肠道时间和pH均会影响钙的吸收率。钙离子以饱和的主动方式被小肠前端吸收（主要在十二指肠和空肠前段），该过程需要钙结合蛋白（钙结合蛋白D9 k）和骨化三醇（1,25-二羟维生素D_3）的参与。当摄入的钙过低时，甲状旁腺素（PTH）还会参与调控钙的吸收。在能量和维生素D的作用下，先激活Ca^{2+}-Mg^{2+}-ATPase，然后再促进钙结合蛋白D9 k或其他的钙结合蛋白（钙调蛋白）将Ca从基底外侧或浆膜转移到细胞外液中。钙在细胞间的转运也通过这种方式，但不受PTH的调控。当摄入的钙过量时，空肠和回肠通过被动运输或细胞旁路的方式进行吸收。果寡糖、菊糖和其他不可消化的多糖可能会促进钙的细胞旁路转运（Suzuki和Hana，2004）。

◆ 3.2 钙的转运和代谢

血液中的钙大约有一半是以游离或离子形式存在的，其余的钙主要与血清白蛋白和前白蛋白结合，还有极少量形成硫酸盐、磷酸盐或柠檬酸盐。哺乳动物的血浆钙浓度受到PTH和骨化三醇（1,25-二羟维生素D_3）的严密调控（Schröder和Breves，2007）。因此，血浆钙浓度不能准确反映机体的钙营养状况。钙离子主要存在于线粒体、内质网和细胞核等细胞器内，同时也分布在淋巴液、血液和其他体液中。在某些激素和神经递质的作用下，钙可以进入细胞液，再通过直接作用或与钙结合蛋白结合，影响腺苷酸环化酶、钙依赖蛋白激酶、糖原合酶、肌球蛋白激酶、一氧化氮合酶、丙酮酸脱氢酶、丙酮酸激酶、丙酮酸羧化酶、磷酸化酶激酶等的活性。

钙是维持骨骼结构的完整性所必需的物质，但磷、氟、镁、钠和锶等其他矿物元素也是骨骼的重要组分成分。羟磷灰石$[Ca_{10}(PO4)_6(OH)_2]$的羟基是晶格结构，能与骨蛋白质结合。骨蛋白质主要包括胶原蛋白、骨黏连蛋白、骨桥蛋白、骨唾液蛋白、骨钙蛋白（骨Gla或BGP）、基质Gla蛋白（MGP）等。骨细胞主要包括成骨细胞、骨细胞和破骨细胞。成骨细胞构建骨骼的支架。骨细胞由成骨细胞转变而来，存在于骨基质中。破骨细胞的功能是重吸收或破坏骨骼。成骨细胞起源于骨髓，在PTH、骨化三醇和雌激素的刺激下可以分泌胶原蛋白和其他骨蛋白。当钙的摄入量不足时，破骨细胞在PTH、骨化三醇和降钙素的作用下被活化，以提高血液钙浓度。

大多数谷物类饲料的钙含量都很低，而鱼粉、肉粉和肉骨粉等动物性副产品的钙含量非常高，要比谷物类饲料高几百倍。更重要的是，动物对谷物类饲料和植物性蛋白饲料中钙的利用率受到植酸含量的影响。饲粮中添加植酸酶可以提高猪对钙的利用率（从仔猪到出栏），但并不能改善妊娠母猪对钙的利用率（Kemme等，1997a）。植酸酶也能提高泌乳母猪对钙的利用率，但其提高幅度与泌乳阶段相关（Kemme等，1997b）。

4 磷

◆ 4.1 磷的存在形式

体内磷的含量仅次于钙，存在无机磷和有机磷（与蛋白质、脂类和糖类结合）两种形式。猪乳中少部分的磷以无机磷形式存在，但大多数的磷以有机磷形式存在，这与肌肉组织中磷的存在形式相同。谷物类和豆类饲料中，因为与植酸（或肌醇六磷酸）（图7-1）结合的磷不能被猪利用，所以这些饲料中磷的利用率很低；但由于小麦含有较高的植酸酶活性，所以小麦型饲粮的磷利用率相对较高，而且也可以通过添加外源性植酸

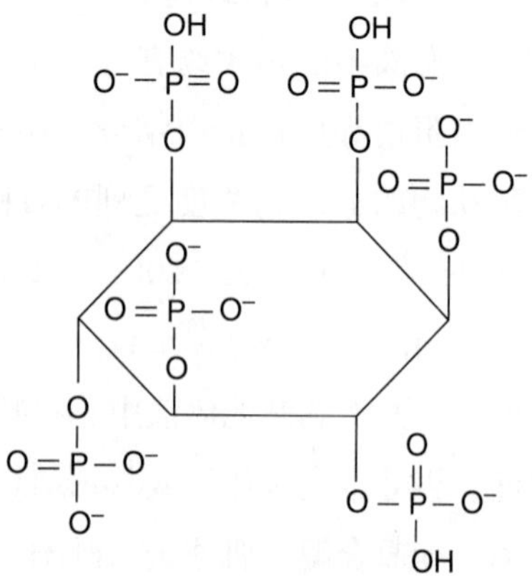

图7-1 植 酸

酶来提高饲粮磷的利用率。玉米、高粱、小麦、大麦、燕麦、豆粕等饲料原料中有60%～70%的磷是以植酸磷形式存在（Nelson等，1968）。

◆ 4.2 磷的需要量

因为猪肌肉组织的磷含量非常稳定，所以与几十年前未经选育的猪相比，生长速度快和瘦肉率高的猪对磷的需要量更高。可见，猪对磷的需要量由基因型和生理阶段或发育阶段决定。瘦肉率中等或偏低的猪对磷的需要量要比瘦肉率高的猪低，这已被前人的研究结果证实（Carter和Cromwell，1998a，1998b）。他们还发现，用重组猪生长激素（pST）处理，在提高瘦肉沉积量、降低采食量的同时，还会增加磷的需要量。很早以前，人们就发现骨骼形成所需的磷水平要比动物生长所需的高，这也在pST处理试验中得到证实（Carter和Cromwell，1998a，1998b）。

◆ 4.3 磷的吸收

磷主要通过两种吸收机制并以无机磷的形式在十二指肠和空肠前段被吸收。当饲粮磷含量较低时，少量的磷通过饱和的载体介导的主动转运机制被吸收，该过程需要钠离子和维生素D（骨化三醇）的参与；但大多数的磷通过剂量依赖性的被动扩散机制被吸收。维生素D可以促进磷的吸收，而镁、铝和钙则会抑制磷的吸收。矿物元素间的这种相互作用通常被用来治疗肾病患者。饲粮中过量的镁会与磷酸结合形成磷酸镁[$Mg_3(PO_4)_2$]，使磷和镁都不能被机体吸收，进而降低磷的吸收。反过来，若胃肠道内的镁含量偏低时，磷的吸收率就会升高。因此，在采用非常规饲料原料配制饲粮时，要特别注意不同矿物元素间的相互拮抗（如镁和磷之间的拮抗作用）。

◆ 4.4 体内的磷

血液中大部分的磷是以有机形式存在的，主要参与脂蛋白中磷脂的构成。而无机磷主要是以HPO_4^{2-}和$H_2PO_4^-$存在，极少量以PO_4^{3-}形式存在。血浆或血清只有经过脱脂处理后，才会含有无机磷。因为血清磷的含量受时间点、动物年龄、饲粮、激素、肾功能、骨骼、磷酸酯代谢等因素的影响，所以在测定磷的利用率时，要考虑这些因素的干扰。

骨骼中的磷含量超过机体总量的85%，剩余的磷存在于细胞外液（如血液）和软组织中，具有广泛的生理功能。磷在中间代谢中的功能是形成ATP高能磷酸键、肌酸磷酸和尿苷酸等。磷还在DNA和RNA合成中起到结构支架的作用，或通过形成cAMP发挥第二信使的功能。此外，细胞膜上含有磷脂，细胞内的主要缓冲液也是磷酸盐。

◆ 4.5 磷的排出

体内大部分的磷是以无机形式从尿中排出的。因为尿磷含量受到饲粮磷水平的影响，所以在饲粮配制时，不宜添加过量的磷，否则会增加体内磷的排出。不能被肠道吸收的磷（主要是植酸磷）会通过粪便排出体外。根据Coffey等（1994）的研究结果，有些磷源可被家禽利用，但不能被猪利用。尽管它们都是单胃动物，但是哺乳动物与家禽的胃肠道存在很大差异。

Pettey等（2006）为了避免不同来源的能量、蛋白质和植酸磷的影响，使用高消化性的半纯合饲粮来饲喂25～100 kg猪，发现每增加1 kg的体重，内源磷的损失大约增加1.6 mg。采用豆粕作为蛋白源，生长猪的粪中内源磷的损失约为NRC（1998）推荐的磷需要量的8.1%（Ajakaiye等，2003）。饲粮中不添加植酸酶时，豆粕中约有51%的磷可被仔猪吸收利用。此外，低植酸大豆生产的豆粕可以降低磷的排出，且添加植酸酶后，还能进一步减少磷的排出（Powers等，2006）。Mahan（1982）在断奶仔猪玉米-豆粕型饲粮中添加0.8%钙和0.68%磷时，骨灰分含量最高，由此估测出断奶仔猪对有效磷的需要量为0.35%。

与饲喂10 g/d的磷相比，饲喂15 g/d的磷显著提高了初产母猪后代的平均体重（Nimmo等，1981b）。该研究小组还发现，在生长和妊娠阶段给初产母猪饲喂0.65%钙和0.5%磷时，23头母猪中有7头不能站立，而在试验过程中被淘汰（Nimmo等，1981a）。实际生产中，由于养猪生产者不能给妊娠期的初产母猪饲喂适宜水平的钙和磷，母猪也会出现类似的现象（Hill，未发表的数据）。

5 钙和磷

◆ 5.1 钙磷比

与饲粮钙磷比例相比，饲粮钙水平对钙利用率的影响可能更大。增加饲粮中磷的水平，会降低尿钙含量，增加体内钙沉积。增加饲粮钙水平而磷含量保持不变，会降低氮的沉积率（Vipperman等，1974）。因此，降低饲粮钙磷比，可以增加钙和磷的沉积率，同时还会提高ADG、ADFI和料重比（Qian等，1996）。当磷的饲喂水平达到或超过NRC（1998）的推荐量时，饲粮钙磷比为2：1，可提高骨骼强度（Hall等，1991）。Cera和Mahan（1998）对生长育肥猪的试验发现，饲粮钙磷比为0.65：0.50时，生长猪的增重最大；而饲粮钙磷比为0.52：0.40时，育肥猪的增重最大。同时他们还发现，随着饲粮钙磷比例的增加，肱骨灰分、轴厚度以及股骨弯曲强度均会增加。

在生长阶段，提高饲粮钙水平对后备母猪的繁殖性能和使用年限没有显著影响（Kornegay等，1985）。Mahan和Newton（1995）研究表明，当配种的后备母猪体重达到135 kg时，体内磷、钾、钠及铜的含量基本保持不变，钙的含量明显增加；但对于不配种的后备母猪，从9月龄到24月龄，体内钙、磷、镁、钾及钠的含量呈线性增加。由此可得出，产了3胎的母猪体内矿物元素含量（钙、磷、镁、钾、钠、锌、铜等）会低于初产母猪。

近年来，养猪户和营养学家已经开始将母猪按胎次分开饲养。通常把初产母猪进行单体饲养和管理，以便根据个体需要调控营养供给与管理；有时，第2胎的母猪也进行单体饲养与管理。Maxson和Mahan（1986）先给妊娠母猪饲喂14%蛋白、0.8%钙和0.6%磷的饲粮，到泌乳阶段再饲喂钙磷比为1.3：1、磷含量0.5%～0.9%的饲粮，发现动物生产性能不受影响。乳中钙和磷的含量随着母猪胎次的增加而提高，但泌乳第7天与第21天相比，血清和乳中钙、磷及镁的含量没有显著差异。与初产母猪相比，产了2胎的母猪肋骨和椎骨的灰分含量要低些。以上研究结果说明，要准确评价某种矿物元素需要量，还需要按同一标准的动物数量和试验技术进行试验。而且对于母猪矿物元素需要量的评价来说，不仅要考虑繁殖阶段、窝产仔数和胎次，还要考虑骨骼组织和检测指标（如灰分、矿物元素含量、长度、弯曲和断裂强度），因为并不是所有的检测指标都会得出相似的结果。此外，还应注意猪在不同生理阶段对各种营养的需要量也不同，这与基因表达差异有关。然而，目前的猪营养研究方法还不能检测到这些细微的变化。

◆ 5.2 钙源与营养

猪对不同钙源的利用率不尽相同。Walker等（1993）以肋骨指标来评价钙利用率，发现与以$CaCO_3$为钙源相比，以苜蓿为钙源的饲粮降低了肋骨的钙含量、密度和灰分，该试验结果可能还受到饲粮是否添加植酸酶和钙磷比例不同的影响。Koch和Mahan（1985）研究表明，猪从哺乳到育肥，饲粮钙磷比例大于1.3：1降低了生长性能，并使许多骨骼指标发生改变，特别是在饲粮磷水平偏低时，这些变化更为明显。Reinhart和Mahan（1986）也发现，提高饲粮磷水平可以增加骨灰分的含量，但当钙磷比超过2：1就会降低猪的生长性能。

降低饲粮的磷水平同时提高钙磷比，会降低血清磷含量（Koch和Mahan，1986）。血清碱性磷酸酶有许多同工酶，过去人们都用它们来反映骨的功能。提高饲粮的钙磷比，可使血清碱性磷酸酶的活性升高；而提高饲粮磷水平，却会降低血清碱性磷酸酶的活性力。虽然某些特殊的血清碱性磷酸酶可以用来评价骨骼健康状况，但难以用于生长育肥猪的评价。

通过营养手段无法改变由遗传决定的性状。Barczewski等（1999）给后备母猪饲喂NRC推荐的钙水平，而磷水平提高到NRC的150%，观察3个胎次发现矿物元素或早期能量的水平不会影响所产母仔的数量。他们还得出，前脚弯曲角度是反映产3胎后的母猪肢体稳健的最佳指标。Cera和Mahan（1998）对生长育肥猪的试验发现，随着饲粮钙磷比例的提高，肱骨灰分、轴厚度以及股骨最大弯曲度均会增加，但骨骼结构的稳定性不受影响。然而，在饲粮营养水平不能满足动物正常的需要，或者某些营养组分严重失衡的时候，骨骼的结构也会出现不良变化。Fernández（1995）利用模型构建发现，矿物元素对猪骨骼生长发育的调控规律与对肌肉生长的不一致，骨骼的增长不受矿物元素摄入量的影响。事实上，提高钙和磷的摄入量，还会降低骨重吸收，从而抑制骨骼正常的生长发育。

◆ 5.3 植酸酶

早在1968年，Nelson等就发现饲粮添加克隆植酸酶对家禽起作用，但是Shurson等（1984）在猪饲粮中添加酵母源（酿酒酵母）植酸酶并没有观测到相似结果。之后，Simons等（1990）报道饲粮添加无花果曲霉生产的植酸酶（1 000 U/kg，低温下稳定），使仔猪的磷消化率提高了约50%。

迄今为止，人们对市场上出售的不同植酸酶可以释放的磷量仍存在争议，而这还必须考虑饲料组成的影响。植酸酶的活性与释放出的磷量并不是线性关系，而且pH和温度不是对所有植酸酶都有相似的作用。植酸中的磷含量一般为植酸×0.282。

继植酸酶的早期研究（Han等，1989；Zyla等，1989）后，Jongbloed等（1992）通过在十二指肠和末端回肠安装T形瘘管来研究玉米-豆粕型饲粮中磷的消化率，发现饲粮添加1 500 U/kg的黑曲霉源植酸酶，磷的消化率提高到27%，但他们在回肠食糜中未能检测到植酸酶活性。该实验室也发现，植酸酶可以提高氮、钙和磷的沉积率，而降低这些养分随粪便的排出量（Mroz等，1994）。Pallauf等（1992）研究也表明，添加1 000 U/kg的植酸酶可使饲粮磷的添加量降低0.2%，粪便磷的排出量降低52%。饲粮中分别添加250 U/kg、500 U/kg和1 000 U/kg黑曲霉源植酸酶，可使猪的生长速度和骨强度呈线性增加（Cromwell等，1993）。

虽然植酸酶可提高锌的表观吸收率（Höhler等，1991），但是Lantzsch等（1988）却首先发现，给猪饲喂大麦或玉米时，添加100ppm锌可以降低小肠内的植酸酶活性；而饲喂小麦时，添加100ppm锌并不影响小肠内的植酸酶活性。植酸酶还可以改善体内的锌平衡，这在随后的保育期配合饲粮中添加高锌的试验中获得证实（Martinez等，2002）。研究表明，在含60ppm锌的饲粮中分别添加500 U/kg或1 000 U/kg植酸酶可以提高镁和锌的表观吸收率及血浆锌含量，但是锰的表观吸收率不受到影响（Pallauf等，1992）。Lei等（1993）以硫酸锌形式在饲粮中补充30ppm的锌，使猪处于锌的负平衡，再添加植酸酶可提高血浆锌含量。同样，在不添加锌的饲粮中添加植酸酶也可提高血浆锌含量，但当饲粮锌水平增至100ppm后，添加植酸酶对血浆锌含量没有明显影响。Adeola等（1995）认为，只有在饲粮中同时添加植酸酶和锌且锌以硫酸锌形式添加时，才能改善体内的锌平衡。

给生长育肥猪饲喂低磷饲粮，且不添加植酸酶，可使平均日增重降低18%；但添加250 U/kg或500 U/kg植酸酶后，猪的生长性能得以恢复（Harper等，1997）。所以他们得出，500 U/kg植酸酶所释放的磷相当于以磷酸氢钙或磷酸二氢钙形式添加的0.87～0.96 g磷。与单独饲喂植酸酶相比，饲喂用植酸酶预处理的饲粮可以降低猪的磷排出（Liu等，1997）。研究还发现，添加黑曲霉源植酸酶可有效替代高粱饲粮中的磷（O’Ouinn等，1997）。Han等（1997）认为，用微生物源或植物源的植酸酶替代生长育肥猪饲粮中所有的无机磷，从生理学角度来讲是完全有可能的。据估计，添加676 U/kg黑曲霉源植酸酶，可将77%的磷从植酸中释放出来，从而降低粪磷含量（Yi等，1996）。而且，当饲粮钙磷比低于1.2：1时，植酸酶的作用效果更加明显（Qian等，1996）。

最近研究表明，在低植酸玉米或低植酸豆粕饲粮中添加植酸酶可以提高磷的利用率，降低磷的排出（Hill等，2008）。此外，他们还发现，与不添加植酸酶相比，添加植酸酶可以降低猪排出的水溶性磷的总量。

与生长育肥猪相比，给妊娠母猪饲喂黑曲霉源植酸酶对磷吸收率的提高效果要差些；而泌乳母猪饲喂植酸酶后，磷的吸收率比妊娠母猪的高3.4%（Kemme等，1997a）。植酸酶还可以提高母猪钙、镁和磷的全肠道表观消化率，但其作用程度取决于泌乳阶段，母猪胎次的影响甚微（Kemme等，1997b）。

6 镁

猪饲粮配制时很少会出现镁缺乏的问题，因为在豆类、鱼粉和谷物类饲料原料中都含有丰富的镁；而且硬水中也含有一定量的镁，通过饮水可以进入猪的胃肠道。硫酸镁（泻药）、氧化镁、氯化镁等均可作为饲粮添加的镁源。与钙和磷的吸收部位不同，镁的吸收部位主要在空肠末端和回肠。然而，瞬时受体电位阳离子通道（TRPM6）却主要分布在小肠（特别是十二指肠）的刷状缘膜上。细胞内高浓度的镁可以抑制这种离子通道的转运作用。当摄入低水平的镁时，镁主要通过主动运输的方式被肠道吸收，该过程需要瞬时受体电位阳离子通道的参与，这是一个饱和的载体介导的吸收过程。当摄入高水平的镁时，镁主要以简单扩散的方式被肠道吸收。

植酸和某些纤维可以降低镁的吸收，同样大量难以消化的脂类也可以降低镁的吸收。摄入高水平的钙和磷会抑制肠道镁的吸收，进而导致镁的缺乏（Hill，未发表数据）。

血液中的镁通常以游离的形式存在，也可以与蛋白质或其他离子结合形成复合物。尽管镁并不影响骨骼的功能，但是体内大多数的镁都沉积在骨骼中。细胞外液以及肝脏和肾脏等软组织中也存在一定量的镁。细胞内的镁既可以与细胞膜上的磷脂结合，也可以作为核酸和某些酶的组成成分。镁还可以作为催化剂，参与碳水化合物、脂类、核酸、蛋白质等物质的代谢。被肠道吸收的镁主要通过肾脏排出体外，也有少量的内源性镁通过粪便排出体外。

当饲喂含0.8%钙、0.6%磷、1 800 U/kg维生素D_3和小于225ppm镁的饲粮时，仔猪会出现镁缺乏症（Miller等，1965）。该试验采用的是半纯合饲粮，以体内钙和磷的最大沉积率为评价指标，仔猪对镁的需要量为325ppm（Miller等，1965）。饲粮蛋白质水平可能会影响动物对镁的需要量（Hendricks等，1970）。妊娠期和泌乳期的母猪对镁的需要量可能会更高，但关于现代选育的猪品系在各个不同生理阶段对镁的确切需要量，迄今未见相关的研究。

携带氟烷基因的育肥猪容易产生PSE肉，而人们曾试图通过饲粮添加镁来改善PSE肉的品质。最近Apple等（2005）发现，虽然饲粮添加镁对猪肉品质没有改善作用，但可以改善屠宰后猪肉的糖酵解和pH。

7 电解质

钠、氯和钾统称为电解质，在调节血液pH中发挥着重要作用。饲粮一般不会出现钾或氯的缺乏，但任何一种电解质的过量摄入都会引发动物健康问题。

◆ 7.1 钠

由于氯化钠中含有40%的钠，因此为了满足动物对钠的需要，最常用的方式就是在饲粮中添加食盐。绝大多数的钠都能被机体吸收（≥95%），只有少量随粪便排出体外。钠主要在结肠通过三种方式吸收：钠离子/葡萄糖转运方式、钠离子与氯离子偶联转运方式、钠的生电吸收方式。血液中的钠以游离形式转运，同时血液中钠、氯和钾的含量均受到严密的调控。

◆ 7.2 氯

氯离子是细胞外液中含量最丰富的一种阴离子，它的负电荷可以中和钠离子的正电荷。氯离子的吸收基本上在小肠内完成，主要是随着钠转运产生的电位梯度被动吸收。此外，钠离子/氯离子偶联转运系统也能将氯离子转入肠黏膜细胞。氯离子是肠黏膜氯离子分泌系统唯一产生的离子，也是胃肠道（特别是胃）内必不可少的一种离子，主要通过肾脏排出体外。

◆ 7.3 钾

体内的钾大多数以阳离子的形式存在细胞内。钾主要在小肠被吸收，但也可以与钠一样被结肠细胞吸收。它主要通过扩散方式或K^+/H^+-ATPase泵被吸收。除了作为电解质、调节酸碱平衡外，钾还是心脏和平滑肌收缩所必需的离子。猪缺乏钾的最早症状是食欲降低（Jensen等，1961）。研究证实，以生长性能为评价指标，仔猪对钾的需要量为0.26%；进一步研究玉米-豆粕型饲粮下不同钾源对断奶仔猪的相对生物学利用率，发现相对于乙酸钾，碳酸钾和碳酸氢钾的生物学利用率分别为103%和107%（Combs等，1985）。同时该研究小组还发现，尿钾含量是衡量不同钾源间相对生物学利用率的最可靠指标。

◆ 7.4 食盐

食盐是影响饲料味道的主要成分，同时在维持渗透压平衡、调节酸碱平衡和水代谢等方面具有极其重要作用。猪饲粮中食盐的添加量一般为5 g/kg，但若饲粮中使用了动物性蛋白饲料，可以适当降低食盐的添加量。Meyer等（1950）采用纯合饲粮的试验得出，生长猪对钠的需要量为0.8～1.1 g/kg DM，这与Hagsten和Perry（1976）采用玉米-豆粕型饲粮的试验结果一致。随着动物年龄的增长，钠的需要量会不断降低。英国科学家（ARC，1981）推荐，体重为5 kg的猪对钠的需要量为1.25 g/kg DM；而对于90 kg猪，钠的需要量仅为0.33 g/kg DM。

由于猪饲粮中的钠、氯和钾通常都是以食盐形式添加，因此目前相关的研究资料非常少。生长猪饲粮中添加0.4% $CaCl_2$和2.22% $Na_2P_3O_{10}$，使平均日增重、平均日采食量和增重/耗料比均出现降低，同时血浆氯离子含量升高，血液pH降低（Yen等，1981）。以上研究表明，猪饲粮中添加适量的电解质非常重要。血液氯离子含量过高，超出了其缓冲能力，可能导致代谢性的酸中毒（Yen等，1981）。Golz和Crenshaw（1990）也发现，当饲粮钾的含量为0.10%时，氯的含量从0.03%提高到0.57%，使体内钠、钾和氯含量都发生变化，导致钾与氯的失衡、体增重降低。当饲粮钠的水平降低到0.03%，平均日增重和增重/耗料比也降低；但当饲粮氯的含量增加到0.57%和钾的含量为0.10%时，平均日采食量和增重/耗料比就会升高，而提高饲粮钠的水平，并不能改善生长性能（Honeyfield等，1985）。当饲粮氯的水平超过0.08%时，血液尿素氮含量显著升高。通常，钠和氯的推荐比例为1：1。通过添加碳酸氢钠或碳酸氢钾来改变猪饲粮的电解质成分，短期内不会影响钠的消化率及镁、钙和氯的平衡（Patience等，1987）。Littledike和Goff（1987）还发现，改变猪饲粮中钾和氯的含量会干扰体内酸碱平衡。猪对钾的需要量为3～4 g/kg DM（Underwood和Suttle，1999）。

8 矿物元素之间的相互作用

不同矿物元素之间的相互作用广泛存在于自然界，包括猪的胃肠道内。然而，矿物元素相互作用的程度取决于动物对矿物元素的需要量及矿物元素的供应量（Hill和Link，2009a）。动物学家认为，饲料原料或饮水中过量的矿物元素是不能被机体利用的。尽管实际情况可能也是如此，但是过量的矿物元素可能还会影响胃肠道内其他矿物元素的吸收。比如，过量的钙会影响铁、镁、磷，或其他矿物元素的吸收利用。饮水中硫含量过高不仅会降低动物的饮水量，而且会改变细胞对钼和铜的利用率。为了阐明矿物元素间的相互作用及其对生物学利用率的影响，人们还需要进一步开展更多、更深入的研究。

9 铁

◆ 9.1 体内的铁

体内大多数的铁存在于血红蛋白中，用来运输氧。其余小部分的铁存在于肌红蛋白（肌肉细胞内）和含铁酶中，或贮存在肝脏中。尽管自然界有很多种铁离子存在形式，但是在动物体内和饲料中只有三价铁离子（Fe^{3+}）和二价铁离子（Fe^{2+}）。血红素铁主要分布于血液（如红细胞）或肌肉组织（如肉粉）中。胃和小肠的蛋白酶有助于球蛋白释放血红素铁，而血红素铁可以被肠道刷状缘细胞完全吸收。

◆ 9.2 铁的吸收

饲料中的铁在胃酸、胃肠道消化酶的作用下被释放出来，形成非血红素铁。胃内的铁是以可溶性的三价铁离子形式存在。但随着小肠pH的升高、越来越接近中性的条件下，三价铁离子会形成不溶性的氢氧化铁而被沉淀下来。二价铁离子在碱性pH条件下仍是可溶性的，但可被氧化成三价铁离子。肠道刷状缘上的还原酶，可以还原维生素C，也可以还原铁离子和铜离子。二价铁离子或其他二价金属离子（如锌、锰、铜）可以通过二价金属转运载体1（DMT1）或其他转运载体，穿过刷状缘转运到肠道黏膜细胞内。目前尚不清楚三价铁离子是否可被肠道吸收，但当三价铁离子与配体或螯合体结合后，形成的复合物很可能还是可溶性的，不会沉淀下来。整合素（一种膜蛋白）可促进三价铁离子和锌离子穿过刷状缘。添加植酸和高水平的钙、磷、锌、锰等微量元素都可能会抑制铁的吸收。

像铁和铜这样的矿物元素既容易被氧化，也容易被还原，很有可能造成机体的氧化损伤。所以，它们在体内不能以游离或离子形式存在，但是可以与氨基酸或蛋白质结合。在肠管闭合前口服铁剂，新生仔猪容易出现铁中毒（甚至死亡）；但通过母猪饲粮的添加提高初乳中维生素E和硒的含量，新生仔猪哺乳后再口服铁剂，就不易出现铁中毒（Loudenslager等，1986）。高铁诱导的心肌坏死还会引起心电图异常，而注射维生素E可以进行预防（Tollerz，1973）。当铁被肠道（主要是十二指肠）黏膜细胞吸收后，它可以进入血液循环、被肠道细胞利用，或伴随肠上皮细胞的脱落一起排出体外。可见，铁被细胞吸收后并不代表它就能发挥生物学功能。细胞内的铁通常与氨基酸（如半胱氨酸和组胺）或蛋白质（如mobilferrin）结合来穿过肠黏膜细胞。在穿过基底侧膜时，铁需要与膜铁转运蛋白结合后，再通过转铁蛋白的运输进入血液。铁在与转铁蛋白结合之前发生的氧化过程需要两种铜蛋白（亚铁氧化酶和血浆铜蓝蛋白）的参与。

没有进入血液循环的铁与脱铁蛋白（apoferritin）结合，以短期储存。而进入血液循环的铁主要以铁蛋白的形式，长期储存在肝脏、骨髓和脾脏中。铁蛋白会不断地分解与合成，使机体能够充分利用铁。由于血清铁蛋白浓度会随着铁储存组织中铁蛋白含量的变化而变化，因此血清铁蛋白浓度可以用来反映机体铁储存情况。血铁黄素是另外一种铁的贮存蛋白。

由于新生仔猪完全依靠母乳来提供营养物质，而母乳大概只能满足仔猪铁需要量的5%～10%（约1 mg/d），因此哺乳仔猪经常会出现贫血症状。不论是采用肌内注射还是腹腔注射，注射到体内的铁都不能完全被吸收；而且仔猪缺铁时，更容易感染疾病（Osborne和Davis，1968）。目前，通常以葡聚糖铁的形式给仔猪注射200 mg铁。然而，Miller等（1981）发现，肌内注射铁胆盐或铁、铜与柠檬酸胆碱的混合物后，都能显著促进血红蛋白的再生。贫血仔猪对铁的吸收率通常要高于铁营养充足的仔猪。与新生仔猪相比，给出生3 d的仔猪注射蛋氨酸铁的效果更好（Kegley等，2002）。同时注射维生素E和铁或在注射铁的前一天注射维生素E，都不能改善仔猪的生长性能和血红蛋白含量（Hill等，1999）。此外，与只在出生第1天注射铁相比，在出生第1天和第14天两次注射铁既不能改善仔猪的生长性能和血液指标（Hill等，1999），也不能增强仔猪的体液免疫（Bruininx等，2000）。研究表明，母猪的叶酸营养状况对于新生仔猪的铁营养非常重要（O’Connor等，1989）。关于其他营养素对新生仔猪注射铁营养的影响，目前的研究还比较缺乏。

◆ 9.3 铁过量

过量的非血红素铁会降低肠道锌的吸收，而维生素A缺乏可以导致铁在体内聚积。铁缺乏会降低体内硒水平，同时会抑制谷胱甘肽过氧化物酶的合成，降低其活性。研究表明，给仔猪饲喂5 000 ppm的铁会引发佝偻病（O' Donovan等，1963）。

◆ 9.4 饲料中的铁

最近分析发现，含铁量超过500 ppm的饲料原料包括鱼粉、血粉、血细胞、磷酸二氢钙、磷酸氢钙、面包、甜菜渣、大豆皮、羽毛粉、肉骨粉、硫酸铜、脱氟磷酸盐、石粉、氧化锌、硫酸锌等（Rincker等，2005；Kerr等，2008）。但是这些原料中有多少铁可以被机体利用，目前仍不清楚。Rincker等（2004）对哺乳仔猪的研究发现，保育料中含80 ppm的铁不能满足仔猪的生长需要。随着母猪胎次的增加，仔猪肝脏铁的含量会不断降低（Hill等，1983b）。

◆ 9.5 铁的代谢

与高铁组（500ppm）相比，饲喂适量的铁（100ppm）提高了猪铁调素（hepcidin）的表达；而与铁正常组（100ppm）相比，饲喂含20ppm铁的基础饲粮提高了DMT1和ZIP14（溶质载体家族成员14）的相对表达量（Hansen等，2009）。与低铁组（20ppm）和铁正常组相比，饲喂500ppm铁降低了猪肝脏中锰的含量，但低铁组十二指肠中锰的含量显著高于高铁组和铁正常组（Hansen等，2009）。Rincker等（2005）研究表明，与新生仔猪或13日龄仔猪都不注射铁相比，出生后第2天注射铁的仔猪到13日龄时的体重和肝脏中锰的含量均显著提高。以上两个试验结果清楚阐述了铁与锰的相互关系，同时也表明这两种矿物元素在吸收进入肠黏膜细胞时可能存在竞争关系。

Rincker等（2005）以新生仔猪和保育猪为试验对象，发现铁调节蛋白（IRP）的结合能力可影响铁的储存，也可通过转录后表达的调控途径影响转运蛋白的表达（图7-2）。因此，当机体铁缺乏时，血液中IRP活性、铁结合能力和铁蛋白含量都会出现增加，这也意味着开始消耗机体储备的铁；机体储备的铁被耗竭之后，才会引起血液中血红蛋白含量和血细胞比容的改变。通过检测铁代谢相关蛋白的变化，有助于理解各种动物（包括猪）对铁的需要量及铁与这些蛋白的相互关系。

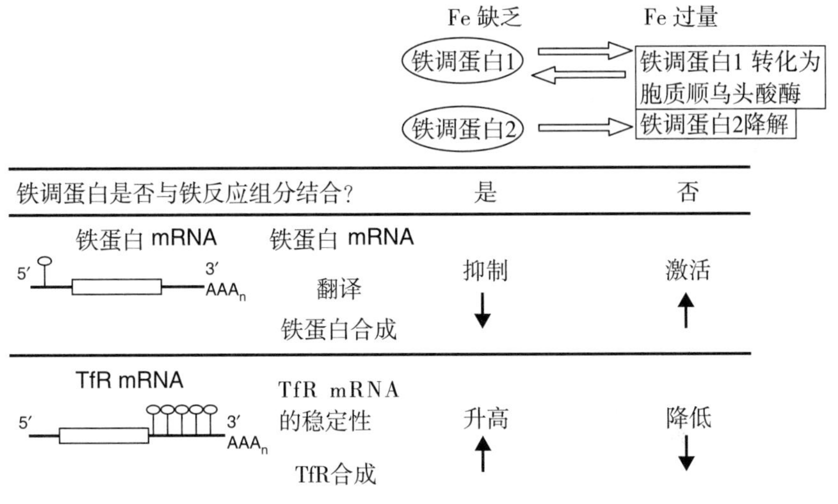

图7-2 铁调控蛋白与铁转运蛋白和储存蛋白的关系

（资料来源于Einstein等，2000）

尽管猪的遗传背景和饲养管理水平发生了巨大变化，但自从Pickett等（1960）推荐猪对铁的需要量为80 ppm以后，NRC（1998）推荐的铁需要量一直没有发生变化。最近，Rincker等（2004）研究表明，现代养猪生产中，饲粮中至少需要添加100 ppm高利用率的铁，以满足猪生长和肌肉沉积的需要。

10 锌

◆ 10.1 锌的吸收

锌广泛分布在机体的各个部位，是许多酶类的组成成分。锌既可以通过与氨基酸结合来维持酶的结构稳定，也可以参与化学反应过程。此外，锌还作为转录酶（如锌指蛋白）的组成部分，参与细胞内和细胞间的核信号传导以及基因转录调控过程（O’Halloran，1993）。饲料中的锌需要从氨基酸、植酸和其他复合物中游离出来，才能被机体吸收。因此，与铁的吸收一样，胃酸和胃肠道消化酶对锌的吸收十分重要。锌转运蛋白ZIP4是将锌转运穿过刷状缘的主要载体。DMT1蛋白也可以转运锌，但不如ZIP4那样重要。与其他许多微量元素一样，当饲粮锌的水平很高时，锌通过被动扩散的方式吸收。

◆ 10.2 锌转运载体

锌在细胞内的转入和转出过程需要四种锌转运载体的参与，以避免锌在细胞内的过多积累，同时又能满足细胞的需要。ZnT1是一种锌转出载体，主要位于小肠细胞和肾脏细胞的质膜上；而ZnT4是另外一种锌转出载体，位于乳腺细胞和脑细胞的质膜上。锌转运蛋白ZnT2参与肠道、睾丸和肾脏的细胞内囊泡对锌的吸收，而ZnT3参与神经细胞和睾丸细胞对锌的吸收。关于这些锌转运蛋白的研究主要集中在实验动物上，但ZnT1最早是在喂高锌的猪上发现的（Martinez-Montemayor等，2008）。Zip1-4是另外一种锌转运载体家族（Kelleher和Lonnerdal，2003）。Zip3转运蛋白是乳腺细胞吸收锌所必需的载体，它受催乳素的调控（Kelleher和Lönnerdal，2005）。

◆ 10.3 锌的代谢（或缺乏与过量）

Pekas（1996）研究发现，静脉注射Zn^{65} 1 h后，90%的Zn^{65}被转出血液，最后有75%～90%的Zn^{65}随粪便排出体外。采用稳定同位素Zn^{70}的研究发现，与饲粮锌含量正常组相比，饲喂锌缺乏的饲粮可以提高猪对锌的吸收，以补偿饲粮锌含量的不足（Serfass等，1996）。研究发现，饲粮组成和猪的遗传性能都会影响锌的新陈代谢。与饲喂锌临界缺乏的饲粮相比，猪采食含足够锌的饲粮后，肝脏锌含量增加了239%。由于机体的代谢性补偿，且玉米-豆粕型饲粮本身含有30～35 ppm的锌，因此生长育肥猪使用玉米-豆粕型饲粮进行饲喂，不会出现明显的锌缺乏症（Hill和Miller，1983）。然而，最近Howdyshell（2008）在保育料中添加高剂量的矿物元素但不补充锌，发现仔猪出现角化不全皮肤病，且全期生长性能受到抑制。Liptrap等（1970）还发现，与同窝出生的阉猪相比，后备母猪和公猪对锌的需要量更高。后备母猪和公猪采食含20～30 ppm锌的饲粮后出现角化不全症状，而阉猪未出现这种情况。

然而，Brink等（1959）发现，以硫酸锌的形式添加，使饲粮锌含量高达4 000 ppm时猪会出现锌中毒。Hill等（1983b）给猪饲喂含5 000 ppm锌（来源于氧化锌）的饲粮后，猪也表现出生长抑制和铜缺乏。尽管关于锌对母猪生殖性能影响的研究资料非常少，但人们都很清楚，在母猪饲粮中必须补充锌，且母猪玉米-豆粕型饲粮的锌含量不能低于50 ppm（Hill和Miller，1983）。但给母猪饲喂含5 000 ppm锌的饲粮后，仔猪断奶时的体重变轻，断奶仔猪数也变少（Hill等，1983b）。

◆ 10.4 高锌

使用高锌的作用主要是降低断奶仔猪的腹泻率，但其作用机制目前仍不清楚（Huang等，1999）。Zhang等（2001）以小鼠为试验模型，发现只有氧化锌（而不是硫酸锌或蛋氨酸锌）可以防预猪痢疾短螺旋体的感染。Jensen-Waern等（1998）发现，以氧化锌的形式添加高剂量的锌对肠道大肠杆菌或肠球菌的数量没有影响，同时嗜中性粒细胞的功能也未受到影响。但也有研究发现，在保育猪非常健康的条件下，以氧化锌的形式添加2 000 ~ 3 000 ppm的锌不会影响仔猪生长、采食量和饲料转化效率（Hill等，2000）。高锌促生长的作用机制可能是改善了肠道形态结构，从而提高了养分的吸收。Carlson等（1998）在保育阶段给早期断奶仔猪或正常断奶仔猪饲喂3 000 ppm的锌（氧化锌形式）后，肠道绒毛高度增加，隐窝深度降低；而且肝脏金属硫蛋白含量也显著提高。金属硫蛋白是一种含硫量非常高的小分子蛋白，可以与阳离子结合。与以硫酸锌形式添加250 ppm的锌相比，以氧化锌形式添加3 000 ppm的锌改善了患有传染性胃肠炎的仔猪的生长性能，促进了肠道健康的恢复。最近，Martinez-Montemayor等（2008）研究表明，保育猪饲喂高锌饲粮使其基因表达发生了变化。而且，饲喂氧化锌形式的高锌还能维持肠道菌群的稳定，保持大肠菌群的多样性，但这种作用在断奶后只能持续2周（Katouli等，1999）。然而，Namkung等（2006）却报道了相反的研究结果，饲喂高锌和高铜的饲粮降低了仔猪回肠菌群的多样性和血液皮质醇的含量。

关于其他锌源是否具有类似氧化锌的功能，目前已经进行了很多相关的对比研究。然而，即使在试验设计、动物日龄、检测指标都基本相同的情况下，也没有任何一个试验表明其他锌源具有类似氧化锌的作用，可以达到改善动物生长、采食量、饲料转化效率等的同等效果。同时饲喂高锌（以氧化锌形式添加2 000 ~ 3 000 ppm的锌）和高铜（以硫酸铜形式添加250 ppm的铜）的试验发现，两者在改善动物生长性能方面没有叠加效应（Smith等，1997；Hill等，2000）。Rincker等（2005）采用代谢笼收集每头猪每天的粪便和尿液，发现饲喂高锌饲粮的前10～14 d，猪体内可吸收大量的锌，粪便排出的锌非常少；但从第10～14天开始，粪便排出的锌量显著增加。

11 铜

初生哺乳动物的肝铜和体铜的含量要比其他生理阶段都高。初乳中铜、锌和铁的含量要比常乳（泌乳期的前3周）的高（Hill等，1983c）。铜是许多酶类和蛋白质必需的组分，主要参与氧和电子的转运过程、氧化还原反应、抗氧化应激过程。人们最熟知的含铜酶和含铜蛋白包括铜蓝蛋白、细胞色素C氧化酶、赖氨酰氧化酶、多巴胺-β-羟化酶、超氧化物歧化酶等。目前对铜的转运和贮存的研究多数是采用先天性缺陷模型或半纯合饲粮。即使玉米和豆粕中可被吸收利用的铜量很低，也很难诱导机体出现铜的缺乏。因为猪会为了满足最低的生命活动要求，不再生长，以维持铜的基本需要。因此，我们不能观测到明显的铜缺乏症，但即使动物不表现出明显的缺乏症状，其健康状况也会受到一定影响（Myers，1976；Hill等，1983a）。Okonkwo等（1979）曾试图确定青年猪对铜的需要量，但在纯合饲粮中补充铜对猪生长性能的影响甚微。Hill等（1983a）通过母猪饲粮添加5 000 ppm的锌来诱导新生仔猪的铜缺乏，发现仔猪对铜的需要量在5～10 ppm范围内。但对于猪的整个生长周期，目前铜的需要量还只是一个估计值，并非来自试验的结果。从初乳和初生仔猪肝脏中高含量的铜可以看出，即使母猪摄入的铜不足，铜也会优先转移给胎儿和初生仔猪（Ehnis等，1996）。

◆ 11.1 铜的吸收

与铁一样，铜通常与氨基酸或蛋白质结合，以防止在氧化/还原反应中沉淀下来。尽管饲粮中铜主要以二价铜离子的形式存在，但也有少量的一价铜离子（亚铜离子）。而且与铁和锌一样，在胃酸、胃蛋白酶及小肠蛋白酶的作用下，铜从复合物中释放出来，并以被动和主动转运方式主要在小肠前段被吸收。Ctr1是铜的主要转运载体（Hill和Link，2009b），但DMT1也可以转运少量的铜。与铁一样，铜价变化会显著影响它与转运蛋白、功能蛋白、存储蛋白的结合能力。动物会通过增加铜的吸收来适应低铜饲粮，但植酸会降低铜的吸收率。

◆ 11.2 铜与其他矿物元素的互作

铜与锌、铁的相互作用具有重要的实际意义。饲喂高锌饲粮的猪会引发铜的缺乏。由于MT优先与铜结合，因此锌诱导合成的MT会与铜结合。而且铜和MT一起，随肠道细胞脱落被排出体外。给母猪饲喂5 000 ppm锌，提高了仔猪肾脏的铜含量，却降低了仔猪肝脏、心脏、胰脏、动脉、睾丸的铜含量（Hill等，1983b）。Carlson等（1996）从断奶到上市给猪饲喂3 000 ppm的锌，降低了体内铜含量；同时在屠宰脱毛过程中，有些猪出现肋骨断裂。这可能是因为高锌抑制了铜的利用率，导致赖氨酰氧化酶活性的降低，而赖氨酰氧化酶是维持骨骼和软骨的健康所必需的。提高铁的摄入量也会降低铜的吸收，这可能是因为两者在胃肠道吸收或其他代谢过程中存在相互竞争，比如高铁会降低红细胞Cu/Zn超氧化物歧化酶活性（Barclay等，1991）。当Cu：Mo：S的比例发生改变，铜会与四硫钼酸盐结合形成复合物，进而阻止铜转入细胞，最终的结果是血浆铜含量正常，而细胞却缺铜。

◆ 11.3 铜的转运

铜被转运载体Ctr1转入细胞后，再经过伴侣蛋白的传递，用以酶的合成或储存在细胞内。伴侣蛋白主要有Atox1、CCS、Cox17以及ATP7A、ATP7B等膜ATP酶相关的蛋白（Hill和Link，2009b）。由于CCS蛋白可传递铜用以合成Cu/Zn SOD，因此它可能是目前衡量体内铜营养状况的最敏感指标，并且猪体内也存在这种伴侣蛋白（Hill和Link，未发表的数据）。

铜蓝蛋白（Cp）是血液中主要的铜转运载体，并参与铁的氧化/还原反应过程。动物体内含有多种铜蓝蛋白。在2日龄，仔猪体内只有CpⅡ；但到14日龄，体内CpⅡ含量保持不变，而CpⅠ含量不断升高，使血液Cp和铜的含量显著升高（Milne和Matrone，1970）。然而，Cp并不是衡量体内铜营养状况的理想指标，因为它是一种急性应激反应蛋白，其合成还受到雌激素的影响。

◆ 11.4 高铜

养猪生产中常常使用高铜饲粮来提高猪的生长性能。早在1948年，Braude就发现，采用铜管饲养的猪的生长速度要比不使用铜管的猪快。然而，高铜在猪饲粮中的应用及其具体作用机制目前仍不是十分清楚，还需要深入研究。Kornegay等（1989）发现，注射铜对猪免疫功能没有明显的影响；但该实验室在1994年的研究表明，注射铜可促进血清促有丝分裂作用（Zhou等，1994）。Moore等（1969）发现，铜可以改变脂肪酸在猪背脂的甘油三酯上的分布位点。Yen和Bienaber（1993）发现，铜降低了门静脉NH_3的净吸收率，但对氧的吸收没有影响。研究表明，铜与锌的相互作用会影响粪便的颜色和黏度（Hill等，2000）。Yen和Pond（1993）发现，铜对小肠的重量没有影响。但Shurson等（1990）发现铜可以促进肠道细胞的周转，而Zhao（2007）发现铜可降低肠道的绒毛高度。诸多研究发现，饲喂125 ~ 250 ppm的Cu可促进仔猪的生长，但其确切的作用机制目前还不清楚。很多研究还发现，其他一些铜源具有硫酸铜类似的效果。但也有很多试验表明，高铜并不能促进仔猪的生长，即使添加250 ppm的硫酸盐形式的铜对仔猪生长也没有促进作用。这些与高锌使用效果类似，因为高铜的使用效果还受到猪的遗传背景、健康状况、管理水平、饲养设施等诸多因素的影响。

◆ 11.5 外界环境中的铜

土壤学家发现，铜是非常难溶性的矿物元素，既可与土壤中其他矿物元素和有机物相互作用，也可以形成沉淀。需要注意的是，土壤铜的最大来源不是猪粪，而是在苹果、桃子、柠檬果、葡萄、蛇麻子、蔬菜等种植过程中使用的杀菌剂。尽管污泥含铜可能会产生一定的毒性，但污泥含镍和镉的毒性更强。Tiller和Merry（1981）认为，污泥的铜含量与猪粪差不多，约为750 ppm。无论铜的污染源是什么，铜污染水的危害最大，而且绵羊易发生铜中毒。

12 锰

哺乳动物体内锰的含量比较少，而且机体满足繁殖性能所需要的锰量要高于满足生长所需要的锰量。由于谷物类和植物性蛋白饲料中含有一定量的锰，因此猪饲喂玉米-豆粕型饲粮一般不会出现锰缺乏。然而，饲粮中植酸、铁、铜或钙的含量会影响锰的吸收利用。采用原子吸收光谱法可以分析锰的含量，但这种方法存在自身的缺点。例如，组织的锰含量很低，采用原子吸收光谱法很难准确测定。另外，饲料和样品处理过程中经常会用到含有锰铁的不锈钢仪器设备，而锰铁又会影响到这种方法对锰测定的准确性。

转运载体DMT1可能参与锰的主动转运过程，但谷物型饲粮的锰含量较高，而机体的锰含量低；因此一般情况下，肠道锰的吸收率都很低。研究发现，在小肠前段，二价锰离子可以转变为三价锰离子。血液中的二价锰离子既可通过游离形式运输，也可与白蛋白和球蛋白等血蛋白结合。若二价锰离子被铜蓝蛋白氧化成三价锰离子，就先由转铁蛋白进行运输，然后再通过转铁蛋白受体被机体利用。

锰分布于机体各个组织，没有主要的储存器官。但研究发现，锰在肾脏、胰脏、肝脏和骨骼中的含量最高。与其他微量元素一样，锰是机体必不可少的元素，可作为酶的激活剂或者酶的组分。需要锰的酶类可分为转移酶、水解酶、裂解酶、合成酶及氧化-还原相关的酶。线粒体Mn-SOD是反映机体锰营养状况的敏感指标。而其他很多需要锰的酶类可用镁替代。关于植酸酶、转基因玉米、转基因豆粕、猪遗传改良等这些因素是否会影响猪各生理阶段对锰的需要量，目前国内外还未见报道。硫酸锰是猪利用率最高的无机锰源，而对于不影响其他微量元素消化利用的有机锰源在猪上的生物学利用率如何，目前尚未进行相关研究。

13 硒

由于动物对硒的需要量非常低，因此它被认为是一种超微量元素。在美国，因为FDA严格限制了硒在猪饲粮中的添加量，所以猪对硒的需要量从未制定过。硒是构成岩石的组成成分，世界各地都含有不同数量的硒。动物机体从缺乏到中毒，对硒的摄入量变化范围非常窄。美国规定的猪饲粮中硒的最大添加量是根据大量试验结果来确定的，以确保FDA允许添加的最高水平既能满足动物的需求，又不会引起动物中毒。

早在20世纪50年代，人们就发现了硒的基本作用；但直到70年代，人们才开始认识硒在体内的代谢功能，至今人们仍在继续研究。最近Beck（2007）发现，一种良性病毒在缺少硒的条件下会发生突变，毒性也变强。以上说明，研究硒的代谢功能对于人类和畜禽来说都具有重要意义。

饲料中的硒是以硒代蛋氨酸或硒代半胱氨酸的形式存在的，这两者都是含硒的氨基酸。由于机体是以氨基酸的形式来吸收和利用硒，因此相对于无机硒源，有机硒源的利用率更高。而且，硒代蛋氨酸必须转化成硒代半胱氨酸，才能用于机体蛋白质的合成。亚硒酸盐、硒化物、硒酸盐等均是饲粮添加的无机硒源。硒在血液中的转运是通过密度脂蛋白（α-球蛋白和β-球蛋白）及硒蛋白P的含硫基团来完成的。

硒在动物体内的利用情况因组织、品种、硒营养状况的不同而异。体内含有四种谷胱甘肽过氧化物酶（GPx），它们都含有四个硒原子（硒代半胱氨酸形式），且均能催化相同的化学反应。GPx1分布在肝脏、肾脏、红细胞和其他组织中，GPx2分布在胃肠道和肝脏中，GPx3分布于血浆、肾脏、甲状腺中，GPx4分布于细胞膜上。GPx利用谷胱甘肽（GSH）来清除体内的过氧化氢，而GSH必须在谷胱甘肽还原酶的催化作用下不断生成。

体内还存在三种碘甲酰原氨酸-5'-脱碘酶，这些酶都含有硒代半胱氨酸，可以把甲状腺素5或5'位的碘脱去。碘甲酰原氨酸-5'-脱碘酶1和碘甲酰原氨酸-5'-脱碘酶2都可将T4转化成T3，而碘甲酰原氨酸-5'-脱碘酶3可将T3转化成为T2。硫氧还蛋白还原酶是另外一种含硒代半胱氨酸的酶，该酶通过FAD催化NADPH将硫氧还蛋白上的双硫键还原。还有一些已知功能的含硒蛋白包括硒代磷酸盐合成酶、硒代蛋白P、硒代蛋白W、蛋氨酸-R-硫氧化物还原酶。

许多含硒蛋白都是机体抗氧化系统的重要组成部分，而抗氧化系统就是利用很多含矿物元素的酶或需矿物元素来催化的酶发挥作用。一般来说，硒的缺乏会导致机体抗氧化能力降低，细胞膜受损，进而降低动物的生长性能和生殖性能。最近研究表明，添加有机硒可提高母猪的产仔数，但其具体的作用机制仍不清楚，有可能是通过增强胚胎的抗氧化能力来实现的。研究发现，有机硒可提高初乳和常乳中的硒含量（Quesnel等，2008）。维生素E通常与硒一起参与机体的抗氧化过程，但是维生素E不能完全替代硒。

众所周知，不同实验室对GPx的分析结果差异很大。这主要是因为GPx很不稳定，对温度非常敏感，而且样品存储时间对GPx活性的影响也很大。Stowe和Miller（1985）认为，与红细胞GPx1相比，GPx3能更好地反映机体的硒营养状况。饲喂高硒的猪要比饲喂低硒的猪生长快。

Mahan等（1985）给保育猪饲喂不同水平的硒（0、0.3、0.5、1.0 ppm，亚硒酸钠形式），发现随着硒水平的升高，粪和尿中硒的含量每周都在增加。他们总结认为，硒的表观消化率大约是70%，且硒在体内的沉积量随饲粮中硒水平的增高而不断增加。Parsons等（1985）发现，饲粮中添加核黄素并不影响硒的吸收，但可通过降低尿中硒的含量来提高硒在体内的沉积量。机体可能会不断摄入硒，并将它贮存在肌肉组织中，直到贮存了足够的硒。肌肉组织中硒的最高含量约为0.08 ppm（以湿重为基础）。以上研究表明，除硒源外，硒的添加水平和其他饲粮组分都会影响硒的沉积量。

14 铬

研究表明，铬并不符合必需微量元素所具备的条件。体内含有且具有一定功能的矿物元素，并不能说明它是动物发挥正常生理功能所必需的。正如Nielsen（1984）的定义，一种必需矿物元素必须具备的条件包括：缺乏时，动物会出现典型的缺乏症状或代谢功能异常；当补充之后，缺乏症即可消失，代谢功能恢复正常。要准确测定饲料和组织中的铬含量非常困难。很多人都误认为，采用电感耦合等离子体的测定方法可以准确测定所有的矿物元素含量，但这完全是不可能的。Anderson等（1997）发现，测定铬含量时，除了要用特制手套小心收集样品外，还需要使用石墨炉原子吸收光谱仪和适当的有机标准品；而且，在样品铬含量非常低的情况下，可能还要通过添加标准品来测定。目前关于铬适宜添加量的研究，各试验结果差异很大，可能是由于饲料铬的检测不准确、添加之前动物体内的铬营养状况不同或者二者均存在。人们可以推测得到，即使饲喂相同剂量、采用相同指标进行衡量，采食足量铬的动物出现的变化规律也会与采食铬缺乏的动物不同。

Mertz（1975）在人上的研究表明，铬具有增强葡萄糖代谢的功能，被认为是“葡萄糖耐受因子”的组成部分。但是，自30多年前被提出以来，这个所谓的“葡萄糖耐受因子”一直都没有被纯化出来，人们对它的特性也不清楚（Mertz，1975）。Anderson等（1997）将两批玉米-豆粕型饲粮（分别含2 790 ng/g和2 587 ng/g的铬）混合后，再添加0.3 μg的铬进行饲喂。这个试验说明，饲粮本身的铬含量差异很容易掩盖添加铬的效果。

有些研究发现，铬具有增加肌肉的生长、减少脂肪的沉积、降低应激反应、改善繁殖性能的功能；但不同研究的结果差异很大，甚至是同一实验室的研究结果有时也相互矛盾（Page等，1993；Boleman等，1995；Lindemann等，1995；Mooney和Cromwell，1995；Ward等，1997；Baldi等，1999）。目前，还没有任何一个研究小组根据其代谢功能，提出它是一种必需微量元素。

15 碘

碘通常在矿物元素章节中进行介绍。虽然碘以离子形式（I^-）发挥生理功能，但是它不是一种金属元素，所以在这里不再讨论。

作者：Gretchen M. Hill

译者：易丹

参考文献

Adeola, O., B.V. Lawrence, A. L. Sutton, and T. R. Cline. 1995. Phytase-induced changes in mineral utilization in zinc-supplemented diets for pigs. J. Anim. Sci. 73:3384-3391.

Ajakaiye, A., M. Z. Fan, T. Archbold, R. R. Hacker, C. W. Forsberg, and J. P. Phillips. 2003. Determination of true digestive utilization of phosphorus and the endogenous phosphorus outputs associated with soybean meal for growing pigs. J. Anim. Sci. 81:2766-2775.

Anderson, R. A., N. A. Bryden, C. M. Evock-Clover, and N. C. Steele. 1997. Beneficial effects of chromium on glucose and lipid variables in control and somatotropin-treated pigs are associated with increased tissue chromium and altered tissue copper, iron, and zinc. J. Anim. Sci. 75:657-661.

Apple, J. K., E. B. Kegley, C. V. Maxwell, Jr., L. K. Rakes, D. Galloway, and T. J. Wistuba. 2005. Effects of dietary magnesium and short-duration transportation on stress response, postmortem muscle metabolism, and meat quality of finishing swine. J. Anim. Sci. 83:1633-1645.

ARC. 1981. Nutrient Requirements of Pigs. Commonwealth Agricultural Bureaux, Farnham Royal, UK.

Baldi, A., V. Bontempo, V. Dell'Orto, F. Cheli, F. Fantuz, and G. Savoini. 1999. Effects of dietary chromium-yeast in weaningstressed piglets. Can. J. Anim. Sci. 79:369-374.

Barclay, S. M., P. J. Aggett, D. J. Lloyd, and P. Duffy. 1991. Reduced erythrocyte superoxide dismutase activity in low birth weight infants given iron supplements. Pediatr. Res. 29:297-301.

Barczewski, R. A., E. T. Kornegay, D. R. Notter, H. P. Veit, and M. E. Wright. 1990. Effects of feeding restricted energy and elevated calcium and phosphorus during growth on gait characteristics of culled sows and those surviving three parities. J. Anim. Sci. 68:3046-3055.

Beck, M. A. 2007. Selenium and vitamin E status: Impact on viral pathogenicity. J. Nutr. 137:1338-1340.

Boleman, S. L., S. J. Boleman, T. D. Bidner, L. L. Southern, T. L. Ward, J. E. Pontif, and M. M. Pike. 1995. Effect of chromium picolinate on growth, body composition, and tissue accretion in pigs. J. Anim. Sci. 73:2033-2042.

Braude, R. 1948. The supplementary feeding of pigs. Vet. Rec. 60:70.

Brink, M. F., D. E. Becker, S. W. Terrill, and A. H. Jensen. 1959. Zinc toxicity in the weanling pig. J. Anim. Sci. 18:836-842.

Bronner, F. 1997. Calcium, in Handbook of Nutritionally Essential Mineral Elements. B. L. O'Dell and R. A. Sunde, eds. Marcel Dekker, Inc., New York.

Bruininx, E. M. A. M., J. W. G. M. Swinkels, H. K. Parmentier, C. W. J. Jetten, J. L. Gentry, and J. W. Schrama. 2000. Effects of an additional iron injection on growth and humoral immunity of weanling pigs. Livest. Prod. Sci. 67:31-39.

Carlson, M. S., G. M. Hill, and J. E. Link. 1999. Early- and traditionally weaned nursery pigs benefit from phase-feeding pharmacological concentrations of zinc oxide: Effect on metallothionein and mineral concentrations. J. Anim. Sci. 77:1199-1207.

Carlson, M. S., S. L. Hoover, G. M. Hill, J. E. Link, and J. R. Turk. 1998. Effect of pharmacological zinc on intestinal metallothionein concentration and morphology in nursery pig. J. Anim. Sci. 76(Suppl. 2):53. (Abstr.)

Carter, S. D., and G. L. Cromwell. 1998a. Influence of porcine somatotropin on the phosphorus requirement of finishing pigs: I. Performance and bone characteristics. J. Anim. Sci. 76:584-595.

Carter, S. D., and G. L. Cromwell. 1998b. Influence of porcine somatotropin on the phosphorus requirement of finishing pigs: II. Carcass characteristics, tissue accretion rates, and chemical composition of the ham. J. Anim. Sci. 76:596-605.

Cera, K. R., and D. C. Mahan. 1988. Effect of dietary calcium and phosphorus level sequences on performance, structural soundness and bone characteristics of growing-finishing swine. J. Anim. Sci. 66:1598-1605.

Coffey, R. D., K. W. Mooney, G. L. Cromwell, and D. K. Aaron. 1994. Biological availability of phosphorus in defluorinated phosphates with different phosphorus solubilities in neutral ammonium citrate for chicks and pigs. J. Anim. Sci. 72:2653-2660.

Combs, N. R., E. R. Miller, and P. K. Ku. 1985. Development of an assay to determine the bioavailability of potassium in feedstuffs for the young pig. J. Anim. Sci. 60:709-714.

Cromwell, G. L., T. S. Stahly, R. D. Coffey, H. J. Monegue, and J. H. Randolph. 1993. Efficacy of phytase in improving the bioavailability of phosphorus in soybean meal and corn-soybean meal diets for pigs. J. Anim. Sci. 71:1831-1840.

Ehnis, L. R., G. M. Hill, J. E. Link, J. B. Barber, and D. R. Hawkins. 1996. Impact of physiological state and breed on long-term copper status of cattle. J. Anim. Sci. 74(Suppl. 1): 77. (Abstr.)

Fernández, J. A. 1995. Calcium and phosphorus metabolism in growing pigs. III A model resolution. Livest. Prod. Sci. 41:255-261.

Golz, D.I., and T. D. Crenshaw. 1990. Interrelationships of dietary sodium, potassium and chloride on growth in young swine. J. Anim. Sci. 68:2736-2747.

Hagsten, I., and T. W. Perry. 1976. Evaluation of dietary salt levels for swine. 1. Effect on gain, water consumption and efficiency of feed conversion. J. Anim. Sci. 42:187−1190.

Hall, D.D., G. L. Cromwell, and T. S. Stahly. 1991. Effects of dietary calcium, phosphorus, calcium: Phosphorus ratio and vitamin K on performance, bone strength and blood clotting status of pigs. J. Anim. Sci. 69:646−655.

Han, Y. M., F. Yang, A. G. Zhou, E. R. Miller, P. K. Ku, M. G. Hogberg, and X. G. Lei. 1997. Supplemental phytases of microbial and cereal sources improve dietary phytate phosphorus utilization by pigs from weaning through finishing. J. Anim. Sci. 75:1017−1025.

Han, Y. W. 1989. Use of microbial phytase in improving the feed quality of soya bean meal. Anim. Feed Sci. Technol. 24:345−350.

Hansen, S. L., N. Trakooljul, H. C. Liu, A. J. Moeser, and J. W. Spears. 2009. Iron transporters are differentially regulated by dietary iron, and modifications are associated with changes in manganese metabolism in young pigs. J. Nutr. 139:1474−1479.

Harper, A. F., E. T. Kornegay, and T. C. Schell. 1997. Phytase supplementation of low-phosphorus growing-finishing pig diets improves performance, phosphorus digestibility, and bone mineralization and reduces phosphorus excretion. J. Anim. Sci. 75:3174−3186.

Hendricks, D. G., E. R. Miller, D. E. Ullrey, J. A. Hoefer, and R.W. Luecke. 1970. Effect of source and level of protein on mineral utilization by the baby pig. J. Nutr. 100:235−240.

Hill, G. M., G. L. Cromwell, T. D. Crenshaw, C. R. Dove, R. C. Ewan, D. A. Knabe, A. J. Lewis, et al. 2000. Growth promotion effects and plasma changes from feeding high dietary concentrations of zinc and copper to weanling pigs (regional study). J. Anim. Sci. 78:1010−1016.

Hill, G. M., P. K. Ku, E. R. Miller, D. E. Ullrey, T. A. Losty, and B. L. O'Dell. 1983a. A copper deficiency in neonatal pigs induced by a high zinc maternal diet. J. Nutr. 113:867−872.

Hill, G. M., and J. E. Link. 2009a. Trace mineral interactions, known, unknown and not used. J. Anim. Sci. 87(E-Suppl. 2):370. (Abstr.)

Hill, G. M., and J. E. Link. 2009b. Transporters in the absorption and utilization of zinc and copper. J. Anim. Sci. 87:E85–E90.

Hill, G. M., J. E. Link, L. Meyer, and K. L. Fritsche. 1999. Effect of vitamin E and selenium on iron utilization in neonatal pigs. J. Anim. Sci. 77:1762−1768.

Hill, G. M., J. E. Link, M. J. Rincker, D. L. Kirkpatrick, M. L. Gibson, and K. Karges. 2008. Utilization of distillers dried grains with solubles and phytase in sow lactation diets to meet the phosphorus requirement of the sow and reduce fecal phosphorus concentration. J. Anim. Sci. 86:112−118.

Hill, G. M., and E. R. Miller. 1983. Effect of dietary zinc levels on the growth and development of the gilt. J. Anim. Sci. 57:106−113.

Hill, G. M., E. R. Miller, and H. D. Stowe. 1983b. Effect of dietary zinc levels on health and productivity of gilts and sows through two parities. J. Anim. Sci. 57:114–122.

Hill, G. M., E. R. Miller, P. A. Whetter, and D. E. Ullrey. 1983c. Concentration of minerals in tissues of pigs from dams fed different levels of dietary zinc. J. Anim. Sci. 57:130–138.

Hinson, R., A. Sutton, A. Schinckel, B. Richert, G. M. Hill, and J. Link. 2009. Effect of feeding a reduced crude protein and phosphorus diet on compartmental and whole body mineral masses of pigs reared under commercial settings. J. Anim. Sci. 84(Suppl. 2): 25. (Abstr.)

Höhler, D., J. Pallauf, and G. Rimbach. 1991. Effect of supplementing microbial phytase on trace element utilization in growing pigs. Umweltaspektr der Tierproduktion 103:475–480.

Honeyfield, D. C., J. A. Froseth, and R. J. Barke. 1985. Dietary sodium and chloride levels for growing–finishing pigs. J. Anim. Sci. 60:691–698.

Howdyshell, R. E. 2008. Evaluating the Efficacy of Dietary Trace Mineral Levels on Physiological Responses and Tissue Mineral Composition in Weanling and Grower Pigs. The OhioStateUniversity, Columbus, OH.

Huang, S. X., M. McFall, A. C. Cegielski, and R. N. Kirkwood. 1999. Dietary zinc supplementation on *Escherichia coli* septicemia in weaned pigs. Swine Health Prod. 7:109–111.

Jensen–Waern, M., L. Melin, R. Lindberg, A. Johannisson, L. Petersson, and P. Wallgren. 1998. Dietary zinc oxide in weaned pigs–effects on performance, tissue concentrations, morphology, neutrophil functions and faecal microflora. Res. Vet. Sci.64:225–231.

Jensen, A. H., S. W. Terrill, and D. E. Becker. 1961. Response of the young pig to levels of dietary potassium. J. Anim. Sci. 20:464–467.

Jongbloed, A. W., Z. Mroz, and P. A. Kemme. 1992. The effect of supplementary Aspergillus niger phytase in diets for pigs on concentration and apparent digestibility of dry matter, total phosphorus, and phytic acid in different sections of the alimentary tract. J. Anim. Sci. 70:1159–1168.

Katouli, M., L. Melin, M. Jensen–Waern, P.Wallgren, and R.Möllby. 1999. The effect of zinc oxide supplementation on the stability of the intestinal flora with special reference to composition of coliforms in weaned pigs. J. Appl. Microb. 87:564–573.

Kegley, E. B., J. W. Spears, W. L. Flowers, and W. D. Schoenherr. 2002. Iron methionine as a source of iron for the neonatal pig. Nutr. Res. 22:1209–1217.

Kelleher, S. L., and B. Lonnerdal. 2003. Zn transporter levels and localization change throughout lactation in rat mammary gland and are regulated by Zn in mammary cells. J. Nutr. 133:3378–3385.

Kelleher, S. L., and B. Lönnerdal. 2005. Zip3 plays a major role in zinc uptake into mammary epithelial cells and is regulated by prolactin. Am. J. Physiol. Cell Physiol. 288:C1042–C1047.

Kemme, P. A., A.W. Jongbloed, Z. Mroz, and A. C. Beynen. 1997a. The efficacy of

Aspergillus niger phytase in rendering phytate phosphorus available for absorption in pigs is influenced by pig physiological status. J. Anim. Sci. 75:2129–2138.

Kemme, P. A., J. S. Radcliffe, A. W. Jongbloed, and Z. Mroz. 1997b. The effects of sow parity on digestibility of proximate components and minerals during lactation as influenced by diet and microbial phytase supplementation. J. Anim. Sci. 75:2147–2153.

Kerr, B. J., C. J. Ziemer, T. E. Weber, S. L. Trabue, B. L. Bearson, G. C. Shurson, and M. H. Whitney. 2008. Comparative sulfur analysis using thermal combustion or inductively coupled plasma methodology and mineral composition of common livestock feedstuffs. J. Anim. Sci. 86:2377–2384.

Koch, M. E., and D. C. Mahan. 1985. Biological characteristics for assessing low phosphorus intake in growing swine. J. Anim. Sci. 60:699–708.

Koch, M. E., and D. C. Mahan. 1986. Biological characteristics for assessing low phosphorus intake in finishing swine. J. Anim. Sci. 62:163–172.

Kornegay, E. T., B. G. Diggs, O. M. Hale, D. L. Handlin, J. P. Hitchcock, and R. A. Barczewski. 1985. Reproductive performance of sows fed elevated calcium and phosphorus levels during growth and development. J. Anim. Sci. 61:1460–1466.

Kornegay, E. T., P. H. G. van Heugten, M. D. Lindemann, and D. J. Blodgett. 1989. Effects of biotin and high copper levels on performance and immune response of weanling pigs. J. Anim. Sci. 67:1471–1477.

Lantzsch, H. J., S. E. Scheuermann, and K. H. Menke. 1988. Influence of various phytase sources on the P, Ca, and Zn metabolism of young pigs at different dietary Zn levels. J. Anim. Physiol. Anim. Nutr. 60:146–157.

Lei, X., P. K. Ku, E. R. Miller, D. E. Ullrey, and M. T. Yokoyama. 1993. Supplemental microbial phytase improves bioavailability of dietary zinc to weanling pigs. J. Nutr. 123:1117–1123.

Lindemann, M. D., C. M. Wood, A. F. Harper, E. T. Kornegay, and R. A. Anderson. 1995. Dietary chromium picolinate additions improve gain:feed and carcass characteristics in growing–finishing pigs and increase litter size in reproducing sows. J. Anim. Sci. 73:457–465.

Liptrap, D. O., E. R. Miller, D. E. Ullrey, D. L. Whitenack, B. L. Schoepke, and R. W. Luecke. 1970. Sex influence on the zinc requirement of developing swine. J. Anim. Sci. 30:736–741.

Littledike, E. T., and J. Goff. 1987. Interactions of calcium, phosphorus, magnesium and vitamin D that influence their status in domestic meat animals. J. Anim. Sci. 65:1727–1743.

Liu, J., D. W. Bollinger, D. R. Ledoux, M. R. Ellersieck, and T. L. Veum. 1997. Soaking increases the efficacy of supplemental microbial phytase in a low–phosphorus corn–soybean meal diet for growing pigs. J. Anim. Sci. 75:1292–1298.

Loudenslager, M. J., P. K. Ku, P. A. Whetter, D. E. Ullrey, C. K. Whitehair, H. D. Stowe, and E. R. Miller. 1986. Importance of diet of dam and colostrum to the biological antioxidant status and parenteral iron tolerance of the pig. J. Anim. Sci. 63:1905–1914.

Mahan, D. C. 1982. Dietary calcium and phosphorus levels for weanling swine. J. Anim. Sci. 54:559–564.

Mahan, D. C. 1985. Effect of inorganic selenium supplementation on selenium retention in postweaning swine. J. Anim. Sci. 61:173–178.

Mahan, D. C., and E. A. Newton. 1995. Effect of initial breeding weight on macro- and micromineral composition over a three-parity period using a high-producing sow genotype. J. Anim. Sci. 73:151–158.

Martinez-Montemayor, M. M., G. M. Hill, N. E. Raney, V. D. Rilington, R. J. Tempelman, J. E. Link, C. P.Wilkinson, et al. 2008. Gene expression profiling in hepatic tissue of newly weaned pigs fed pharmacological zinc and phytase supplemented diets. BMC Genomics 9:421.

Martinez, M. M., G. M. Hill, J. E. Link, J. G. Greene, and D.D. Driksna. 2002. Effects of pharmacological concentrations of zinc oxide and phytase on zinc excretion and performance in the nursery pig. J. Anim. Sci. 80(Suppl. 2):73. (Abstr.)

Maxson, P. F., and D. C. Mahan. 1986. Dietary calcium and phosphorus for lactating swine at high and average production levels. J. Anim. Sci. 63:1163–1172.

Mertz, W. 1975. Effects and metabolism of glucose tolerance factor. Nutr. Rev. 33:129–135.

Meyer, J. H., R. H. Grummer, R. H. Phillips, and G. Bohstedt. 1950. Sodium, chlorine, and potassium requirements of growing pigs. J. Anim. Sci. 9:300–306.

Miller, E. R., M. J. Parsons, D. E. Ullrey, and P. K. Ku. 1981. Bioavailability of iron from ferric choline citrate and a ferric copper cobalt choline citrate complex for young pigs. J. Anim. Sci. 52:783–787.

Miller, E. R., D. E. Ullrey, C. L. Zutaut, J. A. Hoefer, and R. W. Luecke. 1965. Mineral balance studies with the baby pig: effects of dietary vitamin D_2 level upon calcium, phosphorus and magnesium balance. J. Nutr. 85:255–259.

Milne, D. B., and G. Matrone. 1970. Forms of ceruloplasmin in developing piglets. Biochem. Biophys. Acta 212:43–49.

Mooney, K. W., and G. L. Cromwell. 1995. Effects of dietary chromium picolinate supplementation on growth, carcass characteristics, and accretion rates of carcass tissues in growing-finishing swine. J. Anim. Sci. 73:3351–3357.

Moore, J. H., W. W. Christie, R. Braude, and K. G. Mitchell. 1969. The effect of dietary copper on the fatty acid composition and physical properties of pig adipose tissues. Br. J. Nutr. 23:281–287.

Mroz, Z., A. W. Jongbloed, and P. A. Kemme. 1994. Apparent digestibility and retention of nutrients bound to phytate complexes as influenced by microbial phytase and feeding regimen in pigs. J. Anim. Sci. 72:126–132.

Myers, G. 1976. A Copper Deficiency in Young Growing Swine. PurdueUniversity, West Lafayette, IN.

Namkung, H., J. Gong, H. Yu, and C. F. M. de Lange. 2006. Effect of pharmacological intakes of zinc and copper on growth performance, circulating cytokines and gut microbiota of newly weaned piglets challenged with coliform lipopolysaccharides. Can. J. Anim. Sci. 86:511–522.

Nelson, T. S., T. R. Shieh, R. J. Wodzinski, and J. H. Ware. 1968. The availability of phytate phosphorus in soya bean meal before and after treatment with a mold phytase. Poult. Sci. 47:1842–1848.

Nielsen, F. H. 1984. Ultratrace elements in nutrition. Annu. Rev. Nutr. 4:21–41.

Nimmo, R. D., E. R. Peo, Jr., J. D. Crenshaw, B. D. Moser, and A. J. Lewis. 1981a. Effect of level of dietary calcium–phosphorus during growth and gestation on calcium–phosphorus balance and reproductive performance of first litter sows. J. Anim. Sci. 52:1343–1349.

Nimmo, R. D., E. R. Peo, Jr., B. D. Moser, and A. J. Lewis. 1981b. Effect of level of dietary calcium–phosphorus during growth and gestation on performance, blood and bone parameters of swine. J. Anim. Sci. 52:1330–1342.

NRC. 1998. Nutrient Requirements of Swine. 10th Rev. ed. National Academies Press, Washington, DC.

O'Connor, D. L., M. F. Picciano, M. A. Ross, and R. A. Easter. 1989. Iron and folate utilization in reproducing swine and their progeny. J. Nutr. 119:1984–1991.

O'Dell, B. L., and R. A. Sunde. 1997. Pages 2–4 in Handbook of Nutritionally Essential Mineral Elements. B. L. O'Dell and R. A. Sunde, eds. Marcel Dekker, Inc., New York.

O'Donovan, P. B., R. A. Pickett, M. P. Plumlee, and W. M. Beeson. 1963. Iron toxicity in the young pig. J. Anim. Sci. 22:1075–1080.

O'Halloran, T. V. 1993. Transition metals in control of gene expression. Science 261:715–725.

O'Quinn, P. R., D. A. Knabe, and E. J. Gregg. 1997. Efficacy of Natuphos in sorghum–based diets of finishing swine. J. Anim. Sci. 75:1299–1307.

Okonkwo, A. C., P. K. Ku, E. R. Miller, K. K. Kaehey, and D. E. Ullrey. 1979. Copper requirement of baby pigs fed purified diets. J. Nutr. 109:939–948.

Osborne, J. C., and J. W. Davis. 1968. Increased susceptibility to bacterial endotoxin of pigs with iron–deficiency anemia. J. Am. Vet. Med. Assoc. 152:1630–1632.

Page, T. G., L. L. Southern, T. L. Ward, and D. L. Thompson, Jr. 1993. Effect of chromium picolinate on growth and serum and carcass traits of growing-finishing pigs. J. Anim. Sci. 71:656-662.

Pallauf, J., D. H¨ohler, and G. Rimbach. 1992. Effect of microbial phytase supplementation to a maize-soya-diet on apparent absorption of Mg, Fe, Cu, Mn, and Zn and parameters of Zn-status in piglets. J. Anim. Physiol. Anim. Nutr. 68:1-9.

Parsons, M. J., P. K. Ku, D. E. Ullrey, H. D. Stowe, P. A. Whetter, and E. R. Miller. 1985. Effects of riboflavin supplementation and selenium source on selenium metabolism in the young pig. J. Anim. Sci. 60:451-461.

Patience, J. F., R. E. Austic, and R. D. Boyd. 1987. Effect of dietary supplements of sodium or potassium bicarbonate on short-term macromineral balance in swine. J. Anim. Sci. 64:1079-1085.

Pekas, J. C. 1966. Zinc 65 metabolism: Gastrointestinal secretion by the pig. Am. J. Physiol. 211:407-413.

Pettey, L. A., G. L. Cromwell, and M. D. Lindemann. 2006. Estimation of endogenous phosphorus loss in growing and finishing pigs fed semi-purified diets. J. Anim. Sci. 84:618-626.

Pickett, R. A., M. P. Plumlee, W. H. Smith, and W. M. Beeson. 1960. Oral iron requirement of the early-weaned pig. J. Anim. Sci. 19:1284. (Abstr.)

Powers, W. J., E. R. Fritz, W. Fehr, and R. Angel. 2006. Total and water-soluble phosphorus excretion from swine fed low-phytate soybeans. J. Anim. Sci. 84:1907-1915.

Qian, H., E. T. Kornegay, and D. E. Conner, Jr. 1996. Adverse effects of wide calcium:phosphorus ratios on supplemental phytase efficacy for weanling pigs fed two dietary phosphorus levels. J. Anim. Sci. 74:1288-1297.

Quesnel, H., A. Renaudin, N. LeFloc h, C. Jondreville, M. C. Père, J. A. Taylor-Pickard, and J. Le Dividich. 2008. Effect of organic and inorganic selenium sources in sow diets on colostrum production and piglet response to a poor sanitary environment after weaning. Animal 2:859-866.

Reinhart, G. A., and D. C. Mahan. 1986. Effect of various calcium:phosphorus ratios at low and high dietary phosphorus for starter, grower and finishing swine. J. Anim. Sci. 63:457-466.

Rincker, M. J., G. M. Hill, J. E. Link, A. M. Meyer, and J. E. Rowntree. 2005. Effects of dietary zinc and iron supplementation on mineral excretion, body composition, and mineral status of nursery pigs. J. Anim. Sci. 83:2762-2774.

Rincker, M. J., G. M. Hill, J. E. Link, and J. E. Rowntree. 2004. Effects of dietary iron supplementation on growth performance, hematological status, and whole-body mineral concentrations of nursery pigs. J. Anim. Sci. 82:3189-3197.

Schröder, B., and G. Breves. 2007. Mechanisms and regulation of calcium absorption from the gastrointestinal tract in pigs and ruminants: comparative aspects with special emphasis on hypocalcemia in dairy cows. Anim. Health Res. Rev. 7:31–41.

Serfass, R. E., Y. Y. Fang, and M. E. Wastney. 1996. Zinc kinetics in weaned piglets fed marginal zinc intake: Compartmental analysis of stable isotopic data. J. Trace Elem. Exp. Med. 9:73–86.

Shurson, G. C., P. K. Ku, and E. R. Miller. 1984. Evaluation of a yeast phytase product for improving phytate phosphorus bioavailability in swine diets. J. Anim. Sci. 59(Suppl. 1):106. (Abstr.)

Shurson, G. C., P. K. Ku, G. L. Waxler, M. T. Yokoyama, and E. R. Miller. 1990. Physiological relationships between microbiological status and dietary copper levels in the pig. J. Anim. Sci. 68:1061–1071.

Simons, P. C., H. A. Versteegh, A. W. Jongbloed, P. A. Kemme, P. Slump, K. D. Bos, M. G. Wolters, et al. 1990. Improvement of phosphorus availability by microbial phytase in broilers and pigs. Br. J. Nutr. 64:525–540.

Smith, J. W., II, M. D. Tokach, R. D. Goodband, J. L. Nelssen, and B. T. Richert. 1997. Effects of the interrelationship between zinc oxide and copper sulfate on growth performance of early-weaned pigs. J. Anim. Sci. 75:1861–1866.

Stanger, B. R., G. M. Hill, J. E. Link, J. R. Turk, M. P. Carlson, and D. W. Rozeboom. 1998. Effect of high Zn diets on TGE-challenged early-weaned pigs. J. Anim. Sci. 76(Suppl. 2):53. (Abstr.)

Stowe, H. D., and E. R. Miller. 1985. Genetic predisposition of pigs to hypo- and hyperselenemia. J. Anim. Sci. 60:200–211.

Suzuki, T., and H. Hana. 2004. Various nondigestible saccharides open a paracellular calcium signaling in human intestinal Caco-2 cells. J. Nutr. 134:1935–1941.

Tiller, K. G., and R. H. Merry. 1981. Copper pollution of agricultural soils. Pages 119–137 in *Copper in Soils and Plants*. J. F. Loneragan, D. Robson and R. D. Graham, eds. Academic Press, Sydney, Australia.

Tollerz, G. 1973. Vitamin E, selenium and some related compounds and tolerance towards iron in piglets. Acta Agric. Scand. Suppl. 19:184–191.

Underwood, E. J., and N. F. Suttle. 1999. The Mineral Nutrition of Livestock, 3rd ed. CABI Publishing, New York.

Vipperman, P. E., Jr., E. R. Peo, Jr., and P. J. Cunningham. 1974. Effect of dietary calcium and phosphorus level upon calcium, phosphorus and nitrogen balance in swine. J. Anim. Sci. 38:758–765.

Walker, G. L., D. M. Danielson, E. R. Peo, Jr., and R. F. Mumm. 1993. Effect of calcium source, dietary calcium concentration, and gestation phase on various bone characteristics in gestating gilts. J. Anim. Sci. 71:3003–3010.

Ward, T. L., L. L. Southern, and T. D. Bidner. 1997. Interactive effects of dietary chromium tripicolinate and crude protein level in growing-finishing pigs provided inadequate and adequate pen space. J. Anim. Sci. 75:1001–1008.

Yen, J. T., and J. A. Nienaber. 1993. Effects of high–copper feeding on portal ammonia absorption and on oxygen consumption by portal vein–drained organs and by the whole animal in growing pigs. J. Anim. Sci. 71:2157–2163.

Yen, J. T., and W. G. Pond. 1993. Effects of carbadox, copper, or Yucca shidigera extract on growth performance and visceral weight of young pigs. J. Anim. Sci. 71:2140–2146.

Yen, J. T., W. G. Pond, and R. L. Prior. 1981. Calcium chloride as a regulator of feed intake and weight gain in pigs. J. Anim. Sci. 52:778–782.

Yi, Z., E. T. Kornegay, V. Ravindran, M. D. Lindemann, and J. H. Wilson. 1996. Effectiveness of *Natuphos phytase* in improving the bioavailabilities of phosphorus and other nutrients in soybean meal–based semipurified diets for young pigs. J. Anim. Sci. 74:1601–1611.

Zhang, P., M. P. Carlson, N. R. Schneider, and G. E. Duhamel. 2001. Mineral prophylactic concentration of dietary zinc compounds in a mouse model of swine disentery. Anim. Health Res. Rev. 2:67–74.

Zhao, J., A. F. Harper, M. J. Estienne, K. E.Webb, Jr., A. P. McElroy, and D. M. Denbow. 2007. Growth performance and intestinal morphology responses in early weaned pigs to supplementation of antibiotic–free diets with an organic copper complex and spray–dried plasma protein in sanitary and nonsanitary environments. J. Anim. Sci. 85:1302–1310.

Zhou, W., E. T. Kornegay, M. D. Lindemann, J. W. Swinkels, M. K. Welten, and E. A. Wong. 1994. Stimulation of growth by intravenous injection of copper in weanling pigs. J. Anim. Sci. 72:2395–2403.

Zyla, K., M. Kujawski, and J. Koreleski. 1989. Dephosphorylation of phytate compounds by means of acid phosphatase from Aspergillus niger. J. Sci. Food Agric. 49:315–324.

第8章

猪的营养与消化道健康

1 导 语

为了使猪的生长性能和健康状况达到最佳水平，在制订管理措施和营养策略时，应充分考虑炎性反应对胃肠道（GI）功能的影响。胃肠道免疫系统包含了超过70%的机体免疫细胞，它的激活将直接引起机体产生的一类特异性细胞和信号分子的损失，并导致胃肠道功能的降低。因此，在干净环境中饲养的猪，因病原体诱发的炎性反应较少，比恶劣环境中饲养的猪采食更多、长速更快、氮营养素转化成肌肉的效率也更高（Williams等，1997a，1997b）。虽然免疫系统可随时应对病原体，或在高危条件或确认有病原体入侵时可能被激活，但通过饲粮措施避免免疫系统的过度激活，仍是提高动物生产效率的重要途径。

在讨论炎性反应对胃肠道功能的影响时，应注意到胃肠道炎性反应会影响猪的食欲及生长的内分泌调节。患病动物不采食，免疫系统与中枢神经系统的相互作用是其根本原因。炎性反应细胞因子白介素（IL）-1β和肿瘤坏死因子（TNF）作用于中枢神经系统，导致动物采食量下降（Kent等，1992；Plata-Salaman，1994；Warren等，1997），同时引起瘦素表达的增加并从脂肪细胞中释放出来（Grunfeld等，1996；Johnson等，1998；Sarraf等，1997）。炎性反应刺激也会影响生长和能量分配的调节。例如，炎性反应引起的胰岛素样生长因子（IGF-1）浓度的下降，可解释动物采食量不下降而生长受阻的机制（Hevener等，1997）。炎性反应对采食和生长调节的影响，可部分解释肠道炎性反应如何影响动物生长。然而，本章内容并不是详细阐述这些复杂的内分泌网络，而是集中论述炎性反应与胃肠道功能。

2 炎性反应与胃肠道功能

炎性反应是胃肠道功能的一个基本面。健康的胃肠道处于一种“可控”的炎性反应的稳定状态，胃肠道内高密度的细菌种群、饲粮抗原和毒素保持相对稳定。尽管大多数人可能认为，常规条件下临床上表现健康的猪其小肠中并不存在炎性反应，但与无菌猪相比，普通猪肠道内的炎性细胞因子表达、免疫细胞浸润、淋巴样滤泡组织和淋巴集结明显增加（Shirkey 等，2006）。因此，与“正常”共生菌群相关的胃肠道免疫机制的激活，对宿主的能量消耗和宿主消化和吸收营养素的能力存在重要影响。当存在明显的肠道感染情况下，肠道的炎性反应加剧（Coopersmith等，2002），宿主肠道的能量消耗和消化功能会受到进一步的影响。在下一节中，我们分析了炎性反应对肠道的细胞更新速率、消化与吸收功能、屏障完整性、黏液分泌、肠道血流及其运动的影响。

◆ 2.1 肠上皮细胞的更新

即使在没有明显炎性反应情况下，肠上皮细胞也呈现为一个快速的更新过程，即隐窝细胞增殖后，沿着绒毛轴逐渐成熟，最后从绒毛顶端脱落，更新周期一般为3 ~ 5 d（Potten等，1997）。肠上皮细胞的更新是先天性防御的一种重要机制，其受肠道内菌群定植的显著影响。与共生菌群定植相关的胃肠道免疫机制的激活，是决定上皮细胞死亡率和增殖率的关键因素。这些差异可以通过显微镜观察到，与无菌猪相比，定植有肠道菌群的猪其绒毛高度低2倍，而绒毛隐窝深度变得更深（Shirkey等，2006）。另有研究表明，与定植肠道菌群的猪相比，无菌猪的肠上皮细胞增殖速度降低2倍（Miniats和Valli，1973）。

普通动物的肠绒毛较无菌动物短，表明共生菌群对肠道细胞的凋亡和脱落发挥着重要作用。在不发生炎性反应情况下，老化、损伤或不需要的细胞通过细胞凋亡或程序性细胞死亡的方式被清除。尽管炎性细胞因子可以诱导细胞凋亡，但这一过程可避免诱发更严重的炎性反应，即伴随细胞碎片的释放，并发生细胞核固缩、破碎、胞质出现空泡和细胞皱缩等一系列的步骤（Kerr等，1972）。与细胞凋亡不同，细胞坏死是细胞受到直接损伤时产生的一种病理反应，即受损细胞的细胞膜通透性增加、细胞破裂，导致周围组织发生炎性反应。细胞凋亡在调节肠上皮细胞数量中的作用尚不清楚。最初对肠道组织的观察结果表明，肠上皮细胞的数量主要是通过绒毛顶端细胞的脱落这一机制进行调控（Tsai等，1997；Tsubouchi和Leblond，1979）。这与最近的研究结果不同。最近的研究发现，在绒毛轴上存在大量的细胞凋亡，这可解释每天都有大量细胞沿着绒毛轴向上迁移的原因（Hall等，1994）。细胞凋亡并不是随机分布的，而是集中于细胞迁移路径的末端（Hall等，1994）。然而，细胞凋亡并不是只限于绒毛细胞，可增殖的隐窝细胞也存在低水平的细胞凋亡，遗传毒性损伤可增加这种细胞凋亡的发生（Merritt等，1994；Potten等，1990）。

细菌可通过外源途径和内源途径（线粒体途径）诱发细胞凋亡。外源途径作为炎性反应的一部分，可被TNF家族死亡配体激活。肠细胞基底外侧膜可表达死亡配体、Fas配体（FasL）和TNF的死亡受体（Strater和Moller，2000）。由肠道细菌定植（Shirkey等，2006；Willing和van Kessel，2007）和严重感染（Wesche-Soldato等，2005）引起的炎性反应，可使FasL和TNF的表达量升高，与死亡受体的结合可引起细胞内的级联反应，包括启动因子和执行半胱氨酸蛋白酶的激活，并最终上调细胞核及其他细胞组分降解调节酶的表达（Gupta，2003）。在猪的胃肠道中，当脱离基底膜、缺乏生长因子、缺氧、氧化应激以及其他化学因素介导的细胞膜、蛋白质和DNA变化从而引起细胞损伤时（Gupta，2003），细胞的损伤超过其修复能力，可通过多个途径激活内源途径。由内源途径引起的细胞凋亡既可能与执行半胱氨酸蛋白酶的激活有关，也可能无关（Gupta，2003）。与无菌猪相比，普通猪死亡配体（包括FasL和TNF）的诱导表达与细胞凋亡活性的升高相一致（Willing和van Kessel，2007）。此外，由饲粮和内源性底物发酵产生的一些微生物产物，如脱氧胆酸、氨、吲哚类物质和生物胺也具有毒性，也会引起细胞凋亡（Bakke和 Midtvedt，1970；Hughes等，2008；Jourd’heuil等，1997；Mather和Rottenberg，2001；McGarr等，2005；Suzuki等，2002）。另外，这些产物的毒素作用还没有定论（Leschelle等，2002）。微生物发酵碳水化合物生成的短链脂肪酸可作为肠上皮细胞增殖和凋亡的调节因子（von Engelhardt等，1998；Wesche-Soldato等，2005）。有趣的是，机体其他部位的炎性反应也会影响小肠的细胞凋亡。例如，肺部感染绿脓杆菌后，可诱导肠道细胞的凋亡活性（Coopersmith等，2002）。

在某种程度上，为了补偿从绒毛上不断脱落的上皮细胞，宿主增强了肠道细胞的增殖活力，表现为增殖细胞核抗原表达量升高和肠隐窝加深（Willing和van Kessel，2007）。当肠道发生炎性反应时，刺激隐窝细胞增殖的一些调节因子（包括胰高血糖素样肽2，GLP-2）开始表达，这有助于降低肠道的通透性（Burrin等，2003）。由共生菌群引起的肠黏膜表层Toll样受体的激活，对肠上皮细胞的内稳态和受损肠道的修复具有重要作用（Rakoff-Nahoum等，2004）。尽管在某些肠道炎性反应及由绿脓杆菌诱发的败血症情况下肠隐窝细胞会出现增殖，但此时细胞增殖速率减慢，细胞更新基本停滞（Coopersmith等，2003）。肠道的自我修复能力取决于养分的利用率。例如，当存在脂多糖引起的内毒素血症时，补充谷氨酰胺可以提高肠细胞的增殖率，促进肠黏膜的修复（Sukhotnik等，2007）。由于炎性反应引起的细胞更新会给机体带来极大的代谢消耗，因此宿主将通过肠道内高水平表达的抗炎性细胞因子来减小这种影响，进而维持肠道内环境的平衡（Sydora 等，2003）。

尽管肠上皮细胞的脱落和更新可减少病原菌定植的机会（Potten和Loeffler，1990），但也会产生巨大的代谢损耗。氨基酸作为内分泌物合成的组成原料和能量来源，每天的蛋白质合成率会超过100%（Nyachoti等，2000），回肠的内源性蛋白质损失量为10.5 ~ 17.1 g/kg干物质摄入量（Jansman等，2002）。

◆ 2.2 消化与吸收功能

与炎性反应有关的肠上皮细胞更新速率的变化将影响肠黏膜对养分的消化和吸收等功能。在无菌猪肠道内接种正常菌群，将会引起氨肽酶N（APN）、乳糖酶-根皮苷水解酶（lactase phloryzin hydrolase，LPH）等刷状缘酶活性的降低

（Willing和Kessel，2009）。以APN为例，宿主可通过增强APN基因的表达来弥补该酶活性的降低，虽然这并不能使酶活水平恢复到无菌动物的水平。无菌猪感染鼠伤寒沙门氏菌后，其空肠和回肠中γ-谷酰转肽酶（GGT）的活性会急剧降低；但是对乳糖酶、蔗糖酶、葡萄糖糖化酶、碱性磷酸酶和二肽基肽酶IV等其他刷状缘酶的活性却没有影响（Trebichavsky等，2002）。由此可知，在炎性反应过程中，酶活的变化是特异性的，而非普遍性的。

在断奶等应激情况下，炎性反应会降低与消化和吸收相关基因的表达，进而对营养状况产生危害。断奶引起的炎性反应也会引起消化功能受损。例如，在断奶后0～2 d，IL-1β、IL-6和TNF的水平显著增加，同时蔗糖酶的活性则明显降低（Pie等，2004）。在断奶期间，铁的营养状况对仔猪健康十分重要（Zhao等，2002），断奶引起的炎性反应会降低铁吸收相关基因的表达，从而导致仔猪出现缺铁现象（Yasuda等，2009）。给断奶仔猪补充菊糖可以改善其体内的铁贮备状况（Yasuda等，2006）。添加菊糖后仔猪铁储备状况得到改善，同时铁代谢相关基因的表达增强，这正好与炎性反应基因表达下降相吻合（Yasuda等，2009）。除了影响常规养分的吸收，炎性反应还严重影响初生仔猪对初乳中免疫球蛋白（Ig）的吸收。健康和炎性小肠组织间的二维凝胶电泳图谱存在明显的差别，表明炎性小肠组织无法吸收初乳中的Ig并转运到小肠上皮细胞中（Danielsen等，2006）。不只是肠道炎性反应可引起消化酶活性的改变。在腹腔注射脂多糖诱导的败血症模型中，可观察到果糖吸收的减少，黏膜-浆膜间跨上皮流量发生改变，以及刷状缘膜囊泡的摄取（Garcia-Herrera等，2008）。这种摄取的减少，与GLUT5蛋白水平的降低、TNF缓解LPS诱导的败血症对GLUT5表达和抵御果糖摄取的影响作用被阻断有关（Garcia-Herrera等，2008）。

◆ 2.3 血流量、神经传递和运动性

炎性反应常常导致宿主能量的再分配，使其不仅用于消化还用于其他功能。研究表明，注射LPS引发的全身性炎性反应，可显著降低黏膜组织的氧压力和微血管血红蛋白的氧饱和度（Maier等，2009）。患败血症的动物，其舌下和肠道部位的微血管密度显著降低（>50%），平均红细胞流速也明显下降（Verdant等，2009）。尽管在感染期间血液分流到其他部位，对动物清除感染并避免死亡

十分重要，但是胃肠道血流量的减少会影响养分的消化和吸收效率。

炎性反应对肠道功能的重要影响，包括改变肠道神经的神经生理学、神经化学和形态学特征，而这最终会导致肠道功能的变化（Mawe等，2009）。后超极化（AH）神经元受炎性反应的影响极为明显，AH神经元负责肠道蠕动、黏膜分泌和血管舒张等肠道的重要功能（Blackshaw等，2007；Brookes，2001；Furness等，2004）。在用三硝基苯磺酸（TNBS）诱发的肠道炎症模型中，研究人员发现发生炎性反应的结肠，其推进运动速率下降，而用环氧合酶2（COX-2）抑制剂能够缓解这一症状。发生肠道炎性反应时，可经常观察到神经元的过度兴奋。电生理学研究表明，通过化学方法（TNBS）和线虫（旋毛线虫）诱导的肠道炎性反应，常引起肠肌层和黏膜下丛神经元的兴奋性增强（Linden等，2004；Lomax等，2005）。肠道神经被切除后，炎性反应会造成长期的影响。感染旋毛线虫35 d后，空肠的电生理信号仍在发生变化，并且在炎性反应消除后AH神经元依然保持过度兴奋（Chen等，2007）。许多研究报道，食糜通过无菌动物小肠的速率更低（Abrams和Bishop，1967；Sacquet等，1971）。相对于普通大鼠，无菌大鼠小肠中移行动力复合物运动减缓并受时空的限制，使食糜的通过速率降低（Falk等，1998）。

◆ 2.4 肠道屏障功能、黏液和上皮细胞紧密连接

肠道屏障功能通常是指胃肠道上皮细胞阻止肠腔内有害物质吸收的能力，这些有害物质包括饲粮和微生物源的化学成分以及微生物病原体。肠道屏障功能可以被认为是由与黏液分泌有关的外在因子和肠上皮细胞的内在因子（如连接细胞的蛋白质-蛋白质网络结构）所组成（van der Klis和Jansman，2002）。分泌的黏液在肠上皮细胞表面形成一层“非流动层”，作为阻止有害物质和病原微生物扩散的物理屏障，同时也作为肠上皮细胞糖基化跨膜蛋白上可用的微生物黏附位点（Deplancke和Gaskins，2001）。在这个黏液层中，也可能存在较高浓度的由上皮细胞分泌的抗菌蛋白和免疫球蛋白。肠上皮层的屏障功能通常与肠道的通透性有关，并受一系列因素的影响，包括前面所述的上皮细胞的更新、黏液的分泌、紧密连接的构成等（Groschwitz和Hogan，2009）。炎性反应对影响肠道屏障通透性的因素也有显著的影响（Madara，1990）。

黏液的黏稠胶体特性主要与肠壁分泌的一种称作黏蛋白的糖蛋白有关。黏蛋白由一些密切相关的分子组成，这些分子具有一个共同的基本结构，但在不同种类动物、不同组织器官之间差异很大（Lamont，1992）。黏蛋白由一种叫杯状细胞的特定肠上皮细胞合成（Dharmani等，2009），并具有响应特定刺激的结构性（Forstner，1995）和加速性的（McCool等，1995；Plaisancie等，1997）分泌机制。肠上皮细胞分泌的黏蛋白的分子量很高，可达2×10^6 Da，含有高比例的O-连接碳水化合物、50%～80%的干物质（Montagne等，2004）。膜结合的黏蛋白也存在于肠上皮细胞的基顶膜，可作为上皮黏液的覆盖层。膜结合黏蛋白与分泌黏蛋白有相似的特性，但对其功能的了解还不多。黏蛋白分子的蛋白核心存在富含苏氨酸和丝氨酸的结构域，可为寡糖链提供黏附位点。蛋白核心也能够抵抗蛋白质水解酶的降解，因为在半胱氨酸残基中含有大量的分子内二硫键（Lidell等，2006）。糖基化作用使黏蛋白具有了大量的常规特性，比如抗蛋白酶、源于唾液酸和硫酸根离子的高电荷密度及系水力（Moncada和Chadee，2002）。

存在21种编码黏蛋白型糖蛋白的不同基因，且同属于MUC基因家族（Dharmani等，2009）。黏蛋白基因MUC2、MUC3和MUC6主要在十二指肠表达，而MUC1、MUC2、MUC3、MUC4、MUC6、MUC17和MUC20则在整个小肠中表达。黏蛋白基因MUC1、MUC2、MUC3、MUC4、MUC11、MUC12、MUC13、MUC17和MUC20在结肠中的表达水平各异（Andrianifahanana等，2006）。肠道中黏蛋白的表达受复杂的环境因素（如饲粮）、免疫细胞、上皮细胞及肠道菌群的精密调节。黏蛋白的从头合成可以在转录、转录后和翻译后水平上进行调节。在黏蛋白基因调节的遗传方面，MUC基因存在特定的、唯一的启动子序列，组织特异性的差异表达，以及转录因子的调节（Theodoropoulos和Carraway，2007）。炎性细胞因子对黏蛋白分泌的调节作用会得到详尽的阐述，但需要指出的是，现有研究已表明在各类细胞中存在多种生长因子、分化因子和菌源性因子，对黏蛋白基因的表达具有调节作用（Andrianifahanana等，2006）。

黏液胶体可被微生物作为养分和能量来源及定植位点所利用。微生物能够结合在肠道黏蛋白碳水化合物上，以避免自身被排出体外，这些定植在黏液中的微生物也可以阻止病原微生物定植于黏液层（Deplancke和Gaskins，2001）。这种定植还允许它们利用潜在的分子信号通路。因此，这对宿主共生菌和病原菌调控黏液的合成、黏蛋白的分泌及其组成十分有益。实际上我们已经发现，与单一菌悉菌猪相比，隔离饲养的普通猪的MUC1、MUC2和MUC13基因的表达均明显上调（Malik，2009）。读者可以直接从近年来的一些综述性论文中，了解微生物的作用机理以及微生物产物影响黏蛋白基因表达、分泌和糖基化作用方面的详细讨论。

细胞因子既是免疫细胞的产物，也是炎性反应等病理生理过程的调节枢纽。Ⅰ型细胞因子，包括IL-2、干扰素-γ（IFNγ）、IL-12和TNF，与细胞免疫应答的形成有关。Ⅱ型细胞因子，包括IL-4、IL-5、IL-6、IL-9、IL-10和IL-13，可促使体液免疫反应的产生（Lucey，1996）。目前已证实，各种Ⅰ型细胞因子对不同类型细胞中的膜结合型MUC基因具有调节作用，这些黏蛋白基因表达的异常或上调主要与恶性疾病相关。IFNγ与一系列的生物学过程有关，包括免疫应答的调控、细胞凋亡及卵巢癌变MUC1基因表达的上调（Boehm等，1997；Clark等，1994）。TNF-α则可以诱导人类DiFi直肠癌细胞的分化，并上调MUC1基因的表达（Novotnysmith等，1993）。最近的研究表明，TNF-α还可以激活鼻腔上皮细胞MUC1基因的表达，这可能是人类上呼吸道系统炎性疾病的发病机理（Shirasaki等，2003）。IL-1β和TNF-α是与黏蛋白分泌诱导并导致各种炎性疾病有关的Ⅰ型细胞因子。除了单个细胞因子的作用外，细胞因子之间也可相互作用，并协同上调黏蛋白的表达。Ⅱ型细胞因子作为实验动物杯状细胞组织转化的重要调节因子，在体内和体外条件下对黏蛋白的表达均具有刺激作用（Andrianifahanana等，2006）。关于呼吸道系统中Ⅱ型细胞因子对分泌型黏蛋白的调控作用已有了广泛研究，尤其是发现IL-13与炎性反应的表现型（包括呼吸道过敏、嗜酸性白细胞浸润和杯状细胞组织转化）有关（Wills-Karp等，1998）。相对于Ⅰ型细胞因子对分泌型黏蛋白的调节作用，人们对Ⅰ型和Ⅱ型细胞因子分别对膜结合型黏蛋白和分泌型黏蛋白的影响进行了更为广泛的研究。

紧接着肠壁黏液层的下方是一个单层的上皮细胞，作为半通透性屏障，可吸收养分、电解质和水分，但不能吸收有害物质、抗原和微生物（Groschwitz和Hogan，2009）。穿过该屏障可经由跨细胞途径和细胞旁路途径，而跨细胞途径穿过主要是选择性机制介导的可溶物转运。细胞旁路转运主要与肠道屏障功能有关，该途径通过渗透或扩散作用介导溶质非选择性穿过（Turner，2009）。细胞旁路转运受上皮细胞间复杂的跨膜蛋白网络调控，包括细胞桥粒、黏着连接和紧密连接。细胞桥粒和黏着连接（黏附带）是细胞间最为重要的机械连接，而紧密连接（闭合蛋白）则在上皮细胞的顶缘周围形成一种环状结构，负责调节细胞旁路转运（Groschwitz和Hogan，2009）。紧密连接由四个跨膜蛋白家族（occludin、claudins、结合黏附分子和tricellulin）构成，通过细胞间隙的嗜同种和异嗜性相互作用联系在一起。细胞内结构域与由一系列胞质骨架、衔接蛋白和信号蛋白介导的细胞极性、细胞增殖、细胞分化及细胞迁移相互影响，共同作用（Groschwitz和Hogan，2009）。人类连接蛋白中的claudin蛋白家族包含24个基因成员。Claudins含有两个细胞外环，用于家庭成员之间的异嗜性相互作用，以形成选择性离子通道。Occludins蛋白家族成员将胞内Claudins结构域连接到细胞骨架上（Forster，2008）。Tricellulin是细胞紧密连接的一个重要构件，借助于此三个细胞的边缘可以接触到一起（Westphal等，2010）。

虽然微生物的定植通过加强紧密连接可以形成一个更强大的屏障，但实际上仔猪在刚出生时其肠上皮细胞间的紧密连接就已开始形成。与无菌小鼠相比，带菌小鼠中富含脯氨酸蛋白-2（sprr2a）的表达上调100倍以上（Hooper和Gordon，2001），sprr2a是角化细胞被膜和紧密连接中桥连蛋白的一个重要组成部分。此外，小鼠（Cario等，2004）和猪（Danielsen等，2007）中闭锁小带1（occludens1）的表达受细菌识别而上调。

由微生物病原体及其诱发的炎性反应一般与紧密连接功能失调和肠道屏障功能降低有关（Madara，1990）。已有研究证明，有几种病原体可以通过其分泌的毒素或胞膜黏附作用直接影响肠上皮细胞的紧密连接，从而导致细胞损伤，诱发细胞凋亡，或者通过破坏紧密连接蛋白复合体的稳定性而影响肠上皮细胞的紧密连接。例如，致病性大肠埃希氏杆菌附着到肠上皮单层后，导致其屏障功能降低，紧密连接蛋白复合体出现异常（Muza-Moons等，2004）。另外，宿主炎性反应也可以导致肠道屏障功能的改变。紧密连接的促炎细胞因子调节剂主要包括TNF和IFNγ，已证明其可以直接下调occludin的表达（Zolotarevsky等，2002），重组紧密连接蛋白，并降低屏障功能（Mankertz等，2000）。此外，除了对紧密连接蛋白的直接影响外，促炎细胞因子尤其是TNF和IL-13诱导的细胞凋亡，可以导致细胞凋亡部位屏障功能的显著下降（Schulzke等，2006）。相反，抗炎细胞因子IL-10可以阻断由IFNγ诱导的单层上皮通透性的增强（Madsen等，1997），而IL-10基因敲除小鼠在自发性慢性肠道炎症产生之前，肠道的通透性就已经增加（Madsen等，1999）。

3 营养的意义与策略

在商业动物生产体系中，消除所有的炎性刺激是一种既不可行也是不能期望的选择。但是，有一些营养策略可以用于减少由炎性反应带来的损害，或是限制炎性反应通路的诱发。在本节中，我们将考虑氨基酸、脂肪酸、氧化锌和饲粮策略的作用，通过调节微生物的组成来控制肠道炎性反应和功能障碍。

◆ 3.1 氨基酸

传统上，氨基酸营养主要集中于明确每种必需氨基酸满足动物在特定阶段（断奶仔猪、泌乳母猪等）实现最大生产性能（如生长速度、产奶量等）的最低日需要量。总之，就是重点考虑补充氨基酸以满足蛋白质合成的需要。在某些情况下，为了应对炎性刺激，氨基酸的需要量可能会增加；然而，我们的讨论将仅限于前面提到的为了满足动物生产性能需要而补充必需氨基酸和非必需氨基酸，已有证据表明这对免疫反应或炎性应答功能有益。读者可以在第4章了解有关猪氨基酸需要量的详细论述。

在过去的大概十年时间里，因精氨酸和谷氨酰胺在营养代谢和免疫应答中表现出各种各样的功能，所以这两种氨基酸备受大家关注（Wu等，2007；Wu等，2009；Ziegler等，2003）。这些氨基酸在许多代谢途径，包括嘌呤、嘧啶、多胺、尿素、一氧化氮和谷胱甘肽（GSH）合成中，发挥着关键的调节作用和前体物作用，所有这些物质对肠道健康和屏障功能均有明显的作用（Wu等，2007）。另外，谷氨酰胺既是猪肠上皮细胞的主要能量底物，也是合成N-乙酰氨基葡糖-6-磷酸的必需成分，其中N-乙酰氨基葡糖-6-磷酸是糖蛋白（包括黏蛋白）合成的重要前体物（Wu等，2007）。精氨酸和谷氨酰胺似乎可以上调哺乳动物雷帕霉素靶蛋白（mTOR）的表达，mTOR是肌肉和肠道中蛋白质合成（Fumarola等，2005）和肠上皮细胞迁移（Rhoads等，2006）的重要调节因子。谷氨酰胺还可以保护结肠上皮细胞免受细胞因子诱导的细胞凋亡（Evans等，2003）。

一些添加试验已经证实了这些氨基酸的各种功能性作用。在断奶仔猪饲粮中补充1.0%的谷氨酰胺可缓解空肠黏膜的萎缩（Wu等，1996）。最近的研究表明，补充0.5%～1.0%的精氨酸可有效控制LPS诱导的仔猪肠道炎性反应，包括降低炎性细胞因子的表达和绒毛细胞的凋亡率（Liu等，2008）。其他研究表明，补充精氨酸可以减轻猪的免疫应答反应，并降低环磷酰胺引起的免疫抑制作用（Han等，2009）。Hernandez等（2009）最近研究发现，饲粮中补充精氨酸后，断奶仔猪的生产性能得到明显提高。另外，给感染小球隐孢子虫的新生仔猪口服高剂量的精氨酸，可引发前列腺素介导的腹泻（Gookin等，2008）。

最近研究人员利用右旋糖酐硫酸钠诱导的猪结肠炎模型，探讨了饲粮补充半胱氨酸的潜在抗炎作用（Kim等，2009）。半胱氨酸对重要的抗氧化剂——谷胱甘肽（GSH）的合成速率具有限制作用，在啮齿动物结肠炎模型中，半胱氨酸和含有半胱氨酸的化合物可以促进结肠黏蛋白的分泌，并抑制炎性细胞因子的表达。Kim等（2009）报道，随着促炎性细胞因子表达的降低，结肠组织学结构得到明显改善。

◆ 3.2 脂肪酸

目前对猪的脂肪酸营养关注得较少，然而谈到炎性反应，脂肪酸被认为是猪肠炎模型中调节免疫应答的潜在的具有重要药理作用的营养物质（Bassaganya-Riera和Honte-cillas，2006）。脂肪酸的总量及其组成决定了饲粮脂类对机体免疫系统的影响（Calder，1998）。过量摄入饱和脂肪酸可诱发炎性反应，而n-3多不饱和脂肪酸（PUFA）已被证明具有抗炎特性。谈到炎性反应，大家感兴趣的脂肪酸主要包括共轭亚油酸（CLA）、n-3多不饱和脂肪酸二十碳五烯酸（EPA）、二十二碳六烯酸（DHA）和n-6多不饱和脂肪酸花生四烯酸，其中花生四烯酸是前列腺素和白细胞三烯等炎性类花生酸类物质的前体物。共轭亚油酸和n-3多不饱和脂肪酸已在猪化学性结肠炎模型中进行过试验。结果表明，共轭亚油酸可以降低结肠炎的严重程度，这与其诱导增殖物激活受体-γ（PPAR-γ）表达、下调肿瘤坏死因子TNF表达有关（Bassaganya-Riera和Hontecillas，2006）。由于给PPAR-γ基因敲除的结肠炎模型小鼠补充CLA并没有取得积极效果，说明

CLA的有益作用是通过PPAR-γ诱导的（Bassaganya-Riera等，2004a）。反之，单独使用n-3 PUFA或联合CLA使用n-3 PUFA，结果导致结肠炎的发生更快。虽然n-3 PUFA有害地阻断了CLA诱导的PPAR-γ基因的表达，但它可通过诱导PPAR-δ的表达加快结肠炎的康复。研究发现，CLA可缓解细菌性结肠炎的炎性反应（Bassaganya-Riera等，2004b）。同样，这也与CLA诱导PPARγ表达并降低IFNγ表达有关。

对肠炎疾病的研究结果，促进了生产中添加共轭亚油酸的研究。与未补充CLA母猪生产的仔猪相比，饲粮添加CLA母猪生产的仔猪，断奶后经肠毒性大肠杆菌（ETEC）攻毒，其肠炎的发生明显减少，血清IgG和IgA明显升高（Patterson，2008）。虽然从妊娠中期至断奶期给母猪补充共轭亚油酸对断奶仔猪似乎有免疫刺激携带效应，但在保育饲粮中添加共轭亚油酸并未发现对仔猪健康的有益作用。

◆ 3.3 氧化锌

随着欧洲禁用促生长用抗生素以及北美使用抗生素的压力日益增加，氧化锌（ZnO）已被确定为一种潜在的免疫调节剂和控制炎性反应的一种手段。大量的研究报道一致认为，ZnO可减少断奶仔猪的腹泻，并促进其生长。ZnO的促生长作用与其降低肠道炎性反应和功能紊乱密切相关，但对其作用方式还不完全了解。降低炎性反应的意义，包括改善肠道功能，如增加绒毛长度、增加空肠和回肠刷状缘酶的活性以及增加大肠黏蛋白和杯状细胞数量（Hedemann等，2006；Slade等，2011）。

回肠微生物菌群在很大程度上受ZnO添加的影响（Vahjen，2010），这是ZnO影响炎性反应的一种作用机制。以前的研究推测，ZnO的降低腹泻作用与其对致病性大肠杆菌的抑制作用有关；然而，添加ZnO还可增加柠檬酸杆菌属、肠杆菌属、奈瑟氏菌属和不动杆菌属等肠细菌的定植水平，可能提示这并非ZnO的直接抗菌作用。添加ZnO后，干细胞因子的表达降低、肥大细胞在下游募集以及组胺的释放，这为解释其对大肠杆菌没有直接抗菌作用而仍然可以减少腹泻提供了一个潜在的作用机制。Sargeant等（2010）最近报道，饲粮中添加ZnO可减轻动物感染ETEC后的炎性反应。他们认为，ETEC的延迟定植可能是MUC4表达降低的结果，而MUC4是一个针对ETEC K88的受体。

无论作用机制如何，空肠的蛋白质组学分析表明，ZnO可降低氧化型谷胱甘肽和活性caspase-3的水平，表明ZnO可以提高氧化还原状态并降低细胞凋亡活性（Feng等，2010）。据报道，添加ZnO的猪，其肠道细菌向肠系膜淋巴结的移位水平降低（Broom等，2006），是细胞凋亡和氧化应激程度降低的原因。但目前尚不清楚，ZnO是如何降低细菌移位的。给断奶仔猪补充ZnO后，其肠道IGF-1水平增加（Li et等，2006）。IGF-1是肠道发育和结构完整性的重要调节因子（Burrin等，1996；Carey和Alexander，1999；Herman等，2004）。然而，炎性反应时IGF-1的水平下降（Hevener等，1997），从而减轻肠道炎性反应，而不是阻止炎性反应。无论IGF-1水平的提高是肠道炎性反应减轻的原因还是结果，IGF-1均代表着ZnO提高动物增重和饲粮效率的一种措施。

Molist等（2011）最近的研究表明，为了促进肠道健康，考虑ZnO和饲粮的相互作用是十分重要的。他们发现，虽然麦麸和ZnO均可独自减少大肠杆菌的定植，但当两者同时使用时，大肠杆菌的定植数量反而增加（Molist等，2011）。这种有害交互效应的机制目前尚不清楚。有机锌替代品，包括甘氨酸锌，也被证实对动物的生产性能有促进作用（Feng等，2010）。已有研究表明，提高有机锌的生物利用率（Carlson和Case，2002；Ward等，1996），可使其在低添加量条件下就发挥有益作用，进而减少锌排放对环境的污染（Wang等，2009）。将来关于锌源对宿主影响的比较研究，也可加深对其作用方式的理解。

◆ 3.4 营养与肠道微生物群落

动物（包括猪）胃肠道内定植大量不同种类的微生物（Hill等，2002；Leser等，2002），对胃肠道健康（Flint等，2007；McGarr等，2005）和动物生产性能的作用越来越被大家所认识。肠道菌群沿单胃动物肠道的延伸而发生变化，从近端到远端区域菌群的密度和种类一般会不断增加（Richards等，2005）。猪胃和十二指肠内每克食糜含有的微生物数量为$10^3 \sim 10^5$ cfu，到回肠逐渐提高到$10^8 \sim 10^9$ cfu，而到了盲肠和结肠则达到$10^{10} \sim 10^{12}$ cfu最大密度。由于饲粮成分不仅为动物还可以为微生物提供最基本的营养，因此饲粮组成是决定猪肠道微生物菌群结构的主要因素之一（Hill等，2005；Pieper等，2008）。另外，因菌群中各种微生物的水解谱和营养需要差异很大，所以饲粮组成的改变可以有利于特定微生物群落。尤其是一些非消化性养分，如饲粮纤维、抗性淀粉和非消化性蛋白质，可以避开宿主的消化和吸收过程，充当微生物代谢的主要底物。

小肠末端（回肠）和后肠中纤维和蛋白质在该处微生物菌群共同作用下进行发酵。碳水化合物多聚体首先被酶解成单体（单糖），然后才能被转运到细胞内发生代谢。单糖发酵主要被降解为短链脂肪酸（SCFAs），包括乳酸、乙酸、丙酸和丁酸，以及二氧化碳和氢气。在回肠中主要产生乳酸，因为回肠中含有大量的产乳酸杆菌；而在后肠中主要产生乙酸、丙酸和丁酸，因为后肠内梭菌属细菌及其代谢的共生作用占优势；而乳酸可以被进一步代谢为丁酸或丙酸（Flint等，2007）。一般认为，SCFAs对宿主是有益的，可以为肠上皮细胞和全身提供额外的能量来源，并以不确定的形式抑制某些“不受欢迎”细菌的生长，刺激上皮细胞的增殖，促进紧密连接的形成，并至少部分地通过受体介导的机制来抑制炎性反应和遗传毒性（Blaut和Clavel，2007；Brown等，2003；Peng等，2009）。

同样，蛋白质在胞内转运和代谢之前，首先被微生物菌群的胞外酶水解为肽和氨基酸（Gaskins，2001）。氨基酸可以直接用于微生物蛋白质的合成，或发酵成各种产物。氨基酸经脱氨基作用生成氨，并利用碳架形成可供能的SCFA，包括支链脂肪酸。在上皮细胞表面，微生物脲酶也可将尿素降解为氨。当血浆尿素浓度较高时，这一降解过程增强。相反，氨基酸脱羧基会生成胺，包括组胺、腐胺和尸胺。芳香族氨基酸代谢也可产生许多酚类和吲哚化合物（Blaut和Clavel，2007；Macfarlane和Macfarlane，2006）。含硫氨基酸代谢，有助于产生大量的含硫发酵产物，包括硫化氢和二甲基三硫化物（Geypens等，1997）。与碳水化合物代谢产生的发酵产物不同，蛋白质发酵产物主要与上皮细胞的毒性反应和致癌有关（Blaut和Clavel，2007；Gaskins，2001；McGarr等，2005），包括破坏肠道上皮细胞的屏障功能。

饲粮中含有多种可发酵碳水化合物，包括菊粉、抗性淀粉、谷物纤维和甜菜浆，它们可增加猪肠道中SCFA含量、降低蛋白质发酵产物的量（Awati等，2006；Htoo等，2007；Nyachoti等，2006；Smiricky-Tjardes等，2003；Williams等，2001），其中的原因可能是饲粮中的可发酵碳水化合物通过减少蛋白质发酵供能，促进微生物菌体蛋白的合成，抑制了蛋白质发酵产生有毒性的终产物

（Awati等，2006；Bhandari等，2009；Bikker等，2006；Htoo等，2007；Jeaurond等，2008；Nyachoti等，2006）。虽然生长性能对饲粮调控的反应变异很大，但研究结果支持，随着各种来自可发酵碳水化合物的发酵性纤维的增加，有害的蛋白质发酵产物降低。在人和猪上也已证实，可发酵碳水化合物可以促进胃肠道的健康和屏障功能（Nofrarias等，2007；Topping和Clifton，2001）。同样，当饲粮中含有较高水平的可发酵蛋白质时，蛋白质的发酵产物会增加（Bikker等，2006；Geypens等，1997），结肠上皮细胞的损伤和增殖也增加（Govers等，1993）。

4 结 语

胃肠道免疫机制激活与共生微生物区系组成及相应发酵产物变化，以及因肠道病原体定植引起的炎症过度反应有关，它对猪养分的消化和利用具有重要的影响。胃肠道黏膜表面的主要生理变化包括免疫细胞代谢活性增加和上皮细胞更新加快，后者影响养分消化酶和转运蛋白，肠蠕动、血流以及屏障功能（包括黏液分泌和上皮细胞紧密连接蛋白）增强。这些代谢变化对肠道的抗病能力、养分的消化吸收以及生长所需的必需养分都有影响，因此这也关乎可持续性猪营养生产实践理论的发展。

许多营养策略正在逐步形成，人们制订这些营养策略的目的或是维护良好的屏障功能，或是避免肠道上皮暴露于对屏障功能具有损害作用的化合物中。这两种营养策略均是尽量减少与厌食症和营养素从瘦肉组织流失有关的免疫激活和潜在的炎性反应。这些策略不只是依赖营养物质提供必需营养素的价值，更主要的是依赖于其功能特性，以满足被激活的炎性反应的额外需求，为增强猪的健康、提高猪肉生产的养分利用率提供更多机会。

作者：Benjamin P. Willing,
Gita Malik和Andrew
G. Van Kessel
译者：游金明

参考文献

Abrams, G. D., and J. E. Bishop. 1967. Effect of normal microbial flora on gastrointestinal motility. Proc. Soc. Exp. Biol. Med.126:301–304.

Andrianifahanana, M., N. Moniaux, and S. K. Batra. 2006. Regulation of mucin expression: Mechanistic aspects and implications for cancer and inflammatory diseases. Biochim. Biophys. Acta–Rev. Cancer 1765:189–222.

Awati, A., B. A. Williams, M. W. Bosch, W. J. J. Gerrits, and M. W. A. Verstegen. 2006. Effect of inclusion of fermentable carbohydrates in the diet on fermentation end−product profile in feces of weanling piglets. J. Anim. Sci. 84:2133–2140.

Bakke, O. M., and T. Midtvedt. 1970. Influence of germfree status on excretion of simple phenols of possible significance in tumor prmotion. Experientia 26:519.

Bassaganya−Riera, J., and R. Hontecillas. 2006. CLA and n−3 PUFA differentially modulate clinical activity and colonic PPAR responsive gene expression in a pig model of experimental IBD. Clin. Nutr. 25:454–465.

Bassaganya−Riera, J., J. King, and R. Hontecillas. 2004a. Health benefits of conjugated linoleic acid: Lessons from pig models in biomedical research. Eur. J. Lipid Sci. Technol. 106:856–861.

Bassaganya−Riera, J., K. Reynolds, S. Martino−Catt, Y. Z. Cui, L. Hennighausen, F. Gonzalez, J. Rohrer, A. U. Benninghoff, and R.Hontecillas. 2004b. Activation of PPAR gamma and delta by conjugated linoleic acid mediates protection from experimental inflammatory bowel disease. Gastroenterol. 127:777–791.

Bhandari, S. K., C. M. Nyachoti, and D. O. Krause. 2009. Raw potato starch in weaned pig diets and its influence on postweaning scours and the molecular microbial ecology of the digestive tract. J. Anim. Sci. 87:984–993.

Bikker, P., A. Dirkzwager, J. Fledderus, P. Trevisi, I. le Huerou−Luron, J. P. Lalles, and A. Awati. 2006. The effect of dietary protein and fermentable carbohydrates levels on growth performance and intestinal characteristics in newly weaned piglets. J. Anim.Sci. 84:3337–3345.

Blackshaw, L. A., Brookes S. J. H., Grundy D., Schemann M. (2007) Sensory transmission in the gastrointestinal tract. Neurogastroenterol.Motil. 19:1–19.

Blaut M., and T. Clavel. 2007. Metabolic diversity of the intestinal microbiota: Implications for health and disease. J. Nutr.137:S751–S755.

Boehm, C. M., M. C. Mulder, R. Zennadi, M. Notter, A. Schmitt-Graeff, O. J. Finn, J. Taylor-Papadimitriou, et al. 1997.Carbohydrate recognition on MUC1-expressing targets enhances cytotoxicity of a T cell subpopulation. Scand. J. Immunol.46:27–34.

Brookes, S. J. H. 2001. Classes of enteric nerve cells in the guinea-pig small intestine. Anat. Rec. 262:58–70.

Broom, L. J., H. M. Miller, K. G. Kerr, and J.S. Knapp. 2006. Effects of zinc oxide and Enterococcus faecium SF68 dietary supplementation on the performance, intestinal microbiota and immune status of weaned piglets. Res. Vet. Sci. 80:45–54.

Brown, A. J., S. M. Goldsworthy, A. A. Barnes, M. M. Eilert, L. Tcheang, D. Daniels, A. I. Muir, et al. 2003. The orphan G protein-coupled receptors GPR41 and GPR43 are activated by propionate and other short chain carboxylic acids. J. Biol.Chem. 278:11312–11319.

Burrin, D. G., B. Stoll, and X. Guan. 2003. Glucagon-like peptide 2 function in domestic animals. Domest. Anim. Endocrinol. 24:103–122.

Burrin, D. G., T. J.Wester, T. A. Davis, S. Amick, J. P. Heath. 1996. Orally administered IGF-I increases intestinal mucosal growth in formula-fed neonatal pigs. Am. J. Physiol.-Reg. Integr. Compar. Physiol. 270:R1085–R1091.

Calder, P. C. 1998. Dietary fatty acids and the immune system. Nutr. Rev. 56:S70–S83.

Carey, H. V., and A. N. Alexander. 1999. Oral IGF-I enhances nutrient and electrolyte absorption in neonatal piglet intestine. Am.J. Physiol. 277:G619–G625.

Cario, E., G. Gerken, and D. K. Podolsky. 2004. Toll-like receptor 2 enhances ZO-1-associated intestinal epithelial barrier integrity via protein kinase C. Gastroenterol. 127:224–238.

Carlson, M. S., and C. L. Case. 2002. Effect of feeding organic and inorganic sources of additional zinc on growth performance and zinc balance in nursery pigs. J. Anim. Sci. 80:1917–1924.

Chen, Z. X., Z. Suntres, J. Palmer, J. Guzman, A. Javed, J. J. Xue, J. G. Yu, et al. 2007. Cyclic AMP signaling contributes to neural plasticity and hyperexcitability in AH sensory neurons following intestinal Trichinella spiralis-induced inflammation. Int. J. Parasitol. 37:743–761.

Clark, S., M. A. McGuckin, T. Hurst, and B. G.Ward. 1994. Effect Of interferon-gamma and Tnf-Alpha on Muc1 mucin expression in ovarian-carcinoma cell-lines. Disease Markers 12:43–50.

Coopersmith, C. M., K. C. Chang, P. E. Swanson, K. W. Tinsley, P.E. Stromberg, T. G. Buchman, I. E. Karl, and R. S. Hotchkiss. 2002. Overexpression of Bcl-2 in the intestinal epithelium improves survival in septic mice. Crit. Care Med. 30:195–201.

Coopersmith, C. M., P. E. Stromberg, C. G. Davis, W. M. Dunne, D. M. Amiot, I. E. Karl, R. S. Hotchkiss, and T. G. Buchman. 2003. Sepsis from *Pseudomonas aeruginosa* pneumonia decreases intestinal proliferation and induces gut epithelial cell cycle arrest. Crit. Care Med.

31:1630–1637.

Danielsen, M., H. Hornshoj, R. H. Siggers, B. B. Jensen, A. G. van Kessel, and E. Bendixen. 2007. Effects of bacterial colonization on the porcine intestinal proteome. J. Proteome Res. 6:2596–604.

Danielsen, M., T. Thymann, B. B. Jensen, O. N. Jensen, P. T. Sangild, and E. Bendixen. 2006. Proteome profiles of mucosal immunoglobulin uptake in inflamed porcine gut. Proteomics. 6:6588–6596.

Deplancke, B., and H. R. Gaskins. 2001. Microbial modulation of innate defense: goblet cells and the intestinal mucus layer. Am.J. Clin. Nutr. 73:1131S–1141S.

Dharmani, P., V. Srivastava, V. Kissoon-Singh, and K. Chadee. 2009. Role of intestinal mucins in innate host defense mechanisms against poathogens. J. Innate Immun. 1:123–135.

Evans, M. E., D. P. Jones, and T. R. Ziegler. 2003. Glutamine prevents cytokine-induced apoptosis in human colonic epithelial cells. J. Nutr. 133:3065–3071.

Falk, P. G., L. V. Hooper, T. Midtvedt, and J. I. Gordon. 1998. Creating and maintaining the gastrointestinal ecosystem: What we know and need to know from gnotobiology. Microbiol. Molecul. Biol. Rev. 62:1157–1170.

Feng, J., Y. Wang, J. W. Tang, M. Q. Ma, and J. Feng. 2010. Dietary zinc glycine chelate on growth performance, tissue mineral concentrations, and serum enzyme activity in weanling piglets. Biol. Trace Elem. Res. 133:325–334.

Flint, H. J., S. H. Duncan, K. P. Scott, and P. Louis. 2007. Interactions and competition within the microbial community of the human colon: Links between diet and health. Environ. Microbiol. 9:1101–1111.

Forster, C. 2008. Tight junctions and the modulation of barrier function in disease. Histochem. Cell Biol. 130:55–70.

Forstner, G. 1995. Signal-transduction, packaging and secretion of mucins. Annu. Rev. Physiol. 57:585–605.

Fumarola, C., S. La Monica, and G. G. Guidotti. 2005. Amino acid signaling through the mammalian target of rapamycin (mTOR) pathway: Role of glutamine and of cell shrinkage. J. Cell. Physiol. 204:155–165.

Furness, J. B., C. Jones, K. Nurgali, and N. Clerc. 2004. Intrinsic primary afferent neurons and nerve circuits within the intestine.Prog. Neurobiol. 72:143–164.

Garcia-Herrera, J., M. C. Marca, E. Brot-Laroche, N. Guillen, S. Acin, M. A. Navarro, J. Osada, and M. J. Rodriguez-Yoldi. 2008.Protein kinases, TNF-alpha, and proteasome contribute in the inhibition of fructose intestinal transport by sepsis *in vivo*. Am.J. Physiol. Gastrointest. Liver Physiol. 294:G155–G164.

Gaskins, H. R. 2001. Intestinal bacteria and their influence on swine growth. Pages

585–608 in *Swine Nutrition*. A. J. Lewis and L. L. Southern, eds. CRC Press, Boca Raton, FL.
Geypens, B., D. Claus, P. Evenepoel, M. Hiele, B. Maes, M. Peeters, P. Rutgeerts, and Y. Ghoos. 1997. Influence of dietary protein supplements on the formation of bacterial metabolites in the colon. Gut 41:70–76.

Gookin, J. L., D. M. Foster, M. R. Coccaro, and S. H. Stauffer. 2008. Oral delivery of L-arginine stimulates prostaglandin-dependent secretory diarrhea in *Cryptosporidium parvum*-infected neonatal piglets. J. Pediatr. Gastroenterol. Nutr. 46:139–146.

Govers, M. J., J. A. Lapre, H. T. De Vries, and R. van der Meer. 1993. Dietary soybean protein compared with casein damages colonic epithelium and stimulates colonic epithelium and stimulates colonic epithelial proliferation in rats. J. Nutr. 123:1709.

Groschwitz, K. R., and S. P. Hogan. 2009. Intestinal barrier function: molecular regulation and disease pathogenesis. J. Allergy Clin. Immunol. 124:3–20.

Grunfeld, C., C. Zhao, J. Fuller, A. Pollock, A. Moser, J. Friedman, and K. R. Feingold. 1996. Endotoxin and cytokines induce expression of leptin, the ob gene product, in hamsters - A role for leptin in the anorexia of infection. J. Clin. Invest. 97:2152–2157.

Gupta, S. 2003. Molecular signaling in death receptor and mitochondrial pathways of apoptosis (Review). Int. J. Oncol. 22:15–20.

Hall, P. A., P. J. Coates, B. Ansari, and D. Hopwood. 1994. Regulation of cell number in the mammalian gastrointestinal-tract—the importance of apoptosis. J. Cell Sci. 107:3569–3577.

Han, J.,Y. Liu,W. Fan, J. Chao,Y. Hou,Y.Yin, H. Zhu, et al. 2009. Dietary l-arginine supplementation alleviates immunosuppression induced by cyclophosphamide in weaned pigs. Amino Acids 37:643.

Hedemann, M. S., B. B. Jensen, and H. D. Poulsen. 2006. Influence of dietary zinc and copper on digestive enzyme activity and intestinal morphology in weaned pigs. J. Anim. Sci. 84:3310–3320.

Herman, A. C., E. M. Carlisle, J. B. Paxton, and P. V. Gordon. 2004. Insulin-like growth factor-I governs submucosal growth and thickness in the newborn mouse ileum. Pediatr. Res. 55:507–13.

Hernandez, A., C. F. Hansen, B. P. Mullan, and J. R. Pluske. 2009. L-arginine supplementation of milk liquid or dry diets fed to pigs after weaning has a positive effect on production in the first three weeks after weaning at 21 days of age. Anim. Feed Sci. Tech. 154:102–111.

Hevener,W., G.W. Almond, J. D. Armstrong, and R. G. Richards. 1997. Effects of acute endotoxemia on serum somatotropin and insulin-like growth factor I concentrations in prepubertal gilts. Am. J. Vet. Res. 58:1010–1013.

Hill, J. E., S. M. Hemmingsen, B. G. Goldade, T. J. Dumonceaux, J. Klassen, R. T. Zijlstra,

S. H. Goh, and A. G. van Kessel. 2005. Comparison of ileum microflora of pigs fed corn-, wheat-, or barley-based diets by chaperonin-60 sequencing and quantitative PCR. Appl. Environ. Microbiol. 71:867–875.

Hill, J. E., R. P. Seipp, M. Betts, L. Hawkins, A. G. van Kessel, W. L. Crosby, and S. M. Hemmingsen. 2002. Extensive profiling of a complex microbial community by high-throughput sequencing. Appl. Environ. Microbiol. 68:3055–3066.

Hooper, L. V., and J. I. Gordon. 2001. Commensal host-bacterial relationships in the gut. Sci. 292:1115–1118.

Htoo, J. K., B. A. Araiza, W. C. Sauer, M. Rademacher, Y. Zhang, M. Cervantes, and R. T. Zijlstra. 2007. Effect of dietary protein content on ileal amino acid digestibility, growth performance, and formation of microbial metabolites in ileal and cecal digesta of early-weaned pigs. J. Anim. Sci. 85:3303–3312.

Hughes, R., M. J. Kurth, V. McGilligan, H. McGlynn, and I. Rowland. 2008. Effect of colonic bacterial metabolites on Caco-2 cell paracellular permeability *in vitro*. Nutr. Cancer 60:259–266.

Jansman, A. J. M., W. Smink, P. van Leeuwen, and M. Rademacher. 2002. Evaluation through literature data of the amount and amino acid composition of basal endogenous crude protein at the terminal ileum of pigs. Anim. Feed Sci. Tech. 98:49–60.

Jeaurond, E. A., M. Rademacher, J. R. Pluske, C. H. Zhu, and C. F. M. de Lange. 2008. Impact of feeding fermentable proteins and carbohydrates on growth performance, gut health and gastrointestinal function of newly weaned pigs. Can. J. Anim. Sci.88:271–281.

Johnson, R. W., B. N. Finck, K. W. Kelley, and R. Dantzer. 1998. *In vivo* and *in vitro* evidence for the involvement of tumor necrosis factor-alpha in the induction of leptin by lipopolysaccharide. Endocrinol. 139:2278–2283.

Jourd'heuil, D., Z. Morise, E. M. Conner, and M. B. Grisham. 1997. Oxidants, transcription factors, and intestinal inflammation. J. Clin. Gastroenterol. 25:S61–S72.

Kent, S., R. M. Bluthe, R. Dantzer, A. J. Hardwick, K. W. Kelley, N. J. Rothwell, and J. L. vannice. 1992. Different receptor mechanisms mediate the pyrogenic and behavioral-effects of interleukin-1. Proc. Natl. Acad. Sci. 89:9117–9120.

Kerr, J. F. R., A. H. Wyllie, and A. R. Currie. 1972. Apoptosis—basic biological phenomenon with wide-ranging implications in tissue kinetics. Br. J. Cancer 26:239–257.

Kim, C. J., J. Kovacs-Nolan, C. Yang, T. Archbold, M. Z. Fan, and Y. Mine. 2009. L-cysteine supplementation attenuates local inflammation and restores gut homeostasis in a porcine model of colitis. Biochim. Biophys. Acta-Gen. Subj. 1790:1161–1169.

Lamont, J. T. 1992. Mucus—the front-line of intestinal mucosal defense, neuro-immuno-physiology of the gastrointestinal Mucosa.Pages 190–201 in Implications for Inflammatory

Diseases. New York Academy of Sciences, New York.

Leschelle, X., V. Robert, S. Delpal, B. Mouill'e, C. Mayeur, P. Martel, and F. Blachier. 2002. Isolation of pig colonic crypts for cytotoxic assay of luminal compounds: Effects of hydrogen sulfide, ammonia, and deoxycholic acid. Cell Biol. Toxicol.18:193.

Leser, T. D., J. Z. Amenuvor, T. K. Jensen, R. H. Lindecrona, M. Boye, and K. Moller. 2002. Culture-independent analysis of gut bacteria: The pig gastrointestinal tract microbiota revisited. Appl. Environ. Microbiol. 68:673–90.

Li, D. F., X. L. Li, J. D. Yin, X. J. Chen, J. J. Zang, and X. Zhou. 2006. Dietary supplementation with zinc oxide increases IGF-I and IGF-I receptor gene expression in the small intestine of weanling piglets. J. Nutr. 136:1786–1791.

Lidell, M. E., D. M. Moncada, K. Chadee, and G. C. Hansson. 2006. Entamoeba histallytica cysteine proteases cleave the MUC2 mucin in its C-terminal domain and dissolve the protective colonic mucus gel. Proc. Natl. Acad. Sci. 103:9298–9303.

Linden, D. R., K. A. Sharkey, W. Ho, and G. M. Mawe. 2004. Cyclooxygenase-2 contributes to dysmotility and enhanced excitability of myenteric AH neurones in the inflamed guinea pig distal colon. J. Physiol. 557:191–205.

Liu, Y., J. J. Huang, Y. Q. Hou, H. L. Zhu, S. J. Zhao, B. Y. Ding, Y. L. Yin, et al. 2008. Dietary arginine supplementation alleviates intestinal mucosal disruption induced by *Escherichia coli* lipopolysaccharide in weaned pigs. Br. J. Nutr. 100:552–560.

Lomax, A. E., G. M. Mawe, and K. A. Sharkey. 2005. Synaptic facilitation and enhanced neuronal excitability in the submucosal plexus during experimental colitis in guinea-pig. J. Physiol. 564:863–875.

Lucey, D. R., M. Clerici, and G. M. Shearer. 1996. Type 1 and type 2 cytokine dysregulation in human infectious, neoplastic, and inflammatory diseases. Clin. Microbiol. Rev. 9:532–562.

Macfarlane, S., and G. T. Macfarlane. 2006. Composition and metabolic activities of bacterial biofilms colonizing food residues in the human gut. Appl. Environ. Microbiol. 72:6204–6211.

Madara, J. L. 1990 Pathobiology of the intestinal epithelial barrier. Am. J. Pathol. 137:1273–1281.

Madsen, K. L., S. A. Lewis, M. M. Tavernini, J. Hibbard, and R. N. Fedorak. 1997. Interleukin 10 prevents cytokine-induced disruption of T84 monolayer barrier integrity and limits chloride secretion. Gastroenterol. 113:151–159.

Madsen, K. L., D. Malfair, D. Gray, J. S. Doyle, L. D. Jewell, and R. N. Fedorak. 1999. Interleukin-10 gene-deficient mice develop a primary intestinal permeability defect in response to enteric microflora. Inflamm. Bowel Dis. 5:262–270.

Maier, S.,W. Pajk, H. Ulmer, H. Hausdorfer, C. Torgersen, J. Klocker,W. Hasibeder, and H. Knotzer. 2009. Epoprostenol improves mucosal tissue oxygen tension in an acute endotoxemic pig model. Shock 31:104–110.

Malik, G. 2009. Effect of cereal type and commensal bacteria on availability of methionine sources and intestinal physiology in pigs. Dept. of Animal and Poultry Science, Univ. of Saskatchewan, Saskatoon, Canada.

Mankertz, J., S. Tavalali, H. Schmitz, A. Mankertz, E. O. Riecken, M. Fromm, and J. D. Schulzke. 2000. Expression from the human occludin promoter is affected by tumor necrosis factor alpha and interferon gamma. J. Cell Sci. 113:2085–2090.

Mather, M., and H. Rottenberg. 2001 Polycations induce the release of soluble intermembrane mitochondrial proteins. Biochim.Biophys. Acta 1503:357–368.

Mawe, G. M., D. S. Strong, and K. A. Sharkey. 2009. Plasticity of enteric nerve functions in the inflamed and postinflamed gut.Neurogastroenterol. Motility 18:21:481–491.

McCool, D. J., J. F. Forstner, and G. G. Forstner. 1995. Regulated and unregulated pathways for Muc2 mucin secretion in human colonic Ls180 adenocarcinoma cells are distinct. Biochem. J. 312:125–133.

McGarr, S. E., J. M. Ridlon, and P. B. Hylemon. 2005. Diet, anaerobic bacterial metabolism, and colon cancer—A review of the literature. J. Clin. Gastroenterol. 39:98–109.

Merritt, A. J., C. S. Potten, C. J. Kemp, J. A. Hic*km*an, A. Balmain, D. P. Lane, and P. A. Hall. 1994. The role of P53 in spontaneous and radiation-induced apoptosis in the gastrointestinal-tract of normal and P53-deficient mice. Cancer Res. 54:614–617.

Miniats, O. P., and V. E. Valli. 1973. The gastrointestinal tract of gnotobiotic pigs. Pages 575–583 in Germfree Research: Biologic Effect of Gnotobiotic Environments. J. B. Heneghan, ed. Academic Press, New York.

Molist, F., R. G. Hermes, A. G. de Segura, S. M. Martin-Orue, J. Gasa, E. G. Manzanilla, and J. F. Perez. 2011. Effect and interaction between wheat bran and zinc oxide on productive performance and intestinal health in post-weaning piglets. Br. J. Nutr. 105:1592–1600.

Moncada, D. M., and K. Chadee. 2002. Production, structure, and function of gastrointestinal mucins Pages 57–59 in Infections of the Gastrointestinal Tract. M. J. Blaser, ed. Lippincott Williams and Wilkins, Philidelphia.

Montagne, L., C. Piel, and J. P. Lalles. 2004. Effect of diet on mucin kinetics and composition: Nutrition and health implications .Nutr. Rev. 62:105–114.

Muza-Moons, M. M., E. E. Schneeberger, and G. A. Hecht. 2004. Enteropathogenic *Escherichia coli* infection leads to appearance of aberrant tight junctions strands in the lateral membrane of intestinal epithelial cells. Cell Microbiol. 6:783–793.

Nofrarias, M., D. Martinez-Puig, J. Pujols, N. Majo, and J. F. Perez. 2007. Long-term intake of resistant starch improves colonic mucosal integrity and reduces gut apoptosis and blood immune cells. Nutr. 23:861–870.

Novotnysmith, C. L., M. A. Zorbas, A. M. McIsaac, T. Irimura T., B. M. Boman, L. C. Yeoman, and G. E. Gallick. 1993. Down–Modulation of epidermal growth–factor receptor accompanies Tnf–induced differentiation of the Difi human adenocarcinoma cell–line toward a goblet–like phenotype. J. Cell. Physiol. 157:253–262.

Nyachoti, C. M., C. F. de Lange, B. W. McBride, S. Leeson, and V. M. Gabert. 2000. Endogenous gut nitrogen losses in growing pigs are not caused by increased protein synthesis rates in the small intestine. J. Nutr. 130:566–572.

Nyachoti, C. M., F. O. Omogbenigun, M. Rademacher, and G. Blank. 2006. Performance responses and indicators of gastrointestinal health in early–weaned pigs fed low–protein amino acid–supplemented diets. J. Anim. Sci. 84:125–134.

Ou, D., D. Li, Y. Cao, X. Li, J. Yin, S. Qiao, and G. Wu. 2007. Dietary supplementation with zinc oxide decreases expression of the stem cell factor in the small intestine of weanling pigs. J. Nutr. Biochem. 18:820–826.

Patterson, R., M. L. Connor, D. O. Krause, and C. M. Nyachoti. 2008. Response of piglets weaned from sows fed diets supplemented with conjugated linoleic acid (CLA) to an *Escherichia coli* K88(+) oral challenge. Animal 2:1303–1311.

Peng, L., Z. R. Li, R. S. Green, I. R. Holzman, and J. Lin. 2009. Butyrate enhances the intestinal barrier by facilitating tight junction assembly via activation of AMP–activated protein kinase in Caco–2 cell monolayers. J. Nutr. 139:1619–1625.

Pie S., J. P. Lalles, F. Blazy, J. Laffitte, B. Seve, and I. P. Oswald. 2004. Weaning is associated with an upregulation of expression of inflammatory cytokines in the intestine of piglets. J. Nutr. 134:641–647.

Pieper, R., R. Jha, B. Rossnagel, A. G. van Kessel, W. B. Souffrant, and P. Leterme. 2008. Effect of barley and oat cultivars with different carbohydrate compositions on the intestinal bacterial communities in weaned piglets. FEMS Microbiol. Ecol. 66:556–566.

Plaisancie, P., A. Bosshard, J. C. Meslin, and J. C. Cuber. 1997. Colonic mucin discharge by a cholinergic agonist, prostaglandins,and peptide YY in the isolated vascularly perfused rat colon. Digestion 58:168–175.

Plata–Salaman, C. R. 1994. Meal patterns in response to the intracerebroventricular administration of interleukin–1 beta in rats.Physiol. Behav. 55:727–33.

Potten, C. S., C. Booth, and D. M. Pritchard. 1997. The intestinal epithelial stem cell: the mucosal governor. Int. J. Exp. Pathol.78:219–243.

Potten, C. S., and M. Loeffler. 1990. Stem–Cells—attributes, cycles, spirals, pitfalls and uncertainties –lessons for and from the crypt. Development 110:1001–1020.

Potten, C. S., G. Owen, and S. A. Roberts. 1990. The Temporal and spatial changes in cell-proliferation within the irradiated crypts of the murine small-intestine. Int. J. Radiat. Biol. 57:185–199.

Rakoff-Nahoum, S., J. Paglino, F. Eslami-Varzaneh, S. Edberg, and R. Medzhitov. 2004. Recognition of commensal microflora by toll-like receptors is required for intestinal homeostasis. Cell 118:229–241.

Rhoads, J. M., X. Niu, J. Odle, and L. M. Graves. 2006. Role of mTOR signaling in intestinal cell migration. Am. J. Physiol. Gastrointest. Liver Physiol. 291:G510–517.

Richards, J. D., J. Gong, and C. F. M. de Lange. 2005. The gastrointestinal microbiota and its role in monogastric nutrition and health with an emphasis on pigs: Current understanding, possible modulations, and new technologies for ecological studies. Can. J. Anim. Sci. 85:421–435.

Sacquet, E., P. Raibaud, and J. Garnier. 1971. Comparative study of microbial flora of stomach, small intestine and caecum in holoxenic (conventional) rats and in rats with artificial self-filling blind loops or bile duct fistulas. Ann. Inst. Pasteur. (Paris) 120:501–524.

Sargeant, H. R., M. A. Shaw, M. AbuOun, J. W. Collins, M. J. Woodward, R. M. La Ragione, and H. M. Miller. 2010. The metabolic impact of zinc oxide on porcine intestinal cells and enterotoxigenic *Escherichia coli* K88. Livest. Sci. 133: 45–48.

Sarraf, P., R. C. Frederich, E. M. Turner, G. Ma, N. T. Jaskowiak, D. J. Rivet, J. S. Flier, et al. 1997. Multiple cytokines and acute inflammation raise mouse leptin levels: Potential role in inflammatory anorexia. J. Exp. Med. 185:171–175.

Schulzke, J. D., C. Bojarski, S. Zeissig, F. Heller, A. H. Gitter, and M. Fromm. 2006. Disrupted barrier function through epithelial cell apoptosis. Pages 288–299 in Inflammatory Bowel Disease: Genetics, Barrier Function, Immunologic Mechanisms, and Microbial Pathways. Blackwell Publishing, Oxford, U.K.

Shirasaki, H., E. Kanaizumi, K. Watanabe, N. Konno, J. Sato, S. I. Narita, and T. Himi. 2003. Tumor necrosis factor increases MUC1 mRNA in cultured human nasal epithelial cells. Acta Oto-Laryngologica 123:524–531.

Shirkey, T. W., R. H. Siggers, B. G. Goldade, J. K. Marshall, M. D. Drew, B. Laarveld, and A. G. van Kessel. 2006. Effects of commensal bacteria on intestinal morphology and expression of proinflammatory cytokines in the gnotobiotic pig. Exp. Biol. Med. (Maywood) 231:1333–1345.

Slade R. D., I. Kyriazakis, S. M. Carroll, F. H. Reynolds, I. J. Wellock, L. J. Broom, and H. M. Miller. 2011. Effect of rearing environment and dietary zinc oxide on the response of group-housed weaned pigs to enterotoxigenic *Escherichia coli* O149 challenge. Animal 5:1170–1178.

Smiricky-Tjardes, M. R., E. A. Flickinger, C. M. Grieshop, L. L. Bauer, M. R. Murphy, and G. C. Fahey, Jr. 2003 *In vitro* fermentation characteristics of selected oligosaccharides by swine fecal microflora. J. Anim. Sci. 81:2505–2514.

Steinert, P. M., and L. N. Marekov. 1999. Initiation of assembly of the cell envelope barrier structure of stratified squamous epithelia. Mol. Biol. Cell 10:4247–4261.

Strater, J., and P. Moller. 2000. Expression and function of death receptors and their natural ligands in the intestine. Ann. NY Acad.Sci. 915:162–170.

Sukhotnik, I., M. Agam, R. Shamir, N. Shehadeh, M. Lurie, A. G. Coran, E. Shiloni, and J. Mogilner. 2007. Oral glutamine prevents gut mucosal injury and improves mucosal recovery following lipopolysaccharide endotoxemia in a rat. J. Surg. Res.143:379–384.

Suzuki, H., A. Yanaka, T. Shibahara, H. Matsui, A. Nakahara, N. Tanaka, H. Muto, et al. 2002. Ammonia-induced apoptosis is accelerated at higher pH in gastric surface mucous cells. Am. J. Physiol. Gastrointest. Liver Physiol. 283:G986–G995.

Sydora, B. C., M. M. Tavernini, A. Wessler, L. D. Jewell, and R. N. Fedorak. 2003. Lack of interleukin-10 leads to intestinal inflammation, independent of the time at which luminal microbial colonization occurs. Inflamm. Bowel Diseases 9:87–97.

Theodoropoulos, G., and K. L. Carraway. 2007. Molecular signaling in the regulation of mucins. J. Cell. Biochem. 102:1103–1116.

Topping, D. L., and P.M. Clifton. 2001. Short-chain fatty acids and human colonic function: Roles of resistant starch and nonstarch polysaccharides. Physiol. Rev. 81:1031–1064.

Trebichavsky, I., H. Kozakova, and H. Splichal. 2002. Plasma lipopolysaccharide level and enterocyte brush border enzymes in gnotobiotic piglets infected with *Salmonella typhimurium.* Vet. Med. (Czech.) 47:280–294.

Tsai, C. H., M. Hill, S. L. Asa, P. L. Brubaker, and D. J. Drucker. 1997. Intestinal growth-promoting properties of glucagon-like peptide-2 in mice. Am. J. Physiol. 273:E77–E84.

Tsubouchi, S., and C. P. Leblond. 1979. Migration and turnover of entero-endocrine and caveolated cells in the epithelium of the descending colon, as shown by autoradiography after continuous infusion of thymidine-H-3 into mice. Am. J. Anat. 156:431–451.

Turner, J. R. 2009. Intestinal mucosal barrier function in health and disease. Nat. Rev. Immunol. 9:799–809.

Vahjen, W., R. Pieper, and J. Zentek. 2010. Bar-coded pyrosequencing of 16S rRNA gene amplicons reveals changes in ileal porcine bacterial communities due to high dietary zinc intake. Appl. Environ. Microbiol. 76:6689–6691.

van der Klis, J. D., and A. J. M. Jansman. 2002. Optimising nutrient digestion, absorption and gut barrier function in monogastrics: reality or illusion? Pages 15–36 in Nutrition and Health of the Gastrointestinal Tract. M. C. Blok, H. A. Vahl, L. de Langwe, A. E. van de Braak, G. Hemke, and M Hessing, eds. Wageningen Academic Publishers, The Netherlands.

Verdant, C. L., D. De Backer, A. Bruhn, C. M. Clausi, F. H. Su, Z. Wang, H. Rodriguez, A.

R. Pries, and J. L. Vincent. 2009. Evaluation of sublingual and gut mucosal microcirculation in sepsis: A quantitative analysis. Crit. Care Med. 37:2875–2881.

von Engelhardt, W., J. Bartels, S. Kirschberger, H. D. M. Z. Duttingdorf, and R. Busche. 1998. Role of short-chain fatty acids in the hind gut. Vet. Quart. 20:S52–S59.

Wang, J. J., X. Q.Wang, D. Y. Ou, J. D. Yin, and G. Y.Wu. 2009. Proteomic analysis reveals altered expression of proteins related to glutathione metabolism and apoptosis in the small intestine of zinc oxide-supplemented piglets. Amino Acids 37:209–218.

Ward, T., G. Asche, G. Louis, and D. Pollman. 1996. Zinc-methionine improves growth performance of starter pigs. J. Anim. Sci. 74(Suppl.1):182. (Abstr.)

Warren, E. J., B. N. Finck, S. Arkins, K. W. Kelley, R.W. Scamurra, M. P. Murtaugh, and R.W. Johnson. 1997. Coincidental changes in behavior and plasma cortisol in unrestrained pigs after intracerebroventricular injection of tumor necrosis factor-alpha. Endocrinol. 138:2365–2371.

Wesche-Soldato, D. E., J. L. Lomas-Neira, M. Perl, L. Jones, C. S. Chung, and A. Ayala. 2005. The role and regulation of apoptosis in sepsis. J. Endotoxin Res. 11:375–382.

Westphal, J. K., M. J. Dorfel, S. M. Krug, J. D. Cording, J. Piontek, I. E. Blasig, R. Tauber, et al. 2010. Tricellulin forms homomeric and heteromeric tight junctional complexes. Cell. Mol. Life Sci. 67:2057–2068.

Williams, B. A., M.W. A. Verstegen, and S. Tamminga. 2001. Fermentation in the large intestine of single-stomached animals and its relationship to animal health. Nutr. Res. Rev. 14:207–227.

Williams, B. A., M. W. A. Verstegen, and S. Tamminga. 2001. Fermentation in the monogastric large intestine: Its relation to animal health. Nutr. Res. Rev 14:207–227.

Williams, N. H., T. S. Stahly, and D. R. Zimmerman. 1997a. Effect of chronic immune system activation on body nitrogen retention, partial efficiency of lysine utilization, and lysine needs of pigs. J. Anim. Sci. 75:2472–2480.

Williams, N. H., T. S. Stahly, and D. R. Zimmerman. 1997b. Effect of level of chronic immune system activation on the growth and dietary lysine needs of pigs fed from 6 to 112 kg. J. Anim. Sci. 75:2481–2496.

Willing, B. P., and A. G. van Kessel. 2007. Enterocyte proliferation and apoptosis in the caudal small intestine is influenced by the composition of colonizing commensal bacteria in the neonatal gnotobiotic pig. J. Anim. Sci. 85:3256–3266.

Willing, B. P., and A. G. van Kessel. 2009. Intestinal microbiota differentially affect brush border enzyme activity and gene expression in the neonatal gnotobiotic pig. J. Anim. Physiol. Anim. Nutr. 93:586–595.

Wills-Karp, M., J. Luyimbazi, X. Y. Xu, B. Schofield, T. Y. Neben, C. L. Karp, and D. D.

Donaldson. 1998. Interleukin-13: Central mediator of allergic asthma. Science 282:2258–2261.

Wu, G. Y., F. W. Bazer, T. A. Davis, L. A. Jaeger, G. A. Johnson, S. W. Kim, D. A. Knabe, et al. 2007. Important roles for the arginine family of amino acids in swine nutrition and production. Livest. Sci. 112:8–22.

Wu, G. Y., F. W. Bazer, T. A. Davis, S. W. Kim, P. Li, J. M. Rhoads, M. C. Satterfield, et al. 2009. Arginine metabolism and nutrition in growth, health and disease. Amino Acids 37:153–168.

Wu, G. Y., S. A. Meier, and D. A. Knabe. 1996. Dietary glutamine supplementation prevents jejunal atrophy in weaned pigs. J. Nutr. 126:2578–2584.

Yasuda, K., H. D. Dawson, E. V. Wasmuth, C. A. Roneker, C. Chen, J. E. Urban, R. M. Welch, et al. 2009. Supplemental dietary inulin influences expression of iron and inflammation related genes in young pigs. J. Nutr. 139:2018–2023.

Yasuda, K., K. R. Roneker, D. D. Miller, R. M.Welch, and X. G. Lei. 2006. Supplemental dietary inulin affects the bioavailability of iron in corn and soybean meal to young pigs. J. Nutr. 136:3033–3038.

Zhao, J. M., D. F. Li, X. S. Piao, W. J. Yang, and F. L. Wang. 2002. Effects of vitamin C supplementation on performance, iron status and immune function of weaned piglets. Arch. Anim. Nutr.-Arch. Fur Tierernahrung 56:33–40.

Ziegler, T. R., M. E. Evans, C. Fernandez-Estivariz C., and D. P. Jones. 2003. Trophic and cytoprotective nutrition for intestinal adaptation, mucosal repair, and barrier function. Ann. Rev. Nutr. 23:229–261.

Zolotarevsky, Y., G. Hecht, A. Koutsouris, D. E. Gonzalez, C. Quan, J. Tom, R. J. Mrsny, and J. R. Turner. 2002. A membranepermeant peptide that inhibits MLC kinase restores barrier function *in vitro* models of intestinal disease. Gastroenterol. 123:163–172.

第9章
饲料配方和饲喂程序

1 导 语

在绝大多数养猪企业中，饲料成本占养猪总成本的比例最大。因此，提高饲料利用效率对养猪业至关重要。在商业化养猪生产中，设计饲料配方和制订饲喂策略的主要目标是获得最大的利润，而这并不一定意味着要实现猪的最佳生产性能。设计出既能获得最佳生产性能、成本又最低的饲料配方，是一项十分复杂而艰巨的任务。这需要我们掌握基础营养学、动物对能量和营养素的需要量、能够提供能量和营养素的饲料原料、环境因素等方面的基础知识。最低成本配方和原料使用限量取决于原料的适口性、可消化性、有效性、毒性、与其他饲料原料的相容性以及其他因素等，所有这些因素构成了最佳饲料配方不可或缺的部分。

为了在养猪生产中盈利，就有必要采用适当的饲养程序，以确保这些最优化的饲料得到有效利用。最佳饲喂策略是给猪提供接近满足但不超过猪需要量的能量和必需营养素，这样才能够获得更大的利润。此外，这样也可以减少未被利用的能量和营养素的排泄，也将对当前自觉环保意识社会的形成起到正面作用。本章的主要目的是简要综述如何优化饲料配方和饲喂程序，为健康、可持续养猪业的发展提供依据。

2 饲料配方

◆ 2.1 饲料配方的目的

2.1.1 营养水平

设计猪饲料配方的目的很多，但首要目的是满足动物的营养需要。在绝大多数参考文献（ARC，1981；NRC，1998）中，一般首先考虑使用动物的生长、年龄和体重阶段来描述其营养需要量。简而言之，我们可以根据维持、生产和非维持功能所需的营养素来确定动物的营养需要量。对于很多营养素，我们还需要考虑内源损失的需要。这绝对与维持和生产需要不同，因为内源损失是不可避免的，如与消化相关的氨基酸的内源损失。在任何情况下，我们都必须清楚并确定配方软件中涉及的营养成分的最低需要量，以确定如何满足动物对这些营养素的需要。

我们可以将维持需要和生产功能需要进一步划分为瘦肉生长、脂肪生长、胎儿生长、产奶和一般生理活动的需要。ARC（1981）和NRC（1998）等，对这种划分方式进行了模型分析。从配方设计角度看，使用这些模型可以提高配方的精准性，但这是相对的而非绝对的。例如，多种动物的能量摄入量通常用如下公式表示：

能量摄入量=维持能量需要+生产能量需要

对于刚断奶的仔猪来讲，配方师必须考虑如何让仔猪更顺利地由采食母乳向采食以谷物为主的干饲料过渡。刚断奶的仔猪需要快速适应新的环境、新的饲料和饮水来源、新的社会秩序以及与母猪分离等其他影响。由于存在已知的饲料过敏反应（如大豆蛋白的致敏性；Li等，1991；Kim等，2010）、消化酶的不足、断奶带来的心理应激和潜在的疾病威胁等问题，因此这一适应过程是很重要的。配方师可采用多种不同的技术来解决动物的应激问题，如设置特定原料中营养素的最小用量和最大用量，所有这些限制都会使饲料的最终价格受到显著影响。

2.1.2 采购支持

大多数的饲料配方师使用多种饲料原料、采用最低成本原则设计饲料配方，但并不是所有原料都是解决问题必需的。配制最低成本的饲料配方是家畜生产或饲料生产经营者的根本利益所在，而其真正的价值产出是积累了配方经验。企业单位中饲料配方的作用是多方面的，既要考虑满足动物的需要，又要考虑采购因素。学者和学术领域对采购支持的了解较少，主要是因为其过分强调营养水平和营养需要成分。企业单位经常要求配方师确定原料价格采购点、相对预估值、量化节约成本及其他一些财务指标。多数采购机构需要不依赖于配方软件而作出决策，因此需要了解一些简单的工具或与制定采购决策相关的知识，而配方师却经常是提供这些数据的熟练工。

原则上，商业化饲料厂和家畜养殖业主的饲料配方无较大差别，但实际生产中其差异很大。在家畜养殖业主自己的饲料加工车间里，他们需要配制的饲料数量很少，而且他们所使用的饲料原料种类也不多。事实上，两种配方师更愿意使用2～3种主要原料，提供大部分能量（玉米或小麦）和蛋白质或氨基酸（豆粕）。一般来讲，这些原料在绝大多数地方都可以很容易地随时大量获得。是否使用这些常规原料的替代品，常取决于时间（收获季节）、产地及饲料加工厂是否靠近食品或饲料加工企业等因素。例如，位于面粉加工厂附近的养殖场最可能以低价获得小麦次粉，而位于偏远农村地区的养殖场就很难获得这样的原料。同样，大麦在其产地是玉米和小麦的良好替代品，但仅限于收获季节，且数量有限。配方师需要帮助采购人员评估价格最合理的替代原料。

除了给采购人员提供价格和采购工具外，配方师还能通过使用营养价值表或配方技术和程序，来显著影响采购决策。饲料原料的营养成分会因其加工工艺和种植等情况引起的价格变化而变化。这种原料中能量和营养素水平的变化会使动物的生长性能水平明显低于预期，尤其是在配方中能量和营养素不足的情况下。配方师有多种办法来应对由于原料组分变异造成的配方不能满足关键营养素需要的风险，这种配方常被称为随机配方。随机配方综合考虑了原料的变异、达到特定能量和营养素水平的概率两方面的知识。从实践的角度讲，配方师既可解决单一原料能量和营养素规格的变异，也可解决配方时间（真正的随机配方）的差异。Roush等（2007）认为，与调整配方的原料组成相比，实时随机配方更好，但是如果考虑真实概率，这种观点还是值得商榷的。事实上，与只考虑饲料养分总变异的方法相比，使用给单一成分赋予变异的方法可以在采购之前更准确地估计原料的营养成分含量。

2.1.3 其他目的

除了满足动物维持和生产需要、降低生产成本、提供采购支持以外，在配方设计过程中配方师还要考虑一些其他目的。为了实现上述目的，配方师在配方设计过程中引入一些限制因素，有利于确定管理要素的成本。配方设计的其他限制因素和目的可能包括：

- 环境方面的法规要求配方师强制限制氮和磷等营养素。
- 原料中大量的霉菌毒素或其他毒素，或者大豆蛋白等原料的成本，迫使配方师对原料加以限制。
- 特定功能参数，如制粒质量、流散性、容重或者水分含量，都是需要考虑的限制因素。

◆ 2.2 原料营养价值表的发展

商业化饲料配方软件可为配方师提供一些来自于NRC（1998）、《Feedstuffs》杂志或其他出版物的基础营养价值表。配方师最好利用当地的原料数据，建立一个基础营养价值表，这样在配方设计时可为他们提供更多的机会。理论上讲，配方师应掌握一定数量的实测数据，以便建立每种原料的变异、能量和营养素含量的均值。软件用户要能辨别出加工原料中相对稳定的营养素含量，但其含量常与产品的入境渠道和加工方法密切相关。值得注意的是，即使供应商提供的两种原料成分相同或十分相似，在配方中也不能作为同一种原料。换言之，多种相似的原料在配方中应该分开评价。

多数营养素并不能被动物完全利用，而且不同成分的利用率也不同。因此，在众多原料中，对于那些能量和营养具有重要经济价值，而且已获得足够的代表性数据并建立了可靠利用率的原料，推荐使用可利用能量和营养素数据。配方师必须承认，与总的能量和营养素相比，使用可利用能量和可利用营养素是一种进步，并且其大方向几乎总是正确的。简单来说，这增加了原料的利用率，但其应用风险未明显增加。能量利用率确实存在递减现象。尽管一般认为，猪的理想能量体系是净能体系，但应该考虑到确定净能值所需要的资料数量和质量是有限的，因此还不能大幅度改进“修正”代谢能体系。尽管净能体系代表了改进的方向，但相关信息的缺乏使我们不能得到利用净能体系的好处，而且使我们更可能陷入没有改进效果的风险中。

◆ 2.3 配方方法学

2.3.1 最低成本法

多数饲料配方师和软件供应商致力于设计最低成本的饲料配方。配方师通常提出一组限制条件，确保满足氨基酸、主要矿物质和能量等关键营养素的需要。根据提供的原料和营养素，软件解出一系列联立方程，给出最优成本解决方案。获得一个可行性方案需要注意，当限制条件最少时，通常会获得一系列成本最低而又可行的配方；当增加限制条件时，成本也会相应增加。

设计配方时，联合使用多种配方方法才能实现最低单位增重成本。一般情况下，很难找到具有这种功能的商业化软件，但某些公司为了满足配方师的要求，可以提供定制的程序。为了预测一个成本函数，通常可将所得结果与具体的营养素和能量变量进行绑定。一般来讲，这项技术可不断优化配方，但并不能用于未来生产成绩的预测。

2.3.2 对软件的主要关注

对于配方系统的技术参数，关键要考虑的是营养素的限量。例如，设定氨基酸的最佳比例使氨基酸的比例得以保持。在多种配方系统中，会将大多数的重要营养素与配方中的能量浓度进行配比。由于大部分动物可根据能量浓度调节采食量，因此这种软件的应用效果很好。一般来讲，将营养素按与能量浓度的比例配给代表了配方设计中的大多数和方向性的改进。

系数设定与配比不一定相同。系数设定通常将营养素设定为一个具体数值，而配比只是保持营养素的比例而不考虑营养素的水平。例如，如果我们设定赖氨酸与能量的系数是每1 500千卡是1.5，我们就可以确定当能量是1 500千卡时赖氨酸就是1.5，但是如果能量是1 600千卡时，并不表示赖氨酸就是1.6。与营养素配比相比，了解如何使用营养系数设定是更重要的，因为软件对二者的处理方式不同。

配方师经常会根据饲料厂的机械系统来决定生产最小用量。例如，个别饲料厂小规模生产1t饲料时，不能准确称量少于1.8 kg的原料。因此，就可设定

1.8 kg为生产的最小用量，在配方中要么不使用该原料，要么该原料的使用量在1.8 kg以上。

配方过程中软件工程师如何处理约数是一个关键的问题。因为这与植酸酶等高效原料显著相关，小量的舍入对营养素水平的影响很大。

3 饲喂程序

◆ 3.1 饲喂程序的原理

本部分内容将探讨不同年龄或生理阶段猪的常规饲养程序。详细的猪营养需要在其他章节中已有讨论（见第18章）。在养猪生产中，设计饲养程序的主要目的包括：① 提供营养素，以满足最佳生产性能的需要；② 获得最大的经济效益；③ 通过提高营养素的利用效率，使得营养素的排泄降到最低。制订合理饲养程序的基本原则，是根据猪的生长阶段或生理状态进行分阶段饲喂。

◆ 3.2 根据不同年龄或生理状态分阶段饲喂

3.2.1 母猪

泌乳母猪饲养管理的主要目的，是提高产奶量和尽量降低体组织的损失。然而，实现这些目标的主要障碍，就是母猪的自由采食量不足。因此，泌乳母猪的饲喂程序就应该考虑提高自由采食量和增加营养素的利用效率。妊娠期母猪体况会影响泌乳期的自由采食量，这一点已有详细论述（Williams，1998；Kim和Easter，2003）。妊娠期采食过多会导致母猪肥胖，其泌乳期采食量比正常体重母猪的要少。这提示，自由采食量的降低与胰岛素抵抗的增加以及血液中胰岛素、瘦素和胃饥饿素（ghrelin）水平的改变有关（Weldon等，1994；Papadopoulos等，2009）。

为了预防母猪分娩时肥胖，通常在妊娠期进行限饲。妊娠母猪通常在定位栏中单个饲养，目的是控制能量摄入进行单独饲喂。最近社会上对妊娠定位栏的禁用比较关注，这对设法控制妊娠母猪能量摄入的养猪经营者来说是个挑战。小栏群养，每圈可同时饲养5～20头母猪。这种小圈的食槽能进行局部分割，可以单独饲喂，因此可以控制能量摄入。然而，重要的是要根据母猪体况进行分组。

如果设备成本的增加不成问题，使用电子饲喂器对群养妊娠母猪进行饲喂就是一种很好的方式。

饲喂泌乳母猪应该考虑增加其对能量和营养素的摄入。一般的饲喂程序应该是，在分娩后的3～5 d逐步增加饲料的喂给量，从第6天开始经常喂给新鲜饲料（每天2～4次）以刺激泌乳母猪的食欲。在炎热和潮湿气候条件下饲喂泌乳母猪，由于热应激的存在，因此维持母猪食欲是一个挑战。在这种气候条件下，通过把喂料时间改为早晨早些时候和晚上晚些时候，可以增加母猪的采食量；也可以考虑给泌乳母猪提供高能、高营养浓度的饲料。母猪采食量不足时，增加能量和营养素浓度有助于防止泌乳期出现严重的分解代谢状态。

计算泌乳母猪的能量和营养素需要量时，需要考虑的关键因素包括：① 用于产奶的能量和营养素；② 用于乳腺组织合成的能量和营养素；③ 用于母体生长和维持的能量和营养素。乳腺产乳和乳腺实质组织生长的需要，会影响泌乳母猪对能量和营养素的需要（Kim等，1999a，1999b）。当采食量不足时，母猪会动用其体组织为乳腺提供营养素。

如果不能控制泌乳母猪的自由采食量，体组织的损失可根据产奶时能量和营养素的输出量来预测。因此，如果母猪产奶量和窝产仔数相似，能量和营养素对采食量不同母猪的作用不同。由于来自母体组织动用的氨基酸与用于乳汁合成的氨基酸组分不同，因此营养素的摄入量会因母体组织动用的数量而改变饲粮理想氨基酸模式（Kim等，2001；表9-1）。青年母猪自由采食量不足，不能满足母体生长所需的能量和营养素；而成年母猪自由采食量充足，即使对于它们而言，母体生长所需的能量和营养素有限。考虑到青年母猪和成年母猪对营养素需要的质量和数量均不同，可以对泌乳母猪进行分胎次饲喂。青年母猪采食量不足的同时还需要供应母体生长的需要，而成年母猪采食量充足且母体生长需要也不显著，因此与成年母猪相比，青年母猪饲粮中应该含有更高的能量和营养浓度。

表9-1 泌乳母猪的赖氨酸为基础的理想蛋白质模式和限制性氨基酸顺序[1]

体重损失（kg）[2]	75～80	33～45	12～15	6～8	0	0～7
动员数量（%）[3]	50	40	20	5	0	ND[4]
理想氨基酸模式（占赖氨酸的百分比）						
赖氨酸	100	100	100	100	100	100
苏氨酸	75	69	63	60	59	62
缬氨酸	78	78	78	77	77	85
亮氨酸	128	123	118	115	115	114
异亮氨酸	60	59	59	59	59	56
精氨酸	22	38	59	69	72	56
限制性氨基酸顺序[5]						
第一	苏氨酸	赖氨酸	赖氨酸	赖氨酸	赖氨酸	赖氨酸
第二	赖氨酸	苏氨酸	苏氨酸	缬氨酸	缬氨酸	缬氨酸
第三	缬氨酸	缬氨酸	缬氨酸	苏氨酸	苏氨酸	苏氨酸

[1]资料来源于Kim等，2009。

[2]指泌乳期21d的母猪体重损失，根据Kim等（2001a）测定的蛋白质损失量和体组织成分估计。

[3]指母猪泌乳中来自体组织蛋白质分解代谢产生的氨基酸百分含量。

[4]NRC（1998）估计值没有考虑组织蛋白质的动员（ND代表无资料）。

[5]假设在泌乳期间饲喂典型的玉米-豆粕型饲粮（0.9%的赖氨酸）。

高产母猪需要负担14～18个胎儿的生长。胎儿生长主要是在妊娠70 d以后（McPherson等，2004）。妊娠70 d后，胎儿组织中蛋白质的沉积量至少增加19倍，而脂肪沉积在整个妊娠期保持稳定（Kim等，2009）。另外，母猪还需要支持乳腺的生长，乳腺生长主要是发生在妊娠70 d以后（Ji等，2006）。妊娠70 d后，乳腺组织中蛋白质的沉积量至少增加24倍，而脂肪沉积在整个妊娠期相当稳定（Kim等，2009；表9-2）。由于不同类型组织中蛋白质的沉积速率不同，因此妊娠前期和妊娠后期母猪的饲粮理想氨基酸模式也不同（Kim等，2009；表9-3）。饲喂妊娠母猪可施行分阶段饲喂，在妊娠前期饲喂低蛋白质饲粮、妊娠后期饲喂高蛋白质饲粮。应该考虑到，妊娠后期母猪对蛋白质的需要量显著增加，因此需要改变理想氨基酸的比例。分阶段饲喂不同蛋白质浓度的饲粮，可以限制饲粮的能量，而不影响妊娠后期母猪的营养需要量。

表9-2 母体增重和维持的氨基酸需要量（g/d）[1,2]

氨基酸	0 ~ 70 d			70 d 至分娩		
	总量	维持	增重	总量	维持	增重
赖氨酸	6.41	1.64	4.77	8.06	1.78	6.28
苏氨酸	5.19	2.48	2.71	6.78	2.69	4.09
色氨酸	0.93	0.43	0.50	1.17	0.46	0.71
蛋氨酸	1.60	0.46	1.14	2.02	0.50	1.52
缬氨酸	4.12	1.10	3.02	4.66	1.19	3.47
亮氨酸	5.58	1.15	4.43	6.23	1.25	4.98
异亮氨酸	3.80	1.23	2.57	4.68	1.34	3.34
精氨酸	5.77	1.23	4.54	7.96	1.34	6.62

[1]资料来源于Kim等，2009。

[2]配种时母猪的平均体重是160 kg，妊娠70 d体重是195 kg，妊娠114 d体重是220 kg（Ji等，2005）；色氨酸和蛋氨酸是育肥猪的需要量（Mahan和Shields，1998）。

表9-3 妊娠不同胎儿数母猪的赖氨酸为基础的理想蛋白质模式[1]

胎儿数	天数[2]	赖氨酸	苏氨酸	色氨酸	蛋氨酸	缬氨酸	亮氨酸	异亮氨酸	精氨酸
6	0 ~ 70	1.00	0.80	0.15	0.25	0.65	0.88	0.59	0.90
	70 ~ 114	1.00	0.73	0.15	0.26	0.65	0.92	0.56	0.95
8	0 ~ 70	1.00	0.80	0.15	0.25	0.65	0.88	0.59	0.90
	70 ~ 114	1.00	0.72	0.16	0.27	0.66	0.93	0.56	0.96
10	0 ~ 70	1.00	0.80	0.15	0.25	0.65	0.88	0.59	0.90
	70 ~ 114	1.00	0.72	0.16	0.27	0.66	0.94	0.56	0.97
12	0 ~ 70	1.00	0.79	0.15	0.25	0.65	0.88	0.59	0.90
	70 ~ 114	1.00	0.71	0.16	0.27	0.66	0.95	0.56	0.97

（续）

胎儿数	天数[2]	赖氨酸	苏氨酸	色氨酸	蛋氨酸	缬氨酸	亮氨酸	异亮氨酸	精氨酸
14	0 ~ 70	1.00	0.79	0.15	0.25	0.65	0.88	0.59	0.90
	70 ~ 114	1.00	0.71	0.16	0.27	0.66	0.96	0.55	0.98
16	0 ~ 70	1.00	0.79	0.15	0.25	0.65	0.88	0.59	0.90
	70 ~ 114	1.00	0.70	0.16	0.27	0.67	0.97	0.55	0.98
18	0 ~ 70	1.00	0.79	0.15	0.25	0.65	0.89	0.59	0.90
	70 ~ 114	1.00	0.70	0.16	0.28	0.67	0.97	0.55	0.99

[1]资料来源于Kim等，2010。

[2]妊娠天数。

3.2.2 哺乳仔猪

在美国的典型养猪场，母猪为新生仔猪哺乳的时间是14 ~ 28 d。在这么短的哺乳期内，仔猪每天增重150 ~ 250 g，如果不饲喂教槽料，母乳是唯一的能量和营养素来源，用以满足仔猪快速生长的需要。仔猪出生时，其体内贮存的能量和营养素很少，必须采食充足的母乳才能确保哺乳仔猪的快速生长。母猪的初乳和常乳可为仔猪提供必需的营养素（Lin等，2009；Mavromichalis等，2006）。然而，如果母猪的体况较差和/或采食量较低，乳腺就无法给哺乳仔猪提供高质量的营养素（Kim等，1999a，1999b）。而且已有研究表明，母猪泌乳量是保证哺乳仔猪正常生长的限制因素（Aherne，1980；Zijlstra等，1996）。如果母猪泌乳量不足，饲喂开食料或教槽料是一种保证哺乳仔猪正常生长的实用方法。

3.2.3 保育猪

在养猪生产过程中，仔猪通常在14 ~ 28日龄断奶。因为饲粮突然从母乳变成固体饲料、仔猪经历了与母猪分离和环境的改变，刚断奶的仔猪会经历一个应激期（Maxwell和Carter，2000）。由于断奶应激，仔猪会经常出现断奶后生长阻滞。由于胃酸和胰腺消化酶分泌不足，刚断奶的仔猪对结构复杂的饲料组分消化能力有限，突然改变饲粮类型会导致腹泻。因此，在刚断奶仔猪的饲粮中要首先使用消化率高的原料，然后逐步过渡到玉米和豆粕等常规原料，这一点是很重要的。由于与猪乳的结构特性相似，因此在刚断奶仔猪的饲粮中经常使用干燥的乳清、乳糖和乳清浓缩蛋白等乳制品副产物（Tokach等，1989；Mahan等，2004；Cromwell等，2008）。与结构复杂并可能含有抗营养成分的植物性蛋白质相比，血浆蛋白、血粉、鱼粉、肉粉和肉骨粉等动物源性饲料原料的消化率高，是刚断奶仔猪的优质蛋白质原料（de Rodas等，1995；Kim和Easter，2001；Adedokun和Adeola，2005）。然而，加工后的植物蛋白质产品含有水解营养素，而不含有抗营养因子，因此也可给刚断奶的仔猪饲喂（Kim等，2003，2010；González-Vega等，2011；Goebel和Stein，2011）。

分阶段饲喂程序也可很好地用于饲喂保育猪。保育猪通常分3 ~ 4个阶段进行饲喂，即从断奶开始至9或10周龄，猪的体重可达到22 ~ 25 kg。保育饲粮的第一个阶段可以使用乳制品副产物和动物源性蛋白质，第二至最后一个阶段可以逐步增加饲粮中玉米和豆粕的用量。

3.2.4 **育肥猪**

在商业化猪场中，通常在9 ~ 10周龄、体重达到22 ~ 25 kg时，保育猪会被转移到育肥舍（或生长育肥舍）。这一阶段猪的食欲好，采食量和生长通常不成问题。而脂肪沉积增加和饲料转化效率降低（增重/饲料）是该阶段养猪业者面临的最大问题。

由于阉公猪和后备母猪的生长速度和瘦肉增长潜力不同，因此在养猪生产中经常采用分性别饲养措施。阉公猪比同龄的后备母猪摄入的能量更多，生长更快。阉公猪的脂肪增长也比后备母猪快。如果将阉公猪和后备母猪饲养在同一个圈中，采食相同能量和营养素组成的饲粮，后备母猪会比阉公猪更瘦。因为阉公猪和后备母猪对能量和营养素的需要量不同，上市时其体重差异就很大。因此，合理的饲养方式是将阉公猪和后备母猪分栏饲养，并提供不同的饲料。与后备母猪相比，可以给阉公猪饲喂蛋白质浓度相对低的饲粮。因此，采用分性别饲喂方式更有利于增加阉公猪的瘦肉生长、后备母猪的增重和猪群的均匀度。

随着猪的生长，其采食量增加，脂肪沉积也加快。为了促进瘦肉增长，应该设计氨基酸平衡（如理想蛋白质）饲粮，这也是蛋白质沉积所需的。如果任何一种氨基酸不足（如限制性氨基酸），蛋白质合成就会受阻。因此，平衡用于肌肉蛋白质合成的饲粮氨基酸对于增加瘦肉增长是很重要的。已有研究报道了理想蛋白质模式，并对其特征进行了分析（Wang和Fuller，1989；Chung和Baker，

1991)。在饲料配方中引入理想蛋白质的概念，能降低环境中氮的排泄和饲料成本，有利于养猪生产。表9-4归纳了生长猪的理想蛋白质模式。L-赖氨酸和L-苏氨酸是猪饲料中通常添加的氨基酸，偶尔也会添加L-色氨酸和DL-蛋氨酸。

表9-4　生长猪的赖氨酸为基础的理想蛋白质模式[1]

项目	体重(kg)					
	3 ~ 5	5 ~ 10	10 ~ 20	20 ~ 50	50 ~ 80	80 ~ 120
赖氨酸	1.00	1.00	1.00	1.00	1.00	1.00
苏氨酸	0.63	0.62	0.62	0.63	0.65	0.65
色氨酸	0.18	0.18	0.18	0.18	0.18	0.19
蛋氨酸	0.27	0.27	0.27	0.27	0.27	0.27
缬氨酸	0.68	0.68	0.68	0.67	0.68	0.67
亮氨酸	1.01	1.01	1.01	1.00	1.02	0.98
异亮氨酸	0.54	0.55	0.54	0.54	0.56	0.56
苯丙氨酸	0.60	0.60	0.60	0.59	0.61	0.60
组氨酸	0.32	0.32	0.32	0.31	0.32	0.31
精氨酸	0.40	0.41	0.42	0.40	0.36	0.31

[1]资料来源于NRC，1998。

在美国的典型养猪场中，常给猪饲喂干饲料。猪可以自由采食料槽中的干饲料，而饮水是通过单独的饮水器提供的。然而，液体饲喂系统或湿喂系统能增加采食量。在液体饲喂系统中，经常将干料与水在料盘中混合成浆状，其中干物质的含量是20%～30%。液体饲喂时也可以使用来源于食品、乳品加工产物或液体发酵产品等液体成分作为饲料。由于采食量增加，液体饲喂能提高猪的增重(Gonyou和Lou，2000；de Lang等，2006)，但并没有提高饲料转化效率。液体饲喂的猪，其采食量更大，相对于干饲料饲喂时脂肪沉积更多。当采用液体饲喂时，料槽会残存废料，此时面临的最大问题是预防腐败和霉变。

已有研究表明，在饲料中添加抗生素能提高猪的瘦肉率。因此，在养猪生产中经常使用抗生素作为生长促进剂。然而，为了防止猪肉中抗生素的残留，在上市前一段时间在猪饲料中要停止添加抗生素，抗生素的停用时间取决于抗生素的类型。

饲料的粒度大小影响猪对饲料中营养素的消化。以粉料形式饲喂时，玉米的粉碎粒度通常为600～900 μm。如果玉米被粉碎得更细，在350～450 μm时，玉米的营养素消化率会提高。然而，当玉米粉碎过细时，最好对饲料进行制粒，以防止料槽和料仓内出现结块现象。

4 结 语

在设计猪的饲料配方时，应该考虑猪的生长阶段、年龄和体重等因素对能量和营养素需要量的影响。可以根据维持和生产（或增重）需要，计算总的营养素需要量。绝大多数配方师使用多种原料设计最低成本饲料配方。企业设计饲料配方的目标是多方面的，既要满足猪的营养需要，又要为采购人员提供技术支持等。绝大多数配方师和软件供应商致力于将配方中单位饲料的生产成本降至最低。在配方中营养素的限量是需要重点考虑的因素，应将大多数的主要营养素按照能量浓度进行配比。

制订饲喂程序的主要目的，不仅能为实现最佳生产性能满足需要提供营养素，而且能尽量扩大盈利，并尽量减少营养素的排泄。制订合理饲喂程序的基本原则是根据生长阶段和生理状态进行分阶段饲喂。对于母猪来讲，配方师应该考虑增加泌乳期营养素的摄入以满足大量泌乳的需要。然而，应该控制妊娠母猪对营养素尤其是能量的摄入，防止分娩时出现肥胖。对于断奶仔猪来讲，配方师必须考虑如何让仔猪由哺乳向采食主要由谷物组成的干饲料顺利过渡。对于育肥猪，由于养猪企业中育肥猪消耗了大部分的饲料，因此配方的主要目的是提高饲料转化效率。

作者：Sung Woo Kim和
Jeffrey A. Hansen
译者：乔家运

参考文献

Adedokun, S. A., and O. Adeola. 2005. Metabolizable energy value of meat and bone meal for pigs. J. Anim. Sci. 83:2519–2526.

Aherne, F. X. 1980. Management and nutrition of the newly weaned pig. Page 55 in University of Illinois Pork Industry Conference. Urbana, IL.

ARC. 1981. *The Nutrient Requirements of Pigs*. Commonwealth Agricultural Bureaux, Farnham Royal, UK.

Chung, T. K., and D. H. Baker. 1991. A chemically defined diet for maximal growth of pigs. J. Nutr. 121:979–982.

Cromwell, G. L., G. L. Allee, and D. C. Mahan. 2008. Assessment of lactose level in the mid– to late–nursery phase on performance of weanling pigs. J. Anim. Sci. 86:127–133.

de Lange, C. F. M., C. H. Zhu, S. Niven, D. Columbus, and D. Woods. 2006. Swine liquid feeding: Nutritional considerations. Pages 37–50 in Proc. 27th Western Nutr. Conf. Dept. of Animal Science, University of Manitoba, Winnipeg, MB, Canada.

de Rodas, B. Z., K. S. Sohn, C. V. Maxwell, and L. J. Spicer. 1995. Plasma protein for pigs weaned at 19 to 24 days of age: Effect on performance and plasma insulin–like growth factor I, growth hormone, insulin, and glucose concentrations. J. Anim. Sci. 73:3657–3665.

Goebel, K. P., and H. H. Stein. 2011. Phosphorus digestibility and energy concentration of enzyme–treated and conventional soybean meal fed to weanling pigs. J. Anim. Sci. 89:764–772.

Gonyou, H. W., and Z. Lou. 2000. Effects of eating space and availability of water in feeders on productivity and eating behavior of grower–finisher pigs. J. Anim. Sci. 78:865–870.

González–Vega, J. C., B. G. Kim, J. K. Htoo, A. Lemme, and H. H. Stein. 2011. Amino acid digestibility in heated soybean meal fed to growing pigs. J. Anim. Sci. 89:3617–3625.

Ji, F., W. L. Hurley, and S. W. Kim. 2006. Characterization of mammary gland development in pregnant gilts.J. Anim. Sci. 84:579–587.

Kim, S. W. 2010. Recent advances in sow nutrition. Revista Brasileira de Zootecnia 39:303–310.

Kim, S. W., and R. A. Easter. 2001. Nutritional value of fish meals in the diet for young pigs. J. Anim. Sci. 79:1829–1839.

Kim, S. W., and R. A. Easter. 2003. Amino acid utilization for reproduction in sows. Pages 203–222 in Amino Acids in Animal Nutrition. J. P. F. D'Mello, ed. CABI Publishing,

Wallingford, UK.

Kim, S. W., W. L. Hurley, I. K. Han, and R. A. Easter. 1999a. Changes in tissue composition associated with mammary gland growth during lactation in the sow. J. Anim. Sci. 77:2510–2516.

Kim, S. W., W. L. Hurley, I. K. Han, H. H. Stein, and R. A. Easter. 1999b. Effect of nutrient intake on mammary gland growth in lactating sows. J. Anim. Sci. 77:3304–3315.

Kim, S. W., D. H. Baker, and R. A. Easter. 2001. Dynamic ideal protein and limiting amino acids for lactating sows: Impact of amino acid mobilization. J. Anim. Sci. 79:2356–2366.

Kim, S. W., D. L. Knabe, K. J. Hong, and R. A. Easter. 2003. Use of carbohydrases in corn–soybean meal–based nursery diets. J. Anim. Sci. 81:2496–2504.

Kim, S. W., W. L. Hurley, G. Wu, and F. Ji. 2009. Ideal amino acid balance for sows during gestation and lactation. J. Anim. Sci. 87:E123–E132.

Kim, S. W., E. van Heugten, F. Ji, C. H. Lee, and R. D. Mateo. 2010. Fermented soybean meal as a vegetable protein source for nursery pigs: I. Effects on growth performance of nursery pigs. J. Anim. Sci. 88:214–224.

Li, D. F., J. L. Nelssen, P. G. Reddy, F. Blecha, R. D. Klemm, D. W. Giesting, J. D. Hancock, et al. 1991. Measuring suitability of soybean products for early–weaned pigs with immunological criteria. J. Anim. Sci. 69:3299–3307.

Lin, C., D. C. Mahan, G.Wu, and S.W. Kim. 2009. Protein digestibility of colostrums by neonatal pigs. Livest. Sci. 121:182–186.

Mahan, D. C., N. D. Fastinger, and J. C. Peters. 2004. Effects of diet complexity and dietary lactose levels during three starter phases on postweaning pig performance. J. Anim. Sci. 82:2790–2797.

Mavromichalis, I., T. M. Parr, V. M. Gabert, and D. H. Baker. 2001. True ileal digestibility of amino acids in sow's milk for 17–day–old pigs. J. Anim. Sci. 79: 707–713.

Maxwell, C. V., and S. D. Carter. 2000. Feeding the weaned pig. Pages 691–716 in *Swine Nutrition*. 2nd ed. A. J. Lewis and L. L. Southern, eds. CRC Press, Boca Raton, FL.

McPherson, R. L., F. Ji, G.Wu, and S.W. Kim. 2004. Fetal growth and compositional changes of fetal tissues in the pigs. J. Anim. Sci. 82:2534–2540.

NRC. 1998. Nutrient Requirements of Swine. 10th rev. ed. National Academies Press, Washington, DC.

Papadopoulos, G. A., D. G. D. Maes, S.van Weyenberg, T. A. T. G. van Kempen, J. Buyse, and G. P. J. Janssens. 2008. Peripartal feeding strategy with different n–6:n–3 ratios in sows: Effects on sows' performance, inflammatory and periparturient metabolic parameters. Br. J. Nutr. 101:348–357.

Roush, W. B., J Purswell, and S. L Branton. 2007. An adjustable nutrient margin of safety comparison using linear and stochastic programming in an excel spreadsheet. J. Appl. Poult. Res. 16:514–520.

Tokach, M. D., J. L. Nelssen, and G. L. Allee. 1989. Effect of protein and (or) carbohydrate fractions of dried whey on performance and nutrient digestibility of early weaned pigs. J. Anim. Sci. 67:1307–1312.

Wang, T. C., and M. F. Fuller. 1989. The optimum dietary amino acid pattern for growing pigs. 1. Experiments by amino acid deletion. Br. J. Nutr. 62:77–89.

Weldon,W. C., A. J. Lewis, G. F. Louis, J. L. Kovar, and P. S. Miller. 1994. Postpartum hypophagia in primiparous sows: II. Effects of feeding level during gestation and exogenous insulin on lactation feed intake, glucose tolerance, and epinephrine–stimulated release of nonesterified fatty acids and glucose. J. Anim. Sci. 72:395–403.

Williams, I. H. 1998. Nutritional effects during lactation and during the interval from weaning to estrus. Pages 159–182 in The Lactating Sow. M. W. A. Verstegen and P.S. Moughan, eds. Wageningen University Press, Wageningen, The Netherlands.

Zijlstra, R. T., K. Y. Whang, R. A. Easter, and J. Odle. 1996. Effect of feeding a milk replacer to early–weaned pigs on growth, body composition, and small intestinal morphology, compared with suckled littermates. J. Anim. Sci. 74:2948–2959.

第10章

猪饲粮中的非常规原料

1 导 语

非常规原料对养猪业的可持续发展具有重要作用。首先，为了实现经济的可持续发展，使用非常规原料，尤其是副产品已成为控制饲料成本快速上升的重要手段（Zijlstra和Beltranena，2009）。随着新型工业（尤其是生物燃料工业）对饲料谷物需求量的增大，谷物价格将不断提高。短期内，采用适当的风险管理策略（包括采用现代饲料质量评价体系），使用非常规原料可以有效控制养猪生产中的饲料成本。其次，为了农业的可持续发展，某些具有独特农艺学性状的非常规作物可能发挥重要的作用（Miller等，2002）。例如，由于水资源的日益缺乏，具有抗旱特性的高粱和黑小麦可部分替代不耐旱的玉米和小麦等传统饲料谷物。而黑小麦的种植投入成本要比小麦低14%（David-Knight和Weightman，2008）。豆类作物（非油料豆类作物）氮利用效率高，至少可以部分替代饲料谷物，其应用也不断增加。同时，农作物的轮作还可以减轻作物病虫害压力（Krupinsky等，2002）。农业可持续发展取得成功的重要标志是非常规原料市场的形成。当这些新作物质量指标达不到一级市场要求时，就可以进入非常规原料市场。最后，从社会和环境可持续发展的角度来说，利用副产品作为猪饲料有助于解决人猪争粮的矛盾（Nonhebel，2004）。如果将食品、生物燃料以及生物技术产业产生的不可食残渣转化为高质量的动物蛋白，就可以减轻这些行业对环境的影响。尽管在一些地区已经有效地使用食品行业的副产品来生产猪肉，从而减少了养猪业对饲料谷物的依赖，但是在超市每种食品的背后，都至少有1个有用的副产品被忽视了。猪作为一种杂食性动物（Stevens和Hume，1995），适于摄入各种饲料原料，因此，在养猪业中使用非常规原料对畜牧业的可持续发展具有重要意义。

在养猪业中使用非常规饲料原料并不是新鲜事。在传统的养猪生产中，猪的饲养规模小，也不重视生长速度。给猪饲喂的饲料实际上就是我们目前所说的非常规原料，例如剩菜剩饭（Pond和Lei，2001）。在全球（尤其是在亚洲）的小规模养猪生产中，这种传统做法仍十分普遍（Chen，2009）。在现代养猪生产中，要求猪的生长速度快，所生产的猪肉安全、质量稳定，具有市场竞争力。这要求饲料谷物和一些蛋白原料能稳定供应（Pond和Lei，2001）。鉴于此，在北美的商业化猪场，只有当普通谷物饲料和蛋白类原料的价格上涨时，中采用非常规饲料才认为是有利的。

目前使用副产品存在一定的限制因素。首先，副产品无法稳定供应是个大问题。在过去的五年里，北美的养猪业中，只有玉米酒精糟（DDGS）这一非常规原料达到了商品化供应的程度（Patience等，2007）。从世界范围来说，很少地区在商业养猪生产中依靠副产品作为主要饲料原料。然而，在欧洲一些土地面积较小的国家（如荷兰），历史上就严重依赖大量的非常规饲料原料（FEFAC，2005）。非常规饲料原料在有机猪肉生产中很受重视（Partanen等，2006），因为使用非常规原料可以避免使用含有抗除草剂或其他转基因成分的大豆和玉米，或者可以满足使用本土化有机饲料原料的需要。其次，采用非常规饲料原料可能会对猪生长性能的稳定性和猪肉品质带来一定的风险。应用现代饲料评价技术可将风险控制在一定程度。本章介绍了三类非常规饲料原料：① 传统农作物改良产品；② 非常规农作物；③ 来自于生物燃料、食品工业的副产品和粮食加工副产品。有关非常规饲料原料前人已经进行了详细的综述（Thacker和Kirkwood，1990；Chiba，2001），本章将重点介绍最近十年新的进展。

2 饲料配方与风险管理

◆ 2.1 营养

在猪饲粮中采用非常规饲料原料，在改善养猪生产经济效益的同时，也带来了风险，因此必须进行适当的风险管理。风险因素有多种：营养方面，如养分变异性和常量养分含量较大范围的变化；化学方面，如化学残留物；生物学方面，如霉菌毒素和抗营养因子；以及其他一些对猪肉质量具有潜在负面影响的因素（De Lange，2000；Smits和Sijtsma，2007）。

一些欧盟国家（如荷兰）严重依赖于非常规饲料原料（FEFAC，2005）。非常规原料的使用，在扩大饲料原料来源的同时，也会导致常量养分，尤其是非淀粉多糖（NSP）和蛋白质的变化幅度更大。毫无疑问，选择不同的能量评价体系会改变饲料的相对能值（Noblet等，1993）。在能量评估方面，消化能（DE）和代谢能（ME）体系高估了能量对维持和生长的贡献（Black，1995），而净能（NE）体系评价提供了更为精准的原料能值（Whittemore，1997）。目前，已有诸多饲料营养价值表被公布（CVB，2007；Sauvant，2004）。自1970年以来，荷兰的饲料行业已经采用了净能体系（CVB，1993），这在一定程度上克服了原料来源多样化所带来的风险。

在不同国家、不同研究人员之间，所采用的能量评估方法的不同会影响到研究结果。通常情况下，一些新的副产品（如玉米DDGS、小麦DDGS）作为被测试原料，配制等DE或等ME 的饲粮，用梯度法评价其有效能值。毫无疑问，这种方法会导致生产性能下降（Roth-Maier等，2004；Friesen等，2006；Thacker，2006；Whitney等，2006；Widyaratne和Zijlstra，2007）。其原因是采用等DE或ME为基础配制饲粮时，使用高纤维或高蛋白原料会降低饲粮的NE值。但最终结果是人们会将生长性能的下降归因于这种被测原料，而不是所采用的饲料能量和氨基酸质量评价体系。

在欧洲，人们认为获得非常规原料准确的NE预测值至关重要，这可以确保采用非常规饲料原料或副产品时，动物能取得相似的生产性能（Smits和Sijtsma，2007）。然而，事实上，采用NE体系配制猪饲粮来评定非常规原料的试验研究或发表的科学文献很少。仅有的一个例子是，在法国的一项研究中，用

NE和可消化氨基酸配制饲粮，当双低菜籽粕用量增加到18%时，对生长育肥猪生产性能也没有影响（Albar等，2001）。在北美，当饲料中使用一种非常规饲料原料（如无单宁蚕豆）时，如果采用NE体系配制饲粮，可以减少非常规原料对生产性能的影响（Zijlstra等，2008）。因此，采用合适的饲料能量评价体系对成功使用新的饲料原料至关重要。

与饲料原料相关的最大风险是农作物固有的营养成分含量发生改变，其原因是农作物受农艺、天气、收割时间和贮藏条件的影响导致基因表达发生改变。对于副产品和次级品，加工工艺是导致营养成分变化的另一个重要因素（Zijlstra等，2001）。例如，在猪饲粮中使用DDGS的主要风险之一是质量不稳定，尤其是第一限制性氨基酸（赖氨酸），因为烘干处理可对赖氨酸造成破坏（ Zijlstra和Beltranena，2008）。饲料热处理过度造成蛋白质破坏早已为人熟知（van Barneveld等，1994），DDGS中赖氨酸不同程度的破坏也已经被确认（Fontaine等，2007）。除了热损伤，在提取油料种子中的油时，采用不同的提取技术（溶剂抽提、螺杆压榨、冷压），也可能导致饼粕中含有不同程度的油脂残留，因此其能值存在很大差异。

◆ 2.2 其他风险

使用非常规原料的另一个风险是化学残留，尤其是一些来源不明或有害的残留物质。最糟糕的情况是在饲料中使用了被聚氯联苯（polychlorinated biphenyls，PCB）/二噁英污染的原料（Bernard等，2002；Covaci等，2008）。目前已证实猪可通过饲料接触到少量的这类残留物（Glynn等，2009）。PCB等残留物能够沉积在猪肉中（Hoogenboom，2004），给消费者健康带来很大的隐患。最近发生的一个事件是三聚氰胺污染的原料进入了宠物食品和人的食品，这可能是通过预先污染了猪饲料而引起的（González等，2009）。这些事件表明，在饲料业中严格执行危害分析与关键控制点（hazard analysis and critical control point，HACCP）等预防措施和即时召回程序的重要性（den Hartog，2003）。新的副产品，如粗甘油也可能含有残留物，因此也应仔细监测。具体来说，粗甘油可能含有残留的甲醇，甲醇残留水平高时，可能引起代谢性酸中毒、呕吐、失明或胃肠道的问题（Kerr等，2007）。

霉菌毒素在自然条件下可在农作物中产生，因此也同样存留在其副产品中。一些霉菌毒素对发酵和干燥处理抵抗力较强，因此不易灭活。事实上，在一些粮食加工（如乙醇生产）过程中，由于淀粉的去除，呕吐毒素（mycotoxin deoxynivalenol，DON）在副产品DDGS（Schaafsma等，2009）中的含量会增加3倍。和原料相比，除DON外，DDGS中其他一些毒素，如黄曲霉毒素、伏马毒素和玉米赤霉烯酮含量也增加了（Wu和Munkvold，2008）。虽然一些研究表明，DDGS中霉菌毒素污染可能并不是一个普遍现象（Zhang等，2009），但也应克服毒素含量偶尔升高这一风险，因为DON即使在低浓度下也可能会严重影响生长和繁殖性能（House等，2002；Dänicke等，2004）。了解原料谷物的地理位置来源等相关信息，并与谷物生长与收获时期的温度、湿度等农艺条件相结合十分重要，因为这直接关系到乙醇生产中的粮食及其副产品DDGS中DON的含量（Schaafsma等，2001）。

3 农作物

◆ 3.1 谷物

传统上，养猪业的竞争力取决于大规模、低成本的饲料谷物生产，这在北美尤为明显。因为当地的农艺条件不同，其粮食产品标准会有所不同。一些非常规谷物品种已被开发，以提高产量或可消化养分浓度。

3.1.1 玉米

玉米是全球谷物的标准，是美国与拉丁美洲商业化养猪生产中的基础。马齿形杂交黄玉米的营养价值已经被进行过详细描述（Sauber和Owens，2001）。最近十年，经过育种改良，一些特殊的非常规玉米品种，如低植酸（Spencer等，2000；Veum等，2001；Hill等，2009）、抗除草剂（Hyuan等，2004）、抗根虫（Hyun等，2005；Stein等，2009）、生长周期短（Opapeju等，2006）、含植酸酶（Nyannor等，2007）、高油和高氨基酸（Pedersen等，2007）的玉米品种已被开发出来。这些品种不仅产量提高，而且在营养价值及其他农艺性状方面均得到改善。

3.1.2 小杂粮

在某些国家或地区，大麦、高粱和小麦等小杂粮也是养猪业中重要的饲料来源。例如，在加拿大西部和澳大利亚，猪的主要饲料原料是大麦和小麦，而在墨西哥，猪的主要饲料原料是高粱。尽管通常认为，大麦和小麦的DE相比玉米变异更大（Fairbairn等，1999；Zijlstra等，1999），但这些谷物的营养价值也已有详尽的描述（Sauber和Owens，2001）。酶制剂的应用可减少小麦品质变异对断奶仔猪生产性能的影响（Cadogan等，2003）。和玉米的育种类似，人们已培育出一些具有独特属性的大麦和高粱品种，如低植酸大麦（Veum等，2002；Htoo等，2007a，2007b），具有特殊淀粉结构的高粱和大麦等（Shelton等，2002；Bird等，2004）。然而，这方面的进展远远落后于玉米。尽管将小杂粮的育种改良重点放在了增产上，但近几十年来，玉米的增产幅度远远高于小杂粮（Alston等，2009）。在高温和多雨地区，玉米的产量远远高于大麦和小麦，这也是美国小杂粮的生产被玉米大量取代的主要原因。

3.1.3 黑小麦

水和氮（nitrogen，N）素是谷物丰产的关键。在加拿大西部和澳大利亚等半干旱地区，干旱和供水问题一直是个困扰农业生产的难题。增强玉米和小杂粮的耐旱性和N利用效率，可能是增加粮食产量的一种有效途径。在农作物生长条件贫瘠的地区，要使每公顷土地所产谷物满足最大猪肉生产量的需要，种植高产低投入的谷物是一个重要的解决方法（Davis-Knight和Weightman，2008）。黑小麦是小麦和黑麦的杂交品种（Radecki和Miller，1990）。在生产条件贫瘠的地区，与种植小麦相比，种植黑小麦可以提高作物产量（McLeod等，2001），而投入可以降低14%（Davis-Knight和Weightman，2008）。早期的研究表明，与玉米相比，使用黑小麦会降低仔猪生长性能（Hale和Utley，1985）。因此，传统上认为，饲喂黑小麦会降低猪的生产性能。然而，现代黑小麦品种的胰蛋白酶抑制剂含量低，其对猪的适口性也不存在问题（Radecki和Miller 1990）。事实上，用含66.5%小麦或66.5%黑小麦的饲粮饲喂断奶仔猪，猪的生长性能没有差异（Beltranena等，2008；图10-1）。

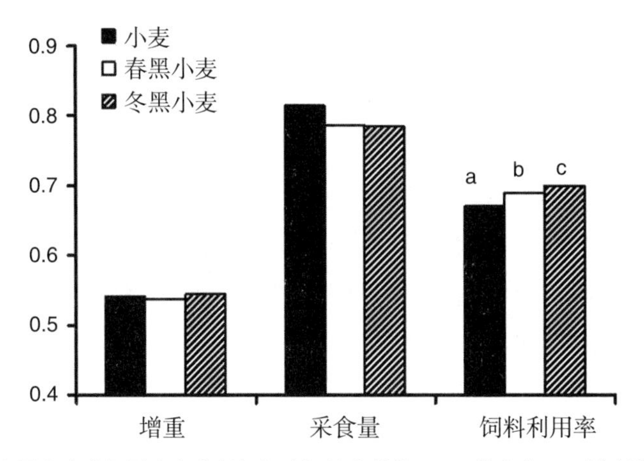

图10-1　春黑小麦或冬黑小麦在断奶仔猪饲粮中替代66.5%的小麦，不影响增重（kg/d；SEM=0.013）和采食量（kg/d；SEM=0.022），但提高了饲料利用率（$P<0.05$；SEM= 0.01）

（资料来源于Beltranena等，2008）

◆ 3.2　豆类

非油料豆类作物N素利用效率高，因此，在目前的气候条件下，种植非油料豆类作物对农业可持续发展具有重要意义。根瘤菌与豆科作物的根是共生关系，可以固定大气中的N元素，从而减少对化肥的需求，甚至会降低作物轮作系统中下一季的谷物与油料作物对化肥的需求。用作饲料的豆类作物包括紫花豌豆、扁豆、鹰嘴豆、白咖啡豆（Rochfort和Panozzo，2007）。与油料作物相比，非油料豆类作物蛋白含量低，淀粉含量高。因此，非油料豆类用作饲料具有双重作用：提供能量和蛋白质。传统认为，豆类种子含有不同数量的抗营养因子，会阻碍营养物质的消化、吸收和利用（Huisman和Jansman，1991）。目前，通过作物育种已经大幅降低了一些豆类种子（如紫花豌豆和蚕豆）的抗营养因子含量（Clarke和Wiseman，2000）。然而，要灭活其他作物（如大豆和菜豆）中的抗营养因子需要通过合适的加工处理（van der Poel 等，1990）。豆类种子中的植物源生物活性物质对人类健康有益（Rochfort和Panozzo，2007），然而，对于猪的有益作用还没有得到证实。

3.2.1 豌豆

在欧洲和加拿大一些地区，豌豆是猪饲粮中重要的能量和蛋白质来源，但是其供应受限于人类食品市场需求。由于现代豌豆品种的抗营养因子含量较低，因此可以在育肥猪饲粮中无限量使用（Gunawardena等，2007）。一般来说，豌豆的消化能值低于玉米，氨基酸的利用率低于豆粕（Mariscal-Landín等，2002；Sauvant等，2004；Stein等，2004）。豌豆淀粉在回肠末端的消化率为90%（Fledderus等，2003；Stein等，2007），通过挤压膨化可以进一步提高（Mariscal-Landín等，2002；Stein等，2007）。不同的豌豆蛋白成分在胃肠道的消化率不同（Le Guen，2007）。这意味着并非所有的豌豆蛋白都能被很好地消化，因而豌豆消化率还有待进一步提高。豌豆的NSP和低聚糖具有独特的发酵特性（Leterme等，1998）。与其他豆科类相似，植酸阻碍了豌豆中磷的利用（Stein等，2007）。

研究表明，给生长育肥猪饲喂含有豌豆的饲粮，在其生产性能、胴体性状（Stein等，2004；2006；Gunawardena等，2007）和肉品质（Gunawardena等，2007）等方面能达到和以豆粕作为主要蛋白源的饲粮一样的效果。豌豆在断奶仔猪上的效果还不确定。研究表明，在断奶仔猪第三阶段饲粮中豌豆的添加量达到18%时，不影响生长性能（Stein等，2004）；但当豌豆添加量达到30%时，其生长性能线性下降（Friesen等，2006）。综上所述，如果采用现代饲料评价技术对能量、氨基酸和磷含量进行准确评定，豌豆就可在生长育肥猪或断奶仔猪后期饲粮中无限量添加，但在断奶仔猪早期饲粮中，添加量应控制在20%以下。

3.2.2 无单宁蚕豆

蚕豆是加拿大西部和欧洲北部部分地区新兴的豆类作物。雨量充沛时，籽实产量和固氮量高于豌豆（Strydhorst等，2008）。因此，从生态学角度来说，与豌豆相比，无单宁蚕豆对维持环境的可持续性具有更重要意义。通过作物育种，无单宁蚕豆中单宁含量低于1%。在猪上的消化试验结果表明，无单宁蚕豆能被猪很好地消化（van der Poel等，1992a），其净能含量和豌豆相似，标准回肠可消化（standardized ileal digestible，SID）氨基酸含量高于豌豆（Zijlstra等，2008）。研究表明，在保育后期饲粮中用无单宁蚕豆完全替代豆粕，不会降低仔猪生长性能（Beltranena等，2009；图10-2）。在生长育肥猪饲粮中，只要净能和SID氨基酸平衡，用无单宁蚕豆完全取代豆粕，不影响猪生长性能、胴体评定和肉品质（Gunawardena等，2007）。

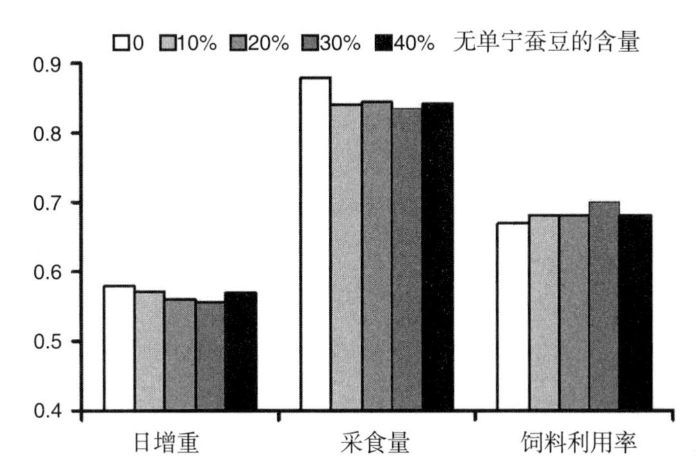

图10-2　在断奶仔猪饲粮中用无单宁蚕豆代替豆粕对日增重（kg/d；SEM=0.011）、采食量（kg/d；SEM=0.019）和饲料利用率（SEM=0.01）无影响

（资料来源于Beltranena等，2009）

3.2.3 羽扇豆

羽扇豆是澳大利亚和欧洲的特产，主要有白羽扇豆、狭叶羽扇豆和黄羽扇豆三种。猪能够很好地利用狭叶羽扇豆和黄羽扇豆，但对白羽扇豆的利用效率较低，其原因尚不清楚（van Barneveld，1999；Písaříková和Zralý，2009）。甜羽扇豆（狭叶的）生物碱含量较低，能量和氨基酸消化率较高（Kim等，2008，2009）。尽管甜羽扇豆营养品质可能存在变异（Kim等，2009），但其在猪饲粮中大量添加也不影响生产性能（van Barneveld 等，1999）。甜羽扇豆可能还含有改变猪脂肪代谢的生物活性物质（Martins等，2005）。在断奶仔猪饲粮中，甜羽扇豆的建议添加量可高达15%（Kim等，2008）。

3.2.4 其他豆类

除以上豆类作物外，还有多种其他豆类作物，如鹰嘴豆、扁豆、菜豆等。一般情况下，应确保这些豆类作物的抗营养因子（如胰蛋白酶抑制剂、凝集素和单宁等）含量较低，或经过加工处理使其降低到动物可耐受的水平（van der Poel等，1990；Rubio，2005）。有了这些保障措施，这些豆类也可以作为一种优质的非常规饲料使用（Mustafa等，2000）。

◆ 3.3 油料作物

大豆、油菜籽和亚麻籽等油料籽实都可以用作猪的饲料原料。但是，通常情况下，若在猪饲粮中使用这些作物籽实，会使饲料价格过高，因而主要是其副产品作为蛋白源使用。一些达不到特定级别的油料籽实也可能会用在猪饲粮中。但也有利用亚麻籽生产高ω-3脂肪酸猪肉的情况，从经济可持续发展的角度来讲，这可能是一个例外。将亚麻籽与豌豆共挤压膨化能提高营养物质（如脂肪酸）消化率（Htoo等，2008；图10-3）。ω-3脂肪酸可以有效地被结合进入猪的脂肪，提高熏猪肉、火腿和其他肉制品的附加值，从而增加这些产品进入食品市场的机会（Musella等，2009）。

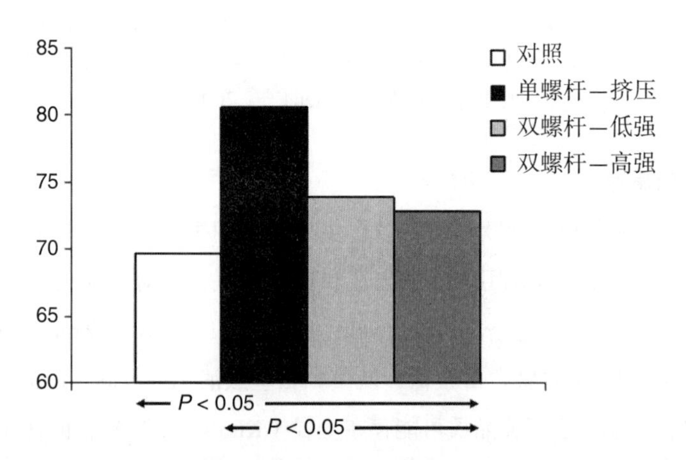

图10-3　采用单/双螺杆及不同膨化强度对亚麻籽和豌豆共膨化时对全肠道能量消化率的影响（SEM=1.9）

（资料来源于Htoo等，2008）

4 副产品

作物籽实的各种组分可用通过一系列技术进行分离，以供人类食品、生物制品和饲料生产应用（Zijlstra等，2004；Zijlstra和Beltranena，2007；Vasanthan和Temelli，2008）。传统上，根据组分的物理特性将作物籽实进行干法（无溶剂）分离，提取其中高价值的重要成分用于人类食品。例如，油的提取使用压榨、磨碎、过滤等方法，而蛋白质和淀粉分离采用风选分离。生产出的副产物可以用作饲料。这种干分离技术的优点是可以连续进行而不是分批处理，加工成本也比较低，并且不存在溶剂残留或浆液（Hemery等，2007）。干法分离的缺点是成分分离不彻底，产品的性能可能达不到用于人类食品或生物工业所需的较高品质要求。因而，依据组分化学特性，使用水、酸、碱、盐或有机溶剂分离高价值成分的湿法分离加工技术逐渐发展起来（Vasanthan和Temelli，2008）。湿分离的优点包括可使高价值组分达到更高的纯度。此外，可溶性抗营养因子（如硫代葡萄糖苷、酚）可以在浆液中被洗脱，改变pH或添加酶可分解副产品中的植酸，使植酸磷转化为有效磷（Drew，2004）。但是，湿法分离加工成本较高，因为主产品和副产品要长时间运输、长期贮存或应用于干饲料，必须要进行干燥处理。虽然烘干有利于抗营养因子的灭活，增加矿物质的可利用性，但也可能损害副产品中的蛋白质，从而降低营养价值。喷雾干燥法比传统的干燥方法昂贵，但可避免蛋白质受损，从而保持蛋白质的营养和功能特性。猪饲粮中使用副产品可以降低饲料成本，促进养猪业经济的可持续发展，因而副产品成为越来越有吸引力的非常规饲料（Jha等，2010；Zijlstra等，2010）。副产品一般含有较高的NSP，因此NSP降解酶可以被用来提高养分消化率（Zijlstra等，2009b）。有关副产品加工的问题不在此详细讨论。

液体饲喂系统允许在猪饲料中使用湿的副产品，可完全避免干燥及相关的能源成本。因此，可以认为液体饲喂更环保、更有利于养猪经济的可持续发展。前提是猪场附近须有一个副产品加工厂，否则运输会成为一个问题。液体饲喂可通过浸泡改变饲料特性，以提高营养物质的消化率（Choct等，2004；Niven等，2007）；或者通过发酵，改善肠道健康和生长性能（Scholten等，1999）。但是，对有些已经过发酵的副产物来说，这种改善的概率就很小了。

◆ 4.1 生物燃料行业

矿物燃料是人类活动的主要能源。由于多种原因，当前用生物柴油和乙醇等可再生能源代替矿物燃料的日益迫切，由此产生的大量DDGS、菜籽饼粕和粗甘油成为可利用的非常规饲料原料。然而，这些副产品营养品质的变异性是一个需要关注的主要问题（Zijlstra和Beltranena，2008）。在全球围绕粮食供应的讨论中，有关谷物在畜禽饲料和生物燃料中的应用，受到了相当大的关注（Blaxter，1983；Avery，2006；Dale，2008）。生物燃料行业与畜牧业和食品工业直接竞争粮食供应，从而提高了当地粮食价格。相应地，生物燃料和食品工业所产生的副产品，也可在畜禽饲料中应用。如果决定生产生物燃料，那么就要建立副产物市场。这样，从成本方面考虑，在猪饲粮中使用生物燃料副产品可能对养猪生产者有很大吸引力（Lammers等，2010）。

4.1.1 DDGS

玉米DDGS作为非常规饲料已经成为全球性商品。最近，对玉米DDGS在猪上的营养价值已经进行了综述（Stein和Shurson，2009）。简要地说，由于淀粉等糖分发酵为乙醇，导致DDGS中其他常量营养素和矿物质含量升高。以玉米为例，玉米中油的含量比小麦高。由于粗脂肪含量的增加，玉米DDGS可以达到和玉米接近的消化能（DE）和代谢能（ME）含量，其蛋白含量也比玉米更高。因此，玉米DDGS可同时作为能量和氨基酸来源，成为猪饲粮中一个有吸引力的饲料原料。Xu等（2010a）研究表明，在育肥猪饲粮中添加高达30%的玉米DDGS，不改变猪的生长性能。然而，玉米DDGS的添加并不总能保持一致的生产性能（Stein和Shurson，2009）。这种差异既可能与由于发酵、干燥和可溶物比例不同导致的玉米DDGS的质量变化有关（Zijlstra和Beltranena，2008），也可能与饲粮能量、常量营养成分和氨基酸模式有关。Xu等（2010a）研究发现，在能量与氨基酸平衡的玉米-豆粕型饲粮中，随着玉米DDGS添加量的增加，胴体瘦肉率和背膘厚未发生变化。然而，玉米DDGS添加量的增加，会导致饲粮纤维和不饱和脂肪酸含量的增加，从而导致屠宰率降低，胴体中不饱和脂肪酸含量增加（Xu等，2010a）。纤维可增加肠道重量（Jørgensen等，1996），多不饱和脂肪酸可直接结合进胴体脂肪（Averette Gatlin等，2002）。为了减少玉米DDGS对猪肉脂肪硬度的负面影响，可以在屠宰前3周从配方中撤出玉米DDGS（Xu等，2010b；Beltranena等，2010；Beltranena和Zijlstra，2010）。

在小麦和其他小杂粮中，小麦的脂肪含量低于玉米，导致小麦DDGS能量比小麦籽粒低得多（Widyaratne和Zijlstra，2008，2009）。与玉米DDGS相比，小麦DDGS作为蛋白质来源比作为能量来源更有价值。玉米DDGS和小麦DDGS

的P含量和消化率也高于原粮。最初，在加拿大的研究发现，小麦DDGS对猪的饲养效果并不理想。即使将饲粮DE和SID氨基酸水平调整到一致（Widyaratne和Zijlstra，2007），添加100 g/kg或更高水平的小麦DDGS也会降低猪的生产性能（Thacker，2006）。这可能是因为用于这些研究的小麦DDGS在干燥过程中被过度加热（Zijlstra和Beltranena，2008）。近年来，随着发酵和干燥工艺的改善，乙醇加工厂可以生产出质量更好的小麦DDGS。实际上，在饲粮中添加15%的小麦DDGS对断奶仔猪生产性能影响较小（Avelar等，2010）。此外，商业养猪中，在育肥猪饲粮中可以添加高达30%的小麦DDGS，但是，当添加量再进一步提高时，就会降低猪的生产性能（Beltranena和Zijlstra，2010）。酶可以提高小麦DDGS（Yanez等，2009）和含有小麦DDGS的饲粮的消化率（Emiola等，2009）。

4.1.2 粗甘油

粗甘油可作为猪的能量来源。含有0～20%粗甘油（来自于大豆油）的饲粮，其全肠道能量表观总消化率为89%～92%。这表明，粗甘油可以被生长猪很好地消化（Lammers等，2008a）。生产生物柴油时，从油料籽实或动物脂肪中提取油，用甲醇水解，NaOH或KOH做催化剂，从而得到甲酯（生物柴油）和粗甘油（Kerr等，2007）。生产1L生物柴油的同时可以生产79 g粗甘油。大规模生产生物柴油开始于欧洲，尤其是德国。在饲粮中用粗甘油（来自于菜籽油）代替10%的大麦，生长猪的平均日增重（ADG）可增加8%（Kijora和Kupsch，1996）。在美国，用粗甘油（来自于大豆油）取代6%的玉米，可提高保育猪的日增重（Groesbeck等，2008）。这些结果表明，在质量相等的基础上，粗甘油可代替5%～10%的谷物饲料（Lammers等，2008b；Zijlstra等，2009a；图10-4）。

使用粗甘油也存在隐患。例如，粗甘油可能含有加工处理后残留的杂质，如甲醇和NaCl。用作饲料的粗甘油，其甲醇含量应不超过150ppm，因为较高水平的甲醇可能引起代谢酸中毒、呕吐、失明或胃肠道问题（Kerr等，2007）。较高水平的NaCl可能限制甘油的添加量，以避免饲粮中的钠和氯超出推荐水平。此外，甘油是一种黏稠的凝胶，可能会导致饲料混合和流动出现问题（Kerr等，2007）。但是，甘油可增加颗粒耐久性，降低电流值、电动机负载，并提高制粒机的生产效率（Groesbeck等，2008）。

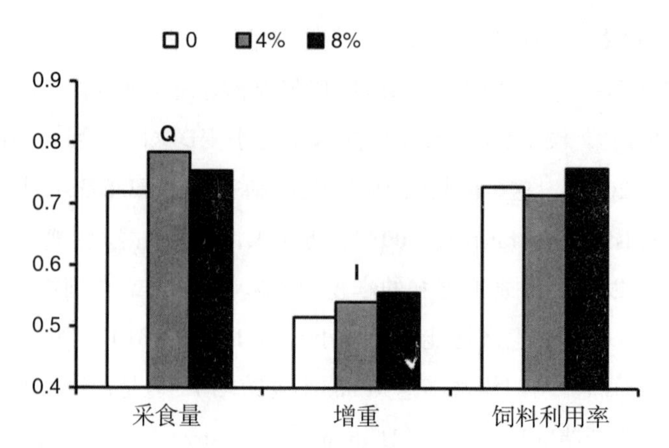

图10-4 在小麦基础饲粮中添加粗甘油，使断奶仔猪采食量呈二次曲线增加（kg/d）（Q；*P*<0.05；SEM＝0.02），日增重线性增加（kg/d）（I；*P*<0.10；SEM＝0.01）

（资料来源于Zijlstra等，2009）

◆ 4.2 食品工业

在超市里每一种食品背后，都应该至少有一种副产品。这些副产品涵盖范围广，如甜菜渣、菜籽粕、柑橘渣、乳清、面包渣、面粉副产品、肉骨粉等。在畜牧业中，可将这些低价值的副产品转化为高品质的动物性蛋白。

4.2.1 油籽粕

尽管生物柴油和生物制品变得日益重要，但油籽粕类产品主要还是来自人类压榨食用油的副产物。在2008—2009年，菜籽油是仅次于大豆油和棕榈油的世界第三大植物油。菜籽粕也是世界上第二大的蛋白来源，大约相当于豆粕产量的1/5（USDA，2010）。在北美和欧洲，大部分油菜籽的硫代葡萄糖苷和芥酸含量较低，故被称为双低油菜籽。因此，普通油菜籽和双低油菜籽之间的营养品质差别较大（本文以下部分油菜籽/粕均指双低油菜籽/粕）。

在油菜籽中，油占籽实的45%，是最有价值的部分。溶剂抽提、螺杆压榨和低温压榨可分别生产得到菜籽毛油及溶剂抽提菜籽粕、螺杆压榨菜籽粕和低温压榨菜籽饼。这些产品都可以作为非常规饲料原料（Leming和Lember，2005）。在实际应用中，溶剂抽提菜籽粕在育肥猪饲粮中的用量限制在15%以下，尽管建议的最大用量为25%（Canola Council of Canada，2009）。在生长育肥猪饲粮中用菜籽粕替代50%豆粕，猪的生长性能和胴体品质比饲喂全豆粕饲粮的猪差（Shelton等，2001）。其原因可能是菜籽粕饲粮中的ME比豆粕饲粮低得多。与豆粕相比，菜籽粕可消化纤维和粗蛋白含量较低，导致其可利用能量和氨基酸含量较低，这是饲粮中限制使用菜籽粕的主要原因（Bell，1993）。

由于溶剂抽提出油率高（>95%），因此大多数油脂厂都是采用溶剂抽提方法，但这也导致菜籽粕DE含量较低（Spragg和Mailer，2007）。螺杆压榨的出油率为75%，因此螺杆压榨菜籽粕含有10%～15%的残油（Leming和Lember，2005），从而比溶剂抽提的油菜粕有更高的消化能值和较低的可消化氨基酸含量（Woyengo，2010；Seneviratne等，2010）。与溶剂提取菜籽粕相比，螺杆压榨菜籽粕含有较高的能量含量。这对能量需求较高的生长育肥猪而言，是更好的非常规原料，尽管残留的硫代葡萄糖苷可能稍微降低采食量（Seneviratne等，2010）。使用螺旋压榨机生产的菜籽饼含18%～20%的残油（Schöne等，2002）。饲粮添加15%的菜籽饼会降低采食量和体增重，残留的硫代葡萄糖苷可能是一个诱因（Schöne等，2002）。每千克猪饲粮中，硫代葡萄糖苷最高限量为2 mmoL/kg（Schöne等，1997）。

亚麻粕是亚麻压榨行业的副产物。根据油的提取工艺，溶剂抽提亚麻粕含有3%～7%的残油（Batterham等，1991；Bell和Keith，1993；Farmer和Petit，2009），螺杆压榨亚麻粕含有13%的油（Eastwood等，2009）。因为残油量较低，所以亚麻粕在母猪饲粮中添加不影响母猪血浆和奶中的脂肪酸组成（Parmer和Petit，2009），然而，饲喂亚麻籽或亚麻籽油可以增加奶中α-亚麻酸含量。由于残油量较高，饲喂螺杆压榨亚麻粕会增加背膘和眼肌中α-亚麻酸含量（Eastwood等，2009）。在生长育肥猪饲粮中，螺杆压榨亚麻粕的添加量可达15%（Eastwood等，2009）。

4.2.2 小麦副产品

小麦通过干法磨碎把大部分淀粉分离出去，就可以得到人食用的面粉，而剩余的就是小麦副产品（Holden和Zimmerman，1991）。小麦副产品包括麦麸、粗粉、次粉、筛下物、小麦加工副产物等。小麦磨粉前分离出的杂质统称为小麦筛下物，通常包括不完善麦粒、混杂种子和其他污染物。通常小麦筛下物粗纤维含量低于7%，并且破碎或瘪小麦含量高于35%（Audren等，2002）。

麦麸来自于在商业面粉加工过程中清洗干净的小麦外种皮，含有12%的粗纤维（AAFCO，1988）。小麦次粉是覆盖在小麦籽粒胚乳的外种皮以内的部分（Huang等，1999），通常含有5%～10%的粗纤维和15%～20%的粗蛋白（CP）。粗麦粉包括绝大部分完整的麸皮和胚芽颗粒，至少含有15%的CP（O'Hearn和Easter，1983）。小麦加工副产物由粗糠、次粉、筛下物和粗麦粉组成（AAFCO，1988），含有约9.5%的粗纤维（Dale，1996）。这些副产品的营养组成变异很大，而且同一类副产品的营养组成也变异很大（Cromwell等，2000）。

小麦副产物组成上差别很大。例如，小麦次粉的变异是因为糠和胚乳比例不同（Huang等，1999）。小麦副产品中性洗涤纤维（NDF）含量和氨基酸消化率呈负相关（Huang等，2001）。麦麸的不可溶性纤维增加食糜黏度，从而减少养分消化率（Sakata和Saito，2007）。使用木聚糖酶可以提高小麦副产品的养分消化率（Nortey等，2007，2008）。

4.2.3 甜菜渣

甜菜渣为甜菜加工制糖后的副产品，含有较高的NSP，特别是果胶（Spagnuolo等，1999）。尽管甜菜渣发酵速度比其他快速发酵的NSP（如菊糖）低很多，但仍能够很好地被猪发酵（Awati等，2006）。可发酵NSP（如甜菜渣）会干扰养分消化。然而，从可持续发展的目标来说，这些可发酵NSP的一些功能性特性，在改变N排泄模式、改善肠道健康和动物福利三个方面已引起了人们的极大关注。如甜菜渣等饲料含有可发酵NSP，微生物可与N结合形成菌体蛋白，从而将尿氮转变为粪氮（Bindelle等，2009），而燕麦壳等主要含有非发酵性NSP则没有这种作用（Zervas和Zijlstra，2002b；图10-5）。N排泄模式的转变及粪便pH的降低，可使猪粪便中氨的排放减少（Canh等，1998），但臭味排放可能不会降低（Payeur等，2002）。高蛋白饲粮中加入甜菜渣能刺激消化道发育和肠道健康，但在低蛋白饲粮中将会损害肠道健康。最后，尽管与淀粉相比，可发酵NSP的可利用能值较低，但这可以通过改变猪的行为，减少其活动量来弥补（Schrama等，1998）。在饲粮中添加45%的甜菜渣可提高妊娠母猪的采食量，对繁殖性能也没有负面影响（van der Peet-Schwering等，2004）。

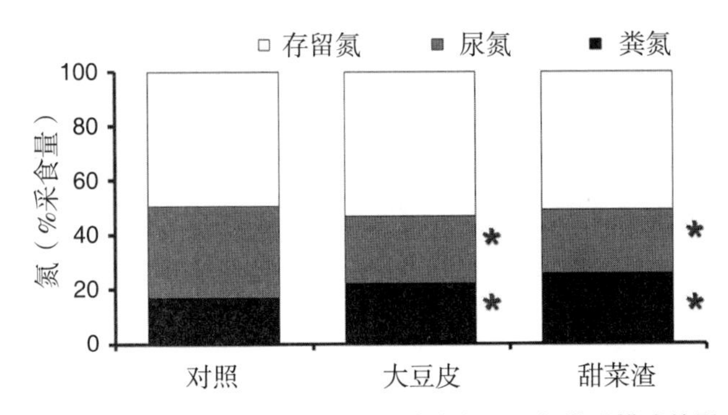

图10-5 大豆皮和甜菜渣中可发酵纤维对生长猪粪氮和尿氮排泄模式的影响（$P<0.05$）

（资料来源于Zervas和Zijlstra，2002）

4.2.4 其他副产品

除了油菜、亚麻、小麦和甜菜的副产品，还存在很多其他的副产品。前人已对这些副产品进行了详细阐述（Thacker和Kirkwood，1990；Chiba，2001；Myer和Brendemuhl，2001；Sauber和Owens，2001）。以下是最新收集到的一些的研究成果。

如果采用溶剂抽提，脂肪含量低（2%）、纤维素含量高（54%NDF）、CP和淀粉含量分别为21%和14%（Weber等，2010）。在营养平衡饲粮中，玉米胚芽粕的添加量可以高达40%（Harbach等，2007）。尽管玉米胚芽粕的确切发酵模式还不甚明了，但确实可以很好地被猪利用（Weber等，2010）。

对于可持续发展的家禽业来说，羽毛粉等是主要的废弃物，应像生猪屠宰的下脚料一样妥善处理。处理方法包括热加工消除病原菌，水解以改善氨基酸消化率。给猪饲喂含有10%水解羽毛粉的基础饲粮，并补充所需的氨基酸，猪的体增重及瘦肉沉积与饲喂玉米-豆粕型饲粮的一样（Divakala等，2009）。

宠物食品副产品泛指不符合特定质量规范、在处理过程中被损坏、或者已经被分发到零售商店而在有效期内不能出售的不合格宠物食品。如果得到正常的监管和批准，将宠物食品的副产品用于断奶仔猪的饲粮中，可以有效提供蛋白质和脂肪（Jablonski等，2006）。

◆ 4.3 分离

分离是保持谷物或其他原料各个组分的营养和功能特性不变或不降低的一种处理方法。采取特定的干法或湿法可以将谷物产品分离成不同的组分，这可能会获得新的高附加值的谷物原料产品，有利于满足对营养要求高的畜禽的需要（Zijlstra等，2004；Zijlstra和Beltranena，2007）。利用空气对豆类籽实进行分离可能就是一个很好的例证，因为豆类种子含有淀粉和蛋白质两部分，可以在气流中很好地分离。具体来说，脱壳的豌豆通过精细研磨和气流分离成两部分：轻的、精细的（豌豆蛋白浓缩物）和重的、粗糙的（主要是淀粉）（可用于猪饲料）（Wu和Nichols，2005）。因此，在湿法分离浓缩蛋白或分离蛋白的基础上可以进一步分离得到浓缩蛋白和浓缩淀粉。在人的食品方面，油脂分离具有悠久传统。另外，经过脱皮后的纤维组分，采用现代分离技术，也可以提取出具有特殊功能的纤维产品。

4.3.1 蛋白组分

大豆浓缩蛋白约含60%的CP，而大豆分离蛋白约含90%的CP。作为一个可持续的非常规原料，在豆类作物中，传统上就常将豌豆各组分进行分离应用（Bramsnaes和Olsen，1979）。利用空气分离也可以从无单宁蚕豆中有效分离出高纯度的浓缩蛋白（Gunawardena等，2010a）。利用空气分离的豆类浓缩蛋白，是仔猪营养中某些特殊蛋白源最具吸引力的替代品（Valencia等，2008；Gunawardena等，2010b；图10-6）。然而，由于其他豆类浓缩蛋白的养分消化率可能低于大豆浓缩蛋白，因此，在试验验证其效果前，应准确测定其养分消化率和可消化养分组成（Valencia等，2008）。豌豆分离蛋白可采用湿法分离生产，有可能被用来替代喷雾干燥血浆蛋白粉。豌豆分离蛋白极易消化，部分原因是由于抗营养因子被有效地去除（Le Guen等，1993a，1993b）。然而，豌豆分离蛋白只具有类似血浆蛋白粉的营养特性，但不具有血浆蛋白的其他功能特性。因此，豌豆分离蛋白必须与卵黄抗体（从超免蛋鸡上获得，含特异性抗肠毒素大肠杆菌K88的抗体）混合，来达到控制细菌感染的作用（Owusu-Asiedu等，2003a，2003b）。

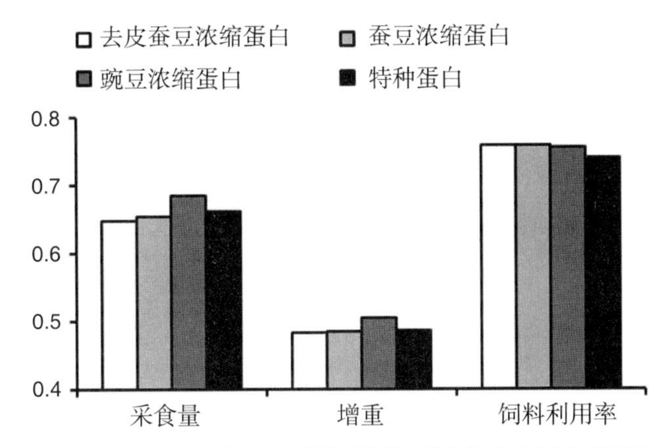

图10-6 饲粮添加16%～17.5%的不同来源浓缩蛋白替代特殊蛋白源对断奶仔猪的采食量（kg/d；SEM＝0.016）、增重（kg/d；SEM＝0.012）和饲料利用效率（SEM＝0.01）无影响

（资料来源于Gunawardena等，2010b）

4.3.2 淀粉组分

淀粉是猪的主要能量来源。90%的淀粉供人食用，很少用于商品猪饲料。在猪营养中应用分离淀粉，是用于研究淀粉化学对血糖波动的影响（van Kempen等，2010；图10-7）。浓缩淀粉可以采用干法或湿法分离生产。利用空气分离生产的易消化豌豆淀粉和无单宁蚕豆淀粉可作为仔猪的饲料原料（Gunawardena等，2010b）。挤压膨化可提高浓缩蚕豆淀粉的养分利用率（Wierenga等，2008）。然而，煮熟的大米可能是仔猪更喜欢的淀粉来源。采取湿分离法从大麦或燕麦提取β-葡聚糖时也可以产生浓缩淀粉，猪对其消化率很高（Johnson等，2006）。在仔猪饲粮中添加生土豆淀粉可以作为抗性淀粉的来源（Bhandari等，2009）。抗性淀粉源具有益生元的活性，利用饲料原料的这一特性可以代替饲料添加剂，降低抗生素的使用。此外，抗性淀粉发酵与可发酵纤维类似，可以使挥发性含氮化合物的排泄降低，从而减少养猪场异味排放（Willig等，2005）。最后，抗性淀粉还可减少粪臭素的形成，因而，在一定程度上，有助于解决养猪生产中为了动物福利而不阉割公猪对肉质风味造成影响的问题（Losel和Claus，2005）。

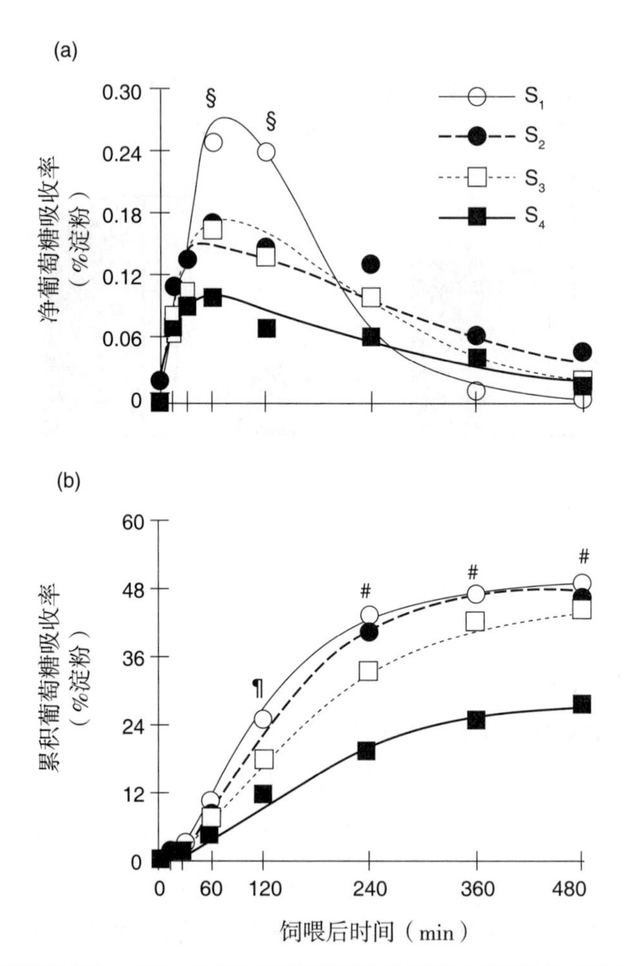

图10-7　分别给猪饲喂含S_1至S_4（从快速到慢速消化四种淀粉）的饲粮对门脉系统葡萄糖吸收率的影响

利用改进的 Chapman–Richards 模型，根据不同时间点的观测值来预测门脉葡萄糖动力学（R_2=0.96）；图a表示门脉葡萄糖净吸收率；图b表示累积门脉葡萄糖吸收率；§表示S_1和S_2；S_3和S_4分别差异显著，¶表示S_1与S_4差异显著；#表示S_1，S_2，S_3和S_4相比差异显著（P<0.05；n=4）。图a中SE范围从0.01到0.03，图b中SE范围从0.02%到5.22%。（资料来源于van Kempen等，2010）

4.3.3 纤维组分

通常认为脱皮有利于提高谷物、豆类或油料籽实对猪的营养价值。脱壳籽实比完整籽实具有更高的能值和营养物质消化率。例如，脱皮蚕豆、扁豆、大麦、燕麦、玉米等谷物和脱皮油菜籽都比相应的带壳籽实有更高的营养物质消化率（van der Poel，1992b；Kracht，2004；Hennig等，2006；Moeser等，2002）。外皮中的NSP在猪消化道内通常不易溶解，因此发酵特性较差（Williams等，2005）。饲喂抗营养因子含量低的谷壳NSP，能增加饲料容积，因此可增加限饲妊娠母猪的饱感，减少其刻板行为（Matte等，1994；Holt等，2006）。但是，来自同类谷物的不同样品，其纤维的发酵特性也可能不同，易发酵纤维会产生益生作用（Pieper等，2009）。湿法分离可以提高NSP的黏度和益生作用。例如，用湿法从大麦或燕麦分离β-葡聚糖时，可以生产具有较高发酵特性和特定体外黏度的β-葡聚糖浓缩液，其黏度取决于β-葡聚糖链的长度。这些组分有益生元活性（Metzler-Zebeli等，2010），并能影响血糖波动（Hooda等，2010）。

4.3.4 脂肪组分

油从油料籽实中提取，经纯化后，作为人类食品具有较高的价值。初步分离得到的粗植物油，对胃肠道发育不成熟的仔猪具有较高的价值，因为植物油比动物性脂肪有更高的粗脂肪消化率（Duran-Montge等，2007）。然而，从价格或生产方面考虑，一般不允许在饲料中添加过高水平的液态植物油，因为其可能会阻碍物料的流动。饱和动物脂肪（如牛油）是生长育肥猪更经济的能量来源，而且不像玉米DDGS因含有大量不饱和脂肪酸（如亚油酸）而使猪脂肪变软（Stein和Shurson，2009）。饲喂亚麻籽油尽管会提高猪肉中ω-3脂肪酸水平（Health Canada，2004），但可能会降低猪脂肪硬度，除非和共轭亚油酸一起使用（Dugan等，2004）。在妊娠母猪和哺乳母猪饲粮中添加亚麻油，会使猪体组织和奶中α-亚麻酸水平升高，从而增加哺乳仔猪，尤其是刚出生仔猪体内的α-亚麻酸含量（Boudry等，2009）。α-亚麻酸可改善仔猪肠道屏障（Boudry等，2009）和免疫力（Farmer等，2010），从而改善其健康状况。最后，饲粮添加不饱和脂肪会增加维生素E的需要量。维生素E是一种生物膜抗氧化剂（Schaefer等，1995），因此，在含不饱和脂肪酸的饲粮中推荐增加维生素E添加量。

5 结 语

在可持续养猪业中，经济性、循环农业、社会接受度和环境影响是关键要素。猪作为杂食性动物能有效地把非常规饲料转化为猪肉，因此非常规原料在构成可持续养猪业的关键要素中发挥重要作用。传统上认为，应用非常规原料仅仅是为了降低生产成本。但是，最近一些经过改良的非常规农作物已被用来降低猪肉生产对环境的影响。非常规饲料可能具有独特的功能，如可以调控动物的健康、行为、养分排泄模式，甚至是猪肉品质。但是使用非常规原料也带来一些问题。首先，使用副产品会使基础饲料的常量养分含量的变异增加，远超出了谷物本身的变异情况。因此，对非常规原料的能量、氨基酸、磷含量及利用率或消化率等方面的评价就非常重要，所选择的饲料评价体系也同样重要。其次，副产物可能含有化学残留物、霉菌毒素，会降低自由采食量并影响繁殖性能。最后，副产品使用可能会降低胴体性状和肉品质。纤维含量高的副产品可降低屠宰率，一些油含量高的副产品可提供不饱和脂肪酸，能使猪脂肪变软。总之，使用非常规原料可以降低每单位猪肉生产的饲料成本，但同时，也可能会影响到成本效率、生长性能预测、动物健康、合理的环境影响和预期的胴体性状和肉品质等生产目标的实现。

作者：Ruurd T. Zijlstra和
Eduardo Beltranena
译者：李振田

参考文献

AAFCO. 1988. Official Publication. Assoc. Am. Feed Control Off., Charleston, WV.

Albar, J., J. Chauvel, and R. Granier. 2001. Incidence du taux de tourteau de colza sur les performances de post-sevrage et d'engraissement [Effects of the level of rapeseed meal on performances in the post-weaning and the growing/finishing periods]. J. Rech. Porc. 33:197–203.

Alston, J. M., J. M. Beddow, and P. G. Pardey. 2009. Agricultural research, productivity, and food prices in the long run. Science 325:1209–1210.

Audren, G. P., H. L. Classen, K. V. Schwean, and V. Racz. 2002. Nutritional value of wheat screenings for broiler chickens. Can. J. Anim. Sci. 82:393–398.

Avelar, E., R. Jha, E. Beltranena, M. Cervantes, A. Morales, and R. T. Zijlstra. 2010. The effect of feeding wheat distiller's dried grain with solubles on growth performance and nutrient digestibility in weaned pigs. Anim. Feed Sci. Technol. 160:73–77.

Averette Gatlin, L., M. T. See, J. A. Hansen, D. Sutton, and J. Odle. 2002. The effects of dietary fat sources, levels, and feeding intervals on pork fatty acid composition. J. Anim. Sci. 80:1606–1615.

Avery, D. 2006. Biofuels, Food, or Wildlife? The Massive Land Costs of U.S. Ethanol. Competitive Enterprise Institute, Washington, DC.

Awati, A., B. A. Williams, M. W. Bosch, Y. C. Li, and M. W. A. Verstegen. 2006. Use of the *in vitro* cumulative gas production technique for pigs: An examination of alterations in fermentation products and substrate losses at various time points. J. Anim. Sci. 84:1110–1118.

Batterham, E., L. Andersen, D. Baigent, and A. Green. 1991. Evaluation of meals from LinolaTM low-linolenic acid linseed and conventional linseed as protein sources for growing pigs. Anim. Feed Sci. Technol. 35:181–190.

Bell, J. M. 1993. Factors affecting the nutritional value of canola meal: A review. Can. J. Anim. Sci. 73:679–697.

Bell, J. M., and M. Keith. 1993. Nutritional evaluation of linseed meals from flax with yellow or brown hulls, using mice and pigs. Anim. Feed Sci. Technol. 43:1–18.

Beltranena, E., D. F. Salmon, L. A. Goonewardene, and R. T. Zijlstra. 2008. Triticale as a replacement for wheat in diets for weaned pigs. Can. J. Anim. Sci. 88:631–635.

Beltranena, E., S. Hooda, and R. T. Zijlstra. 2009. Zero-tannin faba bean as a replacement for soybean meal in diets for starter pigs. Can. J. Anim. Sci. 89:489–492.

Beltranena, E., and R. T. Zijlstra. 2010. Research update: Alternative feedstuffs—DDGS. Pages 167–175 in Advances in Pork Production, Proc. Banff Pork Seminar. Vol. 21. R. O. Ball, ed. University of Alberta, Edmonton, Alberta, Canada.

Beltranena, E., M. Dugan, J. Aalhus, M. Young, N. Campbell, M. Oryschak, and R. T. Zijlstra. 2010. Withdrawal of corn DDGS from finisher diets: Effects on bac*kf*at and belly quality. Page Abstract 27 in Advances in Pork Production, Proc. Banff Pork Seminar. Vol. 21. R. O. Ball, ed. University of Alberta, Edmonton, Alberta, Canada.

Bernard, A., F. Broeckaert, G. de Poorter, A. de Cock, C. Hermans, C. Saegerman, and G. Houins. 2002. The Belgian PCB/dioxin incident: Analysis of the food chain contamination and health risk evaluation. Environ. Res. 88:1–18.

Bhandari, S. K., C. M. Nyachoti, and D. O. Krause. 2009. Raw potato starch in weaned pig diets and its influence on postweaning scours and the molecular microbial ecology of the digestive tract. J. Anim. Sci. 87:984–993.

Bindelle, J., A. Buldgen, M. Delacollette, J. Wavreille, R. Agneessens, J. P. Destain, and P. Leterme. 2009. Influence of source and concentrations of dietary fiber on *in vivo* nitrogen excretion pathways in pigs as reflected by *in vitro* fermentation and nitrogen incorporation by fecal bacteria. J. Anim. Sci. 87:583–593.

Bird, A. R., M. Jackson, R. A. King, D. A. Davies, S. Usher, and D. L. Topping. 2004. A novel high-amylose barley cultivar (Hordeum vulgare var. Himalaya 292) lowers plasma cholesterol and alters indices of large-bowel fermentation in pigs. Br. J. Nutr. 92:607–615.

Black, J. L. 1995. Modelling energy metabolism in the pig—critical evaluation of a simple reference model. Pages 87–102 in Modelling Growth in the Pig. P. J. Moughan, M. W. A. Verstegen, and M. Visser-Reyneveld, eds. Wageningen Pers, Wageningen, The Netherlands.

Blaxter, K. 1983. Animal agriculture in a global context. J. Anim. Sci. 56:972–978.

Boudry, G., V. Douard, J. Mourot, J. P. Lallés, and I. Le Huërou-Luron. 2009. Linseed oil in the maternal diet during gestation and lactation modifies fatty acid composition, mucosal architecture, and mast cell regulation of the ileal barrier in piglets. J. Nutr. 139:1110–1117.

Bramsnaes, F., and H. S. Olsen. 1979. Development of field pea and faba bean proteins. J. Am. Oil Chem. Soc. 56:450–454.

Cadogan, D. J., M. Choct, and R. G. Campbell. 2003. Effects of storage time and exogenous xylanase supplementation of new season wheats on the performance of young male pigs. Can. J. Anim. Sci. 83:105–112.

Canh, T. T., A. L. Sutton, A. J. Aarnink, M. W. A. Verstegen, J. W. Schrama, and G. C. Bakker. 1998. Dietary carbohydrates alter the fecal composition and pH and the ammonia emission from slurry of growing pigs. J. Anim. Sci. 76:1887–1895.

Canola Council of Canada. 2009. Canola Meal Feed Industry Guide. 4th ed. Canola

Council of Canada, Winnipeg, Manitoba,Canada. Chen, D. 2009. Nutrition and feed strategies for sustainable swine production in China. Front. Agric. China 3:471–477.

Chiba, L. I. 2001. Protein supplements. Pages 803–837 in Swine Nutrition. 2nd ed. A. J. Lewis and L. L. Southern, eds. CRC Press, Boca Raton, FL.

Choct, M., E. A. D. Selby, D. J. Cadogan, and R. G. Campbell. 2004. Effect of liquid to feed ratio, steeping time, and enzyme supplementation on the performance of weaner pigs. Aust. J. Agric. Res. 55:247–252.

Clarke, E. J., and J. Wiseman. 2000. Developments in plant breeding for improved nutritional quality of soya beans II. Anti-nutritional factors. J. Agric. Sci. (Cambridge) 134:125–136.

Covaci, A., S. Voorspoels, P. Schepens, P. Jorens, R. Blust, and H. Neels. 2008. The Belgian PCB/dioxin crisis—8 years later: Anoverview. Environ. Toxicol. Pharmacol. 25:164–170.

Cromwell, G. L., T. R. Cline, J. D. Crenshaw, T. D. Crenshaw, R. A. Easter, R. C. Ewan, C. R. Hamilton, et al. 2000. Variability among sources and laboratories in analyses of wheat middlings. NCR-42 Committee on Swine Nutrition. J. Anim. Sci. 78:2652–2658.

CVB [Centraal Veevoeder Bureau (Central Feedstuff Bureau)]. 1993. Net Energy of Feedstuffs for Swine. CVB Report No. 7. CVB,Lelystad, The Netherlands.

CVB [Centraal Veevoeder Bureau (Central Feedstuff Bureau)]. 2007. Veevoedertabel (Table of feeding value of animal feed ingredients). CVB, Lelystad, The Netherlands.

Dale, N. 1996. The metabolizable energy of wheat by-products. J. Appl. Poult. Res. 5:105–108.

Dale, B. 2008. Biofuels: Thinking clearly about the issues. J. Agric. Food Chem. 56:3885–3891.

Dänicke, S., H. Valenta, F. Klobasa, S. Döll, M. Ganter, and G. Flachowsky. 2004. Effects of graded levels of Fusarium toxin contaminated wheat in diets for fattening pigs on growth performance, nutrient digestibility, deoxynivalenol balance and clinical serum characteristics. Arch. Anim. Nutr. 58:1–17.

Davis-Knight, H. R., and R. M. Weightman. 2008. The potential of triticale as a low input cereal for bioethanol production. Project Report 434. Home-Grown Cereals Authority, Kenilworth, UK.

De Lange, C. F. M. 2000. Overview of determinants of the nutritional value of feed ingredients. Pages 17–32 in Feed Evaluation. Principles and Practice. P. J. Moughan, M. W. A. Verstegen, and M. I. Visser-Reyneveld, eds. Wageningen Pers, Wageningen, The Netherlands.

Den Hartog, J. 2003. Feed for Food: HACCP in the animal feed industry. Food Control 14:95–99.

Divakala, K. C., L. I. Chiba, R. B. Kamalakar, S. P. Rodning, E. G. Welles, K. A. Cummins, J. Swann, et al. 2009. Amino acid supplementation of hydrolyzed feather meal diets for finisher pigs. J.

Anim Sci. 87:1270–1281.

Drew, M. 2004. Canola protein concentrate as a feed ingredient for salmonid fish in Advances en Nutrición Acuícola VII. L. E. Cruz Suárez, D. Ricque Marie, M. G. Nieto López, D. Villarreal, U. Scholz, and M. González, eds. Memorias del Simposium Internacional de Nutrición Acuícola. November 16–19, 2004. Hermosillo, Sonora, México.

Dugan, M. E. R., J. L. Aalhus, and J. K. G Kramer. 2004. Conjugated linoleic acid pork research. Am. J. Clin. Nutr. 79:1212S–1216S.

Duran−Montgé, P., R. Lizardo, D. Torrallardona, and E. Esteve−Garcia. 2007. Fat and fatty acid digestibility of different fat sources in growing pigs. Livest. Sci. 109:66–69.

Eastwood, L., P. R. Kish, A. D. Beaulieu, and P. Leterme. 2009. Nutritional value of flaxseed meal for swine and its effects on the fatty acid profile of the carcass. J. Anim. Sci. 87:3607–3619.

Emiola, I. A., F. O. Opapeju, B.A. Slominski, and C. M. Nyachoti. 2009. Growth performance and nutrient digestibility in pigs fed wheat distillers dried grains with solubles−based diets supplemented with a multicarbohydrase enzyme. J. Anim. Sci. 87:2315–2322.

Fairbairn, S. L., J. F. Patience, H. L. Classen, and R. T. Zijlstra. 1999. The energy content of barley fed to growing pigs: characterizing the nature of its variability and developing prediction equations for its estimation. J. Anim. Sci. 77:1502–1512.

Farmer, C., and H. V. Petit. 2009. Effects of dietary supplementation with different forms of flax in late−gestation and lactation on fatty acid profiles in sows and their piglets. J. Anim. Sci. 87:2600–2613.

Farmer, C., A. Giguère, and M. Lessard. 2010. Dietary supplementation with different forms of flax in late gestation and lactation: Effects on sow and litter performances, endocrinology, and immune response. J. Anim. Sci. 88:225–237.

FEFAC. 2005. Feed and Food Statistical Yearbook 2005. Euro. Feed Manufac. Fed., Brussels, Belgium.

Fledderus, J., P. Bikker, and R. E. Weurding. 2003. *In vitro* assay to estimate kinetics of starch digestion in the small intestine of pigs. Pages 4–6 in Proc. 9th Int. Symp. Digest. Physiol. Pigs. Vol. 2. R. O. Ball, ed. University of Alberta, Edmonton, AB, Canada.

Fontaine, J., U. Zimmer, P. J. Moughan, and S. M. Rutherford. 2007. Effect of heat damage in an autoclave on the reactive lysine contents of soy products and corn distillers dried grains with solubles. Use of the results to check on lysine damage in common qualities of these ingredients. J. Agric. Food Chem. 55:10737–10743.

Friesen, M. J., E. Kiarie, and C. M. Nyachoti. 2006. Response of nursery pigs to diets with increasing levels of raw peas. Can. J. Anim. Sci. 86:531–533.

Glynn, A., M. Aune, I. Nilsson, P. O. Darnerud, E. H. Ankarberg, A. Bignert, and I. Nordlander. 2009. Declining levels of PCB, HCB and p,p −DDE in adipose tissue from food producing bovines and swine in Sweden 1991–2004. Chemosphere 74:1457–1462.

González, J., B. Puschner, V. Pérez, M. C. Ferreras, L. Delgado, M. Munõz, C. Pérez, et al. 2009. Nephrotoxicosis in Iberian piglets subsequent to exposure to melamine and derivatives in Spain between 2003 and 2006. J. Vet. Diagn. Invest. 21: 558–563.

Groesbeck, C. N., L. J. McKinney, J. M. DeRouchey, M. D. Tokach, R. D. Goodband, S. S. Dritz, J. L. Nelssen, A. W. Duttlinger, A. C. Fahrenholz, and K. C.Behnke, 2008. Effect of crude glycerol on pellet mill production and nursery pig growth performance. J. Anim. Sci. 86:2228–2236.

Gunawardena, C., W. Robertson, M. Young, R. T. Zijlstra, and E. Beltranena. 2007. Zero–tannin fababean, field pea and soybean meal as dietary protein sources for growing–finishing pigs. J. Anim. Sci. 85(Suppl. 2):42. (Abstr.)

Gunawardena, C. K., R. T. Zijlstra, and E. Beltranena. 2010a. Characterization of the nutritional value of air–classified protein and starch fractions of field pea and zero–tannin faba bean in grower pigs. J. Anim. Sci. 88:660–670.

Gunawardena, C. K., R. T. Zijlstra, L. A. Goonewardene, and E. Beltranena. 2010b. Protein and starch concentrates of air–classified field pea and zero tannin faba bean for weaned pigs. J. Anim. Sci. 88:2627–2636.

Hale, O. M., and P. R. Utley. 1985. Value of Beagle 82 triticale as a substitute for corn and soybean meal in the diet of pigs.J. Anim. Sci. 60:1272–1279.

Harbach, A. P. R., M. C. R. da Costa, A. L. Soares, A. M. Bridi, M. Shimokomaki, C. A. da Silva, and E. I. Ida. 2007. Dietary corn germ containing phytic acid prevents pork meat lipid oxidation while maintaining normal animal growth performance. Food Chem. 100:1630–1633.

Health Canada. 2004. Novel food information on: Omega–3 enhanced pork and products derived therefrom. Accessed Mar. 9,2010. http://www.hc–sc.gc.ca/fn–an/gmf–agm/appro/dd109v3–eng.php.

Hemery, Y., X. Rouau, V. Lullien–Pellerin, C. Barron, and J. Abecassis. 2007. Dry processes to develop wheat fractions and products with enhanced nutritional quality. J. Cereal Sci. 46:327–347.

Hennig, U., S. Kuhla, W. B. Souffrant, A. Tuchscherer, and C. C. Metges. 2006. Effect of partial dehulling of two– and six–row barley varieties on precaecal digestibility of amino acids in pigs. Arch. Anim. Nutr. 60:205–17.

Hermes, R. G., F. Molist, M. Ywazaki, M. Nofrarías, A. Gomez de Segura, J. Gasa, J. F. Pérez. 2009. Effect of dietary level of protein and fiber on the productive performance and health status of piglets. J. Anim. Sci. 87:3569–3577.

Hill, B. E., A. L. Sutton, and B. T. Richert. 2009. Effects of low–phytic acid corn, low–phytic acid soybean meal, and phytase on nutrient digestibility and excretion in growing pigs. J. Anim.

Sci. 87:1518–1527.

Holden, P. J., and D. R. Zimmerman. 1991. Utilization of cereal grain by-products in feeding swine. Pages 585–593 in Swine Nutrition. E. R. Miller, D. E. Ullrey, and A. J. Lewis, eds. Butterworth-Heinmann, Boston, MA.Holt, J. P., L. J. Johnston, S. K. Baidoo, and G. C. Shurson. 2006. Effects of a high-fiber diet and frequent feeding on behavior,reproductive performance, and nutrient digestibility in gestating sows. J. Anim. Sci. 84:946–955.

Hooda, S., J. J. Matte, T. Vasanthan, and R. T. Zijlstra. 2010. Dietary purified oat β -glucan reduces peak glucose absorption and portal insulin release in portal-vein catheterized grower pigs. Livest. Sci. 134:15–17.

Hoogenboom, L. A. P., C. A. Kan, T. F. H. Bovee, G. van der Weg, C. Onstenk, and W. A. Traag. 2004. Residues of dioxins and PCBs in fat of growing pigs and broilers fed contaminated feed. Chemosphere 57:35–42.

House, J. D., D. Abramson, G. H. Crow, and C. M. Nyachoti. 2002. Feed intake, growth and carcass parameters of swine consuming diets containing low levels of deoxynivalenol from naturally contaminated barley. Can. J. Anim. Sci. 82:559–565.

Htoo, J. K., W. C. Sauer, Y. Zhang, M. Cervantes, S. F. Liao, B. A. Araiza, A. Morales, and N. Torrentera. 2007a. The effect of feeding low-phytate barley-soybean meal diets differing in protein content to growing pigs on the excretion of phosphorus and nitrogen. J. Anim. Sci. 85:700–705.

Htoo, J. K., W. C. Sauer, J. L. Yánẽz, M. Cervantes, Y. Zhang, J. H. Helm, and R. T. Zijlstra. 2007b. Effect of low-phytate barley or phytase supplementation to a barley-soybean meal diet on phosphorus retention and excretion by grower pigs. J. Anim. Sci. 85:2941–2948.

Htoo, J. K., X. Meng, J. F. Patience, M. E. R. Dugan, and R. T. Zijlstra. 2008. Effects of co-extrusion of flaxseed and field pea on the digestibility of energy, ether extract, fatty acids, protein, and amino acids in grower-finisher pigs. J. Anim. Sci. 86:2942–2951.

Huang, S. X., W. C. Sauer, B. Marty, and R. T. Hardin. 1999. Amino acid digestibilities in different samples of wheat shorts for growing pigs. J Anim Sci. 77:2469–2477.

Huang, S. X., W. C. Sauer, and B. Marty. 2001. Ileal digestibilities of neutral detergent fiber, crude protein, and amino acids associated with neutral detergent fiber in wheat shorts for growing pigs. J. Anim. Sci. 79:2388–2396.

Huisman J., and A. J. M. Jansman. 1991. Dietary effects and some analytical aspects of antinutritional factors in peas (*Pisum sativum*), common beans (*Phaseolus vulgaris*) and soybeans (*Glycine max* L.) in monogastric farm animals. A literature review. Nutr. Abstr. Rev. 61:901–921.

Hyun, Y., G. E. Bressner, M. Ellis, A. J. Lewis, R. Fischer, E. P. Stanisiewski, and G. F. Hartnell. 2004. Performance of growing–finishing pigs fed diets containing Roundup Ready corn (event nk603), a nontransgenic genetically similar corn, or conventional corn lines. J. Anim. Sci. 82:571–580.

Hyun, Y., G. E. Bressner, R. L. Fischer, P. S. Miller, M. Ellis, B. A. Peterson, E. P. Stanisiewski, and G. F. Hartnell.2005. Performance of growing-finishing pigs fed diets containing YieldGard Rootworm corn (MON 863), a nontransgenic genetically similar corn, or conventional corn hybrids. J. Anim. Sci. 83:1581–1590.

Jablonski, E. A., R. D. Jones, and M. J. Azain. 2006. Evaluation of pet food by-product as an alternative feedstuff in weanling pig diets. J. Anim. Sci. 84:221–228.

Jha, R., J. K. Htoo, M. G. Young, E. Beltranena, and R. T. Zijlstra. 2010. Effects of co-products inclusion on growth performance and carcass characteristics of grower–finisher pigs. J. Anim. Sci. 88(Suppl. 2):553–554. (Abstr.)

Johnson, I. R., T. Vasanthan, F. Temelli, and R. T. Zijlstra. 2007. Chemical characterization and digestible nutrient content of barley and oat starch ingredients derived via wet fractionation for grower pigs. J. Anim. Sci. 85(Suppl. 2):43. (Abstr.)

Jørgensen, H., X. Q. Zhao, and B. O. Eggum. 1996. The influence of dietary fibre and environmental temperature on the development of the gastrointestinal tract, digestibility, degree of fermentation in the hind-gut and energy metabolism in pigs. Br. J. Nutr. 75:365–378.

Ju´arez, M., M. E. R. Dugan, N. Aldai, J. L. Aalhus, J. F. Patience, R. T. Zijlstra, and A. D. Beaulieu. 2010. Feeding co-extruded flaxseed to pigs: Effects of duration and feeding level on growth performance and bac*kf*at fatty acid composition of grower-finisher pigs. Meat Sci. 84:578–584.

Kerr, B. J., W. A. Dozier, III, and K. Bregendahl. 2007. Nutritional value of crude glycerine for nonruminants. Pages 6–18 in Proc. 23rd Carolina Swine Nutr. Conf., Raleigh, NC.

Kijora, C., and R. D. Kupsch. 1996. Evaluation of technical glycerols from "biodiesel" production as a feed component in fattening of pigs. Fett Lipid 98:240–245.

Kijora, C., R. D. Kupsch, H. Bergner, C. Wenk, and A. L. Prabucki. 1997. Comparative investigation on the utilization of glycerol,free fatty acids, free fatty acids in combination with glycerol and vegetable oil in fattening of pigs. J. Anim. Physiol. Anim. Nutr. 77:127–138.

Kim, J. C., J. R. Pluske, and B. P. Mullan. 2008. Nutritive value of yellow lupins (Lupinus luteus L.) for weaner pigs. Austr. J. Exp. Agr. 48:1225–1231.

Kim, J. C., B. P. Mullan, J. M. Heo, C. F. Hansen, and J. R. Pluske. 2009. Decreasing dietary particle size of lupins increases apparent ileal amino acid digestibility and alters fermentation characteristics in the gastrointestinal tract of pigs. Br. J. Nutr.102:350–360.

Kim, J. C., B. P. Mullan, J. M. Heo, A. Hernandez, and J. R. Pluske. 2009. Variation in digestible energy content of Australian sweet lupins (*Lupinus angustifolius* L.) and the development of prediction equations for its estimation. J. Anim. Sci. 87:2565–2573.

Kracht, W., S. Dänicke, H. Kluge, K. Keller, W. Matzke, U. Hennig, and W. Schumann. 2004. Effect of dehulling of rapeseed on feed value and nutrient digestibility of rape products in pigs. Arch. Anim. Nutr. 58:389–404.

Krupinsky, J. M., K. L. Bailey, M. P. McMullen, B. D. Gossen, and T. K. Turkington. 2002. Managing plant disease risk in diversified cropping systems. Agron. J. 94:198–209.

Lammers, P. J., B. J. Kerr, T. E. Weber, W. A. Dozier, III, M. T. Kidd, K. Bregendahl, and M. S. Honeyman. 2008a. Digestible and metabolizable energy of crude glycerol for growing pigs. J. Anim. Sci. 86:602–608.

Lammers, P.J., B. J. Kerr, T. E. Weber, K. Bregendahl, S. M. Lonergan, K. J. Prusa, D. U. Ahn, et al. 2008b. Growth performance, carcass characteristics, meat quality, and tissue histology of growing pigs fed crude glycerin–supplemented diets. J. Anim. Sci. 86:2962–2970.

Lammers, P. J., M. D. Kenealy, J. B. Kliebenstein, J. D. Harmon, M. J. Helmers, and M. S. Honeyman. 2010. Nonsolar energy use and one–hundred–year global warming potential of Iowa swine feedstuffs and feeding strategies. J. Anim. Sci. 88:1204–1212.

Lauridsen, C., S. Hojsgaard, and M. T. Sorensen. 1999. Influence of dietary rapeseed oil, vitamin E, and copper on the performance and the antioxidative and oxidative status of pigs. J. Anim. Sci. 77:906–916.

Le Gall, M., L. Quillien, B. Séve, J. Guéguen, and J. P. Lallés. 2007. Weaned piglets display low gastrointestinal digestion of pea (*Pisum sativum* L.) lectin and pea albumin 2. J. Anim. Sci. 85:2972–2981.

Le Guen, M. P., J. Huisman, J. Gueguen, G. Beelen, and M. W. A. Verstegen. 1995a. Effects of a concentrate of pea antinutritional factors on pea protein digestibility in piglets. Livest. Prod. Sci. 44:157–167.

Le Guen, M. P., J. Huisman, and M. W. A. Verstegen. 1995b. Partition of the amino acids in ileal digesta from piglets fed pea protein diets. Livest. Prod. Sci. 44:169–178.

Leming, R., and A. Lember. 2005. Chemical composition of expeller–extracted and cold–pressed canola meal. Agraarteadus 16:103–109.

Leterme, P., E. Froidmont, F. Rossi, and A. Théwis. 1998. The high water–holding capacity of pea inner fibers affects the ileal flow of endogenous amino acids in pigs. J. Agric. Food Chem. 46:1927–1934.

Lösel, D., and R. Claus. 2005. Dose–dependent effects of resistant potato starch in the diet on intestinal skatole formation and adipose tissue accumulation in the pig. J. Vet. Med. A Physiol. Pathol. Clin. Med. 52:209–212.

Mariscal–Landín, G., Y. Lebreton, and B. Sève. 2002. Apparent and standardised true ileal digestibility of protein and amino acids from faba bean, lupin and pea, provided as whole seeds, dehulled or extruded in pig diets. Anim. Feed Sci. Technol. 97:183–198.

Martins, J. M., M. Riottot, M. C. de Abreu, A. M. Viegas–Crespo, M. J. Lanc¸a, J. A. Almeida, J. B. Freire, and O. P. Bento. 2005. Cholesterol–lowering effects of dietary blue lupin

(*Lupinus angustifolius* L.) in intact and ileorectal anastomosed pigs. J. Lipid Res. 46:1539–1547.

Matte, J. J., S. Robert, C. L. Girard, C.L., C. Farmer, and G. P. Martineau. 1994. Effect of bulky diets based on wheat bran or oat hulls on reproductive performance of sows during their first two parities. J. Anim. Sci. 72:1754–1760.

McLeod, J. G., W. H. Pfeiffer, R. M. DePauw, and J. M. Clarke. 2001. Registration of "AC Ultima" spring triticale. Crop Sci. 41:924–925.

Metzler–Zebeli, B. U., S. Hooda, R. T. Zijlstra, R. Mosenthin, and M. G. Gänzle. 2010a. Dietary supplementation of viscous and fermentable non–starch polysaccharides (NSP) modulates microbial fermentation in pigs. Livest. Sci. 133:95–97.

Metzler–Zebeli, B. U., S. Hooda, R. Pieper, R. T. Zijlstra, A. G. van Kessel, R. Mosenthin, and M. G. Gänzle. 2010b. Non–starch polysaccharides modulate bacterial microbiota, pathways for butyrate production, and abundance of pathogenic Escherichia coli in the gastrointestinal tract of pigs. Appl. Environ. Microbiol. 76:3692–3701.

Miller, P. R., B. G. McConkey, G. W. Clayton, S. A. Brandt, J. A. Staricka, A. M. Johnston, G. P. Lafond, et al. 2002. Pulse crop adaptation in the Northern Great Plains. Agron. J. 94:261–272.

Moeser, A. J., I. B. Kim, E. van Heugten, and T. A. van Kempen. 2002. The nutritional value of degermed, dehulled corn for pigs and its impact on the gastrointestinal tract and nutrient excretion. J. Anim. Sci. 80:2629–2638.

Musella, M., S. Cannata, R. Rossi, J. Mourot, P. Baldini, and C. Corino. 2009. Omega–3 polyunsaturated fatty acid from extruded linseed influences the fatty acid composition and sensory characteristics of dry–cured ham from heavy pigs. J. Anim Sci.87:3578–3588.

Mustafa, A., P. A. Thacker, J. J. McKinnon, D. A. Christensen, and V. J. Racz. 2000. Nutritional value of feed grade chickpeas for ruminants and pigs. J. Sci. Food Agr. 80:1581–1588.

Myer, R. O., and J. H. Brendemuhl. 2001. Miscellaneous feedstuffs. Pages 839–864 in Swine Nutrition. 2nd ed. A. J. Lewis and L. L. Southern, eds. CRC Press, Boca Raton, FL.

Niven, S. J., C. Zhu, D. Columbus, J. R. Pluske, and C. F. M. de Lange. 2007. Impact of controlled fermentation and steeping of high moisture corn on its nutritional value for pigs. Livest. Sci. 109:166–169.

Noblet, J., H. Fortune, C. Dupire, S. Dubois. 1993. Digestible, metabolisable and net energy value of 13 feedstuffs for growing pigs: Effect of energy system. Anim. Feed Sci. Technol. 42:131–149.

Nonhebel, S. 2004. On resource use in food production systems: The value of livestock as rest–stream upgrading system. Ecol. Econ. 48:221–230.

Nortey, T. N., J. F. Patience, P. H. Simmins, N. L. Trottier, and R. T. Zijlstra. 2007. Effects of individual or combined xylanase and phytase supplementation on energy, amino acid, and phosphorus digestibility and growth performance of grower pigs fed wheat–based diets containing wheat millrun. J. Anim. Sci. 85:1432–1443.

Nortey, T. N., J. F. Patience, J. S. Sands, N. L. Trottier, and R. T. Zijlstra. 2008. Effects of xylanase supplementation on digestibility and digestible content of energy, amino acids, phosphorus, and calcium in wheat by-products from dry milling in grower pigs. J. Anim. Sci. 86:3450–3464.

Nyannor, E. K. D., P. Williams, M. R. Bedford, and O. Adeola. 2007. Corn expressing an Escherichia coli-derived phytase gene: A proof-of-concept nutritional study in pigs. J. Anim. Sci. 85:1946–1952.

O'Hearn, V. L., and R. A. Easter. 1983. Evaluation of wheat middlings for swine diets. Nutr. Rep. Int. 28:403–411.

Opapeju, F. O., C. M. Nyachoti, J. D. House, H. Weiler, and H. D. Sapirstein. 2006. Growth performance and carcass characteristics of pigs fed short-season corn hybrids. J. Anim. Sci. 84:2779–2786.

Owusu-Asiedu, A., C. M. Nyachoti, S. K. Baidoo, R. R. Marquardt, and X. Yang. 2003a. Response of early-weaned pigs to an enterotoxigenic *Escherichia coli* (K88) challenge when fed diets containing spray-dried porcine plasma or pea protein isolate plus egg yolk antibody. J. Anim. Sci. 81: 1781–1789.

Owusu-Asiedu, A., C. M. Nyachoti, and R. R. Marquardt. 2003b. Response of early-weaned pigs to an enterotoxigenic *Escherichia coli* (K88) challenge when fed diets containing spray-dried porcine plasma or pea protein isolate plus egg yolk antibody, zinc oxide, fumaric acid, or antibiotic. J. Anim. Sci. 81:1790–1798.

Parera, N., R. P. L´azaro, M. P. Serrano, D. G. Valencia, and G. G. Mateos. 2010. Influence of the inclusion of cooked cereals and pea starch in diets based on soy or pea protein concentrate on nutrient digestibility and performance of young pigs. J. Anim. Sci. 88:671–679.

Partanen, K., H. Siljander-Rasi, and T. Alaviuhkola. 2006. Feeding weaned piglets and growing-finishing pigs with diets based on mainly home-grown organic feedstuffs. Agric. Food Sci. 15:89–105.

Patience, J. F., P. Leterme, A. D. Beaulieu, and R. T. Zijlstra. 2007. Utilization in swine diets of distillers dried grains with solubles derived from corn or wheat used in ethanol production. Pages 89–102 in Biofuels: Implications for the Feed Industry, J. Doppenberg and P. van der Aar, eds. Wageningen Academic Press, Wageningen, The Netherlands.

Payeur, M., S. P. Lemay, S. Godbout, L. Chénard, R. T. Zijlstra, E. M. Barber, and C. Laguë. 2002. Impact of combining a low protein diet including fermentable carbohydrates and oil sprinkling on odour and dust emissions of swine barns. July 28–31, 2002 ASAE Ann. Int. Mtg./ CIGR XVth World Congress. Chicago, IL.

Pedersen, C., M. G. Boersma, and H. H. Stein. 2007. Energy and nutrient digestibility in

NutriDense corn and other cereal grains fed to growing pigs. J. Anim. Sci. 85:2473–2483.

Pieper, R., J. Bindelle, B. Rossnagel, A. van Kessel, and P. Leterme. 2009. Effect of carbohydrate composition in barley and oat cultivars on microbial ecophysiology and proliferation of Salmonella enterica in an *in vitro* model of the porcine gastrointestinal tract. Appl. Environ. Microbiol. 75:7006–7016.

Písaříková, B., and Z. Zral′y. 2009. Nutritional value of lupine in the diets for pigs (a review). Acta Vet. Brno. 78:399–409.

Pond, W. G., and X. G. Lei. 2001. Of pigs and people. Pages 3–18 in Swine Nutrition. 2nd ed. A. J. Lewis and L. L. Southern, ed. CRC Press, Boca Raton, FL.

Rochfort, S., and J. Panozzo. 2007. Phytochemicals for health, the role of pulses. J. Agric. Food Chem. 55:7981–7994.

Roth−Maier, D. A., B. M. B¨ohmer, and F. X. Roth. 2004. Effects of feeding canola meal and sweet lupin (L. luteus, L. angustifolius) in amino acid balanced diets on growth performance and carcass characteristics of growing–finishing pigs. Anim. Res. 53:21–34.

Rubio, L. A. 2005. Ileal digestibility of raw and autoclaved kidney−bean (*Phaseolus vulgaris*) seed meals in cannulated pigs. Anim. Sci. 81:125–133.

Sakata, T., and M. Saito. 2007. Insoluble dietary fiber of wheat bran increased viscosity of pig whole cecal contents *in vitro*. J. Nutr. Sci. Vitaminol. 53:380–381.

Salmon, D. F. 2004. Production of triticale on the Canadian Prairies. M. Mergoum and H. G′omez MacPherson, eds. Triticale Improvement and Production. Plant Production and Protection Paper 179. FAO. Rome, Italy, pp 154.

Sauber, T. E., and F. N. Owens. 2001. Cereal grains and by−products for swine. Pages 785–802 in Swine Nutrition. 2nd ed. A. J. Lewis and L. L. Southern, eds. CRC Press, Boca Raton, FL.

Sauvant, D., J. M. Perez, and G. Tran. 2004. Tables of composition and nutritional value of feed materials: Pigs, poultry, cattle, sheep, goats, rabbits, horses, fish. Wageningen Academic Publishers, Wageningen, The Netherlands and INRA Editions, Versailles, France.

Schaafsma, A. W., L. Tamburic−Ilinic, J. D. Miller, and D. C. Hooker. 2001. Agronomic considerations for reducing deoxynivalenol in wheat grain. Can. J. Plant Pathol. 23:279–285.

Schaafsma, A. W; V. Limay−Rios, D. E. Paul, and D. J. Miller. 2009. Mycotoxins in fuel ethanol co−products derived from maize: a mass balance for deoxynivalenol. J. Sci. Food Agricul. 89:1574–1580.

Schaefer, D. M., Q. Liu, C. Faustman, and M. Yin. 1995. Supranutritional administration of vitamins E and C improves oxidative stability of beef. J. Nutr. 125:S1792–S1798.

Scholten, R. H. J., C. M. C. van der Peet−Schwering, M. W. A. Verstegen, L. A. den Hartog, J. W. Schrama, and P. C. Vesseur. 1999. Fermented co−products and fermented compound diets for pigs: A review. Anim. Feed Sci. Technol. 82:1–19.

Schöne, F., B. Rudolph, U. Kirchheim, and G. Knapp. 1997. Counteracting the negative effects of rapeseed and rapeseed press cake in pig diets. Br. J. Nutr. 78:947–962.

Schöne, F., F. Tischendorf, U. Kirchheim, W. Reichardt, and J. Bargholz. 2002. Effects of high fat rapeseed press cake on growth, carcass, meat quality and body fat composition of leaner and fatter pig crossbreeds. Anim. Sci. 74:285–297.

Schrama, J. W., M. W. Bosch, M. W. A. Verstegen, A. H. Vorselaars, J. Haaksma, and M. J. Heetkamp. 1998. The energetic value of nonstarch polysaccharides in relation to physical activity in group-housed, growing pigs. J. Anim. Sci. 76:3016–3023.

Seneviratne, R. W., M. G. Young, E. Beltranena, L. A. Goonewardene, R. W. Newkirk, and R. T. Zijlstra. 2010. The nutritional value of expeller-pressed canola meal for grower–finisher pigs. J. Anim. Sci. 88:2073–2083.

Shelton, J. L., M. D. Hemann, R. M. Strode, G. L. Brashear, M. Ellis, F. K. McKeith, T. D. Bidner, and L. L. Southern. 2001.

Effect of different protein sources on growth and carcass traits in growing–finishing pigs. J. Anim. Sci. 79:2428–2435.

Shelton, J. L., J. O. Matthews, L. L. Southern, A. D. Higbie, T. D. Bidner, J. M. Fernandez, and J. E. Pontif. 2004. Effect of nonwaxy and waxy sorghum on growth, carcass traits, and glucose and insulin kinetics of growing–finishing barrows and gilts. J. Anim. Sci. 82:1699–1706.

Smits, C., and R. Sijtsma. 2007. A decision tree for co-product utilization. Pages 213–221 in Advances in Pork Production, Proc.

Banff Pork Seminar. Vol. 18, R. O. Ball and R. T. Zijlstra, eds. University of Alberta, Edmonton, Alberta, Canada.

Spagnuolo, M., C. Crecchio, M. D. Pizzigallo, and P. Ruggiero. 1999. Fractionation of sugar beet pulp into pectin, cellulose, and arabinose by arabinases combined with ultrafiltration. Biotechnol. Bioeng. 64:685–691.

Spencer, J. D., G. L. Allee, and T. E. Sauber. 2000. Growing–finishing performance and carcass characteristics of pigs fed normal and genetically modified low-phytate corn. J. Anim. Sci. 78:1529–1536.

Spragg, J., and R. Mailer. 2007. Canola meal value chain quality improvement. A final report prepared for AOF and Pork CRC.JCS Solutions Pty Ltd. Victoria, Australia.

Stein, H. H., G. Benzoni, R. A. Bohlke, and D. N. Peters. 2004. Assessment of the feeding value of South Dakota-grown field peas (*Pisum sativum* L.) for growing pigs. J. Anim. Sci. 82:2568–2578.

Stein, H. H., A. K. R. Everts, K. K. Sweeter, D. N. Peters, R. J. Maddock, D. M. Wulf, and C. Pedersen. 2006. The influence of dietary field peas (*Pisum sativum* L.) on pig performance, carcass quality, and the palatability of pork. J. Anim. Sci. 84:3110–3117.

Stein, H. H., M. G. Boersma, and C. Pedersen. 2006. Apparent and true total tract digestibility of phosphorus in field peas (*Pisum sativum* L.) by growing pigs. Can. J. Anim. Sci. 86:523–525.

Stein, H. H., and R. A. Bohlke. 2007. The effects of thermal treatment of field peas (*Pisum sativum* L.) on nutrient and energy digestibility by growing pigs. J. Anim. Sci. 85:1424–1431.

Stein, H. H., and G. C. Shurson. 2009. Board-invited review: The use and application of distillers dried grains with solubles in swine diets. J. Anim. Sci. 87:1292–1303.

Stein, H. H., D. W. Rice, B. L. Smith, M. A. Hinds, T. E. Sauber, C. Pedersen, D. M. Wulf, and D. N. Peters. 2009. Evaluation of corn grain with the genetically modified input trait DAS-59122-7 fed to growing–finishing pigs. J. Anim. Sci. 87:1254–1260.

Stevens, C. E., and I. D. Hume. 1995. Comparative physiology of the vertebrate digestive system. 2nd Ed. Cambridge University Press. Cambridge, UK.

Strydhorst, S. M., J. R. King, K. J. Lopetinsky, and K. N. Harker. 2008. Weed interference, pulse species, and plant density effects on rotational benefits. Weed Sci. 56:249–258.

Thacker, P. A., and R. N. Kirkwood. 1990. Nontraditional feed sources for use in swine production. Butterworth, Stoneham,MA.

Thacker, P. A. 2006. Nutrient digestibility, performance and carcass traits of growing–finishing pigs fed diets containing dried wheat distiller's grains with solubles. Can. J. Anim. Sci. 86:527–529.

USDA. 2010. Oilseeds: World Markets and Trade. Circular Series FOP 2-10, United States Department of Agriculture, Foreign Agriculture Service. Accessed Mar 9, 2010. http://www.fas.usda.gov/psdonline/circulars/oilseeds.pdf.

Thompson, M. E., M. R. Lewin-Smith, V.F. Kalasinsky, K. M. Pizzolato, M. L. Fleetwood, M. R. McElhaney, and T. O. Johnson. 2008. Characterization of melamine-containing and calcium oxalate crystals in three dogs with suspected pet food-induced nephrotoxicosis. Vet. Pathol. 45:417–426.

Valencia, D. G., M. P. Serrano, C. Centeno, R. Lázaro, and G. G. Mateos. 2008. Pea protein as a substitute of soya bean protein in diets for young pigs: Effects on productivity and digestive traits. 118:1–10.

van Barneveld, R. J., E. S. Batterham, D. C. Skingle, and B. W. Norton. 1995. The effect of heat on amino acids for growing pigs. 4. Nitrogen balance and urine, serum and plasma composition of growing pigs fed on raw or heat-treated field peas (*Pisum sativum*). Br. J. Nutr. 73:259–273.

van Barneveld, R. J. 1999. Understanding the nutritional chemistry of lupin (*Lupinus* spp.) seed to improve livestock production efficiency. Nutr. Res. Rev. 12:203–230.

van der Peet-Schwering, C. M., B. Kemp, J. G. Plagge, P. F. Vereijken, L. A. den Hartog, H. A. Spoolder, and M. W. A. Verstegen. 2004. Performance and individual feed intake characteristics of group-housed sows fed a nonstarch polysaccharides diet adlibitum during gestation over three parities. J. Anim. Sci. 82:1246–1257.

van der Poel, T. F. B., D. J. van Zuilichem, and M. G. van Oort. 1990. Thermal inactivation of lectins and trypsin inhibitor activity during steam processing of dry beans (*Phaseolus vulgaris*) and effects on protein quality. J. Sci. Food Agr. 53: 215–228.

van der Poel, A. F. B., L. M. W. Delleart, A. van Norel, and J. P. F. G. Helsper. 1992a. The digestibility in piglets of faba bean (*Vicia faba* L.) as affected by breeding towards the absence of condensed tannins. Br. J. Nutr. 68:793–800.

van der Poel, A. F. B., S. Gravendeel, D. J. van Kleef, A. J. M. Jansman, and B. Kemp. 1992b. Tannin-containing faba beans (*Vicia faba* L.): Effects of methods of processing on ileal digestibility of protein and starch for growing pigs. Anim. Feed Sci. Technol. 36:205–214.

van Kempen, T. A. T. G., P. R. Regmi, J. J. Matte, and R. T. Zijlstra. 2010. *In vitro* starch digestion kinetics, corrected for estimated gastric emptying, predicts portal glucose appearance in pigs. J. Nutr. 140:1227–1233.

Vasanthan, T., and F. Temelli. 2008. Grain fractionation technologies for cereal beta-glucan concentration. Food Res. Int. 41:876–881.

Veum, T. L., D. R. Ledoux, D. W. Bollinger, V. Raboy, and A. Cook. 2002. Low-phytic acid barley improves calcium and phosphorus utilization and growth performance in growing pigs. J. Anim. Sci. 80:2663–2670.

Veum, T. L., D. R. Ledoux, V. Raboy, and D. S. Ertl. 2001. Low-phytic acid corn improves nutrient utilization for growing pigs. J. Anim. Sci. 79:2873–2880.

Weber, T. E., S. L. Trabue, C. J. Ziemer, and B. J. Kerr. 2010. Evaluation of elevated dietary corn fiber from corn germ meal in growing female pigs. J. Anim Sci. 88:192–201.

Wierenga, K. T., E. Beltranena, J. L. Yáñẽz, and R. T. Zijlstra. 2008. Starch and energy digestibility in weaned pigs fed extruded zero-tannin faba bean starch and wheat as an energy source. Can. J. Anim. Sci. 88:65–69.

Whitney, M. H., G. C. Shurson, L. J. Johnston, D. M. Wulf, and B. C. Shanks. 2006. Growth performance and carcass characteristics of grower–finisher pigs fed high-quality corn distillers dried grain with solubles originating from a modern Midwestern ethanol plant. J. Anim. Sci. 84:3356–3363.

Whittemore, C. T. 1997. An analysis of methods for the utilisation of net energy concepts to improve the accuracy of feed evaluation in diets for pigs. Anim. Feed Sci. Technol. 68:89–99.

Widyaratne, G. P., and R. T. Zijlstra. 2007. Nutritional value of wheat and corn distiller's dried grain with solubles: Digestibility and digestible contents of energy, amino acids and phosphorus, nutrient excretion and growth performance of grower–finisher pigs. Can. J. Anim. Sci. 87:103–114.

Widyaratne, G. P., and R. T. Zijlstra. 2008. Nutritional value of wheat and corn distiller's dried grain with solubles: Digestibility and digestible contents of energy, amino acids and phosphorus, nutrient excretion and growth performance of grower–finisher pigs. Can. J. Anim. Sci. 88:515–516. (Erratum).

Williams, B. A., M. W. Bosch, H. Boer, M. W. A. Verstegen, and S. Tamminga. 2005. An *in vitro* batch culture method to assess potential fermentability of feed ingredients for monogastric diets. Anim. Feed Sci. Technol. 123–124:445–462.

Willig, S., D. L¨osel, and R. Claus 2005. Effects of resistant potato starch on odor emission from feces in swine production units. J. Agric. Food Chem. 53:1173–1178.

Woyengo, T. A., E. Kiarie, and C. M. Nyachoti. 2010. Energy and amino acid utilization in expeller−extracted canola meal fed to growing pigs. J. Anim. Sci. 88:1433–1441.

Wu, Y. V., and N. Nichols. 2005. Fine grinding and air classification of field peas. Cereal Chem. 82:341–344.

Wu, F., and G. P. Munkvold. 2008. Mycotoxins in ethanol co−products: modeling economic impacts on the livestock industry and management strategies. J. Agric. Food Chem. 56:3900–3911.

Xu, G., S. K. Baidoo, L. J. Johnston, D. Bibus, J. E. Cannon, and G. C. Shurson. 2010a. Effects of feeding diets containing increasing levels of corn distillers dried grains with solubles (DDGS) to grower–finisher pigs on growth performance, carcass composition, and pork fat quality. J. Anim. Sci. 88:1398–1410.

Xu, G., S. K. Baidoo, L. J. Johnston, D. Bibus, J. E. Cannon, and G. C. Shurson. 2010b. The effects of feeding diets containing corn distillers dried grains with solubles (DDGS), and DDGS withdrawal period, on growth performance and pork quality in grower–finisher pigs. J. Anim. Sci. 88:1388–1397.

Yanez, J., E. Beltranena, and R. T. Zijlstra. 2009. Effect of phytase and xylanase supplementation or particle size on nutrient digestibility of diets containing distillers dried grains with solubles (DDGS) co−fermented from wheat and corn in cannulated grower pigs. J. Anim. Sci. 87(E−Suppl. 3):52. (Abstr.)

Zervas, S., and R. T. Zijlstra. 2002a. Effects of dietary protein and oathull fiber on nitrogen excretion patterns and postprandial plasma urea profiles in grower pigs. J. Anim. Sci. 80:3238–3246.

Zervas, S., and R. T. Zijlstra. 2002b. Effects of dietary protein and fermentable fiber on nitrogen excretion patterns and plasma urea in grower pigs. J. Anim. Sci. 80:3247–3256.

Zhang, Y., J. Caupert, P. M. Imerman, J. L. Richard, and G. C. Shurson. 2009. The occurrence and concentration of mycotoxins in U.S. distillers dried grains with soluble. J. Agric. Food Chem. 57:9828–9837.

Zijlstra, R. T., E. D. Ekpe, M. N. Casano, J. F. Patience. 2001. Variation in nutritional value of western Canadian feed ingredients for pigs. Pages 12–24 in Proc. 22nd Western Nutr. Conf., Saskatoon, SK, Canada.

Zijlstra, R. T., and R. L. Payne. 2007. Net energy system for pigs. Pages 80–90 in Manipulating Pig Production XI, J. E. Patterson and J. A. Barker, eds. Australasian Pig Science Association, Werribee, Vic, Australia.

Zijlstra, R. T., and E. Beltranena. 2007. New frontier in processing: ingredient fractionation. Pages 216–222 in Manipulating Pig Production XI. J. E. Patterson and J. A. Barker, eds. Australasian Pig Science Association, Werribee, Vic, Australia.

Zijlstra, R. T., and E. Beltranena. 2008. Variability of quality in biofuel co–products. Pages 313–326 in Recent Advances in Animal Nutrition—2008. P. C. Garnsworthy and J. Wiseman, eds. Nottingham Academic Press, Nottingham, UK.

Zijlstra, R. T., and E. Beltranena. 2009. Regaining competitiveness: alternative feedstuffs for swine. Pages 237–243 in Advances in Pork Production Proc. Banff Pork Seminar. Vol. 20, R. O. Ball, ed. University of Alberta, Edmonton, AB, Canada.

Zijlstra, R. T., C. F. M. de Lange, and J. F. Patience. 1999. Nutritional value of wheat for growing pigs: Chemical composition and digestible energy content. Can. J. Anim. Sci. 79:187–194.

Zijlstra, R. T., A. G. van Kessel, and M. D. Drew. 2004. Ingredient fractionation: the value of value–added processing for animal nutrition. The worth of the sum of parts versus the whole. Pages 41–53 in Proc. 25th Western Nutr. Conf., Saskatoon, SK, Canada. Zijlstra, R. T., K. Lopetinsky, and E. Beltranena. 2008. The nutritional value of zero–tannin faba bean for grower–finisher pigs. Can. J. Anim. Sci. 88:293–302.

Zijlstra, R. T., K. Menjivar, E. Lawrence, and E. Beltranena. 2009a. The effect of crude glycerol on growth performance and nutrient digestibility in weaned pigs. Can. J. Anim. Sci. 89:85–89.

Zijlstra, R. T., E. Beltranena, C. M. Nyachoti, and S. W. Kim. 2009b. Phytase and NSP–degrading enzymes for alternative feed ingredients. J. Anim. Sci. 87(E–Suppl. 2):187. (Abstr.)

Zijlstra, R. T., R. Jha, M. G. Young, J. F. Patience, E. Beltranena, and J. K. Htoo. 2010. Effects of dietary crude protein and inclusion of co–products on growth performance and carcass characteristics of grower–finisher pigs. J. Anim. Sci. 88(Suppl. 2):554.(Abstr.)

第11章 猪的纤维素营养

1 导 语

纤维可被定义为不能被动物内源性消化酶消化但对动物和人类具有一定生理作用的碳水化合物或植物木质素。由于纤维素是自然界中含量最丰富的碳水化合物，因此寻找能利用饲粮中纤维素和其他纤维成分或非淀粉多糖（NSP）的方法，对未来养猪生产的可持续发展至关重要。

近年来，随着淀粉和油类在生物燃料工业上的应用不断增加，许多加工产品或副产品可作为饲料原料用于养猪生产。但遗憾的是，这些非常规饲料原料中的纤维含量较高。猪虽然可以从纤维中获得能量，但只有经过胃肠道微生物发酵后以挥发性脂肪酸（VFA）的形式才能被吸收，从而为猪供能。

尽管可溶性纤维容易被发酵，但一般情况下，纤维尤其是不可溶性纤维并不能被猪很好地利用。纤维含量越高，饲粮能量的总消化率就越低。另外，纤维也会降低氨基酸、脂类和一些矿物质的消化率。因此，纤维方面的理论知识和实用知识或者纤维的利用，抑或两个方面，不仅对纤维自身的利用问题，而且还对非常规饲料资源在养猪生产中的有效利用产生很大的影响。本章内容旨在简要概述纤维及其在猪上的应用，以期有助于养猪生产的可持续性发展。

2 饲粮纤维的定义

饲粮纤维的定义很多，但其中大部分要么是将饲粮纤维定义为一组用分析方法所测定的化合物，要么是被认定为一组具有特殊生理功能的化合物（IOM，2001）。在19世纪，Weende法将纤维定义为经酸和碱处理后的不溶性有机残留物（Mertens，2003）。这部分饲粮成分被作为饲粮纤维的公认定义，但其对动物并没有实际价值（AACC，2001）。

后来，两位研究者以不同的方式指出，这一不可消化残留物可以提高人类的健康（Kritchevsky，1988）。Denis Burkit报道，食用“高残留物食物”的人很少患肠癌；Hugh Trowell认为，摄入大量的不可消化残留物有助于保护发展中国家的人群免遭缺血性心脏病（Burkitt等，1972；Kritchevsky，1988；Carpenter，2003）。这些结论引发了人们对饲粮纤维的兴趣。但是，很明显，饲粮纤维是一种由多种成分组成的具有多种生理功能的化学混合物，因此很难对其进行准确定义（Carpenter，2003）。

目前认可的对饲粮纤维的精确定义，应包括纤维的生理作用（IOM，2006）。因此，其定义的一个重要方面，就是饲粮纤维是由不能被动物内源性消化酶消化的碳水化合物组成（AACC，2001；IOM，2006）。定义中的这一条内容很重要，但难以对其进行测定（Englyst等，2007）。目前常用的饲粮纤维的定义（AACC，2001）包括以下几个方面：① 饲粮中不可消化的部分；② 源于碳水化合物或木质素；③ 是植物的一部分；④ 对人类具有促进排便、降低血液胆固醇或葡萄糖等生理作用。

IOM将饲粮纤维的定义分为三部分（包括饲粮纤维、功能性纤维和总纤维）。饲粮纤维由植物中固有的、完整的不可消化性碳水化合物和木质素组成。功能性纤维由对人类具有有益生理功能的可分离的不可消化性碳水化合物组成。总纤维则是饲粮纤维和功能型纤维之和（IOM，2006）。

NSP的概念与饲粮纤维相关，但不能涵盖饲粮纤维的所有组分（Elia和Cummings，2007）。例如，NSP不包括寡糖和木质素，而在AACC（2001）和IOM

（2006）对饲粮纤维的定义中则包括寡糖和木质素。由于饲粮纤维不仅仅限于NSP或植物细胞壁成分，因此使用NSP的概念不能对饲料原料中的纤维进行准确描述。

确切定义饲粮纤维，对标注人类食品中纤维的含量很重要。在猪饲粮中，清晰描述对动物具有营养和生理功能的纤维组分以及明确可以提供饲粮成分能值的组分都很重要。同样，找到能够准确测定动物饲料及其原料中饲粮纤维含量的分析方法也很重要。

3 动物饲料原料中纤维的分析

目前已有很多方法，可以用于测定人类食品、动物饲料及其原料中纤维的含量。所有的方法均包括两个基本的步骤：饲粮中碳水化合物和其他非纤维成分（如蛋白质和脂肪）的消化和不可消化残留物的定量测定。消化过程一般采用化学试剂（如酸、碱和洗涤剂）或酶（如淀粉酶、淀粉葡萄糖苷酶和蛋白酶）。不可消化残留物的测定可以通过称量残留物的重量（重量分析法），或使用气相-液相色谱或高效液相色谱法测定残留物的化学组成。目前，有一些更新的方法用于研究植物细胞壁中NSP的组成和结构，及其与它们在肠道中降解的关系（Guillon等，2006）。这些方法包括拉曼显微光谱法、傅里叶变换红外光谱法（FT-IR）、免疫标记法、荧光法、质谱分析法及其他有关的分析方法（Guillon等，2006）。

◆ 3.1 粗纤维测定法

这是德国Weende农业试验站建立的一种粗略分析饲料原料的化学重量分析方法（Grieshop等，2001）。该方法将碳水化合物分成无氮浸出物和粗纤维两部分。粗纤维是饲料样品经1.25%硫酸和1.25%氢氧化钠消化后的残留物（Cho等，1997；Furda，2001）。在确定这些试验步骤时，我们只知道消化包含酸和碱处理过程，但却不清楚其中的关键酶（Mertens，2003）。粗纤维的测定步骤很粗放，也可重复，但是这种方法测定出的粗纤维与AACC和IOM定义的饲粮纤维无关（Mertens，2003），因为纤维素（40%～100%）、半纤维素（15%～20%）和木质素（5%～90%）的回收是不完全的（Grieshop等，2001；Mertens，2003）。然而，该方法仍然用于调节猪饲料中粗纤维的最大保证水平（AAFCO，2008）。

◆ 3.2 洗涤纤维测定法

洗涤纤维测定法是一种将纤维含量分析值与其生理特性经验性地联系在一起的化学重量法（van Soest等，1991）。该方法是由van Soest（1963）提出，将饲粮纤维分为中性洗涤纤维（NDF）、酸性洗涤纤维（ADF）和酸性洗涤木质素（Robertson和Horvath，2001）。虽然该方法是在粗纤维测定法基础上改进的一种方法，但该方法没有将果胶、黏胶、树胶和β-葡聚糖等可溶性饲粮纤维包含在内（Grieshop等，2001）。相对于富含可溶性饲粮纤维的大豆壳和甜菜碱等饲料原料，富含不可溶性纤维的玉米和玉米酒糟（DDGS）（Johnston等，2003）等谷物在进行洗涤纤维测定时，其可溶性饲粮纤维成分的回收量较少。洗涤法还存在其他一些问题，如样品中的淀粉和蛋白质可能会污染洗涤残留物，从而降低了测定的稳定性和可重复性（Mertens，2003）。

◆ 3.3 总饲粮纤维测定法

Prosky分析法是测定饲粮总纤维含量的一种方法（TDF；Method 985.29；AOAC，2006），改进后可用于测定可溶性饲粮纤维和不可溶性饲粮纤维（Method 991.43；AOAC，2006）。TDF法是先使用酶（如淀粉酶、葡萄糖淀粉酶和蛋白酶）模拟小肠的消化，然后将残留物进行称重（Prosky等，1984）。该残留物也可用于不可消化蛋白质和灰分的分析。TDF测定法虽然比粗纤维测定法和洗涤纤维测定法更耗时，且重复性差，但其测定值更能代表饲粮纤维的概念（Mertens，2003）。还需要进一步改进TDF测定法，能够用于测定低分子量的不可消化性碳水化合物，并矫正由不可消化性残留物带来的干扰（Gordon等，2007）。

◆ 3.4 酶化学分析法

有两种常用的酶化学分析方法，它们结合了不可消化残留物中糖的化学测定过程中酶的初始消化步骤（Theander和Åman，1979；Englyst等，1982）。Uppsala法是将耐淀粉酶多糖、糖醛酸和硫酸木素之和计为饲粮纤维含量（AOAC，2006）。AOAC法（994.13）的消化步骤采用的酶是热稳定性α-淀粉酶和淀粉葡萄糖苷酶（AOAC，2006）。消化后的残留物可被80%的乙醇分为可溶性部分和不可溶性部分。析出的中性糖以气相-液相色谱法和糖醛酸层析法测定为多羟糖醇乙酸酯衍生物（Theander和Åman，1979）。由Englyst等（1982）建立的NSP测定法与Uppsala测定法相似，但NSP测法测定的纤维值不包括木质素和抗性淀粉（Grieshop等，2001）。

◆ 3.5 饲粮纤维的差值推算法

为了实用目的，饲料原料中不可消化养分的含量可通过饲料原料中有机残留物（OR）之和进行计算。计算公式如下：

有机残留物=干物质（DM）-（灰分+淀粉+糖+粗蛋白+粗脂肪）

该公式假设所有的淀粉和糖均可被小肠消化和吸收，而除淀粉和糖外的其他碳水化合物均不能被哺乳动物的消化酶消化，因而归属于纤维（Noblet等，1994；de Lange，2008）。

◆ 3.6 饲粮纤维测定方法的比较

任何一种单独的纤维分析方法都无法精确测定饲粮纤维定义中所包含的所有碳水化合物（NRC，2007）。TDF法被认为是一种能够测定饲粮纤维中最多碳水化合物的方法。但是，TDF测定值中不一定包括低聚果糖等低聚糖和多聚果糖（NRC，2007）。

4 饲粮纤维的生理特性

饲粮纤维区别于可消化多糖的特有属性，受纤维的化学成分和物理结构的影响。与人和动物营养相关的理化特性包括溶解性、系水力、持水力、黏度和可发酵性。饲粮纤维的这些理化特性可提高人类的健康，但同时会降低动物的生产效率。

◆ 4.1 溶解性

饲粮纤维分为可溶性纤维和不可溶性纤维（Cho等，1997）。饲粮纤维的溶解性不仅指饲粮纤维溶于水的能力（Oakenfull，2001），还可以被定义为溶于稀酸、稀碱、缓冲液或模拟胃肠道内消化酶溶液的能力（Cho等，1997）。通过酶消化后，利用乙醇沉淀法可将可溶性纤维从总纤维中分离出来（Cho等，1997）。

饲粮纤维的溶解性，在很大程度上受组成饲粮纤维的两个或多个单糖之间连接键的影响（Oakenfull，2001）。这种连接键为饲粮纤维的水化特性提供了物理结构。纤维素中葡萄糖之间的β-（1-4）糖苷键使其形成规则的晶体结构。该结构可以阻止水分子的进入，从而使得纤维素不可溶（Oakenfull，2001）。但是，β-葡聚糖中β-（1-3）支链的存在，使其不能形成与纤维素类似的规则的晶体结构，这使得β-葡聚糖成为可溶性纤维（Oakenfull，2001）。

饲粮纤维的溶解性不能反映出碳水化合物的组成、物理结构和聚合度，但由于可溶性纤维和不可溶性纤维对人类健康和动物生产的生理作用和整体贡献不同，因此纤维的溶解性显得非常重要。可溶性纤维可增加食糜的黏性，从而减缓人和狗采食后胰岛素和血糖的增加（Dikeman和Fahey，2006）；而不可溶性饲粮纤维则会加快食糜通过胃肠道的速率，增加排便量（Chesson，2006）。

◆ 4.2 持水力和系水力

纤维的生理特性受纤维与水之间相互作用的影响。纤维可通过不同的机制与水结合，如离子间相互作用、氢键结合和毛细管吸附作用（Chaplin，2003）。由于与水结合的机制不同，因此可溶性纤维与不可溶性纤维结合水的能力也不同（Oakenfull，2001）。与水的结合强度和结合数量很大程度上取决于纤维的形态结构和组成。因此，不同来源的纤维，其结合水的强度和数量存在很大的差异（Cadden，1987；Chaplin，2003）。

饲粮纤维的持水能力有不同的表达方式。持水力（water-holding capacity，WHC）常常是指在没有任何外力作用下纤维结合水的数量，而系水力（water-binding capacity，WBC）或更好的概念“保水力”（water-retention capacity）是指受到外力作用后在含水纤维内部所保持的水分的数量（Robertson等，2000）。然而，在文献中这些术语一般可以通用（Ang，1991；Leterme等，1998；Chaplin，2003）。

有多种方法可用于测定纤维的持水能力。持水力可用过滤法（Chaplin，2003）或鲍曼仪器分析法（Auffret等，1994）测定。系水力可用离心法、减压法或使用透析管浸泡在模拟肠道内容物中进行测定（Stephen和Cummings，1979；Cadden，1987；Chaplin，2003）。这些不同的测定方法可评价纤维结合水的不同机制。因此，纤维系水力的测量结果取决于系水力的测定方法。一项欧洲的协作研究，推荐了评价纤维系水力和其他水化特性的标准方法（Robertson等，2000）。该方法是基于离心技术，但为了尽可能减少样品的损失需调整样品重量和离心速度，样品损失会影响测定结果（Robertson 等，2000）。

由于纤维的膨胀特性与其系水力呈正相关（Auffret等，1993），因此系水力是一种合适的容积测定方法（Kyriazakis和Emmans，1995）。可溶性纤维的系水力通常高于不可溶性纤维（Auffret等，1994；Robertson等，2000）。纤维素和木质素的持水力一般较低，而半纤维素的持水力一般较高（Shelton和Lee，2000）。半纤维素的单糖组成包括阿拉伯糖和木糖，与持水力呈正相关（Holloway和Greig，1984）。

◆ 4.3 黏度

黏度是指饲粮纤维尤其是可溶性饲粮纤维在溶液中增稠或形成胶体的能力（Dikeman和Fahey，2006）。不可溶性纤维尽管可通过其吸收水分能力影响其黏度，但通常与黏度无关（Takahashi等，2009）。

饲粮纤维引起的黏度通常受饲粮纤维水平的影响，但这种影响并不是线性的（Dikeman和Fahey，2006）。可溶性饲粮纤维含量较低时，溶液中的分子处于离散状态，可自由流动；但当达到临界浓度时，分子运动受到限制，饲粮纤维分子就会发生物理凝聚（Oakenfull，2001）。此时，含可溶性饲粮纤维溶液的黏度随果胶浓度的增加而迅速增加（Buraczewska等，2007）。饲粮纤维在溶液中的黏度测定取决于液体的剪切率或搅拌速率（Oakenfull，2001）。剪切率越大，黏度的测定结果就越低（Dikeman和Fahey，2006）。在很多研究中，黏度的测定只使用一种剪切率，但由于不同的剪切率得到的黏度值不同，因此无法对溶液或食糜中的黏度值进行比较（Dikeman和Fahey，2006）。为了克服这种局限，针对不同的饲粮纤维，推荐采用不同的剪切率进行黏度测定（Dikeman和Fahey，2006）。

溶液或食糜中饲粮纤维的黏度也会受分子量和颗粒大小的影响。含量相同时，由高分子量瓜尔胶制备的溶液比低分子量瓜尔胶溶液的黏性更大（Dikeman和Fahey，2006），猪盲肠内容物中较大颗粒产生的表观黏度比小颗粒产生的表观黏度更大（Takahashi和Sakata，2002）。

◆ 4.4 阳离子结合能力

饲粮纤维也可以结合矿物元素和有机分子（Oakenfull，2001）。游离羧基和糖醛酸（离子基团）可吸附到金属离子上。纤维和矿物元素之间的吸附作用可阻止Ca^{2+}、Mg^{2+}和Zn^{2+}等矿物元素的吸收（Cho等，1997）。饲粮纤维中的部分化合物与矿物质结合为植酸盐，木质素和其他结合成分也会影响矿物质的吸收（Kritchevsky，1988；Adlercreutz等，2006）。饲粮纤维也可结合到胆酸等有机分子上（Scheneeman，1998），木质素是饲粮纤维中结合能力最强的物质之一（Kritchevsky，1988）。

◆ 4.5 发酵

饲粮纤维被微生物发酵的程度取决于其接触到的后肠中微生物的数量（Oakenfull，2001）。溶解度和系水力对饲粮纤维发酵率的影响很大。饲粮纤维吸收水分后变得膨胀，可增加多糖与微生物接触的表面积（Canibe和Bach Knudsen，2001）。由于可溶性纤维的系水力比不可溶性纤维高，其膨胀程度相应地更大，因此可溶性纤维比不可溶性纤维的发酵速度更快（Auffret等，1993，1994；Oakenfull，2001）。可溶性饲粮纤维的发酵主要发生在近端结肠，而不可溶性纤维的发酵将一直持续到远端结肠（Cho，1997）。

增加粪便量是饲粮纤维发酵的一个主要功能（Stephen和Cummings，1979）。可发酵碳水化合物能够促进微生物的生长，而这又通过提高粪便中微生物的总量来增加粪便的排泄量（Cho等，1997）。不可发酵饲粮纤维的残留也增加了粪便的排泄量（Stephen和Cummings，1979）。因此，无论是由可溶性纤维还是不可溶性纤维组成的饲粮纤维，粪便排泄量的增加均可归因于粪便中微生物数量或不可降解纤维残留物数量的增加（Cho等，1997）。为了达到通便的目的，官方指南推荐使用饲粮纤维粗粉（Jenkins等，1999）。但是，无论使用麦麸粗粉还是细粉，动物的粪便排出量均是相近的；不过，与麦麸细粉相比，使用麦麸粗粉时肠道的蠕动频率更高（Jenkins等，1999）。然而，麦麸细粉比麦麸粗粉的发酵程度更高，因此给动物饲喂麦麸细粉，其肠道中丁酸的含量更高（Jenkins等，1999）。这些结果说明，颗粒大小会影响饲粮纤维的发酵特性和通便效果。

纤维发酵的主要产物是乙酸、丙酸、二氧化碳、甲烷和氢气，其中每种

VFA的浓度取决于饲粮纤维的化学结构和物理结构（Lunn和Buttriss，2007）。乙酸是含量最丰富的一种VFA，大约占后肠产生的总短链脂肪酸的60%，而丙酸和丁酸的量相对较少（Lunn和Buttriss，2007）。

5 饲粮纤维消化率的定性

◆ 5.1 消化

消化是营养素在吸收前被胃肠道分泌至肠腔中的消化酶进行化学降解的过程（Tso和Crissinger，2000）。这些酶主要由口腔腺细胞、胃腺细胞、胰腺外分泌细胞和刷状缘小肠腺分泌（Johnson，2001）。哺乳动物的消化酶可水解少量的淀粉和麦芽低聚糖中的α-（1-4）糖苷键、淀粉和糊精中的α-（1-6）糖苷键、蔗糖中的β-（1-2）糖苷键和乳糖中的β-（1-4）糖苷键等连接键。其他一些连接键，如纤维素中的β-（1-4）糖苷键，并不能被哺乳动物的内源酶水解，需要由发酵过程中产生的细菌酶才能水解（Tso和Crissinger，2000）。营养素的消化和吸收主要发生于小肠，而发酵部分发生在小肠末端，主要集中于大肠。

◆ 5.2 VFA的生成

肠道内环境要求微生物在无氧条件下才能够生存。有三种类型的微生物可以在无氧环境中生存，即厌氧光能利用菌、厌氧呼吸菌（硫酸盐还原菌、产甲烷菌和产乙酸菌）和可发酵微生物（White，2000；Müller，2008）。发酵是一个储能过程，即在氧化还原反应中形成的电子被转移到部分底物，由此产生能量。在这一过程中，底物只有部分被氧化，且只有少量的能量用于微生物的生长（Müller，2008）。

饲粮纤维在猪肠道发酵过程中，微生物将多糖降解为小分子多糖或组分单糖（Müller，2008）。解聚反应通常伴随着少量的化学反应（如水解、氧化、磷酸化和裂解）。单体被吸收进入菌体细胞，并用于中枢代谢途径（White，2000）。发酵过程中，己糖氧化（糖酵解EMP途径或ED途径）和戊糖氧化（磷酸戊糖途径）均可形成丙酮酸，随后被氧化为乙酸、丙酸或丁酸（图11-1、图11-2和图11-3）。

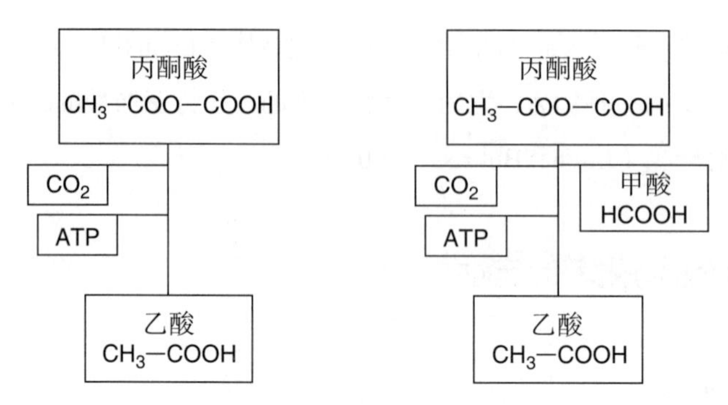

图11-1　发酵时由丙酮酸转化为乙酸的两种途径

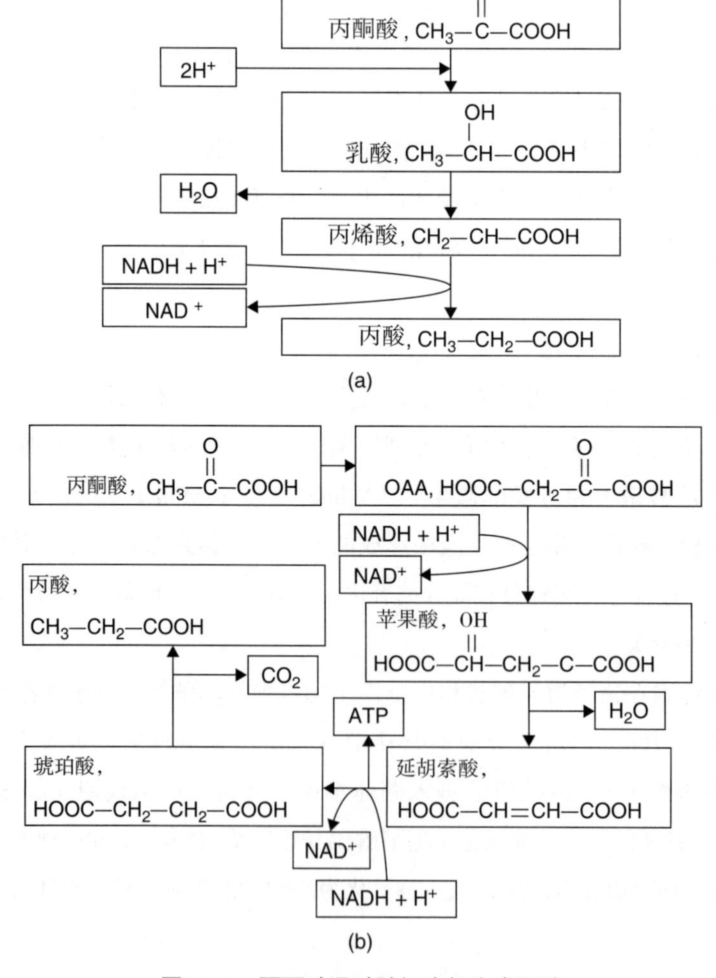

图11-2　丙酮酸通过随机途径生成丙酸

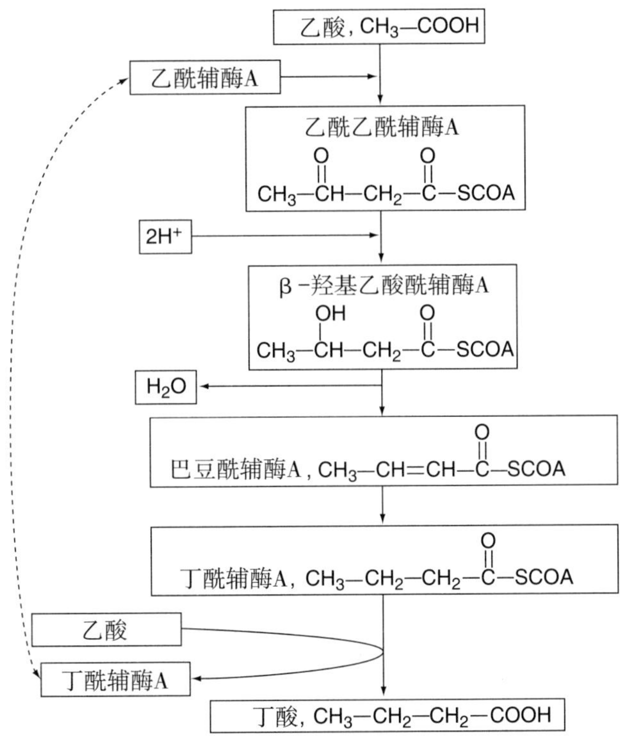

图11-3　由乙酸合成丁酸

◆ 5.3 VFA的吸收

由肠道微生物产生的VFA从微生物细胞中排出进入肠腔。其他微生物可利用这些产物作为底物产生次级产物（厌氧食物链）。然而，猪可吸收一些VFA，这些VFA可改善机体的能量状态。猪大肠中VFA的吸收是一个高效的过程（Barcroft等，1944）。在生长猪盲肠中灌注VFA，其中不到1%的VFA可通过粪便排出体外（Jørgensen等，1997）。VFA的吸收被认为是通过三种机制完成的：①VFA离子的扩散，②阴离子交换（Wong等，2006）③转运载体介导的吸收（Kirat和Kato，2006）。VFA离子的扩散可能只是少量吸收的一种形式，因为在生理pH条件下，肠腔中只有1%的VFA可被离子化（Cook和Sellin，1998）。如果是阴离子交换，VFA被肠上皮细胞摄取，同时HCO_3^-被释放到肠腔中（Cook和Sellin，1998）。最近的研究表明，还存在VFA的主动转运方式。VFA的主动转运载体属于单羧酸酯家族，MCT1是存在于猪肠道上的转运载体（Welter和Claus，2008）。在人结肠细胞中表达的另一种转运载体是钠结合型单羧酸酯载体或

SLC5A8，其可能参与VFA的吸收，尤其是丁酸的吸收（Thangaraju等，2008）。在猪的肠细胞中可检测到MCT1转运载体，但不清楚SLC5A8是否也存在于猪的结肠细胞中。

VFA的吸收也可促进饲粮中其他营养素的吸收。水分和钠离子是随同VFA一起被吸收的（Yen，2001）。植物木酚素属于双酚化合物，与内源性类固醇激素类似，也可与VFA协同转运（BachKnudsen等，2006）。菊粉可提高贫血仔猪对玉米-豆粕型饲粮中铁的生物学利用率（Yasuda等，2006）。目前尚不清楚，菊粉促进铁的吸收是否是通过提高VFA的产生，进而提高铁的吸收；或是否是通过VFA降低了肠腔的pH，进而增加了铁的溶解度；或者是因为VFA提高了铁转运载体的表达（Tako等，2008）。

◆ 5.4 VFA的代谢

VFA通过三种途径进行代谢：① 被结肠细胞用作能源；② 丙酸被肝脏用于糖异生；③ 被脂肪和肌肉组织代谢（Wong等，2006）。所有VFA的氧化始于与辅酶A结合后被激活（如乙酰CoA），之后进入中枢代谢途径（图 11-4）。乙酸被转化为乙酰CoA，丙酸被转化为琥珀酰辅酶A，而丁酸被转化为乙酰乙酰辅酶A（Nelson和Cox，2008）。

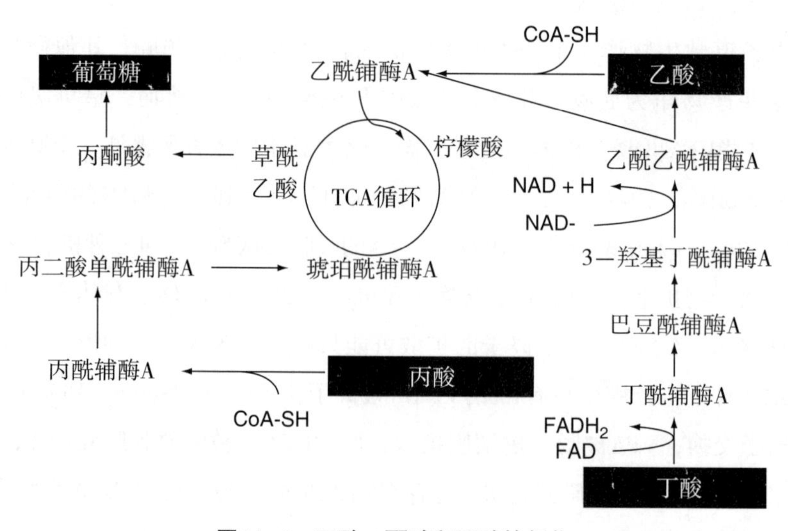

图11-4　乙酸、丙酸和丁酸的氧化

门静脉血中VFA的含量和摩尔比与小肠食糜中的不同，提示VFA在肠细胞中发生了代谢（Argenzio和Southworth，1974；Marsono等，1993）。肠道内容物中VFA的摩尔比通常是65：25：10（乙酸：丙酸：丁酸）。而经肝脏循环后，相应的比例为90：10：0，说明VFA在肠细胞和肝脏中发生了选择性代谢（Robertson，2007）。

人的结肠细胞将吸收丁酸的70%～90%代谢为CO_2和酮体，因此丁酸可作为谷氨酰胺的替代能源（Kritchevsky，1988；Elia和Cummings，2007）。丁酸不仅可作为能源物质，而且还可以调节细胞的增殖和分化，这反过来又有助于预防结肠癌和其他疾病（Cook和Sellin，1999；Wong等，2006）。

大多数乙酸和丙酸未被代谢就离开肠道进入肝脏，丙酸在肝脏中被代谢进入糖异生途径（Wong等，2006）。丙酸代谢可抑制羟甲基戊二酰（HMG）CoA还原酶，进而抑制了胆固醇的合成（Wong等，2006）。有研究认为，大多数乙酸被转移到脂肪组织和骨骼肌中，用于脂肪酸的合成或ATP合成过程中的氧化作用（Elia和Cummings，2007）。

对于饲喂高水平纤维饲粮的猪，其后肠以VFA形式吸收的能量占其吸收总能量的67%～74%，来自于VFA的能量为猪提供了7.1%～17.6%的总可利用能量（Anguita等，2006）。在盲肠中灌注VFA，其提供能量的82%可被机体保留（Jørgensen等，1997）。

VFA对脂肪酸代谢和脂肪分配的影响，是目前研究的热点（Robertson，2007）。丙酸可改变脂肪组织的脂解作用、脂肪细胞的大小和分化以及体脂的分配。另外，VFA也可激活PPARγ、乙酰辅酶A羧化酶和脂肪酸合成酶（Lee和Hosser，2002）。

6 影响饲粮纤维能值的因素

◆ 6.1 猪自身的因素

猪的日龄和品种会影响纤维的消化率。与生长育肥猪相比，母猪能更多地消化饲粮纤维（Le Goff和Noblet，2001），这可以解释为什么母猪肠道中食糜的通过速度更慢（Grieshop等，2001）。母猪微生物区系消化纤维的固有能力可能并不是一个重要因素（Le Goff等，2003）。

梅山猪对纤维的消化能力比西方品种猪更强（Fevriere等，1988；Kemp等，1991）。其他地方如Mukota（津巴布韦）、Mong Cai（越南）和Kunekune（新西兰）的品种猪，其消化饲粮纤维的能力比西方商品猪更强（Ndindana等，2002；Len等，2006；Morel等，2006）。由于瘦肉型生长猪大肠中纤维素分解菌的含量更高，因此其消化的能量比长速慢的猪更多（Varel等，1988）。但是，在其他的一些研究中，并没有发现梅山猪的纤维消化能力比西方品种猪更强（Yen等，2004）。目前，尚没有关于地方品种猪比西方品种猪能够更好地消化哪部分饲粮纤维（不可溶或可溶）的研究。

◆ 6.2 饲粮自身的因素

有多种方法可用于提高纤维性饲料原料中饲粮纤维的微生物降解作用，从而提高饲料的能值。这些方法包括物理法（粉碎、加热、辐射和植物组分的物理分离）和化学法（水解和氧化剂）。氢氧化钠可将大麦秸秆的瘤胃有机物（OM）消化率从52%提高至76%，将其他作物秸秆的干物质消化率提高22%（Fahey等，1993）。但使用NaOH的缺点是，它可能泄漏到土壤中造成污染。无水NH_3、NH_4OH、热氨及尿素已被用于处理纤维性原料，但是用这些方法处理后其消化率的提高并没有使用NaOH处理的效果明显。在对32个作物秸秆进行处理的试验中发现，反刍动物对干物质的消化率提高了15%（Fahey等，1993）。其他化学药品如$Ca(OH)_2$和KOH也曾被用于处理纤维性作物秸秆，但多数试验都是在反刍动物上开展的，因此并不知道在猪上是否也具有相似的结果。

使用臭氧等氧化剂处理纤维，其体外干物质消化率可从44%提高到67%。但地面上的臭氧也是一种污染源，因此需要控制臭氧向环境中泄露（Fahey等，1993）。过氧化氢可将纤维素的瘤胃表观消化率从56.5%提高至85.7%（Kerley等，1985）。二氧化硫可将干物质的体外消化率提高80%，但是动物可能无法接受经过量硫处理的饲料（Fahey等，1993）。

7 发酵能量的分配

◆ 7.1 饲粮纤维的消化率

不同的饲料原料、不同类型的纤维之间，饲粮纤维的发酵不同（Bindelle等，2008）。51个消化试验的数据显示，饲喂高水平纤维的饲料原料时，猪对饲粮纤维的回肠表观消化率（AID）为10%~62%（Back Knudsen和Jørgensen，2001）。大麦纤维素的全肠道表观消化率（ATTD）为23%~65%，小麦和小麦副产品纤维素的全消化道表观消化率分别为24%和64%，黑麦和黑麦片分别为10%和84%，小麦、玉米和燕麦的麸皮和壳分别为13%和42%。DDGS中TDF的全消化道表观消化率平均为47.5%，原料的来源不同其消化率在29.3%~57%范围内变化（Urriola等，2009）。可溶性饲粮纤维的ATTD（92%）大于不可溶性饲粮纤维的ATTD（41.3%）（Urriola等，2009）。

◆ 7.2 每克可发酵纤维产生的VFA量

每克可发酵纤维产生的VFA数量，因发酵纤维类型的不同而异。大豆中的棉籽糖和水苏糖等α-半乳糖苷，发酵可产生更多的气体（甲烷和氢气），易引起胀气，但在大肠中发酵产生的VFA比纤维素和半纤维素发酵产生的VFA更少（Liener，1994）。乙酸、丙酸和丁酸在产生的VFA中的含量最高，因此是大多数试验报道的仅有的VFA。这些VFA的相对产量随发酵底物不同而略有差异（Topping和Clifton，2001），但在实际应用中，乙酸、丙酸和丁酸之间的比例可以假定为恒定（de Lange，2008）。然而，支链氨基酸发酵产生的是支链VFA（异丁酸、异戊酸和戊酸）。因此，支链VFA的产量取决于支链氨基酸的发酵程度。在大多数情况下，这三种支链VFA的产量不到总VFA产量的5%。

◆ 7.3 吸收和代谢的每摩尔VFA产生的ATP摩尔数

动物氧化每摩尔VFA产生的ATP摩尔数分别为：乙酸，10 mol ATP；丙酸，18 mol ATP；丁酸，28 mol ATP。每摩尔ATP产生的能量与这三种VFA均相似，平均约为8.4×10^4J/mol ATP（Blaxter，1989）。该能值与利用其他营养素产生的ATP获得的能量也相近。

8 纤维对能量和养分消化率的负面影响

◆ 8.1 对能量消化率的影响

通过增加饲粮中麸皮的用量（0～40%）提高粗纤维的含量，可逐步降低全肠道能量消化率（Wilfart等，2007）。饲粮能量消化率的降低与干物质和有机物消化率的降低相关（Wilfart等，2007）。通过添加麸皮、玉米麸、大豆皮和甜菜渣的混合物提高饲粮纤维含量后，猪全肠道能量消化率也降低，同时碳水化合物的消化率也相应降低（Le Gall等，2009）。据测算，NDF含量每增加1%，能量消化率就降低1%（Le Gall等，2009）。由于甜菜碱的全肠道消化率比大豆皮高，因此纤维的溶解度会影响能量的消化率（Mroz等，2000）。饲粮纤维中存在的木质素也会降低能量的消化率（Wenk，2001）。能量消化率降低的原因包括：结合到不易消化的纤维源细胞壁成分中的CP和碳水化合物替代了可消化的CP和碳水化合物（如淀粉），纤维的理化特性影响了饲粮营养素的消化和吸收过程，以及纤维影响了胃肠道的生理功能（Le Gall等，2009）。

◆ 8.2 对氨基酸消化率的影响

在大豆和玉米淀粉基础饲粮中添加7.5%的柑橘果胶后，CP和AA的AID分别降低8.2～28.7个百分点（Mosenthin等，1994）。在小麦、玉米和豆粕型基础饲粮中添加4%或8%的苹果胶后，CP和AA的标准回肠消化率（SID）也有所降低（Buraczewska等，2007）。当向大豆分离蛋白和玉米淀粉型基础饲粮中添加由小麦麸加工而成的纯化NDF时，猪的回肠氮消化率呈线性下降（Schulze等，1994）。添加15%的小麦纯化NDF，AA的AID也降低了2～5.5个百分点，Cys、Ala和Gly的AID则分别降低了18、16和12个百分点（Lenis等，1996）。向豆粕-玉米淀粉型基础饲粮中添加梯度水平（3%～9%）的大豆皮，进而使NDF含量从2.72%增加到4.16%时，大部分AA的AID和SID呈线性或二次下降（Dilger等，2004）。但是，当向仔猪豆粕-玉米淀粉型基础饲粮中添加10%的纤维素和大麦秸秆时，除Leu和Gly外，其他AA的AID均没有降低（Sauer等，1991）。当向仔猪豆粕-玉米淀粉型基础饲粮中添加梯度水平（4.3%～13.3%）的纤维素粉时，CP和AA的AID也没有降低（Li 等，1994）。相反，在饲粮中添加羧甲基纤维素反而会提高CP和AA的SID（Larsen等，1994；Bartelt等，2002；Fledderus等，2007）。不溶性和发酵性差的纤维（如纤维素），可通过其持水特性影响CP的消化率，而羧甲基纤维素和果胶等可溶性纤维则通过自身的黏稠特性实现其对CP消化率的影响。

饲粮纤维可通过影响消化过程，降低CP吸收或增加内源性CP和AA的损失而降低CP和AA的利用效率（Mosenthin等，1994）。Schulze等（1995）研究发现，给猪饲喂20%的麦麸纯化NDF，其回肠氮流量增加，其中59%为内源氮。向无氮饲粮中添加梯度水平的豌豆内源纤维，回肠N流量指数增加，这与饲粮WHC的增加有关（Leterme等，1998）。回肠上皮细胞的更新也随着黏蛋白和微生物的线性增加而呈指数增加（Leterme等，1998）。饲粮中添加黏性不可发酵纤维（羟甲基纤维素）时，黏蛋白的分泌和内源氮的损失也会增加，但回肠微生物的数量并没有变化（Bartelt等，2002；Piel等，2004）。然而，Owusu-Asieda等（2006）研究发现，用黏性可发酵纤维瓜尔豆胶饲喂猪时，回肠中某些微生物的数量有所增加。相反，在饲粮中添加3.31%～16.5%的纤维素（一种不可溶的、发酵性差的纤维），并不能引起内源性CP和AA损失的增加，这可能是由于添加纤维素时，并未引起饲粮CP和AA的AID降低（Li等，1994）。饲粮纤维的水平和来源是影响内源CP和AA损失的重要因素（Sauer和Ozimek，1986），而纤维素水平只有在超出某一阈值时才可能降低AA的AID（Li等，1994）。

纤维对胰腺分泌和酶活的影响，同样也受纤维理化特性的调控。与玉米淀粉型、酪蛋白型和纤维素型基础饲粮相比，大麦或小麦型基础饲粮可增加胆汁和胰液的分泌，但不影响消化酶的分泌量（Low，1989）。然而，当向等氮、等能饲粮中分别添加400 g的麸皮时，糜蛋白酶和胰蛋白酶的分泌量比饲粮中不添加麸皮时更大（Langlois等，1986）。相反，在SBM型基础饲粮中添加果胶，既不会提高胰腺的分泌，也不影响胰蛋白酶和糜蛋白酶的分泌量和酶活（Mosenthin等，1994）。然而，饲粮羟甲基纤维素会降低胃蛋白酶的活性，但不影响胰蛋白酶和糜蛋白酶的活性（Larsen等，1993）。

◆ 8.3 对黏蛋白产生和能量内源损失的影响

胃肠道杯状细胞分泌的黏蛋白，是一种高分子量的糖蛋白，可润滑上皮表面，保护肠道免受物理磨损、化学侵蚀和病原微生物的吸附，这些均可能损害肠道的健康（Forstner和Forstner，1994；Tanabe等，2006）。黏蛋白对营养素的消化和吸收也发挥着重要的作用，黏蛋白分泌的改变可能会引起肠道中饲粮营养素和内源分子吸收动力学的变化（Tanabe等，2006）。

有些研究表明，饲粮纤维可增加黏蛋白的分泌。在猪的无氮饲粮中梯度添加小麦秸秆、玉米穗和木纤维素后，小肠黏蛋白中氨基糖（葡糖胺和半乳糖胺）的浓度呈线性增加（Mariscal-Landín等，1995）。饲喂羟甲基纤维素（一种具有黏性但不可发酵的纤维），也可增加黏蛋白的浓度及其在回肠末端的排出量，并增加了断奶仔猪小肠中每根绒毛上回肠杯状细胞的总数量（Piel等，2005）。在纯合饲粮中添加5%的柑橘纤维也可大幅增加小肠黏蛋白的分泌（Satchithanandam等，1990）。

在胃中，纤维的结构特性和可发酵性不影响黏蛋白的分泌；但在盲肠中，低聚果糖和甜菜碱的可发酵性能增加黏蛋白的分泌（Tanabe等，2006）。Libao-Mercado等（2007）报道了相似的研究结果，即在饲粮中添加果胶可增加结肠黏蛋白的分泌和黏膜粗蛋白的合成，但对空肠没有影响。

黏蛋白分子是由连接有碳水化合物侧链上的蛋白质骨架组成的，其有1/2的结构域含有一个由Pro、Ser和Thr组成的蛋白质骨架。这个区域可以抵抗蛋白酶的消化，因为其中80%的蛋白质骨架受寡糖的保护，这些寡糖包括海藻糖、半乳糖、N-乙酰氨基半乳糖、N-乙酰葡糖胺和唾液酸（Montagne等，2004）。由于黏蛋白中存在抗蛋白酶消化结构域，黏蛋白很难被消化，因此大幅增加了回肠食糜中内源CP和碳水化合物的数量（Lee等，1988；Lien等，1997）。

回肠食糜中存在的内源性CP和AA大部分来源于胰腺酶、上皮细胞、菌体和黏蛋白，而内源性碳水化合物主要来源于黏蛋白（Lien等，1997；Miner-Williams等，2009）。在回肠末端前无法被重吸收的内源性CP和AA，可被猪的后肠微生物利用（Souffrant等，1993；Libao-Mercado等，2009）。粪中含有很少的黏蛋白，这进一步说明黏蛋白在后肠已被完全发酵（Lien等，2001）。

◆ 8.4 对其他营养素利用的影响

8.4.1 碳水化合物

在大麦型基础饲粮中添加麦麸，不影响淀粉的消化率（Högberg和Lindberg，2004）；在谷物类基础饲粮中添加20%和40%的麦麸，也不影响淀粉的消化率（Wilfart等，2007）。99%的淀粉是在小肠中被消化，在粪便中几乎检测不到（Högberg和Lindberg，2004；Wilfart 等，2007）。相反，以类似的配方添加梯度水平的麦麸、米糠、大豆皮或甜菜碱，使饲粮纤维从12%提高至38%后，碳水化合物的ATTD降低（Le Gall等，2009）。在饲粮中添加瓜尔豆胶后，空肠对葡萄糖的吸收降低了50%（Rainbird等，1984）。Nunes和Malmlof（1992）、Owusu-Asiedu等（2006）也报道到了类似的结果，即瓜尔豆胶而不是纤维素可减少猪的血浆葡萄糖浓度。瓜尔豆胶引起的食糜黏性降低了肠腔中葡萄糖向上皮细胞扩散的速率，从而降低了葡萄糖的吸收（Rainbird等，1984；Kritchevsky，1988）。

8.4.2 脂类

与对照组饲粮相比，在谷物类基础饲粮中添加20%或40%的麸皮可使无氮浸出物的ATTD降低7%～12%（Wilfart等，2007）。在基础饲粮中添加甜菜碱，也会降低脂肪的AID和ATTD，但添加麸皮不会降低脂肪的AID和ATTD（Graham等，1986）。相反，将黑小麦、小麦和麸皮混合物作为纤维源添加到谷物类基础饲粮中，脂肪的AID和ATTD显著高于对照组（Högberg和 Lindberg，2004）。这一发现表明，含不同来源纤维饲粮的溶解度会影响脂肪的消化率，因为将麦麸、米糠、大豆皮和甜菜碱混合物梯度添加到低水平纤维饲粮中时，尽管增加了饲粮中总纤维的含量，但脂肪的ATTD并没有受到影响（Le Gall等，2009）。饲粮纤维的添加水平也会影响脂质的消化率，因为将椰子粕、大豆皮或甜菜碱梯度添加到低水平纤维对照饲粮中时，脂肪的消化率会降低（Canh等，1998）。

8.4.3 矿物质

由多糖构成的饲粮纤维能够结合矿物质，但有关饲粮纤维影响矿物质消化率的研究结果并不一致。饲粮中添加6%的纤维素，降低了Ca、P、Mg和K的表观吸收率。与玉米-豆粕型饲粮相比，母猪采食含有玉米穗和麸皮或燕麦和燕麦壳的高水平纤维饲粮后，其每单位摄入的矿物质中血清Ca、P、Cu和Zn的浓度明显降低（Girard等，1995）。相反，在饲粮中添加燕麦壳、大豆皮和苜蓿草粉，并不影响Ca、P、Zn或Mn的总肠道消化率（Moore等，1988）。同样，在猪饲粮中添加6%的菊粉，对Ca、P、Mg和Zn的AID和ATTD也没有影响（vanhoof和De Schrijver，1996）。另外，尽管在低水平纤维饲粮中添加20%和40%的麸皮对灰分AID也没有影响，但在饲粮中添加高水平的麸皮会降低灰分的AID（Wilfart等，2007）。

◆ 8.5 饲粮纤维对氮排泄和粪便特性的影响

饲粮纤维对猪氮排泄的主要影响就是将尿氮排泄物转变为粪氮排泄物。纤维的存在增强了猪后肠微生物的发酵，这样一来，饲粮和内源蛋白质发酵产生的氨就被用于微生物的代谢和生长（Zervas和Zijlstra，2002）。因此，由血液吸收用于肝脏尿素合成的可利用氨的浓度总体下降（Mroz等，2000；Zervas和Zijlstra，2002），结果造成尿氮排泄量减少。从尿氮排泄到粪氮排泄的转变取决于饲粮纤维的水平，因为增加甜菜碱的含量可线性降低尿氮排泄物转变成粪氮排泄物的比例（Bindelle等，2009）。尿氮转变为粪氮的程度同样也受纤维来源的影响，因为大麦型基础饲粮的尿氮排泄物转变为粪氮排泄物的比例小于玉米-小麦型基础饲粮，含甜菜碱饲粮的尿氮排泄物转变为粪氮排泄物的比例小于甘薯粉饲粮（Canh等，1997；Leek等，2007）。与纤维素等不可发酵纤维相比，果胶和马铃薯淀粉等可发酵纤维对尿氮排泄向粪氮排泄的转变有更大的影响（Pastuszewska等，2000）。用燕麦壳梯度替代甜菜碱，同样可以提高尿氮排泄向粪氮排泄转变的比例（Bindelle等，2009）。尿氮排泄的减少，有利于减轻养猪生产体系中氨排放带来的环境问题（Aarnink和Verstegen，2007）。

提高饲粮纤维的含量，可线性增加猪每天的排粪量（Moeser和van Kempen，2002）。然而，随着纤维采食量的增加，粪便的干物质量却相应地减少。这表明，水对饲喂高纤维饲粮猪的排粪量起了很大的作用（Canh 等，1998）。粪便的pH随饲粮纤维的添加而降低，饲喂大豆皮和甜菜碱饲粮的猪，其粪便pH低于饲喂不含大豆皮和甜菜碱对照饲粮的猪（Mroz等，2000）。饲喂含22%大豆皮NDF的饲粮，猪粪便的pH低于饲喂含6%和12%NDF饲粮的猪（Moeser和van Kempen，2002）。猪粪便pH的降低，归因于粪便中存在高浓度的VFA（Canh等，

1998）。而VFA的浓度取决于纤维的水平和来源（Canh等，1998）。

9 结 语

纤维包含植物性饲料原料中不能被动物内源酶消化的碳水化合物和果胶，这会对动物和人类产生生理作用。纤维可采用不同的方法进行分析，但饲粮总纤维分析法可比较准确地反映饲料原料中纤维的含量，此分析法包括可溶性纤维和不可溶性纤维的含量。

由于燃料工业中淀粉和油的应用不断增加，因此目前许多猪饲料原料中含有较高的纤维。但是猪不能很好地利用纤维，饲粮中纤维的含量越高，干物质和有机物的消化率就越低。只有当纤维被胃肠道微生物发酵后，动物和人类才能从吸收的VFA中获得纤维能量。乙酸、丙酸和丁酸是纤维发酵产生的含量最多的三种VFA，它们可改善猪的能量状态。

可溶性纤维容易被发酵，其总肠道消失率可达90%以上。但是，不可溶性纤维较难被发酵，其总肠道消失率通常不超过50%。因此，增加饲粮中纤维的含量会降低饲粮总能的消化率。纤维也会降低AA的消化率，因为饲喂高纤维饲粮后，猪的内源性AA损失会增加。纤维可引起黏蛋白分泌量的增加，这也是AA内源损失增加的原因之一。高纤维饲粮也会降低脂类和矿物质的总肠道消化率，但饲粮纤维可以将尿氮排泄转变为粪氮排泄，因而有助于减少氨的合成。

作者：Pedro E. Urriola,
Sarah K. Cervantes-
Pahm和Hans H.
Stein
译者：游金明

参考文献

AACC. 2001. The Definition of Dietary Fiber. AACC Report. Am. Assoc. Cereal Chem. 46:112–126.

AAFCO. 2008. Association of American Feed Control Officials: Official Publication. The Association, Springfield, IL.

Aarnink, A. J. A., and M. W. A. Verstegen. 2007. Nutrition, key factor to reduce environmental load from pig production. Livest. Sci. 109:194–203.

Adlercreutz, H., J. Penalvo, S. M. Heinonen, and A. Linko-Parvinen. 2006. Lignans and other co-passengers. Pages 199–218 in Dietary Fiber Components and Functions. H. Salovaara, F. Gates, and M. Tenkanen eds. Wageningen Academic Publishers. Wageningen, Netherlands.

Ang, J. F. 1991. Water retention capacity and viscosity effect of powdered cellulose. J. Food Sci. 56:1682–1684.

Anguita, M., N. Canibe, J. F. Pérez, and B. B. Jensen. 2006. Influence of the amount of dietary fiber on the available energy from hindgut fermentation in growing pigs: Use of cannulated pigs and *in vitro* fermentation. J. Anim. Sci. 84:2766–2778.

AOAC. 2006. Official Methods of Analysis, 18th ed. Assoc. of. Anal. Chemists, Arlington, VA.

Argenzio, R. A., and M. Southworth. 1974. Sites of organic acid production and absorption in gastrointestinal tract of the pig. Am. J. Physiol. 228:454–460.

Auffret, A., J. L. Barry, and J. F. Thibault. 1993. Effect of chemical treatments of sugar beet fibre on their physico-chemical properties and on their *in vitro* fermentation. J. Sci. Food Agric. 61:195–203.

Auffret, A., M. C. Ralet, F. Guillon, J. L. Barry, and J. F. Thibault. 1994. Effect of grinding and experimental conditions on the measurement of hydration properties of dietary fibers, Lebensmitteln Wiss. Technol. 27:166–172.

Bach Knudsen, K. E., and H. Jørgensen. 2001. Intestinal degradation of dietary carbohydrates—from birth to maturity. Pages 109–120 in Digestive Physiology of Pigs. J. E. Lindberg, and B. Ogle eds. Cabi Publishing New York, NY.

Bach Knudsen, K. E., A. Serena, H. Jørgensen, J. L. Peñalvo, and H. Adlercreutz. 2006. Rye and other natural cercal fiber enhance the production and plasma concentration of enterolactone and butyrate. Pages 219–233 in Dietary Fiber Components and Functions. H. Salovaara, F. Gates, and M. Tenkanen eds. Wageningen Academic Publishers. Wageningen, Netherlands.

Barcroft, J., R. A. McAnally, and A. T. Phillipson. 1944. Absroption of VFA from the alimentary tract of the sheep and other animals. J. Exp. Biol. 20:120–132.

Bartelt, J., A. Jadamus, F. Wiese, E. Swiech, L. Buraczewska, and O. Simon. 2002. Apparent precaecal digestibility of nutrients and level of endogenous nitrogen in digesta of the small intestine of growing pigs as affected by various digesta viscosities. Arch. Tierernaehr. 56:93–107.

Bindelle, J., A. Buldgen, M. Delacollette, J. Wavreille, R. Agneessens, J. P. Destain, and P. Leterme. 2009. Influence of source and concentrations of dietary fiber on *in vivo* nitrogen excretion pathways in pigs as reflected by *in vitro* fermentation and nitrogen incorporation by fecal bacteria. J. Anim. Sci. 87:583–593.

Blaxter, K. L. 1989. Energy Metabolism in Animals and Man. Cambrige University Press. Cambridge, UK.

Buraczewska, L., E. Świ ech, A. Tuśnio, M. Taciak, M. Ceregrzyn, and W. Korczyński. 2007. The effect of pectin in amino acid digestibility and digesta viscosity, motility and morphology of the small intestine, and on N–balance and performance of young pigs. Livest. Sci. 109:53–56.

Burkitt, D. P., A. R. P. Walker, N. S. Painter. 1972. Effect of dietary fiber on stools and transit–times, and its role in the causation of disease. The Lancet 30:1408–1411.

Cadden, A. 1987. Comparative effects of particle size reduction on physical structure and water binding properties of several plant fibers. J. Food Sci. 52:1595–1599.

Canh, T. T., M. W. Verstegen, A. J. Aarnink, and J. W. Schrama. 1997. Influence of dietary factors on nitrogen partitioning and composition of urine and feces of fattening pigs. J. Anim. Sci. 75:700–706.

Canh, T. T., A. L. Sutton, A. J. Aarnink, M. W. Verstegen, J. W. Schrama, and G. C. Bakker. 1998. Dietary carbohydrates alter the fecal composition and pH and the ammonia emission from slurry of growing pigs. J. Anim. Sci. 76:1887–1895.

Canibe, N., and K. E. Bach Knudsen. 2001. Degradation and physicochemical changes of barley and pea fibre along the gastrointestinal tract of pigs. J. Sci. Food Agric. 82:27–39.

Carpenter, K. J. 2003. A short story of nutritional science: Part 4 (1945—1985). J. Nutr. 133:3331–3342.

Chaplin, M. F. 2003. Fibre and water binding. Proc. Nutr. Soc. 62:223–227.

Chesson, A. 2006. Dietary fiber. Pages 629–663 in Food Polysaccharides and Their Applications. 2nd ed. A. M. Alistair, G. O. Philips, and P. A. Williams, eds. CRC Press, Boca Raton, FL.

Cho, S., J. W. de Vries, and L. Prosky. 1997. Dietary fiber analysis and applications. AOAC Intl., Gaithersburg, MD.

Cook, S. I., and J. H. Sellin. 1998. Review article: Short chain fatty acids in health and disease. Aliment. Pharmacol. Ther. 12:499–507.

de Lange, C. F. M. 2008. Efficiency of utilization of energy from protein and fiber in the pig—a case for NE systems. Pages 58–72 in Swine Nutr. Conf. Indianapolis, IN.

Dikeman, C. L., and G. C. Fahey Jr. 2006. Viscosity as related to dietary fiber. A review. Crit. Rev. Food Sci. Nutr. 45:649–663.

Dilger, R. N., J. S. Sands, D. Ragland, and O. Adeola. 2004. Digestibility of nitrogen and amino acids in soybean meal with added soyhulls. J. Anim. Sci. 82:715–724.

Elia, M., and J. H. Cummings. 2007. Physiological aspects of energy metabolism and gastrointestinal effects of carbohydrates. Eur. J. Clin. Nutr. 61(Suppl 1):S40–74.

Englyst, H. N., H. S. Wiggins, and J. H. Cummings. 1982. Determination of the non-starch polysaccharides in plant foods by gas–liquid chromatography of constituent sugars as alditol acetates. Analyst 107:307–318.

Englyst, K. N., S. Liu, and H. N. Englyst. 2007. Nutritional characterization and measurement of dietary fiber carbohydrates. Eur. J. Clin. Nutr. 61(Suppl 1):S19–39.

Fahey, G. C., Jr., L. D. Bourquin, E. C. Titgemeyer, and D. G. Atwell. 1993. Postharvest treatment of fibrous feedstuffs to improve their nutritive value. Pages 715–766 in Forage CellWall Structure and Digestibility. H. G. Jung, D. R. Buxton, D. R. Hatfield, and J. Ralph, eds. Am. Soc. Agron., Madison, WI.

Fevriere, C., D. Bourdon, A. Aumaitre, J. Peiniau., Y. Lebreton, Y. Jaguelin, N. Meziere, and A. Blanchard. 1988. Digestive capacity of the Chinese pig—effect of dietary fibre on digestibility and intestinal and pancreatic enzymes. Pages 172–179 in Proc. of IV Intl. Seminar of Digest. Physiol. in Pigs, Jablona, Poland.

Fledderus, J., P. Bikker, and J.W. Kluess. 2007. Increasing diet viscosity using carboxymethylcellulose in weaned piglets stimulates protein digestibility. Livest. Sci. 109:89–92.

Forstner, J. F., and G. G. Forstner. 1994. Gastrointestinal mucus. Page 1255 in Physiology of the Gastrointestinal Tract. L. R. Johnson, ed. Raven Press, New York, NY.

Furda, I. 2001. The crude fiber method. Pages 11–112 in Dietary Fiber in Human Nutrition. 3rd ed. G. A. Spiller, ed. CRC Press, Boca Raton, FL.

Girard, C. L., S. Robert, J. J. Matte, C. Farmer, G. P. Martineaub. 1995. Influence of high fibre diets given to gestating sows on serum concentrations of micronutrients. Livest. Prod. Sci. 43:15–26.

Gordon, D. T., B. V. McCleaary, and T. Sontag-Strohm. 2007. Summary of dietary fibre methods workshop. Pages 323–338 in Dietary Fiber Components and Functions. H. Salovaara, F. Gates, and M. Tenkanen eds. Wageningen Academic Publishers, Wageningen, Netherlands.

Graham, H., K. Hesselman, and P. Åman. 1986. The influence of wheat bran and sugar-beet pulp on the digestibility of dietary components in a cereal-based pig diet. J. Nutr. 116:242–251.

Grieshop, C. M., D. E. Reese, and G. C. Fahey Jr. 2001. Non-starch polysaccharides and oligosaccharides in swine nutrition. Pages 107–130 in Swine Nutrition. A. J. Lewis, and L. L.

Southern eds. CRC Press, Boca Raton, FL.

Guillon, F., L. Saulnier, P. Robert, J. F. Thibault, and M. Champ. 2006. Chemical structure and function of cell wall through cereal grains and vegetable samples. Pages 31–64 in Dietary Fiber Components and Functions. Salovaara, H., F. Gates, and M. Tenkanen eds. Wageningen Academic Publishers, Wageningen, Netherlands.

Högberg, A., and J. E. Lindberg. 2004. Influence of cereal non-starch polysaccharides on digestion site and gut environment in growing pigs. Livest. Prod. Sci. 87:121–130.

Holloway, W. D., and R. I. Greig. 1984. Water holding capacity of hemicelluloses from fruits, vegetables and wheat bran. J. Food Sci. 49:1632–1633.

IOM. 2001. Institute of Medicine. Proposed definition of dietary fiber. Pages 1–64 in Dietary Reference Intakes.National Academies Press, Washington, DC.

IOM. 2006. Institute of Medicine. Dietary, functional, and total dietary fiber. Pages 340–421 in Dietary Reference Intakes. National Academies Press, Washington, DC.

Jenkins, D. J. A., C. W. C. Kendall, V. Vuksan, L. S. A. Augustin, Y. Li, B. Lee, C. C. Mehling, et al. 1999. The effect of wheat bran particle size on laxation andcolonic fermentation. J. Amer. Coll. Nutr. 18:339–345.

Johnson, L. R. 2001. Digestion and absorption. Page 120 in Gastrointestinal Physiology. L. R. Johnson, ed. Mosby, St Louis, MO.

Johnston, L. J., S. Noll, A. Renteria, and G. C. Shurson. 2003. Feeding by-products high in concentration of fiber to nonruminants, in 3rd Natl. Symp. Alternative Feeds for Livest. Poult., Kansas City, MO.

Jørgensen, H., T. Larsen, X. Q. Zhao, and B. O. Eggum. 1997. The energy value of short-chain fatty acids infused into the cecum of pigs. Br. J. Nutr. 77:745–756.

Kemp, B., L. A. den Hartog, J. J. Klok, and T. Zandstra. 1991. The digestibility of nutrients, energy, and nitrogen in the Meishan and Duroc Landrase pig. J. Anim. Physiol. Anim. Nutr. 65:263–266.

Kerley, M. S., G. C. Fahey, Jr., L. L. Berger, J. M. Gould, and F. L. Baker. 1985. Alkaline hydrogen peroxide treatment unlocks energy in agricultural by-products. Science 230:820.

Kirat, D., and S. Kato. 2006. Monocarboxylase transporter 1 (MCT1) mediates transport of short chain fatty acids in bovine cecum. Exp. Physiol. 91:835–844.

Kritchevsky, D. 1988. Dietary fiber. Ann. Rev. Nutr. 8:301–328.

Kyriazakis, I., and G. C. Emmans. 1995. The voluntary feed intake of pigs given feeds based on wheat bran, dried citrus pulp and grass meal, in relation to measurements of feed bulk. Br. J. Nutr. 73:191–207.

Langlois, A., T. Corring, and J. A. Chayvialle. 1986. Effets de la consommation de son de blé sur la sécretion pancréatique exocrine et la teneur plasmatique de quelques peptides régulateurs chez le porc. Reprod. Nutr. Dev. 26:1178.

Larsen, F. M., M. N. Wilson, and P. J. Moughan. 1994. Dietary fiber viscosity and amino acid digestibility, proteolytic digestive enzyme activity and digestive organ weights in growing rats. J. Nutr. 124:833–841.

Larsen, F. M., P. J. Moughan, and M. N. Wilson. 1993. Dietary fiber viscosity and endogenous protein excretion at the terminal ileum of growing rats. J. Nutr. 123:1898–1904.

Le Gall, M., M. Warpechowski, Y. Jaguelin–Peyraud, and J. Noblet. 2009. Influence of dietary fibre level and pelleting on the digestibility of energy and nutrients in growing pigs and adult sows. Animal 3:352–359.

Le Goff, G., and J. Noblet. 2001. Comparative digestibility of dietary energy and nutrients in growing pigs and adult sows. J. Anim.Sci. 79:2418–2427.

Le Goff, G., J. Noblet, C. Cherbut. 2003. Intrinsic ability of the microbial flora to ferment dietary fiber at different growth stages of pigs. Livest. Prod. Sci. 81:75–87.

Lee, S. H., and K. L. Hossner. 2002. Coordinate regulation of the adipose tissue gene expression by propionate. J. Anim. Sci.80: 2840–2849.

Lee, S. P., J. F. Nicholls, A. M. Roberton, H. Z. Park. 1988. Effect of pepsin on partially purified pig gastric mucus and purified mucin. Biochem. Cell Biol. 66:367–373.

Leek, A. B. G., J. J. Callan, P. Reilly, V. E. Beattie, and J. V. O'Doherty. 2007. Apparent component digestibility and manure ammonia emission in finishing pigs fed diets based on barley, maize or wheat prepared without or with exogenous non–starch polysaccharide enzymes. Anim. Feed Sci. Technol. 135:86–99.

Len, N. T., J. E. Lindberg, and B. Ogle. 2006. Digestibility and nitrogen retention of diets containing different levels of fibre in local (Mong Cai), F1 (Mong Cai × Yorkshire) and exotic (Landrace × Yorkshire) growing pigs in Vietnam. J. Anim. Physiol. Anim. Nutr. 91:297–303.

Lenis, N. P., P. Bikker, J. van der Meulen, J. T. M. van Diepen, J. G. M. Bakker, and A. W. Jongbloed. 1996. Effect of dietary neutral detergent fiber on ileal digestibility and portal flux of nitrogen and amino acids and on nitrogen utilization in growing pigs. J. Anim. Sci. 74:2687–2699.

Leterme, P., E. Froidmont, F. Rossi, and A. Théwis. 1998. The high water–holding capacity of pea inner fibers affects the ileal flow of endogenous amino acids in pigs. J. Agric. Food Chem. 46:1927–1934.

Li, S., W. C. Sauer, and R. T. Hardin. 1994. Effect of dietary fiber level on amino acid digestibility in young pigs. Can. J. Anim.Sci. 74:1649–1656.

Libao–Mercado, A. J. O., C. L. Zhu, J. P. Cant, H. Lapierre, J. N. Thibault, B. Séve, M. F. Fuller, and C. F. M. de Lange. 2009. Dietary and endogenous amino acids are the main contributors to microbial protein in the upper gut of normally nourished pigs. J. Nutr. 139:1088–1094.

Libao–Mercado, A. J., C. L. Zhu, M. F. Fuller, M. Rademacher, B. Séve, and C. F. M. de Lange.

2007. Effect of feeding fermentable fiber on synthesis of total and mucosal protein in the intestine of the growing pig. Livest. Sci. 109:125–128.

Lien, K. A., W. C. Sauer, and M. Fenton. 1997. Mucin output in ileal digesta of pigs fed a protein–free diet. Z. Ernahrungswissenschaft. 36:182–190.

Lien, K.A., W. C. Sauer, and H. E. He. 2001. Dietary influences on the secretion into and degradation of mucin in the digestive tract of monogastric animals and humans. J. Anim. Feed Sci. 10:223–245.

Liener, I. E. 1994. Implications of antinutritional components in soybean foods. Crit. Rev. Food Sci. Nutr. 34:31–67.

Low, A. G. 1989. Secretory response of the pig gut to non–starch polysaccharides. Anim. Feed Sci. Technol. 23:55–65.

Lunn, J., and J. L. Buttriss. 2007. Carbohydrates and dietary fibre. Nutr. Bull. 32:21–64.

Mariscal–Landín, G., B. Séve, Y. Colléaux, and Y. Lebreton. 1995. Endogenous amino nitrogen collected from pigs with end–to–end ileorectal anastomosis is affected by the method of estimation and altered by dietary fiber. J. Nutr. 125:136–146.

Marsono, Y., R. J. Illman, J. M. Clarke, R. P. Trimble, and D. L. Topping. 1993. Plasma lipids and large bowel volatile fatty acids in pigs fed white rice, brown rice and rice bran. Br. J. Nutr. 70:503–513.

Mertens, D. R. 2003. Challenges in measuring insoluble dietary fiber. J. Anim. Sci. 81:3233–3249.

Miner–Williams, W., P. J. Moughan, and M. F. Fuller. 2009. Endogenous components of digesta protein from the terminal ileum of pigs fed a casein–based diet. J. Agric. Food Chem. 57:2072–2078.

Moeser, A. J., and T. A. T. G. van Kempen. 2002. Dietary fibre level and enzyme inclusion affect nutrient digestibility and excreta characteristics in grower pigs. J. Sci. Food Agric. 82:1606–1613.

Montagne, L., C. Piel, J. P. Lallés. 2004. Effect of diet on mucin kinetics and composition: Nutrition and health implications. Nutr. Rev. 62:105–114.

Moore, R. J., E. T. Kornegay, R. L. Grayson, and M. D. Lindemann. 1988. Growth, nutrient utilization and intestinal morphology of pigs fed high–fiber diets. J. Anim. Sci. 66:1570–1579.

Morel, P. C. H., T. S. Lee, and P. J. Moughan. 2006. Effect of feeding level, liveweight and genotype on the apparent faecal digestibility of energy and organic matter in the growing pig. Anim. Feed. Sci. Technol. 126:63–74.

Mosenthin, R., W. C. Sauer, and F. Ahrens. 1994. Dietary pectin's effect on ileal and fecal amino acid digestibility and exocrine pancreatic secretions in growing pigs. J. Nutr. 124:1222–1229.

Mroz, Z., A. J. Moeser, K. Vreman, J. T. van Diepen, T. van Kempen, T. T. Canh, and A. W. Jongbloed. 2000. Effects of dietary carbohydrates and buffering capacity on nutrient digestibility and manure characteristics in finishing pigs. J. Anim. Sci. 78:3096–3106.

Müller, V. 2008. Bacterial Fermentation. Pages 1–8 in Encyclopedia of Life Sciences. John Wiley & Sons, Ltd: Chichester. doi: 10.1002/9780470015902.a0001415.pub2

Ndindana, W., K. Dzama, P N. B. Ndiweni, S. M. Maswaure, and M. Chimonyo. 2002. Digestibility of high fibre diets and performance of growing Zimbabwean indigenous Mukota pigs and exotic Large White pigs fed maize based diets with graded levels of maize cobs. Anim. Feed Sci. Technol. 97:199–208.

Nelson, D. L., and M. M. Cox. 2008. Carbohydrates and glycobiology Pages 235–241 in Lehninger Principles of Biochemistry. 5th ed. W. H. Freeman New York, NY.

Noblet, J., H. Fortune, X. S. Shi, and S. Dubois. 1994. Prediction of net energy value of feeds for growing pigs. J. Anim. Sci. 72:344–354.

NRC. 2007. Nutrient Requirement of Horses. National Academies Press. Washington, DC.

Nunes, C. S., and K. Malmlöf. 1992. Effects of guar gum and cellulose on glucose absorption, hormonal release and hepatic metabolism in the pig. Br. J. Nutr. 68:693–700.

Oakenfull, D. 2001. Physical chemistry of dietary fiber. Pages 33–47 in Dietary Fiber in Human Nutrition. 3rd ed. G. A. Spiller, ed. CRC Press, Boca Raton, FL.

Owusu–Asiedu, A., J. F. Patience, B. Laarveld, A. G. van Kessel, P. H. Simmins, and R. T. Zijlstra. 2006. Effects of guar gum and cellulose on digesta passage rate, ileal microbial populations, energy and protein digestibility, and performance of grower pigs. J. Anim. Sci. 84:843–852.

Pastuszewska, B., Kowalczyk, and J. Ochtabinska, 2000. Dietary carbohydrates affect caecal fermentation and modify nitrogen excretion patterns in rats. Arch. Anim. Nutr. 53:207–225.

Piel, C., L. Montagne, B. Séve, and J. P. Lallés. 2005. Increasing digesta viscosity using carboxymethylcellulose in weaned piglets stimulates ileal goblet cell numbers and maturation. J. Nutr. 135:86–91.

Prosky, L., G. N. Asp, I. Furda, J. W. de Vries, T. F. Schweizer, and B. F. Harland. 1984. Determination of total dietary fiber in foods, food products, and total diets: Interlaboratory study. J. Assoc. Off. Anal. Chem. 67:1044–1052.

Rainbird, A. L., A. G. Low, and T. Zebrowska. 1984. Effect of guar gum on glucose and water–absorption from isolated loops of jejunum in conscious growing pigs. Br. J. Nutr. 52:489–498.

Robertson, M. D. 2007. Metabolic cross talk between the colon and the periphery: Implications for insulin sensitivity. Proc. Nutr. Soc. 66:351–361.

Robertson, J. B., and P. J. Horvath. 2001. Detergent analysis of foods. Page 63 in Dietary Fiber in Human Nutrition. 3rd ed. G. A.

Spiller, ed. CRC Press, Boca Raton, FL. Robertson, J. A., F. D. de Monredon, P. Dysseler, F. Guillon, R. Amado, and J. F. Thibault. 2000. Hydration properties of dietary fibre and resistant starch: A European collaborative study. Lebensmittel–Wissenschaft und –Technol. 33:72–79.

Satchithanandam, S., M. Vargofcak-Apker, R. J. Calvert, A. R. Leeds, and M. M. Cassidy. 1990. Alteration of Gastrointestinal Mucin by Fiber Feeding in Rats. J. Nutr. 120:1179–1184.

Sauer, W. C., and L. Ozimek. 1986. Digestibility of amino acids in swine: Results and their practical applications. A review. Livest.Prod. Sci. 15:367–388.

Sauer, W. C., R. Mosenthin, F. Ahrens, and L. A. den Hartog. 1991. The effect of source of fiber on ileal and fecal amino acid digestibility and bacterial nitrogen excretion in growing pigs. J. Anim. Sci. 69:4070–4077.

Scheneeman, B. O. 1998. Dietary fiber and gastrointestinal function. Nut. Res. 18:625–632.

Schulze, H., P. van Leeuwen, M.W. Verstegen, J. Huisman,W. B. Souffrant, and F. Ahrens. 1994. Effect of level of dietary neutral detergent fiber on ileal apparent digestibility and ileal nitrogen losses in pigs. J. Anim. Sci. 72:2362–2368.

Schulze, H., P. van Leeuwen, M. W. Verstegen, and J. W. van den Berg. 1995. Dietary level and source of neutral detergent fiber and ileal endogenous nitrogen flow in pigs. J. Anim. Sci. 73:441–448.

Shelton, D. R., and W. J. Lee. 2000. Cereal carbohydrates. Page 385–416 in Handbook of Cereal Science and Technology. 2nd ed. K. Kulp and J. G. Ponte, eds. Marcel Dekker, New York, NY.

Souffrant,W. B., A. Rerat, J. P. Laplace, B. Darcy-Vrillon, R. Kohler, T. Corring, and G. Gebhart. 1993. Exogenous and endogenous contributions to nitrogen fluxes in the digestive tract of pigs fed a casein diet. III. Recycling of endogenous nitrogen. Reprod. Nutr. Dev. 33:373.

Stephen, A. S., and J. H. Cummings. 1979. Water-holding by dietary fibre *in vitro* and its relationship to faecal output in man. Gut 20:722–729.

Takahashi, T., and T. Sakata. 2002. Large particles increase viscosity and yield stress of pig cecal contents without changing basic viscoelastic properties. J. Nutr. 132:1026–1030.

Takahashi, T., Y. Furuichi, T. Mizuno, M. Kato, A. Tabara, Y. Kawada, Y. Hirano, et al. 2009. Water-holding capacity of insoluble fibre decreases free water and elevates digesta viscosity in the rat. J. Sci. Food Agric. 89:245–250.

Tako, E. R. P. Glahn, R. M. Welch, X. Lei, K. Yasuda, and D. D. Miller. 2008. Dietary inulin affects the expression of intestinal enterocyte iron transporters, receptors, and storage protein and alters the microbiota in the pig intestine. Br. J. Nutr. 99:472– 480.

Tanabe, H., H. Ito, K. Sugiyama, S. Kiriyama, T. Morita. 2006. Dietary indigestible components exert different regional effects on luminal mucin secretion through their bulk-forming property and fermentability. Biosci. Biotechnol. Biochem. 70:1188– 1194.

Thangaraju, M., G. Cresci, S. Itagaki, J. Mellinger, D. D. Browing, F. G. Berger, P. D. Prasad, and V. Ganapathy. 2008. Sodiumcoupled transport of the short chain fatty acid butyrate by SLC5A8 and its relevance to colon cancer. J. Gastrointest. Surg.12:1773–1782.

Theander, O., and P. Aman. 1979. The chemistry, morphology, and analysis of dietary fiber components. P 215–244 in Dietary Fiber Chemistry and Nutrition. G. E. Inglett, and S. I. Falkehag, eds. Academic Press, New York, NY.

Topping, D. L., and P. M. Clifton. 2001. Short chain fatty acids and human colonic functions: Roles of resistant starch and non-starch polysaccharides. Physiol. Rev. 81:1031–1064.

Tso, P., and K. Crissinger. 2000. Overview of digestion and absorption. Pages 75–106 in Biochemical and Physiological Aspects of Human Nutrition. M. H. Stipanuk, ed. Saunders. Philadelphia, PA.

Urriola, P. E., G. C. Shurson, H. H. Stein. 2009. Digestibility of dietary fiber in distillers co-products fed to growing pigs. J. Anim. Sci. 87(Suppl. 3):145. (Abstr.)

vanhoof, K., and R. De Schrijver. 1996. Availability of minerals in rats and pigs fed non-purified diets containing inulin. Nutr. Res. 16:1017–1022.

van Soest, P. J., J. B. Robertson, and B. A. Lewis. 1991. Symposium: Carbohydrate methology, metabolism, and nutritional implications in dairy cattle. Methods for dietary fiber, neutral dietary fiber, and nonstarch polyssacharides in relation to animal nutrition. J. Dairy Sci. 74:3583–3597.

Varel, V. H., H. G. Jung, and W. G. Pond. 1988. Effects of dietary fiber of young genetically lean, obese, and contemporary pigs: Rate of passage, digestibility, and microbial data. J. Anim. Sci. 66:707–712.

Welter, H., and R. Claus. 2008. Expression of the monocarboxylate transporter 1 (MCT1) in cells of the porcine intestine. Cell Biol. Int. 32:638–645.

Wenk, C. 2001. The role of dietary fibre in the digestive physiology of the pig. Anim. Feed Sci. Technol. 90:21–33.

White, D. The Physiology and Biochemistry of Prokaryotes. 2nd ed. Oxford University Press. New York, NY.

Wilfart, A., L. Montagne, H. Simmins, J. van Milgen, and J. Noblet. 2007. Sites of nutrient digestion in growing pigs: Effects of dietary fiber. J. Anim. Sci. 85:976–983.

Wong, J. M., R. de Souza, C. W. Kendall, A. Emam, and D. J. Jenkins. 2006. Colonic health: Fermentation and short chain fatty acids. J Clin Gastroenterol. 40:235–243.

Yasuda, K., K. R. Roneker, D. D. Miller, R. M. Welch, and X. G. Lei. 2006. Supplemental dietary inulin affects bioavailability of iron present in corn and soybean meal to young pigs. J. Nutr. 136:3033–3038.

Yen, J. T. 2001. Anatomy of the digestive system and nutritional physiology. Pages 31–63 in Swine Nutrition. A. J. Lewis and L.

第12章

酶制剂及其在猪饲粮中的应用

1 导 语

本书对饲粮中添加酶制剂的概念有了很好的说明。因为在实际生产中需要给动物提供营养充足的饲粮，这样会造成大量未被消化利用的养分排泄到环境中，同时造成饲料资源耗竭。例如，通过添加磷酸盐来满足动物对磷的需要，会造成大量的磷排出体外，并引起磷资源的不断消耗。由于饲料资源是有限的，且植物能有效利用动物排泄物的能力也是有限度的，因此目前为了满足动物需要所采用的饲养方法是不可持续的。

使用外源酶制剂能降低养猪业对非再生饲料资源的依赖，并减少养分排泄。同时，使用外源酶制剂还可以提高生长速度（每千克增重所消耗的养分更少），降低排泄物的处理费用（养分排泄量更少），减少矿物质（如磷酸氢钙）的添加量，最终降低养猪生产的成本。

为了阐述添加酶制剂在本书中所发挥的重要作用，本章首先介绍与猪消化过程有关的低效性，然后再讨论酶制剂在养猪生产中的应用及其对生长和养分利用的影响，最后探讨酶制剂在猪营养中的应用前景。

2 猪消化过程的简要概述

猪的消化过程从口腔开始，但口腔的消化作用并不很明显，因为口腔内只分泌非常少量的唾液α-淀粉酶（Corring，1980）。猪的胰外分泌腺可以分泌大量消化酶，包括碳水化合物酶、蛋白酶、脂肪酶和核酸酶（Cranwell，1995），这些酶用于消化碳水化合物、蛋白质和脂肪。在刷状缘酶或在刷状缘膜上消化酶的作用下，将胰酶消化产物最后降解成简单的糖类（单体单元）。

延长食糜在大肠内停留的时间有利于微生物发酵。大肠微生物产生的酶能降解动物摄入的纤维（Yen，2001）。挥发性脂肪酸（VFA）是碳水化合物和蛋白质微生物发酵的最重要产物，能为成年猪提供高达30%的维持能量需要（Rérat等，1987）和菌体氨基酸（AA）。

一般情况下，猪能很好消化碳水化合物、蛋白质和脂类，因为它本身能分泌大量的消化酶，可将这些复杂的化合物降解成各种结构单元。尽管如此，养分的消化吸收效率并不能达到100%（图12-1），所以饲粮中的部分养分最终会被排出体外。饲粮组成、动物生长发育阶段、采食量等诸多因素均会影响猪对养分和能量的利用效率，因此在可持续的养猪生产中必须考虑到养分的利用效率。

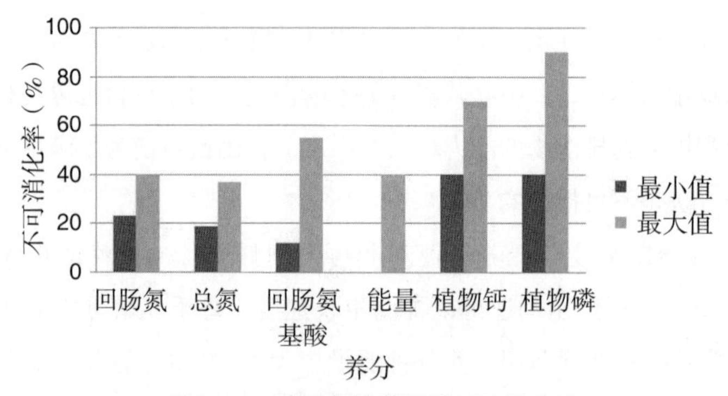

图12-1 猪对部分养分的不可消化率

能量消化率的数据是指蛋白质、脂肪和碳水化合物在猪体内被转化为能量的消化率，不可消化率（%）是由100%减去消化率（%）计算而来。（资料来源于Kornegay，1985；Crenshaw，2001；Noblet，2001；Han等，2003）

如图12-1所示，在不添加外源酶制剂时，植物中不可消化的钙和磷可高达90%，而氮和能量的消化率相对较高（Kornegay，1985；Crenshaw，2001；Noblet，2001；Han等，2003）。除饲料成本外，不可消化养分所涉及的费用还包括这些养分被排放到环境中的处理费用。假设猪的增重/耗料比为0.3，仅将磷的添加量降低0.1%，在猪整个生长周期磷的排放量能减少约20%（Knowlton等，1994）。这些潜在的不可利用的营养物质过量供给，会造成养猪经济成本的增加，而添加外源酶制剂可以降低养分的不可消化率。因此，通过提高养分利用率来减少饲料用量、并降低单位时间或单位增重所排出的粪便量的方法是可行的。

3 猪饲料中的抗营养因子

猪饲粮中大量使用的谷物类、豆类和油籽类饲料，是饲粮中非淀粉多糖（NSP）和植酸盐的主要来源。黑麦、小麦和黑小麦中的NSP主要是阿拉伯木聚糖，而燕麦和大麦中的NSP主要是β-葡聚糖（Bach Knudsen和Hansen，1991）。谷物类饲料中的纤维造成的不利影响似乎比豆类饲料的更大（Fernandez和Jorgensen，1986），但从数量来看，植酸盐可能是谷物类饲料中最重要的抗营养因子。虽然大豆的植酸含量是很多谷物类饲料的2倍，但与豆类饲料相比，谷物类饲料的植酸磷占总磷的比例更高，有些甚至高达70%（Eeckhout和Paepe，1994）。

NSP会对有效能的利用产生负面影响，其原因主要包括：首先，猪自身不能分泌一些降解复杂细胞壁结构的酶类；其次，增加饲粮纤维含量，会降低养分浓度，增加动物采食量，并提高食糜通过速度，最终导致养分和能量利用率的降低（Kass等，1980）。另外，NSP还会对蛋白质和脂肪的消化率产生负面影响（Shi和Noblet，1993）。

可溶性NSP的化学特性决定了其更容易发生分子“交联”和形成胶状物，进而抑制养分消化，减缓食糜通过速度（Härkönen等，1997）。食糜黏性对猪的负面作用并不如家禽那样明显（Bartelt等，2002），因为食糜在猪肠道内的滞留时间更长，猪肠道微生物对食糜的自然发酵作用也更强，且食糜水分含量相对较低（Bedford和Schulze，1998）。但增加猪饲粮的NSP浓度，会使小肠内的微生物菌群数量增加（Bartelt等，2002），进而增强宿主与小肠微生物对养分的竞争，最终导致宿主可利用养分的减少及新陈代谢需要的增加（Just等，1979；Yin等，2000）。

高NSP饲粮中含有大量不可消化的养分，可能是造成氨基酸（AA）、蛋白质、脂类和矿物质消化率降低的主要原因（Graham等，1998；Torre等，1991；Myrie等，2008）。这仅仅是因为它们对营养物质的物理性包裹阻碍了养分的消化吸收（Bedford等，1992）。而NSP的另外一些副作用是由纤维造成的脂肪和氮的内源损失增加（Goff和Noblet，2001；Bartelt等，2002），或是由NSP引起的菌群数量增加产生的间接影响（Yin等，2000）。

尽管NSP对猪矿物质利用有不利影响，但其作用方式目前仍不清楚。有些学者提出，离子型纤维与矿物质间的相互作用会降低矿物质的利用效率（Torre等，1991）。也有学者认为，NSP可能通过阻碍养分与肠道吸收绒毛间的接触来降低矿物质的利用，而植酸也在其中发挥着重要作用。

植酸，又称肌醇1,2,3,4,5,6-六磷酸（IP6），以盐形式存在，被称为植酸盐。几乎所有的植物性饲料都含有植酸。植酸的化学结构及与其他矿物质和养分的结合位置如图12-2所示。在谷物类饲料中，玉米的植酸磷含量最高；豆粕和葵花籽粕的植酸磷含量可分别高达69%和80%（Weremko等，1997；Steiner等，2007）。

植酸最主要的危害作用是能结合六个磷酸基，非反刍动物由于自身不能分泌植酸酶，故无法利用植酸磷中的磷。在猪消化道pH变化范围内，植酸盐可以带有1～6个负电荷，这使它可与多种二价阳离子形成不溶性的配体-金属复合物，造成磷和结合后的矿物质不能被猪利用（Bebot-Brigand等，1999）。而在猪饲粮中添加无机形式的磷和钙，不仅会增加饲料成本，还会提高不可利用的钙和磷的排放量。

此外，植酸还能直接与蛋白质结合形成复合物，也可以与钙、蛋白质一起形成复合物（Cheryan，1980）。小麦比玉米更容易形成这类复合物（Champagne，1988；Selle等，2003）。植酸盐可以通过与消化酶的直接结合，或与消化酶活化所必需的离子结合，来降低胰蛋白酶（Singh和Krikorian，1982）和羧肽酶A（Martin和Evans，1989）等消化酶的功能。

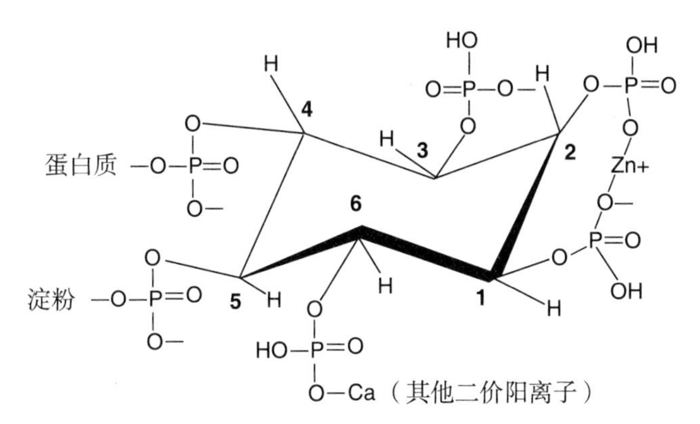

图12-2 植酸与一价或二价阳离子、淀粉、蛋白质结合的化学结构

植酸的6个碳原子上都含有磷酸基，其中5个在轴向位置，1个在赤道位置（碳原子2）

4 猪肠道微生物作用的局限性

肠道微生物有助于提高猪对植物性饲料的利用能力，包括那些NSP含量高的饲料。到盲肠末端，豌豆中有高达84%的NSP可以被消化，小麦中也有高达65%的NSP可以被消化（Goodlad和Mathers，1991）。猪对有机物的总消化率平均约有17%是由大肠完成（Wilfart等，2007），这个数值在13%（低NDF饲料）到32%（高NDF饲料）范围内变化（Shi和Noblet，1993）。尽管所有被消化的NSP不可能全部被猪代谢利用，但大量NSP能在猪消化道内被降解，这进一步说明NSP对猪的影响比对家禽的要更小。

虽然大量NSP在母猪大肠内被降解，但Shi和Noblet（1993）却发现，由NSP降解产物提供的能量可以忽略不计，有时甚至是负值。这是由于微生物发酵产生的VFA利用效率非常低，而且纤维还会影响其他养分的利用。因此，将养分消化利用的位点转移到消化道前段对机体更有益处，这也是使用外源酶制剂的主要优点之一。

5 酶制剂在猪营养中的应用

植物性饲料中不同的抗营养组分可以使用各种不同的外源酶制剂。植酸酶是最为熟知的一种外源酶制剂，且使用的历史时间最长；而NSP水解酶是近来从20世纪90年代开始才被广泛使用。表12-1列出了一些在猪营养中普遍使用的外源酶制剂特性，可以清楚看到植酸酶是一种研究最为透彻的外源酶制剂。

表12-1 猪营养中广泛使用的酶制剂特性[3]

来源	特性	缺点
植酸酶		
真菌 曲霉属	1. 从肌醇环的第 3 个碳原子开始去磷酸化 2. 抗胰蛋白酶 3. 不确定对三磷酸肌醇或者含磷酸基更少的肌醇有去磷酸化作用 4. 终产物是单磷酸肌醇（C2 上带有一个磷酸基） 5. 适宜 pH 范围较窄	1. 对胃蛋白酶敏感 2. 对低分子量肌醇的水解活性较低
[1]细菌 大肠杆菌 芽孢杆菌	1. 从肌醇环的第 6 个碳原子开始去磷酸化（大肠杆菌） 2. 抗胃蛋白酶 3. 适宜 pH 范围较宽，酶活性在消化道中保留的时间更长 4. 耐热性更强 5. 可以释放所有的磷酸基	对胰蛋白酶敏感
植物源植酸酶	1. 从肌醇环的第 6 个碳原子开始去磷酸化 2. 花粉源植酸酶从肌醇环的第 5 个碳原子开始去磷酸化	1. 与真菌和细菌源植酸酶相比，对蛋白酶的耐受性较低 2. 耐热性差；制粒过程中酶活性更易失活 3. 酶解作用的 pH 范围较窄；在消化道转运过程中更易失活 4. 对三磷酸肌醇或者含磷酸基更少肌醇的去磷酸化能力较低

（续）

来源	特性	缺点
非淀粉多糖水解酶[2]		
真菌 曲霉属 细菌 芽孢杆菌 植物源	1. 对 NSP 的特异性强 2. 微生物源木聚糖酶在消化道中对钝化有耐受作用	NSP 水解酶在猪上的应用效果通常不如植酸酶那么显著
淀粉酶		
真菌 曲霉属 细菌 芽孢杆菌	较胰淀粉酶有更高的淀粉水解活性	

[1]绝大多数特性仅适用于大肠杆菌源植酸酶。

[2]NSP水解酶的更详细资料参考Henrissat和Bairoch（1993）；表中未列出的其他酶类包括微生物源的蛋白酶和脂肪酶，主要来源于芽孢杆菌。

植酸酶是肌醇六磷酸水解酶，能水解植酸。植酸酶可分为植物源和微生物源两大类，不同来源的植酸酶均有不同的特性和酶活性。如图12-3所示，植酸酶可以分步水解植酸，首先生成较低分子量的肌醇（肌醇五磷酸和肌醇四磷酸），最终水解为三磷酸肌醇、二磷酸肌醇和单磷酸肌醇（Frølich，1990；Nagashima等，1999）。一般来说，越低分子量的肌醇（三磷酸肌醇或者磷酸基更少的肌醇）对植酸酶活性的耐受性就越强，因此在消化道内植酸水解的终产物并不一定是无磷酸肌醇。

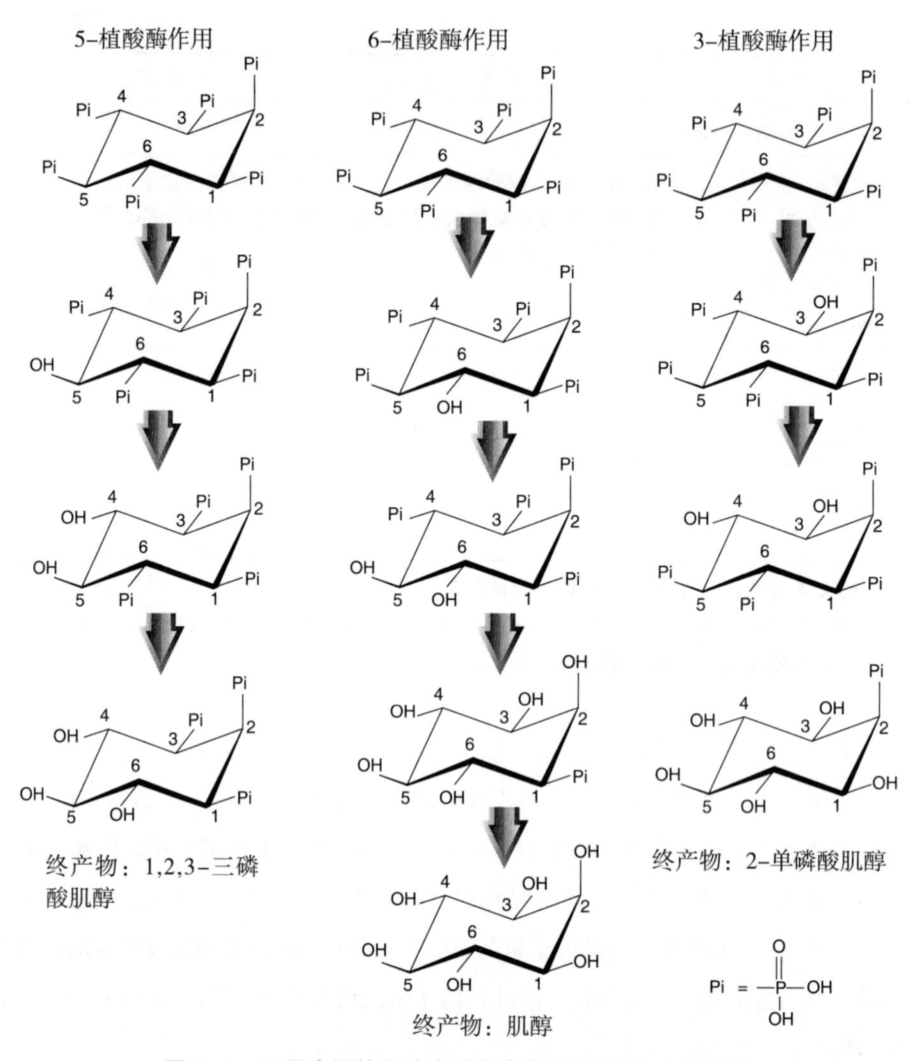

图12-3　不同来源的植酸酶对肌醇磷酸盐的分步水解过程

在酶解过程中，四磷酸肌醇和五磷酸肌醇迅速生成，而含磷酸基更少的肌醇生成缓慢，说明低分子量的肌醇对植酸酶的水解作用有更强的耐受性。尽管从化学理论上推理，存在生成无磷酸肌醇的可能性，但在消化道内生成无磷酸肌醇的可能性微乎其微。（资料来源于Irving和Cosgrove，1972；Barrientos等，1994；Dvořáková等，2000；Nakano等，2000；Bohn等，2007）

不同来源的植酸酶在水解耐受性更强的低分子量肌醇（三磷酸肌醇或者磷

酸基更少的肌醇）方面的能力不尽相同，且对蛋白酶钝化的耐受力也不同。例如，与烟曲霉源植酸酶相比，黑曲霉源植酸酶去除三磷酸肌醇和二磷酸肌醇中磷酸基的功能更弱（Wyss等，1999）；与黑曲霉源植酸酶相比，从猪结肠分离到的大肠杆菌分泌的植酸酶对胃蛋白酶降解的耐受力更强（Rodriguez等，1999）。植物花粉源植酸酶从肌醇环的第5个碳原子开始去磷酸化，水解终产物是三磷酸肌醇（图12-3；Irving和Cosgrove，1972；Barrientos等，1994；Dvořáková等，2000；Nakano等，2000；Bohn等，2007）。Wyss等（1999）研究发现，真菌或细菌源植酸酶不能将植酸水解成肌醇2-单磷酸。因此，在动物饲粮中使用不同特性的植酸酶组合，可能使其作用得到最大程度的发挥。

虽然猪饲粮中使用的外源性植酸酶主要是植物源植酸酶，但其对猪的作用很有限。黑麦、小麦和大麦含有较高活性的植酸酶，而玉米、燕麦和豆科植物中的植酸酶活性非常低（Ravindran等，1995；Steiner等，2007）。由于植物源植酸酶发挥其活性的适宜pH范围较窄，因此它们在消化道内发挥的酶解作用就较低。此外，植物源植酸酶发挥活性的适宜温度临界值较低，它们在制粒等过程进行热处理时很容易失活（Phillippy，1999）。尽管如此，在使用小麦和黑麦进行饲喂且添加外源性植酸酶的实际作用甚微的情况下，植物源植酸酶依然很重要。

植酸酶的作用效果受到饲粮中钙与总磷比例的影响（Lei等，1994）。钙可能通过与植酸结合形成稳定的钙-植酸盐复合物来影响植酸酶的作用效果（Fisher，1992）。一些研究表明，当钙与总磷的比例降低时，添加植酸酶能增加磷的吸收（Liu等，2000；Adeola等，2006）。Liu等（2000）提出，当钙与总磷的比例为1：1时，植酸酶活性最高。

与植酸不同，由于NSP是一类由不同的组分和化学键组成的复合物，因此需要很多种NSP酶才能将它们进行水解。值得注意的是，NSP水解酶对猪的作用效果并不像植酸酶的那么一致。

木聚糖内切酶属于糖基水解酶家族，分布在第10或第11家族（Henrissat和Bairoch，1993）。与第11家族木聚糖内切酶相比，第10家族木聚糖内切酶的分子量更高，结构更为复杂，导致其底物特异性更低。纤维素酶和木聚糖酶都有一个催化结构域和一个或多个非催化结构域（Fontes等，2004）。非催化结构域可促进酶与底物的接触，并延长接触时间（Reilly，1981；Tervilä-Wilo等，1996）。

还有细菌和真菌源的糖基水解酶（Henrissat和Bairoch，1993）。微生物源木聚糖酶对猪消化道钝化作用的耐受力非常大，而β-葡聚糖酶的耐受力却较小（Inborr等，1999）。小麦、大麦和黑麦中含有较高的内源性木聚糖内切酶活性（Courtin和Delcour，2002）。Dornez等（2006）指出，谷物中内源性木聚糖酶总活性的90%来自微生物。消化道中栖居的微生物可能是糖基水解酶的来源，这些酶类可能会降低外源酶制剂的作用（Inborr等，1999）。其他NSP水解酶还包括纤维素酶、甘露聚糖酶、果胶酶和半乳糖苷酶。

尽管非反刍动物自身能分泌淀粉酶，但有时在猪饲粮中还会添加微生物源淀粉酶。Planchote等（1995）研究表明，与芽孢杆菌或曲霉菌源淀粉酶相比，猪胰腺分泌的淀粉酶对纯淀粉颗粒的水解作用更低。在猪饲粮中也会单独添加蛋白酶，或与其他酶混合添加（Caine等，1997；Olukosi等，2007a）。但蛋白酶对猪的作用效果并不一致（Caine等，1997；Dierick等，2004）。

针对饲粮中不同的抗营养因子，使用不同的酶进行组合添加，有时会比单独添加这些单体酶的效果更好，因为不同酶之间可能会协同互作。例如，添加碳水化合物酶水解细胞壁后，能使植酸酶与植酸盐得到更充分的接触，进而提高植酸酶的作用（Parkkonen等，1997；Simon，1998）。

◆ 5.1 酶制剂在仔猪上的应用

仔猪本身需要大量的养分，但其消化道发育尚不成熟，这使得它们成为研究外源酶制剂的理想对象。由于仔猪的养分消化吸收率低，被排泄的养分就更多，而且抗营养因子对仔猪的负面作用更加明显，因此可以推测，在仔猪上使用外源酶制剂的作用效果要优于年龄更大的猪。如图12-4所示，Shelton等（2005）发现，添加植酸酶仅仅提高了仔猪生长保育阶段的日增重；但外源酶制剂对仔猪的作用效果受到诸多因素的影响，并不都是表现日龄依赖性（Shelton等，2005；Jendza等，2005；Brana等，2006）。关于植酸酶和NSP水解酶在仔猪饲粮中的应用已开展了大量研究，在此仅介绍少数代表性研究。

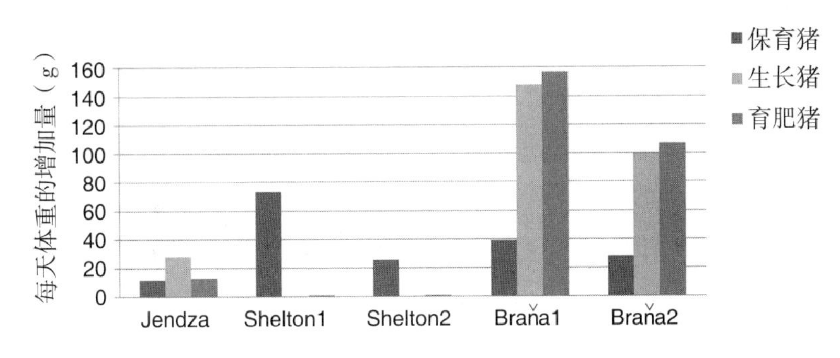

图12-4 植酸酶促进仔猪和生长育肥猪的生长作用

促生长作用是根据植酸酶添加组的日增重减去对照组的日增重之差计算得到的。在所有的研究当中，如果添加多个植酸酶水平，是将对照组和最高水平植酸酶组的生长性能进行比较。Shelton1：添加矿物质预混料的对照组与添加500 FTU/kg的植酸酶组进行比较（植酸酶不影响生长育肥猪的日增重）。Shelton2：不添加矿物质预混料组与植酸酶组进行比较（植酸酶不影响生长育肥猪的日增重）。Brañ a1：负对照组与添加大肠杆菌源植酸酶组进行比较。Brañ a2：负对照组与黑曲霉源植酸酶组进行比较。

低磷饲粮中添加黑曲霉源植酸酶（Lei等，1993；Han等，1997）、大肠杆菌源植酸酶（Sands等，2001；Jendza等，2005）或隔孢伏革菌源植酸酶（Adeola等，2006），均能提高仔猪的生长性能。植酸酶的促生长作用是由增加其他矿物元素或氨基酸的释放所引起的，但锌（Adeola等，1995）、铜（Adeola，1995）或赖氨酸（Selle等，2003）的作用甚微。

Young等（1993）的研究表明，饲粮中添加植酸酶对猪生长性能的提高幅度，与补充0.17%无机磷的作用效果相当。Han等（1997）在猪玉米-豆粕型饲粮中添加植酸酶而不添加无机磷，发现该处理组的增重与添加足量的无机磷组相似。他们认为，若在仔猪饲粮中使用植酸酶，就可以不添加无机磷；但需注意的是，他们研究所用的不是无磷饲粮。Han等（1997）试验饲粮的总磷水平与其他很多试验中所采用的低磷饲粮的总磷水平基本相同（0.34%）。

添加植酸酶在提高猪生长性能的同时，也能增加养分消化率，降低养分排泄。如表12-2所示，植酸酶对磷消化率的提高幅度最大，超过基础饲粮组的1倍（Jendza等，2005）。植酸酶对钙消化率的提高幅度仅次于磷（Lei等，1993；Jendza等，2005；Olukosi等，2007a），而对其他养分消化率的提高幅度一般都很低，或者几乎没有提高，即使对蛋白质缺乏的饲粮也是如此（Liao等，2005；Sands等，

2009）。在常规饲料或低植酸盐含量的饲粮中添加植酸酶均能降低磷的排泄量，其降低量有时高达40%（Han等，1997；Sands等，2001；Baxter等，2003）。低磷饲粮中添加植酸酶能通过增加磷的沉积而提高骨骼矿化（Augspurger等，2003；Pagano等，2007），但不会影响骨骼组织化学特性，这与使用磷充足饲粮的作用效果类似（Pagano等，2007）。

表12-2　断奶仔猪饲粮中添加酶制剂对肠道养分和能量总消化率的影响[1]

干物质	氮	磷	钙	能量	参考文献
植酸酶					
N/A	N/A	48.7	17.9	N/A	Lei 等，1993
N/A	3.75	27.8	0.7	N/A	Han 等，1997
0.68	0.17	105.9	19.4	N/A	Jendza 等，2005
0.12	−1.12	46.7	28.6	−1.9	Olukosi 等，2007a
非淀粉多糖水解酶					
2.1	6.7	N/A	N/A	0.1	Li 等，1996[2]
1.2	1.0	N/A	N/A	1.9	Li 等，1996[3]
1.0	0.7	N/A	N/A	0.7	Yin 等，2000[4]
2.6	1.4	26.1	5.6	0.4	Olukosi 等，2007a[5]

[1]养分消化率的提高率是根据[（加酶饲粮养分消化率−对照饲粮养分消化率）]/ 对照饲粮养分消化率×100的公式计算得到。

[2]裸大麦型饲粮中添加β−葡聚糖酶。

[3]小麦型饲粮中添加β−葡聚糖酶。

[4]小麦型饲粮中添加木聚糖酶。

[5]玉米−小麦麸型饲粮中添加木聚糖酶、淀粉酶和蛋白酶。

外源性NSP酶有时也能改善猪的生长性能，提高养分利用率；但与植酸酶相比，外源性NSP酶对磷和钙利用率的提高幅度普遍偏低（表12−2）。有些研究表明，添加NSP酶能改善生长性能（Bedford等，1992；Diebold等，2004；Fan等，2009），但其他一些研究却没有得到类似结果（Bedford等，1992；Mavromichalis等，2000；Olukosi等，2007b）。造成以上结果差异的原因可能是各试验添加的酶制剂在提高养分利用率方面的功能不同，最终造成对基础饲粮养分利用率的改善效果也不同。

例如，Bedford等（1992）发现木聚糖酶和β-葡聚糖酶都不能提高黑麦或大麦的淀粉消化率，但β-葡聚糖酶能降解豆粕中的β-葡聚糖，进而提高氮的消化率。因此可以推出，木聚糖酶不能改善生长性能，但β-葡聚糖酶却可以提高生长性能。在养分充足条件下，添加β-葡聚糖酶能提高猪对大麦型基础饲粮中蛋白质、氨基酸和能量的回肠消化率，但对小麦型基础饲粮没有这种作用（Li等，1996）。Yin等（2000）的研究表明，小麦型基础饲粮中添加木聚糖酶能提高干物质、粗蛋白质和能量的回肠消化率，提高幅度约4%；同时还能小幅度地提高部分必需氨基酸和非必需氨基酸的回肠消化率。在养分充足的小麦型基础饲粮中添加木聚糖酶对养分消化率的作用效果并不一致（Mavromichalis等，2000）。Olukosi等（2007b）报道，在黑麦和小麦型基础饲粮中添加木聚糖酶，其添加量最高达到32 000IU/kg时，猪对不同饲粮养分消化率的影响也不同，并给出了能够明显影响猪生长的木聚糖酶最低添加量。

有些研究表明NSP水解酶不发挥作用，可能是因为NSP在猪后肠内被微生物发酵。Högberg和Lindberg（2004）研究表明，给仔猪饲喂NSP含量高的饲粮时，β-葡聚糖的回肠和盲肠消化率都很高（72%～93%），所以额外添加酶制剂对消化率进一步提升的空间就很有限。然而，NSP水解酶既可以通过降低养分和氨基酸的内源损失来改善养分利用率（Yin等，2000），也可以通过降解细胞壁，将养分消化利用的位置转移到消化道更前端（Li等，1996），增加肠道中段消化酶的降解作用，进而改善养分利用率。

复合酶制剂的使用效果并不一定优于单一酶制剂。在养分充足的小麦型基础饲粮中，添加木聚糖酶和磷脂酶能提高生长性能，但对养分和氨基酸的回肠消化率和全肠道消化率均无明显影响（Diebold等，2004）。Olukosi等（2007b）研究表明，将植酸酶与木聚糖酶、淀粉酶和蛋白酶复合酶组合添加，能提高消化能和磷的消化率。但总体而言，这种组合添加的效果并不比单独使用植酸酶的好。

根据饲粮中的限制性养分有针对性地使用酶制剂，对于断奶仔猪而言非常重要，因为只有这样才能使酶制剂的作用得到最大程度的发挥。绝大多数研究表明，植酸酶能提高钙和磷的利用率，因此可以降低基础饲粮中这两种物质的添加量，进而减少无机磷的使用量和养分排泄量，这完全符合可持续性养猪生产的要求。酶制剂在提高能量、氮和氨基酸利用率方面的作用效果很少一致，因此在推荐使用时需要特别注意。NSP酶对生长和养分利用的影响不一致，导致很难对其使用效果得出明确的结论。然而，针对谷物中特定的NSP使用专门的NSP酶，其作用效果可能会很好（如在大麦型饲粮中添加β-葡聚糖酶，在黑麦型饲粮中添加木聚糖酶）。

◆ 5.2 酶制剂在生长育肥猪上的应用

因为商品猪的生长育肥期要比保育期长得多，所以猪在生长育肥阶段摄入的养分总量非常多。若在这个阶段的饲粮中添加酶制剂，对减少养分的总摄入量会产生重要影响。事实上，在生长育肥阶段使用酶制剂对猪增重的影响效果要比在保育阶段的更显著，如图12-4所示。

有些研究表明，植酸酶能提高猪的生长性能（Cromwell等，1993；Matsui等，2000），而另外一些研究则没有得到类似结果（Olukosi等，2007b；Woyengo等，2008）。部分原因可能与这些研究中所使用的谷物种类不同有关，因为在小麦饲粮中添加植酸酶的效果不明显（可能是由于小麦中内源植酸酶活性很高）。对于玉米型基础饲粮，在钙和磷的添加量分别降低37%和42%（Adeola等，1998），或者不添加无机磷（Han等，1998）的情况下，添加植酸酶完全能达到常规饲粮的效果。Cromwell等（1993）和Matsui等（2000）报道，在低磷饲粮中添加植酸酶能提高骨骼的抗断强度和矿化程度。

表12-3给出了植酸酶、NSP水解酶或者二者构成的复合酶对生长育肥猪养分和能量消化率的影响。正如在仔猪上的应用一样，在生长育肥猪上使用植酸酶或植酸酶与NSP酶的复合酶对提高磷和钙利用的作用，各研究结果都非常的一致（Traylor等，2001；Paditz等，2004；Olukosi等，2007b）。有些研究表明，植酸酶可使磷利用率提高20%～40%（Cromwell等，1993；Mroz等，1994）。还有研究报道，植酸酶也能促进氨基酸和其他养分的利用（Morz等，1994；Traylor等，2001；Nortey等，2007），但在其他一些研究中未发现这种作用（Paditz等，2004；Olukosi等，2007b）。造成以上差异的原因是各试验所使用的谷物类饲料不同以及基础饲粮的养分和氨基酸消化率本身就存在差异。与那些养分消化率高的基础饲粮相比（Traylor等，2001；Paditz等，2004），在那些养分消化率较低的基础饲粮中添加酶

制剂的作用效果更加明显（Nortey等，2007；Emiola等，2009）。但基础饲粮的磷和钙消化率的高低对植酸酶作用的影响却非常小（Olukosi等，2007b；表12-4）。

表12-3 断奶仔猪饲粮中添加酶制剂对养分和能量消化率的提高比率[1]（%）

干物质	氮	磷	钙	能量	参考文献
植酸酶					
2.2	2.8	82.0	9.9	—	Morz 等，1994
—	-0.2	37.0	8.2	—	Traylor 等，2001
2.9	—	5.4	1.3	5.9	Nortey 等，2007（回肠）
1.7	—	14.6	10.6	1.7	Nortey 等，2007（总）
—	0	9.0	2.3	-0.9	Olukosi 等，2007b
1.8	2.1	26.4	21.0	—	Woyengo 等，2008
NSP 水解酶					
6.3	1.7	—	—	—	Bartelt 等，2002[6]
6.2	—	13.6	-3.2	6.3	Nortey 等，2007（回肠）[2]
2.4	—	4.9	2.1	2.8	Nortey 等，2007（总）[2]
—	0.1	-1.2	0	0	Olukosi 等，2007b[3]
2.4	3.1	4.5	17.6	—	Woyengo 等，2008[5]
5.6	3.0	20.4	38.0	5.1	Emiola 等，2009[4]
NSP 水解酶与植酸酶的复合酶					
4.2	—	18.3	-11.1	6.5	Nortey 等，2007（回肠）
3.5	—	32.9	6.9	4.0	Nortey 等，2007（总）
—	1.3	13.4	5.2	0.1	Olukosi 等，2007b
1.8	2.3	21.9	21.8	—	Woyengo 等，2008

[1]消化率的提高比率是根据[（加酶组的养分消化率-对照组的养分消化率）]/ 对照组的养分消化率×100公式计算得到。

[2]小麦型饲粮中（含小麦糠）添加木聚糖酶。

[3]小麦型饲粮中添加木聚糖酶、淀粉酶和蛋白酶。

[4]大麦-玉米型饲粮中（含30%小麦DDGS）添加木聚糖酶、β-葡聚糖酶和纤维素酶。

[5]小麦型饲粮中添加木聚糖酶。

[6]燕麦-小麦型饲粮中添加木聚糖酶。

表12-4 基础饲粮的磷和粗蛋白含量及酶制剂对猪养分和氨基酸消化率的影响

基础饲粮中的养分含量（%）		消化率（%）						参考文献
		磷		CP		氨基酸（平均值）		
总磷	CP	基础	加酶	基础	加酶	基础	加酶	
0.36	17.0	29.4	53.5[1]	83.3	85.6	77.1	78.2[2]	Morz 等，1994
0.42	13.2	49.8	64.3[1]	82.5	82.2	86.3	87.2[2]	Traylor 等，2001
0.29	20.8	22.2	37.3[1]	84.4	84.2	N/A[4]	N/A	Paditz 等，2004
0.63	20.4	N/A	N/A	76.0	76.8	78.8	79.4	Liao 等，2005
0.70	20.3	N/A	N/A	72.0	74.7[2]	75.9[2]	78.7[2]	Liao 等，2005
0.62	N/A	42.9	45.9[1]	N/A	N/A	74.9	79.1[1]	Nortey 等，2007
0.47	17.9	68.1	74.2[1]	91.2[3]	91.2[3]	N/A	N/A	Olukosi 等，2007b
0.32	14.0	23.7	38.9[1]	77.3[3]	76.1[3]	70.1	70.1	Sands 等，2009
0.49	14.0	21.5	44.1[1]	75.7[3]	73.7[3]	71.7	71.6	Sands 等，2009
0.35	17.6	N/A	N/A	68.6	75.2[1]	N/A	N/A	Emiola 等，2009
0.45	18.6	38.3	46.1	70.7	72.8	69.9	75.5[2]	Emiola 等，2009

[1]消化率大幅提高。
[2]消化率小幅提高。
[3]氮消化率。
[4]N/A表示无数据。

关于NSP水解酶对生长育肥猪生长性能的影响，不同研究结果差异很大，大多数的试验结果是完全没有影响（Thacker等，1991；Barrera等，2004；Olukosi等，2007a；Emiola等，2009）。有研究报道，NSP水解酶能提高养分利用，但提高幅度通常很小（Graham等，1988；Bartelt等，2002；Barrera等，2004），或者根本没有提高（Olukosi等，2007b）。如表12-4所示，NSP酶的作用有时还因基础饲粮的养分消化率不同而异。

不同NSP水解酶组合使用（Emiola等，2009）或进一步与植酸酶组合使用（Olukosi等，2007b；Woyengo等，2008），并不一定会提高生长性能和养分利用率。对于高NSP含量的小麦型基础饲粮，植酸酶比木聚糖酶能提高更多种氨基酸回肠消化率（Nortey等，2007）。在小麦型基础饲粮中，单独添加植酸酶不会影响能量沉积，但同时添加植酸酶和木聚糖酶可使能量和磷的沉积比单独添加植酸酶组提高16%（Olukosi等，2007b）。Woyengo等（2008）研究表明，在消化率相对较低的基础饲粮中，同时添加植酸酶和木聚糖酶能提高氨基酸消化率。

诸多研究表明，生长育肥猪饲粮中植酸酶比NSP水解酶可能更有应用前景。这也许是因为在能量与养分摄入量限制了猪生长性能的时候，添加NSP水解酶也不一定能提高能量利用率，增加可消化养分的摄入量；而添加植酸酶基本上都能提高饲粮磷的利用率。虽然在提高养分利用方面，饲粮添加NSP酶与植酸酶的复合酶的效果并不一定优于单独添加植酸酶，但添加这种复合酶的效果很可能要比单独添加NSP酶更好。

◆ 5.3 酶制剂在繁殖母猪上的应用

关于母猪饲粮添加外源酶制剂的研究相对较少，这可能因为母猪比仔猪和生长育肥猪能更好地利用纤维饲料（Shi和Noblet，1993），或者是母猪在妊娠阶段需要控制体增重。但有些研究表明，在妊娠期和泌乳期添加植酸酶对母猪有益处。Czech和Grela（2004）报道，添加植酸酶能增加母猪血液中的铁和其他矿物质浓度、血红蛋白含量及白细胞数量。还有研究报道，植酸酶能提高饲料养分尤其是矿物质的消化率（Kemme等，2007；Baidoo等，2003；Jongbloed等，2004；Manner和Simon，2006），并降低粪中养分的排泄（Baidoo等，2003；Hill等，2008）。

植酸酶对母猪的作用效果还与母猪繁殖阶段有关。Nyachoti等（2006）研究表明，与妊娠前期相比，妊娠后期添加植酸酶更能提高母猪养分消化率。与妊娠期相比，泌乳期添加植酸酶能使母猪获得更多的可消化磷（Jongbloed等，2004），这也与妊娠期饲粮纤维含量高的作用有关。

植酸酶通常不会影响妊娠期母猪的体重或体重损失（Kemme等，1997；Baidoo等，2003）、仔猪出生窝重（Jongbloed等，2004；Liesegang等，2005；Manner和Simon，2006）和仔猪增重（Jongbloed等，2004；Lyberg等，

2006）。根据仔猪增重计算，妊娠期添加植酸酶能增加母猪泌乳量（Manner和Simon等，2006），但不影响乳中钙和磷的浓度（Liesegang等，2005）。但值得注意的是，大多数试验使用的低磷饲粮对母猪繁殖性能都没有负面影响，因此植酸酶的添加效果不明显也就不足为奇。

与植酸酶相比，NSP水解酶在母猪饲粮中的应用研究更少。有一项研究表明，在妊娠期，玉米-豆粕型饲粮中添加蛋白酶和纤维素酶或木聚糖酶对母猪生产性能都没有显著影响；但在泌乳期，添加木聚糖酶能提高回肠和全消化道中干物质和氮的消化率（Souza等，2007）。如果基础饲粮的氨基酸消化率较高，酶制剂发挥作用的空间就非常有限。

那些添加酶制剂不能提高母猪产仔性能的研究表明，即使在不添加酶制剂的情况下，母猪都能获得充足营养的供应。在这种情况下，可以进一步降低基础饲粮的养分含量，之后再添加酶制剂，就可能获得更好的效果。

6 酶制剂在猪营养中的应用前景

目前关于酶制剂的多数研究表明，为了合理有效地使用酶制剂，需要清楚地了解饲料中的限制性物质及能降解这些限制性物质的相应酶制剂。总体来说，NSP酶的应用效果不是很明显，而植酸酶的应用效果则很稳定。而且，在营养供给充足的饲粮中添加酶制剂会造成浪费，达不到预期的使用效果，事实上还会进一步恶化养分排泄的问题。

然而，除了对磷利用率的影响外，植酸酶对能量、氮和AA利用的作用仍受到一些质疑。有些人认为，植酸酶可通过降解植酸、抑制它与淀粉或蛋白质的结合来促进能量的利用率（Yoon等，1983）。但植酸酶对能量利用率的作用效果并不稳定（Liao等，2005；Adeola等，2006）。Kies等（2005）认为，植酸酶会促进能量依赖型的矿物质转运，这可能会抵消它对能量利用的改善效果。因为植酸酶通常都是在低磷饲粮中使用，所以可以用这种观点去解释；但对于磷和钙过量的饲粮使用植酸酶，无法用这种观点来解释。采用净能作为衡量酶制剂作用的标准可能更为恰当（Olukosi等，2008）。Ketaren等（1993）早期研究发现，植酸酶能增加猪的能量沉积，而对消化能没有显著影响。

植酸可以与蛋白质结合，因此会降低蛋白质和氨基酸的利用率（Cheryan，1980）。但关于植酸酶在提高氨基酸利用率方面的研究，各试验结果并不一致（Adeola和Sands，2003）。究其原因可能与饲料原料组成密切相关（Liao等，2005），基础饲粮的氨基酸消化率、氨基酸缺乏程度、内源植酸酶含量和动物年龄等也都是造成这种差异的因素，但目前还不清楚这些影响因素的重要程度，是哪一种因素更重要还是所有这些影响因素都很重要。

人们未来可能会更关注外源酶制剂的保健功能，以及酶制剂与基因表达调控之间的关系。Högberg和Lindberg（2004）发现，在NSP含量高的仔猪饲粮中添加NSP水解酶能增加回肠乳酸的摩尔浓度。提示NSP水解酶能影响仔猪健康状况，因为乳酸菌能通过抑制致病菌的生长来促进肠道健康（Pluske等，2001）。最近Kiarie等（2007）也发现NSP水解产物具有提高仔猪健康状况的潜能。另外，关于营养与基因的关系已在小鼠上开展了大量研究。但更重要的是必须弄清酶制剂的作用机理，与此同时还应该更加关注，酶制剂究竟如何影响不同基因的表达及怎样才能降低养分排泄？这无疑将促进可持续性养猪生产的发展。

7 结 语

饲粮添加酶制剂可以促进猪生长，提高养分利用效率，并降低养分排泄。这些酶制剂主要作用于植酸盐、淀粉、蛋白质和NSP。植酸酶虽可提高生长性能、增加磷的利用率，但并不一定能提高其他养分的利用。尽管在猪上使用NSP水解酶有一定的作用效果，尤其是仔猪，但是NSP水解酶对生长和养分利用的作用效果并不稳定。酶制剂在猪营养中的未来研究方向主要包括两方面：一是研究酶制剂改善机体健康的功能，二是探索酶制剂对基因表达的调控机制。本章探讨了如何使用外源酶制剂才能符合可持续性养猪生产的要求。

作者：Oluyinka A. Olukosi
和OlayiwolaAdeola
译者：乔家运

参考文献

Adeola, O. 1995. Digestive utilization of minerals by weanling pigs fed copper- or phytase-supplemented diets. Can. J. Anim. Sci. 75:603-610.

Adeola, O., and J. S. Sands. 2003. Does supplemental dietary microbial phytase improve amino acid utilization? A perspective that it does not. J. Anim Sci. 81:78-85.

Adeola, O., B.V. Lawrence, A. L. Sutton, and T. R. Cline. 1995. Phytase-induced changes in mineral utilization in zinc-supplemented diets for pigs. J. Anim. Sci. 73:3384-3391.

Adeola, O., J. I. Orban, D. Ragland, T. R. Cline, and A. L. Sutton. 1998. Phytase and cholecalciferol supplementation of low-calcium and low-phosphorus diets for pigs. Can. J. Anim. Sci. 78:307-313.

Adeola, O., O. A. Olukosi, J. A. Jendza, R. N. Dilger, and M. R. Bedford. 2006. Responses of growing pigs to *Peniophora lycii*- and *Escherichia coli*-derived phytases or varying ratios of calcium to total-phosphorus. Anim. Sci. 82:637-644.

Augspurger, N. R., D. M. Webel, X. G. Lei, and D. H. Baker. 2003. Efficacy of an *E. coli* phytase expressed in yeast for releasing phytate-bound phosphorus in young chicks and pigs. J. Anim. Sci. 81:474-483.

Bach Knudsen, K. E., and I. Hansen. 1991. Gastrointestinal implications in pigs of wheat and oat fractions. 1. Digestibility and bulking properties of polysaccharides and other major constituents. Br. J. Nutr. 65:217-232.

Baidoo, S. K., Q. M. Yang, and R. D.Walker. 2003. Effects of phytase on apparent digestibility of organic phosphorus and nutrients in maize-soya bean meal based diets for sows. Anim. Feed. Sci. Technol. 104:133-141.

Barrera, M., M. Cervantes, W. C. Sauer, A. B. Araiza, N. Torrentera, and M. Cervantes. 2004. Ileal amino acid digestibility and performance of growing pigs fed wheat-based diets supplemented with xylanase. J. Anim. Sci. 82:1997-2003.

Barrientos, L., J.J. Scott, and P.P.N. Murthy. 1994. Specificity of hydrolysis of phytic acid by alkaline phytase from lily pollen. Plant Physiol. 106:1489-1495.

Bartelt, J., A. Jadamus, F. Wiese, E. Swiech, L. Buraczewska, and O. Simon. 2002. Apparent precaecal digestibility of nutrients and level of endogenous nitrogen in digesta of the small intestine of growing pigs as affected by various digesta viscosities. Archiv. Anim. Nutr. 56:93-107.

Baxter, C. A., B. C. Joern, D. Ragland, J. S. Sands, and O. Adeola. 2003. Phytase, high-available-phosphorus corn, and storage effects on phosphorus levels in pig excreta. J. Environ. Qual. 32:1481-1489.

Bebot-Brigand, A., C. Dange, N. Fauconnier, and C. Gérard. 1999. ^{31}P NMR, potentiometric and spectrophotometric studies of phytic acid ionization and complexation towards Co^{2+}, Ni^{2+}, Cu^{2+}, Zn^{2+}, and Cd^{2+}. J. Inorg. Biochem. 75:71-78.

Bedford, M. R., and H. Schulze. 1998. Exogenous enzymes for pigs and poultry. Nutr. Res. Rev. 11:91-114.

Bedford, M. R., J. F. Patience, H. L. Classen, and J. Inborr. 1992. The effect of dietary enzyme supplementation of rye- and barley-based diets on digestion and subsequent performance in weanling pigs. Can. J. Anim. Sci. 72:97-105.

Bohn, L., L. Josefsen, A. S. Meyer, and S. K. Rasmussen. 2007. Quantitative analysis of phytate globoids isolated from wheat bran and characterization of their sequential dephosphorylation by wheat phytase. J. Agric. Food Chem. 55:7547-7552.

Brana, D. V.,M. Ellis, E. O. Castaneda, J. S. Sands, and D. H. Baker. 2006. Effect of a novel phytase on growth performance, bone ash, and mineral digestibility in nursery and grower-finisher pigs. J. Anim. Sci. 84:1839-1849.

Caine, W. R., W. C. Sauer, S. Tamminga, M. W. A. Verstegen, and H. Schulze. 1997. Apparent ileal digestibilities of amino acids in newly weaned pigs fed diets with protease-treated soybean meal. J. Anim. Sci. 75:2962-2969.

Champagne, E. T. 1988. Effects of pH on mineral-phytate, protein-mineral-phytate and mineral-fiber interactions. Possible consequences of atrophic gastritis on mineral bioavailability from high fiber diets. J. Am. Coll. Nutr. 7:499-508.

Cheryan, M. 1980. Phytic acid interactions in food systems. CRC Crit. Rev. Food. Sci. Nutr. 13:297-335.

Corring, T. 1980. Endogenous secretions in the pig. Pages 136-150 in Current Concepts of Digestion and Absorption in Pigs. A. G. Low, and I. G. Partridge, eds. National Institute for Research in Dairying, Reading, UK.

Courtin, C. M., and J. A. Delcour. 2002. Arabinoxylans and endoxylanase in wheat flour bread-making. J. Cereal Sci. 35:225-243.

Cranwell, P. D. 1995. Development of the neonatal gut and enzyme systems. Pages 99–154 in The Neonatal Pig, Development and Survival. M. A. Varley, ed. CAB International, Wallingford, Oxfordshire, U.K.

Crenshaw, T. D. 2001. Calcium, phosphorus, vitamin D, and vitamin K in swine nutrition. Pages 187-212 in Swine Nutrition. A. J. Lewis and L. L. Southern, eds. CRC Press, Boca Raton, FL.

Cromwell, G. L., T. S. Stahly, R. D. Coffey, H. J. Monegue, and J. H. Randolph. 1993. Efficacy of phytase in improving the bioavailability of phosphorus in soybean meal and corn-soybean meal diets for pigs. J. Anim. Sci. 71:1831–1840.

Czech, A., and E. R. Grela. 2004. Biochemical and haematological blood parameters of sows during pregnancy and lactation fed the diet with different source and activity of phytase. Anim. Feed Sci. Technol. 116:211–223.

Diebold, G., R. Mosenthin, H. P. Piepho, and W. C. Sauer. 2004. Effect of supplementation of xylanase and phospholipase to a wheat-based diet for weanling pigs on nutrient digestibility and concentrations of microbial metabolites in ileal digesta and feces. J. Anim. Sci. 82:2647–2656.

Dierick, N. J. Decuypere, K. Molly, and E. vanderbeke. 2004. Microbial protease addition to a soybean meal diet for weaned piglets: Effect on performance, digestion, gut flora and gut function. Pages 229–234 in Recent advances of research in Antinutritional Factors in Legume Seeds and Oilseeds. Proc. 4th Int. Workshop on Antinutritional Factors in Legume Seeds and Oilseeds. EAAP Publication No. 11o, Toledo, Spain.

Dornez, E., I. J. Joye, K. Gebruers, J. A. Delcour, and C. M. Courtin. 2006. Wheat-kernel associated endoxylanases consist of a majority of microbial and a minority of wheat endogenous endoxylanases. J. Agric. Food Chem. 54:4028–4034.

Dvořáková, J., J. Kopecký, V. Havliček, V. and Kren. 2000. Formation of *myo*-inositol phosphates by Aspergillus niger 3-phytase. Folia Microbio. 45:128–132.

Eeckhout, W., and M. D. Paepe. 1994. Total phosphorus, phytate-phosphorus and phytase activity in plant feedstuffs. Anim. Feed Sci. Technol. 47:19–29.

Emiola, I. A., F. O. Opapeju, B. A. Slominski, and C. M. Nyachoti. 2009. Growth performance and nutrient digestibility in pigs fed wheat distillers dried grain with soluble-based diets supplemented with a multicarbohydrase enzyme. J. Anim. Sci. 87:2315–2322.

Fan, C. L., X. Y. Han, Z. R. Xu, L. J. Wang, and L. R. Shi. 2009. Effects of β-glucanase and xylanase supplementation on gastrointestinal digestive enzyme activities of weaned piglets fed a barley-based diet. J. Anim. Physiol. Anim. Nutr. 93:271–276.

Fernandez, J. A., J. M. and Jorgensen. 1986. Digestibility and absorption of nutrients as affected by fiber content in the diet of the pig. Livest. Prod. Sci. 15:53–71.

Fisher, H. 1992. Low-calcium diet enhanced phytate-phosphorus availability. Nutr. Rev. 50:170–171.

Fontes, C. M. G. A., P. I. P. Fonte, T. C. Reis, M. C. Soares, L. T. Gama, F. M. V. Dias, and L. M. A. Ferreira. 2004. A family 6 carbohydrate-binding module potentiates the efficiency of a recombinant xylanase used to supplement cereal-based diets for poultry. Br. Poult. Sci. 45:648–656.

Frølich,W. 1990. Chelating properties of dietary fiber and phytate. The role for mineral availability. Pages 83–93 in Developments in Dietary Fiber. Physiological, Physicochemical and Analytical Aspects. I. Furda, and C. J. Brine, eds. New Plenum Press, New York, NY.

Goff, G. L., and J. Noblet. 2001. Comparative total tract digestibility of dietary energy and nutrients in growing pigs and adult sows. J. Anim. Sci. 79:2418–2427.

Goodlad, J. S., and J. M. Mathers. 1991. Digestion by pigs of non–starch polysaccharides in wheat and raw peas (*Pisium sativum*) fed in mixed diets. Br. J. Nutr. 65:259–270.

Graham, H., K. Hesselman, and P. Aman. 1986. The influence of wheat bran and sugar beet pulp on the digestibility of dietary components of a cereal–based pig diet. J. Nutr. 116:242–251.

Graham, H.,W. lowgren, D. Pettersson, and P. Aman. 1988. Effect of enzyme supplementation on digestion of a barley/pollard–based pig diet. Nutr. Rep. Int. 38:1073–1079.

Han, Y. M., F. Yang, A. G. Zhou, E. R. Miller, P. K. Ku, M. G. Hogberg, and X. G. Lei. 1997. Supplemental phytases of microbial and cereal sources improve dietary phytate phosphorus utilization by pigs from weaning through finishing. J. Anim. Sci. 75:1017–1025.

Han, Y. M., K. R. Roneker, W. G. Pond, and X. G. Lei. 1998. Adding wheat middlings, microbial phytase, and citric acid to corn–soybean meal diets for growing pigs may replace inorganic phosphorus supplementation. J. Anim. Sci. 76:2649–2656.

Han, Y., D. B. Wilson, and X. G. Lei. 1999. Expression of *Aspergillus niger* phytase gene (phy A) in *Saccharomyces cerevisae*. Appl. Environ. Microbiol. 65:1915–1918.

Han, Y. K., I. H. Kim, J. W. Hong, and O.S. Kwon. 2003. Nutrient digestibility of nutrient in plant protein feedstuffs for finishing pigs. Asian Austral J. Anim. Sci. 16:1020–1024.

Härkönen, H., E. Pessa, T. Suortti, and K. Poutanen. 1997. Distribution and some properties of cell wall polysaccharides in rye milling fractions. J. Cereal Sci. 26:95–104.

Henrissat, B., and A. Bairoch. 1993. Updating the sequence–based classification of glycosyl hydrolases. Biochem. J. 293:781–788.

Hill, G. M., J. E. Link, M.J. Ricnker, D. L. Kirkpatrick, M. L. Gibson, and K. Karges. 2008. Utilization of distillers dried grains with soluble and phytase in sow lactation diets to meet phosphorus requirement of the sow and reduce fecal phosphorus concentration. J. Anim. Sci. 86:112–118.

Högberg, A., and J. E. Lindberg. 2004. Influence of cereal non–starch polysaccharides and enzyme supplementation on digestion site and gut environment in weaned piglets. Anim. Feed Sci. Technol. 116:113–128.

Inborr, J., J. Puhakka, J. G. M. Bakker, and J. van der Meulen. 1999. *β*–Glucanase and xylanase activities in stomach and ileum of growing pigs fed wheat bran based diets with and without enzyme treatment. Archiv. Anim. Nutr. 52:263–274.

Irving, G. C. J., and D. J. Cosgrove. 1972. Inositol phosphate phosphatases of microbial origin: the inositols pentaphosphate products of *Aspergillus ficuum* phytases. J. Bacteriol. 112:434–438.

Jendza, J. A., R. N. Dilger, S. A. Adedokun, J. S. Sands, andO. Adeola. 2005. *Escherichia coli* phytase improves growth performance of starter, grower, and finisher pigs fed phosphorus–deficient diets. J. Anim. Sci. 83:1882–1889.

Jongbloed, A. W., J. T. M. van Diepen, P. A. Kemme, and J. Broz. 2004. Efficacy of microbial phytase on mineral digestibility in diets for gestating and lactating sows. Livest. Prod. Sci. 91:143–155.

Just, A., W. C. Sauer, H. Bech-Anderson, H. Jogersen, and B. O. Eggum. 1979. The influence of the hind gut microflora on the digestibility of protein and amino acids in growing pigs elucidated by addition of antibiotics to different fractions of barley. J. Anim. Physiol. Anim. Nutr. 43:83–91.

Kass, M. L., P. J. van Soest, and W. G. Pond. 1980. Utilization of dietary fiber from alfalfa by growing swine. I. Apparent digestibility of diet components in specific segments of the gastrointestinal tract. J. Anim. Sci. 50:175–191.

Kemme, P. A., J. S. Radcliffe, A. W. Jongbloed, and Z. Mroz. 1997. The effects of sow parity on digestibility of proximate components and minerals during lactation as influenced by diet and microbial phytase supplementation. J. Anim. Sci. 75:2147–2153.

Ketaren, P. P., E. S. Batterham, E. B. Dettmann, and D. J. Farrell. 1993. Phosphorus studies in pigs. Effect of phytase supplementation on the digestibility and availability of phosphorus in soya-bean meal for grower pigs. Br. J. Nutr. 70:289–311.

Kiarie, E. G., B. A. Slominski, D. O. Krause, and C.M. Nyachoti. 2008. Nonstarch polysaccharide hydrolysis products of soybean and canola meal protect against enterotoxigenic *Escherichia coli* in pigs. J. Nutr. 138:502–508.

Kies, A. K., L. H. De Jonge, P. A. Kemme, and A.W. Jongbloed. 2005. Interaction between protein, phytate, and microbial phytase. *In vitro* studies. J. Agric. Food Chem. 54:1753–1758.

Knowlton, K. F. J. S. Radcliffe, C. L. Novak, and D. A. Emerson. 1994. Animal management to reduce phosphorus losses to the environment. J. Anim. Sci. 82:173–195.

Kornegay, E. T. 1985. Calcium and Phosphorus in Animal Nutrition. Pages 1–106 in Calcium and Phosphorus in Animal Nutrition. National Feed Ingredients Association, West Des Moines, IA.

Lei, X. G., P. K. Ku, E. R. Miller, and M. T. Tokoyama. 1993. Supplementing corn-soybean meal diets with microbial phytase linearly improves phytate phosphorus utilization by weanling pigs. J. Anim. Sci. 71:3359–3367.

Lei, X. G., P. K. Ku, E. R. Miller, M. T. Yokoyama, and D. E. Ullrey. 1994. Calcium level affects the efficacy of supplemental microbial phytase in corn-soybean meal diets of weanling pigs. J. Anim. Sci. 72:139–143.

Liao, S. F.,W. C. Sauer, A. K. Kies, Y. C. Zhang, M. Cervantes, and J. M. He. 2005. Effect of phytase supplementation to diets for weanling pigs on the digestibilities of crude protein, amino acids, and energy. J. Anim. Sci. 83:625–233.

Liesegang, A., L. Loch, E. Burgi, and J. Risteli. 2005. Influence of phytase added to a vegetarian diet on bone metabolism in pregnant and lactating sows. J. Anim. Physiol. Anim. Nutr. 89:120–128.

Li, S., W. C. Sauer, S. X. Huang, and V. M. Gabert. 1996. Effect of beta-glucanase supplementation to hulless barley or wheat-soybean meal diets on the digestibilities of energy, protein, beta-glucans, and amino acids in young. J. Anim. Sci. 74:1649-1656.

Liu, J., D.W. Bollinger, D. R. Ledoux, and T. L. Veum. 2000. Effects of dietary calcium:phosphorus ratios on apparent absorption of calcium and phosphorus in the small intestine, cecum, and colon of pigs. J. Anim. Sci. 78:106-109.

Lyberg, K., H. K. Andersson, A. Simonsson, and J. E. Lindberg. 2006. Influence of different phosphorus levels and phytase supplementation in gestation diets on sows performance. J. Anim. Physiol. Anim. Nutr. 91:304-311.

Manner, K., and O. Simon. 2006. Effectiveness of microbial phytases in diets of sows during gestation and lactation. J. Anim. Feed. Sci. 15:199-211.

Martin, C. J., and W. J. Evans. 1989. Phytic acid-enhanced metal ion exchange reactions: The effect on carboxypeptidase A. J. Inorg. Biochem. 35:267-288.

Matsui, T., Y. Nakagawa, A. Tamura, C. Watanabe, K. Fujita, T. Nakajima, and H. Yano. 2000. Efficacy of yeast phytase in improving phosphorus bioavailability in a corn-soybean meal-based diet for growing pigs. J. Anim. Sci. 78:94-99.

Mavromichalis, I., J. D. Hancock, B. W. Senne, T. L. Gugle, G. A. Kennedy, R. H. Hines, and C. L. Wyatt. 2000. Enzyme supplementation and particle size of wheat in diets for nursery and finishing pigs. J. Anim. Sci. 78:3086-3095.

Mroz, Z., A. W. Jongboed, and P. A. Kemme. 1994. Apparent digestibility and retention of nutrients bound to phytate complexes as influenced by microbial phytase and feeding regimen in pigs. J. Anim. Sci. 72:126-132.

Myrie, S. B., R. F. Bertolo, W. C. Sauer, and R. O. Ball. 2008. Effect of common antinutritive factors and fibrous feedstuffs in pig diets on amino acid digestibilities with special emphasis on threonine. J. Anim. Sci. 86:609-619.

Nakano, T., J. Toshio, K. Narita, and T. Hayakawa. 2000. The pathway of dephosphorylation of myo-inositol hexakisphosphate by phytases from wheat bran of *Triticum aestivum* L. cv. Nourin # 61. Biosci. Biotech. Bioch. 64:995-1003.

Nagashima, T., T. Tange, and H. Anazawa. 1999. Dephosphorylation of phytate by using *Aspergillus niger* phytase with a high affinity for phytate. Appl. Environ. Microbiol. 65:4862-4864.

Noblet, J. 2001. Digestive and metabolic utilization of dietary energy in pig feeds: Comparison of energy systems. Pages 161–184 in Recent Developments in Pig Nutrition3. P. C. Garnsworthy and J.Wiseman, eds. NottinghamUniversity Press, Nottingham, UK.

Nortey, T. N., J. F. Patience, P. H. Simmins, N. L. Trottier, and R. T. Zijlstra. 2007. Effects of individual or combined xylanase and phytase supplementation on energy, amino acids, and phosphorus digestibility and growth performance of grower pigs fed wheat-based diets containing wheat millrun. J. Anim. Sci. 85:1432-1443.

Nyachoti, C. M., J. S. Sands, M. L. Connor, and O. Adeola. 2006. Effect of supplementing phytase to corn– or wheatbased gestation and lactation diets on nutrient digestibility and sow and litter performance. Can. J. Anim. Sci. 86:501–510.

Olukosi, O. A., A. J. Cowieson, and O. Adeola. 2008. Energy utilization and growth performance of broilers receiving diets supplemented with enzymes containing carbohydrase or phytase activity individually or in combination. Br. J. Nutr. 99:682–290.

Olukosi, O. A., J. S. Sands, and O. Adeola. 2007a. Supplementation of carbohydrases or phytase individually or in combination to weanling and growing–finishing pigs. J. Anim. Sci. 85:1702–1711.

Olukosi, O. A., M. R. Bedford, and O. Adeola. 2007b. Xylanase in diets for growing pigs and broiler chicks. Can. J. Anim. Sci. 87:227–234.

Paditz, K., H. Kluth, and M. Rodehutscord. 2004. Relationship between graded doses of three microbial phytases and digestible phosphorus in pigs. Anim. Sci. 78:429–438.

Pagano, A. R., K. Yasuda, K. R. Roneker, T. D. Crenshaw, and X. G. Lei. 2007. Supplemental *Escherichia coli* phytase and Strontium enhance bone strength of young pigs fed a phosphorus–adequate diet. J. Nutr. 137:1795–1801.

Parkkonen, T., A. Tervila–Wilo, M. Hopeakoski–Nurminen, A. Morgan, K. Poutanen, and K. Autio. 1997. Changes in wheat microstructure following *in vitro* digestion. Acta Agric. Scand. B – S P. 47:43–47.

Phillippy, B. Q. 1999. Susceptibility of wheat and *Aspergillus niger* phytases to inactivation by gastrointestinal enzymes. J. Agric. Food. Chem. 47:1385–1388.

Planchote, V., P. Colonna, D. J. Gallant, and B. Bouchet. 1995. Extensive degradation of native starch granules by alpha–amylase from *Aspergillus fumigatus*. J. Cereal Sci. 21:163–171.

Pluske, J. R., J. C. Kim, D. E. McDonald, D. W. Pethick, and D. J. Hampson. 2001. Non–starch polysaccharides in the diets of young weaned piglets. Pages 81–112 in The Weaner Pig: Nutrition and Management. M. A. Varley and J. Wiseman, eds. CABI Publishing, Wallingford, UK.

Ravindran, V., W. L. Bryden, and E. T. Kornegay. 1995. Phytates occurrence, bioavailability and implications in poultry nutrition. Poult. Avian Bio. Rev. 6:125–143.

Reilly, P. J. 1981. Xylanases: Structure and function. Pages 111–129 in Trends in the Biology of Fermentation for Fuels and Chemicals. A. Hollaender, ed. Basic Life Sciences, Plenium Press, New York, NY.

Rérat, A. A. 1985. Intestinal absorption of end–products of digestion of carbohydrates and proteins in the pig. Archiv. Tierernäh. 35:461–480.

Rérat, A., M. Fiszlewicz, A. Giusi, and P. Vaugelade. 1987. Influence of meal frequency on postprandial variations in the production and absorption of volatile fatty acids in the digestive tract of conscious pigs. J. Anim. Sci. 64:448–456.

Rodriguez, E., Y. Han, and X. G. Lei. 1999. Cloning, sequencing, and expression of an *Eschericia coli* acid phospaphatase/ phytase gene (appA2) isolated from pig colon. Biochem. Bioph.

Res. Comm. 257:117–123.

Sands, J. S., D. Ragland, C. Baxter, B. C. Joern, T. E. Sauber, and O. Adeola. 2001. Phosphorus bioavailability, growth performance, and nutrient balance in pigs fed high available phosphorus corn and phytase. J. Anim. Sci. 79:2134–2142.

Sands, J. S., D. Ragland, R. N. Dilger, and O. Adeola. 2009. Responses of pigs to *Aspergillus niger* phytase supplementation of low–protein or high–phytin diets. J. Anim. Sci. 87:2581–2589.

Selle, P. H., D. J. Cadogan, and W. L. Bryden. 2003. Effects of phytase supplementation of phosphorus–adequate, lysine–deficient, wheat–based diets on growth performance of weaner pigs. Aust. J. Agric. Res. 54:323–330.

Shelton, J. L., D. W. Dean, L. L. Southern, and T. D. Bidner. 2005. Effect of protein and energy sources and bulk density of diets on growth performance of chicks. Poult. Sci. 84:1547–1554.

Shi, X. S., and J. Noblet. 1993. Contribution of the hindgut to digestion of diets in growing pigs and adult sows: Effect of diet composition. Livest. Prod. Sci. 34:237–252.

Simon, O. 1998. The mode of action of NSP hydrolyzing enzymes in the gastrointestinal tract. J. Anima. Feed Sci. 7:115–123.

Singh, M., and A. D. Krikorian. 1982. Inhibition of trypsin activity *in vitro* by phytate. J. Agric. Food Chem. 30:799–800.

Souza, A. L. P.,M. D. Lindemann, and G. L. Cromwell. 2007. Supplementation of dietary enzymes has varying effects on apparent protein and amino acid digestibility in sows. Livest. Sci. 109:122–124.

Steiner, T., R. Mosenthin, B. Zimmermann, R. Greiner, and S. Roth. 2007. Distribution of phytase activity, total phosphorus and phytate phosphorus in legume seeds, cereals and cereal by–products as influenced by harvest year and cultivar. Anim. Feed Sci. Technol. 133:320–334.

Tervilä–Wilo, A., T. Parkkonen, A. Morgan, M. Hopeakoski–Nurminen, K. Poutanen, P. Heikkinen, and K. Autio. 1996. *In vitro* digestion of wheat microstructure with xylanase and cellulose from *Trichoderma reesei*. J. Cereal Sci. 24:215–225.

Thacker, P. A., G. L. Campbell, and J. GrootWassink. 1991. The effect of enzyme supplementation on the nutritive value of rye–based diets for swine. Can. J. Anim. Sci. 71:489–496.

Torre, M., A. R. Rodriguez, and F. Saura–Calixto. 1991. Effects of dietary fiber and phytic acid on mineral availability. Crit. Rev. Food Sci. Nutr. 1:1–22.

Traylor, S. L., G. L. Cromwell, M. D. Lindemann, and D. A. Knabe. 2001. Effect of level of supplemental phytase on ileal digestibility of amino acids, calcium, and phosphorus in dehulled soybean meal for growing pigs. J. Anim. Sci. 79:2634–2642.

Weremko, D., H. Fandrejewski, and T. Zebrowska. 1997. Bioavailability of phosphorus in feeds of plant origin for pigs – review. Aust. J. Anim. Sci. 10:551–566.

Wilfart, A., L. Montagne, P. H. Simmins, J. van Milgen, and J. Noblet. 2007. Site of nutrient digestion n growing pigs: Effects of dietary fiber. J Anim Sci. 85:976–983.

Woyengo, T. A., J. S. Sands, W. Guenter, and C. M. Nyachoti. 2008. Nutrient digestibility and performance responses of growing pigs fed phytase– and xylanase–supplemented wheat–based diets. J. Anim. Sci. 86:848–857.

Wyss, M., R. Brugger, A. Kronenberger, R. Rémy, M. Tessier, A. Kronenberger, A. Middendorf, et al. 1999. Biochemical characterization of fungal phytases (*myo*–inositol hexakisphosphate phosphohydrolases):catalytic properties. Appl. Environ. Microbiol. 65:367–373.

Yen, J. T. 2001. Digestive system. Pages 390–453 in Biology of the Domestic *Pig*. W. G. Pond and H. J. Mersmann, eds. CornellUniversity Press, Ithaca, NY.

Yin, Y. L., J. D. G. McEvoy, H. Schulze, U. Hennig, W. B. Souffrant, and K. C. McCracken. 2000. Apparent digestibility (ileal and overall) of nutrients and endogenous nitrogen losses in growing pigs fed wheat (va. Soissons) or its by–products without or with xylanase supplementation. Livest. Prod. Sci. 62:119–132.

Yoon, J. H., L. U. Thompson, and D. J. A. Jenkins. 1983. The effect of phytic acid on *in vitro* rate of starch digestibility and blood glucose response. Am. J. Clin. Nutr. 38:835–842.

Young, L. G., M. Leunissen, and J. L. Atkinson. 1993. Addition of microbial phytase to diets of young pigs. J. Anim. Sci. 71:2147–2150.

第13章 猪饲料添加剂

1 导 语

猪饲料中通常会使用很多种饲料添加剂，这些添加剂用于促进猪生长、驱除寄生虫、改善肠道健康、提高饲料能量或养分的消化以及其他作用。饲料添加剂可分为抗菌剂（抗生素和化学药物）、微生物制剂（益生菌）、低聚糖（益生元）、药用微量元素、酸化剂、植物提取物、酶制剂、香味剂、除臭剂、抗氧化剂、黏结剂、流化剂、胴体改良剂和驱虫药。

抗菌剂是猪饲料添加剂中使用最为广泛的添加剂。因为它们可以抑制或阻止微生物的生长，故而被称为抗菌剂。抗菌剂与驱虫药一样被美国食品药品管理局（FDA）认定为药物，故受到FDA的监管。每年《饲料添加剂纲要》都会发布这些药物的允许添加水平、配伍组合、停药期（如果有停药期的话）。其他种类的饲料添加剂的使用准则每年由美国饲料管理协会（通常简称为AAFCO，2012）发布。

由于抗菌剂在养猪生产中使用的广泛性和重要性，本章将对它们进行重点介绍，其他种类的添加剂将作简单讨论，而酶制剂已在第12章进行了详细介绍。

2 抗菌剂

抗菌剂包括抗生素和化学药物。抗生素是由霉菌、酵母菌和其他微生物自然代谢产生的物质，而化学药物则是由化学合成的药物。

◆ 2.1 抗菌剂的使用背景

抗菌剂在20世纪50年代初开始被广泛使用。饲料中添加低剂量（远低于治疗剂量）的抗菌剂能促进动物生长，提高饲料利用效率，降低死亡率和改善繁殖性能；添加中高剂量（预防剂量）的抗菌剂能预防某些疾病的发生，而高剂量（治疗剂量）的添加则可用于某些猪病的治疗。

早在20世纪40年代末，人们就发现了猪饲喂抗生素的有益作用，与此同时维生素B_{12}被发现。Stokstad等（1949）给肉仔鸡饲喂金霉素链霉菌（*Streptomyces aureofaciens*）的发酵产物来研究维生素B_{12}的用量，发现肉仔鸡的生长速度远远超过了维生素B_{12}含量所能产生的效果。随后在猪上的试验也发现类似的促生长效应（Cunha等，1949，1950；Jukes等，1950；Lepley等，1950；Luecke等，1950）。最后，Stokstad和Jukes发现，原始发酵产物中的有效成分是一种抗生素——金霉素。

继金霉素被发现后不久，其他一些抗生素也陆续被发现，且将它们喂给猪都出现类似的促生长作用。短短几年时间之后，市场上就出现饲料级抗生素，且养猪业很乐意在饲养过程中使用它们。到1963年，每年约有1 000 t抗生素用于动物饲料中。到1988年，约有13 000 t 抗生素和化学药物被生产出来，其中有4 650 t 出售到美国（美国国际贸易委员会，1989），大概有一半的抗生素用作饲料添加剂。关于每年抗生素和化学药物的生产、销售、价格的更详细信息可从其他资料中获得（Cromwell，2001，2002）。

目前有15种抗菌剂被FDA批准用于猪饲料（饲料添加剂纲要，2012），包括11种抗生素和4种化学药物。11种抗生素分别是亚甲基双水杨酸杆菌肽（BMD）、杆菌肽锌、班贝霉素、金霉素、林可霉素、新霉素、土霉素、青霉素、硫黏菌素、泰乐菌素和维吉尼亚霉素。4种允许添加的化学药物分别是卡巴多司、洛克沙胂、磺胺二甲嘧啶和磺胺噻唑。另外，有些药物可配伍组合使用（如金霉素-

磺胺二甲嘧啶-青霉素、新霉素-土霉素），而其他药物则不能相互配伍使用。

那些能有效提高猪生长性能的抗菌剂都具有抑制或阻止某些微生物生长的作用。但它们的化学成分、抗菌谱及其吸收和排泄方式却存在很大差异。例如，一些抗菌剂很容易被机体吸收（如磺胺类药物和四环素类药物），而其他抗菌剂大部分不能被机体吸收（如杆菌肽斑驳霉素）。有些抗菌剂比其他抗生素更快速地排出体外（如磺胺噻唑比磺胺二甲嘧啶排出得更快）。虽然不同抗菌剂之间存在这些差异，但抗菌剂的特性与其促进猪生长的功能之间并没有必然的联系。

◆ 2.2 抗菌剂作生长促进剂的效果

诸多研究证实，抗生素具有提高猪生长速度和效率的功能。Hays（1977）、CAST（1981）和Zimmerman（1986）的综述是关于抗生素作用效果的优秀综述的代表。表13-1总结了美国从1950年到1985年25年时间内开展的大量试验。结果显示，使用抗生素饲喂猪特别是仔猪，能显著提高生长速度和饲料利用率。

表13-1 抗生素作猪生长促进剂的作用效果[1]

阶段	对照组	抗生素组	提高幅度（%）
早期阶段（7 ~ 25 kg）			
日增重（kg）	0.39	0.45	16.4
料重比	2.28	2.13	6.9
生长阶段（17 ~ 49 kg）			
日增重（kg）	0.59	0.66	10.6
料重比	2.91	2.78	4.5
生长育肥阶段（24 ~ 89 kg）			
日增重（kg）	0.69	0.72	4.2
料重比	3.30	3.23	2.2

[1]资料来源于1 194个试验、32 555头猪（Hays，1977；Zimmerman，1986）。

尽管一些抗生素已经使用了60多年，但仍有证据显示它们依然有促生长作用。表13-2比较了抗生素最初使用的28年时间（1950—1977年）与随后的8年时间（1978—1985年）内的作用效果，发现从早期28年到之后8年的时间使用抗生素的综合效果并没有出现降低。

表13-2 1950—1977年与1978—1985年抗生素的试验效果

阶段	使用抗生素后的提高幅度（%）	
	1950—1977 年[1]	1978—1985 年[2]
早期阶段		
日增重（kg）	16.1	15.0
料重比	6.9	6.5
生长育肥阶段		
日增重（kg）	4.0	3.6
料重比	2.1	2.4

[1]数据来自657个试验，15 689头猪（Hays，1977）。
[2]数据来自239个试验，11 083头猪（Zimmerman，1986）。

由于大部分有关抗生素作用效果的试验都是在大学进行的，与一般的商业猪场或者农场相比，通常大学的试验环境更清洁、疫病发生率更低、环境应激更少，因此表13-1和表13-2的试验评价结果很可能低于商业猪场的观测数据。表13-3的数据表明，抗生素在农场的使用效果约为在环境清洁的大学或科研院所的2倍。

表13-3 大学试验场与农场使用抗生素的效果试验

地点	试验个数	使用抗生素后的提高幅度（%）	
		日增重	料重比
总结 1[1]			
试验场	128	16.9	7.0
农场	32	28.4	14.5
总结 2[2]			

（续）

地点	试验个数	使用抗生素后的提高幅度（%）	
		日增重	料重比
试验场	9	13.2	4.7
农场	67	25.5	10.0

[1]数据来源于7～26 kg体重的约12 000头猪，使用的抗菌剂为金霉素-磺胺二甲嘧啶-青霉素、泰乐菌素-磺胺二甲嘧啶、四环素和卡巴多司（Hays，1997）。

[2]数据来源于8～20 kg体重的3 321头猪，使用的抗菌剂为金霉素-磺胺二甲嘧啶-青霉素（NCR-89，1984；Maddock，1985）。

抗生素除了具有促生长的功能外，还能减少死亡率和发病率，特别是在仔猪上。对22年时间内开展的大量猪场试验进行总结，发现抗生素使仔猪死亡率降低了一半，而且在发病率高和环境应激大的条件下使仔猪死亡率降低了约5倍（表13-4）。

表13-4　商业猪场使用抗生素对生长性能和死亡率的影响[1]

项目	对照组	抗生素	提高率（%）
总结 1[2]			
日增重（kg）	0.31	0.40	26
料重比	2.48	2.23	10
死亡率	4.3	2.0	—
总结 2（疫病高发水平）[3]			
日增重（kg）	0.30	0.38	4.7
料重比	3.07	2.50	10.0
死亡率	15.6	3.1	—

[1]抗菌剂为金霉素-磺胺二甲嘧啶-青霉素或泰乐菌素-磺胺二甲嘧啶；资料来源于Maddock（1985）。

[2]从1960年到1982年进行的67个试验，涉及1 597头猪。

[3]在疫病高发水平下进行的5个猪场试验，涉及638头猪，体重为8～31 kg。

◆ 2.3 抗菌剂对繁殖性能的影响

研究已证实，在繁殖周期的某些特定阶段，如配种期给母猪饲喂抗生素具有改善繁殖性能的作用。对9项研究结果的总结发现，在配种期饲喂高水平的可吸收的抗生素使妊娠率提高了约7%，并使分娩时的窝产活仔数增加了约0.4头（表13-5）。

表13-5 配种期或哺乳期使用抗菌剂对母猪繁殖性能的影响

项目	对照组	抗生素
配种期使用抗生素[1]		
产仔率（%）[2]	75.2	82.1
窝产活仔数	9.9	10.3
哺乳期使用抗生素[3]		
窝产活仔数	9.9	10.2
窝产断奶仔猪数	8.2	8.6
断奶时仔猪存活率（%）	85.1	86.8
断奶体重（kg）	4.86	4.90

[1]来源于Cromwell（2001，2002）对9个试验、1931头母猪的总结资料。大多数试验中的抗菌剂从配种前1周到配种后2～3周进行饲喂，饲喂剂量为0.5～1.0 g/d。

[2]100×（分娩母猪数/配种母猪数）。

[3]来源于Cromwell（2001，2002）对13个试验、2338头母猪的总结资料。大多数试验在分娩前3～7d开始添加高水平的抗菌剂（110～275 mg/kg），且哺乳期持续添加14～21d。

研究还发现，在分娩时和早期哺乳阶段使用抗生素也有益处。在一定条件下，母猪饲料中添加高水平抗生素降低了无乳症和子宫炎的发生率，而无乳症和子宫炎通常发生在母猪产后不久（Langlois等，1978a）。表13-5的数据表明，分娩前和哺乳期母猪饲料中添加抗生素还可使仔猪存活率和断奶体重得到小幅度的提高。肯塔基大学的试验表明，与之前14年期间常规使用抗生素的母猪相比，母猪在之后13年期间不使用抗生素，显著降低了其繁殖性能（表13-6）。

表13-6 不使用抗生素对母猪繁殖性能的影响[1]

项目	使用抗生素（1963—1972 年）	不使用抗生素（1972—1985 年）
窝数	398	688
受胎率	91.4	82.6
窝产仔数	10.8	10.2
窝产活仔数	9.8	9.3
平均出生重（kg）	1.29	1.38
21d 断奶仔猪数（窝）	8.8	7.5
平均断奶体重（kg）	5.67	5.37
仔猪存活率（%）	89.7	80.9
MMA 发病率[2]	<10	66[3]

[1]试验在肯塔基大学密闭的、无特殊病原的猪群进行。在1963—1972年，抗生素被用在妊娠期饲料、哺乳期饲料、教槽料、生长料和育肥料中。从1972—1985年，不在饲料中添加抗生素，或不用抗生素治病。

[2]MMA=乳腺炎，子宫炎和无乳症。

[3]仅在1972—1975年。

◆ 2.4 抗菌剂的作用机制

抗生素和其他抗菌剂促进猪生长的作用机制目前还不十分清楚。Hays（1978）提出，抗菌剂的可能作用机制可分为三类：代谢机制、营养机制、疫病控制机制。

2.4.1 代谢机制

抗生素可直接影响水和氮的排泄等代谢过程（Braude和Johnson，1953），抑制肝脏线粒体的脂肪酸氧化（Brody等，1954），抑制磷酸化反应和氧化反应（Weinberg，1957），增加蛋白合成率（Hash等，1964；Moser等，1980）。尽管抗生素可以引起这些代谢反应，但当给猪饲喂低剂量抗生素时，组织中的抗生素水平很可能不足以起到促生长作用。此外，代谢机制不能合理解释那些无法吸收的抗生素的作用。

2.4.2 营养机制

某些肠道微生物可以合成一些机体所必需的维生素和氨基酸，而另外一些肠道微生物则会与宿主动物竞争维生素、氨基酸和其他必需养分。因为某些肠道微生物合成的营养物质是动物所必需的，当饲喂抗生素致使这些肠道微生物菌群数量增加后，就能为宿主动物提供更多的可利用营养物质。

使用抗生素也可能降低那些与宿主动物竞争营养物质的微生物数量。早期研究发现，四环素抑制了乳酸菌的生长，而乳酸菌需要利用氨基酸（Kellogg等，1964）。当给猪饲喂氨基酸或维生素临界缺乏的饲粮时，这些竞争性微生物数量的减少可能对动物机体有利。

Henderickx等（1981）和Hedde（1981）发现，维吉霉素使肠道菌群数量增加，导致氨气、挥发性脂肪酸和乳酸的产量降低。由于这些代谢产物代表蛋白质和能量的潜在损失，故他们推断维吉霉素可为宿主动物节省能量和氨基酸。

使用抗生素喂猪发生的另一个显著变化是肠壁变薄和肠重量变轻（Braude等，1955；Taylor和Harrington，1955；Yen等，1985）。由于某些微生物产生的毒素会损害肠组织，致使肠壁增厚。但当饲粮添加抗生素时，肠道氨气（作为肠道刺激物质）的产量减少，进而导致肠壁变薄、肠重量变轻（Visek，1978）。而Catron等（1953）和Henderickx等（1981）提出，肠壁变薄有利于养分吸收。此外，肠道代谢活动非常高，基本上都是为机体产热消耗（Webster，1981；Koong等，1982），故肠重量的降低意味着有更多能量转化用于机体生长，而不是用于维持产热。

如果我们认同某些抗生素可节省能量和蛋白质的观点，那么饲喂抗生素的猪对饲粮氨基酸的需要量会更少一些。Catron等（1952）使用金霉素的试验及Yen等（1979）和Moser等（1980）使用卡巴多司的研究都证实，饲喂抗生素时猪对饲粮蛋白质的需要量降低。

2.4.3 疫病控制机制

关于抗生素促进动物生长的作用机制的各种学说，疫病控制机制是最为广泛接受的。因为猪只会不停地接触外界各种微生物，从而引起不同程度的非特异性、亚临床的疾病；而这些疾病又会反过来降低动物生长性能。抗生素通过抑制有害微生物的生长使猪生长潜能得到最大程度的发挥。

支持疾病控制机制的证据远比支持代谢机制和营养机制的多。其中一个证据就是，抗生素对仔猪的作用比对大猪更为明显（表13-1）。究其原因是，仔猪天然免疫力低下，更容易受到病原微生物的感染。哺乳期间从母猪初乳获得的免疫球蛋白在仔猪血液中不断减少，通常到仔猪断奶时或之后不久达到最低水平（Miller等，1961）。由于幼龄仔猪缺乏合成抗体的能力，故在6～8周龄之前它们的免疫保护力都很低。之后，随着年龄的增长，仔猪血液免疫球蛋白水平不断升高（Miller等，1961），这使它们能更好地应对环境病原微生物或其他有害微生物的侵害。

支持疾病控制机制的另一证据是，抗生素的作用程度受到环境清洁度和动物抗病力的影响。表13-7的数据显示，在肮脏的保育舍内使用抗生素比在清洁环境内使用更为有效（Hays和Speer，1960）。

表13-7 保育舍环境清洁度对仔猪使用抗生素效果的影响[1]

项目	对照组	抗生素组[2]	提高幅度（%）
增重（kg）			
清洁的保育舍	8.1	10.4	28
肮脏的保育舍	4.8	7.5	55
料重比			
清洁的保育舍	1.90	1.74	8
肮脏的保育舍	2.89	2.05	29

[1]数据来自于615头仔猪，初始体重平均为5.7 kg，试验期为4～5周（Hays和Speer，1960）。

[2]使用的抗菌剂为螺旋霉素、金霉素或土霉素。

◆ 2.5 经济效益

养猪户在养殖过程中使用抗生素的经济效益受到诸多因素的影响，包括抗生素的使用效果（即生长速度和生长性能的提高率、繁殖性能的改善幅度、死亡和发病的降低率）、抗生素价格、饲料及其他不可控原料的成本、猪市场价格。Zimmerman（1986）根据平均数值计算出，使用抗菌剂的经济效益是每头上市猪2.64美元。Cromwell（1999，2002）估算，由生猪使用抗生素提高生长和饲料利用率产生的经济效益是每头上市猪2.99 ~ 3.93美元，由母猪使用抗生素产生的经济效益每窝高达7.12美元。然而，提高生产效率的最终受益者是消费者。据估计，1981年在动物生产中使用抗生素，使美国消费者用于购买肉产品的费用每年节约35亿美元（CAST，1981）。而到目前，这种节约费用很有可能会更多。

◆ 2.6 避免药物残留

抗菌剂必须在经《饲料添加剂纲要》（2012）批准允许的配伍组合和添加水平范围内才可使用。而且，砷制剂（对氨基苯胂酸和洛克沙胂）、卡巴多司、新霉素和磺胺类药物（磺胺二甲嘧啶和磺胺噻唑）等一些药物必须在屠宰前规定期间内停止使用，以避免在可食用的组织中残留。

尽管猪肉中很少发现药物残留，但过去发现的最主要残留抗菌剂是磺胺二甲嘧啶。这是由这种药物独特的特性造成的。磺胺类药物特别是磺胺二甲嘧啶的静电作用，使它们容易在饲料粉尘中聚集，进而污染其他饲料。饲料中含低至2 ppm的磺胺二甲嘧啶就会在肝脏组织中残留（Cromwell等，1981）。不正确的饲料混合流程或者不能有效防止含磺胺类药物的饲料污染“干净”饲料，都将会造成加工后的饲料含有足够量的磺胺类药物，进而在动物肝脏和肌肉中发生残留。目前使用的磺胺二甲嘧啶是颗粒状，比过去使用的粉末状药物造成的交叉污染会更小一些（Cromwell等，1982）。

◆ 2.7 抗菌剂的安全性

人们一直质疑动物饲料中使用低剂量抗生素是否会对人类健康产生危害。有些人担心抗生素会在有耐药性的肠原杆菌中蓄积，这些肠原杆菌又能将其耐药性传染给病原菌，进而对公共卫生造成潜在威胁。其中，青霉素和四环素是最受关注的两种抗生素，因为这两种药物同时也在人类医学上使用。

虽然抗生素抵抗质粒（R-质粒）的转移通常发生在体外试验，但目前仍不是很清楚在不同动物之间或在动物与人之间发生抗生素抵抗质粒转移的概率有多大。因为动物上的细菌只有在摄入剂量非常高的情况下才能在人体内定植，即便如此，也只能短暂定植（Smith，1969）。

1987年美国FDA授权国家科学院医学研究所进行了一项专门针对动物饲料中使用青霉素和四环素对人类健康影响的调研及其危害性的量化评估。但该研究机构并未找到在动物饲料中使用这两种抗菌剂会危害人类健康的直接证据（Institute of Medicine，1988）。同样，其他研究机构也得到了相同结果，也就是没有确切的证据表明在动物上使用低剂量的抗菌剂会对人类健康造成影响（National Research Council，1980，1999；Council for Agricultural Science and Technology，1981）。近期研究得出的结论与之前的相同（Hillips等，2004；Institute for Food Technologists，2006）。

在肯塔基大学的研究表明，完全禁止使用低剂量抗菌剂只会对抗菌剂的抵抗水平和方式产生微弱的影响（Langlois等，1978a，1978b，1986）。其中一项研究表明，即使不使用抗菌剂长达13年之久，仍有很大一部分的猪肠道微生物出现四环素抗性。因此，限制或禁止使用抗生素是否会显著影响药物抵抗的微生物种类值得怀疑。20世纪70年代，英国政府采取了限制在处方中使用特定药物的措施，但这既没有减少四环素在英国的使用量，也没有对英国的猪肠原杆菌抗药方式造成影响（Smith，1975，1977；Braude，1978）。最近发现，丹麦在很多年前就禁止使用低剂量抗生素，但猪病问题变得越来越严重，抗生素使用的总量比禁止使用之前的还要多（Ministry of Food，Agriculture和Fisheries，2008）。

对人和动物中细菌耐药性的监测与监管一直都在进行中，但仍不是很清楚人畜交叉传染途径。即使目前人药物抵抗的发生率非常高，但仍没有有力的证据证明使用低剂量的抗生素会改变人药物抵抗的水平和方式（Lorian，1986）。抗菌剂在数以10亿的动物上已经使用了近60年，并没有得到令人信服的证据说明，给动物饲喂低剂量的抗菌剂会直接对人类的健康产生任何不利的影响。

3 微生物制剂

微生物制剂，有时被称为“益生菌”，但更确切叫做“饲用微生物”，由活的、天然存在的、可饲喂动物的微生物组成。饲料中添加微生物制剂的目的在于使肠道微生物菌群更加平衡，从而改善宿主动物的肠道健康和生产性能。猪上常用的微生物制剂包括乳酸杆菌类、肠球菌类（过去称为链球菌）和酵母。目前已有大量关于这些微生物制剂在猪上使用效果的报道（Pollmann，1986，1992；Fuller，1989，1992a；van Belle等，1990；Jonsson和Conway，1992；Stavric和Kornegay，1995）。

迄今有45种不同微生物被认定为“公认安全的添加剂”（GRAS），可以直接用作动物饲用微生物（Anonymous，2008；AAFCO，2012）。这些微生物包括2类真菌，即霉菌（黑曲霉）和酵母（酿酒酵母）；7类革兰氏阳性菌，即肠膜明串珠菌、芽孢杆菌类（5种）、双歧杆菌类（6种）、乳酸杆菌类（14种）、足球菌（3种）、丙酸杆菌（3种）、肠球菌（6种）；以及1类革兰氏阴性菌（拟杆菌属，4种）。其中双歧杆菌类、乳酸杆菌类、肠球菌类和酿酒酵母通常是直接饲喂，而黑曲霉和芽孢杆菌类还可用于产酶。肠膜明串珠菌、足球菌类和丙酸杆菌类主要用作青贮饲料的添加剂。在所有被批准使用的微生物中，除保加利亚乳酸杆菌和嗜热肠球菌是用于制作酸奶的发酵微生物外，其他微生物均属于肠道菌株。

◆ 3.1 微生物制剂的使用背景

回顾历史，对微生物制剂的研究可追溯到20世纪初。Metchnikoff（1903，1908）研究认为，保加利亚人的长寿与他们每天食用的酸奶密切相关，酸奶中的乳酸杆菌可以平衡肠道环境，抑制病原菌的生长，进而提高人的健康状况，延长人的寿命。在20世纪的上半个世纪，嗜酸乳杆菌被广泛用于治疗儿童腹泻（Winkelstein，1956）。

虽然在20世纪初就有一些微生物制剂用于动物饲料中，但直到90年代微生物制剂商品化之后才开始被广泛使用。1974年Parker首先提出“益生素”（probiotics）这个词，是指有利于肠道菌群平衡的微生物和物质（Fuller，1992a）。起初，益生素的定义涵盖了所有喂给动物的活微生物、死微生物或者是发酵副产物（Fox，1988），但Fuller（1989）提出将益生素的定义仅仅限定于活的微生物。目前，美国FDA规定用“饲用微生物”一词来代替“益生素”，并且规定这类添加剂必须是“一类活的、天然存在的微生物”（Pendelton，1992）。饲用微生物被美国FDA和AAFCO归类于“公认安全的添加剂”（GRAS）范畴之内。GRAS的定义是：经过由药理学专家和毒理学专家组成的专家组鉴定的可安全使用的食品和饲料添加剂，而专家组根据食品和饲料添加剂被人摄入后现有的技术资料来进行界定（Pollmann，1986）。

新生仔猪在刚出生时，消化道内基本上不含有微生物。当它们一旦接触外界自然环境，各种各样的微生物就会迅速移植到消化道内。在仔猪健康且无应激的条件下，有益菌群与宿主共生并定植在肠道表面，而那些可能致病的有害菌群被抑制。形成一个稳定的保护性肠道微生物菌群对动物的健康和生长非常重要。当动物处于应激状态时，肠道有益菌群与有害菌群之间的平衡被打乱，会导致动物腹泻、肠胃炎和生产性能的下降。饲用微生物的作用机理是通过摄入大量的有益菌群来抑制病原微生物在肠道内的大量繁殖，以减轻外界应激产生的负面效果。

◆ 3.2 饲用微生物的功能

直接饲喂微生物制剂的促生长效果被认为不如饲喂抗菌剂那样具有稳定的促生长效果。过去有许多声称微生态制剂具有提高生长速度和饲料利用率、改善动物健康水平、降低腹泻和提高繁殖性能的功能的报道，但这些功能并未得到有对照组的试验数据支持（Pollmann，1992），没有像抗菌剂一样用大量有对照组的试验来进行证实。因此，很多微生物制剂的功能是在有限的甚至没有任何试验数据证实的情况下，而只是基于对它们有利作用的主观判断，就开始在养猪业中进行提倡、生产、销售和使用。

1986年Pollmann发表了一篇关于猪饲喂乳酸杆菌的作用效果的综述。他发现，在18个试验中有13个试验表明，断奶仔猪的生长性能在数值上得到了提高。而对7个生长育肥猪试验的总结发现，饲料中添加乳酸杆菌或屎肠球菌并不是都能提高生长性能。同时他还报道，仔猪或母猪饲料中添加枯草芽孢杆菌也不是一定能提高仔猪生长性能或母猪繁殖性能。Nousiainen和Setala（1993）对26个试验的总结发现，有16个试验的饲用微生物在数值上表现出具有促生长的作用（只有2个试验在统计上有显著差异），而有9个试验的饲用微生物表现出具有抑制生长的作用（有2个试验在统计上有显著差异）。Stavric和Kornegay（1995）等其他学者也得出类似的结论，即不同的仔猪和生长育肥猪试验，饲用微生物的作用效果不尽相同。

肯塔基大学对仔猪教槽料中单独添加酿酒酵母、嗜酸乳杆菌、粪链球菌的混合制剂，或与抗菌剂（氯四环素、青霉素和磺胺二甲嘧啶的混合产品）联合使用的作用效果进行了系列评价试验（Cromwell等，1991；表13-8）。这项研究包括5个试验，每个处理29个重复，每个重复6头猪。结果表明，5个试验使用微生物制剂都没有影响猪的生长速度和饲料利用率，但添加抗菌剂都可以提高猪的生长、降低料重比。

表13-8 饲用微生物（DFM）和抗菌剂在断奶仔猪饲粮中的使用效果[1]

项目	不添加组	饲用微生物组[2]	抗菌剂组[2]	联合使用组
日增重（g）[3]	247	237	306	310
日采食量（g）[3]	467	460	540	550
料重比[3]	1.92	1.96	1.77	1.75

[1]综合了5个试验，包括764头4周龄断奶的仔猪（平均体重7.4 kg），每个处理26个重复，每个重复5头/圈，试验为期4周（Cromwell等，1991）。

[2]饲用微生物是由酿酒酵母、嗜酸乳杆菌和粪链球菌混合而成，抗菌剂是由氯四环素、青霉素和磺胺甲嘧啶混合而成。

[3]抗菌药的作用效果（$P<0.001$）。

对于饲用微生物的作用效果，虽然在有对照组的试验中无法得到统一结果，但饲用微生物在实际生产中的使用效果较为一致，尤其是在疫病高发和应激大的条件下。有些有对照组的试验表明，饲用微生物可以降低仔猪腹泻（Hale和Newton，1979；Maeng等，1989；Eigel，1989），但其他有对照组的试验并没有发现这种效果（Wu等，1987；McLeese等，1992；De Cupere等，1992；Apgar等，1993）。

4 低聚糖

还有一些有机物被发现可改变肠道内固有微生物菌群的组成，如低聚糖。甘露寡糖（MOS）是一类在酵母细胞壁中含量丰富的短链甘露糖。这类化合物与某些病原菌的特异位点有很高的亲和力，从而可以阻止这些病原菌吸附在肠上皮细胞上。因此，这些病原菌就会被排出小肠，而有益菌群（如乳酸杆菌）

就能吸附并定植在肠上皮。这个过程通常被称为“竞争排斥”。研究发现，这种作用在家禽上更为明显（Newman，1996），而在猪上报道的相对较少。Davis等（1999）研究发现，饲粮添加MOS提高了断奶仔猪的生长，但作用效果约为使用高剂量硫酸铜的一半，而硫酸铜是一种众所周知的抗菌剂（表13-9）。对29个试验的系统分析发现，MOS可使猪增重提高4.2%，使料重比降低2.3%（Miguel等，2003）。他们还发现，MOS对生长慢的猪的作用效果要比对生长快的猪更为明显。Rozeboom等（2005）饲喂MOS也得到类似的结果。

表13-9 甘露寡糖（MOS）和抗菌剂在断奶仔猪饲粮中的使用效果[1]

项目	对照组	甘露寡糖添加组[2]	铜添加组[3]	联合使用组
日增重（g）	364	4.2	439	452
日采食量（g）	564	596	659	638
料重比	1.56	1.45	1.42	1.37

[1]试验为期38d，仔猪21日龄断奶，平均初始体重5.8 kg，每个处理9个重复，每个重复6头/圈（Davis等，2002）。

[2]饲粮甘露寡糖添加水平为0.2%。

[3]硫酸铜的添加量为185ppm Cu。

低聚糖果寡糖（FOS）是一种末端含葡萄糖基的中短链果糖。而菊糖则是由果寡糖组成的碳水化合物，天然存在于洋姜、菊苣及一些其他植物中。果寡糖在小肠内不能被消化，但可在后肠被微生物发酵。果寡糖与甘露寡糖的作用方式不同，因为果寡糖可以作为双歧杆菌等一些肠道有益菌群的发酵底物，从而抑制有害菌群或者病原菌在肠道内的大量繁殖。采用大肠杆菌对仔猪攻毒的试验发现，与不饲喂果寡糖组相比，饲喂果寡糖减少了肠道内大肠杆菌的数量，且降低了腹泻率和死亡率（Bunce等，1995）。对健康仔猪的试验发现，饲粮添加果寡糖对增重的提高幅度与使用抗生素的作用差不多（Russell等，1996；Howard等，1999；表13-10）。但有些研究表明，给仔猪饲喂果寡糖并没有得到正面效果（Farnworth等，1991，1992，1995；Kornegay等，1992）。另有试验发现，给仔猪饲喂果寡糖改善了猪肠道形态结构，如增加绒毛高度和绒毛/隐窝比例（Spencer等，1997）。

表13-10　果寡糖（FOS）和抗菌剂在断奶仔猪饲粮中的使用效果[1]

项目	对照组	甘露寡糖添加组[2]	卡巴多司添加组[2]	联合使用组
日增重（g）	338	379	380	420
日采食量(g）	534	594	560	622
料重比	1.58	1.57	1.47	1.48

[1]总结了2个试验，每个处理8个重复，每个重复4头猪/圈，仔猪15日龄断奶，平均初始体重为5.5 kg，试验持续28d；资料来源于Russell等（1996）。

[2]每天可提供383 mg果寡糖，卡巴多司在饲粮中的添加量55 mg/kg。

低聚半乳糖（GOS）天然存在于黄豆中，也可用乳糖合成。Smiricky-Tjardes等（2003）发现，饲喂低聚半乳糖影响了养分消化率和肠道菌群的生长。壳寡糖（COS）是从角质素提取得到的。研究发现，仔猪饲喂壳寡糖改变了猪肠道菌群，提高了养分消化率，减少了腹泻并提高了免疫力（Liu等，2008，2010）。

5 药用微量元素

当铜和锌在饲粮中的添加量远远超过猪的营养需要量时，这两种微量元素就表现出抗菌作用。诸多试验表明，高铜和高锌都具有促进猪生长的作用（Cromwell，1997，2001）。还有些研究发现，给母猪饲喂高铜饲粮提高了其繁殖性能（Cromwell等，1993）。

爱荷华州立大学Evvard等（1928）和英国Braude（1945）最早发现，饲喂高铜饲粮具有提高猪生长速度的作用。随后，高铜的促生长作用得到世界各地的大量试验证实。表13-11的数据表明，饲粮高铜（200 ~ 250 ppm）对生长速度和饲料利用率的提高程度与抗生素的饲喂效果差不多。但高铜与抗生素的作用机制可能并不相同，因为同时给仔猪饲喂抗生素和高铜时，其作用效果具有可加性（表13-12）。

关于铜添加剂量的研究表明，饲粮添加200 ~ 250 ppm铜的作用效果最佳，而添加100 ~ 125 ppm铜只能达到最佳作用效果的75% ~ 80%（Cromwell等，1989）。硫酸铜、碳酸铜和氯化铜都是铜的有效添加形式（Wallace，

1967；Cromwell等，1998），而氧化铜和硫化铜则不是（Cromwell等，1989；Cromwell，1997）。赖氨酸铜螯合物（Coffey等，1994；Apgar等，1995）及其他有机铜复合物（Bunch等，1965；Stansbury等，1990）也是铜的有效添加形式。添加超高水平的铜（≥500 ppm）可造成过量的铜在肝脏内蓄积，进而可能引起猪中毒（Cromwell，1997）。

表13-11 硫酸铜对断奶仔猪和生长育肥猪生长性能的影响

阶段	对照组	高铜组[1]	提高幅度（%）
仔猪阶段（8 ~ 20 kg）[2]			
日增重（kg）	0.34	0.38	11.9
料重比	1.87	1.78	4.5
生长阶段（18 ~ 56 kg）[3]			
日增重（kg）	0.67	0.71	6.9
料重比	2.80	2.70	3.6
生长育肥阶段（18 ~ 93 kg）[3]			
日增重（kg）	0.71	0.74	3.1
料重比	3.18	3.10	2.5

[1]除微量元素预混料所提供的6～12ppm铜外，另外添加硫酸铜提供200～250ppm铜。
[2]在1978—1997期间肯塔基大学开展的23个试验的综合结果，涉及1376头3～4周龄断奶的仔猪，试验为期4～5周。
[3]在1970—1980期间肯塔基大学开展的18个试验的综合结果，涉及672头断奶仔猪。

表13-12 抗生素和硫酸铜单独使用或联合使用对断奶仔猪生长性能的影响[1, 2]

项目	对照组	抗生素组	高铜（250ppm）组	联合使用组
日增重（kg）	0.26	0.30	0.31	0.34
日采食量（kg）	0.54	0.58	0.59	0.63
料重比	2.10	1.95	1.91	1.84

[1]在6个不同试验基地开展的14个试验的综合结果，涉及1 700头猪。
[2]资料来源于Beames和Lloyd，1965；Mahan，1980；Stahly等，1980；Edmonds等，1985；Hagen等，1987；Burnell等，1988。

Kulwich等（1953）最早发现，添加高剂量的氧化锌具有降低断奶仔猪腹泻率的作用。随后的研究表明，以氧化锌形式添加1 500～3 000 ppm的锌还具有促进仔猪生长的作用（Hahn和Baker，1993；LeMieux等，1995；Hill等，2000；Hollis等，2004）。表13-13对高剂量氧化锌促生长的试验进行了总结。添加其他形式的锌没有出现氧化锌的促生长作用（Hollis等，2004）。

表13-13　饲粮中高锌对断奶仔猪生长性能的影响

项目	对照组	高锌组[1]	提高幅度（%）
总结 1[2]			
日增重（g）	375	422	12.5
采食量（g）	637	690	8.3
料重比	1.71	1.64	4.1
汇总 2[3]			
日增重（g）	352	392	11.4
采食量（g）	564	609	8.0
料重比	1.59	1.56	1.9

[1]添加氧化锌在总结1中提供3 000ppm的锌、在汇总2中提供2 000ppm或2 500ppm的锌。在总结1中的对照组和高锌饲粮中添加220ppm氯四环素。

[2]12所大学开展的12个试验的综合结果，试验为期4周，使用了678头22日龄仔猪，平均初始体重为6.6 kg（每个处理55个重复）（Hill等，2000）。

[3]12个大学展开的14个试验的综合结果，试验持续4周，使用了402头21日龄仔猪，平均初始体重为6.3 kg（每个处理41个重复）（Hollis等，2005）。

美国FDA没有严格控制猪饲喂高铜和高锌，反而将高铜、高锌归类为“公认安全的添加剂”（GRAS）的范围内。然而，使用高铜或高锌饲粮的养猪户应注意的是，绝大部分铜或锌不能被机体吸收，而是通过粪便排出体外。因此，湖泊和土壤在使用过这些富含铜或锌的肥料后，很可能蓄积有大量的铜或锌。而大量铜或锌的蓄积是否会对自然环境产生负面影响，仍是个有争议的问题。

6 酸化剂

仔猪教槽料中添加1%～3%有机酸（如柠檬酸、富马酸和甲酸等）对早期断奶仔猪有益处（表13-14）。饲粮添加磷酸等无机酸也有相同作用。酸化剂的作用机制目前还不十分清楚，但降低胃肠道前段的pH有助于减少有害菌在仔猪胃和小肠内的增殖。也有人提出，降低胃的pH可提高胃蛋白酶活性。有机酸还被用作高水分含量的谷物的防腐剂或饲料防霉剂（Crenshaw等，1986）。

表13-14 有机酸在断奶仔猪教槽料中的使用效果[1, 2]

项目	对照组	有机酸组	提高率（%）
日增重（g）	302	318	5.3
日采食量（g）	565	553	−2.1
料重比	1.87	1.74	7.0

[1]对10个试验的综合结果：体重为8.9～16.6 kg的697头猪，添加的有机酸是柠檬酸和富马酸。
[2]资料来源于Kirchgessner和Roth，1982；Falkowski和Aherne，1984；Giesting和Easter，1985；Edmonds等，1985；Burnell等，1988。

7 植物源添加剂

植物提取物添加剂（通常被称为植物源抗生素，phytobiotics或botanicals）是一类从植物中提取的复合物，包括各种各样的植物药、香料及其提取产品，主要是精油。近些年，植物源添加剂作为抗生素替代品引起世界各国广泛的关注。Windisch等（2010）对植物源添加剂试验进行了总结，发现很多植物源添加剂虽在体外试验有抗氧化和抗菌的特性，但在动物饲养试验上得到的证据仍相当少。有些研究发现，植物源添加剂添加剂改变了肠道微生物区系，改善了肠道健康。

8 酶制剂

由纤维素酶、半纤维素酶和蛋白酶组成的复合酶制剂有时用作饲料添加剂，以提高饲料中复杂的碳水化合物和蛋白质的消化率。这些复合酶制剂常常用在那些饲料原料更多样化的国家或地区，而在使用典型玉米-豆粕型饲粮的北美洲，复合酶制剂的使用就非常少。有些研究表明使用这些酶制剂是有益的（Wenk，1992）。在用大麦或黑麦作饲料原料的地区，有时会使用β-葡聚糖酶和戊聚糖酶来降解这些原料中含有的β-葡聚糖和戊聚糖（这两种复杂的碳水化合物会干扰其他养分的消化）（Newman等，1980；Li等，1995），但并不一定能改善猪的生长性能（Thacker，1993；Thacker和Baas，1996）。同时，关于在保育仔猪饲粮中添加淀粉酶和蛋白酶对养分消化率的作用效果，各研究结果也不尽相同（Lewis等，1955；Cunningham和Brisson，1957a，1957b；Combs等，1960）。Wenk和Boessinger（1993）及van Hartingsveldt等（1995）对使用饲用酶制剂的资料进行了全面综述。

近些年，植酸酶受到世界各国的广泛关注。植酸酶可从植酸中酶切正磷酸基，而植酸是谷物和油籽饼粕中磷的主要存在形式。添加植酸酶可以显著提高猪对植酸磷的利用率（Simons等，1990；Jongbloed等，1992；Cromwell等，1995），并减少磷在环境中的排放量。更详细的植酸酶资料参考第12章。

9 香味剂

很多种人工合成的香味剂可用于饲料中，其目的是提高饲料适口性或者掩盖饲料中难闻的气味。多数试验表明，当给猪提供选择机会的话，猪会选择性摄入含有香味剂或芳香化合物的饲料，但在没有任何选择的情况下，绝大多数香味剂的作用就微乎其微（Hines，1973；Hines等，1975；Kornegay等，1979；Ogunbameru等，1979）。McLaughlin等（1983）对猪饲料中使用香味剂进行了综述。

10 除臭剂

一些除臭剂被用作饲料添加剂，以达到减少猪粪臭气的目的。例如，丝兰皂甙就是其中的一种，它是丝兰植物提取物，可以抑制脲酶活性，从而减少猪粪中氨气的释放量（Sutton等，1992）。研究还发现，丝兰皂甙能提高断奶仔猪和生长育肥猪的生长性能（Foster，1983；Cromwell等，1985）。其他除臭剂，如某些自然界存在的、干燥的活菌制剂，添加到饲料中也能减少猪粪的臭味。还有些研究发现，沸石减少了猪粪的臭味和氮的排放（Barrington和El Moueddeb，1995）。某些低聚糖也可通过影响后肠微生物来减少猪粪的臭味（Sutton等，1991；Farnworth等，1995）。

11 抗氧化剂

乙氧喹或丁基羟基甲苯等抗氧化剂通常用于含大量不饱和脂肪酸的饲料中，以减少不饱和脂肪酸的氧化损失，同时还可以防止维生素的氧化。

12 黏结剂和流化剂

膨润土等黏土在制粒前添加到饲料中，用来增加饲料的黏结性、防止颗粒料的破碎。还有一些黏土和沸石可通过吸附饲料中的黄曲霉毒素来阻止毒素被吸收，从而防止猪发生黄曲霉毒素中毒（Schell等，1993），但这种作用并没有获得美国FDA的认可。流化剂与黏结剂一样也具有吸附黄曲霉毒素的功能，但使用流化剂的目的是阻止某些饲料原料结块并增加其流动性。如水合硅铝酸钠钙，虽然它防止黄曲霉毒素中毒的作用没有获得美国FDA的认可，但它仍能有效吸附黄曲霉毒素（Lindemann等，1993）。

13 胴体改良剂

研究发现，克伦特罗、西马特罗和盐酸莱克多巴胺等一些β-肾上腺素兴奋剂在饲料中添加时，能增加胴体瘦肉率（Jones等，1985；Moser等，1986；Cromwell等，1988；Watkins等，1990；Bark等，1992）。在这些β-肾上腺素兴奋剂中，盐酸莱克多巴胺是唯一一种被FDA批准在美国使用的添加剂。在一定条件下，甜菜碱和肉碱也具有提高胴体瘦肉率的作用（Odle，1995）。有些研究报道，铬（如吡啶甲酸铬）能提高胴体瘦肉率（Page等，1993；Lindemann等，1995；Mooney和Cromwell，1995），但其他报道并未出现类似结果（Crow和Newcomb，1997；Mooney和Cromwell，1999）。最近研究发现，共轭亚油酸（油酸的衍生物，具有共轭双键的脂肪酸的位置和几何异构体）具有降低体脂和增加瘦肉组织的功能（Pariza，1997；Parrish等，1997）。

14 驱虫剂

这类药物也被称为“除虫剂（dewormers）”，在饲料中添加用于控制体内寄生虫（如蛔虫、肺蠕虫和蛲虫）。美国FDA批准了五种猪上使用的驱虫剂，分别是敌敌畏、芬苯达唑、伊维菌素、盐酸左旋四咪唑和噻吩嘧啶酒石酸（饲料添加剂纲要，2012）。其中，伊维菌素还可有效驱除全身的外部寄生虫（如虱子和螨虫）。

作者：Gary L. Cromwell

译者：杨飞云，黄金秀

参考文献

AAFCO. 2012. Official Publication. Association of American Feed Control Officials, Oxford, IN.

Anonymous. 2008. Direct-Fed Microbial, Enzyme, and Forage Additive Compendium. 9th ed. The Miller Publishing Co., Minnetonka,MN.

Apgar, G. A., E. T. Kornegay, M. D. Lindemann, and D. R. Notter. 1995. Evaluation of copper sulfate and a copper lysine complex as growth promotants for weanling swine. J. Anim. Sci. 73:2640–2646.

Apgar, G. A., E. T. Kornegay, M. D. Lindemann, and C. M. Wood. 1993. The effect of feeding various levels of Bifidobacterium globosum A on the performance, gastrointestinal measurements, and immunity of weanling pigs and on the performance and carcass measurements of growing-finishing pigs. J. Anim. Sci. 71:2173–2179.

Bark, L. J., T. S. Stahly, G. L. Cromwell, and J. Miyat. 1992. Influence of genetic capacity for lean tissue growth on rate and efficiency of tissue accretion in pigs fed ractopamine. J. Anim. Sci. 70:3391–3400.

Barrington, S., and K. El Moueddeb. 1995. Zeolite to control swine manure odours and nitrogen volatilization. Pages 65–68 in Proc. International Livestock Odor Conference. Iowa State University, Ames, IA.

Beames, R. M., and L. E. Lloyd. 1965. Response of pigs and rats to rations supplemented with tylosin and high levels of copper.J. Anim. Sci. 24:1020–1026.

Braude, R. 1945. Some observations on the need for copper in the diet of fattening pigs. J. Agric. Sci. 35:163–167.

Braude, R. 1978. Antibiotics in animal feeds in Great Britain. J. Anim. Sci. 46:1425–1436.

Braude, R., M. F. Coates, M. K. Davies, G. F. Harrison, and K. G. Mitchell. 1955. The effect of aureomycin on the gut of a pig.Br. J. Nutr. 9:363–368.

Braude, R., and B. C. Johnson. 1953. Effect of aureomycin on nitrogen and water metabolism in growing pigs. J. Nutr. 49:505–512.

Brody, T. M., M. R. Hurwitz, and J. A. Bain. 1954. Magnesium and the effect of the tetracycline antibiotics on oxidative processes in mitochondria. Antibiot. Chemother. 4:864–870.

Bunce, T. J., M. D. Howard, M. S. Kerley, G. L. Allee, and L. W. Pace. 1995. Protective effect of fructooligosaccharide (FOS) in prevention of mortality and morbidity from infectious E.

coli K:88 challenge. J. Anim. Sci. 63(Suppl. 1):69. (Abstr.)

Bunch, R. J., J. T. McCall, V. C. Speer, and V. W. Hays. 1965. Copper supplementation for weanling pigs. J. Anim. Sci. 24:995–1000.

Burnell, T.W., G. L. Cromwell, and T. S. Stahly. 1988. Effects of dried whey and copper sulfate on the growth responses to organic acid in diets for weanling pigs. J. Anim. Sci. 66:1100–1108.

Catron, D. V., A. H. Jensen, P. G. Homeyer, H. M. Maddock, and G. C. Ashton. 1952. Re-evaluation of protein requirements of growing-fattening swine as influenced by feeding an antibiotic. J. Anim. Sci. 11:221–232.

Catron, D. V., M. D. Lane, L. Y. Quinn, G. C. Ashton, and H. M. Maddock. 1953. Mode of action of antibiotics in swine nutrition.Antibiot. Chemother. 3:571–577.

Coffey, R. D., G. L. Cromwell, and H. J. Monegue. 1994. Efficacy of a copper-lysine complex as a growth promotant for weanling pigs. J. Anim. Sci. 72:2880–2886.

Combs, G. E., W. L. Alsmeyer, H. D. Wallace, and M. Koger. 1960. Enzyme supplementation of baby pig rations containing different sources of carbohydrate and protein. J. Anim. Sci. 19:932–937.

Council for Agricultural Science and Technology. 1981. Antibiotics in Animal Feeds. Report No. 88. Council for Agricultural Science and Technology, Ames, IA.

Crenshaw, J. D., E. R. Peo, Jr., A. J. Lewis, and N. R. Schneider. 1986. The effects of sorbic acid in high-moisture sorghum grain diets on performance of weanling swine. J. Anim. Sci. 63:831–837.

Cromwell, G. L. 1997. Copper as a nutrient for animals. Pages 177–202 in Handbook of Copper Compounds and Applications. H.W. Richardson, ed. Marcel Dekker, Inc., New York, NY.

Cromwell, G. L. 1999. Subtherapeutic use of antibiotics for swine: Performance, reproductive efficiency and safety issues. Pages 70–87 in Proc. 40th Annual George A. Young Swine Health and Management Conference. Univ. of Nebraska, Lincoln, NE.

Cromwell, G. L. 2001. Antimicrobial and Promicrobial Agents. Pages 401–426 in Swine Nutrition. 2nd ed. A. J. Lewis and L. L.Southern, eds. CRC Press, Boca Raton, FL.

Cromwell, G. L. 2002. Why and how antibiotics are used in swine production. Anim. Biotech. 13:7–27.

Cromwell, G. L., R. D. Coffey, G. R. Parker, H. J. Monegue, and J. H. Randolph. 1995. Efficacy of a recombinant-derived phytase in improving the bioavailability of phosphorus in corn–soybean meal diets for pigs. J. Anim. Sci. 73:2000–2008.

Cromwell, G. L., R. I. Hutagalung, and T. S. Stahly. 1982. Effects of form of sulfamethazine (powder vs. granular) on sulfa carry-over in swine feed. J. Anim. Sci. 55(Suppl. 1): 267. (Abstr.)

Cromwell, G. L., J. D. Kemp, T. S. Stahly, and R. H. Dalrymple. 1988. Effects of dietary level and withdrawal time on the efficacy of cimaterol as a growth repartitioning agent in finishing swine. J. Anim. Sci. 66:2193–2199.

Cromwell, G. L., M. D. Lindemann, H. J. Monegue, D. D. Hall, and D. E. Orr, Jr. 1998. Tribasic copper chloride and copper sulfate as copper sources for weanling pigs. J. Anim. Sci. 76:118–123.

Cromwell, G. L., H. J. Monegue, and T. S. Stahly. 1993. Long-term effects of feeding a high copper diet to sows during gestation and lactation. J. Anim. Sci. 71:2996–3002.

Cromwell, G. L., T. S. Stahly, K. A. Dawson, H. J. Monegue, and K. Newman. 1991. Probiotics and antibacterial agents for weanling pigs. J. Anim. Sci. 69(Suppl. 1):114. (Abstr.)

Cromwell, G. L., T. S. Stahly, and H. J. Monegue. 1985. Efficacy of sarsaponin for weanling and growing-finishing swine housed at two animal densities. J. Anim. Sci. 61 (Suppl. 1):111. (Abstr.)

Cromwell, G. L., T. S. Stahly, and H. J. Monegue. 1989. Effects of source and level of copper on performance and liver copper stores in weanling pigs. J. Anim. Sci. 67:2996–3002.

Cromwell, G. L., T. S. Stahly, H. J. Monegue, E. R. Peo, Jr., B. D. Moser, and A. J. Lewis. 1981. Effects of sulfamethazine vs. sulfathiazole in finishing feed on sulfa residues in swine. J. Anim. Sci. 53(Suppl. 1):95. (Abstr.)

Crow, S. D., and M. D. Newcomb. 1997. Effect of dietary chromium additions along with varying protein levels on growth performance and carcass characteristics. J. Anim. Sci. 74:79. (Abstr.)

Cunha, T. J., J. E. Burnside, D. M. Buschman, R. S. Glasscock, A. M. Pearson, and A. L. Shealy. 1949. Effect of vitamin B12, animal protein factor and soil for pig growth. Arch. Biochem. 23:324–326.

Cunningham, H. M., and G. J. Brisson. 1957a. The effect of amylases on the digestibility of starch by baby pigs. J. Anim. Sci.16:370–376.

Cunningham, H. M., and G. J. Brisson. 1957b. The effect of proteolytic enzymes on the utilization of animal and plant proteins by newborn pigs and the response to predigested protein. J. Anim. Sci. 16:568–573.

Davis, M. E., C. V. Maxwell, D. C. Brown, E. B. Kegley, B. Z. de Rodas, K. G. Friesen, D. H. Hellwig, and R. A. Dvorak. 2002.Effect of dietary mannan oligosaccharides and(or) pharmacological additions of copper sulfate on growth performance and immunocompetence of weanling and growing/finishing pigs. J. Anim. Sci. 80:2887–2894.

De Cupere, F., P. Deprez, D. Demeulenaere, and E. Muylle. 1992. Evaluation of the effect of 3 probiotics on experimental Escherichia coli enterotoxaemia in weaned piglets. J. Vet. Med. B 39:277–284.

Edmonds, M. S., O. A. Izquierdo, and D. H. Baker. 1985. Feed additive studies with newly weaned pigs: Efficacy of supplemental copper, antibiotics and organic acids. J. Anim. Sci. 60:462–469.

Eigel, W. N. 1989. Ability of probiotics to protect weanling pigs against challenge with enterotoxigenic E. coli. Pages 10–19 in Proceedings Chr. Hansen Biosystems Technical Conf. San Antonio, TX.

Evvard, J. M., V. E. Nelson, and W. E. Sewell. 1928. Copper salts in nutrition. Proc. Iowa Acad. Sci. 35:211.

Falkowski, J. F., and F. X. Aherne. 1984. Fumaric and citric acid as feed additives in starter pig nutrition. J. Anim. Sci. 58:935–938.

Farnworth, E. R., N. Dilawri, H. Yamazaki, H.W. Modler, and J. D. Jones. 1991. Studies on the effect of adding Jerusalem artichoke flour to pig milk replacer. Can. J. Anim. Sci. 71:531–536.

Farnworth, E. R., H. W. Modler, J. D. Jones, N. Cave, H. Yamazaki, and A. V. Rao. 1992. Feeding Jerusalem artichoke flour rich in fructooligosaccharides to weanling pigs. Can. J. Anim. Sci. 72:977–980.

Farnworth, E. R., H. W. Modler, and D. A. Mackie. 1995. Adding Jerusalem artichoke (*Helianthus tuberosus* L.) to weanling pig diets and the effect on manure composition and characteristics. Anim. Feed Sci. Tech. 55:153–160.

Feed Additive Compendium. 2012. The Miller Publishing Co., Minnetonka, MN.

Foster, J. R. 1983. Effects of sarsaponin in growing-finishing swine diets. J. Anim. Sci. 57(Suppl. 1):94. (Abstr.)

Fox, S. M. 1988. Probiotics: intestinal inoculants for production animals. Vet. Med. 83:806–830.

Fuller, R. 1992a. History and development of probiotics. Pages 1–8 in Probiotics: The Scientific Basis. R. Fuller, ed. Chapman and Hall, London, UK.

Fuller, R. 1989. Probiotics in man and animals. J. Appl. Bacteriol. 66:365–378.

Giesting, D.W., and R. A. Easter. 1985. Response of starter pigs to supplementation of corn-soybean meal diets with organic acids.J. Anim. Sci. 60:1288–1294.

Hagen, C. D., S. G. Cornelius, R. L. Moser, J. E. Pettigrew, and K. P. Miller. 1987. High levels of copper alone or in combination with antibacterials in weanling pig diets. Nutr. Rep. Int. 35:1083.

Hahn, J. D., and D. H. Baker. 1993. Growth and plasma zinc responses of young pigs fed pharmacologic levels of zinc. J. Anim.Sci. 71:3020–3024.

Hale, O. M., and G. L. Newton. 1979. Effects of a nonviable lactobacillus species fermentation product on performance of pigs. J.Anim. Sci. 48:770–775.

Hash, J. H., M. Wishnick, and P. A. Miller. 1964. On the mode of action of the tetracycline antibiotics in Staphylococcus aureus.J. Biol. Chem. 239:2070–2078.

Hays, V. W. 1977. Effectiveness of Feed Additive Usage of Antibacterial Agents in Swine and Poultry Production. Office of Technology Assessment, U.S. Congress. Washington, DC.

Hays, V. W. 1978. The role of antibiotics in efficient livestock production, in Nutrition and Drug Interrelations. Academic Press,New York, NY.

Hays, V. W., and V. C. Speer. 1960. Effect of spiramycin on growth and feed utilization of young pigs. J. Anim. Sci. 19:938–942.

Hedde, R. D. 1981. Intestinal fermentation in the pig and how it is influenced by age and virginiamycin. Pages 10–20 in Proc. Growth Promotion Mode-of-Action Symp. SmithKline Corp., Philadelphia, PA.

Henderickx, H. K., I. J. Vervaecke, J. A. Decuypere, and N. A. Dierick. 1981. Mode of action of growth promotion drugs. Pages3–9 in Proc. Growth Promotion Mode-of-Action Symp. SmithKline Corp., Philadelphia, PA.

Hill, G. M., G. L. Cromwell, T. D. Crenshaw, C. R. Dove, R. C. Ewan, D. A. Knabe, A. J. Lewis, et al. 2000. Growth promotion effects and plasma changes from feeding high dietary concentrations of zinc and copper to weanling pigs (regional study). J.Anim. Sci. 78:1010–1016.

Hines, R. H. 1973. Feed flavors in swine starter rations. Pages 37–42 in Proc. Swine Industry Day, Kansas State Univ., Manhattan,KS.

Hines, R. H., B. A. Koch, and G. L. Allee. 1975. Attractants for swine starter feed: Aroma vs. taste. Pages 20–23 in Proc. Swine Industry Day, Kansas State Univ., Manhattan, KS.

Hollis, G. R., S. D. Carter, T. R. Cline, T. D. Crenshaw, G. L. Cromwell, G. M. Hill, S. W. Kim, et al. 2005. Effects of replacing pharmacological levels of dietary zinc oxide with lower dietary levels of various organic zinc sources for weanling pigs. J.Anim. Sci. 83:2123–2129.

Howard, M. D., H. Liu, J. D. Spencer, M. S. Kerley, and G. L. Allee. 1999. Incorporation of short-chain fructooligosaccharides and Tylan into diets of early weaned pigs. J. Anim. Sci. 77(Suppl. 1):63. (Abstr.)

Institute for Food Technologists. 2006. Antimicrobial resistance: Implications for the food system.

Accessed Sept.1,2010. http://www.ift.org/Knowledge%20Center/Read%20IFT%20Publications/Science%20Reports/Expert%20Reports/Antimicrobial%20Resistance.aspx?page=viewall.

Institute of Medicine. 1988. Human Health Risks with the Subtherapeutic Use of Penicillin or Tetracycline in Animal Feed. National Academies of Science, National Academies Press, Washington, DC.

Jones, R. W., R. A. Easter, F. K. McKeith, R. H. Dalrymple, H. M. Maddock, and P. J. Bechtel. 1985. Effect of the β-adrenergic agonist cimaterol (CL 263,780) on the growth and carcass characteristics of finishing swine. J. Anim. Sci. 61:905–913.

Jongbloed, A., W. Z. Mroz, and P. A. Kemme. 1992 The effect of supplementary Aspergillus niger phytase in diets for pigs on concentration and apparent digestibility of dry matter, total phosphorus, and phytic acid in different sections of the alimentary tract. J. Anim. Sci. 70:1159–1168.

Jonsson, E., and P. Conway. 1992. Probiotics for pigs. Pages 260–316 in Probiotics: The Scientific Basis. R. Fuller, ed. Chapman and Hall, London, UK.

Jukes, T. H., E. L. R. Stokstad, R. R. Taylor, T. J. Cunha, H. M. Edwards, and G. B. Meadows. 1950. Growth promoting effect of aureomycin on pigs. Arch. Biochem. 26:324–325.

Kellogg, T. F., V. W. Hays, D. V. Catron, L. Y. Quinn, and V. C. Speer. 1964. Effect of level and source of dietary protein on performance and fecal flora of baby pigs. J. Anim. Sci. 23:1089–1094.

Kirchgessner, M., and F. X. Roth. 1982. Fumaric acid as a feed additive in pig nutrition. Pig News and Information 3:259–263.

Koong, L. J., J. A. Neinaber, J. C. Pekas, and J. T. Yen. 1982. Effects of plane of nutrition on organ size and fasting heat production in pigs. J. Nutr. 112:1638–1642.

Kornegay, E. T., S. E. Tinsley, and K. L. Bryant. 1979. Evaluation of rearing systems and feed flavors for pigs weaned at two to three weeks of age. J. Anim. Sci. 48:999–1006.

Kornegay, E. T., C. M. Wood, and L. A. Eng. 1992. Effectiveness and safety of fructooligosaccharides for pigs. J. Anim. Sci.70(Suppl. 1):19. (Abstr.)

Kulwich, R., S. L. Hansard, C. L. Comar, and G. K. Davis. 1953. Copper, molydenum, and zinc interrelationships in rats and swine. Proc. Soc. Exp. Biol. Med. 84:487. (Abstr.)

Langlois, B. E., G. L. Cromwell, and V. W. Hays. 1978a. Influence of chlortetracycline in swine feed on reproductive performance and on incidence and persistence of antibiotic resistant enteric bacteria. J. Anim. Sci. 46:1369–1382.

Langlois, B. E., G. L. Cromwell, and V. W. Hays. 1978b. Influence of type of antibiotic and length of antibiotic feeding period on performance and persistence of antibiotic resistant enteric bacteria in growing–finishing swine. J. Anim. Sci. 46:1383–1396.

Langlois, B. E., K. A. Dawson, G. L. Cromwell, and T. S. Stahly. 1986. Antibiotic resistance in pigs following a 13 year ban. J.Anim. Sci. 62(Suppl. 3):18–32.

LeMieux, F. M., L. V. Ellison, T. L. Ward, L. L. Southern, and T. D. Bidner. 1995. Excess dietary zinc for pigs weaned at 28 days . J. Anim. Sci. 73(Suppl. 1):72. (Abstr.)

Lepley, K. C., D. V. Catron, and C. C. Culbertson. 1950. Dried whole aureomycin mash and meat and bone scraps for growingfattening swine. J. Anim. Sci. 9:608–614.

Lewis, C. J., D. V. Catron, C. H. Liu, V. C. Speer, and G. C. Ashton. 1955. Enzyme supplementation of baby pig diets. J. Agric.Food Chem. 3:1047–1050.

Li, S., W. C. Sauer, R. Mosenthin, and B. Kerr. 1995. Effect of β–glucanase supplementation of cereal-based diets for starter pigs on the apparent digestibilities of dry matter, crude protein, and energy. Anim. Feed Sci. Tech. 59:223–231.

Lindemann, M. D., D. J. Blodgett, E. T. Kornegay, and G. G. Schurig. 1993. Potential ameliorators of aflatoxicosis in weanling/growing swine. J. Anim. Sci. 71:171–178.

Lindemann, M. D., C. M. Wood, A. F. Harper, E. T. Kornegay, and R. A. Anderson. 1995. Dietary chromium picolinate additions improve gain:feed and carcass characteristics in growing-finishing pigs and increase litter size in reproducing sows. J. Anim.Sci. 73:457–465.

Liu, P.,X. S. Piao, S.W. Kim, L.Wang,Y. B. Shen, H. S. Lee, and S.Y. Li. 2008. Effects of chito-oligosaccharide supplementation on the growth performance, nutrient digestibility, intestinal morphology, and fecal shedding of Escherichia coli and Lactobacillus in weaning pigs. J. Anim. Sci. 86:2609–2619.

Liu, P., X. S. Piao, P. A. Thacker, Z. K. Zeng, P. Li, D. Wang, and S. W. Kim. 2010. Chito-oligosaccharide reduces diarrhea incidence and attenuates the immune response of weaned pigs challenged with E. coli K88. J. Anim. Sci. 88:3871–3879.

Lorian, V. 1986. Antibiotic sensitivity patterns of human pathogens in American hospitals. J. Anim. Sci. 62(Suppl. 3):49–55.

Luecke, R. W., W. N. McMillan, and F. Thorp, Jr. 1950. The effect of vitamin B12 animal protein factor and streptomycin on the growth of young pigs. Arch. Biochem. 26:326–327.

Maeng, W. J., C. M. Kim, and H. T. Shin. 1989. Effect of feeding lactic acid bacteria concentrate (LBC, Streptococcus faecium,Cernelle 68) on the growth rate and prevention of scouring in piglet. Korean J. Anim. Sci. 31:318–323.

Mahan, D. C. 1980. Effectiveness of antibacterial compounds and copper for weanling pigs. Report No. 80-2. Ohio Swine Research and Industry. Ohio State Univ., Columbus, OH.

McLaughlin, C. L., C. A. Baile, L. L. Buckholtz, and S. K. Freeman. 1983. Preferred flavors and performance of weanling pigs. J. Anim. Sci. 56:1287–1293.

McLeese, J. M., M. L. Tremblay, J. F. Patience, and G. I. Christison. 1992. Water intake patterns in the weanling pig: effect of water quality, antibiotics and probiotics. Anim. Prod. 54:135–142.

Metchnikoff, E. 1903. The Nature of Man. Studies of Optimistic Philosophy. Heineman, London, UK.

Metchnikoff, E. 1908. Prolongation of Life. G.P. Putnam and Sons, New York, NY.

Miguel, J. C., S. L. Rodriguez-Zas, and J. E. Pettigrew. 2003. Efficacy of Bio-Mos in the nursery pig diet: A meta-analysis of the performance response. J. Anim. Sci. 81(Suppl. 1):49. (Abstr.)

Miller, E. R., D. E. Ullrey, I. Ackerman, D. A. Schmidt, J. A. Hoefer, and R. W. Luecke. 1961. Swine hematology from birth to maturity. I. Serum proteins. J. Anim. Sci. 20:31–35.

Ministry of Food, Agriculture, and Fisheries. 2008. DANMAP. Danish Integrated Antimicrobial Resistance Monitoring and Research Program. Accessed Sept. 1, 2010. http://www.danmap.org/.

Mooney, K. W., and G. L. Cromwell. 1999. Efficacy of chromium picolinate on performance and tissue accretion in pigs with different lean gain potential. J. Anim. Sci. 77:1188–1198.

Mooney, K. W., and G. L. Cromwell. 1995. Effects of dietary chromium picolinate supplementation on growth, carcass characteristics,and accretion rates of carcass tissues in growing–finishing swine. J. Anim. Sci. 75:3351–3357.

Moser, R. L., R. H. Dalrymple, S. G. Cornelius, J. E. Pettigrew, and C. E. Allen. 1986. Effect of cimaterol (CL 263,780) as a repartitioning agent in the diet for finishing pigs. J. Anim. Sci. 62:21–26.

Moser, B. D., E. R. Peo, Jr., and A. J. Lewis. 1980. Effect of carbadox on protein utilization in the baby pig. Nutr. Rep. Int.22:949–956.

NRC. 1980. Effects on Human Health of Subtherapeutic Use of Antimicrobials in Animal Feed. National Academies Press,Washington, DC.

NRC. 1999. The Use of Drugs in Food Animals: Benefits and Risks. National Academies Press, Washington, DC.

NCR–89 Committee on Confinement Management of Swine. 1984. Effect of space allowance and antibiotic feeding on performance of nursery pigs. J. Anim. Sci. 58:801–804.

Newman, K. E. 1996. Nutritional manipulation of the gastrointestinal tract to eliminate salmonella and other pathogens. Pages 37–45 in Biotechnology in the Feed Industry. T. P. Lyons and K. A. Jacques, eds. Nottingham University Press, Nottingham,UK.

Newman, C. W., R. F. Eslick, J. W. Pepper, and A. M. El–Negoumy. 1980. Performance of pigs fed hulless and covered barleys supplemented with or without a bacterial diastase. Nutr. Rep. Int. 22:833–837.

Nousiainen, J., and J. Setala. 1993. Lactic acid bacteria as animal probiotics. Pages 315–356 in Lactic Acid Bacteria. S. Salminen and A. von Wright, eds. Marcel Dekker, New York, NY.

Odle, J. 1995. Betaine and carnitine—Evaluation of performance and carcass effects. Pages 1–14 in Proc. Carolina Swine Nutr.Conf. North Carolina State Univ, Raleigh, NC.

Ogunbameru, B. O., E. T. Kornegay, K. L. Bryant, K. H. Hinkelmann, and J. W. Knight. 1979. Evaluation of a fed flavour in lactation and starter diets to stimulate feed intake of weaned pigs. Nutr. Rep. Int. 20:455–460.

Page, T. G., L. L. Southern, T. L. Ward, and D. L. Thompson, Jr. 1993. Effect of chromium picolinate on growth and serum and carcass traits of growing–finishing pigs. J. Anim. Sci. 71:656–662.

Pariza, M. W. 1997. Conjugated linoleic acid, a newly recognized nutrient. Chem. Ind. 12:464–466.

Parker, R. B. 1974. Probiotics, the other half of the antibiotics story. Anim. Nutr. Health 29:4–8.

Parrish, F. C., Jr., R. L. Thiel, J. C. Sparks, and R. C. Ewan. 1998. Effects of conjugated linoleic acid (CLA) on swine performance and body composition. Pages 187–190 in 1997 Swine Research Report AS-638. Iowa State Univ., Ames, IA.

Pendelton, B. 1992. Challenges of regulation: United States industry perspective. Pages 185–189 in Proceedings of the International Roundtable on Animal Feed Biotechnology—Research and Scientific Regulation. D. A. Leger and S. K. Ho, eds. Agriculture Canada, Ottawa, ON, Canada.

Phillips, I. M., T. Casewell, T. Cox, B. De Groot, C. Friis, R. Jones, C. Nightingale, et al. 2004. Does the use of antibiotics in food animals pose a risk to human health? A critical review of published data. J. Antimicrob. Chemotherapy 53:28–52.

Pollmann, D. S. 1986. Probiotics in pig diets. Pages 193–205 in Recent Advances in Animal Nutrition. W. Haresign and D. J. A.Cole, eds. Butterworths, London, UK.

Pollmann, D. S. 1992. Probiotics in swine diets. Pages 65–74 in Proceedings of the International Roundtable on Animal Feed Biotechnology—Research and Scientific Regulation. D. A. Leger and S. K. Ho, eds. Agriculture Canada, Ottawa, ON, Canada.

Rozeboom, D. W., D. T. Shaw, R. J. Tempelman, J. C. Miguel, J. E. Pettigrew, and A. Connolly. 2005. Effects of mannan oligosaccharide and an antimicrobial product in nursery diets on performance of pigs reared on three different farms. J. Anim.Sci. 83:2637–2644.

Russell, R. J., M. S. Kerley, and G. L. Allee. 1996. Effect of fructooligosaccharides on growth performance of the weaned pig. J.Anim. Sci. 74(Suppl. 1):61. (Abstr.)

Schell, T. C., M. D. Lindemann, E. T. Kornegay, D. J. Blodgett, and J. A. Doerr. 1993. Effectiveness of different types of clay for reducing the detrimental effects of aflatoxin-contaminated diets on performance and serum profiles of weanling pigs. J.Anim. Sci. 71:1226–1231.

Simons, P. C. M., H. A. J. Versteegh, A. W. Jongbloed, P. A. Kemme, P. Slump, K. D. Bos, M. G. E. Wolters, R. F. Beudeker,and G. J. Verschoor. 1990. Improvement of phosphorus availability by microbial phytase in broilers and pigs. Br. J. Nutr.64:525–540.

Smiricky-Tjardes, M. R., C. M. Grieshop, E. A. Flickinger, L. L. Bauer, and G. C. Fahey, Jr. 2003. Dietary galactooligosaccharides affect ileal and total-tract nutrient digestibility, ileal and fecal bacterial concentrations, and ileal fermentative characteristics of growing pigs. J. Anim. Sci. 81:2535–2545.

Smith, H. W. 1969. Transfer of antibiotic resistance from animal and human strains of Escherichia coli to resistant E. coli in the alimentary tract of man. Lancet. 1:1174–1176.

Smith, H. W. 1975. Persistence of tetracycline resistance in pig E. coli. Nature 258:628.

Smith, H. W. 1977. Antibiotic resistance in bacteria and associated problems in farm animals before and after the 1969 Swann Report. Pages 344–357 in Antibiotics and Antibiosis in Agriculture. M. Woodbine, ed. Butterworths, Woburn, MA.

Spencer, J. D., K. J. Touchett, H. Liu, G. L. Allee, M. D. Newcomb, M. S. Kerley, and L. W. Pace. 1997. Effect of spray-dried plasma and fructooligosaccharide on nursery performance and small intestinal morphology of weaned pigs. J. Anim. Sci.75(Suppl. 1):199. (Abstr.)

Stahly, T. S., G. L. Cromwell, and H. J. Monegue. 1980. Effect of single additions and combinations of copper and antibiotics on the performance of weanling pigs. J. Anim. Sci. 51:1347–1351.

Stansbury, W. F., L. F. Tribble, and D. E. Orr, Jr. 1990. Effect of chelated copper sources on performance of nursery and growing pigs. J. Anim. Sci. 68:1318–1322.

Stavric, S., and E. T. Kornegay. 1995. Microbial probiotics for pigs and poultry. Pages 205–231 in Biotechnology in Animal Feeds and Feeding. R. J. Wallace and A. Chesson, eds. VCH Verlagsgesellschaft. Weinheim, Germany.

Stokstad, E. L. R., T. H. Jukes, J. Pierce, A. C. Page, Jr., and A. L. Franklin. 1949. The multiple nature of the animal protein factor.J. Biol. Chem. 180:647–654.

Sutton, A. L., S. R. Goodall, J. A. Patterson, A. G. Mathew, D. T. Kelly, and K. A. Meyerholtz. 1992. Effects of odor control compounds on urease activity in swine manure. J. Anim. Sci. 70(Suppl. 1):160. (Abstr.)

Sutton, A. L., A. G. Mathew, A. B. Scheidt, J. A. Patterson, and D. T. Kelly. 1991. Effect of carbohydrate source and organic acids on intestinal microflora and performance of the weanling pig. Pages 422–427 in Proc. 5th Congress on Digestive Physiology in Pigs. EEAP Pub. No. 54. Pudoc, Wageningen, The Netherlands.

Taylor, J. H., and G. Harrington. 1955. Influence of dietary antibiotic supplements on the visceral weights of pigs. Nature 175:643–644.

Thacker, P. A. 1993. Novel approaches to growth promotion in the pig. Pages 295–306 in Recent Developments in Pig Nutrition. D. J. A. Cole, W. Haresign, and P. C. Garnsworthy, eds. Nottingham University Press, Nottingham, UK.

Thacker, P. A., and F. C. Baas. 1996. Effects of gastric pH on the activity of exogenous pentosanase and the effect of pentosanase supplementation of the diet on the performance of growing-finishing pigs. Anim. Feed Sci. Tech. 63:187–200.

U.S. International Trade Commission. 1989. Synthetic organic chemicals, in USITC Publication No. 2219. U.S. International Trade Commission, Washington, DC.

van Belle, M., E. Teller, and M. Focant. 1990. Probiotics in animal nutrition: A review. Arch. Anim. Nutr. Berlin 40:7:543–567.

van Hartingsveldt, W., M. Hessing, J. P. van der Lugt, and W. A. C. Somers. 1995. The

Second European Symposium on Feed Enzymes. TNO Nutrition and Food Research Institute, Zeist, The Netherlands.

Visek, W. J. 1978. The mode of growth promotion by antibiotics. J. Anim. Sci. 46:1447–1469.

Wallace, H. D. 1967. High Level Copper in Swine Feeding. Int. Copper Res. Assoc., New York, NY.

Watkins, L. E., D. J. Jones, D. H. Mowrey, D. B. Anderson, and E. L. Veenhuizen. 1990. The effects of various levels of ractopamine hydrochloride on the performance and carcass characteristics of finishing swine. J. Anim. Sci. 68:3588–3595.

Webster, A. J. F. 1981. The energetic efficiency of metabolism. Proc. Nutr. Soc. 40:121–128.

Weinberg, E. D. 1957. The mutual effects of antimicrobial compounds and metallic cations. Bacteriol. Rev. 21:46–68.

Wenk, C. 1992. Enzymes in the nutrition of monogastric farm animals. Pages 205–218 in Biotechnology in the Feed Industry. T. P.Lyons, ed. Alltech Tech. Publ., Nicholasville, KY.

Wenk, C., andM. Boessinger. 1993. Enzymes in Animal Nutrition. Zurich, Switzerland: Institut fur Nutztierwissenschaften, Gruppe Ernahrung.

Windisch,W., K. Schedle, C. Plitzner, and A. Kroismayr. 2008. Use of phytogenic products as feed additives for swine and poultry.J. Anim. Sci. 86(E. Suppl.):E140–E148.

Winkelstein, A. 1956. L. acidophilus tables in the therapy of functional intestinal disorders. Am. Pract. Dig. Treatment 7:1637–1639.

Wu, M. C., L. C. Wung, S. Y. Chen, and C. C. Kuo. 1987. Study on the feeding value of Streptococcus faecium M-74 for pigs.

I. Large scale of feeding trial of Streptococcus faecium M-74 on the performance of weaning pigs. Pages 11–12 in Proc. Animal Industry Research Institute. Taiwan Sugar Corp., Chunan Miaoli, Taiwan.

Yen, J. T., J. A. Neinaber, W. G. Pond, and V. H. Varel. 1985. Effect of carbadox on growth, fasting metabolism, thyroid function,and gastrointestinal tract in young pigs. J. Nutr. 115:970–978.

Yen, J. T., T. L. Veum, and R. Lauxen. 1979. Lysine-sparing effect of carbadox in low protein diet for young pigs. J. Anim. Sci.49(Suppl. 1):257. (Abstr.)

Zimmerman, D. R. 1986. Role of subtherapeutic antimicrobials in animal production. J. Anim. Sci. 62(Suppl. 3):6–17.

第14章 饲料中氨基酸、脂肪、碳水化合物的生物学利用率

1 导 语

为了配制具有高利用效率的猪饲料，我们必须了解并掌握猪的能量与营养物质需要量、饲料中能量与营养成分含量，以及饲料中能量与营养物质生物学利用率的相关知识。由于猪并不能对饲料中所有的能量和营养物质进行有效的利用，因此，以饲料中有效的或可消化的能量及营养物质含量来确定猪的营养需要量和配制饲粮，比根据饲料中总的能量和营养含量指标更为准确、更能满足猪的生长需要。但是，目前尚未对每种单一饲料原料的营养价值进行全面的评价；此外，测定饲料中能量及养分生物学利用率的方法也没有完全统一。

对饲料中含有能量的营养素（如蛋白质、脂肪和碳水化合物）的生物学利用率测定十分困难，并且检测成本极高。尽管饲料的消化率与饲料的利用率并不总是相等，但从实用的角度出发，通常以测定饲料中能量和养分的消化率来表示饲料的利用率。根据可消化的能量和养分含量来配制饲粮比根据总的能量和养分含量来配制饲料有了很大的改进。这种基于可消化的能量和营养含量的饲粮配方是发展环境友好型饲料生产的最优策略，对促进养猪生产的可持续发展具有重要

意义。本章将简要综述氨基酸（AA）、脂肪和碳水化合物的生物学利用率的研究，有关矿物质和维生素的生物学利用率研究将在第15章讨论。

2 氨基酸生物学利用率

氨基酸（AA）用于合成动物机体的蛋白质，以满足维持需要和产肉、产奶的生产需要。大部分饲料中都含有氨基酸，但大部分商品型猪饲料中的氨基酸主要是由豆粕或者其他油料饼粕来提供。通常，饲粮中添加合成的赖氨酸、蛋氨酸和苏氨酸用来满足动物饲粮中氨基酸营养平衡的需要。然而，并不是饲粮中所有的氨基酸都能被吸收和利用；此外，不同饲料原料提供可消化氨基酸的能力也是有差异的。因此，评价猪饲粮中饲料原料氨基酸的生物学利用率非常必要。

◆ 2.1 氨基酸的相对生物学利用率

饲粮中只有能够被动物机体吸收并合成体组织蛋白的氨基酸才具有生物学有效性。氨基酸的生物学利用率指饲粮中以一定化学形式被吸收并合成机体蛋白质的氨基酸占饲粮中氨基酸的比例（Batterham，1992；Lewis，1992）。氨基酸的生物学利用率可以用斜率法测定，即以不同浓度梯度的氨基酸饲粮分别饲喂不同组别的动物，通过测定动物整个机体蛋白质沉积量（Bateerham，1992）或氨基酸的氧化（Moehn等，2005）来测定氨基酸的生物学利用率。测试饲粮中的氨基酸的增长效应与标准来源氨基酸的效应的比值，即为氨基酸的相对生物学利用率。

氨基酸利用率的测定必须采用待测氨基酸缺乏的饲粮，即该饲粮中待测氨基酸的所有浓度梯度水平都必须低于实验动物的需要量。斜率法测定氨基酸利用率是在假设动物采食氨基酸的剂量与效应是呈线性关系的基础上建立的。斜率法测定过程繁琐，费用高，且用这种方法一次只能测定一种氨基酸的生物学利用率（Gabert等，2001）。因此，在实际饲料配方中，并不采用通过斜率法测定的氨基酸生物学利用率的数据，而是通过测定氨基酸的消化率，将其等同视为氨基酸的利用率（Stein和Nyachoti，2003）。

◆ 2.2 氨基酸消化率

“氨基酸消化率”不是指动物对氨基酸本身的消化，而是指动物对饲粮蛋白质中肽键（连接氨基酸的化学键）的消化（Fuller，2003）。由于饲粮蛋白质中未消化的氨基酸进入大肠后可以被微生物发酵或代谢，因此动物全肠道的氨基酸消化率值并不能准确地反映动物对氨基酸的吸收（Sauer和Ozimek，1986）。为了避免后肠微生物对氨基酸的作用，对于非反刍动物，通常测定小肠末端氨基酸的消化率（即回肠末端氨基酸消化率），该测定方法更为准确（Sauer和de Lange，1992）。测定回肠末端氨基酸消化率需要收集回肠末端的食糜。目前已有很多收集食糜的技术手段，有研究者对这些技术手段进行了全面的综述（Gbert等，2001；Moughan，2003）。在北美地区，通常选择在猪的回肠末端（回盲瓣前10～15 cm）安装T形瘘管测定氨基酸消化率。这种方法测定结果比较准确，试验之间变异最小。与其他的方法一样，采用T形瘘管收集回肠的部分食糜，而不用收集动物回肠排出的全部食糜，因此需要添加动物不能消化的指示剂来计算氨基酸浓度的变化。通常采用Cr_2O_3做指示剂，此外也应用其他的一些指示剂。回肠末端消化率按公式14-1计算（Stein等，2007）：

$$AID(\%) = (1-[(AAd/AAf) \times (Mf/Md)]) \times 100 \qquad (14-1)$$

式中，AID为表观回肠氨基酸消化率；AAd为回肠食糜中氨基酸浓度，以干物质计（g/kg DM）；AAf为饲料中氨基酸的浓度，以干物质计（g/kg DM）；Mf为饲料中指示剂的浓度，以干物质计（g/kg DM）；Md为回肠食糜中指示剂的浓度，以干物质计（g/kg DM）。

利用这个公式计算的氨基酸消化率为“表观回肠氨基酸消化率”，其结果的计算仅为从摄入的氨基酸的量减去回肠中排出氨基酸的量（Nyachoti等，1997；Mosenthin等，2000；Stein等，2007）。在该公式中，回肠中排出中的氨基酸包括饲粮中未消化的氨基酸和内源性氨基酸。内源性氨基酸是指先被小肠吸收后再以内源性蛋白质的形式排到肠道内的氨基酸。由于回肠内有内源性氨基酸的存在，因此回肠表观氨基酸消化率不能准确地反映饲粮蛋白质的消化率。

◆ 2.3 内源性氨基酸

内源性氨基酸主要来源于动物体内的消化酶、黏蛋白、脱落的上皮细胞、血清蛋白、肽类、游离氨基酸、有机胺类和尿素（Moughan和Scchuttert，1991）。内源性蛋白质主要是由唾液、胃液、胰液、胆汁酸和肠道分泌物组成（Low和Zebrowska，1989；Tamminga等，1995）。其中肠道分泌物占总内源性分泌物的60%以上（Low和Zebrowska，1989），包括脱落的上皮细胞、杯状细胞分泌的黏液素，以及肠上皮细胞分泌的其他糖结合物（Lien等，1997）；唾液、胃液、胰液和胆汁酸各占总内源性分泌物的8%～10%。分泌到动物胃肠道中的内源性氨基酸，有70%～80%在到达回肠末端前被水解和重新吸收（Souffrant等，1993；Krawielitaki等，1994；Fan和Sauer，2002）。未被重新吸收的内源性氨基酸主要是由非结合型胆盐和黏液糖蛋白组成，这些组分在很大程度上可以抵抗蛋白酶的水解作用，因此不能被再吸收利用（Taverner等，1981；Moughan和Schuttert，1991；Lien等，1997）。胆汁酸中90%以上的氨基酸是甘氨酸，黏液糖蛋白中富含脯氨酸、谷氨酸、天冬氨酸、丝氨酸和苏氨酸。肠道内的脯氨酸、甘氨酸、苏氨酸、丝氨酸、天冬氨酸和谷氨酸明显地比其他氨基酸的吸收速度慢（Taverner等，1981）。这几种氨基酸主要是以小肽的形式被吸收，进而在肠上皮细胞内水解。然而由于这种过程很慢，因此导致这些氨基酸的净吸收率低于其他氨基酸（Holmes等，1974）。也有分析认为可能是由于5-羧基吡咯啉还原酶（催化脯氨酸合成的酶）的活性比脯氨酸降解酶和脯氨酸氧化酶的活性强（Mariscal-Land in等，1975），导致脯氨酸和甘氨酸在肠上皮细胞内累积并扩散到肠道内（Gardner，1975）。正是基于这种代谢机制的调节，内源性蛋白质中的脯氨酸、甘氨酸、苏氨酸、丝氨酸、天冬氨酸和谷氨酸含量才相对较高。有关内源性蛋白质氨基酸组成的一些研究结果已经公开发表（Wünsche等，1987；Boisen和Mouhan，1996；Stein等，1999b）。

分泌到肠道内的内源性氨基酸分为基础内源性分泌物和饲粮来源型分泌物（Jansman等，2002；Stein等，2007）。基础内源性氨基酸是指禁食动物肠道内分泌的氨基酸，不包括动物采食干物质后分泌的饲粮来源型的氨基酸，这些内源性氨基酸损失通常是以g/kg DM计。最近的研究表明，内源性损失量（以g/kg DM计）与动物的干物质摄入量有关，干物质摄入量越多，内源性损失就越少。内源性损失减少的原因是由于随着干物质摄入量的增加，基础性内源氨基酸损失量（以g/kg DM计）减少（Moter和Stein，2004）。因此，对生长猪和哺乳母猪，应测定其自由采食条件下的内源性氨基酸损失，而对妊娠母猪，则应在限饲条件下测定其内源性氨基酸的损失，这也和妊娠期母猪的实际生产情况相符（Stein等，1999b）。

基础内源氮的损失通过测定动物采食无氮饲粮后回肠内氨基酸的含量来计算（Stein等，1999b）。尽管由于这种方法是动物处于非正常生理状态时测定的而备受争议，但是通常认为无氮饲粮法对大多数氨基酸内源氮的损失的估测还是比较准确的（Stein等，2007）。无氮饲粮法对甘氨酸和脯氨酸的内源性损失测定有时会产生误差，这两种氨基酸的内源性损失可以用测定肽营养的方法和回归法来替代无氮饲粮法测定（Stein等，2007）。在北美，无氮饲粮方法是通用的方法，基础内源性氨基酸损失量用公式14-2计算（Stein等，2007）。

$$IAA_{end}=AAd \times (Mf/Md) \quad (14-2)$$

式中，IAA_{end}为回肠末端基础内源性氨基酸损失量（mg/kg DMI）；AAd为食糜中待测氨基酸的浓度，以DM计；Mf为饲料中指示剂的浓度，以DM计；Md为食糜中指示剂的浓度，以DM计。

除了基础内源性氨基酸损失外，饲料中很多成分也会导致饲粮来源型的内源性氨基酸的损失，其中纤维素和抗营养因子是两个主要的影响因素（Seve等，1994；Boisen和Moughan，1996；Jansmann等，2002）。不同饲粮来源型内源性氨基酸的损失有很大差异，有些饲粮没有内源性氨基酸损失（纯化饲粮中的组分，如酪蛋白），有的甚至超过基础内源性损失量（像纤维素和抗营养因子含量高的饲料原料，如菜籽粕、次粉等）。关于内源性损失和抗营养因子影响内源性损失的综述已有多篇公开发表（Taminga等，1995；Boisen和Moughan，1996；Nyachoti等，1997；Jansman等，2002）。Jansman等（2002）总结了文献中报道由于内源性蛋白质总量变化的研究成果。

饲粮来源型的内源性氨基酸损失不能直接测定，但可以采用高精氨酸法或者N^{15}同位素稀释技术对一些氨基酸的总内源性损失（包括基础性和饲粮型内源性损失）进行测定。应用这些方法可以估测一些氨基酸的内源性损失量，在假设这些内源性蛋白质组成恒定的前提下，其他氨基酸的内源性损失量可以用已经估测过的氨基酸计算。但是，由于内源性损失的氨基酸组成并不是恒定不变的，用高精氨酸法或N^{15}同位素稀释法估测的内源性氨基酸损失可能会导致结果错误

（Stein等，1999b）。高精氨酸法和N^{15}同位素稀释法测定过程繁琐，检测成本昂贵，因此在实际的饲料营养价值评价中并不经常使用（Stein等，2007）。

◆ 2.4 回肠标准氨基酸消化率

由于猪肠道内有内源性氨基酸的分泌，给准确计算氨基酸的吸收带来很多困难。因为在计算回肠末端表观消化率时，要将这些内源性氨基酸视为回肠氨基酸排出量的一部分。与含高蛋白饲料原料（如豆粕）的饲粮相比，猪采食低蛋白饲料（谷物类饲料等）时，其内源性氨基酸占总肠道氨基酸排出量的比例更大。因此，低蛋白饲粮的氨基酸回肠末端表观消化率的测定值通常偏低（Donkoh和Moughan，1994；Fan等，1994；Stein等，2005）。正是由于对低蛋白饲粮回肠末端表观消化率的低估，因此当饲粮中既含有低蛋白饲料原料，又含有高蛋白饲料原料时，不同饲料原料氨基酸的回肠末端表观消化率的测定值不能累加（Stein等，2005）。但是，如果回肠末端氨基酸消化率的值用内源性氨基酸损失进行校正，这种低估就可以避免。进行校正的回肠标准氨基酸消化率（SID）可以根据Stein等（2005）提出的公式14-3进行计算。

$$SID=[AID+(IAA_{end}/AAf)] \quad (14-3)$$

其中，SID是回肠标准氨基酸的消化率（%），AID和IAA_{end}是分别根据公式14-1和公式14-2计算出来的，AAf是饲料干物质中氨基酸的含量（g/kg DM）。最近的研究表明，即使饲粮中含有低蛋白饲料组分，配合饲料的SID值也具有可加性（Stein等，2005）。由于SID值解决了应用AID值时存在的问题，因此在实际饲粮配方中，使用SID值比用AID值更好。关于SID的计算及依据的研究结果均已公开发表（Mosenthin等，2000；Jansman等，2002；Stein等，2007）。

◆ 2.5 消化率的测定方法

最简单和最容易测定氨基酸消化率的方法简称直接法。用这种方法测定氨基酸消化率时，待测饲料原料提供试验饲粮中所有的氮和氨基酸，并且饲粮中所有氨基酸的AID都可以直接测定。一些试验测定了谷物类饲料AID值，在试验饲粮中，有接近90%的饲料组分是由待测饲料原料提供，其余部分为不含蛋白质的添加剂，如维生素、矿物质、油和蔗糖（Lin等，1987；Green等，1999a；

Pedersen等，2007）。蛋白浓缩饲料的AID也可以用直接法进行测定（Green和Kiener，1989；Fan等，1996；Stein等，1999a）。这种待测饲料通常仅占配合饲料的20%～50%，其他成分为淀粉、玉米油和蔗糖作为无氮能量饲料。直接测定法广泛用于各种饲料组分消化率的测定，文献中报道的一些主要饲料的AID均是由这种方法测定的。

氨基酸消化率也可以用差异法进行测定。差异法测定氨基酸消化率的日粮包括基础饲粮和待测饲粮。基础饲粮由基础的含蛋白质的饲料原料组成，待测饲粮由基础饲粮及待测饲料原料组成（Fan和Saucer，1995a，1995b）。待测饲粮中的粗蛋白质含量不低于14%～16%。当基础饲粮和待测饲料原料消化率值之间没有交互作用时，待测饲料原料的消化率可以采用差异法计算。然而用差异法测定得到的消化率值的准确性与待测饲料原料中每一种氨基酸的贡献有关。氨基酸的贡献越大，其消化率的测定结果越可靠（Fan和Saucer，1995a）。差异法主要用来测定适口性差的饲料原料中氨基酸的消化率，如酵母产品、血液制品等。这些原料的适口性差，在其允许的使用量范围内，无法配制出粗蛋白含量超过10%的饲粮。因此，通常需要先将这些饲料原料与其他适口性较好的饲料原料混合，再用差异法测定其消化率（Mateo和Stein，2007）。

◆ 2.6 氨基酸消化率的体外估测法

饲料原料中氨基酸消化率的测定可以采用体外法替代体内测定法。最常用的体外法是模拟胃和小肠环境条件下的两步酶法。首先，称取少量样品加胃蛋白酶，在pH=2、温度为39℃条件下孵育约2h后，将pH提高至pH=6.8，并在样品溶液中加入胰液素（胰蛋白酶混合物）继续孵育6～8h。其次，将孵育后的样品过滤，用乙醇和丙酮洗涤滤出物后进行干燥。分析滤出物中的粗蛋白质与氨基酸含量，此滤出物中粗蛋白和氨基酸的含量即为饲料原料中没有消化的氨基酸。两步酶法最初是用于估测配合饲料中氨基酸消化率（Boisen和Fernandez，1995）。后来在肉骨粉消化率的测定上得以应用，只是孵育的时间适当延长（Qiao等，2004）。两步酶法测定消化率的局限性是样品量较小，通常仅为0.5 g，经消化后滤出物中仅有很少量的样品没有被消化，很难准确定量分析滤出物中氨基酸的含量。因此有必要增加样品的称样量，或者将几个滤袋中的滤出物收集在一起进行氨基酸的分析。体外法估测饲料原料中的氨基酸消化率还不太常用，当前的研究工作是为了改进体外测定技术，以使其在将来能够被广泛地使用。

◆ 2.7 影响氨基酸消化率的因素

2.7.1 动物的年龄及生理状态

仔猪对奶中的蛋白质有很强的消化能力（Mavromichalis等，2001），而对豆粕中蛋白质氨基酸的AID较低，但是随着日龄的增加消化率会逐渐增加（Wilson和Leibholz，1981；Caine等，1997）。仔猪对植物源性蛋白质饲料消化率低，原因是仔猪早期消化道中蛋白酶的活性低（Moughan，1993）。当采食量相同时，育肥猪和母猪对饲料的消化能力基本相同，但目前还不清楚何时达到最大的消化能力，这也证明了动物的生理状态不会影响氨基酸的消化率（Stein等，2001）。

2.7.2 采食量水平

猪的采食量从接近维持能量需要量水平增加到大约2倍维持需要量水平时，饲料中蛋白质及氨基酸的AID值也逐渐增加（Motor和Stein，2004），但再进一步提高采食量时AID则不再变化（Sauer等，1982；Haydon等，1984；Albin等，2001b；Motor和Stein，2004）。低采食量时猪的AID降低的原因是因为小肠末端的基础性内源氨基酸损失增加。SID是用基础性内源氨基酸损失校正AID计算的，因此SID值随着饲料采食量的增加而线性减少（Motor和Stein，2004）。因此，由于内源性氨基酸损失的影响，饲料采食量水平对SID值的影响与对AID值的影响结果是相反的。正是由于这种差异，生长猪和哺乳母猪的氨基酸消化率是在自由采食条件下测定的，这与商业化猪场生产条件下饲喂情况一致。而测定妊娠期母猪的氨基酸消化率则必须限饲，这也和商业化猪场中妊娠期母猪的生产条件符合。

2.7.3 饲料原料的化学组成

与前面讨论的一样，饲粮中粗蛋白和氨基酸的含量影响氨基酸的回肠末端表观消化率。由于纤维影响内源性氨基酸的损失，因此饲粮的纤维素水平可以降低氨基酸的AID值（Mosenthin等，1994；Schulze等，1994；Lenis等，1996）。如果饲粮中可溶性的纤维能够被利用，则饲粮氨基酸的消化率可能不会一直受不可溶性纤维的影响。

由于饲粮脂肪可以降低食糜的流通速度，从而延长蛋白酶对蛋白的水解时间，因此饲粮脂肪可以提高氨基酸的AID和SID值（Imbeah和Sauer，1991；Li和Sauer，1994；Cervants-Pahm和Stein，2008）。饲粮脂肪水平对内源性氨基酸的损失没有影响（de Lange等，1989）。

饲粮中抗营养因子，如胰蛋白酶抑制剂、植物凝集素和单宁可以降低回肠氨基酸的消化率（Jansman等，1994；le Guen等，1995；Schulze等，1995；Yu等，1996），主要原因是抗营养因子增加了饲粮来源型内源性氨基酸的损失。因此，含有抗营养因子的饲粮或饲料原料的AID和SID值都会降低。

3 脂肪的生物学利用率

脂肪是动物机体组织的重要组成部分，与碳水化合物和蛋白质相比，脂肪能提供更高的能量。因此，猪饲粮中，脂肪对提高饲粮能量水平具有重要的作用。

不同饲料原料中的脂肪含量不同，脂肪酸组成也不一样（表14-1），但所有饲料原料的脂肪主要还是以甘油三酯的形式存在（Stahly，1984）。由于脂肪在胃肠道的水溶性环境中溶解度很低，因此脂肪的消化与其他营养素的消化途径不同。胃肠道内脂肪的消化经过一系列特殊的步骤完成，包括脂肪的乳化、酶水解和乳糜微粒的形成（Bauer等，2005）。脂肪经过消化和吸收后，大部分直接用于合成机体脂肪，小部分在体内氧化以ATP形式释放能量。假设所有吸收的脂肪都具有生物学有效性，都能够被机体利用，那么可以通过测定脂肪的消化率，视其与利用率等同，这与其他营养素的情况一样。饲粮脂肪的消化率与单一脂肪酸的消化率有关，而单一脂肪酸的消化率又无法测定，因此在多数情况下仅测定饲粮总脂肪的消化率。影响脂肪消化率的因素有很多。一般认为，猪对大多数饲料原料的总肠道脂肪消化率为25%～77%（Noblet等，1994）。

表14-1 饲料原料中的粗脂肪及脂肪酸组成

饲料	粗脂肪	脂肪酸（占总脂肪酸的比例，%）					
	（%）	C16∶0	C18∶0	C18∶1	C18∶2	C18∶3	U∶S[1]
玉米[2]	3.7	11.1	1.8	26.9	56.5	1.0	6.52
大麦[2]	1.8	22.2	1.5	12.0	55.4	5.6	2.93
高粱[2]	2.9	13.5	2.3	33.3	33.8	2.6	4.56
麸皮[2]	3.4	17.8	0.8	15.2	56.4	5.9	4.24
膨化全脂大豆[2]	17.9	10.5	3.8	21.7	53.1	7.4	4.68
豆粕（48% CP）[2]	1.9	10.5	3.8	21.7	53.1	7.4	4.68
豆油[3]	100	10.3	3.8	22.8	51.0	6.8	5.64
葵花籽油[3]	100	5.4	3.5	45.3	39.8	0.2	8.47
棕榈油[3]	100	43.5	4.3	36.6	3.1	0.6	0.92
牛油[3]	100	24.9	18.9	36.0	3.1	0.6	0.92
精选白脂膏[3]	100	21.5	14.9	41.1	11.6	0.4	1.45
猪油[3]	100	23.8	13.5	41.2	10.2	1.0	1.44

[1]U∶S：不饱和脂肪酸∶饱和脂肪酸。
[2]资料来源于Sauvant等，2002。
[3]资料来源于NRC，1998。

◆ 3.1 脂肪的理化性质

3.1.1 脂肪酸的化学性质

饲粮脂肪的重要化学性质是脂肪酸的饱和程度以及碳链的长短，其对胃肠道中乳糜微粒的形成和脂肪的溶解度有显著的影响，进而会影响脂肪的消化率。通常情况下，由于不饱和脂肪酸比饱和脂肪酸更容易形成乳糜微粒（Freeman等，1968；Stahly，1984），因此不饱和脂肪酸比饱和脂肪酸更容易被消化（表14-2）。不饱和脂肪酸通过增加饱和脂肪酸乳糜微粒的形成而促进饱和脂肪酸的消化，因此将不饱和脂肪酸与饱和脂肪酸混合可以提高饱和脂肪酸的消化率（Powles等，1993）。脂肪中不饱和脂肪酸与饱和脂肪酸的比例对脂肪消化率有重要影响。Stahly（1984）研究发现，当不饱和脂肪酸和饱和脂肪酸比例大于1.5时，猪饲粮中脂肪的表观消化率为70%～80%；低于1.3时，脂肪的表观消化率会降低。Powles等（1995）也得到了同样的结果，他们发现当饲粮中不饱和脂肪酸和饱和脂肪酸比例低于1.5时，体重12 ~ 90 kg的猪饲粮脂肪的消化能降低。

表14-2 猪的饲粮脂肪总肠道表观消化率（ATTD）[1]

参考文献	体重（kg）	基础饲粮	添加的脂肪		
			脂肪来源	添加量（%）	ATTD（%）
Cera 等，1989	6.1	玉米－豆粕－乳清粉	牛油	8	81.8
			玉米油	8	84.8
			可可油	8	87.3
Li 等，1990	5.6	玉米－豆粕－乳清粉－脱脂奶粉	—	0	40.8
			大豆油	10	80.1
			可可油	10	88.0
			混合油（1 ∶ 1）	10	85.6
Jones 等，1992	5.3	玉米－豆粕－乳清粉－脱脂奶粉	大豆油	10	89.5
			可可油	10	88.8
			牛油	10	80.9
			猪油	10	84.8
Jin 等，1998	5.8	玉米－豆粕－乳清粉	可可油	10	83.4
			玉米油	10	82.3
			大豆油	10	83.7
			牛油	10	79.8
Jørgensen 等，2000	35.0	麦麸－淀粉－蔗糖	—	0	83.4
			鱼油	15	92.8
			菜籽油	15	93.4
			可可油	15	88.4

（续）

参考文献	体重（kg）	基础饲粮	添加的脂肪		
			脂肪来源	添加量（%）	ATTD（%）
Duran-Montgé 等，2007	45 ~ 50	大麦	—	0	29.4
			牛油	10	86.5
			HO 葵花籽油[2]	10	84.7
			葵花籽油	10	85.5
			亚麻籽油	10	85.0
			混合油[3]	10	85.4
Kil，2008	22	玉米－豆粕	—	0	33.2
			大豆油	5	74.2
			大豆油	10	82.4
			精选白脂膏	10	80.5
	84	玉米－豆粕	—	0	49.1
			大豆油	5	73.1
			大豆油	10	82.1
			精选白脂膏	10	81.9

[1]SBM：大豆粕。
[2]HO葵花籽油：高油葵花籽油（油葵籽油）。
[3]混合油脂：5.5%油脂，3.5%葵花籽油，1%亚麻籽油。

脂肪酸碳链的长短也会影响脂肪的消化率。虽然短链脂肪酸（SCFA）主要是饱和脂肪酸（Cera等，1989; Straaup等，2006），但14个碳以下的脂肪酸比长链脂肪酸更容易消化。小肠内环境是水溶性的，短链脂肪酸（SCFA）比长链脂肪酸更容易溶解，并且不用形成乳糜微粒而能直接被上皮细胞吸收（Ramírez等，2001）。此外，短链脂肪酸比长链脂肪酸更容易形成乳糜微粒，也更容易从乳糜微粒中释放出来（Bach和Babayan，1982；Stahly，1984）。在断奶仔猪的试验中发现，可可油的总肠道脂肪表观消化率比玉米油或牛油的消化率高，总肠道的短链脂肪酸消化率超过了90%（Cera等，1989）。而对于生长猪来说，脂肪酸碳链的长短对脂肪消化率的影响还不清楚（表14-2；Jørgensen等，2000）。

脂肪酸在甘油三酯分子上的分布及位置也会影响脂肪的消化率（Bracco，1994）。胰脂肪酶可以专一水解甘油三酯的sn-1和sn-3醚键，生成2-甘油一酯和两个游离的脂肪酸。由于2-甘油一酯比游离脂肪酸更容易形成乳糜微粒，因此在甘油三酯sn-2位置上的脂肪酸更易消化（Ramírez等，2001）。

饲粮中游离脂肪酸（非酯化）的消化率较低，与酯化的脂肪酸相比，游离的脂肪酸很难形成乳糜微粒（Freeman，1984），而更容易于在肠道内皂化形成不溶性的皂化矿物盐类（Ramírez等，2001）。当2-甘油一酯和游离脂肪酸的平衡被打破时，游离脂肪酸的浓度会增加，从而抑制其他脂肪代谢产物形成乳糜微粒（Dierick和Decuypere，2004）。因此，在饲粮中增加游离脂肪酸的浓度会降低断奶仔猪（Swiss和Bayley，1976；Powles等，1994）和生长育肥猪的脂肪消化率（Powles等，1993；Jørgensen和Fernández，2000）。然而，DeRouchey等（2004）研究发现，提高断奶仔猪饲粮中游离脂肪酸水平（由水解的精选白脂膏提供）对脂肪和脂肪酸的消化率均没有影响。

3.1.2 脂肪酸的物理特性

饲粮中的脂肪主要由两部分组成：一部分是饲料组分本身含有的脂肪（结合态脂肪，如玉米和全大豆中的脂肪）；一部分是外源添加的从动物或者植物中提取的油脂。在商品育肥猪的玉米-豆粕型饲粮中，有大约40%的脂肪来源于玉米中结合态的脂肪，60%是外源添加的（Azain，2001）。结合态脂肪中脂肪酸的消化率比外源添加的脂肪中脂肪酸的消化率低，因为结合态脂肪大部分存在于脂肪细胞的细胞膜中，与添加的液态脂肪相比，更不容易被乳化和酶水解（Adams和Jensen，1984；Li等，1990；Duran-Montgé等，2007；Kil，2008）。因此，用物理和化学的方法破坏脂肪细胞壁可以提高结合态脂肪的消化率。制粒可以提高生长育肥猪对玉米及玉米-豆粕型饲粮中脂肪的消化率，但不能提高外源添加脂肪的消化率（Noblet和Champion，2003）。同样，制粒以后，断奶仔猪饲粮中脂肪的表观消化率可以从54.7%提高到70.2%（Xing等，2004）。DDGS中脂肪的表观消化率和真消化率均比玉米的高19%，说明乙醇生产中的发酵或者其他加工处理可以使结合态的脂肪变得更容易消化（Kim等，2009）。

◆ 3.2 影响脂肪消化的饲粮成分

3.2.1 饲粮纤维

饲粮纤维降低了动物对脂肪的消化能力。Stahly（1984）指出，饲粮中粗纤维含量每增加1%，脂肪的表观消化率降低1.3%～1.5%。饲粮纤维使脂肪消化率降低的主要原因是因为纤维加快了肠道食糜的流通速度和降低了肠道内脂肪的溶解性（Stahly，1984）。饲粮纤维也可以增加内源性脂肪的损失，如脱落的上皮细胞、胆汁酸和微生物等，导致脂肪的表观消化率降低（Bach Knudsen等，1991；de Lange，2000）。尽管饲粮中纤维素的含量及性质都会影响脂肪的消化率，但是饲粮中纤维素的性质似乎对脂肪消化率的影响更大。饲粮中添加纯化的纤维素对脂肪消化率没有影响，表明纯化纤维素的简单化学结构对脂肪的消化率没有影响（Kil等，2010）。与此相反，可溶性的饲粮纤维增加了食糜黏度，从而影响脂肪酶的水解作用，与胆汁酸形成乳糜微粒以及黏膜对脂肪的吸收，因此脂肪的消化率降低（Smith和Annison，1996）。

3.2.2 饲粮蛋白质和矿物质

饲粮蛋白质和矿物质可以和动物胃肠道内的脂肪相互作用而影响脂肪消化率。增加饲粮中粗蛋白质的含量，可以提高回肠末端及总肠道的脂肪及脂肪酸的消化率（Just等，1980；Jørgensen等，1992）。饲粮蛋白质对脂肪消化率的特殊影响主要与蛋白质的质量（Frobish等，1970）和饲粮中的脂肪含量有关（Jørgensen等，1992）。饲粮中蛋白质影响脂肪消化率的机制还不清楚，蛋白质对脂肪消化率的促进作用，可能与未消化的蛋白提高乳糜微粒稳定性有关（Meyer等，1976）。

饲粮中的矿物质也会影响脂肪的消化率，因为胃肠道中的钙离子和镁离子容易与长链脂肪酸形成不溶性的脂肪酸盐，直接影响脂肪的消化率（Stahly，1984）。在断奶仔猪和生长猪饲粮中，高含量的钙会降低脂肪的消化率（Jørgensen等，1992；Han和Thacker，2006），但是脂肪酸盐主要是在大肠内形成，因此不会影响小肠内脂肪的吸收（Jørgensen等，1992）。

3.2.3 饲粮中添加剂

脂肪的消化在很大程度上取决于脂肪的乳化及乳糜微粒的形成，因此添加外源性的乳化剂，如卵磷脂、溶血卵磷脂等可以促进脂肪的消化。Jørgensen等

（1992）研究发现，在含有牛油的断奶仔猪饲粮中，添加卵磷脂或溶血卵磷脂可以促进脂肪的消化。添加卵磷脂和溶血卵磷脂饲粮的脂肪消化率分别增加了7.5%和3.0%。一些研究结果证明了饲粮中含有乳化剂可以促进脂肪的消化（Reis de Souza等，1995；Jin等，1998），但也有一些研究认为乳化剂对脂肪的消化没有影响（Soares和Lopex-Bote，2002；Dierick和Decuypere，2004；Xing等，2004）。添加乳化剂的效果不尽相同，其原因可能与饲粮中脂肪来源、添加水平（Jones等，1992；Dierick和Decuypere，2004）及乳化剂种类（Wiel等，1993；Dierick和Decupere，2004）不同有关。

在饲粮中添加外源性脂肪酶可能会提高动物对脂肪的消化，但在生长育肥猪的饲粮中添加外源性脂肪酶并未提高脂肪的消化率（Dierick和Decupere，2004）。抗菌剂（如卡巴氧）可通过抑制微生物的活动而促进猪的脂肪消化（Partanen等，2001；Wang等，2005）。胃肠道中的微生物也会影响脂肪的消化，因为微生物的活动增加，不仅可以促进胆汁酸发生不可逆的分解，导致肠道胆汁酸的活性降低（Smith和Annison，1996），而且还会增加微生物的氢化作用，在小肠内将不饱和脂肪酸转化为消化率低的饱和脂肪酸（Yen，2001）。

◆ 3.3 动物因素对脂肪消化的影响

3.3.1 动物的年龄

断奶仔猪的脂肪消化率很低，刚断奶的仔猪由于突然换成固体饲料导致体内的胰脂肪酶活性降低，从而影响了仔猪对脂肪的消化和吸收能力（Lindemann等，1986）。断奶仔猪对饲粮脂肪的消化能力相对较低，并且对饲料结合态脂肪的消化能力低于植物源性脂肪的消化能力（Cera等，1989；Jones等，1992；Jin等，1998）。这也说明了断奶仔猪对饱和脂肪酸的消化能力低于对不饱和脂肪酸的消化能力。仔猪断奶以后，随着仔猪的生长，猪对脂肪的消化能力逐渐增强。日龄大一些的猪对动物源性脂肪和植物源性脂肪的消化能力似乎没有差异（Cera等，1989；Soares和Lopex-Bote，2002；Straarup等，2006）。例如，生长育肥猪对这两种来源脂肪的消化率没有差异（表14-2；Agunbiade等，1992；Jørgensen和Fernández，2000），40 kg左右的猪就可以完全消化饲粮中的脂肪（Wiseman和Cole，1987；Agunbiade等，1992）。

3.3.2 脂肪的内源性损失

回肠食糜或者粪中的脂肪主要是由饲粮中没有消化的脂肪和动物机体内源性损失的脂肪组成。内源性损失的脂肪包括胆汁酸、脱落的细胞和微生物（Sambrook，1979）。由于有内源性脂肪损失的存在，因此很难合理地解释脂肪的消化率值。与氨基酸相似，脂肪的表观消化率随饲粮脂肪的增加而增加。其原因是，与采食高脂肪含量饲粮的猪相比，采食低脂肪含量饲粮的猪内源性脂肪的损失占总脂肪排出量的比例相对较大（Jørgensen等，1993；Jørgensen和Fernández，2000；Kil等，2010）。由于脂肪真消化率或者标准消化率值是用内源性脂肪损失来校正的，因此脂肪真消化率或标准消化率的结果不受饲粮脂肪摄入量的影响（Kil等，2010）。

◆ 3.4 脂肪消化率测定的方法

3.4.1 脂肪的分析

饲粮、食糜及粪中的脂肪可以用多种方法提取，但是不同的方法测定结果会有差异（Boisen和Verstegen，2000），进而会影响脂肪消化率的计算。不同溶剂对脂肪提取的效率不同，在用乙醚提取前对样品进行酸水解对结合态脂肪的完全提取非常重要。相对于饲粮中的脂肪，测定粪中的脂肪含量时，对样品进行酸水解显得更加重要，因为猪肠道中形成的矿物质复合物和脂肪酸钾盐很难用溶剂提取出其脂肪成分。酸水解可以使这些脂肪酸盐水解而使脂肪游离出来，进而容易被溶剂提取（Just，1982）。测定大豆油的总肠道脂肪表观消化率时，饲粮或粪中脂肪在乙醚提取前未进行酸水解的测定值比进行酸水解时的测定值高7.3%（Agunbiade等，1992）。因此，乙醚提取前的酸水解是必需的，可以防止对脂肪消化率的过高估计（Just，1982）。

3.4.2 回肠与总消化道的脂肪消化率

脂肪的消化率既可以测回肠末端，也可以测定总消化道的脂肪消化率。通常认为回肠末端脂肪的消化率较好地估计了脂肪的生物学利用率，因为大部分脂肪是在回肠末端之前被消化和吸收的（Nordgaard和Mortensen，1995），并且回肠内脂肪的消化率值不受大肠内产生的内源性脂肪的影响（Jørgensen等，2000）。由于猪大肠内微生物对不饱和脂肪酸的氢化作用，因此脂肪酸的消化率应该测定回肠末端的，而不是总肠道的脂肪酸消化率（Jørgensen等，1997；Duran-Montgé等，2007）。猪回肠末端脂肪表观消化率通常比总肠道脂肪消化率值要高，是因为在大肠内可以净合成内源性的脂肪（Shi和Noblet，1993；Reis de Souza等，1995）。然而，当饲粮纤维含量低时，这种差异可以忽略不计（Jørgensen等，2000；Kil等，2010）。因为当猪饲喂低纤维含量的饲粮时，大肠内微生物脂肪的合成量很少。然而当饲粮纤维含量接近于商业饲料中的纤维含量时，总肠道的饲粮纤维的消化率低于回肠消化率（Kil等，2010）。

3.4.3 全收粪法与指示剂法

脂肪消化率可以通过测定饲粮或排泄物中所有的脂肪来进行测定（即全收粪法），或者用定点收粪与不可消化性指示剂的方法进行测定（Adeola，2001）。用Cr_2O_3作为指示剂测定的脂肪消化率，可能会比全收粪法测定的结果偏低（Reis de Souza等，1995）。其原因是Cr_2O_3在胃肠道中容易与脂肪分离（Carlson和Bayley，972）。因此，用脂溶性的指示剂测定脂肪消化率可能是比较好的方法。

4 碳水化合物的生物学利用率

人及非肉食性动物的主要能量来源是食物中植物源性的碳水化合物。碳水化合物主要是根据其化学性质，即碳水化合物的聚合程度、糖苷键的类型及单糖的特性进行分类（Cummings和Stephen，2007）。用这种分类方法，碳水化合物可分为单糖、寡糖和多糖（FAO，1998）。多糖又可以分为两类：一类是由淀粉和糖原组成，另一类是饲粮中的纤维。但是这种分类方法不能够反映饲粮中碳水化合物对人及动物的生理功能或营养作用。为了避免这种分类方法的局限性，饲粮碳水化合物根据以能否被小肠消化（FAO，2003；Cummings和Stephen，2007；Englyst等，2007）或影响血糖水平的能力进行分类（血糖碳水化合物或非血糖碳水化合物；Englyst，2005）。

像其他大多数营养素一样，很少对碳水化合物的生物学利用率进行测定。取而代之的是以测定碳水化合物不同部分的消化率来表示整个碳水化合物的生物学利用率。肠道内排出的内源性碳水化合物的量很少，可以忽略不计。因此，碳水化合物只需测定表观消化率。其表观消化率、标准消化率和真消化率没有差别，这一点与脂肪、氨基酸的消化率是不同的。考虑到碳水化合物在小肠和大肠内都可以被降解，降解的产物又可以在小肠和大肠中被吸收，并能显著改变动物的能量状态，因此碳水化合物在回肠的消化率和总肠道的消化率之间还是有明显的不同。

◆ 4.1 单糖的消化率

一些单糖，像葡萄糖、果糖和半乳糖，通过载体主动转运很容易被小肠吸收（Englyst和Hudson，2000）。其他的单糖，像阿拉伯糖、木糖和甘露糖在食品或者饲料中的含量很低，在小肠内以被动吸收的方式被吸收（Englyst和Hudson，2000；IOM，2001）。大部分单糖被吸收后很快在肝脏进行代谢（Englyst，2005）。葡萄糖、果糖及半乳糖可以进行糖酵解，生成丙酮酸，进而转变成乙酰辅酶A，进入三羧酸循环或者用于脂肪酸的合成。

◆ 4.2 双糖的消化率

很多饲料组分中都含有不同含量的蔗糖（1分子葡萄糖和1分子果糖）和麦芽糖（2分子的葡萄糖）。这两类双糖分别被蔗糖酶和麦芽糖酶水解成相应的单糖后被动物吸收（Englyst和Hudson，2000）。蔗糖酶和麦芽糖酶是刷状缘酶，能分别断裂α-（1～2）和α-（1～4）糖苷键。乳糖是牛奶中存在的一种双糖，包括一个葡萄糖，一个半乳糖，由β-（1～4）糖苷键连接而成，乳糖酶可以水解β-（1～4）糖苷键释放半乳糖和葡萄糖（Englyst和Hudson，2000）。一般认为蔗糖、麦芽糖和乳糖可以在小肠内完全消化，所以在回肠的消化率为100%。这些双糖水解后产生的葡萄糖吸收很快，能使血糖的浓度快速增加。因此，饲粮中的单糖和双糖都被称为血糖碳水化合物，主要是因为这些糖消化后能够快速提高血糖的浓度（Englyst，2005）。

◆ 4.3 寡糖的消化率

饲料原料中天然存在的寡糖主要有三种，即棉籽糖、水苏糖和毛蕊花糖，主要存在于豆类、豆科植物、棉籽和糖蜜中。猪饲粮中4%～7%的寡糖，主要来源于豆粕。棉籽糖是由1分子葡萄糖、1分子果糖和1分子半乳糖组成。水苏糖和毛蕊花糖与棉籽糖结构相似，除含有1分子葡萄糖与1分子果糖外，水苏糖还含有2分子半乳糖，毛蕊花糖还含有3分子的半乳糖。寡糖中的葡萄糖和果糖是以α-（1~2）糖苷键连接的，而水苏糖和毛蕊花糖中的葡萄糖和半乳糖是以α-（1~6）糖苷键连接，半乳糖和半乳糖也是以α-（1~6）糖苷键连接。棉籽糖、水苏糖和毛蕊花糖统称为α-半乳糖苷。α-半乳糖苷中的糖苷键可以被α-半乳糖苷酶水解。由于猪肠道内不能合成α-半乳糖苷酶，因此棉籽糖、水苏糖和毛蕊花糖不能被猪体内的酶消化（Canibe和Bach Knudsen，1997）。有研究指出，猪对半乳糖苷的回肠消化率为50%~80%，可能是因为半乳糖苷在猪小肠内进行发酵而被消化（Bangala-Freire等，1991；Canibe和Bach Knudsen，1997；Smiricky等，2002）。凡是在小肠内不能消化的α-半乳糖苷，在大肠中都可以很快发酵，因此半乳糖苷的总肠道消化率为100%。

另外，这3个半乳糖苷和一些合成的寡糖可作为益生元在猪饲料中添加。这些寡糖主要是可以在小肠内部分发酵为果寡糖和低聚半乳糖（Smiricky-Tjardes等，2003）。此外，猪饲料中还存在果聚糖、酵母果聚糖和甘露寡糖等。在现有的资料中，还没有任何关于寡糖消化率的数据。很明显，所有天然的或合成的寡糖，都可以在小肠里部分被发酵，在大肠里被完全发酵。从能量的角度考虑，寡糖在消化道的哪一段进行发酵并不重要，因为其发酵的最终产物是短链脂肪酸，很容易被肠道吸收。由于寡糖在小肠内发酵以短链脂肪酸的形式被吸收，而不是葡萄糖，因此这些寡糖被称作非血糖碳水化合物（Englyst，2005）。

◆ 4.4 淀粉的消化率

淀粉是大多数猪饲粮中主要的碳水化合物（Wiseman等，2006；Bach Knudsen等，2006）。淀粉是唯一具有天然形成的颗粒结构（淀粉粒）的碳水化合物，是由直链淀粉和支链淀粉组成（BeMiller，2007）。大部分谷物淀粉大约含有25%的直链淀粉和75%的支链淀粉（BeMiller，2007）。直链淀粉呈线型，葡萄糖残基间主要是以α-（1~4）糖苷键连接而成，只有少数直链淀粉可能会有少量分支（侧链），分支点以α-（1~6）糖苷键连接（BeMiller，2007；Commings和Stephen，2007）。支链淀粉有很多分支，其葡萄糖残基之间除了以α-（1~4）糖苷键连接外，其分支点以α-（1~6）糖苷键连接（Commings和

Stephen，2007）。蜡质淀粉主要是由支链淀粉组成（BeMiller，2007）。

人的淀粉消化是从口腔分泌的唾液淀粉酶与食品混合开始的，猪的淀粉消化是从口腔分泌的唾液淀粉酶与饲料混合开始的。食物吞咽后进入到胃里，胃内的低pH导致唾液淀粉酶失活，因此这个消化过程很短（Englyst和Hudson，2000）。淀粉的消化主要在小肠，由胰腺和小肠分泌的α-淀粉酶和异麦芽糖酶水解淀粉生成麦芽糖、麦芽三糖和异麦芽糖（也叫α-糊精）（Gray，1992；Groff和Gropper，2000）。麦芽糖酶是小肠刷状缘酶，将麦芽糖、异麦芽糖水解成葡萄糖；异麦芽糖酶（也叫α-糊精酶）作用于异麦芽糖的α-（1~6）糖苷键，水解后生成2分子的葡萄糖（Groff和Gropper，2000）。淀粉在酶的作用下可以完全被消化，影响淀粉的消化速率及消化程度的因素主要包括以下几个方面：① 淀粉粒的自然晶型或淀粉来源；② 直链淀粉和支链淀粉的比例；③ 淀粉的加工工艺和加工的程度（Cummings等，1997；Englyst和Hudson，2000；Svihus等，2005）。直链淀粉似乎比支链淀粉更能抵抗淀粉酶的消化（Svihus等，2005）。由于有多种因素影响淀粉消化，根据淀粉的消化速率及血液中葡萄糖出现时间的不同，淀粉可以进一步分为快消化淀粉和慢消化淀粉（Englyst等，2007）。

淀粉在小肠的消化率非常高，对大多数谷物籽粒来说，回肠的淀粉消化率高于90%（Bach Knudsen等，2006；Sun等，2006；Wiseman，2006）。豌豆的淀粉回肠消化率比谷物的低（Canibe和Bach Khudsen，1997），红豌豆的淀粉回肠消化率为75%~90%（Sun等，2006；Wiseman，2006）。马铃薯淀粉的消化率比饲料中其他成分的消化率都低，生的马铃薯中淀粉的回肠消化率低于40%（Sun等，2006）。

对于所有的饲料原料，可以通过热处理提高淀粉的消化率。例如，红豌豆中的豌豆淀粉在155℃膨化后，其回肠消化率可以从89.8%提高到95.9%（Stein和Bohlke，2007）。原因是淀粉经过膨化以后，淀粉的糊化度增加，从而提高了淀粉的消化率（Svihus等，2005）。然而也有一些试验表明，制粒对淀粉的消化率没有影响（Svihus等，2005；Stein和Bohlke，2007）。

在小肠内不能被消化而进入大肠的淀粉叫抗性淀粉。抗性淀粉在大肠内发酵后以短链脂肪酸的形式被吸收，因此饲料原料的总肠道淀粉消化率接近100%

（Wiseman，2006；Stein和Bohlke，2007）。从能量利用效率的角度考虑，由于短链脂肪酸的能值低于葡萄糖，因此应尽可能提高小肠内淀粉的消化率。

◆ 4.5 饲粮纤维的消化率

饲粮纤维是指在小肠内不能被消化或者很难被消化，但可以在大肠内发酵的一类碳水化合物（De Vries，2004）。小肠不能消化的碳水化合物称为“不可利用碳水化合物”或“非血糖碳水化合物”（Englyst等，2007）。碳水化合物中的非淀粉多糖、抗性淀粉、非消化性寡糖和糖醇统称为饲粮纤维（Englyst，2005；Englyst等，2007）。植物细胞壁中的大部分碳水化合物（80%～90%）属于非淀粉多糖（Commings等，1997）。与单糖、双糖和淀粉（可利用的淀粉）不同，非淀粉多糖不含有可利用性淀粉特有的α-（1～4）糖苷键（Englyst等，2007）。然而，目前饲粮纤维的定义认为，一些非细胞壁的碳水化合物也具有和细胞壁的碳水化合物类似的生理作用（De Vries，2004）。因此，饲粮纤维包括植物细胞壁中的非淀粉多糖，也包括非细胞壁的非淀粉多糖。

纤维素和半纤维素是细胞壁中常见的非淀粉多糖，木质素也被认为是饲料原料中纤维的组分。食物中纤维素是由葡萄糖单位通过β-（1～4）糖苷键连接而成的线型、无分支的长链（Cummings和Stephen，2007；BeMiller，2007）。由于糖苷键本身的属性，纤维素不能被动物分泌的内源酶消化（BeMiller，2007）。与纤维素不同，半纤维素是支链多糖，是由不同类型的己糖和戊糖组成（Cummings和Stephen，2007）。一年生植物，包括谷物类植物中最常见的半纤维素是木聚糖，木聚糖是以木糖为基本结构单位连接而成的线型或带有较多分支的多糖（BeMiller，2007）。半纤维素在直链和分支点核心结构处有侧链存在，通常由阿拉伯糖、甘露糖、半乳糖和葡萄糖组成（Cummings和Stephen，2007）。一些半纤维素也含有葡萄糖衍生的葡萄糖醛酸、半乳糖衍生的半乳糖醛酸。由于糖醛酸的存在，半纤维素可以与金属离子，如钙、锌等形成盐（Southgate和Spiller，2001；Cummings和Stephen，2007）。由于一些半纤维素是可溶性的，因此在猪大肠中，半纤维素的发酵能力高于纤维素。

果胶、树胶、非消化性寡糖和抗性淀粉不是植物细胞壁的成分，也被认为是饲粮纤维组成。果胶是由半乳糖醛酸通过α-（1～4）糖苷键连接而成的线性聚合物（BeMiller，2007）。果胶也含有鼠李糖、半乳糖和阿拉伯糖侧链（Cummings和Stephen，2007）。树胶是天然的植物性多糖，也可以由发酵获得。植物或灌木物理损伤后的渗出物，以及种子的胚乳中都含有天然的树胶（BeMiller，2007）。渗出物中的树胶为阿拉伯胶，种子胚乳中的树胶叫瓜尔胶。发酵产生的树胶为黄原胶和支链淀粉。

饲粮纤维也可以根据其溶解性来分类，包括不可溶性纤维和可溶性纤维。饲粮不可溶性纤维包括木质素、纤维素和一部分半纤维素，饲粮可溶性纤维包括非细胞壁组分和其余的半纤维素。

饲粮纤维的回肠消化率通常都很低，这表明在猪盲肠前段的发酵是有限的。对大多数饲料原料来说，饲粮总纤维的回肠消化率低于25%（Bach Knudsen和Jørgensen，2001；Urriola等，2010）。饲粮中可溶性纤维的回肠消化率比不可溶性纤维的消化率高（Urriola等，2010）。如果饲料原料中含有的纤维大部分属于可溶性纤维，则饲粮总纤维的回肠消化率将高于50%（Urriola等，2010）。

不同饲料原料及不同类型的饲粮纤维在大肠中的发酵是有差异的，主要受饲料原料中木质素及不可溶性纤维的含量影响（Bindelle等，2008）。玉米副产品中纤维的总肠道表观消化率为40%～60%（Bach Knudsen和Jørgensen，2001；Stein等，2009；Urriola等，2010）。但是，若饲粮中包含的纤维大部分为可溶性纤维，则总肠道的纤维消化率将大于80%（Le Goff等，2002；Urriola，2010）。饲粮纤维的发酵能力和总肠道消化率受制粒、挤压膨化和其他可能的加工工艺的影响（Beltranena等，2009）。目前，这些工艺方法对纤维素利用率影响的研究还很有限。

5 结 语

猪饲料原料中含有能量的营养素（如蛋白质、脂肪和碳水化合物），其生物学利用率的测定难度较大，并且检测费用昂贵。因此，尽管消化率（尤其是蛋白质的消化率）与利用率并不总是相等，但从实用的角度出发，通常测定这些营养物质的消化率，视其等同为利用率。蛋白质、脂肪和可以被酶消化的部分碳水化合物（如双糖和淀粉）先在盲肠之前被消化，然后在小肠中以AA、甘油三酯和单糖的形式被吸收。小肠内不能吸收的蛋白质和脂肪进入大肠后，尽管不能在大肠吸收，但可以被肠道微生物发酵，导致肠道内营养素浓度的改变。因此，AA和脂肪的总肠道消化率值是不准确的，不能真实反应这些养分的消化率，应该测定AA和脂肪的回肠消化率。

在猪小肠内不能被酶消化的碳水化合物可以在大肠内发酵，生成短链脂肪酸被机体吸收，从而给猪提供能量。因此，碳水化合物的消化可分为两部分：在小肠内部分被消化，以单糖形式被吸收；在大肠内被发酵，以短链脂肪酸的形式被吸收。对于蛋白质和脂肪，由于肠道内有内源性物质的排出，表观消化率不能代表饲料中养分的真实消化率，因此，用基础或总的内源性损失的估计值来校正表观消化率值，就可以计算出标准消化率或者真消化率。对于碳水化合物，分泌到肠道内的内源性物质可以忽略不计，表观消化率的数值就代表了饲料中碳水化合物的消化率。

总之，蛋白质和脂肪的生物学利用率分别通过计算标准回肠氨基酸的消化率及脂肪的真消化率来估计。碳水化合物的生物学利用率可以通过测定双糖和淀粉的表观回肠消化率以及饲粮纤维的表观总肠道消化率来估计。

作者：Dong Y. Kil, Sarah K. Cervantes-Pahm和Hans H. Stein

译者：王金荣

参考文献

Adams, K. L., and A. H. Jensen. 1984. Comparative utilization of in-seed fats and the respective extracted fats by the young pig.J. Anim. Sci. 59:1557–1566.

Adeola, O. 2001. Digestion and balance techniques in pigs. Page 903–916 in *Swine Nutrition*. 2nd ed. A. J. Lewis andL. L. Southern, eds. CRC Press, New York, NY.

Agunbiade, J. A., J. Wiseman, and D. J. A. Cole. 1992. Utilization of dietary energy and nutrients from soya bean products bygrowing pigs. Anim. Feed Sci. Technol. 36:303–318.

Albin, D. M., M. R. Smiricky, J. E. Wubben, and V. M. Gabert. 2001a. The effect of dietary level of soybean oil and palm oil onapparent ileal digestibility and postprandial flow patterns of chromium oxide and amino acids in pigs. Can. J. Anim. Sci.81:495–503.

Albin, D. M., J. E. Wubben, M. R. Smiricky, and V. M. Gabert. 2001b. The effect of feed intake on ileal rate of passage andapparent amino acid digestibility determined with or without correction factor in pigs. J. Anim. Sci. 79:1250–1258.

Azain, M. J. 2001. Fat in swine nutrition. Pages 95–105 in Swine Nutrition. 2nd ed. A. J. Lewis and L. L. Southern, eds. CRCPress, New York, NY.

Bach, A. C., and V. K. Babayan. 1982. Medium-chain triglycerides: An update. Am. J. Clin. Nutr. 36:950–962.

Bach Knudsen, K. E., and H. Jorgensen. 2001. Intestinal degradation of dietary carbohydrates—from birth to maturity. Pages109–120 in Digestive Physiology of Pigs. J. E. Lindberg, and B. Ogle eds. Cabi Publishing New York, NY.

Bach Knudsen, K. E., B. B. Jensen, J. O. Andersen, and I. Hansen. 1991. Gastrointestinal implications in pigs of wheat and oatfractions. 2. Microbial activity in the gastrointestinal tract. Br. J. Nutr. 65:233–248.

Bach Knudsen, K. E., H. N. Larke, S. Steenfeldt, M. S. Hedemann, and H. Jorgensen. 2006. *In vivo* methods to study the digestionof starch in pigs and poultry. Anim. Feed Sci. Technol. 130:114–135.

Batterham, E. S. 1992. Availability and utilization of amino acids for growing pigs. Nutr. Res. Rev. 5:1–18.

Bauer, E., S. Jakob, and R. Mosenthin. 2005. Principles of physiology of lipid digestion. Asian-Aust. J. Anim. Sci. 18:282–295.

Beltranena, E., J. Sánchez-Torres, L. Goonewardene, X. Meng, and R. T. Zijlstra. 2009.

Effect of single-or twin-screw extrusion onenergy and amino acid digestibility of wheat or corn distillers dried grains with solubles (DDGS) for growing pigs. J. Anim.Sci. 87(Suppl. 3):166. (Abstr.)

BeMiller, J. 2007. Carbohydrate Chemistry for Food Scientist. 2nd ed. AACC International, Inc., St. Paul, MN.

Bengala-Freire, J., A. Aumaitre, and J. Peiniau. 1991. Effects of feeding raw and extruded peas on ileal digetibility, pancreaticenzymes and plasma glucose and insulin in early weaned pigs. J. Anim. Phys. Anim. Nutr. 65:154–164.

Bindelle, J., A. Buldgen, M. Delacollette, J. Wavreille, R. Agneessens, J. P. Destain, and P. Leterme. 2009. Influence of sourceand concentrations of dietary fiber on *in vivo* nitrogen excretion pathways in pigs as reflected by *in vitro* fermentation andnitrogen incorporation by fecal bacteria. J. Anim. Sci. 87:583–593.

Boisen, S., and J. A. Fernandez. 1995. Prediction of the apparent ileal digestibility of protein and amino acids in feedstuffs andfeed mixtures for pigs by *in vitro* analyses. Anim. Feed Sci. Technol. 51:29–43.

Boisen, S., and P. J. Moughan. 1996. Dietary influences on endogenous ileal protein and amino acid loss in the pig—A review.Acta Agric. Scand., Sect A, Anim. Sci. 46:154–164.

Boisen, S., and M.W. A. Verstegen. 2000. Developments in the measurement of the energy content of feeds and energy utilizationin animals. Page 57–76 in Feed Evaluation— Principles and Practice. P. J. Moughan, M. W. A. Verstegen, and M. I. Visser-Reyneveld, eds. Wageningen Press, Wageningen, The Netherlands.

Bracco, U. 1994. Effect of triglyceride structure on fat absorption. Am. J. Clin. Nutr. 60(Suppl.):S1002–S1009.

Butts, C. A., P. J. Moughan, W. C. Smith, G. W. Reynolds, and D. J. Garrick. 1993. The effect of food dry mater intake on theendogenous ileal amino acid extraction determined under peptide alimentation in the 50 kg liveweight pig. J. Sci. Food Agric.62:235–243.

Caine, W. R., W. C. Sauer, S. Taminga, M. W. A. Verstegen, and H. Schulze. 1997. Apparent ileal digestibilities of amino acid innewly weaned pigs fed diets with protease-treated soybean meal. J. Anim. Sci. 75:2962–2969.

Canibe, N., and K. E. Bach Knudsen. 1997. Digestibility of dried and toasted peas in pigs. 1. Ileal and total tract digestibilities ofcarbohydrates. Anim. Feed Sci. Technol. 64:293–310.

Carlson, W. E., and H. S. Bayley. 1972. Digestion of fat by young pigs: a study of the amounts of fatty acid in the digestive tractusing a fat-soluble indicator of absorption. Br. J. Nutr. 28:339–346.

Cera, K. R., D. C. Mahan, and G. A. Reinhart. 1989. Apparent fat digestibilities and performance responses of postweaning swinefed diets supplemented with coconut oil, corn oil or tallow. J. Anim. Sci. 67:2040–2047.

Cervantes–Pahm, S. K., and H. H. Stein. 2008. Effect of dietary soybean oil and soybean protein concentrate on the concentrationof digestible amino acids in soybean products fed to growing pigs. J. Anim. Sci. 86:1841–1849.

Cummings, J. H., and A. M. Stephen. 2007. Carbohydrate terminology and classification. Eur. J. Clin. Nutr. 61:S5–S18.

Cummings, J. H., M. B. Roberfroid, H. Andersson, C. Barth, A. Ferro–Luzzi, Y. Ghoos, M. Gibney, et al. 1997. A new look atdietary carbohydrate: Chemistry, physiology and health. Eur. J. Clin. Nutr. 51:417–423.

de Lange, C. F. M. 2000. Characterisation of the non–starch polysaccharides. Page 77–92 in Feed Evaluation—Principles andPractice. P. J. Moughan, M. W. A. Verstegen, and M. I. Visser–Reyneveld, eds. Wageningen Press, Wageningen, TheNetherlands.

de Lange, C. F. M., W. C. Sauer, R. Mosenthin, and W. B. Souffrant. 1989. The effect of feeding different protein–free diets on therecovery and amino acid composition of endogenous protein collected from the distal ileum and feces in pigs. J. Anim. Sci.67:746–754.

DeRouchey, J. M., J. D. Hancock, R. H. Hines, C. A. Maloney, D. J. Lee, H. Cao, D. W. Dean, and J. S. Park. 2004. Effectsof rancidity and free fatty acids in choice white grease on growth performance and nutrient digestibility in weanling pigs.J. Anim. Sci. 82:2937–2944.

De Vries, J. W. 2004. Dietary fiber: The influence of definition on analysis and regulation. J. AOAC Int. 87:682–706.

Dierick, N. A., and J. A. Decuypere. 2004. Influence of lipase and/or emulsifier addition on the ileal and faecal nutrient digestibilityin growing pigs fed diets containing 4% animal fat. J. Sci. Food Agric. 84:1443–1450.

Donkoh, A., and P. J. Moughan. 1994. The effect of dietary crude protein content on apparent and true ileal nitrogen and AAdigestibilities. Br. J. Nutr. 72:59–68.

Duran–Montgé, P., R. Lizardo, D. Torrallardona, and E. Esteve–Garcia. 2007. Fat and fatty acid digestibility of different fat sourcesin growing pigs. Livest. Sci. 109:66–69.

Englyst, K. N, and G. J. Hudson. 2000. Carbohydrates. Pages 61–76 in Human Nutrition and Dietetics. 10th ed. J. S. Garrow, W.P. T. James, and A. Ralph, eds. Churchill Livingston, Edinburgh, U. K.

Englyst, K. N., and H. N. Englyst. 2005. Carbohydrate bioavailability. Br. J. Nutr. 94:1–11.

Englyst, K. N., S. Liu, and H. N. Englyst. 2007. Nutritional characterization and measurement of dietary carbohydrates. Eur. J.Clin. Nutr. 61:S19–S39.

Fan, M. Z., and W. C. Sauer. 1995a. Determination of apparent ileal amino acid digestibility in barley and canola meal for pigswith the direct, difference, and regression methods. J. Anim. Sci. 73:2364–2374.

Fan, M. Z., and W. C. Sauer. 1995b. Determination of apparent ileal amino acid digestibility in peas for pigs with the direct,difference, and regression methods. Livest. Prod. Sci. 44:61–72.

Fan, M. Z., and W. C. Sauer. 2002. Determination of true ileal amino acid digestibility and the endogenous amino acid outputsassociated with barley samples for growing and finishing pigs by regression analysis technique. J. Anim. Sci. 80:1593–1605.

Fan, M. Z., W. C. Sauer, and W. M. Gabert. 1996. Variability of apparent ileal amino acid digestibility in canola meal forgrowing–finishing pigs. Can. J. Anim. Sci. 76:563–569.

Fan, M. Z., W. C. Sauer, R. T. Hardin, and K. A. Lien. 1994. Determination of apparent ileal AA digestibility in pigs: Effect ofdietary amino acid level. J. Anim. Sci. 72:2851–2859.

FAO. 1998. Carbohydrates in human nutrition. Accessed Oct. 3, 2009. http://www.fao.org/docrep/W8079E/w8079e00.htm.

FAO. 2003. Food energy—Methods of analysis and conversion factors. Accessed Oct. 3, 2009. ftp://ftp.fao.org/docrep/fao/006/y5022e/y5022e00.pdf.

Freeman, C. P. 1984. The digestion, absorption and transport of fats—non–ruminants. Pages 105–122 in Fats in Animal Nutrition.J. Wiseman. ed. Butterworths, London, U.K.

Freeman, C. P., D. W. Holme, and E. F. Annison. 1968. The determination of the true digestibilities of interesterified fats in youngpigs. Br. J. Nutr. 22:651–660.

Frobish, L. T., V. W. Hays, V. C. Speer, and R. C. Ewan. 1970. Effect of fat source and level on utilization of fat by young pigs.J. Anim. Sci. 30:197–202.

Fuller, M., 2003. Amino acid bioavailability – a brief history. Pages 183–198 in Proc. 9th Intl. Symp. Digest. Physiol. Pigs, Vol.1. Banf, Canada.

Gabert, V. M., H. Jorgensen, and C. M. Nyachoti. 2001. Bioavailability of amino acids in feedstuffs for swine. Pages 151–186 in Swine Nutrition. 2nd ed. A. J. Lewis and L. L. Southern, eds. CRC Press, New York, NY.

Gardner, M. L. G. 1975. Absorption of amino acids and peptides from a complex mixture in the isolated small intestine of the rat.J. Physiol. 253:233–256.

Gray, G. M. 1992. Starch digestion and absorption in nonruminants. J. Nutr. 122:172–177.

Green, S., and T. Kiener. 1989. Digestibilities of nitrogen and amino acids in soybean–meal, sunflower, meat and rapeseed mealsmeasured with pigs and poultry. Anim. Prod. 48:157–179.

Green, S., S. L. Bertrand, M. J. C. Duron, and R. A. Maillard. 1987. Digestibility of amino acids in maize, wheat and barley mealmeasured in pigs with ileo–rectal anatomosis and isolation of the large intestine. J. Sci. Food Agric. 41:29–43.

Groff, J. L., and S. S. Gropper. 2000. Advanced Nutrition and Human Metabolism. 3rd ed. Wadsworth, Belmont, CA.

Han, Y. K., and P. A. Thacker. 2006. Effects of calcium and phosphorus ratio in high zinc diets on performance and nutrientdigestibility in weanling pigs. J. Anim. Vet. Adv. 5:5–9.

Haydon, K. D., D. A. Knabe, and T. D. Tanksley, Jr. 1984. Effect of level of feed intake on nitrogen, amino acid and energydigestibilities measured at the end of the small intestine and over the total digestive tract of growing pigs. J. Anim. Sci.59:717–724.

Holmes, J. H., H. S. Bayley, and P. A. Leadbeater. 1974. Digestion of protein in small and large intestine of the pig. Br. J. Nutr.32:479–489.

Imbeah, M., andW. C. Sauer. 1991. The effect of dietary level of fat on amino acid digestibilities in soybean meal and canola mealand on rate of passage in growing pigs. Livest. Prod. Sci. 29:227–239.

IOM. 2001. Dietary Reference Intakes: Proposed Definition of Dietary Fiber. National Academies Press, Washington, DC.

Jansman, A. J. M., A. A. Frohlich, and R. Marquardt. 1994. Production of proline-rich proteins by parotid glands of rats is enhancedby feeding diets containing tannins from faba beans (*Vicia faba* L.). J. Nutr. 124:249–258.

Jansman, A. J. M., W. Smink, P. van Leeuwen, and M. Rademacher. 2002. Evaluation through literature data of the amount andamino acid composition of basal endogenous crude protein at the terminal ileum of pigs. Anim. Feed Sci. Technol. 98:49–60.

Jin, C. F., J. H. Kim, I. K. Han, H. J. Jung, and C. H. Kwon. 1998. Effects of various fat sources and lecithin on the growthperformance and nutrient utilization in pigs weaned at 21 days of age. Asian-Aust. J. Anim. Sci. 11:176–184.

Jones, D. B., J. D. Hancock, D. L. Harmon, and C. E. Walker. 1992. Effects of exogenous emulsifiers and fat sources on nutrientdigestibility, serum lipids, and growth performance in weanling pigs. J. Anim. Sci. 70:3473–3482.

Jorgensen, H., and J. A. Fernández. 2000. Chemical composition and energy value of different fat sources for growing pigs. ActaAgric. Scand. Sect. A. Anim. Sci. 50:129–136.

Jorgensen, H., V. M. Gabert, M. S. Hedemann, and S. K. Jensen. 2000. Digestion of fat does not differ in growing pigs fed dietscontaining fish oil, rapeseed oil, coconut oil. J. Nutr. 130:852–857.

Jorgensen, H., K. Jakobsen, and B. O. Eggum. 1992. The influence of different protein, fat and mineral levels on the digestibilityof fat and fatty acids measured at the terminal ileum and in faeces of growing pigs. Acta Agric. Scand. Sect. A. Anim. Sci.42:177–184.

Jorgensen, H., K. Jakobsen, and B. O. Eggum. 1993. Determination of endogenous fat and fatty acids at the terminal ileum and onfaeces in growing pigs. Acta Agric. Scand. Sect. A. Anim. Sci. 43:101–106.

Just, A. 1982. The net energy value of crude fat for growth in pigs. Livest. Prod. Sci. 9:501–509.

Just, A., J. O. Andersen, and H. Jorgensen. 1980. The influence of diet composition on the apparent digestibility of crude fat andfatty acids at the terminal ileum and overall in pigs. Z. Tierphyphysiol. Tierernaehr. Futtermittelkde. 44:82–92.

Kil, D. Y. 2008. Digestibility and energetic utilization of lipids by pigs. Ph.D. Diss. Univ. of Illinois, Urbana, IL.

Kil, D. Y., T. E. Sauber, D. B. Jones, and H. H. Stein. 2010. Effect of the form of dietary fat and the concentration of dietary NDF onileal and total tract endogenous losses and apparent and true digestibility of fat by growing pigs. J. Anim. Sci. 88:2959–2967.

Kim, B. G., D. Y. Kil, and H. H. Stein. 2009. Apparent and true ileal digestibility of acid hydrolyzed ether extract in various feedingredients fed to growing pigs. J. Anim. Sci. 87(Suppl. 2):776. (Abstr.)

Krawielitzki, K., F. Kreienbring, T. Zebrowska, R. Schadereit, and J. Kowalczyk. 1994. Estimation of N absorption, secretion, andreabsorption in different intestinal sections of growing pigs using the 15N isotope dilution method. Pages 79–82 in DigestivePhysiology in Pigs, vol. I. EAAP publ. No. 80. W. B. Souffrant and H. Hagemeister, ed. Bad Doberan, Germany.

Le Goff, G., J. van Milgen, and J. Noblet. 2002. Influence of dietary fiber on digestive utilization and rate of passage in growingpigs, finishing pigs, and adult sows. Anim. Sci. 74:503–515.

le Guen, M. P., J. Huismann, J. Gueguen, G. Beelen and M. W. A. Verstegen. 1995. Effects of a concentrate of pea antinutritionalfactors on pea protein digestibility in piglets. Livest. Prod. Sci. 44:157–167.

Lenis, N. P., P. Bikker, J. van der Meulen, J. T. M. van Diepen, J. G. M. Bakker, and A. W. Jongbloed. 1996. Effect of dietaryneutral detergent fiber on ileal digestibility and portal flux of nitrogen and amino acids and on nitrogen utilization in growingpigs. J. Anim. Sci. 74:2687–2699.

Lewis, A. J. 1992. Determination of the amino acid requirements of animals. Pages 67–85 in Modern Methods in Protein Nutritionand Metabolism. S. Nissen, ed. Academic Press Inc., San Diego, CA.

Li, S., andW. C. Sauer. 1994. The effect of dietary fat content on amino acid digestibility in young pigs. J. Anim. Sci. 72:1737–1743.

Li, D. F., R. C. Thaler, J. L. Nelssen, D. L. Harmon, G. L. Allee, and T. L. Weeden. 1990. Effect of fat sources and combinationson starter pig performance, nutrient digestibility and intestinal morphology. J. Anim. Sci. 68:3694–3704.

Lien, K. A., W. A. Sauer, and M. Fenton. 1997. Mucin output in ileal digesta of pigs fed a protein−free diet. Z. Ernährungswiss.36:182–190.

Lin, F. D., D. A. Knabe, and T. D. Tanksley, Jr. 1987. Apparent digestibility of amino acids, gross energy and starch in corn,sorghum, wheat, barley, oat groats and wheat midlings for growing pigs. J. Anim. Sci. 64:1655–1663.

Lindemann, M. D., S. G. Cornelius, S. M. El Kandelgy, R. L. Moser, and J. E. Pettigrew. 1986. Effects of age, weaning and dieton digestive enzyme levels in the piglet. J. Anim. Sci. 62:1298–1307.

Low, A. G., and T. Zebrowska. 1989. Digestion in Pigs. Pages 53–121 in Protein Metabolism in Farm Animals. H. D. Bock, B.O.

Eggum, A. G. Low, O. Simon, and T. Zebrowska, eds. Oxford University Press, Oxford, UK.

Mariscal-Landin, G., B. Seve, Y. Colleaux, and Y. Lebreton. 1995. Endogenous amino nitrogen collected from pigs with end-to-endileorectal anastomosis is affected by the method of estimation and altered by dietary fiber. J. Nutr. 125:136–146.

Mateo, C. D., and H. H. Stein. 2007. Apparent and standardized ileal digestibility of amino acids in yeast extract and spray driedplasma protein by weanling pigs. Can. J. Anim. Sci. 87:381–383.

Mavromichalis, I., T. M. Parr, V. M. Gabert, and D. H. Baker. 2001. True ileal digestibility of amino acids in sow's milk for17-day-old pigs. J. Anim. Sci. 79:707–713.

Meyer, J. H., E. A. Stevenson, and H. D. Watts. 1976. The potential role of protein in the absorption of fat. Gastroenterology70:232–239.

Moehn, S., R. F. P. Bertolo, P. B. Pencharz, and R. O. Ball. 2005. Development of the indicator amino acid oxidation technique todetermine the availability of AA from dietary protein in pigs. J. Nutr. 135:2866–2870.

Mosenthin, R., W. C. Sauer, and F. Ahrens. 1994. Dietary pectin's effect on ileal and fecal amino acid digestibility and exocrinepancreatic secretions in growing pigs. J. Nutr. 124:1222–1229.

Mosenthin, R.,W. C. Sauer, R. Blank, J. Huisman, and M. Z. Fan. 2000. The concept of digestible amino acids in diet formulationfor pigs. Livest. Prod. Sci. 64:265–280.

Moter, V., and H. H. Stein. 2004. Effect of feed intake on endogenous losses and amino acid and energy digestibility by growingpigs. J. Anim. Sci. 82:3518–3525.

Moughan, P. J. 1993. Towards an improved utilization of dietary amino acids by the growing pig. Pages 117–136 in RecentDevelopments in Pig Nutrition 2. D. J. A. Cole, W. Haresign, and P. C. Garnsworthy, eds. Nottingham University Press,Nottingham, UK.

Moughan, P. J. 2003. Amino acid availability: Aspects of chemical analysis and bioassay methodology. Nutr. Res. Rev. 16:127–141.

Moughan, P. J., and G. Schuttert. 1991. Composition of nitrogen-containing fractions in digesta from the distal ileum of pigs fed aprotein-free diet. J. Nutr. 121:1570–1574.

Noblet, J., and M. Champion. 2003. Effect of pelleting and body weight on digestibility of energy and fat of two corns in pigs.J. Anim. Sci. 81(Suppl.):140. (Abstr.)

Noblet, J., H. Fortune, X. S. Shi, and S. Dubois. 1994. Prediction of net energy value of feeds for growing pigs. J. Anim. Sci.72:344–354.

Nordgaard, I., and P. B. Mortensen. 1995. Digestive progresses in the human colon. Nutrition. 11:37–45.

NRC. 1998. Page 141 in *Nutrient Requirements of Swine*. 10th rev. ed. National Academies Press, Washington, DC.

Nyachoti, C. M., C. F. M. de Lange, B. W. McBride, and H. Schulze. 1997. Significance of endogenous gut nitrogen losses in thenutrition of growing pigs: A review. Can. J. Anim. Sci. 77:149–163.

Partanen, K., T. Jalava, J. Valaja, S. Perttila, H. Siljander–Rasi, and H. Lindeberg. 2001. Effect of dietary carbadox or formic acidand fibre level on ileal and faecal nutrient digestibility and microbial metabolite concentrations in ileal digesta of the pig.Anim. Feed. Sci. Technol. 93:137–155.

Pedersen,C., M. G. Boersma, and H. H. Stein. 2007. Energy and nutrient digestibility in NutriDense corn and other cereal grainsfed to growing pigs. J. Anim. Sci. 85:2473–2483.

Powels, J., J.Wiseman, D. J. A. Cole, and B. Hardy. 1993. Effect of chemical structure of fats upon their apparent digestible energyvalue when given to growing/finishing pigs. Anim. Prod. 57:137–146.

Powels, J., J.Wiseman, D. J. A. Cole, and B. Hardy. 1994. Effect of chemical structure of fats upon their apparent digestible energyvalue when given to young pigs. Anim. Prod. 58:411–417.

Powels, J., J. Wiseman, D. J. A. Cole, and S. Jagger. 1995. Prediction of the apparent digestible energy value of fats given to pigs.Anim. Sci. 61:149–154.

Qiao, Y., X. Lin, J. Odle, A. Whittaker, and T. A. T. G. van Kempen. 2004. Refining *in vitro* digestibility assays: Fractionation ofdigestible and indigestible peptides. J. Anim. Sci. 82:1669–1677.

Ramírez, M., L. Amate, and A. Gil. 2001. Absorption and distribution of dietary fatty acids from different sources. Early Hum.Dev. 65(Suppl.):S95–S101.

Reis de Souza, T., J. Peiniau, A. Mounier, and A. Aumaitre. 1995. Effect of addition of tallow and lecithin in the diet of weanlingpigs on the apparent total tract and ileal digestibility of fat and fatty acids. Anim. Feed Sci. Technol. 52:77–91.

Sambrook, I. E. 1979. Studies on digestion and absorption in the intestine of growing pigs. 8. Measurement of the flow of totallipid, acid–detergent fibre and volatile fatty acids. Br. J. Nutr. 42:279–287.

Sauer,W. C., and K. de Lange. 1992. Novel methods for determining protein and AA digestibilities in feedstuffs. Pages 87–120 in Modern Methods in Protein Nutrition and Metabolism. S. Nissen, ed. Academic Press Inc., San Diego, Ca.

Sauer,W. C., and L. Ozimek. 1986. Digestibility of amino acids in swine: Results and their practical applications. A review. Livest.Prod. Sci. 15:367–388.

Sauer, W. C., A. Just, and H. Jorgensen. 1982. The influence of daily feed intake on the apparent digestibility of crude protein,amino acids, calcium and phosphorus at the terminal ileum and overall in pigs. Z. Tierphysiol., Tierernährg. Futtermittelkde.48:177–182.

Sauvant, D., J. M. Perez, and G. Tran. 2004. Tables of Composition and Nutritional Value of Feed Materials: Pig, Poultry,Sheep, Goats, Rabbits, Horses, Fish. Wageningen Academic Publishers, Wageningen, The Netherlands and INRA ed. Paris,France.

Schulze, H., P. van Leuwen, M. W. A. Verstegen, J. Huisman, W. B. Souffrant, and F. Ahrens. 1994. Effect of level of dietaryneutral detergent fiber on ileal apparent digestibility and ileal nitrogen losses in pigs. J. Anim. Sci. 72:2362–2368.

Schulze, H., H. S. Saini, J. Huisman, M. Hessing,W. van den Berg, and M.W. A. Verstegen.1995. Increased nitrogen secretion byinclusion of soya lectin in the diets of pigs. J. Sci. Food Agric. 69:501–510.

Seve, B., G. Mariscal-Landin, Y. Colleaux, and Y. Lebreton. 1994. Ileal endogenous amino acid and amino sugar flows in pigs fedgraded levels of protein or fibre. Pages 35–38 in Digestive Physiology in Pigs, vol. I. EAAP publ. No. 80. W. B. Souffrantand H. Hagemeister, eds. Bad Doberan, Germany.

Shi, X. S., and J. Noblet. 1993. Contribution of the hindgut to digestion of diets in growing pigs and adult sows: Effect of dietcomposition. Livest. Prod. Sci. 34:237–252.

Smiricky, M. R., C. M. Grieshop, D. M. Albin, J. E. Wubben, V. M. Gabert, and G. C. Fahey, Jr. 2002. The influence of soyoligosaccharides on apparent and true ileal amino acid digestibilities and fecal consistency in growing pigs. J. Anim. Sci.80:2433–2441.

Smiricky-Tjardes, M. R., C. M. Grieshop, E. A. Flickinger, L. L. Bauer, and G. C. Fahey, Jr. 2003. Dietary galactooligosaccharidesaffect ileal and total-tract nutrient digestibility, ileal and fecal bacterial concentrations, and ileal fermentative characteristicsof growing pigs. J. Anim. Sci. 81:2535–2545.

Smith, C. H. M., and G. Annison. 1996. Non-starch plant polysaccharides in broiler nutrition—Towards a physiologically validapproach to their determination. World Poult. Sci. J. 52:203–221.

Soares, M., and C. J. Lopex-Bote. 2002. Effects of dietary lecithin and fat unsaturation on nutrient utilization in weaned piglets.Anim. Feed Sci. Technol. 95:169–177.

Souffrant,W. B.,A. Rerat, J. P. Laplace, B. Darcy-Vrillon, R.Köhler, T. Corring, and G.

Gebhard. 1993. Exogenous and endogenouscontributions to nitrogen fluxes in the digestive tract of pigs fed a casein diet. III. Recycling of endogenous nitrogen. Reprod.Nutr. Dev. 33:373–382.

Southgate, D. A. T., and G. A. Spiller. 2001. Polysaccharide food additives that contribute to dietary fiber. Pages 27–31 in CRC Handbook of Dietary Fiber in Human Nutrition. 3rd ed. G. A. Spiller, ed. CRC Press, Boca Raton, FL.

Stahly, T. S. 1984. Use of fats in diets for growing pigs. Pages 313–331 in Fats in Animal Nutrition. J.Wiseman. ed. Butterworths,London, U.K.

Stein, H. H., and R. A. Bohlke. 2007. The effects of thermal treatment of field peas (*Pisum sativum* L.) on nutrient and energydigestibility by growing pigs. J. Anim. Sci. 85:1424–1431.

Stein, H. H., and M. Nyachoti. 2003. Animal effects on ileal amino acid digestiblity. Pages 223–241 in Proc. 9th Intl. Symp. Digest.Physiol. Pigs, Vol. 1. Banf, Canada.

Stein, H. H., S. Aref, and R. A. Easter. 1999a. Comparative protein and amino acid digestibilities in growing pigs and sows.J. Anim. Sci. 77:1169–1179.

Stein, H. H., S. P. Connot, and C. Pedersen. 2009. Energy and nutrient digestibility in four sources of distillers dried grains withsolubles produced from corn grown within a narrow geographical area and fed to growing pigs. Asian–Aust. J. Anim. Sci.22:1016–1025.

Stein, H. H., S. W. Kim, T. T. Nielsen, and R. A. Easter. 2001. Standardized amino acid digestibilities in growing pigs and sows.J. Anim. Sci. 79:2113–2122.

Stein, H. H., C. Pedersen, A. R.Wirth, and R. A. Bohlke. 2005. Additivity of values for apparent and standardized ileal digestibilityof amino acids in mixed diets fed to growing pigs. J. Anim. Sci. 83:2387–2395.

Stein, H. H., B. Seve, M. F. Fuller, P. J. Moughan, and C. F. M. de Lange. 2007. Invited review: Amino acid bioavailability anddigestibility in pig feed ingredients—Terminology and application. J. Anim. Sci. 85:172–180.

Stein, H. H., N. L. Trottier, C. Bellaver, and R. A. Easter. 1999b. The effect of feeding level and physiological status on total flowand amino acid composition of endogenous protein at the distal ileum in swine. J. Anim. Sci. 77:1180–1187.

Straarup, E. M., V. Danielsen, C. E. Hoy, and K. Jakobsen. 2006. Dietary structured lipids for post–weaning piglets: Fat digestibility,nitrogen retention and fatty acid profiles of tissues. J. Anim. Physiol. Anim. Nutr. 90:124–135.

Sun, T., H. N. Larke, H. Jorgensen, and K. E. Bach Knudsen. 2006. The effect of extrusion cooking of different starch sources onthe *in vitro* and *in vivo* digestibility in growing pigs. Anim. Feed Sci. Technol. 131:66–85.

Svihus, B., A. K. Uhlen, and O. M. Harstad. 2005. Effect of starch granule structure, associated components and processing onnutritive value of cereal starch: A review. Anim. Feed Sci. Technol. 122:303–320.

Swiss, L. D., and H. S. Bayley. 1976. Influence of the degree of hydrolysis of beef tallow on its absorption in the young pig. Can.J. Physiol. Pharmacol. 54:719–727.

Tamminga, S., H. Schulze, J. van Bruchem, and J. Huisman. 1995. The nutritional significance of endogenous N-losses along thegastro-intestinal tract of farm animals. Arch. Anim. Nutr. 48:9–22.

Taverner, M. R., I. D. Hume, and D. J. Farrell. 1981. Availability to pigs of amino acids in cereal grains. 1. Endogenous levels ofamino acids in ileal digesta and faeces of pigs given cereal diets. Br. J. Nutr. 46:149–158.

Urriola, P. E. 2010. Digestibility of dietary fiber by growing pigs. Ph.D. diss. Univ. Illinois, Urbana-Champaign, IL.

Urriola, P. E., G. C. Shurson, and H. H. Stein. 2010. Digestibility of dietary fiber in distillers co-products fed to growing pigs.J. Anim. Sci. 88:2373–2381.

Wang, Y., Z. Yuan, H. Zhu, M. Ding, and S. Fan. 2005. Effect of cyadox on growth and nutrient digestibility in weanling pigs.South African J. Anim. Sci. 35:117–125.

Wieland, T. M., X. Lin, and J. Odle. 1993. Utilization of medium-chain triglycerides by neonatal pigs: Effects of emulsificationand dose delivered. J. Anim. Sci. 71:1863–1868.

Wilson, R. H., and J. Leibholz. 1981. Digestion in the pig between 7 and 35 d of age. 3. The digestion of nitrogen in pigs givenmilk and soya-bean proteins. Br. J. Nutr. 45:337–346.

Wiseman, J. 2006. Variations in starch digestibility in non-ruminant animals. Anim. Feed Sci. Technol. 130:66–77.Wiseman, J., and D. J. A. Cole. 1987. The digestible and metabolizable energy of two fat blends for growing pigs as influenced bylevel of inclusion. Anim. Prod. 45:117–122.

Wünsche, J., U. Herrman, M. Meinl, U. Hennig, F. Kreinbring, and P. Zwierz. 1987. influss exogener Faktoren auf die präzäkale Nährstoff—und Aminosäurenresorption, ermittelt an Sweinen mit Ileo-Rektalen-Anastomosen. Arch. Anim. Nutr. 37:745–764.

Xing, J. J., E. van Heugten, D. F. Li, K. J. Touchette, J. A. Coalson, R. L. Odgaard, and J. Odle. 2004. Effects of emulsification,fat encapsulation, and pelleting on weanling pig performance and nutrient digestibility. J. Anim. Sci. 82:2601–2609.

Yen, J. T. 2001. Anatomy of the digestive system and nutritional physiology. Pages 31–63 in Swine Nutrition. 2nd ed. A. J. Lewisand L. L. Southern, eds. CRC Press, New York, NY.

Yu, F., P. J. Moughan, and T. N. Barry. 1996. The effect of cottenseed tannins on the ileal digestibility of amino acids in casein andcottonseed kernel. Br. J. Nutr. 75:683–695.

第15章

饲料中矿物质和维生素的生物学利用率

1 导 语

在商业化的养猪生产中，饲粮配方及饲喂策略的主要目标是获得最大的经济效益，而不是追求动物最佳的生长性能。为了获取最大的经济效益，最佳的饲粮配制应尽可能地满足动物的营养需要量，但又不过量。这种饲粮配制技术可以提高动物对养分的利用，减少未被利用养分的排出，对当前发展环境友好型社会具有积极的意义。最优饲喂策略需要考虑的因素很多，其中重要的就是以饲料中可利用的养分为基础配制饲粮。

不像含能量的营养素一样，对矿物质和维生素生物学利用率的研究资料相对较少。因为很难找到合适的指标来评价维生素和微量元素添加量与效应之间的关系，而且在评价前也很难耗竭动物体内维生素和微量元素的贮存以达到理想的评价效果，所以测定动物对矿物质和维生素利用率很困难。因此，需要进一步测定饲料营养素的真利用率，这对于配制高效的环境友好型饲粮非常必要。本章综述了矿物质和维生素的生物学利用率，这对养猪生产的可持续发展具有重要意义。含能量营养素的生物学利用率在第14章中已进行了介绍。

2 矿物质的生物学利用率

很多学者就矿物质相对生物学利用率进行了优秀的综述（Nelson和Walker，1964；Ammerman和Miller，1972；Peeler，1972；Cantor等，1975a，1975b；Cromwell，1992；Ammerman等，1995）。这些文章中大部分是关于饲料中无机来源的矿物质元素相对生物学利用率的研究内容，而不包括有机来源的矿物质元素。本章还将介绍矿物质元素的真吸收率，在没有猪的直接研究资料时，有些结果是从人、雏鸡或大鼠的试验数据推断而来。与相对吸收率相比，真吸收率可以更准确评估饲料中矿物质元素的吸收利用，因此也显得更为重要。同时，要正确评价矿物质添加剂对酸碱平衡的影响，也需要了解矿物质元素的真吸收效率。

◆ 2.1 钙

人摄入既含有动物性又含有植物性的混合食物时，对钙的真吸收率约为30%（Groff等，1995）。目前关于猪对钙的相对生物利用率的研究资料相对匮乏。以碳酸钙为参照，石粉、贝壳粉、石膏、大理石灰和霰石中钙的相对利用率为100%，而白云石灰岩中钙的相对生物学利用率偏低，为51%～78%（Cromwell等，1989a）。Ross等（1984）研究发现，钙原料的颗粒大小对钙的生物学利用率没有影响。以碳酸钙为参照，干苜蓿粉中的钙生物学利用率仅为21%（Cromwell等，1983）。

Bohlke等（2005）以生长猪为研究对象，测定了普通玉米、低植酸玉米和去皮豆粕中钙的表观回肠消化率（AID）和全肠道表观消化率（ATTD）。结果表明，这几种饲料中钙的AID和ATTD值很相近，普通玉米、低植酸玉米和去皮豆粕中钙的消化率平均分别为49%、70%和49%。这些数值与前期对玉米-豆粕型饲粮中钙消化率的研究结果高度一致（Spencer等，2000；Venum等，2001）。对于猪实用饲粮，绝大多数不仅包括玉米和豆粕中的钙，还包括石粉、磷酸氢钙或者磷酸二钙等来源的无机钙。无机钙源的钙ATTD值要比有机钙源的高，玉米、豆粕、石粉、磷酸氢钙或者磷酸二钙中的钙ATTD值为62%～70%（Stein等，2008；Almeida和Stein，2010）。

家禽对磷酸二钙、磷酸三钙、脱氟磷矿石、葡萄糖酸钙、柠檬酸钙、乳酸钙、硫酸钙和骨粉中钙的相对生物学利用率为100%（Peeler等，1972）。Augspurger和Baker（2004）以雏鸡的骨骼灰分含量为评价指标，以分析级碳酸钙为参照发现，测得的饲料级石粉、柠檬酸钙、柠檬酸-苹果酸钙和贝壳粉中钙的相对生物学利用率没有差异。通常认为肉骨粉和鱼粉的钙利用率与饲料级石粉的相当。

以下办法可以提高动物对钙的吸收效率：① 降低采食量；② 饲粮中添加乳糖；③ 妊娠阶段和哺乳阶段；④ 幼龄动物（Groff等，1995）。有些试验还发现，饲粮中添加植酸酶可以提高钙的消化率（Almeida和Stein，2010）。相反，抑制钙吸收的因素主要有：① 不与食物一起摄入；② 饲粮中的植酸或草酸含量；③ 饲粮中的镁和磷过量；④ 由脂肪代谢障碍导致的脂肪痢。

◆ 2.2 磷

人在摄入杂食性食物时，磷的真吸收效率为70%～90%（Groff等，1995）。Soares（1995）总结了很多关于猪和家禽对无机饲料及有机饲料原料中磷生物利用率的研究。Cromwell（1992）还专门针对磷在猪上的生物学利用率研究进行了综述。相对于商品磷酸一钙［Ca（H_2PO_4）$_2$］，给火鸡饲喂磷酸二钙（商品级）时磷的生物学利用率约为90%，饲喂脱氟磷矿石时磷的生物学利用率约为80%（Waibel等，1984）。饲料级的磷酸二钙实际是Ca（H_2PO_4）$_2$和$CaHPO_4$的混合物，并且混合物中结晶水的数目也会发生变化，如Ca（H_2PO_4）$_2\cdot H_2O$或$CaHPO_4\cdot 2H_2O$（Baker，1989）；因此，只有分析混合物中钙和磷的含量，才能明确混合物中$CaHPO_4$（Ca，23%；P，18.5%）和Ca（H_2PO_4）$_2$（Ca，16%；P，21%）的比例。在进行猪饲粮配方设计时，对于所购买的饲料级磷添加剂，不管是磷酸一钙还是磷酸二钙，其磷的生物学利用率都设定为100%；而对于商品型脱氟磷酸盐，其磷的生物利用率假设为90%（Cromwell，1992）。然而，Coffey（1994）等试验发现，相对于NaH_2PO_4，雏鸡和猪对脱氟磷酸盐中磷的生物学利用率平均约为85%。最近研究发现，生长猪对NaH_2PO_4中磷的ATTD约为92%，对$CaHPO_4$和Ca（H_2PO_4）$_2$中磷的ATTD约为82%（Petersen和Stein，2006）。Stein等（2008）也在生长猪玉米-豆粕型饲粮中添加不同水平的Ca（H_2PO_4）$_2$，发现磷的ATTD平均值为84%。

植酸酶可以提高磷的生物学利用率，尽管一些饲料原料（如小麦和大麦）中含有一定量的植酸酶，但植物源的磷基本上不能被利用（Nelson，1976）。Cromwell（1992）研究发现，相对于$CaHPO_4$，玉米中磷的生物学利用率仅为14%，豆粕中磷的生物学利用率为23%~31%。他还进一步发现，小麦及小麦副产品中磷的相对生物学利用率为29%~49%，稻糠、棉籽粕和花生粕中磷的生物学利用率分别为25%、1%和12%。Cromwell认为，干燥乳清粉、血粉、鱼粉和苜蓿粉中磷的相对生物学利用率接近100%，但肉骨粉中磷的相对生物学利用率只有67%。水分含量高的玉米中的有效磷含量是干燥的黄色马齿玉米的4倍（Cromwell，1992）。

最近试验测定的玉米、豆粕和玉米-豆粕型饲粮中的磷消化率要高于Cromwell（1992）得到的相对生物学利用率。Bolke等（2005）采用去势的生长公猪来估测玉米、低植酸玉米和豆粕中磷的AID和ATTD，发现玉米、低植酸玉米和豆粕中磷的AID值与ATTD值基本相似，分别为29%、56%和38%。其中玉米和豆粕中磷的ATTD值与Stein等（2008）的研究结果较为一致。Stein等（2008）用生长猪估测的玉米-豆粕型饲粮磷的ATTD值为38.4%。DDGS中磷的ATTD要比玉米和豆粕的高，为50%~70%（Pedersen等，2007；Stein等，2009；Almeida和Stein，2010）。高蛋白干酒糟（HPDDG）中磷的ATTD也接近60%（Widmer等，2007）。由于DDGS和HPDDG都是经过发酵的产物，发酵过程中部分植酸键可能被破坏，使磷被释放出来，因此这些原料的磷ATTD值要比玉米和豆粕的都高；而未经发酵的玉米胚芽的磷ATTD值还不到30%（Widmer等，2007）。在不添加植酸酶情况下，生长猪对紫花豌豆的磷ATTD约为55%，但如果饲粮添加500单位植酸酶，该ATTD值就提高到65.9%（Stein等，2006）。

发酵产品中磷的生物学利用率相对较高，因为在酵母、DDGS等这些发酵产品中，大部分磷是以核酸磷的形式存在，主要是RNA。以雏鸡为研究对象，将KH_2PO_4作为参照标准，RNA中磷的生物学利用率约为100%（Burns和Baker，1976）；同样，饲料原料中磷脂形式磷的生物学利用率也是100%（Baker，未发表数据）。

植酸

植酸不仅可以降低磷的生物学利用率，还会影响动物体内钙和其他微量元素的营养状态，但对其作用机制及程度目前仍不是很清楚。各种试验饲粮中使用的植酸盐通常是以植酸、植酸钠或植酸钙的形式添加，而这些化合物不同于与植酸Ca-Mg-K混合盐，后者主要存在于植物源性饲料原料中（Nelson，1967；Erdman，1979）。此外，添加植酸盐可以降低微量元素的利用率，但该作用受到饲粮钙水平的显著影响（O'Dell等，1964；Hendricks等，1969；Bafundo等，1984b）。例如，伊利诺伊州立大学使用典型的玉米-豆粕型肉仔鸡饲粮的试验发现，只有在饲粮钙含量至少超出NRC（1994）推荐量的2倍时，添加1.2%植酸钠才能出现显著的锌拮抗作用（Bafundo等，1984b）。

为了合理解释植物性饲料中植酸对矿物质元素生物学利用率的影响，了解植酸在这些原料中的分布位点就非常重要（Erdman，1979）。玉米是谷类作物中非常特殊的一种原料，因为约90%的植酸都存在于玉米胚乳中。小麦和水稻的胚乳中也含有一定量的植酸，但是种皮部分的植酸含量更高，因此麦麸和稻糠中的植酸含量都较高（Halpin和Baker，1987）。与其他多数油料种子不同，大豆的植酸包含在大豆的蛋白质中，遍布大豆各部位，因此大豆分离蛋白的植酸含量要比豆粕的高（Erdman，1979）。花生、棉籽和葵花籽的植酸仅存在于种子晶体结构和球状体结构中。

饲粮添加微生物源植酸酶或植物源植酸酶均可提高猪和家禽对植酸磷的利用率（Almeida和Stein，2010）。植酸酶还可以显著提高动物对钙、锌、锰和铁的利用率，但不能提高动物对铜或硒的利用率（Aoyagi和Baker，1995）。添加柠檬酸、乳酸和甲酸等有机酸也可以提高植酸磷的利用率，且它们与植酸酶一起添加具有叠加效应。尽管1α-羟基维生素D_3可以显著提高肉仔鸡对植酸磷的利用率，但对猪和产蛋鸡却没有作用（Biehl等，1995；Biehl和Baker，1996；Snow等，2003）。低植酸玉米的有效磷含量比常规玉米的高了近3倍。现代转基因技术已将大肠杆菌载体表达的植酸酶基因转入玉米中（Nyannor等，2007；Nyannor和Adeola，2008），同时还可以培育出唾液分泌植酸酶的转基因猪（Golovan等，2001），这两种先进的方法都可以明显提高饲粮植酸磷的利用率。

◆ 2.3 钠和氯

人们通常认为，钠和氯的真吸收率接近100%（Groff等，1995）。根据动物需要量研究的结果可以得出，NaCl、Na_2SO_4、$NaHCO_3$、乙酸钠和柠檬酸钠中钠的生物学利用率是相同的，而NaCl、KCl、NH_4Cl和$CaCl_2 \cdot 2H_2O$中氯的生物学利用率是相同的。由于钠和氯（也包括钾）主要调节体内酸碱平衡和尿中阴阳离子的排出，故很难找到合适的评价方法测定其生物学利用率。经常在猪饲粮中使用的脱氟磷矿石一般都含有比较高的钠。Miller（1980）以NaCl为参照，估测出脱氟磷中钠的生物学利用率为83%。而硅铝酸盐产品（如沸石）中钠的吸收率与NaCl的差不多（Fethiere等，1988）。

◆ 2.4 镁

相对于分析级MgO中的镁，猪对饲料级的MgO、$MgSO_4$和$MgCO_3$中镁的生物学利用率是100%，而谷类饲料和浓缩料中镁的生物学利用率是50%～60%（Miller，1980）。对大鼠的研究结果也进一步证明，MgO、MgCl、$MgCO_3$、磷酸镁、硫酸镁和硅酸盐中镁的生物学利用率是相等的（Cook，1973）。雏鸡对豆粕和硫酸镁（$MgSO_4 \cdot 7H_2O$）中镁的真吸收率均为60%（Guenter和Sell，1974）。然而，关于猪对植物性饲料和动物性饲料中镁的相对生物学利用率，目前还未见报道。

◆ 2.5 钾

各种钾源饲料中钾的生物学利用率研究数据都很少。由于猪的玉米-豆粕型饲粮中富含钾，因此研究钾生物学利用率主要是学术目的。Peeler（1972）曾预测，相对于氯化钾，碳酸钾、碳酸氢钾、磷酸氢二钾、乙酸钾和柠檬酸钾中钾的生物学利用率是100%。Groff等（1995）也认为，人类摄取的钾中有90%以上都可以被机体吸收。

血钾含量、尿钾含量和钾沉积量都曾是评定钾生物学利用率的反应指标，但可能只有钾沉积量（平衡试验）与钾的摄入量呈线性关系（Combs和Miller，1985）。以乙酸钾为参照标准，在试验误差范围内，应用斜率法评价的碳酸钾、碳酸氢钾、玉米和豆粕中钾的生物学利用率是100%（Combs和Miller，1985；Combs等，1985）。

◆ 2.6 铜

人对杂合食物中铜的真吸收率为25%（高摄入量）到50%（低摄入量）（Groff等，1995）。当饲粮铜含量从缺乏水平升至约250 mg/kg时，在组织（主要是肝脏）中沉积的铜量增加很小（呈曲线增加）。因此在这个变化范围内，铜的生物学利用率很难准确评价，这与锌的生物学利用率的评价情况类似。但当饲粮铜含量超过250 mg/kg后，铜在组织中的沉积量迅速增加，通常呈线性上升。Miller（1980）指出，以$CuSO_4 \cdot 5H_2O$为参比，$CuCl_2$和$CuCO_3$中铜的生物学利用率较高，而CuS的生物学利用率较低。Cromwell等（1989b）对猪的试验发现，CuO几乎完全不能被肠道吸收，这与在鸡上的试验结果类似（Baker等，1991；Aoyagi和Baker，1993a；Bzker和Ammerman，1995a）。Aoyagi和Baker（1993a，1993b，1994）和Aoyagi等（1995）建立了仔鸡饲粮中铜含量高于其需要量（肝铜的沉积量）和低于其需要量（胆囊铜的沉积量）时铜生物学利用率的评价方法，使用这两种方法测定不同的无机铜和饲料原料中铜的相对生物学利用率（RBV），所得的结果较为一致。以分析级$CuSO_4 \cdot 5H_2O$为参比，分析级$CuCl_2$的RBV是145%，饲料级的$CuSO_4 \cdot 5H_2O$、赖氨酸-铜、$Cu_2(OH)_3Cl$和蛋氨酸-铜的RBV为95%～115%，分析级和饲料级的CuO的RBV都是0，分析级的$Cu(OAC) \cdot H_2O$、Cu_2O和$CuCO_3 \cdot Cu(OH)_2$的RBV分别是115%、100%、100%。在动物性和植物性蛋白饲料中铜的RBV，猪肝脏为0，鸡肝脏为115%，家禽副产品粉是90%，牛肝脏是80%，玉米蛋白粉是50%、花生粕和豆渣是45%，棉籽粕和去皮豆粕是40%，大鼠肝脏是20%（Aoyagi等，1993）。

以$CuSO_4 \cdot 5H_2O$为参照，猪从粪便中摄取铜的利用率不超过30%（Izquierdo和Bzker，1986）。当饲粮添加Na_2S（Barber等，1961；Cromwell等，1978）、洛克沙胂（Czarnecki和Baker，1985；Edmonds和Baker，1986）或半胱氨酸、抗坏血酸等其他还原剂时（Baker和Czarnecki-Maulden，1987，Aoyagi和Baker，1994），铜在肠道内的吸收率会显著降低。

◆ 2.7 碘

关于碘的生物学利用率，在所有物种上的研究都非常稀少。事实上，最早在大鼠上的试验主要目的并不是评价碘的生物学利用率。但Miller（1980）提出，相对于NaI，KI、Ca（io_3）$_2$ · $2H_2O$、KIO_3和CuI中碘的生物学利用率是100%。Miller和Ammerman（1995）也发现，乙二胺二氢碘（$C_2H_8N_2 \cdot 2HI$）中碘的RBV不低于100%。因此可以推测，相对于NaI，所有碘酸盐中碘的相对生物利用率是100%。

◆ 2.8 铁

集约化饲养的保育猪（基本接触不到土壤）由于生长速度快、且不能从胎盘和母乳中获取铁，特别容易出现缺铁性贫血。因此在实际生产中，通常在出生后的前几天给新生仔猪注射铁剂。注射100 mg或200 mg右旋糖酐铁后，约有90%以上的铁用于血红蛋白的合成（Braude等，1962；Miller，1980）。在仔猪出生后12 h内（在肠道闭合之前），口服右旋糖酐铁也可以有效促进铁与血红蛋白的结合（Harmon等，1974；Thoren-Tolling，1975；Cornelius和Harmon，1976）。

目前还没有有效的方法来提高母猪胎盘或乳房中的铁转运给仔猪。康奈尔大学的研究表明，无论铁源是通过口服还是注射给母猪，仔猪在出生时铁贮存量的增加及猪乳中铁浓度的提高，都不足以避免仔猪缺铁性贫血的发生（Pond等，1961）。给泌乳母猪饲喂高铁饲粮可增加哺乳仔猪的血红蛋白含量，但研究表明，这主要是因为食入母猪粪中的铁所致，而不只是因为母乳中铁的含量增加。

有关猪对铁的生物学利用率研究表明，以$FeSO_4 \cdot H_2O$为参比，$FeSO_4 \cdot 2H_2O$、$FeSO_4 \cdot 7H_2O$、柠檬酸铁和胆酸亚铁中铁的生物学利用率基本上都是100%，而且Fritz等（1970）对鸡和大鼠的研究结果更进一步证实了这些（Harmon等，1967；Ullrey等，1979；Furugouri和Kawabata，1975）。用鸡和大鼠进行的试验表明，对柠檬酸铁铵、$FeCl_2$、富马酸铁和葡萄糖酸铁中铁的生物学利用率也是100%，而$FeCl_3$、$Fe_2(SO_4)_3$和碳酸铁的生物学利用率很低。氧化铁几乎完全不能被机体利用，碳酸盐中铁的生物学利用率是与碳酸盐的来源有关（Harmon等，1969；Henry和Miller，1995）。

商业用磷酸二钙和脱氟磷矿石中富含铁，含有2.5%～3.0%的$FePO_4 \cdot 2H_2O$（Baker，1989）。这些产品中铁的生物利用率通常为35%～85%（Ammerman和Miller，1972；Kornegay，1972；Deming和Czarnecki-Maulden，1989）。

人类的营养学通常将铁源分为血红素铁或非血红素铁。然而，这种分类法常会产生一些误解，因为一些营养学家会无意识地将所有动物源铁都错认为血红素铁，而实际上，动物源铁一般是由血红素铁（利用率高）、铁蛋白（利用率低）和血铁黄素（利用率极低）组成的混合物（Layrisse等，1975；Bogunjoko等，1983）。这也可以说明，为什么新鲜或干燥的肝脏中都含铁丰富，但其生物学利用率要比禽类副产品及肉粉中铁的利用率都低的原因，主要是肝脏中的铁仅有约10%的是血红素铁，而约90%的铁是非血红素铁（Chausow，1987）。商业用血粉中的铁通常被认为具有较高的利用率，但Miller（1980）对快速干燥的血粉进行测定，发现它的铁生物学利用率只有40%～50%（相对于$FeSO_4 \cdot 7H_2O$），而Chausow和Czarnecki-Maulden（1988a）报道的仅有22%。因此，干燥处理过程可能会影响干燥血粉中铁的生物学利用率。

谷物和菜籽粕中的铁主要以结合态存在，大部分与植酸、纤维素或蛋白质结合形成复合物，因此它们的铁生物学利用率可能要比$FeSO_4 \cdot 7H_2O$的低。但有关这方面的研究目前还十分有限。在没有确切数据的情况下，人们通常认为谷物和菜籽粕中铁的生物学利用率不超过50%（相对于$FeSO_4 \cdot H_2O$）。

Chausow（1987）、Chausow和Czarnecki-Maulden（1988a，1988b）以鸡、猫和狗为实验动物，用$FeSO_4 \cdot 7H_2O$作为参照，研究了不同铁源的生物学利用率。鸡的试验结果表明，铁缺乏降低了血红蛋白的浓度，干燥血粉、肉骨粉、禽类副产品、羽毛粉和鱼粉中铁的相对生物学利用率分别是22%、48%、68%、39%和32%。对于植物性饲料，芝麻粕、稻糠、苜蓿粉、去皮豆粕和黄玉米中铁的相对生物学利用率分别是96%、77%、65%、45%和20%（Czarnecki-Maulden，1988a，1988b）。

Boiling等（1998）以$FeSO_4 \cdot 7H_2O$或$FeSO_4 \cdot H_2O$作为参照，用血红蛋白含量的指标来评价不同铁源的生物学利用率，发现分析纯硫酸铁［$Fe_2(SO_4)_3 \cdot 7H_2O$］的相对生物学利用率仅为37%，而棉籽粕中铁的生物学利用率是56%。该试验还研究了镀锌工业两种副产品的铁生物学利用率，它们都是$FeSO_4 \cdot H_2O$和$ZnSO_4 \cdot H_2O$的混合物，其铁的生物学利用率与硫酸亚铁的相同。

当人摄入既含有动物性成分又含有植物性成分的杂合食物时，其铁的真吸收率约为15%（Groff等，1995）。影响铁吸收的因素有很。现已发现，饲粮中的抗坏血酸、半胱氨酸及柠檬酸和乳酸等有机酸都会加倍提高铁的吸收率。但饲粮中的植酸、草酸和过量锌（同时又有植酸存在）都会使铁的吸收率降低，还不到没有这些拮抗因子时铁吸收率的一半。此外，缺铁的动物对铁的吸收率要比不缺铁的动物高很多。

◆ 2.9 锰

因为饲粮锰缺乏对家禽的影响要比对猪的影响更严重，所以锰生物学利用率的研究基本上都是使用禽类为动物模型（Souther和Baker，1983a，1983b；Black等，1984；Henry，1995；Baker等，1986；Halpin和Baker，1986a，1986b；Halpin等，1986；Halpin和Baker，1987）。由于锰在组织（主要是骨骼）中的沉积量随饲粮锰含量增加呈线性增加，因此可以用组织锰含量来评价其生物学利用率。相对于$MnSO_4 \cdot H_2O$，$MnCl_2$、MnO、$MnCO_3$和MnO_2的生物学利用率分别约为100%、75%、55%和30%（Henry，1995）。蛋白质锰络合物和蛋氨酸锰络合物的生物学利用率与$MnSO_4 \cdot H_2O$相近（Baker等，1986；Fly等，1989）。玉米、豆粕、麦麸和鱼粉中的锰全部不能被鸡和猪利用（Baker等，1986）；稻糠中含有丰富的锰，但利用率极低。尽管过量的锰会降低肠道铁的吸收，但玉米-豆粕型饲粮中过量的铁或钴对锰的利用率影响非常小。饲粮添加过量的磷（无机磷或植酸磷）会显著降低锰的利用率（Baker和Wedekind，1988；Baker和Wedekind，1990a，1990b；Wedekind等，1991a，1991b；Baker和Oduho，1994）。

即使添加像$MnSO_4 \cdot H_2O$这样高利用率的锰源，鸡肠道对锰的吸收率也只有2%～3%（Wedekind等，1991a，1991b）。如果除去饲粮中所有的纤维素及植酸，锰的吸收率可以提高近1倍（Halpin等，1986）。

◆ 2.10 硒

亚硒酸钠（Na_2SeO_3）和硒酸钠（Na_2SeO_4）中硒的生物学利用率很高，且以组织硒沉积量为衡量指标，硒代蛋氨酸相对于Na_2SeO_3的生物学利用率超过100%（Mathias等，1967；Cantor等，1975a，1975b；Mahan和Moxon，1978）。谷物中的硒生物学利用率较高，为60%～80%；苜蓿粉的更高，超过100%；但鱼和禽类的副产品中的硒利用率却很低，只有10%～40%。对家禽的研究表明，以谷胱甘肽过氧化物酶活性为评价指标，豆粕硒的生物学利用率仅有18%；而以渗出性素质预防为评价指标，豆粕硒的生物学利用率则为60%（Cantor等，1975b）。相对于Na_2SeO_3，鸡对动物性饲料中硒的生物学利用率平均为28%，而对植物性饲料中硒的生物学利用率为47%（Wedekind等，1998）。富硒酵母相对于Na_2SeO_3的生物学利用率是159%。Mahan和Parrett（1996）在生长育肥猪上评价了富硒酵母相对于Na_2SeO_3的生物学利用率。结果发现，以体内硒沉积为评价指标，富硒酵母的生物学利用率要高于Na_2SeO_3；但以血清谷胱甘肽含量为评价指标，富硒酵母的生物学利用率却低于Na_2SeO_3。

肠道对硒的吸收率较高，其中猪对摄入硒的吸收率约为63%（Wright和Bell，1966）。研究发现，各种砷化物以及半胱氨酸、蛋氨酸、铜、钨、汞、镉和银都会降低肠道无机硒的吸收率（Baker和Czarnecki-Maulden，1987；Lowry和Baker，1989a）。

◆ 2.11 锌

假设人对植物性食品中锌的吸收率低于10%，对动物性食品中锌的吸收率是30%；在此基础上，通常认为人对杂合食品中锌的吸收率约是20%（Groff等，1995）。猪里脊肉、碎牛肉等可食用肉产品中锌的吸收率都很高，主要是因为这些肉产品富含半胱氨酸（和以谷胱甘肽形式存在的半胱氨酸），可以促进锌的吸收率（Hortin等，1991，1993）。

有关猪对含锌添加剂相对生物学利用率的研究相当匮乏。Miller等（1981）报道，猪对锌粉（含99.3%Zn）的生物学利用率高于分析纯ZnO。以饲料级$ZnSO_4 \cdot H_2O$为参比，猪对饲料级ZnO的生物学利用率仅为56%～68%（Hahn和Baker，1993；Wedekind等，1994）。

很多影响铁吸收的因素同样也会影响锌的吸收。锌摄入量低、饲粮中含有抗坏血酸和半胱氨酸等还原剂，都可以提高锌的利用率；而锌摄入量高、饲粮中含有植酸或者氧化剂，都会降低锌的吸收。人处于应激或创伤或同时遭受应激和创伤（如手术、烧伤）时，也会降低对锌的吸收率（Groff等，1995）。

目前在美国，由于通常使用高剂量的饲料级氧化锌（Zn，2 000～3 000 mg/kg）

来促进断奶仔猪的生长性能（Hahn和Baker，1993；Hill等，2000），因此ZnO的生物学利用率问题越来越受到关注。早期对鸡ZnO生物学利用率的研究表明，试剂级ZnO的生物学利用率与试剂级$ZnSO_4 \cdot 7H_2O$的相等（Edwards，1959）。然而，对肉仔鸡的研究发现，饲料工业最常用的锌源——饲料级ZnO的生物学利用率约是饲料级$ZnSO_4 \cdot H_2O$的50%（Wedekind和Baker，1990c；Wedekind等，1992），但不同来源的饲料级ZnO的生物学利用率差别很大（Edwards和Baker，1999）。

Baker和Ammerman（1995b）总结认为，与分析纯$ZnSO_4 \cdot 7H_2O$相比，$ZnSO_4 \cdot H_2O$、$ZnCO_3$、$ZnCl_2$、分析纯ZnO、蛋氨酸锌和乙酸锌都有较高的生物学利用率，因此这些锌源均可作为锌添加剂在猪饲粮中添加。饲喂缺锌饲粮的动物增重及骨骼锌含量是评价锌生物学利用率的最理想指标（Wedekind等，1992）。而软组织中的锌含量、血浆锌含量和血浆碱性磷酸酶活性与饲粮锌添加水平的回归关系不强，因此这些指标不适合用于评价锌的生物学利用率。

与其他很多微量元素一样，非反刍动物对常规玉米-豆粕型饲粮中锌的利用率极低。事实上，动物对常规饲粮中锌的需要量比无植酸（如鸡蛋白）饲粮高出3～4倍。另外，在植酸和纤维素存在的情况下，过量的钙会降低锌的利用率。过量的锌会加剧铜和铁的缺乏，而过量的铜和铁对锌利用率的影响很小（Southern和Baker，1983c；Bafundo等，1984a）。

大豆制品中锌的利用率很差。相对于$ZnSO_4 \cdot 7H_2O$，豆粕、大豆浓缩蛋白和大豆分离蛋白中锌的生物学利用率分别为34%、18%和25%（Edwards和Baker，2000）。动物性饲料中锌的利用率比植物性饲料的高，但某些动物性饲料可能含有锌的拮抗因子，进而影响锌的利用率（Baker和Halpin，1988）。

◆ 2.12 铬

自从Page等（1993）报道育肥猪饲粮中添加三甲基吡啶铬可以改善猪的胴体品质后，人们开始关注铬在猪营养中的作用。随后，Lindemann等（1995，2004）发现，妊娠期母猪饲粮中添加铬可以增加窝仔数。相对于三甲基吡啶铬，生长猪对丙酸铬、蛋氨酸铬和酵母铬的生物学利用率分别为13.1%、50.5%和22.8%（Lindemann等，2008）。

3 维生素的生物学利用率

关于现代猪配合饲料及预混料中的维生素生物学利用率，人们主要关注两个问题：① 维生素和维生素-矿物质预混料的稳定性，也包括它们在饲粮及添加剂中的稳定性；② 植物性和动物性饲料原料中维生素的利用效率。Wornick（1968）、Zhuge和Klopfenstein（1986）、Baker（1995）以及Baker（2001）的综述可供大家参考，这些文章详细阐述了影响晶体维生素在饲粮及预混料中稳定性的因素。然而，有关猪对饲料原料中维生素生物学利用率的研究，目前还未见报道；即使综合鸡和鼠上的研究来看，也只对少数几种饲料原料中的维生素生物学利用率进行了评价。

维生素和矿物质生物学利用率的评价方法还存在很多缺点。在传统的生长试验过程中，体内贮存的维生素可以防止动物出现明显的缺乏症。即使动物出现了明显的缺乏症，我们在试验中还必须搞清楚一个问题：反应指标（通常是增重）的增加是由补充维生素引起的还是由采食量的增加引起的？因为纯合饲粮的适口性很差，添加某些养分后会提高采食量。与脂溶性维生素相比，水溶性B族维生素在一定范围内与动物生长性能之间呈现较好的剂量效应，因此它们在很多方面更容易进行评价。

在应用合适的生物法测定维生素的最大效价和用外推值法评定维生素生物学利用率时，应考虑以下问题：① 受试动物在预试期必须达到所期望的维生素缺乏状态；② 以维生素作为组成或辅因子的关键酶活性通常不是理想的评价指标，但体增重除外；③ 必须考虑某些维生素合成的前体物质的影响（如蛋氨酸合成胆碱和色氨酸合成烟酸）；④ 使用特异性维生素抑制剂有利于准确评价维生素的生物学利用率。

◆ 3.1 维生素A

根据维生素的命名法则，维生素A是指所有β-紫罗酮的衍生物，具有生物活性的全反式视黄醇（如视黄醇或者维生素A_1），但不包括维生素A源——类胡萝卜素（Anonymous，1979）。全反式视黄醇的酯类都被称为视黄醇酯。

动物组织含有维生素A，而大多数植物只含有维生素A源——类胡萝卜素，它在小肠内分解后形成维生素A。维生素A是以视黄醇的形式在血液中运输，但主要以棕榈酸视黄酯形式在肝脏中贮存。动物对维生素A的吸收率在很大的剂量范围内都保持相对稳定，但类胡萝卜素的吸收率会随剂量增加而降低（Erdman等，1988）。

在饲料及预混料中维生素A酯比维生素A醇更稳定，因为视黄醇中的羟基及其侧链的4个双键很容易发生氧化损坏，但维生素A醇的酯化反应并不能完全阻止维生素A醇的氧化损失。目前商业用维生素A添加剂一般都是“被包被”的酯类（如乙酸酯或棕榈酸酯），同时添加乙氧基喹啉或丁羟基甲苯（BHT）等抗氧化剂进行保护。

预混料或饲料原料中的水分含量会影响维生素A的稳定性。水分会软化维生素A微囊，使氧气更容易渗入微囊内。因此，在外界环境湿度高或有游离的氯化胆碱（易吸湿的）存在时，会加剧对维生素A的破坏。在潮湿的环境中，微量元素也会加速预混料中维生素A的损失。因此，为了保持维生素A最大的活性，预混料应该尽可能保持干燥，且pH要大于5。低pH容易导致全反式维生素A发生异构化反应，形成顺式结构，进而降低其活性；同时还会导致维生素A酯发生去酯化反应而变成醇，进而降低其稳定性（DeRitter，1976）。此外，热处理特别是膨化处理，会降低维生素A的生物学利用率（Baker，2001）。

晶体状β-胡萝卜素在肠道中的吸收率比食物和饲料中含有的β-胡萝卜素高（Rao和Rao，1970）。食物中的β-胡萝卜素部分会与蛋白质结合形成复合物。饲料中的纤维素尤其是果胶，可以降低鸡肠道对β-胡萝卜素的吸收（Erdman等，1986）。

Ullrey（1972）综述了猪对维生素A源生物学利用率方面的资料，并指出猪将胡萝卜素转化为维生素A的能力远远低于鼠。相对于全反式棕榈酸视黄酯来说，猪对玉米中胡萝卜素的生物效价仅为7%～14%。猪对玉米（包括玉米蛋白粉）中胡萝卜素的转化最大也不要超过261 IU/mg维生素A。假设所有的胡萝卜素都是全反式β-胡萝卜素，大鼠将胡萝卜素转化为维生素A的理论值低于1667 IU/mg。玉米胡萝卜素由约50%的叶黄素、25%的β-玉米胡萝卜素和25%的β-类胡萝卜素组成（Ullrey，1972）。维生素A生物学利用率的定量分析比较困难，目前最常用的方法是以维生素A在肝脏内的沉积量为反应指标来评定其生物学利用率（Erdman等，1988；Chung等，1990）。

◆ 3.2 维生素D

维生素D是指具有胆钙化醇生物活性的所有类固醇的总称。胆钙化醇又叫维生素D_3，与麦角钙化醇即维生素D_2不同。商业用维生素D_3通常是喷雾干燥的产品或用明胶包被的微囊型产品（经常与维生素A一起）；一个国际单位的维生素D相当于0.025 μg维生素D_3的活性（Anonymous，1979）。这些产品单独在室温条件下贮存时十分稳定，但在配合饲料和矿物质-维生素预混料中室温贮存4～6个月，维生素D的活性损失达到20%（Baker，2001）。

维生素D的前体物在植物饲料中是钙化醇，而在动物饲料中是7-羟基胆钙化醇，它们必须经过紫外线照射才能分别转化为有活性的维生素D_2或维生素D_3。长期以来，人们都认为维生素D_2和维生素D_3对猪具有相同的生物活性，但Horst等（1982）指出，维生素D_3比维生素D_2的生物活性更高。维生素D_3的羟基化形式有三种，分别是25-OH-D_3、1-α-OH-D_3和1,25（OH）$_2$-D_3，其中1α-OH-D_3的生物活性比维生素D_3高。

◆ 3.3 维生素E

维生素E是一类具有α-生育酚活性的母育酚及三烯生育酚衍生物的统称。自然界存在8种维生素E：α、β、γ、δ-生育酚以及α、β、γ、δ-三烯生育酚。在这8种维生素E中，D-α-生育酚的生物活性最高（Bieri和Mckenna，1981）。1个国际单位（IU）的维生素E相当于1 mg DL-α-生育酚乙酸酯的活性。所有消旋的维生素E（如DL-α-生育酚）的活性约为纯D-α-生育酚的70%，而β-生育酚和γ-生育酚的生物活性仅为α-生育酚的40%和10%（Bieri和Mckenna，1981）。根据Bieri和Mckenna（1981）提出的维生素E活性顺序表，α-三烯生育酚是自然界唯一具有维生素E活性的三烯生育酚，其生物效价约为DL-α-生育酚乙酸酯的25%。

植物性饲料比动物性饲料的维生素E活性更高，特别是植物油富含有活性的维生素E，但玉米和玉米油中γ生育酚的含量是α-生育酚的6倍（Ullrey，1981）。去脂豆粕的维生素E活性非常低。

维生素E很容易被氧化破坏，而且加热、水分、不饱和脂肪酸和微量元素都会加速维生素E的氧化。苜蓿在32℃贮存12周，维生素E的损失达50%～70%；而苜蓿脱水后，维生素E的损失达30%（Livingston等，1968）。用有机酸处理高水分的谷物也会加速维生素E的破坏（Young等，1975，1977，1978）。然而，维生素E即使贮存在弱碱条件下其稳定性也会受到影响。石粉或氧化镁的粉末可以直接与维生素E发生作用，显著降低其生物活性。

◆ 3.4 维生素K

维生素K也属于脂溶性维生素，主要有三种类型：植物产生的叶绿醌（维生素K_1）、微生物合成的甲萘醌（维生素K_2）和人工合成的甲萘醌（维生素K_3）。这三种维生素K都具有生物活性，但在猪饲料中只添加人工合成的水溶性甲萘醌。商业用维生素K_3添加剂包括亚硫酸氢钠甲萘醌（MSB）、亚硫酸氢钠甲萘醌复合物（MSBC）和亚硫酸嘧啶甲萘醌（MPB），它们分别含有52%、33%和45.5%的甲萘醌。维生素K_3添加剂在饲粮和预混料中的稳定性受水分、氯化胆碱、微量元素和碱性条件的负影响。MSBC和MPB在含有胆碱的维生素-微量元素预混料中贮存3个月，生物活性损失约80%；但在不含胆碱的相同预混料中，生物活性的损失将大大减少（Baker，2001）。包被的维生素K_3通常比不包被的维生素K_3更稳定。鸡对MPB的生物学利用率比MSB和MSBC都高（Griminger，1965；Charles和Huston，1972）。Seerley等（1976）研究发现，MPB对猪的作用明显。Oduho等（1993）比较了不同维生素K源亚硫酸烟酰胺甲萘醌（MNB；45.7%甲萘醌，32%烟酰胺）与MPB在仔鸡上的作用。结果发现，根据凝血时间来评价，MNB与MPB中的维生素K活性相等。尽管一些饲料原料（如苜蓿粉）中富含猪可利用的活性维生素K，但猪对这些饲料原料中维生素K生物学利用率的定量研究资料仍非常少（Fritchen等，1971）。

◆ 3.5 生物素

商业用D-生物素没有专门的活性单位，所以通常认为1 g D-生物素等于1 g活性单位。制粒或加热对饲料中生物素活性的影响甚微，但氧化酸败会显著降低生物素的生物学利用率。饲料原料中的生物素大部分是以结合状态存在，即

□-N-生物素酰-L-赖氨酸（生物胞素），它是合成蛋白质的一种成分。晶体状生物素很容易在小肠内被吸收，但是不同生物胞素中的生物素生物学利用率差异很大，取决于它们结合的蛋白质的可消化性（Baker，1995）。抗生物素蛋白是存在于蛋清中的一种糖蛋白，可结合生物素，并使其完全不能被机体利用。适度的加热处理蛋清将会使抗生物素蛋白变性，从而阻止其与生物素的结合。使用生物素缺乏的鸡来评定谷物中生物素的生物学利用率发现，玉米的生物素生物学利用率很高（>100%），而小麦、燕麦和高粱的生物素利用率较低，约为50%（Aderson和Warnick，1970；Frig，1976；Anderson等，1978）。玉米、大麦、高粱和小麦中具有生物活性的生物素含量分别是0.11 mg/kg、0.08 mg/kg、0.09 mg/kg和0.04 mg/kg（Anderson等，1978）。饲料成分表列出的豆粕生物素含量是0.30 mg/kg。对产蛋鸡来说，豆粕中生物素的生物学利用率是100%，肉骨粉中生物素的利用率是86%（Buenrostro和Kratzer等，1984）。由于玉米-豆粕型饲粮含有大量可利用的生物素，因此在生长育肥猪玉米-豆粕型饲粮中添加生物素的作用一般不明显。在某些情况下，母猪饲粮添加生物素可以提高受胎率，缩短断奶与发情的间隔时间，改善猪蹄健康和被毛状况，特别对高产母猪的作用更加明显（Bryant等，1985）。Lewis等（1991）报道，在整个妊娠期和泌乳期内，玉米-豆粕型饲粮添加0.33 mg/kg生物素可以增加断奶仔猪数，但不能改善猪蹄健康。Watkins等（1991）进行了类似试验，发现玉米-豆粕型饲粮添加0.44 mg/kg生物素却对猪没有影响。

生物素是唯一一种可用试验真实评价其生物学价值的维生素。在无生物素的纯合饲粮中添加或不添加晶体状抗生物素蛋白，通过测定生长性能指标来评价生物素的生物学利用率。缺乏生物素的鸡生长速度会因添加20%的玉米补充料而成倍增长，但将同样数量的玉米中添加到含有3.81 mg/kg抗生物素蛋白的饲粮中，其生长速度变化不明显（Anderson等，1978）。用大麦进行的试验也得到了相似的结果（Anderson等，1978）。因此，这些结果都进一步证明，在无生物素的纯合饲粮中添加谷物饲料后，所观察到的生长效应是由谷物饲料自身提供的有效生物素所产生的。

◆ 3.6 胆碱

在动物营养学中，胆碱属于B族维生素的范畴，尽管其需要量远远超出维生素所定义的“微量有机营养素”的量。胆碱主要是在小肠内吸收，是机体所必需的养分，其主要作用包括：①用于磷脂合成；②构成乙酰胆碱的主要成分；③参与高半胱氨酸转化成蛋氨酸的转甲基作用。通过饲喂无胆碱饲粮诱导鸡出现胆碱缺乏症的试验表明，与高半胱氨酸合成蛋氨酸的转甲基作用相比，胆碱优先用于体内磷脂的合成、乙酰胆碱的形成或同时参与这两个过程；因为补充甜菜碱（胆碱氧化后的甲基化产品）不能具有胆碱所产生的生长效应。然而，实用饲粮中含有的胆碱能提供满足最佳生长所需的胆碱需要量的1/2～2/3时，添加化学合成的胆碱与甜菜碱的作用效果一样（Lowry等，1987；Dilger等，2007）。

与禽类不同，哺乳动物的饲粮所需要的胆碱可由过量的蛋氨酸所代替。因为结晶状氯化胆碱（74.6%胆碱）容易吸潮，被认为是在维生素-矿物质预混料中其他维生素的破坏剂，所以胆碱通常单都是独添加，而不是与常用的维生素-矿物质预混料一起添加。粗提植物油（如玉米油和大豆油）中所含的胆碱是与磷脂结合的磷脂酰胆碱，这种形式的胆碱生物学利用率不低于100%（Emmert等，1996）。而精炼植物油通常要经过碱处理和“脱色”过程，磷脂包括与磷脂结合的胆碱在这些过程中基本上完全去除了。

鸡对油料饼粕中胆碱的生物学利用率（相对于结晶氯化胆碱）分别为：豆粕83%（Molitoris和Baker，1976a；Emmert和Baker，1997），花生粕76%，而菜籽粕仅为24%（Emmert和Baker，1997）。饲粮过量的蛋白质会增加鸡对饲粮胆碱的需要量（Molitoris和Baker，1976b）。以最小肝脏脂肪含量为评价指标所需要的饲粮胆碱水平，要比以最大生长速度和效率为评价指标所需要的水平高（Anderson等，1979）。

与以色氨酸为前体物的烟酸一样，对胆碱生物学利用价值的评价十分困难，在猪上基本不可能，因为所有的猪常规饲料原料都含有胆碱和蛋氨酸。即使采用转甲基抑制因子乙硫氨酸或通过氨基乙醇的甲基化抑制剂（如2-氨基-2-甲基-1-丙醇）来抑制蛋氨酸合成胆碱，也很难分清两者的效应，无法准确评价胆碱的生物学利用率（Molitoris和Baker，1976a；Anderson等，1979；Lowry等，1987）。

在生长育肥猪的玉米-豆粕型饲粮中添加胆碱的效果不明显，可能是由于豆粕富含胆碱（NCR-42，1980）。但同样是饲喂玉米-豆粕型饲粮，在妊娠母猪上添加胆碱的效果却比较好（Kornegay和Meacham，1973；Stockland和Blaylock，1974；NCR-42，1976）。玉米-豆粕型饲粮添加维生素对生长期的猪和鸡没有效果，不只局限于胆碱，绝大多数的维生素都是如此。伊利诺伊州立大学没有发表的研究数据表明，玉米-豆粕型饲粮添加烟酸或泛酸也没有效果。但是，在养猪生产过程中出现环境恶劣和应激比较大时，应该在维生素预混料给猪补充一定量的胆碱、烟酸和泛酸，以提高猪的抗应激能力。

◆ 3.7 叶酸

叶酸是指具有叶酸活性的一类化合物的总称，食物中叶酸的存在形式有150多种。叶酸的化学结构是由一个喋啶环、对氨基苯甲酸（PABA）和谷氨酸结合而成。动物细胞不能合成PABA，谷氨酸不能与蝶酸结合（也就是喋啶不能与PABA结合）。因此，非反刍动物饲粮必须补充叶酸。饲料和食品中叶酸大部分是以多聚谷氨酸的形式存在。植物中叶酸以共轭多聚谷氨酸的形式存在，包含7个谷氨酸残基以γ-键连接（主要方式）组成的多肽链。肠道内的蛋白酶不能使谷氨酸残基从这种聚合物中裂解出来，但是肠道内的轭合酶（聚谷氨酸叶酸水解酶）能水解出6个谷氨酸残基，最后只留下1个谷氨酸残基。一般认为，只有单一谷氨酸残基的叶酸才可被肠上皮细胞吸收。被刷缘状吸收的叶酸大部分被还原成四氢叶酸（FH_4），然后再甲基化成N^5-甲基-FH_4，进入血液循环。N^5-甲基-FH_4是血浆中叶酸的主要存在形式。血浆中的N^5-甲基-FH_4主要以与蛋白结合的形式存在。

与硫胺素一样，叶酸在喋啶环上也含有一个游离的氨基，这使得叶酸对热非常敏感，在加热处理时很容易失去活性，特别是在含有乳糖或葡萄糖等还原糖的食品或饲料加热时更容易失活。关于叶酸（或硫胺素）游离的氨基是否可以与吡哆醛或磷酸吡哆醛的游离醛基结合，目前还不清楚。一些大豆及可食用的豆子中存在肠道轭合酶的抑制剂可阻止叶酸的吸收（Krumdieck等，1973；Bailey，1988）。饲料和预混料的长期贮存会导致叶酸活性的损失（Verbeeck，1975）。

在常规的玉米-豆粕型饲粮中添加叶酸对生长猪没有效果，因此这类饲粮通常不用添加叶酸（Easter等，1983）。对于妊娠期和哺乳期母猪，有研究发现添加叶酸可以改善其繁殖性能（Lindeman和Kornegay，1989；Matte等；1992），而其他试验则没有得到类似的结果（Pharazyn和Aherne，1987；Easter等，1983；Harper等，1994）。

◆ 3.8 烟酸

烟酸是一类具有尼克酰胺活性的吡啶-3-羧酸及其衍生物，吡啶-3-羧酸本身确切称为尼克酸（Anonumous，1979）。烟酸是非常稳定的维生素，添加到饲料或预混料中几乎不受加热、氧气、水分或光照的影响。在植物性饲料原料中，大部分烟酸是以结合态存在，主要是尼克酰胺核苷，因此不能被机体利用（Yen等，1977）。Ghosh等（1963）认为，谷物中有85%～90%的烟酸、菜籽粕中40%的烟酸都是以不可利用的结合态存在。而这些植物性饲料原料中结合态的烟酸只有通过碱水解的方式才能被释放出来。然而，肉类和奶制品不含有结合态的烟酸，都是游离的烟酸和烟酰胺。

因为过量的色氨酸可以转化为烟酸，而且所有饲料原料都同时含有色氨酸和烟酸，所以对于烟酸生物学利用率的评价，目前还没有找到比较好的方法。50 mg的色氨酸可以生成1 mg的烟酸（Baker等，1973；Czarnecki等，1983）。给猪饲喂玉米-豆粕型饲粮时，过量的亮氨酸是否会抑制色氨酸或烟酸的利用？是否抑制色氨酸向烟酸的代谢转化过程？这个问题一直存在争议（Anonymous，1986），现有的数据对两种观点都支持。鸡的研究数据表明，过量的亮氨酸对色氨酸转化为烟酸及烟酸的生物学活性都没有影响（Lowry和Baker，1989b）。另外，色氨酸转化为烟酸单核苷酸的过程中有2个代谢反应需要铁的参与。因此，Oduho等（1994）得出，鸡缺铁会降低色氨酸转化为烟酸的效率（重量比例从42：1降低到56：1）。

游离的烟酸或烟酰胺的生物利用率都比较高，且都可以购买到相应的商品。烟酰胺相对于烟酸的生物学利用率约为120%（Baker等，1976；Oduho和Baker，1993）。然而，其他研究也表明，烟酸和烟酰胺对鸡具有相同的生物学利用率（Bao-Ji和Combs，1986；Ruiz和Harms，1988）。

◆ 3.9 泛酸

泛酸的商品形式通常为D-或DL-泛酸钙，仅D型结构的泛酸钙具有生物活性（Staten等，1980）。1 g的D-泛酸钙等于0.92 g泛酸（PA）活性，1 g的DL-泛酸钙等于0.46 g泛酸活性。结晶的泛酸在加热、氧气及光照条件下相对稳定，但在潮湿条件下会快速失活。

饲料原料中的泛酸大多是以辅酶A的形式存在，这种形式在肠道内不能被全部吸收。对鸡的生物学试验表明，玉米和豆粕中的泛酸生物学利用率是100%，而大麦、小麦和高粱中的泛酸生物学利用率仅为60%（Southern和Baker，1981）。饲料加工过程可能会使泛酸的生物学利用率降低，但目前还没有确切的动物试验数据来证实。美国成年人的典型饮食中泛酸的生物学利用率仅为50%，冷藏、罐藏、精炼等处理过程可能会进一步降低泛酸的生物学利用率（Sauberlich，1985）。

◆ 3.10 核黄素

核黄素的稳定性相对较低，遇光、碱及氧气都会降低其生物学活性。饲料中的核黄素主要是以核苷酸辅酶的形式存在，这种形式核黄素的生物学利用率可能低于100%。Chung和Baker（1990）估测，鸡对玉米-豆粕型饲粮中核黄素的生物学利用率是60%（相对于晶体核黄素）。维生素-矿物质预混料中的晶体核黄素的活性会随贮存时间的延长而不断降低，且高温贮存会加剧其活性的损失（Zhuge和Klopenstein，1986）。

Sauberlich（1985）提出，影响食物中核黄素生物学利用率的因素很多，包括饲粮添加过量的四环素、铁、锌、铜、维生素C和咖啡因等都会抑制核黄素的利用。然而，Patel和Baker（1996）对鸡的生长试验发现，在核黄素缺乏的大豆分离蛋白半纯合饲粮中，添加过量的铁（420 mg/kg）、锌（448 mg/kg）、铜（245 mg/kg）、抗坏血酸（1 000 mg/kg）、咖啡因（200 mg/kg）或金霉素（500 mg/kg），都不会降低结晶核黄素的利用率。

◆ 3.11 硫胺素

食品或饲料工业中使用的硫胺素包括盐酸硫胺素（含89%硫胺素）或者硝酸

硫胺素（含92%硫胺素）。这些化合物即使在100℃高温下仍非常稳定，但却很易溶于水（NRC，1987）。1IU的硫胺素活性等于3 μg结晶盐酸硫胺素。硫胺素含有一个游离的氨基，加热过程会发生美拉德反应，使硫胺素的生物活性迅速被破坏。同样，任何有碱处理的加工过程都会导致硫胺素的活性损失。猪饲料原料中的硫胺素大部分是以磷酸化的形式存在，以蛋白质磷酸盐复合物、单磷酸硫胺素、双磷酸硫胺素或三磷酸硫胺素的形式存在。某些未加工的饲料原料（如鱼）含有一定的硫胺素酶，可以破坏饲粮中的硫胺素，因此必须添加硫胺素。尽管硫胺素酶在猫及毛皮动物的营养中具有特殊的作用，但在现代的猪营养中的作用非常小。在鱼粉生产过程中，硫胺素会随水溶性物质一起损失掉，因此鱼粉最终所含的硫胺素根本就没有可利用的价值。同样，经过高温处理的肉粉所含的硫胺素的生物学利用率也非常低。

制粒可能也会导致一些硫胺素的损失。盐酸硫胺素和硝酸硫胺素以预混料的形式在温度40 ℃、相对湿度85%的条件下贮存21 d后，硫胺素活性分别剩下48%和95%（Baker，2001）；但以配合饲料的形式在相同的条件下贮存相同的时间，盐酸硫胺素的活性仅剩下21%，而硝酸硫胺素的活性可以保留97%。因此，硝酸盐形式的硫胺素贮存在预期的高温环境下可能更稳定。谷物和豆粕中富含硫胺素，即使考虑加热或长期贮存所造成的活性损失，在猪实用饲粮中也不需要再添加。

◆ 3.12 维生素B_6

玉米和豆粕都含有丰富的维生素B_6，因此在猪实用饲粮中通常不需要添加晶体形式的维生素B_6。伊利诺伊州立大学的研究表明，玉米和豆粕中维生素B_6的生物学利用率分别约为40%和60%（Yen等，1976）。玉米经过适度的加热（80～120 ℃），可以提高其维生素B_6的生物学利用率，但加热温度过高（160 ℃）则会降低维生素B_6的生物学利用率。玉米中可利用的维生素B_6大多是以吡哆醛和吡哆胺形式存在，比吡哆醇对热更加不稳定（Schroeder，1971）。植物性饲料中的维生素B_6还包括葡萄糖苷吡哆醇和赖氨酸吡哆醛，这两种化合物所含的维生素B_6的生物学利用率最低（Gregory和Kirk，1981；Trumbo等，1988）。虽然玉米和豆粕中的维生素B_6相对于结晶盐酸吡哆醇的生物学利用率偏低，但猪实用饲粮所含的可利用维生素B_6通常都是过量的，因此不需要在这些饲粮中再添加。

维生素B_6在预混料中会损失生物活性，特别是在预混料含有碳酸盐或氧化物形式的矿物添加剂时，其生物学活性损失得更多（Ver Leeck，1975）。高温会加速维生素B_6的生物活性损失。在室温下贮存3个月，维生素B_6生物活性可保持76%，但在37℃条件下贮存3个月，其活性则只有45%（Backer，2001）。颗粒配合饲料中的维生素B_6在室温下贮存3个月，其活性平均损失约20%（Backer，2001）。

◆ 3.13 维生素B_{12}

结晶型氰钴胺素或维生素B_{12}都是可利用的，1个美国药典单位（USP）等于1μg的维生素B_{12}。植物性饲料原料基本上不含维生素B_{12}，它主要存在于动物性蛋白饲料和发酵产品中，其生物学利用率被认为是100%（但未被证明）。

动物性饲料和发酵产品中所含的维生素B_{12}是以甲基钴胺素或腺苷钴胺素结合蛋白的形式存在。猪肠道吸收维生素B_{12}需要“固有的”因子的参与，这与人相同，但与绵羊和马不同。结晶维生素B_{12}在饲料和预混料中都相当稳定。

◆ 3.14 维生素C

因为猪自身能够合成维生素C（抗坏血酸），所以人们对维生素C生物学利用率的研究非常少。尽管如此，在猪纯合饲粮中通常还要添加含有维生素C的预混料，这主要是考虑维生素C具有抗氧化和抗应激的功能。饲粮在贮存过程中会损失大量的维生素C，而用乙基纤维素进行包被，可最大限度地降低其损失。制粒和挤压会大幅度降低饲料或预混料中维生素C的活性（Backer，2001）。目前人们已知，维生素C的损失主要是由氧化造成的，首先是还原型抗坏血酸发生可逆的氧化反应，形成脱氢抗坏血酸；然后再进一步氧化成二酮古洛糖酸，而这个反应是不可逆的。不管抗坏血酸是还原形式还是氧化形式，都具有预防坏血病的功能，但二酮古洛糖酸没有这个功能。抗坏血酸和脱氢抗坏血酸都对热不稳定，特别在Cu、Fe或Zn等微量元素存在时，对热更不稳定。

4 结 语

与含能量营养物质相比，对矿物质和维生素生物学利用率的研究资料相对较少。对于很多矿物质和维生素来说，要准确评价其生物学利用率非常困难。因为缺乏敏感的评价指标，或很难获得适当的动物缺乏模型来达到理想的效应。因此，在研究矿物质和维生素的生物学利用率时，采用正确的评价方法非常重要，需要根据每一种维生素和矿物质的特性，采用不同的评价方法来研究这些化合物的生物学利用率。

通常采用常规的消化试验来评价猪对常量元素Ca和P的生物利用率，而且现有的结果证明，这两种常量元素的回肠消化率和全肠道消化率没有差异。因此，可以通过测定全肠道消化率来估测Ca和P的生物学利用率。但是，目前还没有研究证明，其他矿物质元素的生物学利用率也可以通过消化试验来估测。而对于大多数的矿物质和所有的维生素，通常采用斜率比法来评定其生物学利用率。但在分析这些试验结果时，需要特别注意所使用的生物学利用率的评价标准。

对于某些结合态存在的矿物元素（如Se），很显然其生物学利用率取决于所采用的评价指标，这使得它们的评价结果变得更为复杂。此外，大部分矿物质元素的生物学利用率还受饲粮其他矿物质元素含量的影响，有时也受植酸、草酸等抗营养因子含量的影响。因此，应用矿物质和维生素的生物学利用率的数据时，常常需要考虑这些因素的影响。

作者：David H. Baker和
Hans H. Stein
译者：王金荣

参考文献

Almeida, F. N., and H. H. Stein. 2010. Performance and phosphorus balance of pigs fed diets formulated on the basis of values for standardized total tract digestibility of phosphorus. J. Anim. Sci. 88:2968–2977.

Ammerman, C. B., and S. M. Miller. 1972. Biological availability of minor mineral ions: a review. J. Anim. Sci. 35:681–694.

Ammerman, C. B., D. H. Baker, and A. J. Lewis, eds. 1995. Bioavailability of Nutrients for Animals: Amino Acids, Minerals, and Vitamins. Academic Press, San Diego, CA.

Anderson, J. O., and R. E. Warnick. 1970. Studies of the need for supplemental biotin in chick rations. Poult. Sci. 49:569–578.

Anderson, P. A., D. H. Baker, and S. P. Mistry. 1978. Bioassay determination of the biotin content of corn, barley, sorghum and wheat. J. Anim. Sci. 47:654−659.

Anderson, P. A., D. H. Baker, P. A. Sherry, and J. E. Corbin. 1979. Choline−methionine interrelationship in feline nutrition. J. Anim. Sci. 49:522–527.

Anonymous. 1979. Nomenclature policy: generic descriptors and trivial names for vitamins and related compounds. J. Nutr. 109:8–15.

Anonymous. 1986. Pellagragenic effect of excess leucine. Nutr. Rev. 44:26–27.

Aoyagi, S., and D. H. Baker. 1993a. Bioavailability of copper in analytical−grade and feed−grade inorganic copper sources when fed to provide copper at levels below the chick's requirement. Poult. Sci. 72:1075–1083.

Aoyagi, S., and D. H. Baker. 1993b. Nutritional evaluation of copper−lysine and zinc−lysine complexes for chicks. Poult. Sci. 72:165–171.

Aoyagi, S., and D. H. Baker. 1994. Copper−amino acid complexes are partially protected against inhibitory effects of L−cysteine and L−ascorbate. J. Nutr. 124:388–395.

Aoyagi, S., and D. H. Baker. 1995. Effect of microbial phytase and 1,25−dihydroxycholecalciferol on dietary copper utilization in chicks. Poult. Sci. 74:121–126.

Aoyagi, S., D. H. Baker, and K. J. Wedekind. 1993. Estimates of copper bioavailability from liver of different animal species and from feed ingredients derived from plants and animals. Poult. Sci. 72:1746–1755.

Aoyagi, S., K. M. Hiney, and D. H. Baker. 1995. Copper bioavailability in pork liver and in various animal by−products as determined by chick bioassay. J. Anim. Sci. 73:799–804.

Augspurger, N. R., and D. H. Baker. 2004. Phytase improves dietary calcium utilization in chicks, and oyster shell, carbonate, citrate, and citrate-malate forms of calcium are equally bioavailable. Nutr. Res. 24:293–301.

Bafundo, K. W., D. H. Baker, and P. R. Fitzgerald. 1984a. The iron-zinc interrelationship in the chick as influenced by Eimeria acervulina infection. J. Nutr. 114:1306–1312.

Bafundo, K. W., D. H. Baker, and P. R. Fitzgerald. 1984b. Zinc utilization in the chick as influenced by dietary concentration of calcium and phytate and by *Eimeria acervulina* infection. Poult. Sci. 63:2430–2437.

Bailey, L. B. 1988. Factors affecting folate bioavailability. Food Technol. 42:206–210.

Baker, D. H. 1989. Phosphorus supplements for poultry. Multistate Newsl. Poult. Ext. Res. 1:5–6.

Baker, D. H. 1995. Vitamin bioavailability. Pages 399–431 in Bioavailability of Nutrients for Animals: Amino Acids, Minerals, and Vitamins. C.B. Ammerman, D. H. Baker, and A. J. Lewis, eds. Academic Press, San Diego, CA.

Baker, D. H. 2001. Bioavailability of minerals and vitamins. Pages 357–379 in *Swine Nutrition*. 2nd ed. A. J. Lewis and L. L. Southern, eds. CRC Press LLC, Boca Raton, FL.

Baker, D. H., and C. B. Ammerman. 1995a. Copper bioavailability. Pages 127–157 in Bioavailability of Nutrients for Animals: Amino Acids, Minerals, and Vitamins. C. B. Ammerman, D. H. Baker, and A. J. Lewis, ed. Academic Press, San Diego, CA.

Baker, D. H., and C. B. Ammerman. 1995b. Zinc bioavailability. Pages 367–399 in Bioavailability of Nutrients for Animals: Amino Acids, Minerals, and Vitamins. C. B. Ammerman, D. H. Baker, and A. J. Lewis, eds. Academic Press, San Diego, CA.

Baker,D. H., and G. L. Czarnecki-Maulden. 1987. Pharmacologic role of cysteine in ameliorating or exacerbating mineral toxicities. J. Nutr. 117:1003–1010.

Baker, D. H., and K. M. Halpin. 1988. Zinc antagonizing effects of fish meal, wheat bran and a corn-soybean meal mixture when added to a phytate- and fiber-free casein-dextrose diet. Nutr. Res. 8:213–218.

Baker, D. H., and G.W. Oduho. 1994. Manganese utilization in the chick: Effects of excess phosphorus on chicks fed manganesedeficientdiets. Poult Sci. 73:1162–1165.

Baker, D. H., and K. J. Wedekind. 1988. Manganese utilization in chicks as affected by excess calcium and phosphorus ingestion.Pages 29–34 in Proc. of the Maryland Nutr. Conf.

Baker, D. H., N. K. Allen, and A. J. Kleiss. 1973. Efficiency of tryptophan as a niacin precursor in the chick. J. Anim. Sci.36:299–302.

Baker, D. H., K. M. Halpin, D. E. Laurin, and L. L. Southern. 1986. Manganese for poultry—A review. Pages 1–6 in Proc. of theArkansas Nutr. Conf., Little Rock, AR.

Baker, D. H., J. Odle, M. A. Funk, and T. M. Wieland. 1991. Bioavailability of copper in cupric oxide, cuprous oxide and in acopper–lysine complex. Poult. Sci. 70:177–179.

Baker, D. H., J. T. Yen, A. H. Jensen, R. G. Teeter, E. N. Michel, and J. H. Burns. 1976. Niacin activity in niacinamide and coffee.Nutr. Rep. Int. 14:115–120.

Bao–Ji, C., and G. F. Combs. 1986. Evaluation of biopotencies of nicotinamide and nicotinic acid for broiler chickens. Poult. Sci.65(Suppl. 1):24. (Abstr.)

Barber, R. S., J. D. Bowland, R. Braude, K. G. Mitchell, and J. W. G. Porter. 1961. Copper sulphate and copper sulphide (CuS) assupplements for growing pigs. Br. J. Nutr. 15:189.

Biehl, R. R., and D. H. Baker. 1996. Efficacy of supplemental 1α–hydroxycholecalciferol and microbial phytase for young pigsfed phosphorus– or amino acid–deficient corn–soybean meal diets. J. Anim. Sci. 74:2960–2966.

Biehl, R. R., D. H. Baker, and H. F. DeLuca. 1995. 1α–hydroxylated cholecalciferol compounds act additively with microbialphytase to improve phosphorus, zinc and manganese utilization in soy–based diets fed to chicks. J. Nutr. 125:2407–2416.

Bieri, J. G., and M. C. McKenna. 1981. Expressing dietary values for fat–soluble vitamins: changes in concepts and terminology.Am. J. Clin. Nutr. 34:289–394.

Black, J. R., C. B. Ammerman, P. R. Henry, and R. D. Miles. 1984. Biological availability of manganese sources and effects ofhigh dietary manganese on tissue mineral composition of broiler–type chicks. Poult. Sci. 63:1999–2006.

Black, J. R., C. B. Ammerman, P. R. Henry, and R. D. Miles. 1985. Effect of dietary manganese and age on tissue trace mineralcomposition of broiler–type chicks as a bioassay of manganese sources. Poult. Sci. 64:688–693.

Bogunjoko, F. E., R. J. Neale, and D. A. Ledward. 1983. Availability of iron from chicken meat and liver given to rats. Br. J. Nutr.50:511–520.

Bohlke, R. A., R. C. Thaler, and H. H. Stein. 2005. Calcium, phosphorus, and amino acid digestibility in low–phytate corn, normalcorn, and soybean meal by growing pigs. J. Anim. Sci. 83:2396–2403.

Boling, S. D., H. M. Edwards, III, J. L. Emmert, R. R. Biehl, and D. H. Baker. 1998. Bioavailability of iron in cottonseed meal,ferric sulfate and two ferrous sulfate by–products of the galvanizing industry. Poult. Sci. 77:1388–1392.

Braude, R., A. G. Chamberlain, K. Kotarbinska, and K. G. Mitchell. 1962. The metabolism of iron in piglets given labeled ironeither orally or by injection. Br. J. Nutr. 16:427.

Bryant, K. L. Emmert, E. T. Kornegay, J.W. Knight, H. P. Veit, and D. R. Natter. 1985. Supplemental biotin for swine III. Influenceof supplementation to corn– and wheat–based diets on the incidence and severity of toe lesions, hair and skin characteristicsand structural soundness of sows housed in confinement during four parities. J. Anim. Sci. 60:154–162.

Buerostro, J. L., and F. H. Kratzer. 1984. Use of plasma and egg yolk biotin of white leghorn hens to assess biotin availability fromfeedstuffs. Poult. Sci. 63:1563–1570.

Burns, J. M., and D. H. Baker. 1976. Assessment of the quantity of biologically available phosphorus in yeast RNA and single−cellprotein. Poult. Sci. 55:2447–2455.

Cantor, A. H., M. L. Langevin, T. Noguchi, and M. L. Scott. 1975a. Efficacy of selenium compounds and feedstuffs for preventionof pancreatic fibrosis in chicks. J. Nutr. 105:106–111.

Cantor, A. H., M. L. Scott, and T. Noguchi. 1975b. Biological availability of selenium in feedstuffs and selenium compounds forprevention of exudative diathesis in chicks. J. Nutr. 105:96–105.

Charles, O.W., and T. M. Huston. 1972. The biological activity of vitamin K materials following storage and pelleting. Poult. Sci.51:1421–1427.

Chausow, D. G. 1987. Selected Aspects of Mineral Nutrition of the Cat and Dog with Special Emphasis on Magnesium and Iron.Ph.D. Diss. Univ. of Illinois, Urbana, IL.

Chausow, D. G., and G. L. Czarnecki−Maulden. 1988a. The relative bioavailability of iron from feedstuffs of plant and animalorigin to the chick. Nutr. Res. 8:175–185.

Chausow, D. G., and G. L. Czarnecki−Maulden. 1988b. The relative bioavailability of plant and animal sources of iron to the catand chick. Nutr. Res. 8:1041–1050.

Chung, T. K., and D. H. Baker. 1990. Riboflavin requirement of chicks fed purified amino acid and conventional corn−soybeanmeal diets. Poult. Sci. 69:1357–1363.

Chung, T. K., J.W. Erdman, and D. H. Baker. 1990. Hydrated sodium calcium aluminosilicate: effects on zinc, manganese, vitaminA and riboflavin utilization. Poult. Sci. 69:1364−1370.

Coffey, R. D., K. W. Mooney, G. L. Cromwell, and D. K. Aaron. 1994. Biological availability of phosphorus in defluorinatedphosphates with different phosphorus solubilities in neutral ammonium citrate for chicks and pigs. J. Anim. Sci. 72:2653–2660.

Combs, N. R., and E. R. Miller. 1985. Determination of potassium availability in K_2CO_3, $KHCO_3$, corn and soybean meal for theyoung pig. J. Anim. Sci. 60:715–719.

Combs, N. R., E. R. Miller, and P. K. Ku. 1985. Development of an assay to determine the bioavailability of potassium in feedstuffsfor the young pig. J. Anim. Sci. 60:709–714.

Cook, D. A. 1973. Availability of magnesium: Balance studies in rats with various inorganic magnesium salts. J. Nutr. 103:1365–1370.

Cornelius, S. G., and B. G. Harmon. 1976. Sources of oral iron for neonatal piglets. J. Anim. Sci. 42:1351. (Abstr.)

Cromwell, G. L. 1992. The biological availability of phosphorus in feedstuffs for pigs, Pig News Info. 13:75N–78N.Cromwell, G. L., V. W. Hays, and T. L. Clark. 1978. Effects of copper sulfate, copper sulfide and sodium sulfide on performanceand copper stores of pigs. J. Anim. Sci. 46:692–698.

Cromwell, G. L., T. S. Stahly, and H. J. Monegue. 1983. Bioavailability of the calcium and phosphorus in dehydrated alfalfa mealfor growing pigs. J. Anim. Sci. 57(Suppl. 1):242. (Abstr.)

Cromwell, G. L., R. D. Ross, and T. S. Stahly. 1989a. An evaluation of the requirements and biological availability of calcium andphosphorus for swine. Page 88 in Proc. of the Texas Gulf Nutr. Symp., Raleigh, NC.

Cromwell, G. L., T. S. Stahly, and H. J. Monegue. 1989b. Effects of source and level of copper on performance and liver copperstores in weanling pigs. J. Anim. Sci. 67:2996–3002.

Czarnecki, G. L., and D. H. Baker. 1985. Reduction of liver copper concentration by the organic arsenical, 3–nitro–4–hydroxyphenylarsonic acid. J. Anim. Sci. 60:440–450.

Czarnecki, G. L., K. M. Halpin, and D. H. Baker. 1983. Precursor (amino acid):product (vitamin) interrelationship for growingchicks as illustrated by tryptophan–niacin and methionine–choline. Poult. Sci. 62:371–374.

Deming, J. G., and G. L. Czarnecki–Maulden. 1989. Iron bioavailability in calcium and phosphorus sources. J. Anim. Sci. 67(Suppl.1):253. (Abstr.)

DeRitter, E. 1976. Stability characteristics of vitamins in processed foods. Food Technol. 30:48–53.

Dilger, R. N., T. A. Garrow, and D. H. Baker. 2007. Betaine can partially spare choline in chicks, but only when fed in dietscontaining a minimal level of choline. J. Nutr. 137:2224–2228.

Easter, R. A., P. A. Anderson, E. J. Michel, and J. R. Corley. 1983. Response of gestating gilts and starter, grower and finisherswine to biotin, pyridoxine, folacin and thiamine additions to corn–soybean meal diets. Nutr. Rep. Int. 28:945–954.

Edmonds, M. S., and D. H. Baker. 1986. Toxic effects of supplemental copper and roxarsone when fed alone or in combination toyoung pigs. J. Anim. Sci. 63:533–537.

Edwards, H. M., Jr. 1959. The availability to chicks of zinc in various compounds and ores. J. Nutr. 69:306–308.

Edwards, H. M., III, and D. H. Baker. 1999. Bioavailability of zinc in several sources of zinc oxide, zinc sulfate and zinc metal.J. Anim. Sci. 77:2730–2735.

Edwards, H. M., III, and D. H. Baker. 2000. Zinc bioavailability in soybean meal. J. Anim. Sci. 78:1017–1021.

Emmert, J. L., and D. H. Baker. 1997. A chick bioassay approach for determining the bioavailable choline concentration of normaland overheated soybean meal, canola meal and peanut meal. J. Nutr. 127:745–752.

Emmert, J. L., T. A. Garrow, and D. H. Baker. 1996. Development of an experimental diet for determining bioavailable cholineconcentration, and its application in studies with soybean lecithin. J. Anim. Sci. 74:2738–2744.

Erdman, J. W., Jr. 1979. Oilseed phytates: nutritional implications. J. Am. Oil Chem. Soc.

56:736–741.

Erdman, J. W., Jr., G. C. Fahey, and C. B. White. 1986. Effects of purified dietary fiber sources on β–carotene utilization by thechick. J. Nutr. 116:2415–2423.

Erdman, J.W., Jr., C. L. Poor, and J.M. Dietz. 1988. Processing and dietary effects on the bioavailability of vitamin A, carotenoidsand vitamin E. Food Technol. 42:214–219.

Fethiere, R., R. D. Miles, R. H. Harms, and S. M. Laurent. 1988. Bioavailability of sodium in Ethacal R feed component. Poult.Sci. 67(Suppl. 1):15. (Abstr.)

Fly, A. D., O. A. Izquierdo, K. R. Lowry, and D. H. Baker. 1989. Manganese bioavailability in a Mn–methionine chelate. Nutr.Res. 9:901–910.

Frigg, M. 1976. Bioavailability of biotin in cereals. Poult. Sci. 55:2310–2318.

Fritschen, R. D., O. D. Grace, and E. R. Peo. 1971. Bleeding pig disease, Nebraska Swine Report EC71, 219:22–25.

Fritz, J. C., G. W. Pla, T. Roberts, J. W. Boehne, and E. L. Hove. 1970. Biological availability in animals of iron from commondietary sources. J. Agric. Food Chem. 18:647–651.

Furugouri, K., and A. Kawabata. 1975. Iron absorption in nursing piglets. J. Anim. Sci. 41:1348–1354.

Ghosh, H. P., P. K. Sarkar, and B. C. Guha. 1963. Distribution of the bound form of nicotinic acid in natural materials. J. Nutr.79:451–453.

Golovan, S. P., R. G. Meidinger, A. Ajakaiye, M. Cottrill, M. Z. Wiederkehr, D. J. Barney, C. Plante, et al. 2001. Pigs expressingsalivary phytase produce low–phosphorus manure. Nat. Biotechnol. 19:741–745.

Gregory, J. F., III, and J. R. Kirk. 1981. The bioavailability of vitamin B6 in foods. Nutr. Rev. 39:1–4.

Griminger, P. 1965. Relative vitamin K potency of two water–soluble menadione analogues. Poult. Sci. 44:210–213.

Groff, J. L., S. S. Groper, and S. M. Hunt. 1995. Page 575 in Advanced Nutrition and Human Metabolism. West Publishing, St.Paul, MN.

Guenter,W., and J. L. Sell. 1974. A method for determining true availability of magnesium from foodstuffs using chickens. J. Nutr.104:1446–1457.

Hahn, J. D., and D. H. Baker. 1993. Growth and plasma zinc responses of young pigs fed pharmacologic levels of zinc. J. Anim.Sci. 71:3020–3024.

Halpin, K. M., and D. H. Baker. 1986a. Long–term effects of corn, soybean meal, wheat bran and fish meal on manganese utilizationin the chick. Poult. Sci. 65:1371–1374.

Halpin, K. M., and D. H. Baker. 1986b. Manganese utilization in the chick: Effects of corn, soybean meal, fish meal, wheat branand rice bran on tissue uptake of manganese. Poult. Sci. 65:995–1003.

Halpin, K. M., and D. H. Baker. 1987. Mechanism of the tissue manganese-lowering effect of corn, soybean meal, fish meal, wheatbran and rice bran. Poult. Sci. 66:332–340.

Halpin, K. M., D. G. Chausow, and D. H. Baker. 1986. Efficiency of manganese absorption in chicks fed corn-soy and casein diets.J. Nutr. 116:1747–1751.

Harmon, B. G., D. E. Becker, and A. H. Jensen. 1967. Efficacy of ferric ammonium citrate in preventing anemia in young swine.J. Anim. Sci. 26:1051–1053.

Harmon, B. G., D. E. Hoge, A. H. Jensen, and D. H. Baker. 1969. Efficacy of ferrous carbonate as a hematinic for young swine.J. Anim. Sci. 29:706–710.

Harmon, B. G., S. G. Corneliu, J. Totsch, D. H. Baker, and A. H. Jensen. 1974. Oral iron dextran and iron from steel slats ashematinics for swine. J. Anim. Sci. 39:699–702.

Harper, A. F.,M. D. Lindemann, L. I. Chiba, G. E. Combs, D. L. Handlin, E. T. Kornegay, and L. L. Southern. 1994. An assessmentof dietary folic acid levels during gestation and lactation on reproductive and lactational performance of sows: a cooperativestudy. J. Anim. Sci. 72:2338–2344.

Henricks, D. G., E. R. Miller, D. E. Ullrey, J. A. Hoefer, and R. W. Luecke. 1969. Effect of level of soybean protein andergocalciferol on mineral utilization by the baby pig. J. Anim. Sci. 28:342–348.

Henry, P. R. 1995. Manganese bioavailability. Pages 239–256 in Bioavailability of Nutrients for Animals: Amino Acids, Minerals, and Vitamins. C. B. Ammerman, D. H. Baker, and A. J. Lewis, eds. Academic Press, San Diego, CA.

Henry, P. R., and E. R. Miller. 1995. Iron bioavailability. Pages 169–199 in *B*ioavailability of Nutrients for Animals: Amino Acids, Minerals, and Vitamins. C. B. Ammerman, D. H. Baker, and A. J. Lewis, eds. Academic Press, San Diego, CA.

Hill, G. M., G. L. Cromwell, T. D. Crenshaw, C. R. Dove, R. C. Ewan, D. A. Knabe, A. J. Lewis et al. 2000. Growth promotioneffects and plasma changes from feeding high dietary concentrations of zinc and copper to weanling pigs (regional study).J. Anim. Sci. 78:1010–1016.

Horst, R. L., J. L. Napoli, and E. T. Littledike. 1982. Discrimination in the metabolism of orally dosed ergocalciferol andcholecalciferol by the pig, rat, and chick. Biochem. J. 204: 185–189.

Hortin, A. E., P. J. Bechtel, and D. H. Baker. 1991. Efficacy of pork loin as a source of zinc, and effect of added cysteine on zincbioavailability. J. Food Sci. 56:1505–1508.

Hortin, A. E., G. Oduho, Y. Han, P. J. Bechtel, and D. H. Baker. 1993. Bioavailability of zinc in ground beef. J. Anim. Sci.71:119–123.

Izquierdo, O. A., and D. H. Baker. 1986. Bioavailability of copper in pig feces. Can. J. Anim. Sci. 66:1145–1148.

Kornegay, E. T. 1972. Availability of iron contained in defluorinated phosphate. J. Anim.

Sci. 34:569–572.

Kornegay, E. T., and T. N. Meacham. 1973. Evaluation of supplemental choline for reproducing sows housed in total confinementon concrete or in dirt lots. J. Anim. Sci. 37:506–509.

Krumdieck, C. L., A. J. Newman, and C. E. Butterworth, Jr. 1973. A naturally occurring inhibitor of folic acid conjugase (petroylopolyglutamyl hydrolase) in beans and other pulses. Am. J. Clin. Nutr. 24:460. (Abstr.)

Layrisse, M., C. Martinez-Torres, M. Renzy, and I. Leets. 1975. Ferritin iron absorption in man. Blood 45:689–698.

Lewis, A. J., G. L. Cromwell, and J. E. Pettigrew. 1991. Effects of supplemental biotin during gestation and lactation on reproductiveperformance of sows: A cooperative study. J. Anim. Sci. 69:207–214.

Lindemann, M. D., and E. T. Kornegay. 1989. Folic acid supplementation to diets of gestating-lactating swine over multipleparities. J. Anim. Sci. 67:459–464.

Lindemann, M. D., S. D. Carter, L. I. Chiba, C. R. Dove, F. M. Lemieux, and L. L. Southern. 2004. A regional evaluation ofchromium tripicolinate supplementation of diets to reproducing sows. J. Anim. Sci. 82:2972–2977.

Lindemann, M. D., G. L. Cromwell, H. J. Monegue, and K. W. Purser. 2008. Effect of chromium sources on tissue concentrationof chromium in pigs. J. Anim. Sci. 86:2971–2978.

Lindemann, M. D., C. M. Wood, A. F. Harper, E. T. Kornegay, and R. A. Anderson. 1995. Dietary chromium picolinate additionsimprove gain:feed and carcass characteristics in growing-finishing pigs and increase litter size in reproducing sows. J. Anim.Sci. 73:457–465.

Livingston, A. L., J. W. Nelson, and G. O. Kohler. 1968. Stability of alpha-tocopherol during alfalfa dehydration and storage.J. Agric. Food Chem. 16:492–496.

Lowry, K. R., and D. H. Baker. 1989a. Amelioration of selenium toxicity by arsenicals and cysteine. J. Anim. Sci. 67:959–965.

Lowry, K. R., and D. H. Baker. 1989b. Effect of excess leucine on niacin provided by either tryptophan or niacin. FASEB J.3:A666. (Abstr.)

Lowry, K. R., O. A. Izquierdo, and D. H. Baker. 1987. Efficacy of betaine relative to choline as a dietary methyl donor. Poult. Sci.66(Suppl. 1):135. (Abstr.)

Mahan, D. C., and A. L. Moxon. 1978. Effect of adding inorganic or organic selenium sources to the diets of young swine. J. Anim.Sci. 47:456–466.

Mahan, D. C., and N. A. Parrett. 1996. Evaluating the efficacy of selenium-enriched yeast and sodium selenite on tissue seleniumretention and serum glutathionine peroxidase activity in grower and finisher swine. J. Anim. Sci. 74:2967–2974.

Mathias, M. M., D. E. Hogue, and J. K. Loosli. 1967. The biological value of selenium in bovine milk for the rat and chick. J. Nutr.93:14–20.

Matte, J. J., C. L. Girard, and G. J. Brisson. 1992. The role of folic acid in the nutrition of gestating and lactating primaparoussows. Livest. Prod. Sci. 32:131–138.

Miller, E. R. 1980. Bioavailability of minerals. Pages 144–154 in *Proc. of the Minnesota Nutr. Conf.*, Univ. of Minnesota, St. Paul,MN.

Miller, E. R., and C. B. Ammerman. 1995. Iodine bioavailability. Pages 157–167 in Bioavailability of Nutrients for Animals:Amino Acids, Minerals and Vitamins. C. B. Ammerman, D. H. Baker, and A. J. Lewis, eds. Academic Press, San Diego, CA.

Miller, E. R., P. K. Ku, J. P. Hitchcock, and W. T. Magee. 1981. Availability of zinc from metallic zinc dust for young swine.J. Anim. Sci. 52:312–315.

Molitoris, B. A., and D. H. Baker. 1976a. Assessment of the quantity of biologically available choline in soybean meal. J. Anim.Sci. 42:481–489.

Molitoris, B. A., and D. H. Baker. 1976b. Choline utilization in the chick as influenced by levels of dietary protein and methionine.J. Nutr. 106:412–418.

NCR–42 Committee on Swine Nutrition. 1976. Effect of supplemental choline on reproductive performance of sows: A cooperativeregional study. J. Anim. Sci. 42:1211–1216.

NCR–42 Committee on Swine Nutrition. 1980. Effect of supplemental choline on performance of starting, growing and finishingpigs: A cooperative regional study. J. Anim. Sci. 50:99–102.

Nelson, T. S. 1967. The utilization of phytate phosphorus by poultry. Poult. Sci. 46:862–871.

Nelson, T. S., and A. C. Walker. 1964. The biological evaluation of phosphorus compounds. Poult. Sci. 43:94–98.

NRC. 1987. Vitamin Tolerance of Animals. National Academies Press, Washington, DC.

NRC. 1994. Nutrient Requirements of Poultry. 9th ed. National Academies Press, Washington, DC.

Nyannor, E. K., and O. Adeola. 2008. Corn expressing an Escherichia coli–derived phytase gene: Comparative evaluation study inbroiler chicks. Poult. Sci. 87:2015–2022.

Nyannor, E. K., P. Williams, M. R. Bedford, and O. Adeola. 2007. Corn expressing an *Escherichia coli*–derived phytase gene:proof–of–concept nutritional study in pigs. J. Anim. Sci. 85:1946–1952.

O'Dell, B. L., J. M. Yohe, and J. E. Savage. 1964. Zinc availability in the chick as affected by phytate, calcium and ethylenediaminetetracetate.Poult. Sci. 43:415–419.

Oduho, G., and D. H. Baker. 1993. Quantitative efficacy of niacin sources for the chick: Nicotinic acid, nicotinamide, NAD andtryptophan. J. Nutr. 123:2201–2206.

Oduho, G.W., T. K. Chung, and D. H. Baker. 1993. Menadione nicotinamide bisulfite is a bioactive source of vitamin K and niacinactivity for chicks. J. Nutr. 123:737–743.

Oduho, G., Y. Han, and D. H. Baker. 1994. Iron deficiency reduces the efficacy of tryptophan as a niacin precursor for chicks.J. Nutr. 124:444–450.

Page, T. G., L. L. Southern, T. L. Ward, and D. L. Thompson, Jr. 1993. Effect of chromium picolinate on growth and serum andcarcass traits of growing–finishing pigs. J. Anim. Sci. 71:656–662.

Patel, K., and D. H. Baker. 1996. Supplemental iron, copper, zinc, ascorbate caffeine and chlortetracycline do not affect riboflavinutilization in the chick. Nutr. Res. 16:1943–1952.

Peeler, H. T. 1972. Biological availability of nutrients in feeds: availability of major mineral ions. J. Anim. Sci. 35:695–712.

Pedersen, C., M. G. Boersma, and H. Stein. 2007. Digestibility of energy and phosphorus in 10 samples of distillers dried grainwith solubles fed to growing pigs. J. Anim. Sci. 85:1168–1176.

Petersen, G. I., and H. H. Stein. 2006. Novel procedure for estimating endogenous losses and measuring apparent and truedigestibility of phosphorus by growing pigs. J. Anim. Sci. 84:2126–2132.

Pharazyn, A., and F. X. Aherne. 1987. Folacin requirement of the lactating sow. Pages 16–18 in 66th Annual Feeders Day Report. Univ. of Alberta, Edmonton, AB, Canada.

Pond,W. G., R. S. Lowrey, J. H. Maner, and J. K. Loosli. 1961. Parenteral iron administration to sows during gestation or lactation.J. Anim. Sci. 20:747–752.

Rao, N., and B. S. N. Rao. 1970. Absorption of dietary carotenes in human subjects. Am. J. Clin. Nutr. 23:105–110.

Ross, R. D., G. L. Cromwell, and T. S. Stahly. 1984. Effects of source and particle size on the biological availability of calcium incalcium supplements for growing pigs. J. Anim. Sci. 59:125–134.

Ruiz, N., and R. H. Harms. 1988. Comparison of biopotencies of nicotinic acid and nicotinamide for broiler chickens. Br. Poult.Sci. 29:491–498.

Sauberlich, H. 1985. Bioavailability of vitamins. Prog. Food Nutr. Sci. 9:1–33.

Schroeder, H. A. 1971. Losses of vitamins and trace minerals resulting from processing and preservation of foods. Am. J. Clin.Nutr. 24:562–567.

Seerley, R. W., O. W. Charles, H. C. McCampbell, and S. P. Bertsch. 1976. Efficacy of menadione dimethylpyrimidinol bisulfiteas a source of vitamin K in swine diets. J. Anim. Sci. 42:599–607.

Snow, J. L., M. E. Persia, P. E. Biggs, D. H. Baker, and C. M. Parsons. 2003. 1α–hydroxycholecalciferol has little effect on phytatephosphorus utilization in laying hen diets. Poult. Sci. 82:1792–1796.

Soares, J. H. 1995. Phosphorus bioavailability. Pages 257–294 in Bioavailability of Nutrients for Animals: Amino Acids, Minerals,and Vitamins. C. B. Ammerman, D. H. Baker, and A. J. Lewis, eds. Academic Press, Inc. San Diego, CA.

Southern, L. L., and D. H. Baker. 1981. Bioavailable pantothenic acid in cereal grains and soybean meal. J. Anim. Sci. 53:403–408.

Southern, K. L., and D. H. Baker. 1983a. Eimeria acervulina infection in chicks fed deficient or excess levels of manganese. J. Nutr.113:172–177.

Southern, L. L., and D. H. Baker. 1983b. Excess manganese ingestion in the chick. Poult. Sci. 62:642–646.

Southern, L. L., and D. H. Baker. 1983c. Zinc toxicity, zinc deficiency and zinc–copper interrelationship in *Eimeria acervulinainfectedchicks*. J. Nutr. 113:688–696.

Spencer, J. D., G. L. Allee, and T. E. Sauber. 2000. Phosphorus bioavailability and digestibility of normal and genetically modifiedlow–phytate corn for pigs. J. Anim. Sci. 78:675–681.

Staten, F. E., P. A. Anderson, D. H. Baker, and P. C. Harrison. 1980. The efficacy of DL–pantothenic acid relative to D–pantothenicacid in chicks. Poult. Sci. 59:1664. (Abstr.)

Stein, H. H., M. G. Boersma, and C. Pedersen. 2006. Apparent and true total tract digestibility of phosphorus in field peas (*Pisumsativum* L.) by growing pigs. Can. J. Anim. Sci. 85:523–525.

Stein, H. H., S. P. Connot, and C. Pedersen. 2009. Energy and nutrient digestibility in four sources of distillers dried grains withsolubles produced from corn grown within a narrow geographical area and fed to growing pigs. Asian–Austr. J. Anim. Sci.22:1016–1025.

Stein, H. H., C. T. Kadzere, S. W. Kim, and P. S. Miller. 2008. Influence of dietary phosphorus concentration on the digestibilityof phosphorus in monocalcium phosphate by growing pigs. J. Anim. Sci. 86:1861–1867.

Stockland, W. L., and L. G. Blaylock. 1974. Choline requirement of pregnant sows and gilts under restricted feeding conditions.J. Anim. Sci. 39:1113–1116.

Thoren–Tolling, K. 1975. Studies on the absorption of iron after oral administration in piglets. Acta Vet. Scand. Suppl. 54:1–121.

Trumbo, P. R., J. F. Gregory, and D. B. Sartain. 1988. Incomplete utilization of pyridoxine–beta–glucoside as vitamin B_6 in the rat.J. Nutr. 118:170–175.

Ullrey, D. E. 1972. Biological availability of fat–soluble vitamins: vitamin A and carotene. J. Anim. Sci. 35:648–657.Ullrey, D. E. 1981. Vitamin E for swine. J. Anim. Sci. 53:1039–1056.

Ullrey, D. E., E. R. Miller, J. P. Hitchcock, P. K. Ku, R. L. Covert, J. Hegenauer, and P. Saltman. 1973. Oral ferric citrate vs. ferroussulfate for prevention of baby pig anemia. Mich. Agric. Exp. Sta. Rep. 232:34.

Verbeeck, J. 1975. Vitamin behavior in premixes. Feedstuffs 47:45–48.

Veum, T. L., D. R. Ledoux, V. Raboy, and D. S. Ertl. 2001. Low-phytic acid corn improves nutrient utilization for growing pigs.J. Anim. Sci. 79:2873–2880.

Waibel, P. E., N. A. Nahorniak, H. E. Dziuk, M. M. Walser, and W. G. Olson. 1984. Bioavailability of phosphorus in commercialphosphate supplements for turkeys. Poult. Sci. 63:730–737.

Watkins, K. L., L. L. Southern, and J. E. Miller. 1991. Effect of dietary biotin supplementation on sow reproductive performanceand soundness and pig growth and mortality. J. Anim. Sci. 69:201–206.

Wedekind, K. J., and D. H. Baker. 1990a. Effect of varying calcium and phosphorus level on manganese utilization. Poult. Sci.69:1156–1164.

Wedekind, K. J., and D. H. Baker. 1990b. Manganese utilization in chicks as affected by excess calcium and phosphorus ingestion.Poult. Sci. 69:977–984.

Wedekind, K. J., and D. H. Baker. 1990c. Zinc bioavailability in feed-grade sources of zinc. J. Anim. Sci. 68:684–689.

Wedekind, K. J., M. R. Murphy, and D. H. Baker. 1991a. Manganese turnover as affected by excess phosphorus consumption.J. Nutr. 121:1035–1041.

Wedekind, K. J., E. C. Titgmeyer, A. R. Twardock, and D. H. Baker. 1991b. Phosphorus, but not calcium, affects manganeseabsorption and turnover in chicks. J. Nutr. 121:1776–1786.

Wedekind, K. J., A. E. Hortin, and D. H. Baker. 1992. Methodology for assessing zinc bioavailability: efficacy estimates forzinc-methionine, zinc sulfate and zinc oxide. J. Anim. Sci. 70:178–188.

Wedekind, K. J., A. J. Lewis, M. A. Giesemann, and P. S. Miller. 1994. Bioavailability of zinc from inorganic and organic sourcesfor pigs fed corn-soybean meal diets. J. Anim. Sci. 72:2681–2689.

Wedekind, K. J., R. S. Beyer, and G. F. Combs, Jr. 1998. Is selenium addition necessary in pet foods? FASEB J. A823. (Abstr.)

Widmer, M. R., L. M. McGinnis, and H. H. Stein. 2007. Energy, amino acid, and phosphorus digestibility of high protein distillersdried grain and corn germ fed to growing pigs. J. Anim. Sci. 85:2994–3003.

Wornick, R. C. 1968. The stability of microingredients in animal feed products. Feedstuffs 40:25–28.

Wright, P. L., and M. C. Bell. 1966. Comparative metabolism of selenium and tellurium in sheep and swine. Am. J. Physiol.211:6–10.

Yen, J. T., A. H. Jensen, and D. H. Baker. 1976. Assessment of the concentration of biologically available vitamin B_6 in corn andsoybean meal. J. Anim. Sci. 42:866–870.

Yen, J. T., A. H. Jensen, and D. H. Baker. 1977. Assessment of the availability of niacin in corn, soybeans and soybean meal.J. Anim. Sci. 46:269–278.

Young, L. G., A. Lun, J. Pos, R. P. Forshaw, and D. E. Edmeades. 1975. Vitamin E stability in corn and mixed feed. J. Anim. Sci.40:495–499.

Young, L. G., R. B. Miller, D. E. Edmeades, A. Lun, G. C. Smith, and G. J. King. 1977. Selenium and vitamin E supplementationof high moisture corn diets for swine reproduction. J. Anim. Sci. 45:1051–1060.

Young, L. G., R. B. Miller, D. E. Edmeades, A. Lun, G. C. Smith, and G. J. King. 1978. Influence of method of corn storage andvitamin E and selenium supplementation on pig survival and reproduction. J. Anim. Sci. 47:639–647.

Zhuge, Q., and C. F. Klopfenstein. 1986. Factors affecting storage stability of vitamin A, riboflavin and niacin in a broiler dietpremix. Poult. Sci. 65:987–994.

第16章
猪营养与环境

1 导　语

养猪业是红肉的主要供应者，所提供的猪肉为人类健康、经济和社会活动做出了重要贡献。自有记载的人类文明可追溯到对（包括猪在内）的少数几种动物的驯化家养，动物生产对人类身心的健康、社会和经济的发展均起了重要作用（Diamond，2002；Fan等，2008a）。人们普遍认为，在过去的二十年里，主要是矿物能源的应用和动物生物学的科技进步推动了养猪业朝着工业化和集约化的方向发展。

目前的集约化养猪生产正面临着一些新的可持续发展问题。首先，市场波动、低利润空间以及农村偏远地区养殖户较弱的经济和社会生存能力等风险，均在不断增加（Fan等，2008a）。这问题主要由饲料价格不断上涨引起。人口数量的不断增加提高了对植物性食物的需求，极端天气影响了作物产量以及发达国家生产生物燃料对谷物和油籽用量不断增加，这些因素均导致了饲料价格的上涨。其次，虽然人类对畜产品消费的增加改善了人类的生活质量、提高了人类寿命，但同时也带来了一些严重的健康问题，如肥胖、心血管疾病、Ⅱ型糖尿病、慢性肠炎和结肠直肠癌。尽管人们的生活方式、居住环境和家族遗传史不同，但对畜产品尤其是猪肉等红肉消费量的不断提高，常常被认为与上述健康问题的增加密切相关（Diamond，2002）。如果不能真正理解和解决畜产品消费与这些慢性疾病发生之间存在联系的生物学机制，这些人类健康问题将会成为维持和扩大集约化动物生产的一个主要障碍。

最后，较低的营养物质利用效率以及与此有关的提高饲料产量的生产实践，主要是为了尽量提高生长速度和瘦肉率，以提高养猪生产的利润率。而这些生产实践又会带来一系列的环境问题（NRC，1998，2012），包括甲烷（CH_4）和一氧化二氮（N_2O）等温室气体的排放（Mackie等，1998），引起酸化和异味的氨（NH_3）的排放（Rideout等，2004），来源于猪排泄物的硝酸盐（NO^{3-}）的侵蚀（Jayasundara等，2010）和三价铬（Cr^{3+}）的污染（Blowes，2002；Ellis等，2002），重金属微量矿物离子砷（砷酸盐，AsO_4^{3-}）、镉（Cd^{2+}）、铜（Cu^{2+}）和锌（Zn^{2+}）导致的食物链和生态污染风险的增加（Linden等，1999），为农作物施粪肥带来的过量水溶性磷（P）（Mallin，2000），病原微生物的传播（Hutchison等，2005），抗生素耐药性的产生（Levy，1998；Gorbach，2001；Umber和Bender，2009；Guenther等，2010），以及令人生厌的排泄物臭味（Mackie等，1998；Rideout等，2004；Fan等，2006）和空气污染物（Godbout等，2009；Thorne等，2009；Donham，2010；Schinasi等，2011）的影响。

本章节主要阐述饲粮组分及其应用如何引起上述主要的环境问题，同时还讨论如何从管理和生物技术方面，通过改善饲粮配方，有效缓解目前养猪生产给环境带来的负面影响。

2 饲粮引起的主要环境问题

本质上，动物生产是以动物为生物转换器进行大规模能量转化的过程。饲粮组成是引起负面环境问题的主要来源。养猪生产中饲粮营养素和组分的消化和代谢以及粪便排泄是如何引起的环境问题，以及对这些问题的深入理解是利用营养调控和非饲粮调节策略缓解环境问题必需的。

◆ 2.1 温室气体的排放

众所周知，集约化养猪生产中，大部分的CH_4和N_2O排放来自猪的排泄物，以及冲洗粪池产生的粪浆经厌氧微生物发酵产生的（Mackie等，1998）。其中，来源于饲粮和肠道内源性的未被消化的粪蛋白质、不可消化低聚糖和非淀粉多糖（NSP）的碳（C）是CH_4生物合成的主要碳源（Mackie等，1998；Velthof等，2005）。粪便中的含氮（N）化合物在贮存过程中或施肥到农田后经微生物的硝化-反硝化循环，成为了集约化养猪生产中N_2O生物合成和排放的另外一个重要来源（Mackie等，1998；Jayasundara等，2010）。由于富营养化，因此造成了地表水资源在缺氧条件下主要温室气体N_2O的厌氧生物合成和排放（Naqvi等，2000）。由大量猪粪尿或水溶性磷引起的地表水的富营养化，对主要温室气体的排放起到了间接作用（Forsberg等，2005；Fan等，2008b）。因此，碳、氮和磷的利用率低以及这些营养素在粪尿中的过量排放是集约化养殖生产中温室气体排放的主要原因。

◆ 2.2 氨的排放

在集约化养猪系统中，排泄物中总N损失的75%一般是以NH_3的形式排放的（Mackie等，1998）。猪排泄物中的含氮化合物很少直接以NH_3和铵（NH_4^+）的形式排泄。猪等哺乳类动物的氮代谢终产物大部分通过尿液以尿素的形式排出体外。排泄物中含有很高的细菌脲酶活性，可以降解新鲜粪便中的尿素。粪尿混合后，贮藏过程中猪粪浆中的尿素很容易被细菌脲酶迅速降解为NH_3，这是猪排泄物中NH_3和NH_4^+的主要来源（Mackie等，1998；Le等，2005）。猪粪浆中NH_3和NH_4^+的另外一个来源是，厌氧条件下，饲粮和猪内源蛋白质、肽和氨基酸的细菌氧化脱氨作用（Mackie等，1998）。猪粪浆的pH一般呈碱性，主要是因为其中含有较高水平的Cl^-、K^+和Na^+等离子。这些离子来源于饲粮中添加的食盐和植物性饲料原料，并最终随尿液排出。猪粪浆中较强的碱性环境，有利于NH_3的存在而不是以NH_4^+盐形式存在，最终是以氨的形式从猪粪浆中挥发掉（Mackie等，1998）。

除了污染环境外，NH_3的排放还会大大降低氮肥的利用价值，而且给集约化养猪场及其周边的动物和人类带来NH_3中毒的风险（Mackie等，1998；Schinasi等，2011）。与养猪生产中NH_3排放相关的主要环境问题有三点：① NH_3的排放形成了猪粪的臭味，给当地居民生活带来危害（Mackie等，1998）；

② 猪场NH_3的排放是大气硫酸铵的主要农业来源，容易形成酸雨和地表土壤的酸化（van Breemen等，1982；Rideout等，2004）；③ NH_3排放引起的大气中过量硫酸铵随雨水降落至地面，导致硝酸盐淋溶到地下饮用水资源，引起温室气体N_2O的排放（Mackie等，1998）。因此，猪粪浆中NH_3的排放给局部甚至全球均带来了一些负面的环境影响。

◆ 2.3 硝酸盐淋溶和铬径流

众所周知，硝酸盐是饮水的主要污染源，而土壤表面硝酸盐的过量淋溶是此污染的主要途径。地表土壤硝酸盐的过量累积，源于猪排泄物中NH_3排放引起的大气硫酸铵沉降，以及猪粪经厌氧贮藏和土壤施肥后的含氮化合物经硝化细菌作用产生的硝酸盐（Mackie等，1998；Jayasundara等，2010）。

另外，浅层地下水是主要的饮用水源之一，易受到来自工业、农业和家庭活动的铬离子等污染物的污染（Blowes，2002）。美国环境保护组织（EPA）的环境研究小组2010年开展了一项调查，发现美国35个城市的饮用水中总铬的离子浓度超过了EPA制定的标准（0.1 mg/L）（Yahoo News，2010）。饮用水中总铬离子的浓度包括三价铬（Cr^{3+}）和六价铬（Cr^{6+}）。三价铬在弱酸和中性pH时不易溶于水，因此在环境中相对稳定（Blowes，2002）。六价铬具有致癌性，主要是来源于工业生产（Blowes，2002；Salnikow和Zhitkovich，2008）。鉴于下述几个方面，猪排泄物中Cr^{3+}的过量排放应引起我们的关注。第一，目前铬污染标准（美国EPA制定的水平）中包括Cr^{3+}和Cr^{6+}两种离子形式。第二，环境中的Cr^{3+}有可能被进一步氧化成有毒性的Cr^{6+}（Blowes，2002）。三价铬存在于天然食物和饲料原料中，而且是一种必需的微量矿物元素（NRC，1998，2012），对猪的蛋白质、葡萄糖和脂质代谢均具有重要作用（Sales和Jancík，2011）。在正常养猪生产条件下，猪对Cr^{3+}的需要量可由饲粮中主要的常规能量饲料原料和蛋白质饲料原料提供（NRC，1998，2012）。然而，在养猪生产实践中，为提高瘦肉率和促进代谢，会在饲粮中添加远高于需要水平的吡啶甲酸铬（Sales和Jancík，2011）。因此，在猪生产中，为调节生长和健康状况而在饲粮中添加各种铬添加剂，这无疑会造成排泄物中Cr^{3+}的过量排放，Cr^{3+}因此也会随径流进入地下饮用水资源中。

◆ 2.4 与重金属矿物质有关的食物链与生态问题

众所周知，在主要的重金属微量元素中，动物和人类接触较低剂量的AsO_4^{3-}、Cd^{2+}、Pb^{2+}和Hg^{2+}就可出现中毒现象（NRC，2005）。因此，主要食物链中目前公认的绝大多数重金属污染，包括工业活动中产生的可溶性离子（Zheng等，2007；Zhao等，2010）。集约化畜牧生产带来的主要有毒重金属离子可能是AsO_4^{3-}和Cd^{2+}（Linden等，1999；Jackson等，2003）。蛋白质添加剂、豆粕、双低菜粕和普通菜粕这些主要的饲料原料，以及微量矿物质预混料和无机常量矿物质添加剂（磷酸氢钙和石灰石）等矿物质添加剂，都是饲粮中砷和镉的主要来源（Linden等，1999；Jackson等，2003）。主要饲料原料中的这些有毒重金属元素来源于作物从土壤中的吸收，金属矿井和加工厂附近的这些土壤可能是通过空气沉降而被污染，或者是因回收的城市废物和畜禽排泄物而被污染。在饲料级的矿物质调节剂中，这些有毒重金属元素可能来源于矿物及其化学加工。

在猪排泄物中大量存在而且有可能引起生态系统污染的另外两种微量矿物元素是Cu^{2+}和Zn^{2+}。众所周知，绵羊对铜毒性的耐受力较低（Mendel等，2007）。因此，使用含高铜的猪排泄物作为肥料的土地，收获的饲料和新鲜饲草中也含有较高含量的Cu，这些饲料和饲草对绵羊的毒性风险较大。另外，高浓度的锌可抑制细菌的生长。因此，高锌可通过抑制土壤细菌硝化作用所必需的氨氧化细菌，从而对表层土壤造成污染（Mertens等，2009）。最近的研究表明，排泄物中过量的Cu和Zn容易导致微生物对特定重金属产生抗性（Hölzel等，2012）。猪粪中过量的Cu和Zn来源于各种猪饲粮中过量添加的微量矿物元素（NRC，1998，2012），以及断奶仔猪饲粮中作为生长促进剂使用的药理水平的Cu和Zn（Shelton等，2011）。因此，猪的饲粮是导致猪排泄物中重金属微量元素过量的源头。

◆ 2.5 排泄物中可溶性磷的径流

农田施用猪场等集约化动物生产的排泄物后，排泄物中磷的径流会大大加速地表水资源的富营养化（Mallin，2000）。经典的土壤研究指出，磷的径流是指暴雨后水溶性磷（如吸附于土壤和有机物中的可溶性游离无机磷）沿地表的流失（Sharpley等，1992）。最新的研究表明，施用粪肥中80%的磷随雨水下渗到土壤表层2 cm内，20%进入深层土壤（Vadas等，2007）。这2 cm表层土壤中的磷通过脱吸附作用形成地表径流（Vadas等，2007），径流大小则与雨水大小呈正相关（Sharpley等，2008）。不考虑饲粮类型和植酸酶的使用，猪排泄物中超过30%的总磷是水溶性的，主要由大肠微生物发酵产生（Angel等，2005）。在排泄物贮存过程中以及施肥后在土壤中存在大量的微生物活动，排泄后这些微生物活动会进一步增加表层土壤中水溶性磷的比例。而且水溶性磷试验表明，可溶性磷可作为排泄物和其他生物质中磷径流损失的潜在指示剂（Kleinman等，2007）。

在传统饲养管理条件下，超过50%的饲粮磷通过排泄物排出体外（Fan等，2008b）。相对于氮来讲，作物生长只需要一小部分的磷（Havlin，2004）。因此，在农田中施用富含磷的猪排泄物后，无疑会导致表层土壤中磷的径流，从而加速了地表水的富营养化。

◆ 2.6 病原微生物的传播

整个养猪生产系统中的主要环节，包括生猪饲养、猪肉加工、猪排泄物处理和养殖设施维护等，均是细菌、寄生虫和病毒等各种病原微生物的可能藏身之处（Hutchison等，2005）。早期的相关研究指出，人类与乳头瘤病毒（HPV）有关的肿瘤，如宫颈癌，就与食用猪肉有关（Schneider等，1990）。目前已知，圆环病毒可感染猪和鸟类。近期的比较病毒基因组分析表明，在人类粪便标本中发现了猪圆环病毒基因组，推测是经由食用猪肉传播的（Li等，2010）。而饲粮组分可显著影响肠道和粪便中的微生物区系及其数量（Flickinger等，2003；Dahiya等，2007；Wells等，2010）。在饲粮中添加富含免疫球蛋白的喷雾干燥血浆蛋白粉，可提高断奶仔猪的健康和生长性能，而不会给阴性猪带来病毒传染（Pujols等，2008；Shen等，2011）。然而，最近有报道称，凭经验用灭火法试验性生产的喷雾干燥血浆蛋白粉可能会给养猪业带来生物安全风险，但这与血浆蛋白粉的灭活工艺有关（Patterson等，2010）。

如何通过改变其他饲粮成分来影响养猪生产中动物传染性病原和病毒的致病过程，我们还知之甚少。尽管改变饲粮成分确实可以影响细菌的致病性，但还需要开展更多的研究来了解饲粮成分的变化是如何影响动物传染性病原和病毒的致病性的，这对环境可能会产生一定的影响。

◆ 2.7 抗生素耐药性的产生

养猪生产中抗生素耐药性的产生主要有两种特殊类型，即抗生素诱导的抗生素耐药性（Aminov和Mackie，2007）和对重金属（尤其是Cu和Zn）特异的抗生素耐药性（Fard等，2011）。目前令人担忧的耐抗生素菌株的产生，在某种程度上是由养猪生产中饲料中使用的抗生素添加剂引起的（美国微生物学会，1999）。抗生素最初作为生长促进剂使用，可追溯到20世纪四五十年代（Visek，1978；Cromwell，2001）。据Cromwell（2001）记载，自20世纪60年代饲料级抗生素的平均成本较低后，亚治疗剂量的抗生素在断奶仔猪饲粮中便被广泛使用。

有研究表明，抗生素耐药性基因在"抗生素时代"之前就已经存在了（Aminov和Mackie，2007）。最近的研究表明，抗生素耐药性基因可以通过三种不同的横向基因转移机制在细菌之间进行有效转移，即噬菌体介导的途径、质粒介导途径和基因组移植（Lartigue等，2007；Allen等，2011；Heuer等，2011）。养猪场中产生的携带耐药基因的细菌，可将耐药基因直接传递给养猪工人（Zhang等，2009），或通过食物链传递给消费者（Manges等，2001）。另外，饲粮中约有75%的抗生素不能被家畜吸收和代谢，而直接经由粪便排出体外（Chee-Sanford等，2009）。在猪的排泄物贮存和施肥期间，抗生素耐药菌会进一步进化并扩散到环境和生态系统中（Koike等，2007；Chee-Sanford等，2009）。因家畜养殖场在饲料中添加抗生素产生的耐药性，也正在对人类（Gorbach等，2001）、宠物（Umber和Bender，2009）和野生鸟类（Guenther等，2010）感染性疾病的治疗产生危害。另外，排泄物中过量Cu和Zn引起的对重金属特异的抗生素耐药性与铜绿假单胞菌的耐药性并无关联（Deredjian等，2011）。然而，有另外两项研究表明，Cu和Zn引起的耐药性可引起和增强猪源性细菌的耐药性（Amachawadi等，2010；Cavaco等，2011）。

因此，在猪营养中通过饲粮添加预防剂量的抗生素添加剂以及药理剂量的铜和锌生长促进剂，可促使抗生素耐药性的形成。这也是公众对集约型、现代化养猪业最大的担忧之一。

◆ 2.8 排泄物中主要挥发性臭味化合物的生物合成

猪排泄物中产生的几类挥发性有机化合物会引起猪粪便臭味的方生，包括氨和胺、挥发性短链脂肪酸（VFA）、挥发性硫化物（如硫化氢）、酚类（如对甲酚）以及吲哚类（如甲基吲哚，Mackie等，1998；Rideout等，2004；Le等，2005）。排泄物中的氨主要来源于尿液中的尿素和粪便中蛋白质、小肽和氨基酸的微生物脱氨基作用，而胺则来源于粪便中蛋白质、小肽和氨基酸的微生物脱羧基作用（Mackie等，1998；Le等，2005）。VFA的微生物合成过程已经很清楚。尽管后肠产生的VFA可被猪作为能量物质有效地吸收利用，但向环境中排放的VFA仍被认为是集约化养猪生产引起恶臭的一类挥发性化合物（Mackie等，1998；Le等，2005）。此外，排泄物中的VFA大部分来源于生猪养殖和排泄物贮藏过程中饲粮和内源性未被消化的蛋白质、碳水化合物和脂类碳氢骨架链的微生物发酵（Mackie等，1998；Le等，2005）。

挥发性硫化物是指大量的有机硫化物，包括猪排泄物中含量相对较高的硫化氢和硫醇等简单硫化物（Mackie等，1998；Rideout等，2004；Le等，2005）。挥发性硫化物除了对动物和人的毒性非常大之外，还与猪产生的臭味化合物有关（Mackie等，1998；Le等，2005）。猪饲粮中的无机含硫化合物和有机含硫化合物都是挥发性硫化物生物合成的潜在前体物（Mackie等，1998；Le等，2005）。猪饲粮中的这些化合物通常是无机硫酸根阴离子，与饲粮中的主要饲料原料、微量矿物元素添加剂、含硫氨基酸和小肽（如蛋氨酸、半胱氨酸和胱氨酸）及谷胱甘肽和硫酸盐共轭类固醇等含硫代谢物有关。对甲酚等挥发性酚类化合物可导致猪排泄物臭味的产生。在猪的后肠、排泄物贮存期间以及猪粪肥田后，苯丙氨酸和酪氨酸是厌氧微生物发酵合成酚类化合物的主要前体物（Mackie等，1998；Le等，2005）。

猪排泄物中的挥发性吲哚，尤其是甲基吲哚，形成了猪排泄物的特有臭味（Mackie等，1998；Rideout等，2004；Le等，2005）。值得一提的是，甲基吲哚也是引起猪肉异味的主要物质（Weiler等，2000）。色氨酸（Trp）是猪后肠中或排泄物贮存期间厌氧微生物发酵生成吲哚的主要前体氨基酸（Mackie等，1998；Rideout等，2004；Le等，2005）。在谷物和油菜粕等主要的商品猪饲料原料中，色氨酸的含量通常很低（NRC，1998）。因此Claus和Raab（1999）推测，来源于结肠黏膜的胃肠道内源性蛋白质中，色氨酸的含量较高，是肠腔微生物合成甲基吲哚的主要色氨酸来源。结肠黏膜细胞有丝分裂速率的变化反映了饲粮成分对上皮细胞周转率和寿命的影响，这也与甲基吲哚的微生物合成有关（Claus和Raab，1999）。

因此，饲粮蛋白质和内源性蛋白质损失（受饲粮组分尤其是纤维成分的影响）是猪排泄物中挥发性恶臭化合物产生的根源，厌氧微生物发酵合成了这些化合物。这些恶臭化合物对周边社区的影响，是养猪业必须解决的公众最关心的问题之一。

◆ 2.9 养猪生产带来的空气污染物

除了前面提到的挥发性恶臭化合物外，养猪生产带来的其他空气污染物，包括粉尘（以总颗粒物计量）、内毒素、可培养放线菌、其他类型的细菌和真菌也应引起我们的重视（Thorne等，2009）。集约化、密集型养猪生产带来的空气污染物容易导致养猪场工人的职业卫生问题，增加公众对附近居住环境卫生和安全问题的担忧（Letourneau等，2010）。饲粮结构、饲料加工和动物运输，会受到日常管理、动物健康状况、饲粮、通风设施和猪舍设计的影响，也会影响到集约化养猪场的空气质量。

3 缓解主要环境问题的措施

传统的遗传选择和育种理论与实践促进了全世界各种现代瘦肉型杂交猪的发展，使得瘦肉产量有了突飞猛进的提高，猪上市时间大大缩短。比如，瘦肉型猪从出生到体重达110 ~ 120 kg上市只需要5 ~ 6个月的时间。然而，在集约化养殖条件下，即便提供最适宜的生长环境和营养条件，饲养同样长的时间后一些地方杂交猪种的上市体重只能达到90 kg。

传统的遗传选择和育种，使得瘦肉型猪的饲料转化率和生长速率得到了极大的提高，但是在集约化养殖生产中，氮和能量的总体利用率并没有得到显著提高。其主要原因如下：第一，DNA、RNA和蛋白质等主要含氮聚合物在体内的生物合成和降解需要较高的ATP或能量。动物体内的粗蛋白质（CP）沉积的二级能量效率（kp）很低。例如，生长猪的kp仅为0.47 ~ 0.55，而粗脂肪沉积的二级能量效率（kf）可达到0.67 ~ 0.86（ARC，1981；Fan等，2006）。Rivera-Ferre

等（2005）的研究论证了这一结果。其报道，伊比利亚地方猪比瘦肉型长白猪的骨骼肌蛋白合成速率更高，表明肌肉蛋白质降解率的不同导致了伊比利亚猪的蛋白沉积率较低。第二，Fan等（2006）总结得出，遗传程序化和饲粮变化使得猪机体对氮的利用率从哺乳期（83%）到育成期（51%）急剧下降。在断奶后大部分生长期，猪对氮的利用率均较低。传统的遗传选择和育种计划并不能有效解决该问题，正如传统的遗传干预不能显著改变蛋白质和氨基酸的代谢途径一样。第三，在集约化养殖生产中，饲料效率的提高很大程度上取决于脂肪沉积率的降低。沉积单位脂肪所需要的净能是沉积单位肌肉所需要净能的2倍多。与瘦肉沉积相关的水分含量比脂肪沉积相关的水分含量要多75%，因此，瘦肉沉积所需的饲粮净能较少。

因此，通过将植物饲料转化为脂肪含量低的瘦肉，加速了拥有瘦肉型猪或杂交猪或两者兼有的全球养猪生产系统的生产强度和进度，并形成了较快的周转和集约化养猪生产体系。然而，从生物学角度来看，用于猪生长的CP沉积效率和与之相关的能量利用二级效率并没有得到很大的改善。根本的生物学原因是，过量的气体排放和排泄物中营养素的排放是全球集约化养猪生产带来的主要的环境可持续性问题。

◆ 3.1 饲粮中外源酶的添加

粪便中养分的损失是整个机体养分利用率低的重要组成部分（Fan等，2006）。与猪饲粮中能量和各种养分利用有关的生理生化过程，包括内源酶和外源酶的水解、跨肠上皮细胞的吸收转运，但跨膜转运过程似乎不是限速过程（Weiss等，1998；Fan等，2006）。另外，粪便中代谢性内源养分的损失主要来自于黏膜更新和脱落及其他内源性分泌物，并随饲料原料结构的变化而改变（Fan等，2006）。例如，植物饲料原料中的细胞壁木质素和植酸，则被公认为单胃动物营养中的抗营养因子（Adeola和Cowieson，2011；Ravindran和Son，2011）。

植物细胞壁木质素可包被营养素，而植酸主要与营养素形成络合物。如果饲粮中不添加外源酶，那么植物饲料原料中的木质素和植酸都不能被单胃动物有效利用（Adeola和Cowieson，2011）。饲粮中存在的β-葡聚糖可增加食糜的黏度，降低小肠葡萄糖的转运速率，并抑制门静脉血液循环中葡萄糖浓度峰值和胰岛素的产生（Hooda等，2010）。据报道，小肠的葡萄糖吸收率可影响小肠氨基酸转运蛋白的表达，影响采食后门静脉氨基酸流和胰岛素脉冲的水平（van Der Meulen等，1997；Yin等，2010；Adeola和Coieson，2011；Drew等，2011）。反之，由哺乳动物雷帕霉素靶蛋白（mTOR）信号通路介导的蛋白质翻译调控的猪肌肉蛋白质合成和整个机体氮沉积效率也会受到影响（Yang等，2008；Adeola和Cowieson，2011）。饲粮中β-葡聚糖引起的食糜黏度的增加，使营养素的消化和吸收从空肠近端转移到了空肠末端（Hooda等，2011）。因此，植物细胞壁基质物质和植酸在多方面限制了猪对饲粮营养素的利用效率。

此外，水溶性植物细胞壁木质纤维素聚合物，包括木质素、纤维素和半纤维素，可通过增加猪胃肠道中CP和氨基酸的内源损失而磨平消化道的腔面（Fan和Sauer，2002；Fan等，2006；Myrie等，2008）。植物细胞壁或/和饲料原料中的其他不可消化的NSP成分（如大麦和燕麦中的β-葡聚糖、菜粕和甜菜果酱中的果胶）也经常引起猪消化道中食糜黏度的增加。中等水平到高水平的饲粮果胶，可通过抑制胰蛋白酶活性、增加CP和氨基酸的内源损失，从而降低CP和氨基酸的消化率（Mosenthin等，1994；Myrie等，2008）。然而，有研究发现，中等水平的饲粮果胶可提高猪对磷的消化率和内源磷的排放（Fan等，2004b）。尽管植酸可增加矿物质的内源损失（Davies等，1975；Onyango等，2009；Woyengo等，2009），但对CP和氨基酸内源性损失的影响几乎可以忽略不计，而且对其他单胃动物的影响也不一致（Onyango等，2009；Woyengo等，2009）。因此，植物饲料原料中的NSP和植酸都增加了猪代谢性内源营养素的损失。

自Simons等（1990）对植酸酶在饲料中应用开展的开创性研究以来，单胃动物饲粮中添加植酸酶在全世界得到了广泛的应用。就其工业应用而言，目前植酸酶市场约为3.3亿美元，植酸酶的应用每年为全球饲料工业可节约20亿~30亿美元（Adeola和Cowieson，2011）。通过微生物工业发酵和作物生产（如使用转基因玉米），可表达和获得真菌类和非致病性大肠杆菌来源的外源植酸酶（Simons等，1990；Nyannor等，2007；Adeola和Cowieson，2011）。值得一提的是，大多数的植酸酶添加试验并没有将钙、磷和微量矿物元素的真消化利用和内源损失加以区分。在猪饲粮中添加最适水平的植酸酶，钙、磷和微量矿物元素（如铁和锌）利用率的提高与排泄物中矿物元素排放的减少密切相关（Lei等，1993；

Adeola等，1995；Adeola和Cowieson，2011）。然而，饲粮中添加植酸酶对猪CP、氨基酸和能量消化率的影响似乎并不一致（Adeola和Cowieson，2011）。总而言之，在养猪生产中，饲粮添加或含有最适水平的外源植酸酶可大大提高植物饲料原料中钙、磷和微量矿物元素的消化利用率，降低排泄物中这些矿物元素的排泄量。

α-淀粉酶和半纤维素酶等微生物源的碳水化合物酶成了一类新兴的酶，每年约有2.2亿美元的产值，占全球约40%的市场份额（Adeola和Cowieson，2011）。β-（1,3/1,4）-葡聚糖酶和β-1,4-内切木聚糖酶是其中的两种占主导地位的酶，在行业中得到了广泛推广应用。在大麦型饲粮中添加β-葡聚糖酶可显著提高仔猪的回肠能量、CP和氨基酸消化率（Li等，1996）。已有研究证明，猪采食小麦型饲粮比采食燕麦或大麦型饲粮会产生更多的NH_3和臭味（O'Shea等，2011a）。而其他两项研究却表明，在大麦型饲粮中添加β-葡聚糖酶会增加生长育肥猪排泄物中NH_3和臭味的排放（Gary等，2007；O'Shea等，2010）。这有可能是β-葡聚糖酶的添加将β-葡聚糖的降解从大肠转移到了小肠，从而促进了宿主动物小肠上皮细胞对D-葡萄糖更快、更有效的吸收，而不是大肠中微生物摄取和代谢β-葡聚糖降解产生的D-葡萄糖。此外，燕麦类谷物中β-葡聚糖的链更长，更难溶于水。有研究表明，在燕麦型饲粮中添加外源β-葡聚糖酶对猪排泄物中NH_3的排放没有影响（O'Shea等，2010）。因此，在断奶仔猪的燕麦或大麦型饲粮中添加外源β-葡聚糖酶可能有益于提高碳和氮的消化利用率。然而应当注意的是，在大麦型饲粮中添加β-葡聚糖酶可能会增加生长育肥猪排泄物中NH_3和臭味的排放。

木聚糖酶，主要是β-（1,4）-内切木聚糖酶，为少数半纤维素酶类中的一种，在小麦和玉米型猪饲粮中添加可小幅度地提高能量、CP及一些氨基酸的消化率（He等，2010；Adeola和Cowieson，2011）。在含量较高小麦副产品（30%）的生长猪饲粮中添加，效果更明显（Nortey等，2008）。几项研究报道，在生长育肥猪的玉米或小麦型饲粮中同时添加最适水平的木聚糖酶和植酸酶，对增加生长性能和营养物质消化率存在一致的交互作用（Olukosi等，2007；Woyengo等，2008；Yáñez等，2011）。曾有两项研究考察了玉米-豆粕型饲粮中添加甘露聚糖酶和半乳聚糖酶对生长性能和营养物质消化率的影响，发现在饲粮中添加这两种酶，对断奶仔猪、生长猪和育肥猪没有影响或存在不一致的影响（Petty等，2002；Kim等，2003）。

近年来，有关饲粮中添加外源酶可有效降解木质纤维素方面有了一个概念上的进展，即同时使用纤维素酶、半纤维素酶和植酸酶等多种酶。研究表明，在谷物类饲粮尤其是含有高水平副产品的饲粮中添加这些酶，可改善断奶仔猪、生长猪和育肥猪的生长性能和能量、CP和磷的消化率（Omogbenigun等，2004；Emiola等，2009）。此外，Kiarie等（2007）研究发现，在断奶仔猪的大麦、小麦、豌豆、亚麻籽、菜籽粕和豆粕饲粮中，添加多种外源酶不仅可以改善营养物质的消化率，还能提高回肠末端有益菌乳酸杆菌的数量。由植物细胞壁半纤维素酶酶解释放的单糖主要包括阿拉伯糖、木糖和甘露糖，但这些单糖不是宿主通过主要的肠刷状缘顶端葡萄糖转运载体钠-D-葡萄糖共转运载体-1进行肠道摄取和利用的典型的生理学单糖底物（Yang等，2010；Wright等，2011）。这些释放的阿拉伯糖、木糖和甘露糖很容易被肠道栖生菌利用，从而产生益生作用（Schutte等，1991，1992）。因此，饲粮添加半纤维素酶可提高改善猪肠道健康和排泄物特性等的益生作用。

总之，目前对外源木质纤维素降解酶的研发仍处于初级阶段，即其有效的互补酶、特定酶的活性和酶的热稳定性。例如，用于饲料添加的木质素降解酶在市面上并没有出售。同样，完全降解木聚糖使用的与β-1,4-内切木聚糖酶互补的β-1,4-外切木糖苷酶，在市面上也没有用于饲料添加使用的产品。纤维素是植物细胞壁中含量最高的成分，有限的纤维素酶活性限制了其对纤维素的降解作用。鉴于上述原因，需要指出的是，相对较低的纤维素酶比活性同样也限制了第二代纤维素乙醇作为生物燃料进行商业化生产的可行性（Wilson，2009）。在猪饲料中添加更有效互补的外源性木质纤维素降解酶混合物，就可以更大程度地提高饲粮碳、氮和矿物质的消化利用效率，进一步降低饲养成本，减少对环境的负面影响。

◆ 3.2 低粗蛋白质饲粮配方

猪机体内和主要饲料原料中的含氮化合物主要包括蛋白质、肽和游离氨基酸，以及非蛋白氮（NPN）化合物，如DNA、RNA、核苷酸、核苷和多胺等（属于其他次要形式的NPN），上述物质统称为CP（CP=N × 6.25）。饲料样品中的CP含量在动物营养研究中通常采用凯氏定氮法测定（AOAC，1993；Fan等，2006），目前更多地采用Leco-N分析仪进行测定（Rideout和Fan，2004）。因此，在猪营养中CP和N的利用率是通用的。

从整个机体来看，CP沉积（$CP_{沉积}$）是机体CP合成（$CP_{合成}$）与CP降解（$CP_{降解}$）和CP内源损失（$CP_{内源损失}$）的净差值，Fan等（2006）用以下公式进行描述：

$$CP_{沉积}=CP_{合成}-CP_{降解}-CP_{内源损失}$$

骨骼肌是最大的机体CP储备库。出生后从哺乳到断奶期间，骨骼肌蛋白合成活性急剧下降，且在大部分生长期间都维持在一个相对较低的水平（Reeds等，1993；Fan等，2006；Davis等，2008；Yang，2009），从而增加了出生后氨基酸的氧化和尿氮的排泄，这也是导致养猪生产中氮和能量利用率较低的主要原因之一。

蛋白质降解是影响整个机体CP沉积的显著负面因素（Mulvaney等，1985；Skjaerlund等，1994）。蛋白质的降解途径较为复杂，很多的研究把重点放在了从细胞和分子水平上了解其调节机制（Goll等，2008）。此外，相比对蛋白质合成活性的估测，体内直接测量细胞内的蛋白质降解率是一项更具挑战性的科学问题（Bergen，2008）。泛素-蛋白酶体降解途径是蛋白质降解的一个主要途径，是消耗ATP的过程，每降解1 mol的典型蛋白质需要消耗300 ~ 400 mol的ATP（Benaroudj等，2003）。据估测，用于蛋白质合成的能量消耗约为0.90 kJ/g蛋白质，然而通过化学计算得出用于蛋白质降解的能量消耗约为0.40 kJ/g蛋白质（Fan等，2008a）。目前，对调节猪整个机体、组织和器官中蛋白质降解速率的因素了解较少。

目前普遍认为，细胞外和细胞内的氨基酸浓度受到严格调控，事实上没有大量的L型氨基酸通过尿液排出体外（Newsholme和Leech，1991）。饲粮来源和蛋白质降解返回细胞内氨基酸库的过量氨基酸被有效氧化（Salway和Granner，1999），但不是被循环用于蛋白质、RNA、DNA和其他含氮化合物的生物合成。氮的分解代谢终产物氨主要在肝脏中被转化为尿素，少部分在肠道中被转化为尿素（Bush等，2002；Fan，2003）。需要指出的是，在出生后生长期，饲粮中的一些嘧啶可能被用于了体细胞的RNA合成（Berthold等，1995）；母体和胎儿组织的RNA只能来源于从头合成（Boza等，1996）。因此，饲粮中大部分的核苷、核苷酸、RNA与DNA等NPN可能被降解，随后氮代谢终产物以尿素形式被排出体外。哺乳期N的分解代谢损失较低（9%），断奶后饲喂玉米-豆粕型饲粮时，氮的代谢损失将大大增加（39%～41%）（Fan等，2006）。因此，氮的利用率低下很大程度上归结于出生后遗传和营养因素，这构成了猪粪便中的氮，这就是当前集约化养猪所面临的环境可持续性问题。

猪排泄物中氮最大的来源就是体内氨基酸代谢产生的、主要以尿素形式排泄的尿氮。自20世纪90年代以来，通过饲料原料满足动物对回肠可消化必需氨基酸需要的猪饲粮配方就已得到了广泛认可和应用（NRC，1998；Sauer等，2000）。为了获得猪对饲料原料中氨基酸的回肠真消化率，需要估测出回肠末端内源氨基酸的排出量。而回肠末端内源氨基酸的排出量具有饲料原料特异性，并受测试饲料原料组成的影响（Nyachoti等，1997；Jansman等，2002；Fan等，2006；Stein等，2007）。Stein等（2007）建议，氨基酸回肠标准消化率可通过矫正测定的氨基酸回肠表观消化率和已知的基础回肠内源氨基酸排泄量得出（Jansman等，2002）。Zhe等（2005）研究表明，此方法过于简化，无疑低估了大量饲料原料的真消化率值，从而导致猪饲粮配方和饲养过程中过多的氨基酸氧化和排泄物中氮的排泄。

回归也曾被用于估测猪回肠末端内源氨基酸排出量和饲料原料的氨基酸回肠真消化率（Furuya和Kaji，1992；Fan等，1995；Jansman等，2002）。最近，开始采用替代法估测粪便中的内源磷排泄量和磷的真消化率（Fang等，2007；Fan等，2008b）。替代法是一种成本低、操作简单的方法，可用于估测回肠末端内源氨基酸排泄量和氨基酸回肠真消化率。无论是采用回归分析还是替代法，都需要考虑到待测饲料原料在猪饲粮中所占的可能比例。总之，需要测定氨基酸回肠真消化率，并应用于猪饲粮配方中，以最大限度地减少氨基酸的氧化和排泄物中氮的排放，从而减少养猪生产的负面影响。

采用鱼粉、菜粕和豆粕（SBM）等蛋白质原料配制的猪的常规饲粮，可满足必需氨基酸的推荐水平，断奶仔猪饲粮中含有20%～26%的CP水平，生长猪饲粮中含有16%～18%的CP水平，育肥猪饲粮中含有13%～16%的CP水平（NRC，1998）。这些饲粮除了含有限制性必需氨基酸（如赖氨酸）外，还含有过量的其他氨基酸。因为机体不能贮存蛋白质合成剩余的氨基酸，多余氨基酸在内脏器官、骨骼肌和其他外周组织中被脱氨。另外，氨基酸分解代谢产生的氮会以尿素形式通过尿液排出体外（Voet和Voet，1995）。Windmueller和Spaeth（1975，1976）的经典研究和Stoll等（1998）的最新研究均表明，内脏器官中存在大量的必需氨基酸和非必需氨基酸的首过分解代谢，这反过来促进了尿液中尿素的排泄。Kerr和Easter（1995）计算得出，当给猪饲喂添加了限制性必需氨基酸的玉米-豆粕型饲粮时，饲粮CP水平每降低一个百分点，氮的排泄会降低8%。Deng等（2007）也观察到，氮利用率的大大提高以及排泄物中氮排放量的大大降低。据计算，按照理想蛋白质模式，以回肠真可消化氨基酸供给量为基础，为了节约成本，通过添加限制性必需氨基酸将猪饲粮中的CP水平降低3～3.5个百分点，总粪氮排泄可降低30%左右。

有些研究定量分析了低CP日粮对环境的影响，如温室气体和NH_3的排放以及其他粪臭化合物的分泌与排放。Velthof等（2005）报道，饲喂低CP饲粮可降低排泄物贮藏过程中NH_3和CH_4的排放，减少施肥后土壤中N_2O的排放。NRC（2012）最近总结得出，若以二氧化碳当量作为低CP饲粮对粪污和生产中温室气体排放的最终判定标准，不够完整和严谨。因此，尚需要基于长期实证研究的大规模、累计效应测定，来量化低CP饲粮对集约化养猪生产体系中温室气体排放的影响。

低CP饲粮的确能够降低排泄物中氨的含量（Kerr等，2006；Htoo等，2007a；Ziemer等，2009）和猪粪中NH_3的排放（Canh等，1998a；Sutton等，1999；Otto等，2003；Hayes等，2004；Velthof等，2005；Leek等，2007；Le等，2007b，2008，2009）。然而，低CP饲粮对猪粪臭味的影响结果不一，这依赖于试验是如何开展的以及主要测量指标是什么（NRC，2012）。有些研究表明，配制低CP饲粮可降低生长育肥猪排泄物中臭味的排放和/或臭味强度（Hayes等，2004；Leek等，2007；Le等，2007a，2007b，2008）。Le等（2009）的研究表明，将育肥猪饲粮中的CP水平降低3%，可减少排泄物中主要臭味化合物的浓度，但不影响粪便中臭味的强度和排放。在Otto等（2003）的研究中，经过训练有素的感官小组成员测定，发现将生长猪饲粮中的CP水平从15%降到0，不影响猪粪臭味的产生。需要指出的是，Otto等（2003）的试验是以酪蛋白作为蛋白源的，酪蛋白可被生长猪几乎完全消化（Fan等，2006）。因此，Otto等（2003）的试验中猪采食酪蛋白饲粮后排泄物中的臭味强度，仅能反映粪便中内源CP损失水平的变化。此外，Trabue等（2011）的研究显示，在测量对臭味影响过程中，关键技术和嗅觉测定技术之间的系统差异与人为因素和较大变异有关。在这种情况下，一些证据表明，关键的潜在致病细菌是主要的蛋白质发酵菌（Le等，2005；Bauer等，2006）。饲喂低CP饲料可减少排泄物中产气荚膜杆菌和大肠杆菌等有害菌及其毒素的产生（Le等，2005；Bauer等，2006）。另外，由于蛋白质添加物也是饲粮重金属矿物元素As和Cd的重要来源（Linden等，1999；Jackson等，2003），因此配制低CP饲粮理论上可以减少排泄物中As和Cd的排放。

总之，通过在饲粮中补充限制性必需氨基酸，在NRC（1998）推荐水平基础上将饲粮CP水平降低2%～3%，可有效缓解一些主要的环境问题，包括集约化养猪生产中的NH_3排放、臭味减排、排泄物中病原菌以及重金属As和Cd的排放。

◆ 3.3 低磷饲粮的配制

猪的无机钙和磷添加剂和饲料原料中钙和磷的生物学效价通常采用粪便表观消化率和斜率比法测定（Jongbloed等，1991；Cromwell，1992）。在过去的十年期间，有几项研究已采用回归分析法、无磷饲喂法和替代法，测定了无机钙和磷添加剂和饲料原料中磷的真消化率和粪便中胃肠内源磷的排泄量（Fan等，2001；Petersen和Stein，2006；Fang等，2007；Fan等，2008b）。这些研究表明，磷的表观消化率和效价不稳定，且无机磷添加剂和饲料原料中磷的真生物学效价被大大低估了（Fan等，2001；Petersen和Stein，2006；Fang等，2007）。此外，猪粪便中的内源磷排泄量反映了相当大一部分机体代谢磷的损失和磷的需要量（Schulin-Zeuthen等，2007；Fan等，2008b）。通过饲喂无磷、纯合和半纯合饲粮测定的猪粪便内源磷的排泄量为0.070～0.199 g/kg干物质摄入量，有可能是内源磷的最低水平，反映了猪的粪便内源磷的基础损失（Fan等，2008b；Almeida和Stein，2010）。Fan等（2008b）总结了粪便内源磷的排泄量，发现饲喂不同待测饲料原料时，断奶仔猪、生长猪、育肥猪和母猪粪便内源磷排泄量的变化范围为0.3～1.1 g/kg干物质摄入量。饲喂玉米-豆粕型基础饲粮的生长猪，采用替代法测定时，增加的粪便内源磷排泄量（1.3 g/kg干物质摄入量）（Fan等，2012）。

Almeida和Stein（2010）以玉米和DDGS的磷标准消化率为基础配制饲粮，采用半纯合和无磷饲粮矫正粪磷内源基础排泄量，测定了断奶仔猪的生长性能和磷平衡的变化。显然，使用磷的标准消化率简化了不同饲料原料粪便内源磷排泄量之间的差异，那么必然低估了猪饲粮配方中饲料原料的磷真消化率。因此，粪便内源磷排泄量可能会受到很多因素的影响，如饲料原料的类型、饲粮组成、猪的生长阶段和生理状况。

另外，主要的无机Ca添加剂和饲料原料中Ca的生物学效价也可采用Ca真消化率为指标进行测定（Ajakaiye等，2004；Fan等，2004a）。此外，当采用表观粪便消化率和斜率比法测定时，饲粮Ca：P比和Ca、P需要量取决于总磷和有效磷的量（NRC，1998）。目前的最佳饲粮Ca：P比是采用生长性能为主要测定指标获得的（NRC，1998，2012）。所以，一些研究表明饲粮总Ca：P比对生长性能和骨骼矿化状况的影响受外源植酸酶添加的影响不足为奇（Qian等，1996；L'etourneau-Montminy等，2010）。Wang等（2008c）报道，给断奶仔猪饲喂玉米-豆粕型饲粮时，饲粮Ca：P比变化对N沉积效率没有影响，但是钙沉积效率呈线性增加。然而，Ca：P比值变化对断奶仔猪磷沉积效率和生长性能的影响呈二次曲线，真可消化钙与真可消化磷的比值为（0.96～1）：（1.35～1）时影响最大（Wang等，2008c）。Fan等（2012）观察到，Ca：P比变化对生长性能的影响呈二次曲线，但是对Ca、P沉积效率没有影响。

Stein等（2011）的大型合作研究显示，粪磷表观消化率和钙、磷沉积效率随饲粮钙含量和Ca：P比的增加而线性降低，但对粪钙表观消化率没有影响。机体钙和磷的代谢，包括Ca和P的吸收、骨骼Ca和P的矿化和重吸收、胃肠Ca和P分泌以及肾脏Ca和P的排泄，均受骨三醇、降钙素、甲状旁腺激素、促炎细胞因子和成纤维细胞生长因子23等多种激素和生长因子的调控。因此，饲粮Ca、P的摄入量和饲粮Ca、P比例均会从肠道吸收、内源分泌和尿液排泄等层面影响机体Ca和P的沉积。在以机体Ca、P沉积效率作为主要指标定义生长育肥猪的Ca、P真可消化需要量和最佳真可消化钙/真可消化磷比值时，对这些变化应进行说明。

植物饲料中的大多数磷是植酸磷，可采用一些饲粮措施来提高磷的利用，减少猪排泄物中磷的排放。首先，被广泛采用的方法是通过添加各种外源植酸酶来提高植酸磷的利用，这些植酸酶可通过真菌（Simons等，1990）、非致病性大肠杆菌、酵母（Stahl等，2000）和转基因作物（Nyannor等，2007）进行工业化生产。但是，外源糖酶和植酸酶的热稳定性仍然是饲料制粒过程中的挑战性问题（Gentile等，2002；Kim等，2008）。其次，使用低磷饲料原料配制饲粮，也可提高饲粮磷的消化率（Spencer等，2000；Htoo等，2007b）。

通过使用植酸酶和低磷饲料大大提高植酸磷的消化利用率，对完全不使用无机磷酸盐添加剂来配制满足NRC（1998，2012）推荐的有效磷需要量的低磷饲粮是非常必要的。菜粕和SBM蛋白质添加物中的总磷含量高于大麦、玉米和小麦等谷物类原料（NRC，1998）。同样，另一种配制低磷饲粮的有效方法是使用较少量的蛋白质原料配制低CP饲粮。另外，由于无机磷酸盐和蛋白质添加物均是饲粮重金属As和Cd的主要来源（Linden等，1999；Jackson等，2003），所以添加较少的磷和蛋白质添加物配制低CP饲粮，理论上可以减少排泄物中As和Cd的排放量。

总之，以添加真可消化Ca、P为基础同时外源添加植酸酶和低植酸盐原料配制猪的饲粮，可大大减少排泄物中P以及重金属As和Cd的排放及其对环境的负面影响。此措施还可对环境产生有利的影响，如减少P的径流和地表水的富营养化。

◆ 3.4 低硫和低微量矿物元素饲粮的配制

3.4.1 硫

猪排泄物中的含硫化合物主要来自粪便中不可消化的含硫氨基酸（如饲料中的Met和Cys）、胃肠中内源性蛋白质和谷胱甘肽代谢物以及饲粮来源的含硫氨基酸和硫酸盐经体内完全分解代谢后随尿液排泄的硫酸盐。尽管大多数挥发性含硫化合物都会使猪的排泄物出现异味，但硫化氢和甲硫醇是其中最重要的两种物质，占猪排泄物中总挥发性含硫化合物的70%～97%（Le等，2005）。Bauchart-Thevret等（2011）的体内示踪动力学研究显示，给断奶仔猪饲喂液体代乳料时，胃肠道将饲粮中25%的胱氨酸用于了非氧化生物合成。尽管大多数植物源饲料原料中含硫氨基酸的含量相对偏低，但低CP饲粮时由于降低了总含硫氨基酸的量因此也会导致低S饲粮。一些研究已证明，配制低CP饲粮和低S饲粮可降低排泄物中硫的含量（Le等，2007a；Ziemer等，2009）和猪粪浆中臭味强度，减少挥发性硫化物的排放（Clark等，2005；Le等，2007a）。一般饲粮中添加过量的DL-Met及其类似物DL-2-羟基-4-（甲硫基）丁酸，可以增强抗氧化应激和炎症的主要抗氧化剂谷胱甘肽的生物合成。但在生长猪饲粮中添加超过推荐水平的Met，增加了臭气的排放和粪坑上方空气中的臭味强度（Le等，2007a）。Eriksen等（2010）的研究也证明，在饲粮中添加过量的Met提高了粪污中所有上述提到的五种挥发性硫化物的排放，而且尿液中的硫酸盐是硫的主要存在形式，这说明硫酸盐是体内大量Met分解代谢产生的主要废物。

硫酸盐通常作为阴离子被用于生产大多数的微量矿物添加剂，以降低生产成本和提高微量矿物元素的生物学效价。以微量矿物元素添加剂形式提供的过量硫酸盐容易被吸收进入血液循环，然后大部分进入尿液。小部分未被消化的硫酸盐被循环返回肠道内，通过肠肝循环以粪的形式排出。Armstrong等（2000）的试验表明，给生长育肥猪饲喂添加非硫酸盐形式的铜的低S饲粮，大大降低了粪便臭味的强度。然而，Armstrong等（2004）开展的另一项研究发现，给断奶仔猪饲喂添加非硫酸盐形式的铜的低S饲粮，并不影响粪便的臭味。这两项研究结果存在差异，可能是由以下因素引起的。首先，这两项研究使用了不同生长阶段的猪。其次，饲粮中过量的硫酸盐主要是通过尿液排出，而上述两项研究均没有测定尿样。虽然在这方面尚需要开展更多的研究，但是我们仍可以总结得出，配制低S饲粮可有效降低成本，也是降低猪排泄物中与挥发性硫化物有关的异味排放的有效措施。

3.4.2 微量矿物元素

猪排泄物中大量的Cu、Zn和Cr等微量矿物元素及其对生态系统和饮水资源可能产生的负面影响，主要归结于这些微量矿物元素在机体内的沉积效率较低。正如最新的NRC（2012）所述，猪常规饲粮中微量矿物元素的沉积率只有5%～30%。当在断奶仔猪和生长猪饲粮中添加药理水平的Cu（如200～250 ppm）和Zn（2 000～3 000 ppm）作为生长促进剂时，Cu和Zn的沉积效率将降低至5%～10%，90%～95%的Cu和Zn会随排泄物排出（NRC，2012）。在啮齿目动物上的早期研究也发现，胃肠道中的内源锌循环构成了每日锌需要量的大部分，粪便中内源锌的损失是机体排泄锌的重要部分（Davies和Nightingale，1975；

Flanagan，1984）。而饲粮中存在的植酸盐增加了粪便中锌的总排泄量（Davies和Nightingale，1975；Flanagan，1984）。

另外，吡啶甲酸铬和氯化铬常被用作饲粮铬的添加剂。饲粮中添加铬（如0.2μg）不是为了满足机体对铬的需要，而是作为一种代谢调节物来提高育肥猪的营养物质消化率和瘦肉率（Kornegay等，1997），改善妊娠母猪的体况（Young等，2004）。然而，关于铬在猪体内的沉积效率鲜有报道。啮齿目动物的示踪动力学研究表明，饲粮中添加的吡啶甲酸铬和氯化铬分别仅有1.16%和0.55%的铬被吸收。被吸收的大部分铬不能被进一步代谢，而是很快随尿液排出体外，包括有机的吡啶甲酸铬也是被肠道吸收后通过尿液以完整的形式被排出体外（Kottwitz等，2009）。因此，排泄物中过量的微量矿物元素大部分来源于未被消化的粪便损失，这是由主要饲料原料的生物学利用效率低、粪便中的内源损失以及饲粮中的过量添加造成的。

目前，尚缺少饲料原料和微量矿物添加剂中微量矿物元素的真生物学效价数据。由于存在高水平的体内循环，饲料原料中微量矿物元素的生物学利用效率不能简单地采用动物粪便表观消化率试验进行测定。而稳定性同位素示踪动力学试验对猪等大型动物来说成本太高。采用斜率法测定的饲料原料和微量矿物元素添加剂中的微量元素利用率变异较大。例如，当分别以鸡蛋蛋白和大豆浓缩蛋白为对照时，测得的SBM中Zn的利用率分别为40%和78%（Edwards和Baker，2000）。在配制猪饲粮中微量元素预混料时，通常的做法就是忽略主要饲料原料本身含有的微量元素，这就导致了微量元素的过量添加和在猪粪便中的过量排泄（Creech等，2004；Sutton等，2004）。为了同时测定粪便中内源磷排出量和饲料原料中磷全肠道真消化率而建立的替代法（Fang等，2007），也许适用于微量矿物元素的测定。测定粪便中内源微量矿物元素的排泄量以及饲料原料中微量元素和微量元素添加剂本身的真消化率，将可以真消化率为基础配制猪的饲粮配方，从而尽量减少微量矿物元素的过量添加和过量排泄。

很多饲粮措施均可以提高微量矿物元素的沉积效率，从而尽量减少猪废弃物中微量矿物元素的排泄。目前已有研究表明，在饲粮中添加外源植酸酶可以提高微量矿物元素在猪体内的生物学利用率和沉积率（Lei等，1993；Adeola等，1995）。关于针对Cu和Zn的抗生素耐药性的问题也已有记载。目前已有替代高水平Cu、Zn添加的其他饲粮措施。因此，为了减少养殖排泄物中Cu和Zn的含量，需要重新考虑药理剂量的Cu、Zn作为生长促进剂在猪饲粮中的使用。

尽管饲粮中添加Cr作为代谢调节剂确实可以提高育肥猪和怀孕母猪的营养物质消化率、瘦肉率和代谢状态（Kornegay等，1997；Young等，2004），但是提高的幅度相对较小。猪饲粮中添加的Cr大部分被分泌到排泄物中，可能引起饮用水中总可溶性铬离子超过安全标准。而许多其他有用的饲粮措施也可以提高营养物质消化率和瘦肉率。因此，需要重新考虑在猪饲粮中添加Cr作为代谢调节剂，这样可使排泄物中Cr对饮用水的污染降到最低。

微量矿物元素的体内循环和粪便中的排出量是影响微量矿物元素需要量和排泄量的主要因素。需要进一步量化研究植物细胞壁和植酸盐等饲粮成分对猪体内的微量矿物元素循环、粪便中的排出量以及微量矿物元素需要量的影响。以真可消化微量矿物元素为基础配制满足营养需要的低微量元素饲粮以及植酸酶的使用，可以提高微量矿物元素的沉积效率，从而降低微量元素的排泄，并能最大限度地降低集约化养猪生产对生态系统和饮用水的负面影响。

◆ 3.5 断奶仔猪饲粮中抗菌剂的替代

3.5.1 断奶与抗菌剂

有关仔猪在断奶过程中的生物脆弱性的认识已经很清楚了，即胃酸分泌不足，不能维持较低的pH以应对病原微生物；有效利用谷物-淀粉的消化能力发育不够，消化道中食糜的蛋白质消化效率较低，非免疫性和主动免疫防御功能欠发达（Fan，2003）。有关断奶仔猪的高发病率、高死亡率和生长阻滞的问题已有详实记载（Maxwell和Carter，2001；Pluske等，2002）。最近有报道指出，断奶仔猪小肠刷状缘顶端碱性磷酸酶活性和酶亲和力降低可能与断奶有关（Lackeyram等，2010）。刷状缘顶端的碱性磷酸酶被认为是消化道内必需的非免疫性防御蛋白（Chen等，2010，2011）。可想而知，增强其催化效率和表达，也许可作为改善断奶仔猪肠道健康和营养的生物学标记。

20世纪40年代随着抗生素作为家畜促生长剂的发现，出现了一种简化解决仔猪断奶问题的饲粮方案。随后，在饲粮中添加预防剂量的抗生素被广泛推广使用（Cromwell，2001）。另外，在过去的二十年里，药理水平的Cu（200 ~ 250 ppm）和Zn（2 000～3 000 ppm）也作为一种抗菌促生长剂在断奶仔猪饲粮中使用（NRC，1998，2012）。需要指出的是，使用饲用抗生素和药理水平的Cu、Zn作为促生长剂，导致了完全不同类型的抗生素耐药性问题，前者主要是影响人类和动物健康的医学管理，后者主要是影响环境微生物和生态系统。

要想最终解决与饲用抗生素和药理水平Cu、Zn的应用密切相关的抗生素耐药性问题，就应在包括猪在内的畜禽养殖生产中停止使用饲用抗生素Cu和Zn。早在2006年，欧盟就已经禁止了在家畜养殖过程中使用抗生素作为生长促进剂。同样，2011年，韩国也禁止了抗生素作为生长促进剂的使用。也许在不久的将来，美国和其他国家也会进一步限制甚至取消抗生素在家畜生产中的使用。

目前，在断奶仔猪营养方面，已开展了大量的研究开发抗生素的替代方案（Pluske等，2002；Stein，2002；Pettigrew，2006；Heo等，2012）。这些替代方案是针对断奶仔猪肠道健康和营养的不同方面，可能会为可持续的养猪生产做出巨大贡献，甚至成为必不可少的措施。

3.5.2 **脂质、葡萄糖和乳糖**

断奶使仔猪从采食富含易消化的酪蛋白、乳脂和乳糖的液态奶转换为富含谷物淀粉和植物蛋白源的固体饲料。为了使断奶仔猪消化功能适应这种快速转变，高质量断奶饲粮必须能提供易消化的常量营养素（Stein，2002）。很明显，能量和营养素的高效利用将会对减少该阶段养猪生产对环境的负面影响起到很大的作用。

因为新生仔猪和哺乳仔猪分泌的胰脂肪酶酶活水平和小肠脂肪酸结合蛋白表达都比较高，所以它们对饲粮中油或脂的消化能力很强（Reinhart等，1992；Mubiru和Xu，1998）。早期断奶导致外分泌胰脂肪酶的表达及酶活性会在断奶后的几天内逐渐减少（Cera等，1990；Marion等，2003）。然而，经过适应后，肠腔内总的脂肪酶消化能力以及油和脂的消率能力均增强（Cera等，1988，1990），这也表明饲粮中的油脂对于满足断奶后仔猪机体能量需要非常重要。

另外，饲粮中可被快速消化的糖类也是重要的常量营养素。饲粮中充足的可被快速吸收的D-葡萄糖对确保体内血糖平衡至关重要，以保证正常的生理功能，如为大脑和红细胞提供至关重要的代谢燃料，参与重要的生物合成，刺激胰岛素分泌以维持机体蛋白质的合成，尤其是骨骼肌中的蛋白质合成（Newsholme和Leech，1991；Yang等，2011）。另外，葡萄糖是一种信号营养素，肠道中葡萄糖的快速吸收对细胞中氨基酸转运载体蛋白表达的翻译调控非常重要（Yang等，2008；Roos等，2009；Adeola和Cowieson，2011；Yang等，2011）。因此，葡萄糖的吸收率可影响猪肠腔中氨基酸（蛋白质合成的必需原件）进入门静脉循环的吸收速率，进一步影响机体蛋白质的合成和N的沉积效率（van Der Meulen等，1997；Yang等，2008；Yin等，2010；Drew等，2012）。如果饲粮中没有足够的易消化的碳水化合物提供葡萄糖，快速生长的幼年动物（如断奶仔猪）的机体生长和健康状况以及蛋白质和氨基酸的利用效率将会受到阻碍。

尽管目前已知消化道乳糖酶活性在出生后发育过程中（尤其是断奶过渡期）会降低。但众所周知，结晶乳糖和干乳清粉或奶制品中的乳糖是提高断奶仔猪生长性能和机体氮沉积效率的饲粮碳水化合物的重要来源（Mahan，1992；Nessmith等，1997a；Cromwell等，2008）。Mahan和Newton（1993）进一步研究表明，在断奶后前2周用玉米淀粉替代乳糖或葡萄糖来维持生长性能和机体氮沉积效率是无效的。Nessmith等（1997b）报道，在饲粮中添加梯度水平的结晶乳糖（0、20%和40%），可线性提高早期断奶仔猪第一阶段和第二阶段的生长性能。在一大型合作研究中，Cromwell等（2008）观察到乳糖对断奶仔猪的生长性能有积极影响。而Krause等（1995，1997）没有观察到补充乳糖后回肠和盲肠中黏附乳酸菌计数有所增加，提示了补充乳糖后断奶仔猪生长性能的提高不是因为乳糖的益生素作用。已有研究证明，在饲粮中添加玉米油不能达到添加乳糖改善断奶仔猪生长性能一样的效果（Cera等，1988a），这与脂类不能有效替代断奶仔猪饲粮中易被消化碳水化合物的理论是一致的。

Lackeyram（2012）发现，尽管断奶急剧降低了小肠乳糖酶的最大比活性和整个消化道乳糖酶的消化能力，但整个消化道乳糖酶的消化能力依然很强。并且根据我们的计算，降低的整个消化道乳糖酶消化能力足够维持乳糖的最大消化，Hayhoe等（2012）开展的试验支持了上述结论。因此，断奶仔猪商品饲粮中添加的乳糖对断奶仔猪来说是高度可消化的碳水化合物，可改善机体健康状况，尽可能地降低未被利用的营养素以废物的形式被排泄到体外。

3.5.3 谷类淀粉的利用

已有研究证实，生长育肥猪几乎可以消化大多数谷物中100%的淀粉（Lin等，1987）。然而，断奶仔猪却不能很好地消化大多数谷物中的淀粉。首先，大多数谷物中的淀粉主要存在于半结晶层中（Svihus等，2005），幼龄猪胃酸分泌不足限制了将其完全凝胶化（Svihus等，2005；Wiseman，2006）。蒸煮可有效提高谷物籽实中淀粉的凝胶化（Svihus等，2005）。Pluske等（2002）研究发现，煮熟的大米可提高断奶仔猪的生长性能，减少大肠杆菌病的发生。其次，断奶期间与黏膜消化有关酶（如二糖酶和淀粉消化酶）的活性和消化能力均有显著增强或提高（Fan等，2002），但总唾液和外分泌胰腺α-淀粉酶的消化能力直到生长猪阶段才达到最大值（Fan，2003）。因此，谷物籽实中的淀粉不能被断奶仔猪有效利用。

3.5.4 消化酶和消化能力

Lackeyram（2012）研究发现，断奶仔猪小肠内蔗糖酶、麦芽糖酶和麦芽糖-葡萄糖糖化酶的消化能力远高于哺乳仔猪。尽管蔗糖和麦芽糖是蔗糖酶、异麦芽糖酶和麦芽糖酶及麦芽糖-葡萄糖糖化酶的典型底物，但这些酶与α-淀粉酶互补，可水解淀粉和糊精或/和麦芽糖糊精生成麦芽糖和葡萄糖，同样有利于麦芽糖转化为D-葡萄糖（Nichols等，1998）。据Lackeyram（2012）分析，肠黏膜内的主要双糖酶活性不可能限制淀粉的消化，反而是活性相对较低的黏膜麦芽糖-葡萄糖糖化酶限制了断奶仔猪对谷类籽实中淀粉或/和糊精的最终消化。

新生仔猪和幼龄仔猪具有较高水平的肠刷状缘顶端钠-D-葡萄糖协同转运活性，而这种能力不可能是新生和幼龄哺乳动物消化淀粉和糖过程中的限速步骤（Yang等，2011）。游离D-葡萄糖显然是断奶仔猪饲粮中高度可消化糖类的来源。考虑到蔗糖酶的高消化能力（Lackeyram，2012），蔗糖和麦芽糖都应该成为断奶仔猪的高度可消化的碳水化合物，Naranjo等（2010）的研究结果证实了这一点。

已有明确结论，除了D-葡萄糖外，乳糖、蔗糖、麦芽糖和右旋糖都是可被断奶仔猪快速消化的碳水化合物。饲粮中添加适宜水平的廉价的高度可消化的碳水化合物，对维持断奶仔猪高水平的生长性能、机体氮利用率和健康状况势在必行。而且，这样还可以减少未被利用的营养物质作为废弃物被排出体外，减少甚至消除断奶仔猪饲粮中抗生素的使用，这将大大推动养猪业的可持续发展。

3.5.5 饲粮蛋白质

饲粮蛋白质是另外一个非常重要的常量营养素，尤其是在配制能够改善断奶仔猪生长性能和健康状况的无抗生素添加剂饲粮配方时需要认真考虑（Stein，2002）。由氨肽酶氮等寡肽酶催化的小肠黏膜刷状缘顶端的蛋白质消化对饲粮蛋白质的消化非常重要。Fan等（2002）研究发现，在断奶过渡期的前2周，尽管黏膜氨肽酶-N的最大活性仍然很高，但是它的酶亲和力（例如这种酶的水解效率）相对较低。Pieper等（2012）研究表明，与可发酵纤维相比，可发酵蛋白质可引起仔猪大肠微生态的独特变化。有证据表明，一些潜在的病原菌主要是蛋白质发酵菌（Le等，2005；Bauer等，2006）。梭状芽孢杆菌和大肠杆菌这两种潜在的致病菌亚种可以降解必需氨基酸，包括Trp（Le等，2005）。据记载，甘氨酸可特异性地增加产气荚膜梭菌的增殖和定植，但可减少家禽回肠和盲肠中乳酸菌的数量（Dahiya等，2007）。

饲喂低CP饲粮可减少猪排泄物中产气荚膜梭状芽孢杆菌和大肠杆菌及其分泌的肠毒素（Le等，2005；Bauer等，2006）。Mulder等（2009）研究发现，厚壁菌门尤其是乳酸菌等有益菌群的数量与猪消化道中病原菌的数量存在很强的负相关，并受不同养殖环境的影响。另外两项连续的研究也表明，饲喂低CP饲粮可减少病原菌攻毒的断奶仔猪腹泻的发生（Heo等，2008，2009）。有两种方法可降低残余的CP和氨基酸被用于小肠末端和大肠中的细菌发酵，一种是通过在断奶仔猪饲粮中添加限制性必需晶体氨基酸配制低CP饲粮（Heo等，2008，2009），另一种是使用一些高质量的鱼粉和干乳清粉等高度可消化蛋白源（Stein，2002）。因此，在不使用抗生素条件下，配制低CP饲粮和使用高度可消化蛋白源可有效减少断奶仔猪的肠道疾病、提高其生长性能，也有利于保护环境和养猪生产的可持续性发展。

3.5.6 可供选择的饲料添加剂

目前已研发了一些可供选择的饲料添加剂，并检验了其直接控制断奶仔猪肠腔致病菌的功效。据记载，饲粮添加的喷雾干燥猪或牛血粉、血浆蛋白或免疫球蛋白制剂，可通过直接抑制肠腔病原菌改善断奶仔猪的生长性能和健康状况（Kats等，1994；de Rodas等，1995；Jiang等，2000）。de Rodas等（1995）研究表明，饲粮中添加喷雾干燥猪血浆蛋白粉，可通过提高血液循环中的生长激素和胰岛素浓度改善生长性能。Jiang等（2000）在饲料中添加猪的喷雾干燥血浆蛋白粉，减少了小肠绒毛固有层中固有细胞的密度，从而降低了氨基酸的首过代谢，增加了门静脉中的氨基酸流量。这可能解释了de Rodas等（1995）报道的饲粮添加喷雾干燥猪血浆蛋白为何可以提高血液中生长激素和胰岛素的水平，其原因很可能是通过提高血液循环中的可利用氨基酸实现的。

乳源乳铁蛋白是另外一种具有抗菌活性的生物活性寡肽，同样也可以作为肠道营养生长因子发挥作用；因此，可以有效替代饲用抗生素来改善断奶仔猪的健康状况和生长性能（Tang等，2009）。有几项研究表明，合成乳铁蛋白和重组乳铁蛋白对致病菌具有直接抗菌作用，这是其作用方式之一（Chen等，2006；Yen等，2009）。针对*E. coli* K88^{+}感染制备的卵黄抗体制剂在断奶仔猪中已得到了有效应用（Marquardt等，1999；Pettigrew，2006）。

中草药提取物具有抗菌活性，可以提高肠道健康和生长性能（Kong等，2007；Kim等，2008）。因黏土具有多孔和吸附特性，野生动物可根据其天性或/和后天经验利用黏土对各种抗营养因素和微生物毒素进行解毒（Diamond，1999）。人类通常使用黏土缓解胃痛和止泻，且这些用法由来已久（Diamond，1999）。Song等（2012）研究发现，在饲粮中添加蒙脱石、高岭石和沸石等黏土，虽然没有提高断奶仔猪的生长速率，但缓解了由大肠杆菌引起的断奶仔猪腹泻。这显然是黏土吸附了大肠杆菌及其毒素的结果。

已有大量研究，测定了饲粮中添加各种有机酸对改善断奶仔猪生长性能和健康状况的效果（Gabert和Sauer，1994；Roth和Kirchgessner，1998；Partenan和Mroz，1999；Borysenko，2002）。有机酸主要包括乙酸、丙酸、丁酸、己二酸、苯甲酸、柠檬酸、蚁酸、反丁烯二酸、乳酸、苹果酸、山梨酸、琥珀酸和酒石酸及其盐（Risley等，1992；Roth和Kirchgessner，1998；Kluge等，2006；Walsh等，2007，2012a，2012b）。然而在不同的研究中，饲粮中添加这些有机酸的效果通常不一致，可能是各个研究中使用的有机酸的理化特性（Cherrington等，1991）、吸收和代谢速率、滞留时间以及添加水平不同引起的。然而，在断奶仔猪饲粮中添加有机酸似乎都产生了一定的有益作用。

综上所述，在饲料中添加喷雾干燥猪和牛血浆粉和血浆蛋白粉、乳铁蛋白、卵黄抗体、黏土以及有机酸可以替代抗菌剂，控制肠道病原菌及其毒素，从而改善断奶仔猪生长性能和健康状况。因此，所有这些替代品将有利于未来养猪生产的可持续性发展。

3.5.7 谷氨酰胺

肠腔内来源于饲粮和内源的氨基酸可作为各种生物合成的代谢燃料和前体物质，为肠道黏膜生长提供营养（Burrin和Reeds，1997；Yang等，2008）。肠腔内的氨基酸，尤其是谷氨酰胺和亮氨酸等中性氨基酸，在调节细胞基因表达和其他细胞事件过程中是必不可少的，通过几种已知的级联反应，如mTOR信号通路，增强肠黏膜的增生和肥大（Yang等，2008；Wu等，2011）。已有证据表明，断奶可通过降低氨肽酶-N的亲和力（Fan等，2002）、最大活性和整个肠道中该酶的消化能力（Lackeyram，2012）而降低黏膜的蛋白质消化效率，从而限制了用于断奶仔猪肠黏膜生长的肠腔内游离氨基酸的利用率。然而，小肠刷状缘顶端的Na^+-AA交换蛋白和Na^+-AA协同转运体的表达不受断奶的影响，表明刷状缘顶端氨基酸转运体的功能在断奶过渡期得到了有效的保护（Teng等，2012；Wang等，2012b）。

有研究显示，在饲粮中添加晶体谷氨酰胺，可改善断奶仔猪的生长性能、肠黏膜形态、基因表达、屏障功能和促炎反应（Wu等，1996；Wang等，2008a；Ewaschuk等，2011；Zhong等，2011），节约了体外培养细菌对非必需氨基酸和必需氨基酸的利用（Dai等，2012）。饲粮添加L-丙氨酰-L-谷氨酰胺和甘氨酰-谷氨酰胺（谷氨酰胺的稳定来源），可改善应激情况下幼龄猪的肠上皮细胞的状况或/和机体免疫情况（Haynes等，2009；Jiang等，2009）。因此，饲粮添加晶体谷氨酰胺和谷氨酰胺小肽可有效改善肠道健康、免疫状况和生长性能。

3.5.8 表皮生长因子和β-葡聚糖

肠道营养性生长因子和天然免疫调节剂作为可供选择的饲料添加剂，是已经研发的且经过功效验证的解决断奶仔猪肠黏膜萎缩、生长和健康应激问题的有效饲粮调控措施。除了前面提到的直接抗菌活性，乳铁蛋白也是一种能够促进肠道生长的因子（Wu等，2010）。重组乳铁蛋白目前可通过毕赤酵母和转基因大米生产（Chen等，2004；Lee等，2010）。乳铁蛋白可通过乳铁蛋白特有受体介导的细胞内吞作用被吸收进入肠上皮细胞，乳铁蛋白受体存在于所有日龄猪的刷状缘膜上（Liao等，2007）。Nielsen等（2010）研究表明，小肠固有层中的

T细胞也是乳铁蛋白的效应细胞。有研究显示，在饲粮中添加重组乳铁蛋白可改善断奶仔猪的生长性能、肠道形态和免疫功能（Wang等，2006；Shan等，2007；Tang等，2009）。另外需要注意的是，Kang等（2010）研究发现，口服在食品级细菌乳酸乳球菌中表达的重组表皮生长因子，可改善断奶仔猪的小肠生长和黏膜形态。

β-葡聚糖是一类化学异构的非消化性多糖，是细菌、酵母、真菌以及大麦和燕麦等一些谷物的细胞壁结构成分（Li等，2006；Volman等，2008；Barsanti等，2011）。β-葡聚糖具有高度可发酵性，是黏性可溶性纤维的组成成分，当先天性免疫识别受体Dectin-1被确认为β-葡聚糖的受体时，β-葡聚糖也被认为是一种免疫调节剂（Willment等，2001）。Dectin-1在上皮内淋巴细胞、中性粒细胞、巨噬细胞和树状突细胞等肠道相关免疫系统的各种细胞中均有表达（Goodridge等，2009；Drummond和Brown，2011；Esteban等，2011）。然而，Dectin-1在肠上皮细胞中却没有表达（Volman等，2010）。在所测定的28种NSP产品中，发现微生物和植物来源的β-葡聚糖以及半乳甘露聚糖瓜尔胶可调节由TLR4配体LPS诱导的树状突细胞的细胞因子表达模式（Wismar等，2011）。Chanput等（2012）的研究发现，具有不同分子量和不同分子结构的β-葡聚糖，对巨噬细胞具有不同程度的免疫调节特性。然而，在饲粮中添加低剂量的β-葡聚糖（如0.025%~0.1%），却不能改善断奶仔猪的生长性能和健康状况（Dritz等，1995；Hahn等，2006；Gallois等，2009）。Ewaschuk等（2012）研究表明，给断奶仔猪饲喂富含大麦β-葡聚糖的饲粮，可增强全身免疫功能，但同时也增加了大肠杆菌在肠上皮细胞上的黏附和肠道的通透性。

因此，关于饲粮中添加β-葡聚糖具有正向免疫调节作用的结论尚不能确定。另外，在饲粮中添加重组表皮生长因子或乳铁蛋白替代抗生素，可有效改善断奶仔猪的肠黏膜形态、免疫状况和生长性能。

3.5.9 ω-3脂肪酸

众所周知，断奶仔猪会发生炎性反应（Fan，2003），而ω-3多不饱和脂肪酸具有抗炎效果（Simopoulous，2002）。在母猪饲粮中添加ω-3多不饱和脂肪酸，可改善仔猪断奶过渡期的生长性能、肠道形态和免疫状态（Boudry等，2009；Leonard等，2011a）。Jiang等（1998）观察到，多不饱和脂肪酸可通过上调紧密连接蛋白的表达减少内皮细胞间的通透性。Liu等（2003）研究发现，鱼油多不饱和脂肪酸可通过下调淋巴细胞活性而发挥抗炎作用，这可能是通过影响断奶仔猪细胞内信号通路实现的。在过去的十年里，除了ω-3脂肪酸调节膜的流动性和结构这些基本作用外，ω-3脂肪酸调节生物功能的研究也取得了引人注目的进展。二十碳五烯酸（EPA）和二十二碳六烯酸（DHA）这两种长链ω-3多不饱和脂肪酸，可在酶催化下转化为多种具有生物活性的内分泌物，EPA可使包括免疫细胞在内的多种细胞产生消退素（resolvin）-E1系列、DHA可产生消退素-D系列和保护素（Calder，2012；Zhang和Spite，2012）。消退素和保护素是高效的细胞调节剂，具有抗炎作用，可通过细胞膜表面和细胞内受体缓解炎性反应（Calder，2012）。

Arita等（2005）报道，阿司匹林可加快EPA向消退素-E1的转化，EPA与阿司匹林联合使用可有效治疗化学试剂诱导的结肠炎。Campbell等（2010）研究表明，消退素E1主要通过增强肠上皮细胞碱性磷酸酶的基因表达和酶活，加快鼠的葡聚糖硫酸钠模型结肠炎的消退。可想而知，饲粮添加ω-3多不饱和脂肪酸缓解与断奶有关的小肠和大肠炎症以及健康状况的效果，都可能是通过上调肠道碱性磷酸酶的表达实现的。因此，饲粮添加ω-3多不饱和脂肪酸可有效替代抗生素，改善断奶仔猪的肠黏膜免疫功能和生长性能。

3.5.10 益生菌和益生元

在饲粮中添加有效的益生菌和益生元，是为了通过定植有益菌群而抑制有害菌群，从而改善宿主健康（Gibson和Roberfroid，1995），最终减少或消除抗菌药物的使用。此观点也得到了Mulder等（2009）研究的进一步支持。该研究发现，对断奶仔猪有益的厚壁菌门细菌尤其是乳酸菌数量与有害菌群数量之间存在很强的负相关，并受不同养殖环境的影响。尽管在文献中由于所测产品和试验条件的不同得到的试验结果不尽一致（Callaway等，2008），但是仍有很多报道证明，在饲粮中添加益生菌可控制断奶仔猪的病原菌群，改善肠道局部免疫和生长性能（vanbelle等，1990；Callaway等，2008；Lessard等，2009；Choi等，2011）。

根据Gibson和Roberfroid（1995）以及Gibson等（2004）提出的定义和主要标准，益生元的主要成分是饲粮纤维。然而，并不是所有的饲粮纤维成分都可以作为益生元。因为有些纤维成分不能增加有益菌群，而还有些纤维成分可同时引起有益菌和有害菌的增加。微生物宏基因组方面的最新研究，已阐明了益生元作用的定义和模式（Preidis和Versalovic，2009）。1995年，Gibson和Roberfroid（1995）将低聚果糖列为唯一的益生元。随后益生元的品种大大增加，包括菊苣菊糖、抗性淀粉、半纤维素类成分等其他NSP以及各种不可消化的低聚糖（如低聚半乳糖、葡萄糖低聚糖、甘露低聚糖、异麦芽糖低聚糖、反式乳低聚糖和低聚木糖）、低聚糖（如油料种子粕中提取的棉子糖和水苏糖）、乳果糖、乳蔗糖和拉克替醇等糖醇（Flickinger等，2003；Gibson等，2004；Gallois等，2009；Aachary和Prapulla，2011）。

在饲粮中添加低聚甘露糖可改善断奶仔猪的免疫功能和生长性能（Rozeboom等，2005；Che等，2012）。在饲粮中添加含有海带多糖和墨角藻聚糖的海藻提取物，可减少断奶仔猪结肠中的大肠杆菌数量，提高生长性能（Leonard等，2011b）。在饲粮中添加燕麦壳、马铃薯抗性淀粉和菊粉可改善断奶仔猪的肠道健康或/和生长性能（Kim等，2008；Bhandari等，2009；Hansen等，2010，2011）。然而，在饲粮中添加黏性可溶性纤维——β-葡聚糖、果胶和羧甲基纤维素，可增加肠道内容物的黏度、大肠杆菌在肠道中的定植以及仔猪断奶后发生大肠杆菌病的可能，这就意味着这些黏性纤维成分不是有效的益生元（McDonald等，1999；Hopwood等，2004；Montagne等，2004）。尽管已经证实，一些不可消化的低聚糖和多糖对断奶仔猪具有益生元的作用，但前面提到的黏性可溶性纤维成分并不是益生元。

在过去的十年里，基于细菌对丁酸的生物合成，人们对益生元发挥作用的生物学机制有了很好的认识（Ramsay等，2006；Louis和Flint，2009；Louis等，2010；Metzler-Zebeli等，2010）。根据关键标记酶丁酰-CoA：乙酰CoA-转移酶基因的分布，一些益生元和底物被证实在肠道内可有效诱导有益产丁酸菌。这些益生元包括乳酸、菊糖、抗性淀粉、低聚果糖、低聚木糖以及燕麦麸和壳中的半纤维素阿拉伯木聚糖（Bach Knudsen等，1993；Duncan等，2004；Scott等，2011；Lecerf等，2012）。另外，据记载，丁酸对宿主有很多直接的积极作用。丁酸是肠上皮细胞主要的首选代谢燃料之一，从而可影响细胞的成熟和周转率（Wächtershäuser和Stein，2000；Roy等，2006）。丁酸是一种信号分子，可减少肠上皮隐窝细胞的凋亡，增加肠上皮细胞的分化以及黏蛋白2和肠道碱性磷酸酶等关键功能基因的表达（Hinnebusch等，2003；Kim和Milner，2007；Burger-van Paassen等，2009）。而且，肠道是最大的次级免疫器官，丁酸可直接影响肠道免疫细胞的功能（Schley和Field，2002；Wang等，2008b；Maa等，2010）。

Malo等（2010）研究表明，肠道碱性磷酸酶维持着肠道微生物区系的正常平衡。因此，益生元的效果可以通过维持健康的肠道微生物区系得到进一步的介导，这可通过丁酸加强肠道刷状缘顶端碱性磷酸酶的表达而实现。然而，还需要更多的体内研究来支持该观点，饲粮益生元对断奶仔猪肠道黏膜的保护作用可能通过丁酸参与的各种调控作用而被间接介导。考虑到可供选择的益生元的种类众多，因此在饲粮中添加益生元的经济有效的方式是用来替代饲用抗生素，更好地改善断奶猪的营养和健康。

◆ 3.6 纤维、电解质平衡和酸化剂

3.6.1 氨的排放

在集约化养猪生产中，生长阶段和育肥阶段粪便和尿液的总排泄量非常大，是氨排放和排泄物臭味引起的主要环境问题。生长育肥猪通常采食玉米-豆粕型饲粮，饲粮中的总纤维含量相对较低，限制了大肠中菌体蛋白的合成。而且，饲粮中主要的饲料原料和盐提供的钾、氯和钠等主要电解质的存留效率很低（NRC，1998，2012）。这种过量添加和电解质不平衡，使得过量的电解质从尿液中排出以及尿液和粪污的pH较高。

生长育肥猪对饲粮纤维具有相对较强的发酵能力，通过纤维发酵可获得高达30%的维持代谢能需要量（Varel和Yen，1997）。已有研究表明，在饲粮中添加高水平、易获得、低成本的外源纤维成分，同时维持较好的电解质平衡，可有效减少生长育肥猪粪污中的氨排放。这主要是通过将血液中的尿素从尿液排泄转移到用于大肠微生物的生长，利用高含量的VFA酸化粪便实现（Sutton等，1999）。实用的低成本的外源纤维源包括大麦、燕麦、大豆壳、燕麦麸和小麦麸、小麦次粉、双低菜粕、菜籽粕、螺旋压榨椰子粕、甜菜碱和DDGS。DDGS被认为是容易获得的一种纤维源。Stein和Shurson（2009）报道，DDGS中的总纤维、中性洗涤纤维和酸性洗涤纤维的含量分别为42.1%、25.3%和

9.9%，约是玉米中含量的3倍多。

值得一提的是，在生长育肥猪饲粮中添加相对高水平的这些外源纤维物质是为了给结肠提供足够的能量底物，尽量增加大肠中微生物的生长和微生物蛋白的合成，而不是诱发益生元作用。尽管，富含果胶的甜菜碱容易发酵，但前面提到的其他纤维源由于含有木质素-半纤维素-纤维素混合物而不溶于水，这限制了其可发酵性。纤维素是一种准结晶体聚合结构，在消化道环境中降解木质纤维素的微生物纤维素酶很不容易接触和水解纤维素（Wilson，2011）。一些研究表明，在生长育肥猪饲粮中添加甜菜果浆、大豆壳和压榨椰子粕这些外源纤维成分，可降低尿氮排泄、尿和粪污的pH以及氨的排放（Canh等，1998c；Mroz等，2000；Wang等，2009；Jarret等，2011）。Canh等（1997，1998b）的研究也显示，通过平衡电解质减少饲粮中电解质的总量，可有效降低生长育肥猪的尿和粪污pH，从而减少粪污中氨的排放。

考虑到一些对添加水平的现有报道（Stein和Shurson，2009；Urriola等2010；Jarret等，2011），最重要的是要确保饲粮添加纤维成分不会对生长育肥猪的生长性能产生负面影响。另外，饲粮补充纤维对粪污氨排放的影响可能会随纤维源的不同而异。总之，在饲粮中补充低成本的外源纤维物质，同时配以低水平的平衡电解质，可有效减少生长育肥猪的氨排放。

全血的pH受严格的稳态调节机制调控的，这通过肾脏分泌过量的质子进入血液循环而实现。借助这些生理机制，在饲粮中补充外源有机酸或其盐类可以降低养猪生产中尿液和粪污的pH。Mroz等（2000）观察到，在饲粮中补充苯甲酸钙和有机酸，可提高营养物质的消化率，尿液的pH降低了1.6，也可能减少了粪污的氨排放。2001年van Kempen报道，在饲粮中添加己二酸可使生长猪尿液的pH降低2.2，且粪污中氨的排放减少25%。因此，在饲粮中补充有机酸或其盐类可有效降低生长育肥猪尿液的pH和氨排放。

3.6.2 粪臭素

尽管小肠末端和大肠中Trp的细菌降解导致了主要异味化合物粪臭素的生物合成，但Trp来源及其影响因素可能是调控猪排泄物中粪臭素含量的主要限制因素。根据猪NRC（1998，2012），Trp的需要量和大多数植物饲料原料中的供应量均相对较低，因此饲粮来源的不可消化的Trp可能不是猪肠道中粪臭素生物合成的主要来源。Claus及其合作者（Claus等，1994；Claus和Raab，1999；Mentschel和Claus，2003；Claus等，2007）的研究证明，内源性蛋白质，包括结肠细胞增殖和分化造成的上皮细胞脱落损失，是粪臭素生物合成所需Trp的主要来源。Claus等（2003）的研究也显示，在饲粮中添加土豆抗性淀粉，可降低盲肠pH和结肠细胞凋亡，但增加了盲肠食糜中丁酸的含量，导致血液循环、脂肪组织和粪便中粪臭素含量降低。Willig等（2005）和Lösel等（2006）也得出了相似的结果。而Rideout等（2004）的研究表明，当饲粮中含有5%菊粉时，生长猪粪便中粪臭素的含量减少约50%。

据报道，菊粉、抗性淀粉、低聚果糖、低聚木糖和燕麦麸皮和燕麦壳来源的半纤维素阿拉伯木聚糖是通过诱导肠道中有益菌产丁酸菌生成丁酸的有效底物（Scott等，2011；Lecerf等，2012）。产丁酸菌主要是有益的革兰氏阳性硬壁菌门细菌（Louis和Flint，2009）。然而，饲粮添加低聚木糖以及燕麦麸皮和燕麦壳来源的半纤维素阿拉伯木聚糖抑制生长育肥猪粪臭素生物合成的效果尚有待进一步验证。另外，黏性可溶性纤维成分（如果胶、瓜尔胶和β-葡聚糖等）不是抑制消化道粪臭素生物合成的有效产丁酸底物。壳聚糖是另外一种类型的商品化的黏性可溶性纤维，可降低盲肠中乳酸菌数量和丁酸含量，增加猪排泄物的臭味排放（O'Shea等，2011b）。

因此，尽管黏性可溶性纤维成分不能有效降低主要粪臭味化合物——粪臭素，在饲粮中添加非黏性产丁酸可溶性纤维成分却可有效降低粪便中粪臭素的生物合成和排泄物中病原菌的数量，而这两者正是异味猪肉产生的根源。由于需要在饲粮中添加的水平较高，需要进一步研究来降低饲粮中这些非黏性产丁酸可溶性纤维成分的成本和有效添加水平，从而降低生长育肥猪对环境的影响。

◆ 3.7 管理措施和饲料加工技术

3.7.1 氮的损失

大规模养猪场的液体排泄物储存设施中，以NH_3和其他挥发性异味化合物形式的氮排放，给环境带来一些主要污染问题，如对当地的异味影响、酸雨和对地表和地下水系统的硝酸盐污染（Burton和Beauchamp，1986；Canh等，

1998a）。因此，理想的养猪生产规模应该为小型或中型，使用饲养床-刮除固体粪污方式替代水冲刷清洁系统。这种模式主要适用于劳动成本较低且水资源有限的发展中国家，可以减少集约化养猪生产中NH_3的排放损失。而且，木屑和秸秆覆盖物等多孔生物材料对过滤猪舍空气和降低猪舍和排泄物储存设施排放均有效（Regmi等，2007；Blanes-Vidal等，2009；Chen等，2009）。Lavoie等（2009）研究表明，在猪舍内对液体和固体排泄物进行分离，也可有效控制气味排放、改善猪舍空气质量。

虽然使用硫酸直接处理粪污可有效控制NH_3排放，同时也可为作物生长提供硫酸盐，但硫酸盐经微生物转化生成的挥发性硫化物是一个主要的问题（Eriksen等，2008）。同样，使用磷酸控制粪污pH和NH_3排放，处理后的排泄物肥田后将导致更多的磷径流。另外，Jayasundara等（2010）研究表明，硝酸盐淋溶损失来源于猪排泄物中氮的硝化作用。猪排泄物中的氮经脱氮以N_2O形式形成的气体排放损失和硝酸盐淋溶损失，是肥田后排泄氮损失的主要机制。这提示农田对排泄物中氮的负载量需要优化。再者，无论何种土壤类型，秋天的粪氮损失大于春天（Jayasundara等，2010）。因此，猪舍设计、排泄物处理和农田施肥均需要优化，以尽量减少粪氮的损失。

3.7.2 动物福利与健康

在常规的养猪生产管理过程中，人与动物不恰当的接触，会引起猪的应激，随之产生大量的皮质醇（Hemsworth和Barnett，2000；Fan等，2006；Bregendahl等，2008）。据报道，皮质醇的大量释放可通过影响相关翻译起始因子（Liu等，2004）而减少骨骼肌的蛋白质合成（Fan等，2006；Bregendahl等，2008）。因此，在常规的养猪生产管理过程中，既能够改善动物福利、尽量减少猪的应激的有效措施，也有利于在集约化养猪生产中维持正常水平的骨骼肌蛋白质生长率和氮利用率，这对环境也会产生积极的影响。

在以下两种主要的管理条件下，如果是由于致病菌的存在导致猪的健康状况不佳，通常与炎症和感染有关。第一，研究表明，在养猪生产实践中，新生仔猪生活早期接触各种环境栖生微生物群落对建立正常、健康的微生物区系至关重要（Mulder等，2011；Schmidt等，2011）。因此，尽管在养猪生产中维持良好的卫生条件是没有必要的，成本也是很高的，但是控制不卫生的条件以尽量减少致病菌的传播却非常关键。第二，维持严格高水平的生物安全，对减少致病菌和病毒在集约化养殖体系中传播是必需的。健康状况不良不仅可降低猪的生长性能，也会降低机体的氮利用效率，很大程度上是由于减少了骨骼肌蛋白质合成能力，同时通过促炎细胞因子信号通路增加了内脏器官蛋白质的合成速率（Williams等，1997；Orellana等，2002；Fan等，2006）。另外，已证明细菌和病毒感染均会通过增加猪骨骼肌促炎细胞因子和肌肉生长抑制素（骨骼肌增生和肥大生长的关键负调节因子）的表达（Escobar等，2004），降低蛋白质的沉积率（Gonzalez-Cadavid和Bhasin，2004）。因此，改善卫生条件和猪健康状况的有效措施，既有助于维持养猪生产中正常水平的氮利用率，也可给环境带来正面影响。

3.7.3 饲料加工技术

饲料原料尤其是副产品中营养素的消化利用率以及随之带来的猪排泄物对环境的影响，可通过饲料加工技术得到进一步改善。饲料结构和饲料中存在的NSP、植酸盐、单宁酸、抗原蛋白以及胰蛋白酶和胰凝乳蛋白酶抑制剂等抗营养因子，限制了单胃动物对饲粮营养素的消化率（Bondi和Alumot，1987）。其中的一些抗营养因子的影响，可以通过化学或物理的传统饲料加工技术（如利用溶剂萃取去除单宁酸，利用适当的热处理方法使蛋白酶抑制剂和抗原蛋白变性）和生物处理方法去除（Lacki等，1999）。

Stein和Bohlke（2007）研究发现，经高温热挤压后的豌豆，生长猪对其能量、淀粉、CP和氨基酸的消化率提高。处理条件可影响乙醚浸提物的残余量，冷处理的油菜饼粕含有较多的生长猪可消化的能量，但同样含有较高水平的残余硫苷（Seneviratne等，2011）。Yáñez等（2011）研究表明，DDGS粉碎后，生长猪对其中大部分可测定氨基酸的回肠消化率均增加。对断奶仔猪而言，其胃酸分泌能力有限，摄入的谷物淀粉的糊化作用有限，从而阻碍了淀粉的消化。而微粉化可提高断奶仔猪对裸壳大麦中淀粉和其他营养物质的消化率（Huang等，1998）。采用带有充足水分的热处理，如蒸煮法，可增加淀粉的糊化和消化率以及断奶仔猪的肠道健康和生长性能（Svihus等，2005；Pluske等，2007）。因此，高效的饲料加工技术可有效改善饲粮营养物质的利用率，减少排泄物对环境的负面影响。

◆ 3.8 生物技术措施

3.8.1 碳、氮和磷的利用

生物技术为提高C、N和P的利用率提供了最终的解决方案。与重要营养素相关的可持续发展问题，可以通过改变猪的消化和消化后营养素的利用和代谢途径来解决，同样通过生物技术工程可不同程度上实现农作物的重要性状。

通过在饲粮中添加利用DNA重组技术生产的各种外源酶，已显著提高了饲粮营养素的利用率（Rodriguez等，2000；Cowieson等，2006）。正如上文所述，由于植酸磷相对简单的结构和仅需要单一植酸酶的水解，植酸磷的利用率问题已通过向饲粮中添加大量由酵母、真菌、细菌和农作物生产的外源植酸酶得到了很好的解决。然而，在集约化养猪生产中，仍存在植酸磷利用和磷污染两个关键问题。首先，饲料制粒过程中外源植酸酶的热稳定性仍然是一个挑战性问题（Kim等，2008）。在这种情况下，值得一提的是，转植酸酶基因的猪即“生态猪”的商业化，才是最终的解决办法（Golovan等，2001）。其次，大多数植物饲料原料中总磷的含量过高，其中包括植酸磷。因此，正如Fan等（2008b）所述，在使用植酸酶的同时，仍需努力开发低磷作物，以尽量减少排泄物中磷的过量排放。对CP和淀粉的消化利用来说，小肠黏膜上的寡肽酶和麦芽糖-葡糖淀粉酶对断奶仔猪很可能是限制因素。因此，通过DNA重组技术利用非致病性大肠杆菌和毕赤酵母菌生产这些外源酶，并作为饲粮外源酶添加剂饲喂断奶仔猪，似乎是有前景的。

木质纤维素是植物细胞壁物质中含量最丰富的NSP成分。纤维素是一种准结晶体聚合结构，很难被环境中分离的目前已知的大多数微生物纤维素酶水解（Wilson，2011）。因此，发现和工业生产新一代的微生物木质纤维素降解酶，包括木质素降解酶、半纤维素酶和有效的纤维素酶等外源酶，是进一步提高养猪生产中C、N消化利用率的主要挑战。与此同时，宏基因组测序、编码和表达新型微生物酶基因方面的最新研究进展，为在不久的将来实现这些迫切需要的发明提供了机会（Hess等，2011；Wang等，2012a）。虽然在商业化和公众接受方面还存在一定的问题，另一种长远的生物技术方法就是通过开发转基因猪，在唾液腺中表达这些新型的酶，利用唾液分泌作为载体（Golovan等，2001；Yin等，2006；Yan等，2007）。因此，生物技术有希望进一步提高饲粮营养素中C、N、P的消化利用率，并最终解决集约化养猪生产中主要的可持续性问题。

3.8.2 代谢调节物

骨骼肌蛋白质的合成能力相对较低，却有较高的蛋白质降解率，这是养猪生产中氮和能量利用率低的主要代谢原因（Fan等，2008a）。早在过去的三十年里，常用猪生长激素等肌肉代谢调节物（如猪的生长激素和莱克多巴胺等人工合成的β-肾上腺素能兴奋剂等）调节猪的生长和生产代谢途径，进而提高生产效率（Bergen和Merkel，1991；Bell等，1998；Etherton，1999）。

猪的生长激素在某些国家已被批准用于养猪生产。在生长猪阶段使用猪的外源重组生长激素（每天每千克体重的使用量是100 ~ 150 μg）2 ~ 3个月，可提高瘦肉量和饲料转化效率（Krick等，1993；Etherton，2004），部分原因是由于脂肪组织沉积和采食量的减少（Wang等，1999；Etherton，2004）。而且，外源生长激素可提高CP和氨基酸的沉积率，但是这种效应与CP和氨基酸维持需要量的增加有关（Krick等，1993）。使用猪的外源重组生长激素对瘦肉产量和氮利用率产生的积极作用，主要是通过蛋白质合成途径效应器上调了生长激素-胰岛素-IGF轴，从而提高了骨骼肌的蛋白质合成速率（Wester等，1998；Lewis等，2000；Bush等，2003；Davis等，2004；Yang等，2008）。另外，生长激素的使用降低了肠-肝尿素合成通路中的酶活（Bush等，2002）。但关于使用外源生长激素对肌肉蛋白质降解的影响，研究结果并不一致（vann等，2000；Davis等，2004）。

另外，据Krick等（1993）报道，使用生长激素引起的CP和氨基酸维持需求量的增加，可能是因为内脏器官的蛋白质合成能力增强从而引起胃肠内源氮损失增加造成的（Wester等，1998；Bush等，2003；Davis等，2004）。外源生长激素对内脏的肥大作用与肝脏中核糖体数量和蛋白质合成能力的增加有关（Bush等，2003）。因此可以预知，在养猪生产中，植入外源生长激素是提高瘦肉量和机体氮沉积率的有效生物技术措施。但仍需要进一步定量研究，在猪营养与生产中使用外源生长激素对机体氮利用率和污染问题的影响。

有些国家将合成的β-肾上腺素能兴奋剂，莱克多巴胺盐酸盐（商品名瘦肉精），作为代谢调节物在后期育肥阶段使用（20 ppm）。瘦肉精可以提高饲料转化效率，在很大程度上是由于通过减少脂肪生成、增加脂肪分解，从而使猪育肥后期阶段的脂肪沉积减少实现的（Liu和Mills，1990；Peterla和Scanes，

1990；Mills等，2003）。另外，瘦肉精还可通过调节脂肪和肌肉组织中的线粒体解偶联蛋白使热增耗增加，从而减少脂肪的沉积（Ramsay和Richards，2007）。有研究报道，瘦肉精可提高后期育肥阶段的瘦肉沉积（Yen等，1990；Bark等，1992）。更重要的是，瘦肉精对瘦肉沉积率的增加，是通过选择地增强骨骼肌的蛋白质合成率实现的（Bergen等，1989；Anderson等，1990；Helferich等，1990；Adeola等，1992；Depreux等，2002）。Yen等（1990）研究表明，使用瘦肉精后，一些内脏器官的鲜重降低，提示粪便中内源氮的损失减少。但是我们还未见到关于瘦肉精如何影响猪内脏器官蛋白质合成率和粪便中内源氮损失的任何报道。关于瘦肉精对骨骼肌蛋白降解率的影响也知之甚少。Bergen等（1989）估测出，对照组和瘦肉精组每天的蛋白质降解百分率分别是3.4%和4.9%，然而，氮没有提供与这些估测值相关的变异性。鉴于饲喂瘦肉精可改善瘦肉沉积的事实，那么有理由相信机体的氮利用率也会提高。因此，下一步应该研究瘦肉精对后期育肥猪氮利用率的影响。

3.8.3 转基因策略

在实验室动物上已得到验证的动物生物技术，如通过转基因法下调肌肉生长的负调控因子等，有可能会进一步提高养猪生产中的瘦肉沉积和机体氮利用率。目前已知，肌肉生长抑制素是骨骼肌生长的关键负调节因子（Gonzalez-Cadavid和Bhasin，2004），那么通过转基因技术下调该基因将会增加肌肉蛋白的生长。Yang等（2001）以小鼠为研究对象，发现通过转基因技术过表达肌肉生长抑制素的前结构域，可大大增加肌肉的质量。结节性硬化复合体（TSC），包括TSC1和TSC2，是肿瘤抑制基因，其蛋白产物可抑制其下游调节因子哺乳动物类雷帕霉素靶蛋白（mTOR，是所有蛋白质合成的正向激酶和调节器）（Wan等，2006）。转基因小鼠表达人类的TSC1可形成肌肉萎缩（Wan等，2006）。因此，下调TSC1和TSC2基因可能是提高肌肉生长的另一种转基因技术。细胞内激酶S6K1作为mTOR的关键底物，是营养素和胰岛素/IGF介导细胞生长信号途径的重要组成成分。S6K1基因缺失小鼠的肌肉生长受到了抑制（Aguilar等，2007）。因此，随着对肌肉生长调节分子机制的认识越来越深入，将研发出有效的转基因方法用于促进肌肉生长。同时，需要进一步研究转基因动物的肌肉质量和肌肉蛋白质沉积率是否增加，机体的氮利用率是否同时提高。

尽管氨基酸的分解代谢是保障必需生理功能的一个正常的生理过程，但是氨基酸的过度分解代谢将是导致氮利用率低下的最终原因。Cleveland等（2008）利用小鼠的肝细胞系研究发现，利用RNA干扰技术可以有效下调酵母氨酸途径的赖氨酸分解代谢，从而减少细胞中的赖氨酸分解代谢损失和赖氨酸的需要量。该研究表明，氨基酸分解代谢途径的基因工程是提高养猪生产中机体N利用率的一种有效措施。

4 结 语

在局部乃至全球范围内，集约化养猪生产带来的主要的可持续性环境问题包括主要温室气体、酸性气体和产生异味的氨的排放，硝酸盐和Cr对饮用水资源的污染，排泄物中过量的磷引起的地表水的富营养化，病原微生物传播引起的食品安全和公共健康问题，危及传染性疾病治疗的抗生素耐药的出现，重金属矿物质对食物链和生态系统的污染，以及排泄物的恶臭味和空气污染物对当地社区的影响。

在饲粮中添加外源酶可提高能量和营养素的消化利用，减少胃肠道的内源损失。配制低CP、Ca、P、S和微量矿物元素添加剂的猪饲粮（以真可消化营养素为基础），可减少粪便的排泄量，缓解一些环境问题。同样，在饲粮中添加适宜水平的非黏性可溶性纤维和其他饲料添加剂替代饲用抗生素，可尽量减少抗生素的耐药性，有效减少病原微生物以及挥发性产恶臭味化合物在粪便中的排泄。

综合管理和饲料加工技术使我们更容易解决这些环境问题。因此，应用全面的饲粮和非营养措施可以有效缓解全球集约化养猪生产所面临的环境问题。另外，长远的生物技术措施可以提供最终的解决方案，即完全解决集约化养猪生产中的能量和物质转化效率低的根本生物学问题，从而使环境污染最小化、生产者利润最大化。

作者：Ming Z. Fan
译者：王自蕊

参考文献

Aachary, A. A., and S. G. Prapulla. 2011. Xylooligosaccharides (XOS) as an emerging prebiotics: Microbial synthesis, utilization, structural characterization, bioactive properties, and applications. Compreh. Rev. Food Sci. Food Safety 10:2–16.

Adeola, O., and A. J. Cowieson. 2011. Opportunities and challenges in using exogenous enzymes to improve nonruminant animal production. J. Anim. Sci. 89:3189–3218.

Adeola, O., B.V. Lawrence, A. L. Sutton, and T. R. Cline. 1995. Phytase-induced changes in mineral utilization in zinc-supplemented diets for pigs. J. Anim. Sci. 73:3384–3391.

Adeola, O., R. O. Ball, and L. G. Young. 1992. Porcine skeletal muscle myofibrillar protein synthesis is stimulated by ractopamine. J. Nutr. 122:488–495.

Aguilar, V., S. Alliouachene, A. Sotiropoulos, A. Sobering, Y. Athea, F. Djouadi, S. Miraux, et al. 2007. S6 kinase deletion suppresses muscle growth adaptations to nutrient availability by activating AMP kinase. Cell Metabolism 5:476–487.

Ajakaiye, A., M. Z. Fan, C.W. Forsberg, J. P. Phillips, S. Golovan, R. G. Meidinger, T. Archbold, and R. R. Hacker. 2004. Digestion and absorption of calcium associated with soybean meal is completed by the end of the small intestine in the transgenic phytase Enviropig^{TM}. FASEB J. 18:A526. (Abstr.)

Allen, H. K., T. Looft, D. O. Bayles, S. Humphrey, U. Y. Levine, D. Alt, and T. B. Stanton. 2011. Antibiotics in feed induce prophages in swine fecal microbiomes. MBio: e00260–11.

Almeida, F. N., and H. H. Stein. 2010. Performance and phosphorus balance of pigs fed diets formulated on the basis of values for standardized total tract digestibility of phosphorus. J. Anim. Sci. 88:2968–2977.

Amachawadi, R. G., N. W. Shelton, M. E. Jacob, X. Shi, S. K. Narayanan, L. Zurek, S. S. Dritz, et al. 2010. Occurrence of tcarB, a transferable copper resistance gene, in fecal enterococci of swine. Foodborne Pathog. Dis. 7:1089–1097.

American Academy of Microbiology. 1999. Antimicrobial Resistance: An Ecological Perspective. American Academy of Microbiology, Washington, DC.

Aminov, R. I., and R. I. Mackie. 2007. Evolution and ecology of antibiotic resistance genes. FEMS Microbiol. Lett. 271: 147–161.

Anderson, P. T.,W.G.Helferich, L. C. Parkhill, R. A. Merkel, andW. G. Bergen. 1990. Ractopamine increases total and myofibrillar protein synthesis in cultured rat myotubes. J. Nutr.

120:1677–1683.

Angel, C. R., W. J. Powers, T. J. Applegate, N. M. Tamim, and M. C. Christman. 2005. Influence of phytase on water-soluble phosphorus in poultry and swine manure. J. Environ. Qual. 34:563–571.

AOAC. 1993. Official Methods of Analysis. 15th ed. Assoc. Off. Anal. Chem., Arlington, VA.

ARC. 1981. The Nutrient Requirement of Pigs. Commonwealth Agric. Bureau, Slough, UK.

Arita, M., M. Yoshida, S. Hong, E. Tjonahen, J. N. Glickman, N. A. Petasis, R. S. Blumberg, and C. N. Serhan. 2005. Resolvin E1, an endogenous lipid mediator derived from omega-3 eicosapentaenoic acid, protects against 2,4,6-trinitrobenzene sulfonic acid-induced colitis. Proc. Natl. Acad. Sci. USA. 102:7671–7676.

Armstrong T. A., D. R. Cook, M. M. Ward, C. M. Williams, and J. W. Spears. 2004. Effect of dietary copper source (cupric citrate and cupric sulfate) and concentration on growth performance and fecal copper excretion in weanling pigs. J. Anim. Sci. 82:1234–1240.

Armstrong, T. A., C. M. Williams, J. W. Spears, and S. S. Schiffman. 2000. High dietary copper improves odor characteristics of swine waste. J. Anim. Sci. 78:859–864.

Bach Knudsen, K. E., B. B. Jensen, and I. Hansen. 1993. Oat bran but not a beta-glucan-enriched oat fraction enhances butyrate production in the large intestine of pigs. J. Nutr. 123:1235–1247.

Bark, L. J., T. S. Stahly, G. L. Cromwell, and J. Miyat. 1992. Influence of genetic capacity for lean tissue growth on rate and efficiency of tissue accretion in pigs fed ractopamine. J. Anim. Sci. 70:3391–3400.

Barsanti, L., V. Passarelli, V. Evangelista, A. M. Frassanito, and P. Gualtieri. 2011. Chemistry, physico-chemistry and applications linked to biological activities of β-glucans. Natl. Prod. Rep. 28:457–466.

Bauchart-Thevret, C., J. Cottrell, B. Stoll, and D. G. Burrin. 2011. First-pass splanchnic metabolism of dietary cysteine in weanling pigs. J. Anim. Sci. 89:4093–4099.

Bauer, E., B. A.Williams, H. Smidt, R. Mosenthin, and M.W. A. Verstegen. 2006. Influence of dietary components on development of the microbiota in single-stomached species. Nutr. Res. Rev. 19:63–78.

Bell, A. W., D. E. Bauman, D. H. Beermann, and R. J. Harrell. 1998. Nutriiton, development and efficacy of growth modifiers in livestock species. J. Nutr. 128(Suppl. 2):S360–S363.

Benaroudj, N., P. Zwickl, E. Seemüller, W. Baumeister, and A. L. Goldberg. 2003. ATP hydrolysis by the proteasome regulatory complex PAN serves multiple functions in protein degradation. Mol. Cell 11:69–78.

Bergen, W. G. 2008. Measuring *in vivo* intracellular protein degradation rates in animal system. J. Anim. Sci. 86(E. Suppl.): E3–E12.

Bergen, W. G., and R. A. Merkel. 1991. Body composition of animals treated with partitioning agents: Implications for human health. FASEB J. 5:2951–2957.

Bergen, W. G., S. E. Johnson, D.M. Skjaerlund, A. S. Babiker, N.K. Ames, R. A. Merkel, and D. B. Anderson. 1989. Muscle protein metabolism in finishing pigs fed red ractopamine. J. Anim. Sci. 67:2255–2262.

Berthold, H. K., P. F. Crain, I. Gouni, P. J. Reeds, and P. D. Klein. 1995. Evidence for incorporation of intact dietary pyrimidine (but not purine) nucleotides into hepatic RNA. Proc. Natl. Acad. Sci. USA. 22:10123–10127.

Bhandari, S. K., C. M. Nyachoti, and D. O. Krause. 2009. Raw potato starch in weaned pig diets and its influence on postweaning scours and the molecular microbial ecology of the digestive tract. J. Anim. Sci. 87:984–993.

Blanes-Vidal, V.,M. N. Hansen, and P. Sousa. 2009. Reduction of odor and odorant emissions from slurry stores by means of straw covers. J. Environ. Qual. 38:1518–1527.

Blowes, D. 2002. Tracking hexavalent Cr in groundwater. Sci. 295:2024–2025.

Bondi, A., and E. Alumot. 1987. Anti-nutritive factors in animal feedstuffs and their effects on livestock. Prog. Food Nutr. Sci. 11:115–151.

Borysenko, M. K. 2002. Antibiotics and organic acids in modulation of visceral organ and whole body growth in early-weaned piglets. M.Sc. Thesis, University of Guelph, Guelph, ON, Canada.

Boudry, G., V. Douard, J. Mourot, J. P. Lall'es, and I. Le Huërou-Luron. 2009. Linseed oil in the maternal diet during gestation and lactation modifies fatty acid composition, mucosal architecture, and mast cell regulation of the ileal barrier in piglets. J. Nutr. 139:1110–1117.

Boza, J. J., F. Jahoor, and P. J. Reeds. 1996. Ribonucleic acid nucleotides in maternal and fetal tissues derive almost exclusively from synthesis de novo in pregnant mice. J. Nutr. 126:1749–1758.

Bregendahl, K., X. Yang, L. Liu, J. T. Yen, T. C. Rideout, Y. Shen, G. Werchola, and M. Z. Fan. 2008. Fractional protein synthesis rates are similar when measured by intraperitoneal or intravenous flooding doses of l-[ring-2H5] phenylalanine in combination with a rapid regimen of sampling in piglets. J. Nutr. 138:1976–1981.

Burger-van Paassen, N., A. Vincent, P. J. Puiman, M. van der Sluis, J. Bouma, G. Boehm, J. B. van Goudoever, et al. 2009. The regulation of intestinal mucin MUC2 expression by short-chain fatty acids: Implications for epithelial protection. Biochem. J. 420:211–219.

Burton, D. L., and E. G. Beauchamp. 1986. Nitrogen losses from swine housings. Agric. Wastes 15:59–74.

Bush, J. A., G. Wu, A. Suryawan, H. V. Nguyen, and T. A. Davis. 2002. Somatotropin-induced amino acid conservation in pigs involves differential regulation of liver and gut urea cycle enzyme activity. J. Nutr. 132:59–67.

Bush, J. A., S. R. Kimball, P. M. J. O'Connor, A. Suryawan, R. A. Orellana, H. V. Nguyen, L. S. Jefferson, and T. A. Davis. 2003. Translational control of protein synthesis in muscle and liver of growth hormone-treated pigs. Endocrinol. 144:1273–1283.

Calder, P. C. 2012. Long-chain fatty acids and inflammation. Proc. Nutr. Soc. 28:1–6.

Callaway, T. R., T. S. Edrington, R. C. Anderson, R. B. Harvey, K. J. Genovese, C. N. Kennedy, D. W. Venn, and D. J. Nisbet. 2008. Probiotics, prebiotics and competitive exclusion for prophylaxis against bacterial disease. Anim. Health Res. Rev. 9:217–225.

Campbell, E. L., C. F. MacManus, D. J. Kominsky, S. Keely, L. E. Glover, B. E. Bowers, M. Scully, W. J. Bruyninckx, and S. P. Colgan. 2010. Resolvin E1-induced intestinal alkaline phosphatase promotes resolution of inflammation through LPS detoxification. Proc. Natl. Acad. Sci. USA. 107:14298–14303.

Canh, T. T., A. J. A. Aarnink, J. B. Schutte, A. Sutton, D. J. Langhout, and M. W. A. Verstegen. 1998a. Dietary protein affects nitrogen excretion and ammonia emission from slurry of growing-finishing pigs. Livest. Prod. Sci. 56:181–191.

Canh, T. T., A. J. A. Aarnink, M. W. A. Verstegen, and J. W. Schrama. 1997. Influence of dietary factors on nitrogen partitioning and composition of urine and feces of fattening pigs. J. Anim. Sci. 75:700–706.

Canh, T. T., A. J. A. Aarnink, M. W. A. Verstegen, and J. W. Schrama. 1998b. Influence of dietary factors on the pH and ammonia emission of slurry from growing-finishing pigs. J. Anim. Sci. 76:1123–1130.

Canh, T. T., A. L. Sutton, A. J. A. Aarnink, M.W. A. Verstegen, J.W. Schrama, and G. C. M. Bakker. 1998c. Dietary carbohydrates alter the fecal composition and pH and the ammonia emission from slurry of growing pigs. J. Anim. Sci. 76:1887–1895.

Cavaco, L. M., H. Hasman, and F. M. Aarestrup. 2011. Zinc resistance of *Staphylococcus aureus* of animal origin is strongly associated with methicillin resistance. Vet. Microbiol. 150:344–348.

Cera, K. R., D. C. Mahan, and G. A. Reinhart. 1988. Effects of dietary dried whey and corn oil on weanling pig performance, fat digestibility, and nitrogen utilization. J. Anim. Sci. 66:1438–1345.

Cera, K. R., D. C. Mahan, and G. A. Reinhart. 1990. Effect of weaning, week postweaning and diet composition on pancreatic and small intestinal luminal lipase response in young swine. J. Anim. Sci. 68:384–391.

Chanput, W., M. Reitsma, L. Kleinjans, J. J. Mes, H. F. Savelkoul, and H. J. Wichers. 2012. β-Glucans are involved in immunemodulation of THP-1 macrophages. Mol. Nutr. Food Res. (In press.)

Che, T. M., R.W. Johnson, K.W. Kelley, K. A. Dawson, C. A. Moran, and J. E. Pettigrew. 2012. Effects of mannan oligosaccharide on cytokine secretions by porcine alveolar macrophages and serum cytokine concentrations in nursery pigs. J. Anim. Sci. 90:657–668.

Chee-Sanford, J. C., R. I. Mackie, S. Koike, I. G. Krapac, Y. F. Lin, A. C. Yannarell, S. Maxwell, and R. I. Aminov. 2009. Fate and transport of antibiotic residues and antibiotic resistance genes following land application of manure waste. J. Environ. Qual. 38:1086–1108.

Chen, H. L., C. C. Yen, C. Y. Lu, C. H. Yu, and C. M. Chen. 2006. Synthetic porcine lactoferricin with a 20-residue peptide exhibits antimicrobial activity against *Escherichia coli, Staphylococcus aureus*, and *Candida albicans*. J. Agric. Food Chem. 54:3277–3282.

Chen, H. L., Y. W. Lai, C. C. Yen, Y. Y. Lin, C. Y. Lu, S. H. Yang, T. C. Tsai, Y. J. Lin, C. W. Lin, and C. M. Chen. 2004. Production of recombinant porcine lactoferrin exhibiting antibacterial activity in methylotrophic yeast, *Pichia pastoris*. J. Mol. Microbiol. Biotechnol. 8:141–149.

Chen, K. T., M. S. Malo, A. K. Moss, S. Zeller, P. Johnson, F. Ebrahimi, G. Mostafa, et al. 2010. Identification of specific targets for the gut mucosal defense factor intestinal alkaline phosphatase. Am. J. Physiol. 299:G467–G475.

Chen, K. T., M. S. Malo, L. K. Beasley-Topliffe, K. Poelstra, J. L. Millan, G. Mostafa, S. N. Alam, et al. 2011. A role for intestinal alkaline phosphatase in the maintenance of local gut immunity. Dig. Dis. Sci. 56:1020–1027.

Chen, L., S. Hoff, L. Cai, J. Koziel, and B. Zelle. 2009. Evaluation of wood chip-based biofilters to reduce odor, hydrogen sulfide, and ammonia from swine barn ventilation air. J. Air Waste Manag. Assoc. 59:520–530.

Cherrington, C. A., M. Hinton, G. C. Mead, and I. Chopra. 1991. Organic acids: chemistry, antibacterial activity and practical applications. Adv. Micro. Physiol. 32:87–107.

Choi, J. Y., J. S. Kim, S. L. Ingale, K. H. Kim, P. L. Shinde, I. K. Kwon, and B. J. Chae. 2011. Effect of potential multimicrobe probiotic product processed by high drying temperature and antibiotic on performance of weanling pigs. J. Anim. Sci. 89:1795–1804.

Clark, O. G., B. Morin, Y. Zhang, W. C. Sauer, and J. J. Feddes. 2005. Preliminary investigation of air bubbling and dietary sulfur reduction to mitigate hydrogen sulfide and odor from swine waste. J. Environ. Qual. 34:2018–2023.

Claus, R., and S. Raab. 1999. Influences on skatole formation from tryptophan in the pig colon. Adv. Exp. Med. Biol. 467:679–684.

Claus, R., D. Günthner, and H. Letzguss. 2007. Effects of feeding fat-coated butyrate on mucosal morphology and function in the small intestine of the pig. J. Anim. Physiol. Nutr. 91:312–318.

Claus, R., D. Lösel, M. Lacorn, J. Mentschel, and H. Schenkel. 2003. Effects of butyrate on apoptosis in the pig colon and its consequences for skatole formation and tissue accumulation. J. Anim. Sci. 81:239–248.

Claus, R., U.Weiler, and A. Herzog. 1994. Physiological aspects of androstenone and skatole formation in the boar–A review with experimental data. Meat Sci. 38:289–305.

Cleveland, B. M., A. S. Kiess, and K. P. Blemings. 2008. Alpha–aminoadipate delta–semialdehyde synthase mRNA knockdown reduces the lysine requirement of a murine hepatic cell line. J. Nutr. 138:2143–2147.

Cowieson, A. J., M. Hruby, and E. E. M. Pierson. 2006. Evolving enzyme technology: Impact on commercial poultry nutrition. Nutr. Res. Rev. 19:90–103.

Creech, B. L., J.W. Spears,W. L. Flowers, G. M. Hill, K. E. Lloyd, T. A. Armstrong, and T. E. Engle. 2004. Effect of dietary trace mineral concentration and source (inorganic vs. chelated) on performance, mineral status, and fecal mineral excretion in pigs from weaning through finishing. J. Anim. Sci. 82:2140–2147.

Cromwell, G. L. 1992. The biological availability of phosphorus in feedstuffs for pigs. Pig News Info. 13:75N–78N.

Cromwell, G. L. 2001. Antimicrobial and promicrobial agents. Pages 401–426 in *Swine Nutrition*. 2nd ed. A. J. Lewis and L. L. Southern, eds. CRC Press LLC, Boca Raton, FL.

Cromwell, G. L., G. L. Allee, and D. C. Mahan. 2008. Assessment of lactose level in the mid– to late–nursery phase on performance of weanling pigs. J. Anim. Sci. 86:127–133.

Dahiya, J. P.,D. Hoehler, A. G.van Kessel, and M. D. Drew. 2007. Dietary encapsulated glycine influences *Clostridium perfringens* and *Lactobacilli* growth in the gastrointestinal tract of broiler chickens. J. Nutr. 137:1408–1414.

Dai, Z. L., X. L. Li, P. B. Xi, J. Zhang, G. Wu, and W. Y. Zhu. 2012. L–Glutamine regulates amino acid utilization by intestinal bacteria. Amino Acids. (In press.)

Davies, N. T., and R. Nightingale. 1975. The effects of phytate on intestinal absorption and secretion of zinc, and whole–body retention of Zn, copper, iron and manganese in rats. Br. J. Nutr. 34:243–258.

Davis, T. A., A. Suryawan, R. A. Orellana, H. V. Nguyen, and M. L. Fiorotto. 2008. Postnatal ontogeny of skeletal muscle protein synthesis in pigs. J. Anim. Sci. 86(E. Suppl.):E13–E18.

Davis, T. A., J. A. Bush, R. C. vann, A. Suryawan, S. R. Kimball, and D. G. Burrin. 2004. Somatotropin regulation of protein metabolism in pigs. J. Anim. Sci. 82:E207–E213.

de Rodas, B. Z., K. S. Sohn, C. V. Maxwell, and L. J. Spicer. 1995. Plasma protein for pigs weaned at 19 to 24 days of age: effect on performance and plasma insulin–like growth factor I, growth hormone, insulin, and glucose concentrations. J. Anim. Sci. 73:3657–3665.

Deng, D., R. L. Huang, T. J. Li, G. Y. Wu, M. Y. Xie, Z. R. Tang, P. Kang, et al. 2007.

Nitrogen balance in barrows fed low-protein diets supplemented with essential amino acids. Livest. Sci. 109:220–223.

Depreux, F. F., A. L. Grant, D. B. Anderson, and D. E. Gerrard. 2002. Paylean alters myosin heavy chain isoform content in pig muscle. J. Anim. Sci. 80:1888–1894.

Deredjian, A., C. Colinon, E. Brothier, S. Favre-Bont'e, B. Cournoyer, and S. Nazaret. 2011. Antibiotic and metal resistance among hospital and outdoor strains of *Pseudomonas aeruginosa*. Res. Microbiol. 162:689–700.

Diamond, J. 1999. Evolutionary biology: Dirty eating for healthy living. Nature 400:120–121.

Diamond, J. 2002. Evolution, consequences and future of plant and animal domestication. Nature 418:700–707.

Donham, K. J. 2010. Community and occupational health concerns in pork production: A review. J. Anim. Sci. 88(E. Suppl.):E102–E111.

Drew, M. D., T. C. Schafer, and R. T. Zijlstra. 2012. Glycemic index of starch affects nitrogen retention in grower pigs. J. Anim. Sci. 90:1233–1241.

Dritz, S. S., J. Shi, T. L. Kielian, R. D. Goodband, J. L. Nelssen, M. D. Tokach, M. M. Chengappa, et al. 1995. Influence of dietary beta-glucan on growth performance, nonspecific immunity, and resistance to *Streptococcus suis* infection in weanling pigs.J. Anim. Sci. 73:3341–3350.

Drummond, R. A., and G. D. Brown. 2011. The role of Dectin-1 in the host defence against fungal infections. Curr. Opin. Microbiol. 14:392–399.

Duncan, S. H., P. Louis, and H. J. Flint. 2004. Lactate-utilizing bacteria, isolated from human feces, that produce butyrate as a major fermentation product. Appl. Environ. Microbiol. 70:5810–5817.

Edwards, H. M., III, and D. H. Baker. 2000. Zinc bioavailability in soybean meal. J. Anim. Sci. 78:1017–1021.

Ellis, A. S., T. M. Johnson, and T. D. Bullen. 2002. Chromium isotopes and the fate of hexavalent chromium in the environment. Sci. 295:2060–2062.

Emiola, I. A., F. O. Opapeju, B. A. Slominski, and C. M. Nyachoti. 2009. Growth performance and nutrient digestibility in pigs fed wheat distillers dried grains with solubles-based diets supplemented with a multicarbohydrase enzyme. J. Anim. Sci. 87:2315–2322.

Eriksen, J., A. P. Adamsen, J. V. N ϕr gaard, H. D. Poulsen, B. B. Jensen, and S. O. Petersen. 2010. Emissions of sulfur-containing odorants, ammonia, and methane from pig slurry: effects of dietary methionine and benzoic acid. J. Environ. Qual. 39:1097–10107.

Eriksen, J., P. S ϕr ensen, and L. Elsgaard. 2008. The fate of sulfate in acidified pig slurry during storage and following application to cropped soil. J. Environ. Qual. 37:280–286.

Escobar, J., W. G. van Alstine, D. H. Baker, and R. W. Johnson. 2004. Decreased protein accretion in pigs with viral and bacterial pneumonia is associated with increased myostatin expression in muscle. J. Nutr. 134:3047–3053.

Esteban, A., M. W. Popp, V. K. Vyas, K. Strijbis, H. L. Ploegh, and G. R. Fink. 2011. Fungal recognition is mediated by the association of dectin–1 and galectin–3 in macrophages. Proc. Natl. Acad. Sci. USA. 108:14270–14275.

Etherton, T. D. 1999. Emerging strategies for enhancing growth: is there a biotechnology better than somatotropin? Domes. Anim. Endocrinol. 17:171–179.

Etherton, T. D. 2004. Somatotropic function: the somatomedin hypothesis revisited. J. Anim. Sci. 82 (E. Suppl.):E239–E244.

Ewaschuk, J. B., G. K. Murdoch, I. R. Johnson, K. L. Madsen, and C. J. Field. 2011. Glutamine supplementation improves intestinal barrier function in a weaned piglet model of Escherichia coli infection. Br. J. Nutr. 106:870–877.

Ewaschuk, J. B., I. R. Johnson, K. L. Madsen, T. Vasanthan, R. Ball, and C. J. Field. 2012. Barley–derived beta–glucans increases gut permeability, ex vivo epithelial cell binding to *E. coli*, and naive T–cell proportions in weanling pigs. J Anim. Sci. (In press.)

Fan, M. Z. 2003. Growth and ontogeny of the gastrointestinal tract. Page 31 in Neonatal Pig Gastrointestinal Physiology and Nutrition. R. J. Xu, and P. D. Cranwell, eds. Nottingham University Press, UK.

Fan, M. Z., and W. C. Sauer. 2002. Determination of true ileal amino acid digestibility and the endogenous amino acid outputs associated with barley samples for growing–finishing pigs by the regression analysis technique. J. Anim. Sci. 80:1593–1605.

Fan, M. Z., C. M. F. de Lange, and T. Archbold. 2012. Effects of changes in dietary true digestible Ca to P ratio on growth performance and digestive and post–absorptive Ca and P utilization in grower pigs. J. Anim. Sci. 90.(E. Suppl.) (In press.)

Fan, M. Z., L. I. Chiba, P. D. Matzat, X. Yang, Y. L. Yin, Y. Mine, and H. Stein. 2006. Measuring synthesis rates of nitrogencontaining polymers by using stable isotope tracers. J. Anim. Sci. 84 (E. Suppl.):E79–E93.

Fan, M. Z., O. Adeola, E. K. Asem, and D. King. 2002. Postnatal ontogeny of kinetics of porcine jejunal brush border membranebound alkaline phosphatase, aminopeptidase N, and sucrase activities in pigs. Comp. Biochem. Physiol. Part A. Mol. Integr. Physiol. 132: 599–607.

Fan, M. Z., S.W. Kim, and T. J. Applegate. 2008a. Understanding protein synthesis and degradation and their pathway regulations: An editorial. J. Anim. Sci. 86(E. Suppl.):E1–E2.

Fan, M. Z., T. Archbold, and T. Rennick. 2004a. A moderate dietary level of exogenous cellulose increases the true digestibility and the gastrointestinal endogenous metabolic outputs of calcium associated with soybean meal in the post–weaned piglet. FASEB J. 18:A494.

Fan, M. Z., T. Archbold, and Y. L. Yin. 2004b. A moderate dietary level of exogenous pectin increases true digestibility and the gastrointestinal endogenous metabolic fecal loss of phosphorus associated with soybean meal in the post-weaned piglet. FASEB J. 18:A1306.

Fan, M. Z., T. Archbold, W. C. Sauer, D. Lackeyram, T. Rideout, Y. Gao, C. F. M. de Lange, and R. R. Hacker. 2001. Novel methodology allows simultaneous measurement of true phosphorus digestibility and the gastrointestinal endogenous phosphorus outputs in studies with pigs. J. Nutr. 131:2388–2396.

Fan, M. Z., W. C. Sauer, and V. M. Gabert. 1996. Variability of apparent ileal digestibility of amino acids in different samples of canola meal for growing-finishing pigs. Can. J. Anim. Sci. 76:563–569.

Fan, M. Z., W. C. Sauer, and M. I. McBurney. 1995. Estimation by regression analysis of the endogenous amino acid levels in digesta collected from the distal ileum of pigs. J. Anim. Sci. 73:2319–2328.

Fan, M. Z., Y. Shen, Y. L. Yin, Z. R. Wang, Z. Y. Wang, T. J. Li, T. C. Rideout, et al. 2008b. Methodological considerations for measuring phosphorus utilization in pigs. Page 370 in Mathematical Modelling in Animal Nutrition. J. France, and E. Kebreab, eds. CABI International, Wallingford, Oxon, UK.

Fang, R. J., Y. L. Yin, K. N. Wang, J. H. He, Q. H. Chen, T. J. Li, M. Z. Fan, and G. Wu. 2007. Comparison of the regression analysis technique and the substitution method for the determination of true phosphorus digestibility and faecal endogenous phosphorus losses associated with feed ingredients for growing pigs. Lives. Sci. 109:251–254.

Fard, R. M., M. W. Neuzenroeder, and M. D. Barton. 2011. Antimicrobial and heavy metal resistance in commensal enterococci isolated from pigs. Vet. Micobiol. 148:276–282.

Flanagan, P. R. 1984. A model to produce pure zinc deficiency in rats and its use to demonstrate that dietary phytate increases the excretion of endogenous zinc. J. Nutr. 114:493–502.

Flickinger, E. A., J. van Loo, and G. C. Fahey, Jr. 2003. Nutritional responses to the presence of inulin and oligofructose in the diets of domesticated animals: a review. Crit. Rev. Food Sci. Nutr. 43:19–60.

Forsberg, C. W., S. P. Golovan, A. Ajakaiye, J. P. Phillips, R. G. Meidinger, M. Z. Fan, J. M. Kelly, and R. R. Hacker. 2005. Genetic opportunities to enhance sustainability of pork production in developing countries: a model for food animals. Page 429 in Applications of Gene-based Technologies for Improving Animal Production and Health in Developing Countries. H. P. S. Makkar, and G. J. Viljoen, eds. Springer, Dordrecht, The Netherlands.

Furuya, S., and Y. Kaji. 1992. The effects of feed intake and purified cellulose on the endogenous ileal amino acid flow in growing pigs. Br. J. Nutr. 68:463–472.

Gabert, V. M., andW. C. Sauer. 1994. The effects of supplementing diets for weanling pigs with organic acids. A review. J. Anim. Feed Sci. 3:7–87.

Gallois, M., H. J. Rothkötter, M. Bailey, C. R. Stokes, and I. P. Oswald. 2009. Natural alternatives to in-feed antibiotics in pig production: Can immunomodulators play a role? Animal 3:1644–1661.

Gary, B. P., M. Fogarty, T. P. Curran, M. J. O'Connell, and J. V. O'Doherty. 2007. The effect of cereal type and enzyme addition on pig performance, intestinal microflora, and ammonia and odor emissions. Animal 1:751–757.

Gentile, F., P. Amodeo, F. Febbraio, F. Picaro, A. Motta, S. Formisano, and R. Nucci. 2002. SDS-resistant active and thermostable dimers are obtained from the dissociation of homotetrameric beta-glycosidase from hyperthermophilic *Sulfolobus solfataricus* in SDS. Stabilizing role of the A-C intermonomeric interface. J. Biol. Chem. 277:44050–44060.

Gibson, G. R., and M. B. Roberfroid. 1995. Dietary modulation of the human colonic microbiota: Introducing the concept of prebiotics. J. Nutr. 125:1401–1412.

Gibson, G. R., H. M. Probert, J. V. Loo, R. A. Rastall, and M. B. Roberfroid. 2004. Dietary modulation of the human colonic microbiota: Updating the concept of prebiotics. Nutr. Res. Rev. 17:259–275.

Godbout, S., S. P. Lemay, C. Duchaine, F. Pelletier, J. P. Larouche, M. Belzile, and J. J. Feddes. 2009. Swine production impact on residential ambient air quality. J. Agromed. 14:291–298.

Goll, D. E., G. Neil, S. W. Mares, and V. F. Thompson. 2008. The non-lysosomal Ca^{2+}-dependent protein degradation pathways: and Myofibrillar protein turnover: The calpains and the proteasome. J. Anim. Sci. 86:E19–E35.

Golovan, S., R. D. Meidinger, A. Ajakaiye, M. Cottrill, M. Z. Weiderkehr, C. Plante, J. Pollard, et al. 2001. Enhanced phosphorus digestion and reduced pollution potential by pigs with salivary phytase. Nature Biotechnol. 19:741–745.

Gonzalez-Cadayid, N. F., and S. Bhasin. 2004. Role of myostatin in metabolism. Curr. Opin. Clin. Nutr. Metab. Care 7: 451–457.

Goodridge, H. S., A. J. Wolf, and D. M. Underhill. 2009. Beta-glucan recognition by the innate immune system. Immunol. Rev. 230:38–50.

Gorbach, S. L. 2001. Antimicrobial use in animals: Time to stop. NEJM 345:1202–1203.

Guenther, S., M. Grobbel, A. Lüke-Becker, A. Goedecke, N. D. Friedrich, L. H. Wieler, and C. Ewers. 2010. Antimicrobial resistance profiles of *Escherichia coli* from common European wild bird species. Vet. Microbiol. 144:219–225.

Hahn, T. W., J. D. Lohakare, S. L. Lee, W. K. Moon, and B. J. Chae. 2006. Effects of supplementation of beta-glucans on growth performance, nutrient digestibility, and immunity in weanling pigs. J. Anim. Sci. 84:1422–1428.

Hancock, J. D., and K. C. Behnke. 2001. Use of ingredient and diet processing technologies (grinding, mixing, pelleting, and extrusion) to produce quality feeds for pigs. Pages 469 in *Swine Nutrition*, 2nd ed., A. L. Lewis, and L. L. Southern, eds. CRC Press LLC, NY.

Hansen, C. F., A. Herna'ndez, J. Mansfield, A' . Hidalgo, T. La, N. D. Phillips, D. J. Hampson, and J. R. Pluske. 2011. A high dietary concentration of inulin is necessary to reduce the incidence of swine dysentery in pigs experimentally challenged with *Brachyspira hyodysenteriae*. Br. J. Nutr. 106:1506–1513.

Hansen, C. F., N. D. Phillips, T. La, A. Hernandez, J. Mansfield, J. C. Kim, B. P. Mullan, et al. 2010. Diets containing inulin but not lupins help to prevent swine dysentery in experimentally challenged pigs. J. Anim. Sci. 88:3327–3336.

Havlin, J. L. 2004. Technical basis for quantifying phosphorus transport to surface and groundwaters. J. Anim. Sci. 82(E. Suppl.):E277–E291.

Hayes, E. T., A. B. Leek, T. B. Curran, V. A. Dodd, O. T. Carton, V. E. Beattie, and J. V. O'Doherty. 2004. The influence of diet crude protein level on odor and ammonia emissions from finishing pig houses. Bioresour. Technol. 91:309–315.

Hayhoe, M., T. Archbold, Q.Wang, X. Yang, and M.Z. Fan. 2012. Efficacy of prebiotics on lactose digestibility, whole body protein metabolic status and growth performance in replacing antibiotics in weanling pigs fed corn, soybean meal and dried–whey powder based diets. Page 121 in Book of Abstracts of Int. Symp. Digestive Physiology of Pigs. Digestive Physiology of Pigs Committee, Keystone, CO.

Haynes, T. E., P. Li, X. Li, K. Shimotori, H. Sato, N. E. Flynn, J.Wang, et al. 2009. L–Glutamine or L–alanyl–L–glutamine prevents oxidant– or endotoxin–induced death of neonatal enterocytes. Amino Acids 37:131–142.

He, J., J. Yin, L.Wang, B. Yu, and D. Chen. 2010. Functional characterization of a recombinant xylanase from Pichia pastoris and effect of the enzyme on nutrient digestibility in weaned pigs. Br. J. Nutr. 103:1507–1513.

Helferich,W. G., D. B. Jump, D. B. Anderson, D. M. Skjaerlund, R. A. Merkel, and W. G. Bergen. 1990. Skeletal muscle alpha–actin synthesis is increased pretranslationally in pigs fed the phenethanolamine rectopamine. Endocrinol. 126:3096–3100.

Hemsworth, P. H., and J. L. Barnett. 2000. Human–animal interactions and animal stress. Page 309 in The Biology of Animal Stress—Basic Principles and Implications for Animal Welfare. G. P. Moberg, and J. A., Mench, eds. CAB International, Oxon, UK.

Heo, J. M., F. O. Opapeju, J. R. Pluske, J. C. Kim, D. J. Hampson, and C. M. Nyachoti. 2012. Gastrointestinal health and function in weaned pigs: A review of feeding strategies to control post–weaning diarrhoea without using in–feed antimicrobial compounds. J. Anim. Physiol. Anim. Nutr. (Berl). doi: 10.1111/j.1439–0396.2012.01284.

Heo, J. M., J. C. Kim, C. F. Hansen, B. P. Mullan, D. J. Hampson, and J. R. Pluske. 2008. Effects of feeding low protein diets to piglets on plasma urea nitrogen, faecal ammonia nitrogen, the incidence of diarrhoea and performance after weaning. Arch. Anim. Nutr. 62:343–358.

Heo, J. M., J. C. Kim, C. F. Hansen, B. P. Mullan, D. J. Hampson, and J. R. Pluske. 2009. Feeding a diet with decreased protein content reduces indices of protein fermentation and the incidence of postweaning diarrhea in weaned pigs challenged with an enterotoxigenic strain of Escherichia coli. J. Anim. Sci. 87:2833–2843.

Hess, M., A. Sczyrba, R. Egan, T. W. Kim, H. Chokhawala, G. Schroth, S. Luo, et al. 2011. Metagenomic discovery of biomassdegrading genes and genomes from cow rumen. Science 331:463–467.

Heuer, H., H. Schmitt, and K. Smalla. 2011. Antibiotic resistance gene spread due to manure application on agricultural fields. Curr. Opin. Microbiol. 14:236–243.

Hinnebusch, B. F., J.W. Henderson, A. Siddique, M. S. Malo,W. Zhang, M. A. Abedrapo, and R. A. Hodin. 2003. Transcriptional activation of the enterocyte differentiation marker intestinal alkaline phosphatase is associated with changes in the acetylation state of histone H3 at a specific site within its promoter region *in vitro*. J. Gastrointest. Surg. 7:237–244.

Hölzel, C. S., C. Müller, K. S. Harms, S. Mikolajewski, S. Schäfer, K. Schwaiger, and J. Bauer. 2012. Heavy metals in liquid pig manure in light of bacterial antimicrobial resistance. Environ. Res. 113:21–27.

Hooda, S., J. J. Matte, T. Vasanthan, and R. T. Zijlstra. 2010. Dietary oat β-glucan reduces peak net glucose flux and insulin production and modulates plasma incretin in portal-vein catheterized grower pigs. J. Nutr. 140:1564–1569.

Hopwood, D. E., D. W. Pethick, J. R. Pluske, and D. J. Hampson. 2004. Addition of pearl barley to a rice-based diet for newly weaned piglets increases the viscosity of the intestinal contents, reduces starch digestibility and exacerbates post-weaning colibacillosis. Br. J. Nutr. 92:419–427.

Htoo, J. K., B. A. Araiza, W. C. Sauer, M. Rademacher, Y. Zhang, M. Cervantes, and R. T. Zijlstra. 2007a. Effect of dietary protein content on ileal amino acid digestibility, growth performance, and formation of microbial metabolites in ileal and cecal digesta of early-weaned pigs. J. Anim. Sci. 85:3303–3312.

Htoo, J. K., W. C. Sauer, Y. Zhang, M. Cervantes, S. F. Liao, B. A. Araiza, A. Morales, and N. Torrentera. 2007b. The effect of feeding low-phytate barley-soybean meal diets differing in protein content to growing pigs on the excretion of phosphorus and nitrogen. J. Anim. Sci. 85:700–705.

Huang, S. X., W. C. Sauer, M. Pickard, S. Li, and R. T. Hardin. 1998. Effect of micronization on energy, starch and amino acid digestibility in hulless barley for young pigs.

Can. J. Anim. Sci. 78:81–87.

Hutchison, M. L., L. D. Walters, T. Moore, D. J. Thomas, and S. M. Avery. 2005. Fate of pathogens present in livestock wastes spread onto fescue plots. Appl. Environ. Microbiol. 71:691–696.

Jackson, B. P., P. M. Bertsch, M. L. Cabrera, J. J. Camberato, J. C. Seaman, and C. W. Wood. 2003. Trace element speciation in poultry litter. J. Environ. Qual. 32:535–540.

Jansman, A. J. M., W. Smink, P. V. Leeuwen, and M. Rademacher. 2002. Evaluation through literature data of the amount and amino acid composition of basal endogenous crude protein at the terminal ileum of pigs. Anim. Feed Sci. Technol. 98: 49–60.

Jarret, G., J. Martinez, and J. Y. Dourmad. 2011. Effect of biofuel co-products in pig diets on the excretory patterns of N and C and on subsequent ammonia and methane emissions from pig effluent. Animal 5:622–631.

Jayasundara, J., C. Wagner-Riddle, G. Parkin, J. Lauzon, and M. Z. Fan. 2010. Transformations and losses of swine manure ^{15}N as affected by application timing at two contrasting sites. Can. J. Soil Sci. 90:55–73.

Jiang, R., B. Stoll, M.Z. Fan, and D.G. Burrin. 2000. Feeding porcine plasma protein decreases intestinal growth and cell proliferation in early-weaned pigs. J. Nutr. 130:21–26.

Jiang,W. G., R. P. Bryce, D. F. Horrobin, and R. E. Mansel. 1998. Regulation of tight junction permeability and occludin expression by polyunsaturated fatty acids. Biochem. Biophys. Res. Commun. 244:414–420.

Jiang, Z. Y., L. H. Sun, Y. C. Lin, X. Y. Ma, C. T. Zheng, G. L. Zhou, F. Chen, and S. T. Zou. 2009. Effects of dietary glycylglutamine on growth performance, small intestinal integrity, and immune responses of weaning piglets challenged with lipopolysaccharide. J. Anim. Sci. 87:4050–4056.

Jongbloed, A. W., H. Everts, and P. A. Kemme. 1991. Phosphorous availability and requirements in pigs. Pages 65–80 in Recent Advances in Animal Nutrition. E. R. Heinemann, ed. Butterworth, London, UK.

Kang, P., D. Toms, Y. Yin, Q. Cheung, J. Gong, K. De Lange, and J. Li. 2010. Epidermal growth factor-expressing *Lactococcus lactis* enhances intestinal development of early-weaned pigs. J. Nutr. 140:806–811.

Kats, L. J., J. L. Nelssen, M. D. Tokach, R. D. Goodband, T. L. Weeden, S. S. Dritz, J. A. Hansen, and K. G. Friesen. 1994. The effects of spray-dried blood meal on growth performance of the early-weaned pig. J. Anim. Sci. 72:2860–2869.

Kerr, B. J., and R. A. Easter. 1995. Effect of feeding reduced protein, amino acid-supplemented diets on nitrogen and energy balance in grower pigs. J. Anim. Sci. 73:3000–3008.

Kerr, B. J., C. J. Ziemer, S. L. Trabue, J. D. Crouse, and T. B. Parkin. 2006. Manure composition of swine as affected by dietary protein and cellulose concentrations. J. Anim. Sci. 84:1584–1592.

Kiarie, E., C. M. Nyachoti, B. A. Slominski, and G. Blank. 2007. Growth performance, gastrointestinal microbial activity, and nutrient digestibility in early–weaned pigs fed diets containing flaxseed and carbohydrase enzyme. J. Anim. Sci. 85:2982– 2993.

Kim, J. C., B. P. Mullan, D. J. Hampson, and J. R. Pluske. 2008. Addition of oat hulls to an extruded rice–based diet for weaner pigs ameliorates the incidence of diarrhea and reduces indices of protein fermentation in the gastrointestinal tract. Br. J. Nutr. 99:1217–1225.

Kim, M. S., and X. G. Lei. 2008. Enhancing thermostability of *Escherichia coli* phytase AppA2 by error–prone PCR. Appl. Microbiol. Biotechnol. 79:69–75.

Kim, S. W., D. A. Knabe, K. J. Hong, and R. A. Easter. 2003. Use of carbohydrases in corn–soybean meal–based nursery diets. J. Anim. Sci. 81:2496–2504.

Kim, S. W., M. Z. Fan, and T. J. Applegate. 2008. Nonruminant nutrition symposium on natural phytobiotics for health of young animals and poultry: Mechanisms and application. J. Anim. Sci. 86(E. Suppl.):E138–E139.

Kim, Y. S., and J. A. Milner. 2007. Dietary modulation of colon cancer risk. J. Nutr. 137:2576S–2579S.

Kleinman, P., D. Sullivan, A.Wolf, R. Brandt, Z. Dou, H. Elliott, J. Kovar, et al. 2007. Selection of a water–extractable phosphorus test for manures and biosolids as an indicator of runoff loss potential. J. Environ. Qual. 36:1357–1367.

Kluge, H, J. Broz, and K. Eder. 2006. Effect of benzoic acid on growth performance, nutrient digestibility, nitrogen balance, gastrointestinal microflora and parameters of microbial metabolism in piglets. J. Anim. Physiol. Anim. Nutr. 90: 316–324.

Koike, S., I. G. Krapac, H. D. Oliver, A. C. Yannarell, J. C. Chee–Sanford, R. I. Aminov, and R. I. Mackie. 2007. Monitoring and source tracking of tetracycline resistance genes in lagoons and groundwater adjacent to swine production facilities over a 3–year period. Appl. Environ. Microbiol. 73:4813–4823.

Kong, X. F., G. Wu, Y. P. Liao, Z. P. Hou, H. J. Liu, F. G. Yin, T. J. Li, et al. 2007. Dietary supplementation with Chinese herbal ultra–fine powder enhances cellular and humoral immunity in early–weaned piglets. Lives. Sci. 108:94–98.

Kornegay, E. T., Z. Wang, C. M. Wood, and M. D. Lindemann. 1997. Supplemental chromium picolinate influences nitrogen balance, dry matter digestibility, and carcass traits in growing–finishing pigs. J. Anim. Sci. 75:1319–1323.

Kottwitz, K., N. Laschinsky, R. Fischer, and P. Nielsen. 2009. Absorption, excretion and retention of ^{51}Cr from labelled Cr–(III)– picolinate in rats. Biometals 22:289–295.

Krause, D. O., B. A. White, and R. I. Mackie. 1997. Ribotyping of adherent *Lactobacillus* from weaning pigs: a basis for probiotic selection based on diet and gut compartment. Anaerobe 3:317–325.

Krause, D. O., R. A. Easter, B. A. White, and R. I. Mackie. 1995. Effect of weaning diet on the ecology of adherent *Lactobacilli* in the gastrointestinal tract of the pig. J. Anim. Sci. 73:2347–2354.

Krick, B. J., R. D. Boyd, K. R. Roneker, D. H. Beermann, D. E. Bauman, D. A. Ross, and D. J. Meisinger. 1993. Porcine somatotropin affects the dietary lysine requirement and net lysine utilization for growing pigs. J. Nutr. 123:1913–1922.

Lackeyram, D. 2012. Expression of the Small Intestinal Hydrolases in the Early–Weaned Pig. PhD Diss. Univ. of Guelph, Guelph, ON, Canada.

Lackeyram, D., C.Yang, T. Archbold, K.Swanson, and M. Z. Fan. 2010. Early weaning reduces small intestinal alkaline phosphatase expression in pigs. J. Nutr. 140:461–468.

Lacki, K., and Z. Duvnjak. 1999. A method for the decrease of phenolic content in commercial canola meal using an enzyme preparation secreted by the white–rot fungus trametes versicolor. Biotechnol. Bioeng. 62:422–433.

Lartigue, C., J. I. Glass, N. Alperovich, and R. Pieper. 2007. Genome transplantation in bacteria: change one species to another. Sci. 317:632–638.

Lavoie, J., S. Godbout, S. P. Lemay, and M. Belzile. 2009. Impact of in–barn manure separation on biological air quality in an experimental setup identical to that in swine buildings. J. Agric. Saf. Health. 153:225–240.

Le, P. D., A. J. Aarnink, A. W. Jongbloed, C. M. van der Peet Schwering, N. W. Ogink, and M. W. Verstegen. 2007a. Effects of crystalline amino acid supplementation to the diet on odor from pig manure. J. Anim. Sci. 85:791–801.

Le, P. D., A. J. Aarnink, A. W. Jongbloed, C. M. van der Peet Schwering, N. W. Ogink, and M. W. Verstegen. 2007b. Effects of dietary protein levels on odor from pig manure. Animal 1:734–744.

Le, P. D., A. J. Aarnink, A. W. Jongbloed, C. M. van der Peet Schwering, N. W. Ogink, and M. W. Verstegen. 2008. Interactive effects of dietary crude protein and fermentable carbohydrate levels on odor from pig manure. Livest. Sci. 114:48–61.

Le, P. D., A. J. Aarnink, and A. W. Jongbloed. 2009. Odor and ammonia emission from pig manure as affected by dietary crude protein level. Livest. Sci. 121:267–274.

Le, P. D., A. J. Aarnink, N. W. Ogink, P. M. Becker, and M. W. Verstegen. 2005. Odor from animal production facilities: its relationship to diet. Nutr. Res. Rev. 181:3–30.

Lecerf, J. M., F D'epeint, E. Clerc, Y. Dugenet, C. N. Niamba, L. Rhazi, A. Cayzeele, et al. 2012. Xylo–oligosaccharide (XOS) in combination with inulin modulates both the intestinal environment and immune status in healthy subjects, while XOS alone only shows prebiotic properties. Br. J. Nutr. 23:1–12.

Lee, T. T., C. C. Chang, R. S. Juang, R. B. Chen, H. Y. Yang, L. W. Chu, S. R. Wang, et al. 2010. Porcine lactoferrin expression in transgenic rice and its effects as a feed additive on early weaned piglets. J. Agric. Food Chem. 58:5166–5173.

Leek, A. B., E. T. Hayes, T. P. Curran, J. J. Callan, V. E. Beattie, V. A. Dodd, and J. V. O'Doherty. 2007. The influence of manure composition on emissions of odor and ammonia from finishing pigs fed different concentrations of dietary crude protein. Bioresour. Technol. 98:3431–3439.

Lei, X., P. K. Ku, E. R. Miller, D. E. Ullrey, and M. T. Yokoyama. 1993. Supplemental microbial phytase improves bioavailability of dietary zinc to weanling pigs. J. Nutr. 123:1117–1123.

Leonard, S. G., T. Sweeney, B. Bahar, B. P. Lynch, and J. V. O'Doherty. 2011a. Effect of dietary seaweed extracts and fish oil supplementation in sows on performance, intestinal microflora, intestinal morphology, volatile fatty acid concentrations and immune status of weaned pigs. Br. J. Nutr. 105:549–560.

Leonard, S. G., T. Sweeney, B. Bahar, B. P. Lynch, and J. V. O'Doherty. 2011b. Effects of dietary seaweed extract supplementation in sows and post-weaned pigs on performance, intestinal morphology, intestinal microflora and immune status. Br. J. Nutr. 106:688–699.

Lessard, M., M. Dupuis, N. Gagnon, E. Nadeau, J. J. Matte, J. Goulet, and J. M. Fairbrother. 2009. Administration of *Pediococcus acidilactici* or *Saccharomyces cerevisiae boulardii* modulates development of porcine mucosal immunity and reduces intestinal bacterial translocation after *Escherichia coli* challenge. J. Anim. Sci. 87:922–934.

L'etourneau-Montminy, M. P., A. Narcy, M. Magnin, D. Sauvant, J. F. Bernier, C. Pomar, and C. Jondreville. 2010. Effect of reduced dietary calcium concentration and phytase supplementation on calcium and phosphorus utilization in weanling pigs with modified mineral status. J. Anim. Sci. 88:1706–1717.

L'etourneau, V., B. Nehm'e, A. M'eriaux, D. Mass'e, and C. Duchaine. 2010. Impact of production systems on swine confinement building bioaerosols. J. Occup. Environ. Hyg. 7:94–102.

Levy, S. B. 1998. Multidrug resistance: A sign of the times. NEJM 338:1376–1378.

Lewis, A. J., T. J. Wester, D. G. Burrin, and M. J. Dauncey. 2000. Exogenous growth hormone induces somatotrophic gene expression in neonatal liver and skeletal muscle. Am. J. Physiol. 278:R838–844.

Li, J.,D. F. Li, J. J. Xing, Z. B. Cheng, and C. H. Lai. 2006. Effects of beta-glucan extracted from saccharomyces cerevisiae on growth performance, and immunological and somatotropic responses of pigs challenged with *Escherichia coli* lipopolysaccharide. J. Anim.

Sci. 84:2374–2381.

Li, L, A. Kapoor, B. Slikas, O. S. Bamidele, C. Wang, S. Shaukat, M. A. Masroor, et al. 2010. Multiple diverse circoviruses infect farm animals and are commonly found in human and chimpanzee feces. J. Virol. 84:1674–1682.

Li, S., W. C. Sauer, S. X. Huang, and V. M. Gabert. 1996. Effect of β-glucanase supplementation to hulless barley or wheatsoybean meal based diets on the digestibilities of energy, protein, β-glucans, and amino acids in young pigs. J. Anim. Sci. 74:1649–1656.

Liao, Y., V. Lopez, T. B. Shafizadeh, C. H. Halsted, and B. Lönnerdal. 2007. Cloning of a pig homologue of the human lactoferrin receptor: expression and localization during intestinal maturation in piglets. Comp. Biochem. Physiol. A. Mol. Integr. Physiol. 148:584–590.

Lin, F. D., D. A. Knabe, and T. D. Tanksley, Jr. 1987. Apparent digestibility of amino acids, gross energy and starch in corn, sorghum, wheat, barley, oat groats and wheat middlings for growing pigs. J. Anim. Sci. 64:1655–1663.

Linden, A., I. M. Olsson, and A. Oskarsson. 1999. Cadmium levels in feed components and kidneys of growing-finishing pigs. J. AOAC Int. 82:1288–1297.

Liu, C. Y., and S. E. Mills. 1990. Decreased insulin binding to porcine adipocytes *in vitro* by beta-adrenergic agonists. J. Anim. Sci. 68:1603–1608.

Liu, S., W. C. Willett, and M. J. Stampfer. 2000. A prospective study of dietary glycemic load, carbohydrate intake, and risk of coronary heart disease in US women. Am. J. Clin. Nutr. 71:1455–1461.

Liu, Y., L. Gong, D. Li, Z. Feng, L. Zhao, and T. Dong. 2003. Effects of fish oil on lymphocyte proliferation, cytokine production and intracellular signalling in weanling pigs. Arch. Tierernahr. 57:151–165.

Liu, Z., G. Li, S. R. Kimball, L. A. Jahn, and E. J. Barrett. 2004. Glucocorticoids modulate amino acid-induced translation initiation in human skeletal muscle. Am. J. Physiol. 287:E275–E281.

Liu, Z., T. Uesaka, H.Watanabe, and N. Kato. 2001. High fat diet enhances colonic cell proliferation and carcinogenesis in rats by elevating serum leptin. Int. J. Onc. 19:1009–1014.

Lösel, D., M. Lacorn, D. Büttner, and R. Claus. 2006. Flavor improvement in pork from barrows and gilts via inhibition of intestinal skatole formation with resistant potato starch. J. Agric. Food Chem. 54:5990–5995.

Louis, P., and H. J. Flint. 2009. Diversity, metabolism and microbial ecology of butyrate-producing bacteria from the human large intestine. FEMS Microbiol. Lett. 294:1–8.

Louis, P., P. Young, G. Holtrop, and H. J. Flint. 2010. Diversity of human colonic butyrate-producing bacteria revealed by analysis of the butyryl-CoA:acetate CoA-transferase gene. Environ. Microbiol. 12:304–314.

Maa, M. C., M. Y. Chang, M. Y. Hsieh, Y. J. Chen, C. J. Yang, Z. C. Chen, Y. K. Li, et al. 2010. Butyrate reduced lipopolysaccharide mediated macrophage migration by suppression of Src enhancement and focal adhesion kinase activity. J. Nutr. Biochem. 21:1186–1192.

Mackie, R. I., P. G. Stroot, and V. H. Varel. 1998. Biochemical identification and biological origin of key odor components in livestock waste. J. Anim. Sci. 76:1331–1342.

Mahan, D. C. 1992. Efficacy of dried whey and its lactalbumin and lactose components at two dietary lysine levels on postweaning pig performance and nitrogen balance. J. Anim. Sci. 70:2182–2187.

Mahan, D. C., and E. A. Newton. 1993. Evaluation of feed grains with dried skim milk and added carbohydrate sources on weanling pig performance. J. Anim. Sci. 71:3376–3382.

Mallin, M. 2000. Impacts of industrial animal production on rivers and estuaries. Am. Sci. 88:26–37.

Malo, M. S., S. N. Alam, G. Mostafa, S. J. Zeller, P. V. Johnson, N. Mohammad, K. T. Chen, et al. 2010. Intestinal alkaline phosphatase preserves the normal homeostasis of gut microbiota. Gut 59:1476–1484.

Manges, A. R., J. R. Johnson, B. Foxman, T. T. O'Bryan, K. E. Fullerton, and L.W. Riley. 2001.Widespread distribution of urinary tract infections caused by a multidrug–resistant *Escherichia coli* clonal group. N. Engl. J. Med. 345:1007–1013.

Marion, J., V. Rom'e, G. Savary, F. Thomas, J. Le Dividich, and I. Le Huërou–Luron. 2003.Weaning and feed intake alter pancreatic enzyme activities and corresponding mRNA levels in 7–d–old piglets. J. Nutr. 133:362–368.

Marquardt, R. R., L. Z. Jin, J. W. Kim, L. Fang, A. A. Frohlich, and S. K. Baidoo. 1999. Passive protective effect of egg–yolk antibodies against enterotoxigenic *Escherichia coli* K88$^+$ infection in neonatal and early–weaned piglets. FEMS Immunol. Med. Microbiol. 23:283–288.

Maxwell, C. V., and S. D. Carter. 2001. Feeding the weaned pig. Page 691 in *Swine Nutrition*. 2nd ed. A. J. Lewis and L. L. Southern. CRC Press, Boca Raton, FL.

McDonald, D. E., D. W. Pethick, J. R. Pluske, and D. J. Hampson. 1999. Adverse effects of soluble non–starch polysaccharide (guar gum) on piglet growth and experimental colibacillosis immediately after weaning. Res. Vet. Sci. 67:245–250.

Mendel, M., M. Chlopecka, and N. Dziekan. 2007. Haemolytic crisis in sheep as a result of chronic exposure to copper. Pol. J. Vet. Sci. 10:51–56.

Mentschel, J. and R. Claus. 2003. Increased butyrate formation in the pig colon by feeding raw potato starch leads to a reduction of colonocyte apoptosis and a shift to the stem cell compartment. Metabolism 52:1400–1405.

Mertens, J., K. Broos, S. A. Wakelin, G. A. Kowalchuk, D. Springael, and E. Smolders. 2009. Bacteria, not archaea, restore nitrification in zinc–contaminated soil. ISME J. 3:916–923.

Metzler-Zebeli, B. U., S. Hooda, R. Pieper, R. T. Zijlstra, A. G. van Kessel, R. Mosenthin, and M. G. Gänzle. 2010. Nonstarch polysaccharides modulate bacterial microbiota, pathways for butyrate production, and abundance of pathogenic *Escherichia coli* in the pig gastrointestinal tract. Appl. Environ. Microbiol. 76:3692–3701.

Mills, S. E., M. E. Spurlock, and D. J. Smith. 2003. Beta-adrenergic receptor subtypes that mediate ractopamine stimulation of lipolysis. J. Anim. Sci. 81:662–668.

Mitsuyama, K., A. Toyonaga, and M. Sata. 2002. Intestinal microflora as a therapeutic target in inflammatory bowel disease. J. Gastro. 37(Suppl. 14):73–77.

Montagne, L., F. S. Cavaney, D. J. Hampson, J. P. Lallés, and J. R. Pluske. 2004. Effect of diet composition on postweaning colibacillosis in piglets. J. Anim. Sci. 82:2364–2374.

Mosenthin, R., W. C. Sauer, and F. Ahrens. 1994. Dietary pectin's effect on ileal and fecal amino acid digestibility and exocrine pancreatic secretions in growing pigs. J. Nutr. 124:1222–1229.

Mroz, Z. A., J. Moeser, K. Vreman, J. T. van Diepen, T. van Kempen, T. T. Canh, and A. W. Jongbloed. 2000. Effects of dietary carbohydrates and buffering capacity on nutrient digestibility and manure characteristics in finishing pigs. J. Anim. Sci. 78:3096–3106.

Mubiru, J. N., and R. J. Xu. 1998. Comparison of growth and development of the exocrine pancreas in pigs and rats during the immediate postnatal period. Comp. Biochem. Physiol. A. Mol. Integr. Physiol. 120:699–703.

Mulder, I. E., B. Schmidt, C. R. Stokes, M. Lewis, M. Bailey, R. I. Aminov, J. I. Prosser, et al. 2009. Environmentally-acquired bacteria influence microbial diversity and natural innate immune responses at gut surfaces. BMCBiol. 7:79 doi:10.1186/1741-7007-7-79.

Mulder, I. E., B. Schmidt, M. Lewis, M. Delday, C. R. Stokes,M. Bailey, R. I. Aminov, et al. 2011. Restricting microbial exposure in early life negates the immune benefits associated with gut colonization in environments of high microbial diversity. PLoS One 6:e28279.

Mulvaney, D. R., R. A. Merkel, and W. G. Bergen. 1985. Skeletal muscle protein turnover in young pigs. J. Nutr. 115: 1057–1064.

Myrie, S. B., R. F. Bertolo, W. C. Sauer, and R. O. Ball. 2008. Effect of common antinutritive factors and fibrous feedstuffs in pig diets on amino acid digestibilities with special emphasis on threonine. J. Anim. Sci. 86:609–619.

Naqvi, S. W. A., D. A. Jayakumar, P. V. Narvekar, H. Nalk, V. V. S. S. Sarma, W. D'Souza, S. Joseph, and M. D. George. 2000. Increased marine production of N_2O due to intensifying anoxia on the Indian continental shelf. Nature 408:346–349.

Naranjo, V. D., T. D. Bidner, and L. L. Southern. 2010. Comparison of dried whey permeate and a carbohydrate product in diets for nursery pigs. J. Anim. Sci. 88:1868–1879.

Nessmith, W. B., Jr., J. L. Nelssen, M. D. Tokach, R. D. Goodband, J. R. Bergström, S. S.

Dritz, and B. T. Richert. 1997b. Evaluation of the interrelationships among lactose and protein sources in diets for segregated early–weaned pigs. J. Anim. Sci. 75:3214–3221.

Nessmith,W. B., Jr., J. L. Nelssen, M. D. Tokach, R. D. Goodband, and J. R Bergström. 1997a. Effects of substituting deproteinized whey and(or) crystalline lactose for dried whey on weanling pig performance. J. Anim. Sci. 75:3222–3228.

Newsholme, E. A., and A. R. Leech. 1991. Biochemistry for the Medical Sciences. John Wiley & Sons, New York, NY.

Nichols, B. L. N., J. Eldering, S. Avery, D. Hahn, A. Quaronii, and E. Sterchi. 1998. Human small intestinal maltase–glucoamylase cDNA cloning homology to sucrase–isomaltase. J. Biol. Chem. 273:3076–3081.

Nielsen, S. M., G. H. Hansen, and E. M. Danielsen. 2010. Lactoferrin targets T cells in the small intestine. J. Gastroenterol. 45:1121–1128.

Nortey, T. N., J. F. Patience, J. S. Sands, N. L. Trottier, and R. T. Zijlstra. 2008. Effects of xylanase supplementation on the apparent digestibility and digestible content of energy, amino acids, phosphorus, and calcium in wheat and wheat by–products from dry milling fed to grower pigs. J. Anim. Sci. 86:3450–3464.

NRC. 1998. Nutrient Requirements of Swine. 10th rev. ed. National Academies Press, Washington, DC.

NRC. 2005. Mineral Tolerance of Animals. 2nd ed. National Academies Press, Washington, DC.

NRC. 2012. Nutrient Requirements of Swine. 11th rev. ed. National Academies Press, Washington, DC.

Nyachoti, C. M., C. F. M. de Lange, B. W. McBride, and H. Schulze. 1997. Significance of endogenous gut nitrogen losses in the nutrition of growing pigs: A review. Can. J. Anim. Sci. 77:149–163.

Nyannor, E. K. D., P. Williams, M. R. Bedford, and O. Adeola. 2007. Corn expressing an *Escherichia coli*–derived phytase gene: A proof–of–concept nutritional study in pigs. J. Anim. Sci. 85:1946–1952.

O'Shea, C. J., M. B. Lynch, T. Sweeney, D. A. Gahan, J. J. Callan, and J. V. O'Doherty. 2011a. Comparison of a wheatbased diet supplemented with purified β–glucans, with an oat–based diet on nutrient digestibility, nitrogen utilization, distal gastrointestinal tract composition, and manure odor and ammonia emissions from finishing pigs. J. Anim. Sci. 89:438–447.

O'Shea, C. J., T. Sweeney, M. B. Lynch, D. A. Gahan, J. J. Callan, and J. V. O'Doherty. 2010. Effect of β–glucans contained in barley– and oat–based diets and exogenous enzyme supplementation on gastrointestinal fermentation of finisher pigs and subsequent manure odor and ammonia emissions. J. Anim. Sci. 88:1411–1420.

O'Shea, C. J., T. Sweeney, M. B. Lynch, J. J. Gallan, and J. V. O'Doherty. 2011b. Modifications of selected bacteria and markers of protein fermentation in the distal gastrointestinal tract of pigs upon consumption of chitosan is accompanied by heightened manure odor emissions. J. Anim. Sci. 89:1366–1375.

Olukosi, O. A., J. S. Sands, and O. Adeola. 2007. Supplementation of carbohydrases or phytase individually or in combination to diets for weanling and growing–finishing pigs. J. Anim. Sci. 85:1702–1711.

Omogbenigun, F. O., C. M. Nyachoti, and B. A. Slominski. 2004. Dietary supplementation with multienzyme preparations improves nutrient utilization and growth performance in weaned pigs. J. Anim. Sci. 82:1053–1061.

Onyango, E. M., E. K. Asem, and O. Adeola. 2009. Phytic acid increases mucin and endogenous amino acid losses from the gastrointestinal tract of chickens. Br. J. Nutr. 101:836–842.

Orellana, R. A., P. M. J. O'Connor, H. V. Nguyen, J. A. Bush, A. Suryawan, M. G. Thivierge, M. L. Fiorotto, and T. A. Davis. 2002. Endotoxemia reduces skeletal muscle protein synthesis in neonates. Am. J. Physiol. 283:E909–E916.

Otto, E. R.,M. Yokoyama, S. Hengemuehle, R. D. von Bermuth, T. van Kempen, and N. L. Trottier. 2003. Ammonia, volatile fatty acids, phenolics, and odor offensiveness in manure from growing pigs fed diets reduced in protein concentration. J. Anim Sci. 81:1754–1763.

Partenan, K. H., and Z. Mroz. 1999. Organic acids for performance enhancement in pig diets. Nutr. Res. Rev. 12:117–145.

Patterson, A. R., D. M. Madson, and T. Opriessnig. 2010. Efficacy of experimentally produced spray–dried plasma on infectivity of porcine circovirus type 2. J. Anim. Sci. 88:4078–4085.

Peterla, T. A., and C. G. Scanes. 1990. Effect of beta–adrenergic agonists on lipolysis and lipogenesis by porcine adipose tissue *in vitro*. J. Anim. Sci. 68:1024–1029.

Petersen, G. I., and H. H. Stein. 2006. Novel procedure for estimating endogenous losses and measurement of apparent and true digestibility of phosphorus by growing pigs. J. Anim. Sci. 84:2126–2132.

Pettigrew, J. E. 2006. Reduced use of antibiotic growth promoters in diets fed to weanling pigs: Dietary tools, part 1. Anim. Biotechnol. 17:207–215.

Pettey, L. A., S. D. Carter, B. W. Senne, and J. A. Shriver. 2002. Effects of bata–mannanase addition to corn–soybean meal diets on growth performance, carcass traits, and nutrient digestibility of weanling and growing–finishing pigs. J. Anim. Sci. 80:1012–1019.

Pieper, R., S. Kröger, J. F. Richter, J.Wang, L. Martin, J. Bindelle, J. K. Htoo, et al. 2012. Fermentable fiber ameliorates fermentable protein–induced changes in microbial ecology, but not the mucosal response, in the colon of piglets. J. Nutr. 142:661–667.

Pluske, J. R., D. W. Pethick, D. E. Hopwood, and D. J. Hampson. 2002. Nutritional influences on some major enteric bacterial diseases of pig. Nutr. Res. Rev. 15:333–371.

Pluske, J. R., L. Montagne, F. S. Cavaney, B. P. Mullan, D. W. Pethick, and D. J. Hampson. 2007. Feeding different types of cooked white rice to piglets after weaning influences starch digestion, digesta and fermentation characteristics and the faecal shedding of beta-haemolytic *Escherichia coli*. Br. J. Nutr. 97:298–306.

Preidis, G. A., and J. Versalovic. 2009. Targeting the human microbiome with antibiotics, probiotics, and prebiotics: Gastroenterology enters the metagenomics era. Gastroenterology 136:2015–2031.

Pujols, J., S. L'opez-Soria, J. Segal'e, M. Fort, M. Sibila, R. Rosell, D. Solanes, et al. 2008. Lack of transmission of porcine circovirus type 2 to weanling pigs by feeding them spray-dried porcine plasma. Vet. Rec. 163:536–538.

Qian, H., E. T. Kornegay, and D. E. Conner, Jr. 1996. Adverse effects of wide calcium:phosphorus ratios on supplemental phytase efficacy for weanling pigs fed two dietary phosphorus levels. J. Anim. Sci. 74:1288–1297.

Ramsay, A. G., K. P. Scott, J. C. Martin, M. T. Rincon, H. J. Flint. 2006. Cell-associated alpha-amylases of butyrate-producing Firmicute bacteria from the human colon. Microbiology 152: 3281–3290.

Ramsay, T. G., and M. P. Richards. 2007. Beta-adrenergic regulation of uncoupling protein expression in swine. Comp. Biochem. Physiol. A Mol. Integr. Physiol. 47(2):395–403.

Ravindran, V., and J. H. Son. 2011. Feed enzyme technology: Present status and future developments. Recent Pat. Food Nutr. Agric. 3:102–109.

Reeds, P. J., D. G. Burrin, T. A. Davis, and M. L. Firotto. 1993. Postnatal growth of gut and muscle: Competitors or collaborators. Proc. Nutr. Soc. 52:57–67.

Regmi, S., M. Ongwandee, G. Morrison, M. Fitch, and R. Surampalli. 2007. Effectiveness of porous covers for control of ammonia, reduced sulfur compounds, total hydrocarbons, selected volatile organic compounds, and odor from hog manure storage lagoons. J. Air Waste Manag. Assoc. 57:761–768.

Reinhart, G. A., F. A. Simmen, D. C. Mahan, M. E. White, and K. L. Roehrig. 1992. Intestinal development and fatty acid binding protein activity of newborn pigs fed colostrum or milk. Biol. Neonate 62:155–163.

Rideout, T. C., and M. Z. Fan. 2004. Nutrient utilization in responses to chicory inulin supplementation in studies with pigs. J. Sci. Food Agri. 84:1005–1012.

Rideout, T. C., M. Z. Fan, J. P. Cant, C. Wagner-Riddle, and P. Stonehouse. 2004. Excretion of major odor-causing and acidifying compounds in response to dietary supplementation of chicory inulin in growing-finishing pigs. J. Anim. Sci. 82: 1678–1684.

Risley, C. R., E. T. Kornegay, M. D. Lindemann, C. M. Wood, and W. N. Eigel. 1992. Effect of feeding organic acids on selected intestinal content measurements at varying times postweaning in pigs. J. Anim. Sci. 70:196–206.

Rivera-Ferre, M. G., J. F. Aguilera, and R. Nieto. 2005. Muscle fractional protein synthesis is higher in Iberian than in Landrace growing pigs fed adequate or lysine-deficient diets. J. Nutr. 135:469–478.

Rodriguez, E., E. J. Mullaney, and X. G. Lei. 2000. Expression of the *Aspergillus fumigatus* phytase gene in *Pichia pastoris* and characterization of the recombinant enzyme. Biochem. Biophys. Res. Commun. 268:373–378.

Roos, S., O. Lagerlöf, M. Wennergren, T. L. Powell, and T. Jansson. 2009. Regulation of amino acid transporters by glucose and growth factors in cultured primary human trophoblast cells is mediated by mTOR signaling. Am. J. Physiol. Cell Physiol. 297:C723–C231.

Roth, F. X., and M. Kirchgessner. 1998. Organic acids as feed additives for young pigs: Nutritional and gastrointestinal effects. J. Anim. Feed Sci. 7:25–33.

Roy, C. C., C. L. Kien, L. Bouthillier, and E. Levy. 2006. Short-chain fatty acids: Ready for prime time? Nutr. Clin. Pract. 21:351–366.

Rozeboom, D. W., D. T. Shaw, R. J. Tempelman, J. C. Miguel, J. E. Pettigrew, and A. Connolly. 2005. Effects of mannan oligosaccharide and an antimicrobial product in nursery diets on performance of pigs reared on three different farms. J. Anim. Sci. 83:2637–2644.

Sales, J., and F. Janc'k. 2011. Effects of dietary chromium supplementation on performance, carcass characteristics, and meat quality of growing-finishing swine: A meta-analysis. J. Anim. Sci. 89:4054–4067.

Salnikow, K., and A. Zhitkovich. 2008. Genetic and epigenetic mechanisms in metal carcinogenesis and cocarcinogenesis: nickel, arsenic, and chromium. Chem. Res. Toxicol. 21:28–44.

Salway, J. G. and Granner, D. K. 1999. Metabolism at a Glance. Blackwell Science Ltd, Oxford, UK.

Sauer, W. C., M. Z. Fan, R. Mosenthin, and W. Drochner. 2000. Methods of measuring ileal amino acid digestibility in pigs. Page 279 in Farm Animal Metabolism and Nutrition: Critical Reviews. J. P. F. D'Mello, ed. CAB Int., Wallingford, Oxon, UK.

Schinasi, L., R. A. Horton, V. T. Guidry, S., Wang, S. W. Marshall, and K. B. Morland. 2011. Air pollution, lung function, and physical symptoms in communities near swine feeding operations. Epidemiology 22:208–215.

Schley, P. D., and C. J. Field. 2002. The immune-enhancing effects of dietary fibres and prebiotics. Br. J. Nutr. 87(Suppl. 2):S221–S230.

Schmidt, B., I. E. Mulder, C. C. Musk, R. I. Aminov, M. Lewis, C. R. Stokes, M. Bailey, et al. 2011. Establishment of normal gut microbiota is compromised under excessive hygiene conditions. PLoS One 6:e28284.

Schneider, A., A. Morabia, U. Papendick, and R. Kirchmayr. 1990. Pork intake and human papillomavirus-related disease. Nutr. Cancer 13:209–211.

Schulin-Zeuthen, M., E. Kebreab, W. J. J. Gerrits, S. Lopez, M. Z. Fan, R. S. Dias, and J. France. 2007. Meta-analysis of phosphorus balance data from growing pigs. J. Anim. Sci. 85:1953–1961.

Schutte, J. B., J. de Jong, E. J. van Weerden, and S. Tamminga. 1992. Nutritional implications of L-arabinose in pigs. Br. J. Nutr. 68:195–207.

Schutte, J. B., J. de Jong, R. Polziehn, and M. W. Verstegen. 1991. Nutritional implications of D-xylose in pigs. Br. J. Nutr. 66:83–93.

Scott, K. P., J. C. Martin, C. Chassard, M. Clerget, J. Potrykus, G. Campbell, C. D. Mayer, et al. 2011. Substrate-driven gene expression in Roseburia inulinivorans: Importance of inducible enzymes in the utilization of inulin and starch. Proc. Natl. Acad. Sci. USA. 108(Suppl. 1):4672–4679.

Seneviratne, R. W., E. Beltranena, R. W. Newkirk, L. A. Goonewardene, and R. T. Zijlstra. 2011. Processing conditions affect nutrient digestibility of cold-pressed canola cake for grower pigs. J. Anim. Sci. 89:2452–2461.

Shan, T., Y. Wang, Y. Wang, J. Liu, and Z. Xu. 2007. Effect of dietary lactoferrin on the immune functions and serum iron level of weanling piglets. J. Anim. Sci. 85:2140–2146.

Sharpley, A. N., P. J. Kleinman, A. L. Heathwaite, W. J. Gburek, G. J. Folmar, and J. P. Schmidt. 2008. Phosphorus loss from an agricultural watershed as a function of storm size. J. Environ. Qual. 37:362–368.

Sharpley, A. N., S. J. Smith, and O. R. Jones. 1992. The transport of bioavailable phosphorus in agricultural runoff. J. Environ. Qual. 21:30–35.

Shelton, N. W., M. D. Tokach, J. L. Nelssen, R. D. Goodband, S. S. Dritz, J. M. DeRouchey, and G. M. Hill. 2011. Effects of copper sulfate, tribasic copper chloride, and zinc oxide on weanling pig performance. J. Anim. Sci. 89:2440–2451.

Shen, H. G., S. Schalk, P. G. Halbur, J. M. Campbell, L. E. Russell, and T. Opriessnig. 2011. Commercially produced spray-dried porcine plasma contains increased concentrations of porcine circovirus type 2 DNA but does not transmit porcine circovirus type 2 when fed to naıve pigs. J. Anim. Sci. 89:1930–1938.

Simons, P. C. M., H. A. J. Versteegh, A. W. Jongbloed, P. A. Kemme, P. Slump, K. D. Box, M. G. E. Wolters, et al. 1990. Improvement of phosphorus availability by microbial phytase in broilers and pigs. Br. J. Nutr. 64:525–529.

Simopoulos, A. P. 2002. Omega-3 fatty acids in inflammation and autoimmune diseases. J.

Am. Coll. Nutr. 21:495–505.

Skjaerlund, D. M., D. R. Mulvaney,W. G. Bergen, and R. A. Merkel. 1994. Skeletal muscle growth and protein turnover in neonatal boars and barrows. J. Anim. Sci. 72:315–321.

Song, M., Y. Liu, J. A. Soares, T. M. Che, O. Osuna, C. W. Maddox, and J. E. Pettigrew. 2012. Dietary clays alleviate diarrhea of weaned pigs. J. Anim. Sci. 90:345–360.

Spencer, J. D., G. L. Allee, and T. E. Sauber. 2000. Phosphorus bioavailability and digestibility of normal and genetically modified low–phytate corn for pigs. J. Anim. Sci. 78:675–681.

Stahl, C. H., K. R. Roneker, J. R. Thornton, and X. G. Lei. 2000. A new phytase expressed in yeast effectively improves the bioavailability of phytate phosphorus to weanling pigs. J. Anim. Sci. 78:668–674.

Stein, H. H. 2002. Experience of feeding pigs without antibiotics: A European perspective. Anim. Biotechnol. 13:85–95.

Stein, H. H., and G. C. Shurson. 2009. Board–invited review: the use and application of distillers dried grains with solubles in swine diets. J. Anim. Sci. 87:1292–1303.

Stein, H. H., and R. A. Bohlke. 2007. The effects of thermal treatment of field peas (Pisum sativum L.) on nutrient and energy digestibility by growing pigs. J. Anim. Sci. 85:1424–1431.

Stein, H. H., B. S`eve, M. F. Fuller, P. J. Moughan, and C. F. M. de Lange. 2007. Amino acid bioavailability and digestibility in pig feed ingredients: terminology and application. J. Anim. Sci. 85:172–180.

Stein, H. H., O. Adeola, G. L. Cromwell, S.W. Kim, D. C. Mahan, P. S. Miller. 2011. Concentration of dietary calcium supplied by calcium carbonate does not affect the apparent total tract digestibility of calcium, but decreases digestibility of phosphorus by growing pigs. J. Anim. Sci. 89:2139–2144.

Stoll, B., J. Henry, P. J. Reeds, H. Yu, F. Jahoor, and D. G. Burrin. 1998. Catabolism dominates the first–pass intestinal metabolism of dietary essential amino acids in milk protein–fed piglets. J. Nutr. 128:606–614.

Sutton, A. L., and B. T. Richert. 2004. Nutrition and feed management strategies to reduce nutrient excretions and odors from swine manure. 49:397–404.

Sutton, A. L., K. B. Kephart, M. W. Verstegen, T. T. Canh, and P. J. Hobbs. 1999. Potential for reduction of odorous compounds in swine manure through diet formulation. J. Anim. Sci. 77:430–439.

Svihus, B., A. K. Uhlen, and O. M. Harstad. 2005. Effect of starch granule structure, associated components and processing on nutritive value of cereal starch: A review. Anim. Feed Sci. Technol. 122:303–320.

Tang, Z., Y. Yin, Y. Zhang, R. Huang, Z. Sun, T. Li, W. Chu, et al. 2009. Effects of dietary supplementation with an expressed fusion peptide bovine lactoferricin−lactoferrampin on performance, immune function and intestinal mucosal morphology in piglets weaned at age 21 d. Br. J. Nutr. 101:998–1005.

Teng, X., Q. J.Wang, C. B. Yang, and M. Z. Fan. 2012. Early weaning up−regulates jejunal neutral amino acid exchanger (ASCT2) gene expression in pigs. Page 113 in Book of Abstracts of XII Int. Symp. on Digestive Physiology of Pigs. Digestive Physiology of Pigs Committee, Keystone, CO.

Thorne, P. S., A. C. Ansley, and S. S. Perry. 2009. Concentrations of bioaerosols, odors, and hydrogen sulphide inside and downwind from two types of swine livestock operations. J. Occup. Environ. Hyg. 6:211–220.

Trabue, S., B. Kerr, B. Bearson, and C. Ziemer. 2011. Swine odor analyzed by odor panels and chemical techniques. J. Environ. Qual. 40:1510–1520.

Umber, J. K., and J. B. Bender. 2009. Pets and antimicrobial resistance. Vet. Clin. North Am. Small Anim. Pract. 39:279–292.

Urriola, P. E., H. H. Stein, and G. C. Shurson. 2010. Digestibility of dietary fiber in distillers coproducts fed to growing pigs. J. Anim. Sci. 88:2373–2381.

Vadas, P. A., W. J. Gburek, A. N. Sharpley, P. J. Kleinman, P. A. Moore, Jr., M. L. Cabrera, and R. D. Harmel. 2007. A model for phosphorus transformation and runoff loss for surface−applied manures. J. Environ. Qual. 36:324–332.

van Breemen, N., P. A. Burrough, E. J. Velthorst, H. F. van Dobben, T. de Wit, T. B. Ridder, and H. F. R. Reijnders. 1982. Soil acidification from atmospheric ammonium sulphate in forest canopythroughfall. Nature 299:548–550.

van Der Meulen, J., J. G. M. Bakker, B. Smits, and H. De Visser. 1997. Effect of source of starch on net portal flux of glucose, lactate, volatile fatty acids and amino acids in the pig. Br. J. Nutr. 78:533–544.

van Kempen, T. A. 2001. Dietary aipic acid reduces ammonia emission from swine excreta. J. Anim. Sci. 79:2412–2417.

vanbelle, M., E. Teller, and M. Focant. 1990. Probiotics in animal nutrition: A review. Arch. Anim. Nutr. 40:543–567.

vann, R. C., H. V. Nguyen, P. J. reeds, D. G. Burrin, M. L. Fiorotto, N. C. Steele, D. R.Weaver, and T. A. Davis. 2000. Somatotropin increases protein balance by lwering body protein degradation in fed, growing pigs. Am. J. Physiol. 278:E477–E483.

Varel, V. H., and J. T. Yen. 1997. Microbial perspective on fibre utilization by swine. J. Anim. Sci. 75:2715–2722.

Velthof, G. L., J. A. Nelemans, O. Oenema, and P. J. Kui*km*an. 2005. Gaseous nitrogen and carbon losses from pig manure derived from different diets. J. Envion. Qual. 34:698–706.

Vidovic, M., A. Sadibasic, S. Cupic, and M. Lausevic. 2005. Cd and Zn in atmospheric deposit, soil, wheat, and milk. Environ. Res. 97:26–31.

Visek, W. J. 1978. The mode of growth promotion by antibiotics. J. Anim. Sci. 46:1447–1469.

Voet, D., and J. G. Voet. 1995. *Biochemistry*. 2nd ed. John Wiley & Sons, Inc., New York, NY.

Volman, J. J., J. D. Ramakers, and J. Plat. 2008. Dietary modulation of immune function by beta–glucans. Physiol. Behav. 94:276–284.

Volman, J. J., R. P. Mensink,W. A. Buurman, G. Onning, and J. Plat. 2010. The absence of functional dectin–1 on enterocytes may serve to prevent intestinal damage. Eur. J. Gastroenterol. Hepatol. 22:88–94.

Wächtershäuser, A., and J. Stein. 2000. Rationale for the luminal provision of butyrate in intestinal diseases. Eur. J. Nutr. 39:164–171.

Walsh, M. C., D. M. Sholly, R. B. Hinson, K. L. Saddoris, A. L. Sutton, J. S. Radcliffe, R. Odgaard, et al. 2007. Effects of water and diet acidification with and without antibiotics on weanling pig growth and microbial shedding. J. Anim. Sci. 85:1799–1808.

Walsh, M. C., M. H. Rostagno, G. E. Gardiner, A. L. Sutton, B. T. Richert, and J. S. Radcliffe. 2012a. Controlling *Salmonella* infection in weanling pigs through water delivery of direct–fed microbials or organic acids. Part I: effects on growth performance, microbial populations, and immune status. J. Anim. Sci. 90:261–271.

Walsh, M. C., M. H. Rostagno, G. E. Gardiner, A. L. Sutton, B. T. Richert, and J. S. Radcliffe. 2012b. Controlling *Salmonella* infection in weanling pigs through water delivery of direct–fed microbials or organic acids: Part II. Effects on intestinal histology and active nutrient transport. J. Anim. Sci. 90:2599–2608.

Wan, M., X. Wu, K. L. Guan, M. Han, Y. Zhuang, and T. Xu. 2006. Muscle atrophy in transgenic mice expressing a human TSC1 transgene. FEBS Letters. 580:5621–5627.

Wang, B., A. Morinobu, M. Horiuchi, J. Liu, and S. Kumagai. 2008a. Butyrate inhibits functional differentiation of human monocyte–derived dendritic cells. Cell Immunol. 253:54–58.

Wang, J., L. Chen, P. Li, X. Li, H. Zhou, F. Wang, D. Li, et al. 2008b. Gene expression is altered in piglet small intestine by weaning and dietary glutamine supplementation. J. Nutr. 138:1025–1032.

Wang, W., T. Archbold, M. Kimber, J. Li, J. S. Lam, and M. Z. Fan. 2012a. Porcine gut microbial metagenomic library for mining novel cellulases established from grower pigs fed cellulose–supplemented high–fat diets. J. Anim. Sci. 90. (E. Suppl.) (In press.)

Wang, Y., S. K. Fried, R. N. Petersen, and P. A. Schoknecht. 1999. Somatotropin regulates adipose tissue metabolism in neonatal swine. J. Nutr. 129:139–145.

Wang, Y., T. Shan, Z. Xu, J. Liu, and J. Feng. 2006. Effect of lactoferrin on the growth performance, intestinal morphology, and expression of PR-39 and protegrin-1 genes in weaned piglets. J. Anim. Sci. 84:2636–2641.

Wang, Y., Y. J. Chen, J. H. Cho, J. S. Yoo, Y. Huang, H. J. Kim, S. O. Shin, T. X. Zhou, and I. H. Kim. 2009. Effect of soybean hull supplementation to finishing pigs on the emission of noxious gases from slurry. Anim. Sci. J. 80:316–321.

Wang, Z. X., C. Yang, T. Archbold, M. Hayhoe, K. A. Lien, and M. Z. Fan. 2012b. Expression of the small intestinal Na-neutral amino acid co-transporter B0AT1 (SLC6A19) in early-weaned pigs. Page 84 in Book of Abstracts of XII Int. Symp. On Digestive Physiology of Pigs, Digestive Physiology of Pigs Committee, Keystone, CO.

Wang, Z., T. Archbold, C. Yang, and M. Z. Fan. 2008. Dietary true digestible calcium to phosphorus ratio affects phosphorus utilization and renal sodium and phosphate co-transporter gene expressions in post-weaned pigs. FASEB J. 22:691.

Weiler, U., I. Font, M. Furnols, K. Fischer, H. Kemmer, M. A. Oliver, M. Gispert, et al. 2000. Influence of differences in sensitivity of Spanish and German consumers to perceive androstenone on the acceptance of boar meat differing in skatole and androstenone concentrations. Meat Sci. 54:297–304.

Weiss, S. L., E. A. Lee, and J. Diamond. 1998. Evolutionary matches of enzyme and transporter capacities to dietary substrate loads in the intestinal brush border. Proc. Natl. Acad. Sci. USA 95:217–221.

Wells, J. E.,W. T. Oliver, and J. T. Yen. 2010. The effects of dietary additives on faecal levels of *Lactobacillus* spp., coliforms, and *Escherichia coli*, and faecal prevalence of *Salmonella* spp. and *Campylobacter* spp. in US production nursery swine. J. Appl. Microbiol. 108:306–314.

Wester, T. J., T. A. Davis, M. L. Fiorotto, and D. G. Burrin. 2000. Exogenous growth hormone stimulates somatotropic axis function and growth in neonatal pigs. Am. J. Physiol. 274:E29–E37.

Williams, N. H., T. S. Stahly, and D. R. Zimmerman. 1997. Effect of chronic immune system activation on body nitrogen retention, partial efficiency of lysine utilization and lysine needs of pigs. J. Anim. Sci. 75:2463–2471.

Willig, S., D. Lösel, and R. Claus. 2005. Effects of resistant potato starch on odor emission from feces in swine production units. J. Agric. Food Chem. 53:1173–1178.

Willment, J. A., S. Gordon, and G. D. Brown. 2001. Characterization of the human beta-glucan receptor and its alternatively spliced isoforms. J. Biol. Chem. 276:43818–43823.

Wilson, D. B. 2009. Cellulases and biofuels. Curr. Opin. Biotechnol. 20:295–299.

Wilson, D. B. 2011. Microbial diversity of cellulose hydrolysis. Curr. Opin. Microbiol. 14:259–263.

Windmueller, H. G., and A. E. Spaeth. 1975. Intestinal metabolism of glutamine and glutamate from the lumen as compared to glutamine from blood. Arch. Biochem. Biophy. 171:662–672.

Windmueller, H. G., and A. E. Spaeth. 1976. Metabolism of absorbed aspartate, asparagine, and arginine by rat small intestine *in vivo*. Arch. Biochem. Biophy. 175.670–676.

Wiseman, J. 2006. Variations in starch digestibility in non–ruminants. Anim. Feed Sci. Technol. 130:66–77.

Wismar, R., S. Brix, H. N. Laerke, and H. Frøiaer. 2011. Comparative analysis of a large panel of non–starch polysaccharides reveals structures with selective regulatory properties in dendritic cells. Mol. Nutr. Food Res. 55:443–454.

Woyengo, T. A., A. J. Cowieson, O. Adeola, and C. M. Nyachoti. 2009. Ileal digestibility and endogenous flow of minerals and amino acids: responses to dietary phytic acid in piglets. Br. J. Nutr. 102:428–433.

Woyengo, T. A., J. S. Sands, W. Guenter, and C. M. Nyachoti. 2008. Nutrient digestibility and performance responses of growing pigs fed phytase– and xylanase–supplemented wheat–based diets. J. Anim. Sci. 86:848–857.

Wright, E. M., D. D. Loo, and B. A. Hirayama. 2011. Biology of human sodium glucose transporters. Physiol. Rev. 91:733–794.

Wu, G., F.W. Bazer, G. A. Johnson, D. A. Knabe, R. C. Burghardt, T. E. Spencer, X. L. Li, and J. J.Wang. 2011. Triennial Growth Symposium: Important roles for L–glutamine in swine nutrition and production. J. Anim. Sci. 89:2017–2030.

Wu, G., S. A. Meier, and D. A. Knabe. 1996. Dietary glutamine supplementation prevents jejunal atrophy in weaned pigs. J. Nutr. 126:2578–2584.

Wu, W. Z., X. Q. Wang, G. Y. Wu, S. W. Kim, F. Chen, and J. J. Wang. 2010. Differential composition of proteomes in sow colostrum and milk from anterior and posterior mammary glands. J. Anim. Sci. 88:2657–2664.

Yahoo News. 2010. U.S. water has large amounts of likely carcinogen: Study. Accessed Dec. 19, 2010. http://news. yahoo.com/s/afp/healthusenvironmentpollutionwater.

Yan, X., A. Voutetakis, C. Zhang, B. Hai, C. Zhang, B. J. Baum, and S. Wang. 2007. Sorting of transgenic secretory proteins in miniature pig parotid glands following adenoviral–mediated gene transfer. J. Gene Med. 9:779–787.

Yáñez, J. L., E. Beltranena, M. Cervantes, and R. T. Zijlstra. 2011. Effect of phytase and xylanase supplementation or particle size on nutrient digestibility of diets containing distillers dried grains with solubles cofermented from wheat and corn in ileal–cannulated grower pigs. J. Anim. Sci. 89:113–123.

Yang, X. 2009. Postnatal changes of skeletal muscle protein synthesis in the pig. PhD Diss.

Univ. of Guelph, Guelph, ON, Canada.

Yang, C., D. M. Albin, Z. Wang, B. Stoll, D. Lackeyram, K. C. Swanson, Y. L. Yin, K. A. Tappenden, Y. Mine, R. Y. Yada, D. G. Burrin, and M. Z. Fan. 2011. Apical Na+-D-glucose co-transporter 1 (SGLT1) activity and protein abundance are expressed along the jejunal crypt-villus axis in the neonatal pig. Am. J. Physiol. 300:G60–G70.

Yang, J., T. Ratovitski, J. P. Brady, M. B. Solomon, K. D. Wells, and R. J. Wall. 2001. Expression of myostatin pro domain in muscular transgenic mice. Mol. Reprod. Dev. 60:351–361.

Yang, X., C. Yang, A. Farberman, T. C. Rideout, C. F. M. de Lange, J. France, and M. Z. Fan. 2008. The mTOR-signaling pathway in regulating metabolism and growth. J. Anim. Sci. 86(E. Suppl.):E36–E50.

Yen, C. C., C. Y. Lin, K. Y. Chong, T. C. Tsai, C. J. Shen, M. F. Lin, C. Y. Su, et al. 2009. Lactoferrin as a natural regimen for selective decontamination of the digestive tract: recombinant porcine lactoferrin expressed in the milk of transgenic mice protects neonates from pathogenic challenge in the gastrointestinal tract. J. Infect. Dis. 199:590–598.

Yen, J. T., H. J. Mersmann, D. A. Hill, and W. G. Pond. 1990. Effects of ractopamine on genetically obese and lean pigs. J. Anim. Sci. 68:3705–3712.

Yin, F., Z. Zhang, J. Huang, and Y. Yin. 2010. Digestion of dietary starch affects systemic circulation of amino acids in weaned pigs. Br. J. Nutr. 103:1404–1412.

Yin, H. F., B. L. Fan, B. Yang, Y. F. Liu, J. Luo, X. H. Tian, and N. Li. 2006. Cloning of pig parotid secretory protein gene upstream promoter and the establishment of a transgenic mouse model expressing bacterial phytase for agricultural pollution control. J. Anim. Sci. 84:513–519.

Young, M. G., M. D. Tokach, J. Noblet, F. X. Aherne, S. S. Dritz, R. D. Goodband, J. L. Nelssen, et al. 2004. Influence of Carnichrome on the energy balance of gestating sows. J. Anim Sci. 82:2013–2022.

Zhang, M. J., andM. Spite. 2012. Resolvins: anti-inflammatory and proresolving mediators derived from omega-3 polyunsaturated fatty acids. Annu. Rev. Nutr. 32:203–227.

Zhang, Z. Y., L. J. Ding, and M. Z. Fan. 2009. Resistance patterns and detection of aac(3)-IV gene in apramycin-resistant *Escherichia coli* isolated from farm animals and farm workers in northeastern of China. Res. Vet. Sci. 87:449–454.

Zhao, G., H. Zhou, and Z. Wang. 2010. Concentrations of selected heavy metals in food from four e-waste disassembly localities and daily intake by local residents. J. Environ. Sci. Health A. Tox. Hazard Subst. Environ. Eng. 45:824–835.

Zheng, N., Q. Wang, and D. Zheng. 2007. Health risk of Hg, Pb, Cd, Zn, and Cu to the inhabitants around Huludao Zinc Plant in China via consumption of vegetables. Sci. Total. Environ. 383:81–89.

Zhong, X., X. H. Zhang, X. M. Li, Y. M. Zhou, W. Li, X. X. Huang, L. L. Zhang, and T. Wang. 2011. Intestinal growth and morphology is associated with the increase in heat shock protein 70 expression in weaning piglets through supplementation with glutamine. J. Anim. Sci. 89:3634–3642.

Zhu, C. L., M. Rademacher, and C. F.M. de Lange. 2005. Increasing dietary pectin level reduces utilization of digestible threonine intake, but not lysine intake, for body protein deposition in growing pigs. J. Anim. Sci. 83:1044–1053.

Ziemer, C. J., B. J. Kerr, S. L. Trabue, H. Stein, D. A. Stahl, and S. K. Davidson. 2009. Dietary protein and cellulose effects on chemical and microbial characteristics of swine feces and stored manure. J. Environ. Qual. 38:2138–2146.

第17章

猪营养与肉品质

1 导 语

很多因素都影响猪肉品质，其中遗传、宰前管理和胴体冷却都对猪肉品质发挥重要影响。越来越多的文献表明，营养调控不仅可以抵消不良遗传和饲养管理方面因素给猪肉品质带来的消极影响，而且一些情况下，还可以使饲养管理良好的、具有优良肉质性状基因型的猪表现出更好的肉品质。

猪肉品质通常指肌肉的pH、肉色、质地坚实度、大理石评分或肌内脂肪含量、货架寿命以及冷却排酸猪肉的风味。在美国和国际的消费者心目中，肉品质的定义还包括养猪的生产环境、伦理和动物福利方面。当前猪肉加工业者还倾向于将脂肪的品质（颜色、硬度、组成）、肉的营养组成（如肉中蛋白质、维生素、微量元素的数量和品质）和肉的安全（如有没有致病菌和化学物质残留）纳入到肉品质的定义中。本章主要讨论了营养调控如何影响肉的加工品质（如肉的pH、肉色、肉质地坚实度、系水力等）和食用品质，同时还将重点讨论猪脂肪和鲜五花肉品质的营养调控以及“自然”（户外散养）与传统（全封闭集约化饲养）养猪系统生产的猪肉品质差异。

2 动物宰后机体代谢调控和猪肉品质

动物放血后血液停止循环，肌肉的供氧也随之停止，肌肉的代谢从脂肪有氧代谢转向贮备的肌糖原的无氧代谢。动物宰后肌肉无氧代谢的终产物是乳酸，随着肌肉中乳酸的积累，肌肉pH从7.1 ~ 7.3最终降低到5.4 ~ 5.7。有三种典型的劣质猪肉与猪宰后肌肉pH异常降低有关：①肉色苍白的、质地松软的、汁液渗出的猪肉，即PSE肉；②肉色发暗、质地坚硬、如干柴般的猪肉，即DFD肉；③肉色发红、质地松软、汁液渗出的猪肉，即RSE猪肉。

肌肉中积累了过多的乳酸会使猪宰后1h内，肌肉pH迅速下降至5.8 ~ 6.0甚至以下，肌肉中过高的酸度和随之而来的温度升高导致了肌肉蛋白质变性并最终发展成PSE肉。宰后肌糖原快速降解的原因包括遗传倾向性、宰前应激和交感–肾上腺髓质轴激活或后二者的协同作用。相反，如果肌糖原贮存低，乳酸累积量则打了折扣，导致尸僵过程中肌肉的最终pH早早稳定在6.0以上，就形成了DFD猪肉。宰前各种应激使得猪能量消耗增加，而宰前肌糖原贮备低通常又强化了宰前应激对体内糖原贮备耗尽的效应。

一磷酸腺苷激活蛋白激酶γ–2亚基基因突变的猪，其肌肉中肌糖原含量异常高（Milan等，2000），因此该基因型通常也被称为酸肉基因（RN–）（Monin和Sellier，1985）。肌肉中乳酸过量积累会引起肌肉最终pH降低到肌肉的等电点5.0 ~ 5.3（使肌肉中收缩蛋白的静电荷为零的pH）。虽然RSE猪肉的肉色在等电点时几乎正常，但由于肌肉收缩蛋白在等电点时携带的正负离子数相等，蛋白互相吸引使得肌纤维蛋白吸附的水量下降，引起水分过度丢失，即系水力下降，导致鲜肉的蛋白功能属性下降。

◆ 2.1 屠宰前禁食

调节宰前肌肉中糖原贮量可以显著改善鲜猪肉的肉色和系水力。屠宰前16 ~ 24 h禁食可以有效地减少背最长肌糖原的浓度、升高肌肉初始（45 min）和最终（24 h）的pH，并可以显著改善鲜猪肉的系水力（Partanen等，2007；Sterten等，2009）。但宰前小于16 h的禁食对肌肉糖原贮存、尸僵过程pH下降和系水力则没有改善（Faucitano等，2006）。有趣的是，Bidner等（2004）发现，酸肉基因猪（RN–）宰前禁食30 ~ 60 h不会降低背最长肌糖原水平，这表明单纯屠宰前禁食并不能有效地调节酸肉基因型猪异常高的肌肉糖原贮备。

屠宰前禁食16～48 h可以降低肌肉的亮度（低的L*值），生产出肉色更好的猪肉（Sterten等，2009）；但传统的宰前16～36 h禁食不影响鲜猪肉的红度值（a*）或黄度值（b*）（Faucitano等，2006；Partanen等，2007）。另外屠宰前禁食除了有助于改善肉质，还可以降低运输和待屠宰过程中猪的死亡率，减少因胃肠道被刺破导致的致病菌污染胴体的发生，减少需要抛弃或处理的废弃物（Murray等，2001）。

◆ 2.2 饲喂蔗糖

早期研究发现，屠宰前短时间内补饲含25%～50%蔗糖的饲粮可以升高肌肉糖原含量和胴体肌肉初始pH（Sayre等，1963）。Fernandes等（1979）报道，给育肥猪提供葡萄糖、果糖或蔗糖糖浆溶液也可以提高肌肉糖原含量；而且他们发现猪屠宰前在存放栏中静休时，简单地给猪饮用自来水就足可以升高猪肉的初始pH、最终pH，改善猪肉品质。最近，Camp等（2003）给猪饲喂含0～15%蔗糖的饲粮，发现蔗糖升高了背最长肌的红度值（a*）和黄度值（b*），但是背最长肌的滴水损失率随着蔗糖水平升高而线性升高。短期和长期饲喂蔗糖，很显然可以增加宰前肌肉糖原贮存，但是通过饲喂糖来增加糖原贮备并不等于改善鲜猪肉的品质。

◆ 2.3 降低肌肉糖原含量的饲粮

欧洲最近的研究表明，饲喂高脂（17%～19%）、高蛋白（19%～25%）和低可消化碳水化合物（<5%）的饲粮，可以有效降低猪背最长肌的总糖原，包括糖原前体和高分子糖原的浓度（Rosenvold等，2003；Bee等，2006）。重要的是，饲喂这些降肌糖原饲粮的猪，屠宰后肌肉pH45 min升高，滴水损失下降（Rosenvold等，2001，2002），但对肉色的作用不确定（Rosenvold等，2001，2002；Bee等，2006）。屠宰前给猪饲喂高蛋白、高脂肪和低碳水化合物饲粮对猪肉食用品质影响的报道很少（Rosenvold等，2001；Bee等，2006）。Leheska等（2002）给屠宰前2～14 d的猪饲喂超高蛋白质（33.7% CP）、高脂肪（19.6%粗脂肪）饲粮，没有观察到这个降糖原饲粮对肌肉糖原贮备、肌肉pH下降以及其他鲜猪肉品质的影响；但是这个美国的单一试验中的饲粮纤维含量没有欧洲同类试验中的饲粮纤维含量那么高，这可能是Leheska等（2002）与欧洲研究结果不一致的原因。

◆ 2.4 其他改变宰后胴体代谢的饲粮调控

2.4.1 补充色氨酸

一般认为宰前应激影响肌肉的糖原贮存、最终影响猪肉品质。调节肌肉糖原贮存的一个可能途径是调节猪的应激反应。5-羟色氨是一种调节哺乳动物好斗行为的神经递质，它的产生直接受色氨酸刺激。因此，提高饲粮中的色氨酸水平可以缓解猪的好斗行为（Warner等，1998），同样也可以降低血液中皮质醇和乳酸的浓度（Guzik等，2006）。特别重要的是，Adeola和Ball（1992）研究表明，在育肥猪饲粮中添加0.5%的L-色氨酸也降低了PSE猪肉的发生率。另外，Guzik等（2006）研究表明，屠宰前5 d，猪采食添加0.5%的L-色氨酸可以获得更好的肉色评分和低的L*值。但是在过去十年里的大部分试验，并没有观察到饲粮中添加L-色氨酸能改善猪肉品质的作用（Panella-Riera等，2008，2009）。

2.4.2 补充镁

饲粮中长期（Otten等，1992）和短期（D'Souza等，1999）添加镁都可以有效地减少猪宰前应激反应，降低PSE肉发生率（D'Souza等，1998,2000）。更重要的是，即使宰前1星期给猪饲喂补充镁的饲粮，也可以改善鲜猪肉的系水力，而且与镁源无关（图17-1；Apple，2007）。进一步的研究表明，长期（Apple等，2000）或短期（D'Souza等，2000）饲喂补充镁的饲粮都可以改善鲜猪肉的肉色。

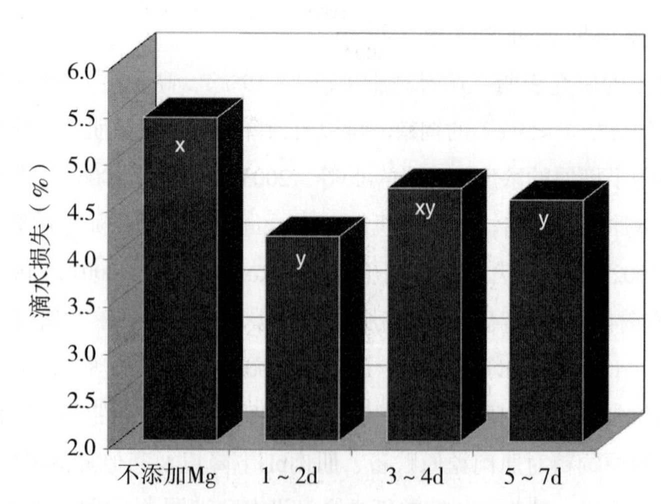

图17-1 屠宰前饲喂补充镁的饲粮的持续时间对背最长肌滴水损失的影响的综合分析

x，y表示差异显著（$P<0.05$）

（改编自Apple, 2007）

2.4.3 补充一水肌氨酸

肌氨酸是肝、肾和胰腺中甘氨酸、精氨酸和蛋氨酸的代谢产物，作为底物，它增加了细胞中磷酸肌酸生成ATP的生物效率。根据人上的研究结果，Berg和Allee（2001）推断猪饲粮中添加一水肌氨酸，可能会通过增加肌肉中磷酸肌酸的水平，节约肌肉中的糖原，减少PSE肉的发生率。但是一水肌氨酸改善猪肉质的作用并不确定、可靠。Young等（2005）报道，屠宰前5 d，饲粮中补充一水肌氨酸提高了背最长肌初始pH。但是其他的研究表明，屠宰前补充肌氨酸并没有改善背最长肌初始和最终pH或鲜猪肉的肉色（James等，2002；Rosenvold等，2007）。事实上，一些研究表明，补充一水肌氨酸升高了猪肉的亮度（L*值升高），降低了猪肉的红度（a*值降低），肉变得更黄（b*值升高）（Stahl等，2001；Young等，2005）。另一方面，有证据表明，育肥猪饲粮中补充肌氨酸能减少背最长肌的滴水损失（James等，2002；Young等，2005）。

3 肌内脂肪含量的饲粮调控

肌内脂肪含量对于消费者感受熟猪肉良好的嫩度、风味和多汁性具有重要作用（Lonergan等，2007）。2.5%～3.0%的肌内脂肪含量是满足美国消费者对熟猪肉品质接受所必需的，而大多数美国猪肉进口伙伴国的消费者则更喜欢肌内脂肪含量至少4%以上的猪肉（NPPC大理石花纹评分4）。但由于美国养猪业者在过去20年饲喂瘦肉型品种的猪，肌内脂肪含量降低到了1.0%（Gil等，2008），因此增加猪肌内脂肪含量或肌肉大理石花纹评分的责任就落在猪营养学家的肩上。

◆ 3.1 饲粮蛋白质和氨基酸对猪肌内脂肪含量的调节

提高饲粮粗蛋白和/或Lys水平被反复证实具有增加猪胴体瘦肉产量（Grandhi和Cliplef，1997）和猪肉的水分含量（Friesen等，1994；Goerl等，1995），但降低了肌内脂肪含量的作用（Goodhand等，1990，1993；Grandhi和Cliplef，1997）。事实上，Goal等（1995）发现饲粮粗蛋白水平从10%升高到22%，背最长肌的肌内脂肪含量下降71.3%。饲粮Lys水平从0.54%升高到1.04%（Frisen等，1994）或从0.8%

升高到1.4%（Johnston等，1993），背最长肌的大理石评分呈线性下降。

另外，降低饲粮粗蛋白或Lys水平是一个有效增加肌内脂肪含量的措施。当生长育肥猪采食粗蛋白水平低的饲粮，肌内脂肪含量升高了66.7%～136.8%（Blanchard等，1999；Cameron等，1999）。但是，猪长期采食蛋白或Lys缺乏的饲粮会对日增重和饲料转化率产生不良影响。有趣的是，育肥阶段的最后5～6星期饲喂Lys缺乏饲粮对猪生产性能没有不利影响，但却提高了背最长肌肌内脂肪含量，只是没达到其他研究报道的相同程度（Cisneros等，1996；Bidner等，2004）。

Cisneros等（1996）考察了育肥猪饲粮中额外添加2.0%的亮氨酸的效果，发现补充亮氨酸提高了猪肉大理石评分（+29.8%），提高了背最长肌肌内脂肪含量（+25.7%）和半腱肌肌内脂肪含量（+18.4%），并且不影响猪的生长性能。Hyun等（2009）给猪饲喂含1.22%或3.22%亮氨酸的饲粮，显著改善了背最长肌大理石评分（+21.9%）和肌内脂肪含量（+41.7%），但饲喂高亮氨酸饲粮猪的育肥阶段的生长速度几乎下降了11%。以上两个例子中，肌内脂肪含量升高直接归因于饲粮亮氨酸水平升高还是因为添加高水平亮氨酸使得Lys吸收降低而引起饲粮氨基酸模式不平衡间接导致的，尚值得推敲。

◆ 3.2 饲粮能量水平与来源对猪肌内脂肪的影响

3.2.1 饲料与能量摄入

育肥猪即使严格限饲，也不会影响肌肉pH（Cameron等，1999；Lebret等，2001）或鲜猪肉的品质（Cameron等，1999；Sterten等，2009）。按自由采食量的75%～80%限饲的试验反复证实，肌内脂肪含量降低8%～27%（Lebret等，2001；Daza等，2007）。有趣的是，降低育肥猪饲粮的能量浓度不改变肌内脂肪的含量或其他鲜猪肉的品质（Lee等，2002），也没有证据表明饲粮中谷物来源会影响大理石花纹的评分（Camp等，2003；Carr等，2005a；Sullivan等，2007）。

3.2.2 油脂

数十年以来，油脂是用来增加猪饲粮能量浓度和改善制粒的，但是饲粮中油脂的水平和/或来源对猪肌内脂肪的作用并不稳定。Miller等（1990）报道，猪饲粮中添加10%的葵花籽油或双低菜籽油，会降低背最长肌大理石花纹评分。Myer等（1992）指出，随着饲粮中双低菜籽油添加水平的提高，大理石花纹评分线性下降。相反，Apple等（2008b）观察到，背最长肌肌内脂肪含量随着饲粮中玉米油含量的增加而增加；给猪采食含5%牛油的饲粮，背最长肌的肌内脂肪含量升高了大约25%（Eggert等，2007）。但是，在大多数的情况下，鲜有研究表明，饲粮中的油脂对猪肉大理石花纹评分（Eggel等，2001；Apple等，2008a）或肌内脂肪含量（Morel等，2006）有影响。

共轭亚油酸（CLA）是亚油酸的同分异构体，是一组在碳位（c8、c10、c9、c11、c10、c10、C12和c13）具有双键（cis/cis、cis/trans和trans/trans）的亚油酸的位置和几何异构体。大多数合成的CLA包括65%的CLA异构体，主要是cis-9/trans-11和trans-10/cis-12异构体。2009年6月CLA面世，商品名为Lutalin（BASF SE，Ludwigschafen，Genmany），用于添加在猪和鸡的饲料中。重要的是猪饲粮中添加CLA，可增加背最长肌大理石花纹评分或肌内脂肪含量（图17-2）。Dugan等（1999）报道，添加CLA，大理石花纹评分提高了11.3%。而且Joo等（2002）、Sun等（2004）和Martin等（2008b）均报道，饲粮中添加CLA，肌内脂肪含量增加幅度最小为12%，最大为44%。Wiegand等（2001）的研究表明，添加0.75% CLA可以使氟烷敏感基因阴性猪、携带者和阳性猪的背最长肌的肌内脂肪含量分别升高17.8%、19.2%和16.6%。

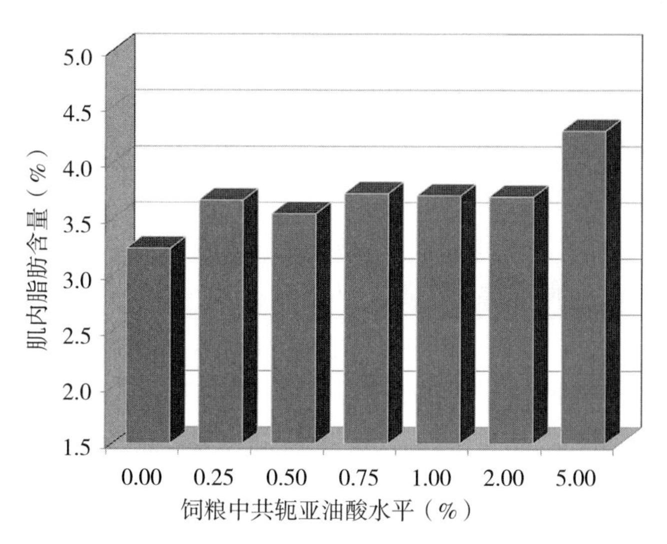

图17-2 饲粮中共轭亚油酸水平（CLA）对背最长肌肌内脂肪（IMF）含量的影响

对有关文献的综合分析表明，随着饲粮中CLA的增加IMF成线性增加（P=0.05）。

3.2.3 补充维生素A

维生素A的衍生物视黄酸参与脂肪细胞的分化和增殖调控（Pairault等，1988），因此视黄酸缺乏可能会直接升高肌内脂肪细胞的增殖和肌内脂肪含量。事实上，给牛饲喂维生素A缺乏饲粮，提高了背最长肌大理石花纹的评分或肌内脂肪含量，而不影响牛的生长性能和胴体组成（Oka等，1998；Gorocica-Buenfil等，2007）。D'Souza等（2003）证明，生长育肥猪采食维生素A缺乏饲粮，肌内脂肪含量提高了54%。而Olivares等（2009）则指出，给猪饲喂含100 000 IU维生素A饲粮，对于肌内脂肪含量倾向高的基因型猪，其肌内脂肪含量确实升高，但高瘦肉率猪的肌内脂肪含量则没有升高。证据表明，维生素A缺乏或超量的饲粮具有增加肌内脂肪/大理石花纹评分的前景，但是维生素A的添加量、饲喂阶段以及与其他饲料原料和饲料添加剂的相互作用在很大程度上还不清楚，特别是什么时候该给生长育肥猪补充维生素A也不清楚。

3.2.4 盐酸莱克多巴胺（RAC）

盐酸莱克多巴胺（Paylean，Elanco Animal Health，Greenfield，IN）（中国禁止使用，编者注）是一种营养重分配的饲料添加物，能改善猪生长性能和胴体瘦肉产量，但不影响鲜猪肉的pH、质地或系水力（Apple等，2007）。尽管有很多报道或传闻说莱克多巴胺可以影响肌肉大理石花纹评分或肌内脂肪含量，如育肥猪饲粮中添加5 mg/kg（Watkins等，1990）、10 mg/kg（Apple等，2008a）或20 mg/kg（Carr等，2009）的莱克多巴胺升高了大理石花纹评分。但事实上，大多数的研究并没有发现莱克多巴胺影响肌肉大理石花纹评分（Carr等，2005b；Patience等，2009；Rincker等，2009）。

4 猪脂肪品质的营养调控

猪肉和脂肪中脂肪酸来源有两个途径：一是从头合成，二是从饲粮中吸收的外源脂肪酸。例如，从头合成通常是利用从玉米和大麦中消化摄入的葡萄糖为底物，因此会增加体脂中的SFA比例，而代价是降低体脂从谷物油脂中吸收来的PUFA的比例（Lampe等，2006）。然而，正如前述，猪饲粮中通常要添加脂肪来增加饲粮的能量浓度和减少饲粮中的谷物，尤其是玉米的比例。

◆ 4.1 饲粮中脂肪的来源和猪脂肪品质

猪饲粮中油脂的品质取决于几个方面因素，包括碘值（IV，油脂的化学不饱和度的指标）和油脂的凝固点、熔点。高饱和度的脂肪源，如牛油、猪油的碘值为30～70 g I/100 g脂肪，凝固点为45～50℃；相反，不饱和程度高的油，如大豆油、菜籽油、玉米油、葵花籽油、红花籽油的碘值超过100 g I/100 g脂肪，凝固点低于30℃，熔点20℃或更低。另外，猪对油脂的消化率很显然随着饱和脂肪酸的比例下降而升高（Averette Gatlin等，2005）。猪脂肪的脂肪酸组成通常是饲粮油脂品质（如脂肪酸组成）的反映。猪采食添加牛油的饲粮，其脂肪倾向含低比例的多不饱和脂肪酸（PUFA）和低的碘值；而采食含植物油的饲粮，猪脂肪中的多不饱和脂肪酸组成将升高，代价是饱和脂肪酸（SFA）和单不饱和脂肪酸（MUFA）组成下降（表17-1）。

表17-1 不同脂肪来源的饲粮与不添加脂肪的对照组中皮下脂肪和背最长肌的脂肪酸组成比例的变化

项 目	牛油[1]	家禽脂肪[1]	大豆油[1]	玉米油[2]	菜籽油[3]	黄油[4]	精选白色油脂[5]
皮下脂肪							
总 SFA	−3.43	−8.33	−12.93	−14.40	−22.25	−0.60	−5.86
软脂酸（16 ： 0）	−4.60	−6.42	−11.70	−12.80	−20.62	−1.67	−5.68
硬脂酸（18 ： 0）	−3.38	−11.79	−15.45	−17.91	−28.36	+1.99	−8.79
总 MUFA	+4.80	+1.69	−11.17	−12.47	+2.93	−2.44	+2.42
油酸（18 ： 1^{cis9}）	+3.86	−0.76	−10.81	−10.41	+4.85	−4.12	+2.51
总 PUFA	−8.94	+10.46	+50.26	+91.50	+86.73	+8.38	+10.73
亚油酸（18 ： 2_{n-6}）	−9.96	+10.43	+46.53	+97.28	+44.44	+6.62	+10.63
亚麻酸（18 ： 3_{n-3}）	−1.37	+13.70	+180.82	+30.19	+162.50	+13.56	+4.76
碘值	−1.35	+5.84	+18.43	+23.72	—	—	—
背最长肌							
总 SFA	−0.98	−2.12	−2.83	−0.78	—	+3.34	−9.71
软脂酸（16 ： 0）	−2.09	−2.26	−3.41	−0.65	—	+2.28	−2.41
硬脂酸（18 ： 0）	+1.05	−1.40	−2.27	−1.26	—	+6.36	−2.49

(续)

项　目	牛油[1]	家禽脂肪[1]	大豆油[1]	玉米油[2]	菜籽油[3]	黄油[4]	精选白色油脂[5]
总 MUFA	−1.45	−2.55	−6.58	−5.70	—	−4.87	−0.33
油酸（18：1^{cis9}）	−0.52	−2.23	−5.35	−2.81	—	−6.85	−0.16
总 PUFA	+13.95	+22.64	+40.93	+31.82	—	+6.86	+16.37
亚油酸（18：2_{n-6}）	+13.67	+24.58	+45.13	+41.60	—	+6.20	+14.65
亚麻酸（18：3_{n-3}）	+17.14	+20.00	+148.57	+18.18	—	+12.31	+5.56
碘值	+3.35	+5.70	+10.83	+5.71	—	—	—

[1]脂肪来源包括5%饲喂基础（Apple等，2009a，2009b）。
[2]脂肪来源包括4%饲喂基础（Apple等，2008c）。
[3]脂肪来源包括5%饲喂基础（Myer等，1992）。
[4]脂肪来源包括4%饲喂基础（Averette Gatlin等，2002）。
[5]脂肪来源包括4%饲喂基础（Engel等，2001）。

尽管食用PUFA显然有益健康，但猪体脂中PUFA比例升高会导致软脂肪的发生。据Wittington等（1986）报道，猪体脂中C18：2_{n-6}脂肪酸含量高于15%就被认为软脂肪，因此饲喂富含C18：2_{n-6}的多不饱和脂肪将导致软脂肪（Miller等，1990；Myer等，1992）和软五花肉的发生（Apple等，2007a，2008b）。软脂肪和软五花肉会引起胴体搬运、处理和切割困难，减少培根肉的产量，造成肉产品没有吸引力、货架期变短，特别是受到美国国内消费者和出口伙伴国消费者的区别对待。研究表明，饲粮中脂肪的碘值从80降到20，五花肉的厚度和硬度显著增加（Averette Gatlin等，2003），因此饲喂动物脂肪不会像饲喂植物油那样严重地降低体脂和五花肉的硬度（Engel等，2001）。有趣的是，Shackelford等（1990）指出，与不添加脂肪的饲粮或添加牛脂肪饲粮的相比，给猪饲喂含葵花籽油、红花籽油、菜籽油的饲粮，其培根肉的松脆度、咀嚼性、含盐性、风味和整体接受度的感官评价得分更低。Teye等（2006）观察到，采食含大豆油的饲粮，猪将产软培根肉和大量低品质的软培根肉切片。

越来越多的证据表明，通过调节饲粮脂肪来源和添加水平，特别是饲粮脂肪来源，在14～35 d就可以改变50%～60%猪体脂的脂肪酸组成，但这种调节作用随着饲喂时间的延长而逐渐变小（Wiseman和Agunbiade，1998）。Apple等（2009a，2009b）报道，给猪饲喂5%大豆油在获得17.4 kg增重的第一个饲养阶段，背最长肌、皮下脂肪和胴体混合样品的脂肪酸组成显著改变，猪体脂的碘值升高了近12个点。Anderson等（1972）发现，猪皮下脂肪中的亚麻酸C18：3的半衰期接近300 d。因此出于节省和提高效率的目的，给生长猪饲喂高水平的油脂，可能在屠宰时给猪脂肪品质造成不可挽回的损失。因而，从育肥后期猪饲粮中去掉所有脂肪或用牛油、氢化脂肪代替饲粮中不饱和油脂对猪脂肪品质会产生戏剧性效果的可行性尚存疑问（Apple等，2009b）。

◆ 4.2 生产生物燃料的副产物

为了降低对化石类燃料的依赖，美国在开发生物燃料方面做出了巨大的努力，在利用玉米生产乙醇方面取得了显著进步，产生了大量的玉米干酒糟（DDGS），可以用来配制猪饲料。DDGS的粗脂肪含量为10%～15%（Rausch和Belyea，2006），而且其中不饱和脂肪酸比例高，因此饲喂DDGS含量高的饲粮会提高猪皮下脂肪的PUFA含量和碘值（Xu等，2008；White等，2009）。鲜五花肉中PUFA的比例随着猪饲粮中DDGS的比例升高而线性升高（Whitney等，2006；Xu等，2008；White等，2009），这会导致五花肉变软、易弯曲和不受欢迎（Whitney等，2006；Weimer等，2008；Widmer等，2008）。Weimer等（2008）报道，随着饲粮中DDGS的比例增加，肉脂分离率也会增加。Xu等（2008）指出，随着饲粮中DDGS的比例升高，虽然DDGS不改变熟培根肉的松脆度、风味和整体可接受度，但培根肉的脂肪和嫩度线性下降了。

任何新的或回收的动植物油脂，在催化剂的作用下均可以与乙醇反应，甲酯化后生成了常说的生物柴油。粗甘油是生物柴油的副产物，同DDGS一样，作为猪饲料的能量来源，引起广泛关注。Mourot等（1994）和Della Casa等（2009）发现，饲粮中添加5%～10%的粗甘油将增加猪背脂中C18：1^{cis9}和所有MUFA的比例，然而Mourot等（1994）和Lammers等（2008）观察到皮下脂肪和肌肉中C18：2_{n-6}的比例下降。而且，给猪饲喂含粗甘油的饲粮，多不饱和脂肪下降导致猪五花肉变得坚实（Schieck等，2009）。

◆ 4.3 共轭亚油酸（CLA）

饲粮中添加CLA通常会增加猪体脂和肌肉中SFA的比例，特别是棕榈酸（C16：0）和硬脂酸（C18：0）在猪体脂（Dugan等，2003；Sun等，2004）和肌肉（Eggert等，2001；Sun等，2004；Martin等，2008b）中的比例。但也有报道称饲粮中添加CLA，体脂和肌肉中C18：1^{cis9}和MUFA的比例没有变化（Eggert等，2001）或者升高了（Joo等，2002）。虽然这些报道互相矛盾，但在CLA对PUFA组成上的作用上却基本一致。除了Thiel-Cooper等（2001）、Averette Gatlin等（2002）和Martin等（2009）报道猪饲粮中添加CLA升高了鲜背最长肌C18：2_{n-6}含量外，绝大多数研究表明，猪饲粮中添加CLA降低（Joo等，2002；Sun等，2004）或者不改变（Eggert等，2001）猪瘦肉和脂肪中PUFA的比例。特别重要的是，SFA含量升高和伴随的PUFA比例降低，造成碘值下降（Eggert等，2001；Averette Gatlin等，2002；Larsen等，2009）和质地更坚实的猪脂肪（Dugan等，2003）和鲜五花肉（Eggert等，2001；Larsen等，2009）。

◆ 4.4 猪脂肪品质的其他营养调控措施

给猪饲喂低蛋白质、低赖氨酸饲粮，增加了鲜背最长肌中SFA、MUFA的比例，降低了PUFA的比例（Wood等，2004；Teye等，2006a）和碘值（Grandhi和Cliplef，1997）。给猪饲喂高油玉米配合的饲粮会增加其体脂中亚油酸（C18：2_{n-6}）和PUFA的比例（Rentfrow等，2003），而给猪饲喂高亚油酸玉米（Della Casa等，2010）或高油酸、高油玉米（Rentfrow等，2003），则显著地提高了鲜猪肉中亚油酸（C18：2_{n-6}）和油酸（C18：1^{cis}）的浓度。另外，Skelley等（1999）观察到，给猪饲喂小麦时由于小麦中的脂肪比玉米的脂肪更硬，但五花肉的坚实度似乎没有受到饲粮中谷物来源的影响（Skelley等，1975；Carr等，2005a）。

当猪采食量下降到自由采食量的70%～85%时，猪脂肪和鲜五花肉的坚实度显著降低（Haydon等，1989）。猪脂肪/五花肉坚实度下降似乎是对限饲引起的猪脂肪和肌肉中C18：1^{cis9}、各种MUFA和C18：2_{n-6}的比例升高的响应（Wood等，1996；Daza等，2007）。Daza等（2007）指出，生长猪阶段限饲会抑制脂肪生成酶类活性，即使随后恢复了自由采食，但猪到了育肥阶段这些酶的活性依然低于正常水平。

有证据表明，育肥猪饲粮中添加200 μg/kg来自吡啶羧酸铬的铬，降低了五花肉中C18：2_{n-6}和C18：3_{n-3}的比例和碘值，但是铬不改变五花肉的厚度和坚实度（Jackson等，2009）。饲粮中添加高水平的铜很经济，常用来做猪的促生长剂，饲粮高水平铜会提高猪脂肪中多不饱和脂肪的比例（Bosi等，2000）。

饲粮中添加莱克多巴胺具有升高猪脂肪中多不饱和脂肪酸和降低饱和脂肪酸的作用（Carr等，2005b；Apple等，2008a），其升高多不饱和脂肪酸的作用在于其能增加C18：2_{n-6}（Carr等，2005b；Apple等，2008a）和C18：3_{n-3}（Apple等，2008a）的沉积。Mill等（1990）报道，由于莱克多巴胺降低了脂肪的从头合成，因此猪皮下脂肪的脂肪酸组成是猪育肥后期饲粮脂肪酸组成的直接反映，特别是给猪饲喂了添加油脂的饲粮。

有趣的是，莱克多巴胺似乎不影响猪腹部肉（五花肉）的坚实度（Carr等，2005b；Apple等，2007a；Scramlin等，2008）、培根的品质（Scramlin等，2008）或培根的适口性（Jeremiah等，1994）。

肉碱是一种维生素类化合物，参与长链脂肪酸穿越线粒体内膜进入线粒体内进行β-氧化的过程，猪饲粮中添加L-肉碱可以提高猪生长效率和胴体瘦肉率（Owen等，2001；Chen等，2008），但不影响鲜猪肉的品质（Apple等，2008b）。Apple等（2008c）报道，猪饲粮中添加L-肉碱降低了背脂中PUFA的比例，升高了背最长肌中MUFA的比例，但不影响猪肉或脂肪的碘值。该书作者据此推测，L-肉碱可能在Δ^9脱饱和酶的作用下，促进C18：2_{n-6}去不饱和生成C18：$1^{cis\,9}$，导致多不饱和脂肪酸下降而单不饱和脂肪酸升高同时发生。

5 猪脂肪和肉色稳定性的营养调控

所有增加猪肉中PUFA含量的饲粮调整都会使猪肉脂肪更易氧化。事实上，猪饲粮中的双低菜籽油（Leskanich等，1997）、鱼油（Leskanich等，1997）、大豆油（Morel等，2006）、亚麻籽油（Morel等，2006）或高油玉米（Guo等，2006），都会升高冷藏中猪肉的TBARS值，因此大量研究聚焦饲粮中添加抗氧化剂，特别是VE，或者通过补充矿物添加剂来刺激内源的抗氧化酶系，来增强猪肉的抗氧化稳定性。

◆ 5.1 维生素E

维生素E（α-生育酚）是使自由基链式反应淬灭的抗氧化物质，以此保护细胞膜的完整性（Morrissey等，1993），以及阻滞脂肪和肌红蛋白的氧化（Faustman等，1989），特别是猪肉在冷藏和零售展示期间。因此在生长育肥猪饲粮中添加超营养量的维生素E，被普遍认为是可以改善猪肉质的营养调控措施。

研究人员反复证实，饲粮中添加100～200 mg/kg的dl-α-生育酚乙酸盐可以有效延迟新鲜、完整的肌肉切块（Monahan等，1994；Boler等，2009）和肉馅（Phillips等，2001；Boler等，2009）、烹饪前猪肉（Guo等，2006）和熟猪肉（Coronado等，2002）的脂质氧化反应。另外，因为脂肪氧化与色素氧化呈正相关，因此在育肥牛饲粮中添加维生素E，不只会延缓肉的变色，还会确实改善鲜牛肉的肉色稳定性（Faustman等，1989）。早期研究表明，猪饲粮中添加dl-α-生育酚乙酸盐也可以改善鲜猪肉的肉色稳定性（Monahan等，1994）。但是大多数的研究没有观察到通过在猪饲粮中添加dl-α-生育酚乙酸盐（Phillips等，2001；Guo等，2006），或其天然存在的立体异构体d-α-生育酚乙酸盐（Boler等，2009）来升高维生素E水平的做法对鲜猪肉的肉色或冷藏期间肉色稳定性有什么好处。

◆ 5.2 维生素C

维生素C是具有抗氧化能力的水溶性维生素，猪通常在肝中利用D-葡萄糖生成足量的维生素C。给即将屠宰的猪皮下注射维生素C可以降低PSE胴体的发生率（Cabadaj等，1983）。不过有研究表明，猪屠宰前4h补饲维生素C会使猪肉变得更红更暗（Peeters等，2006）。但也有研究指出，不论是短期（Ohene-Adjei等，2001；Pion等，2004），还是长期补充维生素C（Eichenberger等，2004；Gebert等，2006）都不影响猪肉肉色和系水力。没有证据表明，猪饲粮中补充维生素C会改善背最长肌脂质的氧化稳定性（Gebert等，2006）。实际上Ohene-Adjei等（2001）和Eichenberger等（2004）都报道给猪饲喂高水平维生素C的日粮，确实升高了背最长肌切块在冷藏期间的TBARS值。饲粮中停止补充维生素C，猪血液中维生素C水平会迅速降回到基线（Pion等，2004）。因此，补充维生素C的时间似乎是发挥维生素C改善肉质作用的关键。

◆ 5.3 矿物元素的补充

5.3.1 硒

硒是内源抗氧化酶谷胱甘肽过氧化物酶的组分，许多研究都表明，猪饲粮中添加亚硒酸钠和酵母硒复合物会增加血清谷胱甘肽过氧化物酶活性（Mahan等，1999；Zhan等，2007）。不过补硒升高谷胱甘肽过氧化物酶活性并不等同于改善了鲜肉的肉色和系水力（Mahan等，1999；Wolter等，1999），或鲜猪肉在贮藏过程中的脂质稳定性（Wolter等，1999；Han和Thacker，2006）。

5.3.2 锰

锰和镁都是二价的金属离子，在一些生物功能中，金属离子转移时，二者可以互换。但锰是过氧化物酶歧化酶（SOD）激活所必需的，以阻断超氧自由基链式反应。因此饲粮中补充锰会降低鲜背最长肌切块的TBARS值（Apple等，2005）。猪采食添加350 mg/kg锰的饲粮，背最长肌在2～4 d零售模拟展示中，肉色较对照组猪的肉色变化更小（Sawyer等，2007）。猪饲粮中补充锰的其他好处还包括提高背最长肌pH和可见肉色评分，降低鲜背最长肌猪肉的L*值（Apple等，2005，2007c）。饲粮中补充锰对于改善猪肉质很有希望，但还需要开展更多的试验，因为添加锰对猪肉适口性的影响还知之甚少，而且到目前为止，所有锰对猪肉质的研究都是由一个团队完成的。

◆ 5.4 育肥猪饲粮中移除维生素和微量元素预混剂的效应

越来越多的猪营养学家认为，大多数生长育肥猪全价饲粮中维生素或矿物元素含量已经满足甚至超过了NRC（1998）推荐量。因此减少饲粮特别是育肥期最后一个月的饲粮中的维生素和矿物元素，将不只降低养猪成本，也将减少磷和其他矿物元素的环境排放量（McGlone，2000）。更何况很少有证据表明，在育肥后期不添加维生素和微量元素会影响鲜猪肉的肉色、大理石花纹、肉的坚实度和剪切力（WBSF值，威纳-伯瑞泽勒剪切力）（Mavromichalis等，1999；Choi

等，2001；Shelton等，2004），唯一的不足是肉在冷冻贮存时TBARS值会升高（Choi等，2001；Hamman等，2001）。

6 熟猪肉适口性的营养调控

尽管鲜猪肉的肉色是影响消费者购买决定的最重要因素，但熟肉的风味将影响消费者是否是会成为再次购买的回头客，所以营养调控能否改善或调节肉的风味至关重要。

◆ 6.1 粗蛋白质/赖氨酸

当育肥猪饲粮中粗蛋白质水平从10%升高到22%，背最长肌熟肉切块的剪切力（WBSF）则升高了近23%（Goerl等，1995）。Goodband等（1990，1993）报道，当育肥猪饲粮赖氨酸（Lys）水平从0.6%上升到1.4%时，背最长肌和半腱肌熟肉块的剪切力（WBSF）也会线性升高。但是Apple等（2004）报道，育肥期饲粮的Lys与能量比值从1.7 g/Mcal（0.56%～0.59% Lys）升高到3.1 g/Mcal（1.02%～1.08% Lys）时，肉的剪切力（WBSF）也会线性升高。Goodband等（1990）也指出，当饲粮Lys水平升高时，肌纤维和整体嫩度感官评价得分降低。而Castell等（1994）报道随着Lys水平升高，猪肉风味评分下降。不过大多数时候，升高猪饲粮中Lys水平不影响熟猪肉的多汁性、风味强度或嫩度评分（Goodband等，1993；Castell等，1994；grandhi和Clilef,1997）。

◆ 6.2 饲粮能量水平和来源

降低生长育肥猪的饲粮能量浓度不影响猪肉风味（Lee等，2002），但是自由采食的猪的背最长肌切块嫩度分别比75%（Cameron等，1999）、80%（Blanchard等，1999）或82%（Elis等，1996）限饲的猪的背最长肌切块嫩度评分更高、剪切力（WBSF值）更低，尽管采食量并不会影响总的和可溶性肌肉胶原蛋白（Wood等，1996；Lebret等，2001）和肌纤维片断化指数（一种评价宰后胴体肌肉蛋白质降解度的指数；Cameron等，1999）。许多研究还表明，受训的品尝人员对自由采食的生长育肥猪的肉风味（Blanchard等，1999；Cameron等，1999）、风味喜好度、多汁性和整体可接受度（Elis等，1996；Cameron等，1999）的评价更高。

猪饲粮中谷物的来源可引起猪肉风味不同。譬如，饲喂小麦的猪比饲喂高粱的风味评分更高（McConnell等，1975），而饲喂33∶67或67∶33的黄玉米和白玉米复合物的猪的背最长肌肉块的多汁性和风味评分要高于分别饲喂黄玉米、白玉米或大麦的猪（Lampe等，2006）。McConnell等（1975）报道，与饲喂高粱相比，饲喂小麦的猪的背最长肌的WBSF更低，嫩度评分更高。Robertson等（1999）指出，饲喂大麦的猪比饲喂玉米或大麦与黑小麦复合物的猪的肉块嫩度感官评价更好。但也有相反的报道，给猪饲喂黄玉米、白玉米、小麦、大麦或黑小麦（Skelley等，1975；Carr等，2005b；Lampe等，2006；Sullivan等，2007），肉的剪切力（WBSF）都没有差异；无论是受过培训的品尝人员（Carr等，2005b；Sullivan等，2007）还是消费者品尝小组（Jeremiah等，1999）都没有觉察到生长猪饲粮、育肥猪饲粮或是生长育肥猪饲粮中不同的谷物来源对猪肉的嫩度、多汁性、风味或整体可接受度有影响。

给猪饲喂双低菜籽油，会使猪肉带上异味或怪味，从而降低熟猪肉的整体可接受度（Miller等，1990；Tikk等，2007）。但是，饲粮油脂来源既不影响肉的剪切力（Miller等，1990；Engel等，2001；Apple等，2008a，2008b），也不影响猪肉的嫩度、多汁性或风味强度的感官评价（Miller等，1990；Engel等，2001；Tikk等，2007）。给猪饲喂DDGS（Whitney等，2006；Widmer等，2008；Xu等，2008）或含甘油的配合饲粮（Lammers等，2008；Della Casa等，2009）都不影响熟的背最长肌肉块的剪切力（WBSF）和风味评分。另外，饲粮中添加CLA不影响猪肉的剪切力（WBSF值）（Dugan等，1999）、风味评分（Dugan等，1999—2003；Wiegand等，2001；Larsen等，2009）、风味品质（Averette Gatlin等，2006）或风味的挥发性组成（Martin等，2008）。

◆ 6.3 代偿性生长

代偿性生长是指猪限饲一段时间后恢复自由采食后发生加速生长的现象。限饲期猪蛋白质降解速度的增加现象在恢复自由采食后也不会降下来，由此Kristensen等（2002）推测，宰前蛋白质降解活性升高将会导致屠宰后肌肉快速嫩化。有趣的是，Kristensen等（2002）和Therkildsen等（2002）都发现，限饲一段时间后恢复自由采食，猪背最长肌中除了Calpastatin以外μ-Calpin和m-Calpin活性都升高了。Therkildsen等（2002）指出，屠宰前自由采食的时间越久，μ-Calpin活性越高。代偿性生长似乎不影响肌肉的总胶原含量，但也有研究表明，背最长肌中可溶性胶原的比例确实会随着限饲后自由采食而升高（Kristensen等，2002，2004；Therkildsen等，2002）。但是，只有真正代偿性生长的猪的肌肉WBSF值和嫩度的感官评分会得到改善（Kristensen等，2002）。也就是说，在那些限饲的时间或程度不足以引起生长速度显著下降的研究中，猪自由采食的时间长短对熟猪肉的风味，特别是嫩度改善作用很小甚至没有（Therkildsen等，2002；Kristensen等，2004；Heyer和Lebret，2007）。

◆ 6.4 维生素D_3

正因为肌肉中钙的浓度与肉嫩度的关系已被确认，所以一般认为增加肌肉中钙的浓度会增加宰后细胞骨架蛋白的钙蛋白酶（Calpin）降解，改善熟肉的嫩度。维生素D_3参与细胞间钙离子的迁移和调节，因此给育肥牛饲喂超营养水平的维生素D_3会升高血和肌肉中钙离子的水平，改善了熟牛肉的嫩度（Swanek等，1999）。但是，尽管给育肥猪饲粮中补充维生素D_3，血浆和肌肉中钙离子的浓度升高了125%（Wiegand等，2002；Lahucky等，2007），但猪肉的WBSF值（Wiegand等，2002；Swigert等，2004；Wilborn等，2004）、嫩度感官评分或者其他所有适口性品质（Swigert等，2004；Wilborn等，2004）都没有改变。有趣的是，有证据表明，给猪饲粮中补充超营养水平的维生素D_3，会改善鲜猪肉的品质，包括提高了肌肉的初始和最终的pH、客观肉色值和背最长肌的红度值a*，降低了亮度值L*和滴水损失（Wilborn等，2004；Swigert等，2004；Lahucky等，2007）。

◆ 6.5 盐酸莱克多巴胺（RAC）

尽管盐酸莱克多巴胺不影响熟猪肉的多汁性或风味（Carr等，2005a，2005b；Patience等，2009；Rincker等，2009），但大多数的研究表明，饲粮中添加少至5 mg/kg的盐酸莱克多巴胺将会提高熟猪肉的剪切力（WBSF值）（Patience等，2009；Rincker等，2009），降低嫩度的感官评价得分（Carr等，2005a，2005b；Patience等，2009）。一般认为，为了使盐酸莱克多巴胺促进猪生长的作用最大化，育肥晚期饲粮中的Lys水平必须达到1.0%（Webster等，2007），因此熟猪肉的剪切力（WBSF值）升高是含盐酸莱克多巴胺饲粮Lys水平升高的结果，而不是饲粮中补充盐酸莱克多巴胺本身的作用。尽管如此，Xiong等（2006）发现，屠宰后2 d、4 d、7 d的猪背最长肌切块的剪切力（WBSF值）高于对照组猪；但宰后10 d、14 d或21 d，猪肉的剪切力（WBSF值）与对照组相比没有差异。这表明，育肥猪饲粮中添加盐酸莱克多巴胺延缓了宰后猪肉嫩化的过程。

7 有机猪肉生产

过去十多年，因为消费者感觉在设施友好的环境中自由放养猪而生产的猪肉更有营养、更有益健康，所以自然放养方式生产的猪肉的小众市场得到了显著发展。很重要的是，这些消费者愿意为享有类似观点的小型家庭养殖场提供的自然猪肉产品支付额外的费用，即使消费者认为这样猪肉的适口性差些。据Honeyman等（2006）报道，美国自然猪肉市场即使每年多达75万头的猪但仍满足不了其国内市场的需求。自然猪肉市场面临着诸多挑战问题，其中生产成本高和为期一年的超长养殖周期是两个最迫切的问题。Stender等（2009）认为，自然猪肉的生产成本为66.50～99.00美元/100磅，而Honeyman等（2006）观察到有机猪肉的生产成本比自然猪肉的生产成本高400%～500%。另外，一年的养殖周期必然遇到冬天户外饲养的问题，寒冷引起断奶前仔猪大量死亡和较高的饲料消耗，都大大降低了生产效率（Bee等，2004；Gentry等，2004）。Honeyman等（2006）指出，猪肉品质对于自然生态养猪市场的可持续至关重要。户内和户外饲养的猪肉初始和最终pH相差无几，但是通过对已发表的33篇文章的综合分析表明，户外饲养的猪的鲜猪肉系水力显著下降（图17-3）。事实上，一些研究表明，与户内养猪相比，户外养猪猪肉滴水损失率升高（Gentry等，2002a，2002b；Galián等，2008）、可榨出水分的百分比升高（Kim等，2009）。综合分析可指出，户外饲养的猪的肉色更深（L*低，P=0.05）、肉色偏黄（P=0.07）（图17-3）。有趣的是，一些研究者报道，户外饲养的猪肉色的红度（a*）值增加了20%或更多（Gentry等，2002b，2004；Hoffman等，2003）。但综合分析表明，户内和户外饲养的猪肉的红度值没有差异。

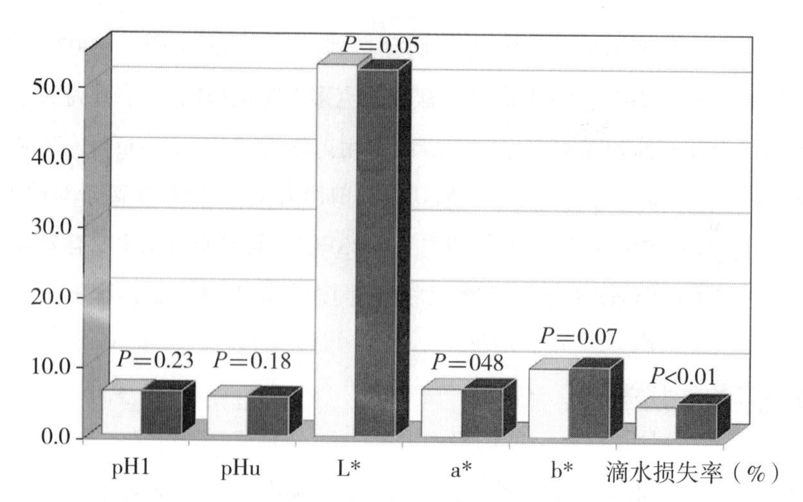

图17-3　综合分析法研究比较了猪舍饲养猪（浅色柱）和生态放养猪（深色柱）的新鲜猪肉的品质（n=33）

品质包括初始肌肉pH（pH1）、最终肌肉pH（pHu）、色差仪评分和滴水损失百分比。

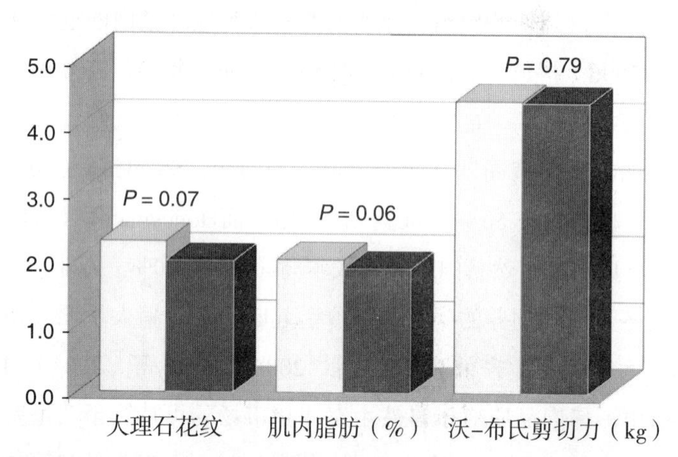

图17-4　综合分析法研究比较了猪舍饲养猪（浅色柱）和生态放养猪（深色柱）的猪肉大理石花纹评分（n=33）、肌内脂肪含量（IMF）和沃-布氏剪切力（WBSF）值

大多数研究和综合分析都表明，传统户内饲养的猪的猪肉大理石花纹（Gentry等，2002a，2004）和肌内脂肪含量（Högberg等，2002；Hoffman等，2003；Bee等，2004）高于同期户外饲养的猪。但是，不论是综合分析（图17-4）还是大量研究都没有发现来源于户外或户内饲养的猪的熟猪肉的剪切力（WBSF值）有什么差别（Hoffman等，2003；Oksbjerg等，2005；Galián

等，2008）。不足为奇的是，户外饲养的猪的肉的多汁性评分低于户内饲养的猪（Gentry等，2002a；Jonsäll等，2002），但感官品尝评定，户内和户外饲养的猪的肉风味、嫩度和整体可接受度相当（Jons äll等，2002；Gentry等，2002a，2002b；Kim等，2009）。

与户内饲养的猪相比，户外饲养的猪的皮下脂肪中，C16：0、C18：0和/或所有饱和脂肪酸的比例通常较低，而C18：2_{n-6}、C18：3_{n-3}或所有多不饱和脂肪酸比例较高（Bee等，2004；Hansen等，2006）。Bee等（2004）、Oksbjerg等（2005）以及González和Tejeda（2007）报道，户外饲养的猪的背最长肌的C16：0的比例下降。Kim等（2009）观察到，自由放养的猪的背最长肌中C18：1^{cis9}降低了11.7%，所有MUFA降低了12.2%，但户外饲养的猪通常只是背最长肌MUFA比例降低了5%的（表17-2）。有趣的是，Högberg等（2002）和Galián等（2008）都观察到，户外饲养的猪的背最长肌的PUFA组成大幅下降，是因为户外饲养的猪的肉中C18：2_{n-6}、C18：3_{n-3}和所有SFA的比例（Oksbjerg等，2005；González和Tejeda，2008；Kim等，2009）似乎都增加了。户外饲养的猪的背最长肌中不饱和脂肪酸升高的原因是与细胞膜组成中有关的极性脂肪（磷脂）部分中的PUFA的含量升高了。如前文指出的，户外饲养的猪的背最长肌中肌内脂肪含量（中性脂肪部分）通常显著低于户内饲养的猪。

表17-2 自然户外饲养猪与猪舍饲养猪的背最长肌肌内脂肪的脂肪酸组成[1,2]

参考文献	16：0	18：0	∑SFA	18：1^{cis9}	∑MUFA	18：2_{n-6}	18：3_{n-6}	∑PUFA
Högberg 等，2001	−1.9	−9.8*	−5.3*	+3.1*	+2.1	+6.9	+19.1*	+1.5
Högberg 等，2002	+3.4	−0.4	+2.5	+5.4*	+7.3*	−30.9*	−47.6*	−30.1*
Hoffman 等，2003	+0.5	−21.7*	−8.3	−5.4	−6.9	+44.8*	+9.9	+33.7*
Bee 等，2004	−1.9*	−0.6	−2.2	−3.3*	−3.6*	+29.1*	+45.2*	+31.3*
	−1.0*	+0.2	−1.2*	−6.7*	−6.8*	+27.2*	+55.6*	+33.1*
	+3.0	+2.2	+2.8	−3.2	−1.9	−7.2	0.0	−6.9
Högberg 等，2004	+0.4	+4.9	+1.8	0.0	−1.4	+4.0	+8.0	+4.0
	−0.8	+2.4	+0.3	−0.7	−0.9	+8.4	+27.6*	+10.1

（续）

参考文献	16∶0	18∶0	∑SFA	$18:1^{cis9}$	∑MUFA	$18:2_{n-6}$	$18:3_{n-6}$	∑PUFA
Oksbjerg 等，2005	−1.3	−1.6	−0.6	−2.9*	−3.5*	+12.5	+13.3	+12.4
	−3.8*	−24.0*	−6.9*	−7.5*	−8.4*	+39.7	−2.2	+38.2*
González 和 Tejeda，2007	−5.8*	−10.6	−7.2*	+3.1	+2.8*	+31.0*	+94.1*	+28.3
Galián 等，2008	+17.4*	−10.6*	+10.7*	−4.2*	−4.1*	−14.7	−3.7	−21.4
Kim 等，2009	−7.8*	−0.6	−6.9*	−11.7*	−12.2*	+54.1*	+31.9	+58.3*

[1]与生态放养猪相比的比例变化，*表明与猪舍饲养猪相比差异显著。

[2]16∶0=软脂酸；18∶0=硬脂酸；∑SFA=所有饱和脂肪酸的总和；$18:1^{cis9}$=油酸；∑MUFA=所有单不饱和脂肪的总和；$18:2_{n-6}$=亚油酸；$18:3_{n-3}$=亚麻酸；∑PUFA=所有多不饱和脂肪酸的总和。

绝大多数关于自然猪肉产品的研究都是在欧洲开展的，美国类似的自然生态养猪的研究较少，在大多数提到的研究中，猪是饲养在苜蓿草地上。因此其他牧草对猪肉品质影响的研究，特别是对熟猪肉的适口性方面，目前还未见到报道。Gentry等（2002b）指出，户外出生和长大的猪的猪肉在零售展示中更容易褪色。此外，自然放养的猪的猪肉PUFA高，但关于其脂肪稳定性的研究几乎没有。动物福利立法很可能迫使猪营养学家和肉品科学家在即将到来的时间里，又重新探讨50年前类似的问题，即如何通过营养调控来改善自然生态养猪的生产效率和猪肉品质。

作者：Jason K. Apple
译者：尹靖东

参考文献

Adeola, O., and R. O. Ball. 1992. Hypothalamic neurotransmitter concentrations and meat quality in stressed pigs offered excess dietary tryptophan and tyrosine. J. Anim. Sci. 70:1888–1894.

Anderson, D. B., R. G. Kauffman, and N. J. Benevenga. 1972. Estimate of fatty acid turnover in porcine adipose tissue. Lipids 7:488–489.

Apple, J. K. 2007. Effects of nutritional modifications on the water-holding capacity of fresh pork: A review. J. Anim. Breeding Gen. 124(Suppl. 1):43–58.

Apple, J. K., C. V. Maxwell, D. C. Brown, K. G. Friesen, R. E. Musser, Z. B. Johnson and T. A. Armstrong. 2004. Effects of dietary lysine and energy density on performance and carcass characteristics of finishing pigs fed ractopamine. J. Anim. Sci. 82:3277–3287.

Apple, J. K., C. V. Maxwell, B. deRodas, H. B. Watson, and Z. B. Johnson. 2000. Effect of magnesium mica on performance and carcass quality of growing–finishing swine. J. Anim. Sci. 78:2135–2143.

Apple, J. K., C. V. Maxwell, D. L. Galloway, C. R. Hamilton, and J. W. S. Yancey. 2009a. Interactive effects of dietary fat source and slaughter weight in growing–finishing swine: I. Growth performance and longissimus muscle fatty acid composition. J. Anim. Sci. 87:1407–1422.

Apple, J. K., C. V. Maxwell, D. L. Galloway, C. R. Hamilton, and J. W. S. Yancey. 2009b. Interactive effects of dietary fat source and slaughter weight in growing–finishing swine: II. Fatty acid composition of subcutaneous fat. J. Anim. Sci. 87:1423–1440.

Apple, J. K., C. V. Maxwell, J. T. Sawyer, B. R. Kutz, L. K. Rakes, M. E. Davis, Z. B. Johnson, et al. 2007a. Interactive effect of ractopamine and dietary fat source on quality characteristics of fresh pork bellies. J. Anim. Sci. 85:2682–2690.

Apple, J. K., C. V. Maxwell, B. R. Kutz, L. K. Rakes, J. T. Sawyer, Z. B. Johnson, T. A. Armstrong, et al. 2008a. Interactive effect of ractopamine and dietary fat source on pork quality characteristics of fresh pork chops during simulated retail display. J. Anim. Sci. 86:2711–2722.

Apple, J. K., P. J. Rincker, F. K. McKeith, S. N. Carr, T. A. Armstrong, and P. D. Matzat. 2007b. Meta-analysis of ractopamine responses in finishing swine. Prof. Anim. Sci. 23:179–196.

Apple, J. K., W. J. Roberts, C. V. Maxwell, L. K. Rakes, K. G. Friesen, and T. M. Fakler. 2007c. Influence of dietary inclusion level of manganese on pork quality during retail display.

Meat Sci. 75:640–647.

Apple, J. K., W. J. Roberts, C. V. Maxwell, C. B. Boger, K. G. Friesen, L. K. Rakes and T. M. Fakler. 2005. Influence of dietary manganese source and supplementation level on pork quality during retail display. J. Muscle Food. 16:207–222.

Apple, J. K., J. T. Sawyer, C. V. Maxwell, J. C. Woodworth, J. W. S. Yancey, and R. E. Musser. 2008b. Effect of L−carnitine supplementation on the performance and pork quality traits of growing−finishing swine fed three levels of corn oil. J. Anim.Sci. 86(E−Suppl. 2):37. (Abstr.)

Apple, J. K., J. T. Sawyer, C. V. Maxwell, J. C. Woodworth, J. W. S. Yancey, and R. E. Musser. 2008c. Effect of L−carnitine supplementation on the fatty acid composition of subcutaneous fat and LM from swine fed three levels of corn oil. J. Anim.Sci. 86(E−Suppl. 2):38. (Abstr.)

Averette Gatlin, L., M. T. See, J. A. Hansen, and J. Odle. 2003. Hydrogenated dietary fat improves pork quality of pigs from two lean genotypes. J. Anim. Sci. 81:1989–1997.

Averette Gatlin, L., M. T. See, D. K. Larick, X. Lin, and J. Odle. 2002. Conjugated linoleic acid in combination with supplemental dietary fat alters pork quality. J. Nutr. 132:3105–3112.

Averette Gatlin, L., M. T. See, and D. K. Larick, and J. Odle. 2006. Descriptive flavor analysis of bacon and pork loin from lean−genotype gilts fed conjugated linoleic acid and supplemental fat. J. Anim. Sci. 84:3381–3386.

Averette Gatlin, L., M. T. See, and J. Odle. 2005. Effects of chemical hydrogenation of supplemental fat on relative apparent lipid digestibility in finishing swine. J. Anim. Sci. 83:1890–1898.

Bee, G., C. Biolley, G. Guex, W. Herzog, S. M. Lonergan, and E. Huff−Lonergan. 2006. Effects of available dietary carbohydrate and preslaughter treatment on glycolytic potential, protein degradation, and quality traits of pig muscles. J. Anim. Sci. 84:191–203.

Bee, G., G. Guex, and W. Herzog. 2004. Free−range rearing of pigs during the winter: adaptations in muscle fiber characteristics and effects on adipose tissue composition and meat quality traits. J. Anim. Sci. 82:1206–1218.

Berg, E. P., and G. L. Allee. 2001. Creatine monohydrate supplemented in swine finishing diets and fresh pork quality: I. A controlled laboratory experiment. J. Anim. Sci. 79:3075–3080.

Bidner, B. S., M. Ellis, D. P. Witte, S. N. Carr, and F. K. McKeith. 2004. Influence of dietary lysine level, pre−slaughter fasting, and rendement napole genotype on fresh pork quality. Meat Sci. 68:53–60.

Blanchard, P. J., M. Ellis, C. C. Warkup, B. Hardy, J. P. Chadwick, and G. A. Deans. 1999. The influence of rate of lean and fat tissue development on pork eating quality. Anim. Sci. 68:477–485.

Boler, D. D., S. R. Gabriel, H. Yang, R. Balsbaugh, D. C. Mahan, M. S. Brewer, F. K.

McKeith, and J. Killefer. 2009. Effect of different dietary levels of natural-source vitamin E in grow-finish pigs on pork quality and shelf life. Meat Sci. 83:723–730.

Bosi, P., J. A. Cacciavillani, L. Casini, D. P. Lo Fiego, M. Marchett, and S. Mattuzzi. 2000. Effects of dietary high-oleic acid sunflower oil, copper and vitamin E levels on the fatty acid composition and the quality of dry cured Parma ham. Meat Sci. 54:119–126.

Cabadaj, R., J. Pleva, and P. Mala. 1983. Vitamin C in the prevention of PSE and DFD meat. Folia Veterinaria 27:81–87.

Cameron, N. D., J. C. Penman, andA. C. Fisken, G. R. Nute, A. M. Perry and J. D.Wood. 1999. Genotype with nutrition interactions for carcass composition and meat quality in pig genotypes selected for components of efficient lean growth rate. Anim. Sci. 69:69–80.

Camp, L. K., L. L. Southern, and T. D. Bidner. 2003. Effect of carbohydrate source on growth performance, carcass traits, and meat quality of growing-finishing pigs. J. Anim. Sci. 81:2488–2495.

Carr, S. N., D. N. Hamilton, K. D. Miller, A. L. Schroeder, D. Fern'andez-Dueñas, J. Killefer, M. Ellis, and F. K. McKeith. 2009. The effect of ractopamine hydrochloride (Paylean R _) on lean carcass yields and pork quality characteristics of heavy pigs fed normal and amino acid fortified diets. Meat Sci. 81:533–539.

Carr, S. N., D. J. Ivers, D. B. Anderson, D. J. Jones, D. H. Mowrey, M. B. England, J. Killefer, et al. 2005a. The effects of ractopamine hydrochloride on lean carcass yields and pork quality characteristics. J. Anim. Sci. 83:2886–2893.

Carr, S. N., P. J. Rincker, J. Killefer, D. H. Baker, M. Ellis, and F. K. McKeith. 2005b. Effects of different cereal grains and ractopamine hydrochloride on performance, carcass characteristics, and fat quality in late-finishing pigs. J. Anim. Sci. 83:223–230.

Castell, A. G., R. L. Cliplef, L. M. Poste-Flynn, and G. Butler. 1994. Performance, carcass and pork characteristics of castrates and gilts self-fed diets differing in protein content and lysine:energy ratio. Can. J. Anim. Sci. 74:519–528.

Chen, Y. J., I. H. Kim, J. H. Cho, J. S. Yoo, Q. Wang, Y. Wang, and Y. Huang. 2008. Evaluation of dietary l-carnitine or garlic powder on growth performance, dry matter and nitrogen digestibilities, blood profiles and meat quality in finishing pigs. Anim. Feed Sci. Technol. 141:141–152.

Choi, S. C., B. J. Chae, and I. K. Han. 2001. Impacts of dietary vitamins and trace minerals on growth and pork quality in finishing pigs. Asian-Aust. J. Anim. Sci. 14:1444–1449.

Cisneros, F., M. Ellis, D. H. Baker, R. A. Easter, and F. K. McKeith. 1996. The influence of short-term feeding of amino acid-deficient deits and high dietary leucine levels on the intramuscular fat content of pig muscle. Anim. Sci. 63:317–322.

Coronado, S. A., G. R. Trout, F. R. Dunshea, N. P. Shah. 2002. Effect of dietary vitamin E, fishmeal and wood and liquid smoke on the oxidative stability of bacon during 16 weeks' frozen storage. Meat Sci. 62:51–60.

Daza, A., A. I. Rey, D. Menoyo, J. M. Bautista, A. Olivares, and C. J. L'opez-Bote. 2007. Effect of level of feed restriction during growth and/or fattening on fatty acid composition and lipogenic enzyme activity in heavy pigs. Anim. Feed Sci. Technol. 138:61–74.

Della Casa, G., D. Bochicchio, V. Faeti, G. Marchetto, E. Poletti, A. Garavaldi, A. Panciroli, and. N. Brogna. 2009. Use of pure glycerol in fattening heavy pigs. Meat Sci. 81:238–244.

Della Casa, G., D. Bochicchio, V. Faeti, G. Marchetto, E. Poletti, A. Rossi, A. Panciroli, et al. 2010. Performance and fat quality of heavy pigs fed maize differing in linoleic acid content. Meat Sci. 84:152–158.

DeVol, D. L., F. K. McKeith, P. J. Bechtel, J. Navakofski, R. D. Shanks, and T. R. Carr. 1988. Variation in composition and palatability traints and relationships between muscle characteristics and palatability in a random sample of pork carcasses. J. Anim. Sci. 66:385–395.

D'Souza, D. N., D.W. Pethick, F. R. Dunshea, J. R. Pluske, and B. P. Mullan. 2003. Nutritional manipulation increases intramuscular fat levels in the Longissimus muscle of female finisher pigs. Aust. J. Agric. Res. 54:745–749.

D'Souza, D. N., R. D. Warner, B. J. Leury, and F. R. Dunshea. 1998. The effect of dietary magnesium aspartate supplementation on pork quality. J. Anim. Sci. 76:104–109.

D'Souza, D. N., R. D. Warner, B. J. Leury, and F. R. Dunshea. 2000. The influence of dietary magnesium supplement type, and supplementation dose and duration, on pork quality and the incidence of PSE pork. Aust. J. Agric. Res. 51:185–189.

Dugan, M. E. R., J. L. Aalhus, L. E. Jeremiah, J. K. G. Kramer, and A. L. Schaefer. 1999. The effects of feeding conjugated linoleic acid on subsequent pork quality. Can. J. Anim. Sci. 79, 45–51.

Dugan, M. E. R., J. L. Aalhus, D. C. Rolland, and L. E. Jeremiah. 2003. Effects of feeding different levels of conjugated linoleic acid and total oil to pigs on subsequent pork quality and palatability. Can. J. Anim. Sci. 83:713–720.

Eggert, J. M., M. A. Belury, A. Kempa-Steczko, S. E. Mills, and A. P. Schinckel. 2001. Effects of conjugated linoleic acid on the belly firmness and fatty acid composition of genetically lean pigs. J. Anim. Sci. 79:2866–2872.

Eggert, J. M., A. L. Grant, and A. P. Schinckel. 2007. Factors affecting fat distribution in pork carcasses. Prof. Anim. Sci. 23:42–53.

Eichenberger, B., H. P. Pfirter, C. Wenk, and S. Gebert. 2004. Influence of dietary vitamin E and C supplementation on vitamin E and C content and thiobarbituric acid reactive substances (TBARS) in different tissues of growing pigs. Arch. Anim. Nutr. 58:195–208.

Ellis, M., A. J. Webb, P. J. Avery, and P. J. Brown. 1996. The influence of terminal sire genotype, sex, slaughter weight, feeding regime and slaughter–house on growth performance and carcass and meat quality in pigs and on the organoleptic properties of fresh pork. Anim. Sci. 62:521–530.

Engel, J. J., J. W. Smith, II, J. A. Unruh, R. D. Goodband, P. R. O'Quinn, M.D. Tokach, and J. L. Nelssen. 2001. Effects of choice white grease or poultry fat on growth performance, carcass leanness, and meat quality characteristics of growing–finishing pigs. J. Anim. Sci. 79:1491–1501.

Faucitano, L., L. Saucier, J. A. Correa, S. Methot, A. Giguere, A. Foury, P. Mormede, and R. Bergeron. 2006. Effect of feed texture, meal frequency and pre–slaughter fasting on carcass and meat quality, and urinary cortisol in pigs. Meat Sci. 74:697–703.

Faustman, C., R. G. Cassens, D. M. Schaefer, D. R. Buege, S. N. Williams, and K. K. Scheller. 1989. Improvement of pigment and lipid stability in Holstein steer beef by dietary supplementation with vitamin E. J. Food Sci. 54:858–862.

Fernandes, T. H., W. C. Smith, M. Ellis, J. B. K. Clark, and D. G. Armstrong. 1979. The administration of sugar solutions to pigs immediately prior to slaughter. 2. Effect on carcass yield, liver weight and muscle quality in commercial pigs. Anim. Prod. 29:223–230.

Friesen, K. G., J. L. Nelssen, R. D. Goodband, M. D. Tokach, J. A. Unruh, D. H. Kropf, and B. J. Kerr. 1994. Influence of dietary lysine on growth and carcass composition of high–lean–growth gilts fed from 34 to 72 kilograms. J. Anim. Sci. 72:1761–1770.

Gali'an, M., A. Poto, M. Santaella, and B. Pelnado. 2008. Effects of the rearing system on the quality traits of the carcass, meat and fat of the Chato Murciano pig. J. Anim. Sci. 79:487–497.

Gebert, S., B. Eichenberger, H. P. Pfirter, and C. Wenk. 2006. Influence of different vitamin C levels on vitamin E and C content and oxidative stability in various tissues and stored m. longissimus dorsi of growing pigs. Meat Sci. 73:362–367.

Gentry, J. G., J. J. McGlone, J. R. Blanton, Jr., and M. F. Miller. 2002a. Alternate housing systems for pigs: influences on growth, composition, and pork quality. J. Anim. Sci. 80:1781–1790.

Gentry, J. G., J. J. McGlone, M. F. Miller, and J. R. Blanton, Jr. 2002b. Diverse birth and rearing environment effects on pig growth and meat quality. J. Anim. Sci. 80:1707–1715.

Gentry, J. G., J. J. McGlone, M. F. Miller, and J. R. Blanton, Jr. 2004. Environmental effects on pig performance, meat quality, and muscle characteristics. J. Anim. Sci. 82:209–217.

Gil, M., M. I. Delday, M. Gispert, M. Fonti Furnols, C. M. Maltin, G. S. Plastow, and R. Klont. 2008. Relationships between biochemical characteristics and meat quality of *Longissimus thoracis* and *Semimembranosus* muscles in five porcine lines. Meat Sci. 80:927–933.

Goerl, K. F., S. J. Eilert, R. W. Mandigo, H.Y. Chen, and P. S. Miller. 1995. Pork characteristics as affected by two populations of swine and six crude protein levels. J. Anim. Sci. 73:3621–3626.

Gonz'alez, E., and J. F. Tejeda. 2007. Effects of dietary incorporation of different antioxidant extracts and free-range rearing on fatty acid composition and lipid oxidation of Iberian pig meat. Anim. 1:1060–1067.

Goodband, R. D., J. L. Nelssen, R. H. Hines, D. H. Kropf, G. R. Stoner, R. C. Thalel, A. J. Lewis, and B. R. Schricker. 1993. Interrelationships between porcine somatotropin and dietary lysine on growth performance and carcass characteristics of finishing swine. J. Anim, Sci. 71:663–672.

Goodband, R. D., J. L. Nelssen, R. H. Hines, D. H. Kropf, R. C. Thaler, B. R. Schricker, G. E. Fitzner, and A. J. Lewis. 1990. The effects of porcine somatotropin and dietary lysine on growth performance and carcass characteristics of finishing swine. J. Anim. Sci. 68:3261–3276.

Gorocica–Buenfil, M. A., F. L. Fluharty, T. Bohn, S. J. Schwartz, and S. C. Loerch. 2007. Effect of low vitamin A diets with high–moisture or dry corn on marbling and adipose tissue fatty acid composition of beef steers. J. Anim. Sci. 85:3355– 3366.

Grandhi, R. R., and R. L. Cliplef. 1997. Effects of selection for lower bac*kf*at, and increased levels of dietary amino acids to digestible energy on growth performance, carcass merit and meat quality in boars, gilts, and barrows. Can. J. Anim. Sci. 77:487–496.

Guo, Q., B. T. Richert, J. R. Burgess, D. M.Webel, and D. E. Orr, M. Blair, A. L. Grant, and D. E. Gerrard. 2006. Effects of dietary vitamin E and fat supplementation on pork quality. J. Anim. Sci. 84:3089–3099.

Guzik, A. C., J. O. Matthews, B. J. Kerr, T. D. Bidner, and L. L. Southern. 2006. Dietary tryptophan effects on plasma and salivary cortisol and meat quality in pigs. J. Anim. Sci. 84:2251–2259.

Hamm, R. 1986. Functional properties of the myofibrillar system and their measurements. Pages 135–199 in Muscle as Food. P. J. Bechtel, ed. Acad. Press Inc., Orlando, FL.

Hamman, L. L., J. G. Gentry, C. B. Ramsey, J. J. McGlone, and M. F. Miller. 2001. The effect of vitamin–mineral nutritional modulation on pork quality of halothane carriers. J. Muscle Food. 12:37–51.

Han, Y. K., and P. A. Thacker. 2006. Effect of l–carnitine, selenium–enriched yeast, jujube fruit and hwangto (red clay) supplementation on performance and carcass measurements of finishing pigs. Asian–Aust. J. Anim. Sci. 19:217–223.

Hansen, L. L., C. Claudi–Magnussen, S. K. Jensen, and H. J. Anderson. 2006. Effect of organic pig production systems on performance and meat quality. Meat Sci. 74:605–615.

Haydon, K. D., T. D. Tanksley, Jr., and D. A. Knabe. 1989. Performance and carcass composition of limit-fed growing-finishing swine. J. Anim. Sci. 67:1916–1925.

Heyer, A., and B. Lebret. 2007. Compensatory growth response in pigs: effects on growth performance, composition of weight gain at carcass and muscle levels, and meat quality. J. Anim. Sci. 85:769–778.

Hoffman, L. C., E. Styger, M. Muller, and T. S. Brand. 2003. The growth and carcass and meat characteristics of pigs raise din a free-range or conventional housing system. S. Afr. J. Anim. Sci. 33:166–175.

Högberg, A., J. Pickova, J. Babol, K. Andersson, and P. C. Dutta. 2002. Muscle lipids, vitamins E and A, and lipid oxidation as affected by diet and RN genotype in female and castrated male Hampshire crossbred pigs. Meat Sci. 60:411–420.

Högberg, A., J. Pickova, P. C. Dutta, J. Babol, and A. C. Bylund. 2001. Effect of rearing system on muscle lipids of gilts and castrated male pigs. Meat Sci. 58:223–229.

Högberg, A., J. Pickova, S. Stern, K. Lundström, and A. C. Bylund. 2004. Fatty acid composition and tocopherol concentrations in muscle of entire male, castrated male and female pigs, reared in an indoor or outdoor housing system. Meat Sci. 68:659–665.

Honeyman, M. S., R. S. Pirog, G. H. Huber, P. J. Lammers, and J. R. Hermann. 2006. The United States pork niche market phenomenon. J. Anim. Sci. 84:2269–2275.

Hyun, Y., M. Ellis, G. Bressner, and D. Baker. 2009. Effect of dietary leucine levels on carcass composition, meat quality, and growth performance in finishing pigs. Acessed Aug., 2009. http://www.livestocktrail.uiuc.edu/uploads/porknet/ paperDisplay.cfm?ContentID=126.

Jackson, A. R., S. Powell, S. L. Johnston, J. O. Matthews, T. D. Bidner, F. R. Valdez, and L. L. Southern. 2009. The effect of chromium as chromium propionate on growth performance, carcass traits, meat quality, and the fatty acid profile of fat from pigs fed no supplemented dietary fat, choice white grease, or tallow. J. Anim. Sci. 87:4032–4041.

James, B. W., R. D. Goodband, J. A. Unruh, M. D. Tokach, J. L. Nelssen, S. S. Dritz, P. R. O'Quinn, and B. S. Andrews. 2002. Effect of creatine monohydrate on finishing pig growth performance, carcass characteristics and meat quality. Anim. Feed Sci. Technol. 96:135–145.

Jeremiah, L. E., R. O. Ball, J. K. Merrill, P. Dick, L. Stobbs, L. L. Gibson, and B. Uttaro. 1994. Effects of feed treatment and gender on the flavour and texture profiles of cured and uncured pork cuts. I. Ractopamine treatment and dietary protein level. Meat Sci. 37:1–20.

Jeremiah, L. E., A. P. Sather, and E. J. Squires. 1999 Gender and diet influences on pork palatability and consumer acceptance. I. Flavor and texture profiles and consumer acceptance. J. Muscle Food. 10:305–316.

Johnston, M. E., J. L. Nelssen, R. D. Goodband, D. H. Kropf, R. H. Hines, and B. R. Schricker. 1993. The effects of porcine somatotropin and dietary lysine on growth performance and carcass characteristics of finishing swine fed to 105 or 127 kilograms. J. Anim. Sci. 71:2986–

2995.

Jonsäll, A., L. Johansson, K. Lundström, K. H. Andersson, A. N. Nilsen, and E. Risvik. 2002. Effects of genotype and rearing system on sensory characteristics and preference for pork (M. longissimus dorsi). Food Qual. Pref. 13:73–80.

Joo, S. T., J. I. Lee, Y. L. Ha, and G. P. Park. 2002. Effects of dietary conjugated linoleic acid on fatty acid composition, lipid oxidation, color, and water–holding capacity of pork loin. J. Anim. Sci. 80:108–112.

Kim, D. H., P. N. Seong, S. H. Cho, J. H. Kim, J. M. Lee, C. Jo, and D. G. Lim. 2009. Fatty acid composition and meat quality of organically reared Korean native black pigs. Livest. Sci. 120:96–102.

Kristensen, L., M. Therkildsen, M. D. Aaslyng, N. Oksbjerg, and P. Ertbjerg. 2004. Compensatory growth improvesmeat tenderness in gilts but not in barrows. J. Anim. Sci. 82:3617–3624.

Kristensen, L., M. Therkildsen, B. Riis, M. T. SÀkensen, N. Oksbjerg, P. P. Purslow, and P. Ertbjerg. 2002. Dietary–inducted changes of muscle growth rate in pigs: Effects on *in vivo* and postmortem muscle proteolysis and meat quality. J. Anim. Sci. 80:2862–2871.

Lahucky, R., I. Bahelka, U. Kuechenmeister, K. Vasickova, K. Nuernberg, K. Ender, and G. Nuernberg. 2007. Effects of dietary supplementation of vitamins D3 and E on quality characteristics of pigs and longissimus muscle antioxidative capacity. Meat Sci. 77:264–268.

Lammers, P. J., B. J. Kerr, T. E. Weber, K. Bregendahl, S. M. Lonergan, K. J. Prusa, D. U. Ahn, et al. 2008. Growth performance, carcass characteristics, meat quality, and tissue histology of growing pigs fed crude glycerin–supplemented diets. J. Anim. Sci. 86:2962–2970.

Lampe, J. F., J. W. Mabry, and T. Baas. 2006. Comparison of grain sources for swine diets and their effect on meat and fat quality traits. J. Anim. Sci. 84:1022–1029.

Larsen, S. T., B. R. Wiegand, F. C. Parrish, J.E. Swan, and J.C. Sparks. 2009. Dietary conjugated linoleic acid changes belly and bacon quality from pigs fed varied lipid sources. J. Anim. Sci. 87:285–295.

Lebret, B., H. Juin, J. Noblet, and M. Bonneau. 2001. The effects of two methods of increasing age at slaughter on carcass and muscle traits and meat sensory quality in pigs. Anim. Sci. 72:87–94.

Lee, C. Y., H. P. Lee, J. H. Jeong, K. H. Baik, S. K. Jin, J. H. Lee, and S. H. Sohnt. 2002. Effects of restricted feeding, lowenergy diet, and implantation of trenbolone acetate plus estradiol on growth, carcass traits, and circulating concentrations of insulin–like growth factor (IGF)–I and IGF–binding protein–3 in finishing barrows. J. Anim. Sci. 80:84–93.

Leheska, J. M., D. M.Wulf, J. A. Clapper, R. C. Thaler, and R. J. Maddock. 2002. Effects of high–protein/low–carbohydrate swine diets during the final finishing phase on pork muscle quality. J. Anim. Sci. 80:137–142.

Leskanich, C. O., K. R. Matthews, C. C.Warkup, R. C. Noble, and M. Hazzledine. 1997. The effect of dietary oil containing (*n*−3) fatty acids on the fatty acid, physiochemical, and organoleptic characteristics of pig meat and fat. J. Anim. Sci. 75:673–683.

Lonergan, S. M., K. J. Stalder, E. Huff–Lonergan, T. J. Knight, R. N. Goodwin, K. J. Prusa, and D. C. Beitz. 2007. Influence of lipid content on pork sensory quality within pH classification. J. Anim. Sci. 85:1074–1079.

Mahan, D. C., T. R. Cline, and B. Richert. 1999. Effects of dietary levels of selenium–enriched yeast and sodium selenite sources fed to growing–finishing pigs on performance, tissue selenium, serum glutathione peroxidase activity, carcass characteristics, and loin quality. J. Anim. Sci. 77:2172–2179.

Martin, D., T. Antequera, E. Muriel, A. I. Andres, and J. Ruiz. 2008a. Oxidative changes of fresh pork loin from pig, caused by dietary conjugated linoleic acid and monounsaturated fatty acids, during refrigerated storage. Food Chem. 111:730–737.

Martin, D., T. Antequera, E. Muriel, A. I. Andres, and J. Ruiz. 2009. Quantitative changes in the fatty acid profile of lipid fractions of fresh loin from pigs as affected by dietary conjugated linoleic acid and monounsaturated fatty acids during refrigerated storage. J. Food Compos. Anal. 22:102–111.

Martin, D., E. Muriel, E. Gonzalez, J. Viguera, and J. Ruiz. 2008b. Effect of dietary conjugated linoleic acid and monounsaturated fatty acids on productive, carcass and meat quality traits of pigs. Livest. Sci. 117:155–164.

Mavromichalis, I., J.D. Hancock, I. H. Kim, B.W. Senne, D. H. Kropf, G. A.Kennedy, R. H. Hines, and K. C. Behnke. 1999. Effects of omitting vitamin and trace mineral premixes and(or) reducing inorganic phosphorus additions on growth performance, carcass characteristics, and muscle quality in finishing pigs. J. Anim. Sci. 77:2700–2708.

McConnell, J. C., G. C. Skelley, D. L. Handlin, and W. E. Johnston. 1975. Corn, wheat, milo and barley with soybean meal or roasted soybeans on feedlot performance, carcass traits and pork acceptability. J. Anim. Sci. 41:1021–1030.

McGlone, J. J. 2000. Deletion of supplemental minerals and vitamins during the late finishing period does not affect pig weight and feed intake. J. Anim. Sci. 78:2797–2800.

Milan, D., J. T. Jeon, C. Looft, V. Amarger, A. Robic, M. Thelander, C. Rogel–Gaillard, et al. 2000. A mutation in PRAKG3 associated with excess glycogen content in pig skeletal muscle. Science 288:1248–1251.

Miller, M. F., S. D. Schackelford, K. D. Hayden, and J. O. Reagan. 1990. Determination of the alteration in fatty acid profiles, sensory characteristics and carcass traits of swine fed elevated levels of monounsaturated fats in the diet. J. Anim. Sci. 68:1624–1631.

Mills, S. E., C. Y. Liu, Y. Gu, and A. P. Schinckel. 1990. Effects of ractopamine on adipose tissue metabolism and insulin binding in finishing hogs. Interaction with genotype and slaughter weight. Domest. Anim. Endocrinol. 7:215–264.

Monahan, F. J., A. Asghar, J. I. Gray, D. J. Buckley, and P. A. Morrissey. 1994. Effect of oxidized dietary lipid and vitamin E on the colour stability of pork chops. Meat Sci. 37:205–215.

Monin, G., and P. Sellier. 1985. Pork of lowtechnological quality with a normal rate of muscle pH fall in the immediate post–mortem period—the case of the Hampshire breed. Meat Sci. 13:49–63.

Morel, P. C. H., J. C. McIntosh, and J. A. M. Janz. 2006. Alteration of the fatty acid profile of pork by dietary manipulation. Asian–Australasain J. Anim. Sci. 19:431–437.

Morrissey, P. A., P. J. Sheehy, and P. Gaynor. 1993. Vitamin E. Int. J. Vit. Nutr. Res. 63:260–264.

Mourot, J.,A. Aumaitre, A. Mounier, P. Peiniaua, and A. C. Françoisb. 1994. Nutritional and physiological effects of dietary glycerol in the growing pigs. Consequences on fatty tissues and post mortem muscular parameters. Livest. Prod. Sci. 38:237–244.

Murray, A., W. Robertson, F. Nattress, and A. Fortin. 2001. Effect of pre–slaughter overnight feed withdrawal on pig carcass and muscle quality. Can. J. Anim. Sci. 81:89–97.

Myer, R. O., J. W. Lamkey, W. R. Walker, J. H. Brendemuhl, and G. E. Combs. 1992. Performance and carcass characteristics of swine when fed diets containing canola oil and copper to alter the unsaturated:saturated ratio of pork fat. J. Anim. Sci. 70:1417–1423.

NRC. 1998. *Nutrient Requirements of Swine*. 10th rev. ed. National Academies Press, Washington, DC.

Ohene–Adjei, S., T. Bertol, Y. Hyun, M. Ellis, S. Brewer, and F. K. McKeith. 2001. The effect of dietary supplemental vitamin E and C on odors and color changes in irradiated pork. J. Anim. Sci. 79(Suppl. 1):443. (Abstr.)

Oka, A., Y. Maruo, T. Miki, T. Yamasaki, and T. Saito. 1998. Influence of vitamin A on the quality of beef from the Tajima strain of Japanese Black cattle. Meat Sci. 48:159–167.

Oksbjerg, N., K. Strudsholm, G. Lindahl, and J. E. Hermansen. 2005. Meat quality of fully or partly outdoor reared pigs in organic production. Acta Agric. Scand. Section A. 55:106–112.

Olivares, A., A. Daza, A. I. Rey, and C. J. Lopez–Bote. 2009. Interactions between genotype, dietary fat saturation and vitamin A concentration on intramuscular fat content and fatty acid composition in pigs. Meat Sci. 82:6–12.

Otten,W., A. Berrer, S. Hartmann, T. Bergerhoff, and H. M. Eichinger. 1992. Effects of magnesium furmarate supplementation on meat quality in pigs. Pages 117–120 in Proc. 38th International Congress of Meat Science and Technology, Clermont–Ferrand, France.

Owen, K. Q., H. Jit, C. V. Maxwell, J. L. Nelssen, R. D. Goodband, M.D. Tokach, G. C.

Tremblay, and S. I. Koo. 2001. Dietary l-carnitine suppresses mitrochondrial branched-chain keto acid dehydrogenase activity and enhances protein accretion and carcass characteristics of swine. J. Anim. Sci. 79:3104–3112.

Pairault, J., A. Quignard-Boulange, I. Dugail, and F. Lasnier. 1988. Differential effects of retinoic acid upon early and late events in adipose conversion of 3T3 preadipocytes. Exp. Cell Res. 177:27–36.

Panella-Riera, N., A. Dalmau, E. Fábrega, M. Font i Furnols, M. Gispert, J. Tibau, J. Soler, et al. 2008. Effect of supplementation with $MgCO_3$ and l-tryptophan on the welfare and on carcass and meat quality of two halothane pig genotypes (NN and nn). Livest. Sci. 115:107–117.

Panella-Rierá, N., A. Velarde, A. Dalmau, E. F`abrega, M. Font i Furnols, M. Gispert, J. Soler, et al. 2009. Effect of magnesium sulfate and l-tryptophan and genotype on the feed intake, behaviour and meat quality of pigs. Livest. Sci. 124:277–287.

Partanen, K., H. Siljander-Rasi, M. Honkavaara, and M. Ruusunen. 2007. Effects of finishing diet and pre-slaughter fasting time on meat quality in crossbred pigs. Agric. Food Sci. 16:245-258.

Patience, J. F., P. Shand, Z. Pietrasik, J. Merrill, G. Vessie, K. A. Ross, and A. D. Beaulieu. 2009. The effect of ractopamine supplementation at 5 ppm of swine finishing diets on growth performance, carcass composition and ultimate pork quality. Can. J.Anim. Sci. 89:53–66.

Peeters, E., B. Driessen, and R. Geers. 2006. Influence of supplemental magnesium, tryptophan, vitamin C, vitamin E, and herbs on stress response and pork quality. J. Anim. Sci. 84:1827–1838.

Phillips, A. L., C. Faustman, M. P. Lynch, K. E. Govoni, T. A. Hoagland, and S. A. Zinn. 2001. Effect of dietary α-tocopherol supplementation on color and lipid stability in pork. Meat Sci. 58:389–393.

Pion, S. J., E. van Heugten, M. T. See, D. K. Larick, and S. Pardue. 2004. Effects of vitamin C supplementation on plasma ascorbic acid and oxalate concentrations and mat quality in swine. J. Anim. Sci. 82:2004–2012.

Rausch, K. D., and R. L. Belyea. 2006. The future of coproducts from corn processing. Appl. Biochem. Biotechnol. 128:47–86.

Rentfrow, G., T. E. Sauber, G. L. Allee, and E. P. Berg. 2003. The influence of diets containing either conventional corn, conventional corn with choice white grease, high oil corn, or high oil high oleic corn on belly/bacon quality. Meat Sci. 64:459–466.

Rincker, P. J., J. Killefer, P. D. Matzat, S. N. Carr, and F. K. McKeith. 2009. The effect of ractopamine and intramuscular fat content on sensory attributes of pork from pigs of similar genetics. J. Muscle Food. 20:79–88.

Robertson, W. M., S. Jaikaran, L. E. Jeremiah, D. F. Salmon, F. X. Aherne, and S. J.

Landry. 1999. Meat quality and palatability attributes of pork from pigs fed corn, hulless barley or triticale based diets. Adv. Pork Prod. 10:35.

Rosenvold, K., H. C. Bertram, and J. F. Young. 2007. Dietary creatine monohydrate has no effect on pork quality of Danish crossbred pigs. Meat Sci. 76:160–164.

Rosenvold, K., B. Ess'en-Gustavsson, and H. J. Andersen. 2003. Dietary manipulation of pro- and macroglycogen in porcine skeletal muscle. J. Anim. Sci. 81:130–134.

Rosenvold, K., H. N. Lærke, S. K. Jensen, A. H. Karlsson, K. Lundström, and H. J. Andersen. 2001. Strategic finishing feeding as a tool in the control of pork quality. Meat Sci. 59:397–406.

Rosenvold, K., H. N. Lærke, S. K. Jensen, A. H. Karlsson, K. Lundström, and H. J. Andersen. 2002. Manipulation of critical quality indicators and attributes in pork through vitamin E supplementation, muscle glycogen reducing finishing feeding and pre–slaughter stress. Meat Sci. 62:485–496.

Sather, A. P., L. E. Jeremiah, and E. J. Squires. 1999. Effects of castration on live performance, carcass yields, and meat Quality of male pigs fed wheat or corn based diets. J. Muscle Food. 10:245–259.

Sawyer, J. T., A.W. Tittor, J. K. Apple, J. B. Morgan, C. V. Maxwell, L. K. Rakes, and T. M. Fakler. 2007. Effects of supplemental manganese on performance of growing–finishing pigs and pork quality during retail display. J. Anim. Sci. 85:1046–1053.

Sayre, R. N., E. J. Briskey, and W. G. Hoekstra. 1963. Effect of excitement, fasting, and sucrose feeding on porcine muscle phosphorylase and post–mortem glycolysis. J. Food Sci. 28:472–477.

Schieck, S. J., L. J. Johnston, G. C. Shurson, and B. J. Kerr. 2009. Evaluation of crude glycerol, a biodiesel co–product, in growing pig diets to support growth and improved pork quality. J. Anim. Sci. 87(E–Suppl. 3):90. (Abstr.)

Scramlin, S. M., S. N. Carr, C. W. Parks, D. M. Fernandez–Duenas, C. M. Leick, F. K. McKeith, and J. Killefer. 2008. Effect of ractopamine level, gender, and duration of ractopamine on belly and bacon quality traits. Meat Sci. 80:1218–1221.

Shackelford, S. D., M. F. Miller, K. D. Haydon, N. V. Lovegren, C. E. Lyon, and J. O. Reagan. 1990. Acceptability of bacon as influenced by the feeding of elevated levels of monounsaturated fats to growing–finishing swine. J. Food Sci. 55:621–624.

Shelton, J. L., L. L. Southern, F. M. LeMieux, T. D. Bidner, and T. G. Page. 2004. Effects of microbial phytase, low calcium and phosphorus, and removing the dietary trace mineral premix on carcass traits, pork quality, plasma metabolites, and tissue mineral content in growing–finishing pigs. J. Anim. Sci. 82:2630–2639.

Skelley, G. C., R. F. Borgman, D. L. Handlin, J. C. Acton, J. C. McConnell, F. B. Wardlaw, and E. J. Evans. 1975. Influence of diet on quality, fatty acids and acceptability of pork. J. Anim. Sci. 41:1298–1304.

Stahl, C. A., G. L. Allee, and E. P. Berg. 2001. Creatine monohydrate supplemented in swine finishing diets and fresh pork quality: II. Commercial applications. J. Anim. Sci. 79:3081–3086.

Stender, D., J. Kliebenstein, R. Ness, J. Mabry, and G. Huber. 2009. Costs, returns, production and financial efficiency of niche pork production in 2008. Accessed Nov. 2009. http://www.valuechains.org/NR/rdonlyres/18362A2A-0540-4E4C-A337- E842501848E0/111967/CostsReturnsProdEffic09212009.pdf.

Sterten, H., T. Frøystein, N. Oksbjerg, A. C. Rehnberg, A. S. Ekker, and N. P. Kjos. 2009. Effect of fasting prior to slaughter on technological and sensory properties of the loin muscle (M. *longissimus dorsi*) of pigs. Meat Sci. 83:351–357.

Sullivan, Z. M., M. S. Honeyman, L. R. Gibson, and K. J. Prusa. 2007. Effect of triticale-based diets on pig performance and pork quality in deep-bedded hoop barns. Meat Sci. 76:428–437.

Sun, D., Z. Zhu, S. Qiao, S. Fan, and D. Li. 2004. Effects of conjugated linoleic acid levels and feeding intervals on performance, carcass traits and fatty acid composition of finishing barrows. Arch. Anim. Nutr. 58:277–286.

Swanek, S. S., J. B. Morgan, F. N. Owens, D. R. Gill, C. A. Stratia, H. G. Dolezal, and F. K. Ray. 1999. Vitamin D3 supplementation of beef steers increases longissimus tenderness. J. Anim. Sci. 77:874–881.

Swigert, K. S., F. K. McKeith, T. C. Carr, M. S. Brewer, andM. Culbertson. 2004. Effects of dietary D3, vitamin E, and magnesium supplementation on pork quality. Meat Sci. 67:81–86.

Teye, G. A., P. R. Sheard, F. M. Whittington, G. R. Nute, A. Stewart, and J. D. Wood. 2006a. Influence of dietary oils and protein level on pork quality. 1. Effects on muscle fatty acid composition, carcass, meat and eating quality. Meat Sci. 73:157–165.

Teye, G. A., J. D. Wood, F. M. Whittington, A. Stewart, and P. R. Sheard. 2006b. Influence of dietary oils and protein level on pork quality. 2. Effects on properties of fat and processing characteristics of bacon and fran*kf*urter-style sausages. Meat Sci. 73:166–177.

Therkildsen, M., B. Riis, A. Karlsson, L. Kristensen, P. Ertbjerg, P. P. Purslow, M. D. Aaslyng, and N. Oksbjerg. 2002. Compensatory growth response in pigs, muscle protein turn-over and meat texture: Effects of restriction/realimentation period. Anim. Sci. 75:367–377.

Thiel-Cooper, R. L., F. C. Parrish, Jr., J. C. Sparks, B. R. Wiegand, and R. C. Ewan. 2001. Conjugated linoleic acid changes swine performance and carcass composition. J. Anim. Sci. 79:1821–1828.

Tikk, K., M. Tikk, M. D. Aaslyng, A. H. Karlsson, G. Lindahl, and A. H. Andersen. 2007.

Significance of fat supplemented diets on pork quality—connections between specific fatty acids and sensory attributes of pork. Meat Sci. 77:275–286.

Warner, R. D., G. A. Eldridge, C. D. Hofmeyr, and J. L. Barnett. 1998. The effect of dietary tryptophan on pig behaviour and meat quality—preliminary results. Anim. Prod. Aust. 22:325.

Watkins, L. E., D. J. Jones, D. H. Mowrey, D. B. Anderson, and E. L. Veenhuizen. 1990. The effect of various levels of ractopamine hydrochloride on the performance and carcass characteristics of finishing swine. J. Anim. Sci. 68:3588–3595.

Webster, M. J., R. D. Goodband, M. D. Tokach, J. L. Nelssen, S. S. Dritz, J. A. Unruh, K. R. Brown, et al. 2007. Interactive effects between ractopamine hydrochloride and dietary lysine on finishing pig growth performance, carcass characteristics, pork quality, and tissue accretion. Prof. Anim. Sci. 23:597–611.

Weimer, D., J. Stevens, A. Schinckel, M. Latour, and B. Richert. 2008. Effects of feeding increasing levels of distillers dried grains with solubles to grow–finish pigs on growth performance and carcass quality. J. Anim. Sci. 86(E-Suppl. 3):85.

White, H. M., B. T. Richert, J. S. Radcliffe, A. P. Schinckel, J. R. Burgess, S. L. Koser, S. S. Donkin, and M. A. Latour. 2009. Feeding conjugated linoleic acid partially recovers carcass quality in pigs fed dried corn distillers grains with solubles. J. Anim. Sci. 87:157–166.

Whitney, M. H., G. C. Shurson, L. J. Johnston, D. M.Wulf, and B. C. Shanks. 2006. Growth performance and carcass characteristics of grower-finisher pigs fed high-quality corn distillers dried grain with solubles originating from modern Midwestern ethanol plant. J. Anim. Sci. 84:3356–3363.

Whittington, F. M., N. J. Prescott, J. D.Wood, and M. Enser. 1986. The effect of dietary linoleic acid on the firmness of bac*kf*at in pigs of 85 kg live weight. J. Sci. Food Agric. 37:753–761.

Widmer, M. R., L. M. McGinnis, D. M. Wulf, and H. H. Stein. 2008. Effects of feeding distillers dried grains with solubles, highprotein distillers grains, and corn germ to growing-finishing pigs on pig performance, carcass quality, and the palatability of pork. J. Anim. Sci. 86:1819–1831.

Wiegand, B. R., F. C. Parrish, Jr., J. E. Swan, S. T. Larsen, and T. J. Baas. 2001. Conjugated linoleic acid improves feed efficiency, decreases subcutaneous fat, and improves certain aspects of meat quality in Stress-Genotype pigs. J. Anim. Sci. 79:2187–2195.

Wiegand, B. R., J. C. Sparks, D. C. Beitz, F. C. Parrish, R. L. Horst, A. H. Trenkle, and R. C. Ewan. 2002. Short-term feeding of vitamin D3 improves color but does not change tenderness of pork-loin chops. J. Anim. Sci. 80:2116–2121.

Wilborn, B. S., C. R. Kerth,W. F. Owsley,W. R. Jones, and L. T. Frobish. 2004. Improving pork quality by feeding supranutritional concentrations of vitamin D3. J. Anim. Sci. 82:218–224.

Wiseman, J., and J. A. Agunbiade. 1998. The influence of changes in dietary fat and oils on fatty acid profiles of carcass fat in finishing pigs. Livest. Prod. Sci. 54:217–227.

Wolter, B., M. Ellis, F. K. McKeith, K. D. Miller, and D. C. Mahan. 1999. Influence of dietary selenium source on growth performance, and carcass and meat quality characteristics in pigs. Can. J. Anim. Sci. 79:119–121.

Wood, J. D., S. N. Brown, G. R. Nute, F. M. Whittington, A. M. Perry, S. P. Johnson, and M. Enser. 1996. Effects of breed, feed level and conditioning time on the tenderness of pork. Meat Sci. 44:105–112.

Wood, J. D., G. R. Nute, R. I. Richardson, O. Southwood, G. Plastow, R. Mansbridge, N. da Costa, and K. C. Chang. 2004. Effects of breed, diet, and muscle on fat deposition and eating quality in pigs. Meat Sci. 67:651–667.

Xiong, Y. L., M. J. Grower, C. Li, C. A. Elmore, G. L. Cromwell, and M. D. Lindemann. 2006. Effect of dietary ractopamine on tenderness and postmortem protein degradation of pork muscle. Meat Sci. 73:600–604.

Xu, G., S. K. Baidoo, L. J. Johnston, J. E. Cannon, and G. C. Shurson. 2008. Effects of adding increasing levels of corn dried distillers grains with solubles (DDGS) to corn−soybean meal diets on pork fat quality of growing−finishing pigs. J. Anim. Sci. 86(E−Suppl. 3):85.

Young, J. F., H. C. Bertram, K. Rosenvold, G. Lindahl, and N. Oksbjerg. 2005. Dietary creatine monohydrate affects quality attributes of Duroc but not Landrace pork. Meat Sci. 70:717–725.

Zhan, X. A., M. Wang, R. Q. Zhao, W. F. Li, and Z. R. Xu. 2007. Effects of different selenium source on selenium distribution,loin quality and antioxidant status in finishing pigs. Anim. Feed Sci. Technol. 132:202–211.

第18章

生长猪和种猪的饲养

1 导 语

本章的主要目的是阐述生长猪和种猪对饲粮的能量和营养物质的需要量。猪的营养需要量是用来满足其维持、生产（如肌肉生长、骨骼生长、产奶和精液生产等）和活动等生物过程的需要量。随着猪的生长，能量和营养物质的维持需要量会随之增加。而且，肌肉生长快或产奶量高的猪需要更多的能量和营养物质，以保证这些功能得到充分的发挥，所以拥有高产潜力的猪对能量和营养物质的需要量，要高于低生产性能的猪。严格地讲，能量和营养物质的需要量就是指满足猪最佳生长和生产性能的需要量。

本章所列出的能量和营养物质的需要量可为养猪生产实际提供合理的指导，这些需要量很大程度上是参考了NRC（1998）和《美国国家猪营养指南》（NSNG，2010），而且本章的作者也在其中贡献颇多。简要来说，本章的营养需要量是在NRC（1998）的基础上，根据现代基因型猪具有较高生长性能潜力进行了矫正。在实际的养猪生产中，各个商业猪场的条件千差万别。饲料混合均匀度误差、天然饲料原料的能值和营养素含量的变异、饲料原料中的抗营养因子含量、其他未被发现的环境或疾病应激，都可能会改变猪的营养需要量。因此，在应用本章所提出的能量和营养物质的需要量时，还应该根据每个猪场的实际情况进行调整，从而使猪达到最佳的生产性能。

2 种 猪

◆ 2.1 后备母猪

在生长发育阶段，给后备母猪提供合理的营养与正确的饲养管理，将为其有一个较长的繁殖寿命和高产的繁殖性能奠定良好基础。目前，有关后备母猪的绝大多数研究集中于生长育肥期的饲养，首要的重点是提高后备母猪的早期发情率。一个成功的后备母猪饲养方案，必须使初产母猪拥有较高的分娩率和较大的窝产仔数。此外，合适的后备母猪营养还要能提高繁殖母猪长时间维持高产的可能性。

2.1.1 后备母猪饲养目标的确定

后备母猪饲养方案的主要难题在于确定合适的生长目标和适宜的营养需要量，以确保其进入繁殖群后具有最大的长期繁殖性能。Rozeboom（2006）提出了五种策略及相应的目标，大体可归纳为两类：① 与商品猪的饲养类似，以最大的瘦肉生长速率为目标，到配种的目标日龄时饲养成为大的母猪；② 控制猪的生长速度，来调节达到首次配种日龄时的体重和体组成（如脂肪、瘦肉和骨骼的发育）。在生长育肥期，后备母猪生长缓慢会导致初情期延迟（Beltranena等，1991），使第1胎的窝产仔数减小（Tummaruk等，2001）。然而，生长速度太快的后备母猪的发情期并不比生长速度合适的猪早（Beltranena等，1991），而且生长速度过快可能还会对后备母猪正常骨骼发育（Williams等，2004；Orth，2007年）和使用寿命（Jorgensen和Sorensen，1998；Johnston等，2007）带来负面影响。可见，后备母猪生长速度过慢和过快对其繁殖性能都不利。母猪从出生到选来配种时的生长速度应控制在600 ~ 800 g /d，以达到最长的繁殖寿命。

初情期的体组成和初配年龄都会影响第1胎的窝产仔数和母猪终生的繁殖性能。对后备母猪生长及其后来的繁殖性能的回顾分析表明，选种时后备母猪的体脂含量与其后来作为繁殖母猪的繁殖寿命呈正相关（Lopez-Serrano等，2000；Stalde等，2005；Johnston等，2007）。这些研究表明，选种时，越肥的后备母猪，进入繁殖母猪群后，其繁殖寿命就越长。但Rozeboom等（1996）在控制试验中发现，改变后备母猪初配期的体组成，并未对这些母猪4胎后的留存率产生影响。同时，Edwards（1998）对多个研究的总结得出，后备母猪进入繁殖群后，通过限制蛋白质水平或提高饲养水平来调节体内脂肪储备，连续观察几个胎次，母猪的繁殖性能都没有得到改善。

为了提高终生繁殖力，后备母猪至少应在135 kg进行初配，但不能超过155 kg。Williams等（2005）发现，当母猪初配体重不足135 kg时，连续前3胎的窝产仔数都下降。初配体重的理想范围应该是足够的体重和体成熟度达到合理平衡。体成熟就是骨骼发育完全、体组织储备丰盈，以达到合适的体尺，保证母猪体内有足够的瘦肉和脂肪储备。Johnston等（2007）的追溯性研究表明，后备母猪过重可能会对其终生生产力产生负面影响。而且，Dourmad等（1994年）还发现，后备母猪体重过大还会减少使用年限。此外，体重和体尺过大还会引起母猪福利问题。因为如果母猪太大，饲养在目前使用的分娩栏和妊娠栏内的舒适度降低。

一些学者建议，为了获得最佳的母猪终生繁殖力，初配期理想的背膘厚应在18～20 mm范围内（Challinor等，1996）。然而，其他一些学者则认为，初配期的背膘厚低于18 mm，母猪才能获得最佳的终生生产力（Stalder等，2005；Johnston，2007）。初配期理想的体脂含量可能与猪品种不同有关。一般来说，在同一个品种内，母猪过瘦或过肥都不好，只有适宜的体脂贮存才能使母猪终生繁殖性能最佳。

2.1.2 后备母猪的营养需要量

表18-1和表18-2所列出的后备母猪营养需要量是以生长速度（600～800 g/d）和初配体重（135～155 kg）为目标建立的。母猪自由采食量的估测值参考了《美国国家猪营养指南》（2010）。然后，再根据NRC（1998）的生长猪模型，利用表中列出的预期采食量和目标生长速度来确定后备母猪的营养需量。在生长的早期和中期阶段，后备母猪的饲喂与商品猪非常相似，但若要求更低的生长速度的情况除外。在整个生长阶段，为了使骨骼的矿化状态达到最佳，需要增加饲粮钙和磷的水平（Nimmo等，1981）。与商品猪相比，后备母猪在最后的生长阶段还需要显著提高饲粮中维生素和微量元素水平。增加这些营养物质的需要量，是为

了让这些母猪更好地发挥其妊娠期的生产性能，而不是变成商品肉猪，所以增加营养水平是为母猪繁殖性能提高到新的水平提供物质基础。这也是商业养猪生产中普遍采用的饲养方法，当后备母猪选作种用后，要换料饲喂，即用种猪专用的维生素-微量元素预混料配制饲粮进行饲喂。表18-2反映了后备母猪从110～140 kg体重阶段对维生素矿物质元素的需求变化。

表18-1　后备母猪能量和氨基酸的需要量（饲喂基础）[1]

项目	体重（kg）			
	20～50	50～80	80～110	110～140
采食量估测值（kg/d）[2]	1.5	2.1	2.5	2.7
饲粮代谢能（kcal/kg）	3 265	3 265	3 265	3 265
代谢能摄入量（Mcal/d）	4.9	6.8	8.2	8.8
预期日增重（g/d）	693	852	925	846
总赖氨酸（%）	0.98	0.84	0.74	0.62
标准回肠可消化氨基酸（%）				
精氨酸	0.33	0.26	0.20	0.13
组氨酸	0.28	0.24	0.21	0.17
异亮氨基酸	0.47	0.41	0.36	0.30
亮氨酸	0.87	0.74	0.64	0.52
赖氨酸	0.86	0.74	0.64	0.53
蛋氨酸	0.23	0.20	0.17	0.14
蛋氨酸+胱氨酸	0.50	0.43	0.38	0.32
苯丙氨酸	0.51	0.44	0.38	0.31
苯丙氨酸+酪氨酸	0.81	0.70	0.61	0.50
苏氨酸	0.55	0.48	0.43	0.36
色氨酸	0.16	0.14	0.12	0.10
缬氨酸	0.58	0.50	0.44	0.36

[1]基于NRC（1998）的生长模型，假设无脂瘦肉日增重为325 g/d。

[2]资料来源于《美国国家猪营养指南》（2010）。

表18-2 后备母猪矿物质和维生素的需要量（饲喂基础）[1]

项目	体重（kg）			
	20 ~ 50	50 ~ 80	80 ~ 110	110 ~ 140
采食量估测值（kg/d）[2]	1.5	2.1	2.5	2.7
饲粮代谢能（kcal/kg）	3 265	3 265	3 265	3 265
代谢能摄入量（mcal/d）	4.9	6.8	8.2	8.8
预期日增重（g/d）	693	852	925	846
矿物质（% 或重量 /kg）[3]				
钙（%）	0.69	0.60	0.56	0.52
总磷（%）	0.69	0.60	0.56	0.52
有效磷（%）	0.31	0.24	0.22	0.19
钠（%）	0.11	0.10	0.10	0.10
氯（%）	0.10	0.08	0.08	0.07
镁（%）	0.04	0.04	0.04	0.04
钾（%）	0.22	0.19	0.17	0.16
铜（mg）	4.1	3.4	3.0	5.0
碘（mg）	0.14	0.14	0.14	0.14
铁（mg）	62.4	48.8	41.0	80.0
锰（mg）	2.3	2.0	1.9	20.0
硒（mg）	0.18	0.15	0.14	0.30
锌（mg）	63.0	53.1	47.6	50.0
每千克饲料中维生素的含量				
维生素 A（IU）	1 440	1 311	1 250	4 000
维生素 D（IU）	166	152	144	200
维生素 E（IU）	11.0	11.0	11.0	44.0
维生素 K（mg）	0.50	0.50	0.50	0.50
生物素（mg）	0.05	0.05	0.05	0.20
胆碱（g）	0.32	0.30	0.29	1.25
叶酸（mg）	0.30	0.30	0.30	1.30
烟酸（平均，mg）	9.4	7.7	6.8	10.0

(续)

项目	体重（kg）			
	20 ~ 50	50 ~ 80	80 ~ 110	110 ~ 140
泛酸（mg）	7.8	7.2	7.0	12.0
核黄素（mg）	2.4	2.1	1.9	3.8
硫胺素（mg）	1.0	1.0	1.0	1.0
维生素 B_6（mg）	1.1	1.0	1.0	1.0
维生素 B_{12}（μg）	9.3	6.2	4.6	15.0

[1]基于NRC（1998）生长模型。

[2]资料来源于《美国国家猪营养指南》（2010）。

[3]后备母猪在最后发育阶段的矿物质和维生素的需要量增加到妊娠母猪的水平。

◆ 2.2 妊娠母猪

妊娠期母猪的饲养目标是要尽量减少胚胎和胎儿的损失，并为母猪分娩和泌乳做好准备。要为母猪分娩和哺乳做好充分的准备，需要控制能量的摄入量及相应的体增重，以避免母猪在妊娠期变得过肥。在妊娠的早期阶段也就是刚刚受孕之后的30 d，首要的饲养目标就是保证最大的胚胎生活率，同时增加分娩时的窝产仔数。在妊娠中期（30 ~ 75 d）的主要饲养目标是保证胎儿发育及母猪营养储备的增加。青年母猪通过不断生长来增加营养贮备；而对于经产母猪，通过恢复上个泌乳期的体重损失来补充营养储备。在妊娠后期（75 ~ 114 d），胎儿继续迅速生长，乳腺发育也非常快，为即将到来的哺乳期做好准备。正确的饲养方案不仅要满足妊娠母猪的营养需求，还需要在合理的成本范围内确保母猪繁殖性能得到延续。

2.2.1 能量和营养物质需要量的确定

母猪每日对能量和营养物质的需要量随体型大小、生产水平、妊娠阶段、健康状况及外界环境的变化而改变。妊娠母猪的能量和营养的需要量基本上可分为三部分：母猪自身的维持需要，胎儿生长的需要及母体增重的需要。每部分都可以单独进行估算，然后再求和，得到母猪每日对能量和其他营养物质的总需要量。

能量和营养素的维持需要主要受母猪体重及其饲养环境的影响。随着体重或年龄的增长，母猪对能量和营养素的维持需要也会不断增加。因此，与体重小的年轻母猪比，体重大的年长母猪需要采食更多的饲料，以满足其基本的维持需要。妊娠母猪的能量维持需要占能量总需求量的75%～85%。一般情况下，母猪体重每增加23 kg，每日代谢能（ME）的需要量就增加约470 kcal；若以玉米-豆粕型饲粮来饲喂，每日需要增加约0.15 kg的饲粮。除体重外，母猪能量的维持需要还受实际饲养环境温度的影响。实际环境温度不一定是温度计的读数，而是母猪自身感受到的温度。当使用垫料提供隔热层时，外界温度降低，母猪感觉也不会那么冷。但在潮湿环境下，由于蒸发散热的作用，母猪感受温度比温度计测定的温度更低。在商业化养殖条件下，人们最关心的是温度低于母猪的等温区或适温区时的情况。环境温度越低，母猪在不动用自身体组织的情况，维持内部体温所需要的能量摄入量和采食量就越高。一般来说，环境温度低于18℃时，每下降5.5℃，单独饲养的妊娠母猪就需要增加360 g的玉米-豆粕型饲粮的摄入量，以满足其维持能量需要。群体饲养和使用垫料有助于母猪保持体温，环境温度降到10～13℃时，母猪才需要提高采食量来维持体温。

在实际生产条件下，很难通过调节采食量影响胚胎的生长及其营养需要。在能量充足的条件下，母猪每日摄入6～10 Mcal ME，胎儿重量的变化则相当小（Noblet等，1990）。同时，饲养水平对胎儿的体成分的影响也很小。然而，在妊娠后期增加母猪脂肪的摄入量，可提高仔猪出生后的成活率。

从理论上讲，在满足母猪维持需要和胎儿生长需要后，“多余的”能量和营养素可用于母体增重。用于母猪自身增重的能量占能量总需要量的15%～25%。母体增重的组成主要由母猪的胎次和饲粮组成决定（Pettigrew和Yang，1997）。同样，理想的母体增重还受母猪年龄的影响（表18-3和表18-4）。未生产过的母猪（产0胎）和初产母猪（产1胎）自身都还在生长，所以与体成熟的年长母猪相比，它们在妊娠期的体增重更高。母猪体重的增加主要是为即将到来的哺乳期提供营养储备，因为产奶需要常常会超过摄入饲粮的营养量。然而，母猪在妊娠期的增重过高会降低其在哺乳期的采食量，进而降低哺乳期的生产性能（Weldon等，1994；Sinclair等，2001），并减少母猪使用寿命（Dourmad等，1994）。

表18-3 妊娠母猪能量和氨基酸的需要量（饲喂基础）[1]

项目	0和1胎[2]		2胎及以上	
窝产仔数（总出生仔数）	10.5	12.5	12.0	14.0
采食量估测值（kg/d）[3]	2.08	2.13	1.85	1.90
总增重估测值（kg）	52	57	36	40
饲粮代谢能（Mcal/kg）	3.3	3.3	3.3	3.3
饲粮中的含量（%）				
总赖氨酸含量	0.68	0.70	0.58	0.60
回肠标准可消化氨基酸[4]				
精氨酸	0.52	0.54	0.45	0.46
组氨酸	0.18	0.18	0.15	0.16
异亮氨酸	0.3	0.34	0.29	0.30
亮氨酸	0.55	0.57	0.47	0.49
赖氨酸	0.58	0.60	0.50	0.52
蛋氨酸	0.16	0.16	0.14	0.14
蛋氨酸+半胱氨酸	0.41	0.42	0.35	0.37
苯丙氨酸	0.34	0.35	0.29	0.30
苯丙氨酸+酪氨酸	0.58	0.60	0.50	0.52
苏氨酸	0.44	0.46	0.38	0.40
色氨酸	0.11	0.11	0.09	0.10
缬氨酸	0.40	0.41	0.34	0.36

[1]资料来源于《美国国家猪营养指南》（2010），所有的饲粮都在适温区限制饲喂。

[2]0表示母猪第一次妊娠。

[3]根据母猪实际体况调整采食量，以获得预期体况或体增重。

[4]SID=标准回肠可消化氨基酸。

表18-4 妊娠母猪矿物质和维生素的需要量（饲喂基础）[1]

养分	%或数量（kg）
矿物质	
钙（%）	0.85

（续）

养分	% 或数量（kg）
总磷（%）	0.70
有效磷（%）	0.45
钠（%）	0.18
氯（%）	0.14
镁（%）	0.05
钾（%）	0.25
铜（mg）	6.00
碘（mg）	0.17
铁（mg）	100
锰（mg）	25
硒（mg）	0.30
锌（mg）	60
每千克饲料中维生素的含量	
维生素 A（IU）	4 600
维生素 D（IU）	230
维生素 E（IU）	50
维生素 K（mg）	0.60
生物素（mg）	0.23
胆碱（g）	1.45
叶酸（mg）	1.50
有效烟酸（mg）	12.00
泛酸（mg）	14.00
核黄素（mg）	4.3
硫胺素（mg）	1.15
维生素 B_6（mg）	1.15
维生素 B_{12}（μg）	18.00

[1]根据NRC（1998）调整，即在NRC（1998）推荐量的基础上增加15%，以满足现代基因型猪生产性能提高的需要。

在妊娠期，胎儿的能量和营养素的供应量，可以影响仔猪出生时的大小和仔猪从出生到上市期间的生长性能。这种现象被称为胎儿印迹，是在早期敏感阶段刺激因素或损害引起的生理“迹象”（Lucas，1991）。妊娠母猪对能量和营养素的摄入量急剧减少或单个胎儿能量和营养素的供给量急剧减少，会引起宫内发育迟缓（IUGR），还可能会导致超低初生重的仔猪（侏儒猪）的出生（Foxcroft等，2006年）。与初生重中等或偏重的仔猪相比，初生重低的仔猪存活率低、生长速度慢、上市时胴体重轻、胴体脂肪含量高、瘦肉率低（Milligan等，2002；Gondret等，2006）。有些学者在提高妊娠期营养的摄入量会降低初生重小的仔猪的发生率的基本假设下，研究了增加妊娠期的营养摄入量对后代初生重和出生后生长性能的影响（Cromwell等，1989；Dwyer等，1994；Wu等，2006），研究表明，给妊娠母猪提供高于推荐量的营养水平，对其后代仔猪的生长不会产生显著的、持久的影响。因此，必须采取更有针对性的手段来减少初生重小的仔猪的发生率。

胎盘是母体营养供应和胎儿发育之间的纽带，所以最近有些学者致力于改善胎盘发育和功能的研究。对于胎儿的成功发育和减少仔猪初生重低的发生率，良好的胎盘发育至关重要（Wu等，2010）。与普通的玉米-豆粕型饲粮相比，从妊娠第30～114 天饲喂含高浓度的精氨酸及其代谢产物（如一氧化氮、多胺）的饲粮，可以改善胎盘的生长和功能，从而增加窝产仔数、提高同窝仔猪初生重的整齐度、降低侏儒猪的比例（Wu等，2010）。精氨酸被认为是一种“功能性氨基酸”。功能性氨基酸可能是必需氨基酸，也可能是非必需氨基酸，在添加量高于或超出维持、生长和繁殖传统需要量的时候，在机体中发挥特殊的生理功能（Wu和Kim，2007）。功能性氨基酸是一个相对较新的概念，未来可能成为妊娠母猪饲粮营养需要的一部分。

妊娠母猪对氨基酸的需求在整个妊娠期并不是保持不变的。Kim等（2009）提出，母猪对氨基酸的需要量及各种氨基酸的比例会随着妊娠阶段的不同而发生变化，这意味着妊娠早期与后期的饲粮的氨基酸模式是不同的。与采用全程饲喂的对照组相比，采用分阶段饲养方法配制的妊娠母猪饲粮，提高了母猪整个妊娠期的体增重，并减少了背膘损失。因为目前大多数商业猪场在整个妊娠期都是用同一料仓饲喂，所以在实际生产上，要采用这种分阶段的营养方案还需要改变储料仓和输送线。

2.2.2 刻板行为

绝大多数养猪生产系统都需要控制母猪在妊娠期的体增重，以防止母猪过肥。在某些地区，高能饲粮价格比较经济，通常要对母猪进行限饲，以控制能量的摄入量。因此，母猪每天都处于饥饿状态，进而促使刻板行为的发生（Lawrence和Terlouw，1993；Brouns等，1994）。Odberg（1978）对刻板行为的定义是指有规律地重复的、无明显目的、无任何功能效果的行为。

虽然提高采食量可以减少母猪的刻板行为（Bergeron等，2000），但体增重也会随之增加，因此这种做法是不切实际的。另外一种可行方法是，通过增加饲粮粗纤维的含量来降低其能量浓度水平，这样就可以增加采食量。一些研究表明，在能量和营养素水平能满足母猪维持、胎儿生长和适度体增重需要的前提下，饲喂含50%～89%粗纤维的饲粮，可以显著减少刻板行为（Brouns等，1994；Bergeron等，2000；Danielsen和Vestergaard，2001）。如果要尽量减少母猪刻板行为的话，妊娠母猪的饲粮需要含有不低于30%NDF（Meunier-Salaun等，2001），可发酵的粗纤维要达到足够高的含量，同时还要确保饲粮含有的能量和其他营养素水平能满足繁殖需要。配制母猪饲粮的基本要求是满足母猪生理功能的营养需求或采食量需求，而更高的要求则是在此基础上还能进一步改善母猪的福利状况（Johnston和Holt，2006）。目前，在北美配制的妊娠母猪饲粮通常都没有考虑到要改善母猪的行为。

◆ 2.3 哺乳母猪

哺乳母猪的饲喂关键是要在尽可能减少体组织动员的条件下提高产奶量，进而最大限度地提高整窝仔猪的生长性能。降低母猪在哺乳期的体组织损失，可能会促进母猪断奶后迅速发情并提高下一窝产仔数。要实现这些目标的基本方法是尽可能地提高母猪采食量，同时饲粮要含有充足的能量和营养素，以满足母猪每天的营养需要量。

2.3.1 营养需要量的确定

哺乳母猪对能量和营养素的需要量取决于母猪体重、产奶量和乳成分，在一定程度上还受饲养环境的影响。正如妊娠母猪一样，哺乳母猪需要充足的能量和营养素来满足维持体重和基本生理功能的需求。与体重小的年轻母猪相比，体重大的年长母猪对能量和营养素的维持需要量更高。因为哺乳母猪通常都是自由采食去努力摄入更多的能量和养分，因而其维持需要很容易得到满足。

与维持需要不同，用于产奶的能量和营养素的需要量很难得到满足。虽然可以通过饲粮供给能量和营养素来满足产奶需要量，但是如果母猪的采食量受到环境、健康状况、品种或饲粮营养浓度不足的抑制，母猪就会进入能量和营养的负平衡，并会动用体贮来满足产奶的能量和营养素需要。母猪的产奶量和乳成分会直接影响产奶所需的能量和营养的需要。乳成分受遗传或饲粮因素的影响很小，在实际饲养方案中不会考虑这些因素的影响。因此在确定产奶需要时，主要考虑的因素是产奶量。在实际生产条件下，不会直接测定母猪产奶量，而是通过测定哺乳仔猪的窝增重来间接推算的。NRC（1998）已建立窝增重与产奶量之间的回归方程，根据这些方程和设定的乳成分标准，可以推测出母猪产奶所需的能量和营养物质（NRC，1998；NSNG，2010）。用于产奶和机体维持的能量和营养素的需要量之和就是哺乳母猪对能量和营养素的总需求量。要实现哺乳母猪营养需要的目标，需要在配制营养充分的饲粮的同时，最大程度地增加母猪采食量。如果哺乳期的采食量低或饲粮营养浓度不够，母猪就会动用体贮来弥补能量和营养素的摄入量不足，还会减少产奶量。

表18-5列出了四种不同情况下哺乳母猪对能量和氨基酸的需要量。表中只选择性地列出了四种情况下的营养需要量，因为提供所有可能情况下母猪的能量和营养素的需要量，超出了本章内容的范围。表18-5列出的营养需要量是根据NRC（1998）提出的哺乳母猪需求预测模型，使用针对NSNG（2010）开发的软件进行计算得来的。与此类似，表18-6列出的矿物质和维生素的需要量，也是根据NRC（1998）的预测模型来计算的，但是考虑到现代基因型母猪的繁殖性能普遍提高，所以在NRC（1998）推荐量的基础上平均增加15%。最近很少有关现代基因型母猪对矿物质和维生素需要量的研究。

表18-5 哺乳母猪能量和氨基酸的需要量（饲喂基础）[1]

项目	第 1 胎	第 1 胎	第 2 胎或以上	第 2 胎或以上
假定的母猪体重变化（kg）	−10	−5	−5	2.5
采食量估测值（kg/d）	5.4	4.6	6.4	5.6
仔猪增重估测值（g/d）	222	180	222	180
窝断奶仔猪数	11	10	12	11
断奶仔猪窝重（kg）	66.0	52.0	73.0	57.0
饲粮代谢能（Mcal/kg）	3.3	3.3	3.3	3.3
代谢能摄入量（Mcal/d）	17.8	15.2	21.1	18.5
饲粮中的含量（%）				
总赖氨酸	1.12	0.95	1.08	0.90
标准回肠可消化氨基酸[2]				
精氨酸	0.52	0.44	0.50	0.42
组氨酸	0.37	0.33	0.36	0.30
异亮氨酸	0.53	0.47	0.51	0.43
亮氨酸	1.09	0.95	1.04	0.88
赖氨酸	0.99	0.83	0.95	0.79
蛋氨酸	0.25	0.22	0.24	0.2
蛋氨酸 + 半胱氨酸	0.45	0.41	0.44	0.37
苯丙氨酸	0.52	0.46	0.50	0.42
苯丙氨酸 + 酪氨酸	0.95	0.84	0.91	0.77
苏氨酸	0.58	0.52	0.56	0.48
色氨酸	0.18	0.16	0.17	0.14
缬氨酸	0.81	0.71	0.78	0.65

[1]资料来源于《美国国家猪营养指南》（2010），母猪在适温区自由采食。

[2]SID=标准回肠可消化氨基酸。

表18-6　哺乳母猪矿物质和维生素的需要量（饲喂基础）[1]

营养物质	% 或数量（kg）
矿物质	
钙（%）	0.85
总磷（%）	0.70
有效磷（%）	0.45
钠（%）	0.25
氯（%）	0.20
镁（%）	0.05
钾（%）	0.25
铜（mg）	6.00
碘（mg）	0.17
铁（mg）	100
锰（mg）	25
硒（mg）	0.30
锌（mg）	60
维生素	
维生素 A（IU）	2 300
维生素 D（IU）	230
维生素 E（IU）	50
维生素 K（mg）	0.60
生物素（mg）	0.23
胆碱（g）	1.15
叶酸（mg）	1.50
烟酸（可利用，mg）	12.00
泛酸（mg）	14.00
核黄素（mg）	4.30
硫胺素（mg）	1.15
维生素 B_6（mg）	1.15
维生素 B_{12}（μg）	18.00

[1]根据NRC（1998）进行调整，即在NRC（1998）推荐量的基础上增加15%，以满足现代基因型猪繁殖性能提高的需求。

2.3.2 围产期的饲养管理

母猪分娩前后饲养管理的关键是如何让母猪从妊娠后期的限制饲喂过渡到哺乳期的自由采食。关于分娩前母猪的合理饲养方法，目前有几种不同的观点。最近，Rozeboom等（2009）用经产母猪评估了常见的三种饲喂方法：①饲喂水平从妊娠到分娩保持不变；②从妊娠第109天至分娩逐步降低饲喂水平；③从妊娠第85天开始到分娩增加饲喂水平。正如预期的那样，他们发现刚分娩后各组母猪体重的差异很小。而且，哺乳期母猪的自由采食量、窝产仔数、断奶前仔猪存活率、仔猪生长速度以及断奶后母猪下一胎的繁殖性能在各组间都没有差异。由于缺少负面影响的数据，可能比较好的饲养方法是，在妊娠后期适度增加母猪采食量，既可以满足胎儿迅速增长的营养需要，也不会给母猪随后的产奶性能带来负面影响。

对于刚刚分娩的母猪，应该逐渐增加其采食量，以便让它们在产后第4天可以自由采食。在哺乳期任一时间限制采食量都会增加母猪体重损失，降低哺乳仔猪生长性能，并损害母猪以后的繁殖性能（Koketsu等，1996）。

◆ 2.4 营养与下一胎的繁殖性能

哺乳期的营养会显著影响母猪下一胎的繁殖性能。哺乳早期或后期限饲，会增加母猪体重和背膘的损失，降低促性腺激素对卵巢功能的促进作用，延长断奶至发情的时间间隔，降低断奶后母猪的排卵率（Zak等，1997）。如果在临近断奶的时候进行限饲，可能会影响配种后胚胎的存活率，从而减少下一胎的窝产仔数（Zak等，1997）。此外，限饲还会抑制卵泡的正常发育和卵母细胞的成熟（Zak等，1997）。

限饲会引起母猪对能量、蛋白质、维生素和矿物质的摄入量减少。其中，蛋白质（氨基酸）摄入量的降低及其引起的体蛋白损失是影响母猪下一胎繁殖性能的最关键因素（King，1987；Clowes等，2003）。由低蛋白饲粮导致的赖氨酸摄入量的降低，会促进肌肉分解，降低泌乳后期促黄体激素的释放，并减少卵泡对卵母细胞正常发育的支持（Yang等，2000a，2000c）。从推理上讲，使用高蛋白质饲粮似乎可以避免出现这种问题。然而，Yang等（2000）研究表明，饲粮蛋白质水平过高，可以降低母猪接下来的窝产仔数，但Tritton等（1996）对此的观点并不一致（Tritton等，1996）。

饲料原料

与生殖周期的其他阶段相比，母猪在哺乳期对能量和营养素的需要最多。因此，哺乳期饲粮必须由高能量、高营养浓度的饲料原料组成。哺乳母猪饲粮中，通常会在饲粮中使用大比例的玉米和其他谷物来提供能量，而通过添加豆粕、菜籽粕及其他油籽粕类等高蛋白饲料原料来作为氨基酸天然来源，还会通过适量添加合成氨基酸来满足哺乳母猪特殊的生理需求。

哺乳期饲粮中添加精选动物油、牛油或大豆油，可以提高饲粮的能量浓度，最终增加母猪的能量摄入量。哺乳期饲粮添加油脂可以提高乳脂率（Pettigrew，1981）和每天产出的乳脂量（Lauridsen和Danielsen，2004），并能提高哺乳仔猪的生长速度（Pettigrew，1981；Lauridsen和Danielsen，2004）。但在实际生产中，人们通常会控制油脂的添加水平，因为添加油脂会增加饲粮在贮存仓和喂料仓内结块的可能性，并且随着其添加量的提高，饲粮油脂酸败的发生率也会增加。

像甜菜渣、小麦秸秆、大豆皮或麦麸这样的粗纤维饲料，由于能值低，一般不用于哺乳母猪饲粮的配制，但也有例外。虽然使用粗纤维饲料会降低母猪的能量摄入量，且违背了泌乳母猪的营养目标，但在分娩前后几天的饲粮中添加粗纤维饲料，可减少便秘的发生。这种饲喂方法虽有助于改善母猪健康状况，但很难改善母猪生产性能。

◆ 2.5 公猪

公猪对整个养猪生产具有重要影响，但可能是因为它们占总猪群的比例非常小，所以人们的关注度相对很少。公猪不仅可以作为遗传改良的父本，还可以影响母猪的分娩率和产仔数（Whitney和Baidoo，2010）。营养是影响种猪繁殖性能和所有猪健康的重要因素。营养状况会影响公猪的性欲、体格健康、使用寿命以及精子的生成量和精液质量。交配方式、成熟的年龄和阶段、膘情、环境条件及射精频率等诸多因素都可能会影响公猪的营养需要量。

对于自然交配的公猪，主要饲喂目标是最大限度地减少成熟体重，以便公猪能与后备母猪及小母猪进行交配。公猪饲喂过量不仅会影响性欲，还可能导致腿部无力和生殖问题、降低使用年限，因此必须进行限制饲喂。而对于人工授精的公猪，主要的营养目标是要尽可能地增加精子的生成和精液质量，同时要确保公猪的健康状态。跛行等引起的福利问题会对公猪产生重要影响，因为这会降低采精时公猪爬跨假母猪的意愿和能力。在人工授精中心，对成熟公猪体重的关注度不如自然交配的公猪；但对于体重过大的公猪，必须考虑工作人员的安全问题。

由于1～2岁的青年种公猪仍会不断生长，因此饲喂方案必须让其保持适度的体增重，为180～250 g/d（Whitney和Baidoo，2010）。青年种公猪的饲养目标是，既要限制能量摄入来实现平缓的生长速度，又要维持较高的氨基酸、维生素和矿物质的摄入量，以保持良好的生育功能和性欲。随着公猪的体重不断增加、年纪慢慢变大，它们的生长速度逐渐下降，用于维持需要的营养所占的比例不断增加。成熟公猪的饲喂要求应该满足自身的维持需要，同时还要保持最佳繁殖性能。公猪的维持需要取决于其体重大小和体况优劣。此外，当公猪处于寒冷气温条件下或其性欲极高时，应该适当增加饲喂量。Kemp等（1989a）估计，在环境温度低于20℃条件下，每降低1℃，需要额外消耗3.8 kcal/kg$^{0.75}$的热量来维持正常体温。

2.5.1 营养对公猪繁殖性能的影响

营养可以影响公猪的性欲、产精量和精液质量。长期的过度限饲会导致体重大幅下降，并导致公猪拒绝采精（Stevermer等，1961）。同样，饲喂低蛋白饲粮，尤其在能量摄入受限的条件下，会降低血液雌二醇-17β的水平，进而降低了公猪的交配意愿（Louiset等，1994a，1994b）。然而，不管是饲喂水平还是营养水平，短期限饲对公猪性欲的影响非常小（Ju等，1985；Kemp等，1989b）。在某些人工授精中心，通过给公猪饲喂高蛋白质水平的饲粮（高于正常需要量的5%～10%）来提高其性欲。但体增重过大会使公猪精神不振，降低其运动的稳定性和平衡性，从而降低其爬跨能力（Westendorf和Richter，1977）。

降低营养水平（正常需要量的50%～70%），也会减少精液量和总精子数（Beeson等，1953；Kemp等，1989b）。而当营养水平增加到正常水平时，精液量和总精子数也恢复到正常。以往的研究表明，公猪可以在短时间内承受非常大的营养水平变化，而不影响精子质量（Stevermer等，1961）。

2.5.2 公猪的营养需要

目前关于种公猪营养需要的研究相当匮乏。表18-7列出的营养需要量是在NRC（1998）的基础上，参考《美国国家猪营养指南》（2010），再根据现代猪种的生产水平进行了调整。实际上，现代青年公猪和成熟公猪的营养需要量是采用NRC（1998）的预测模型，再根据《美国国家猪营养指南》（2010）中的生产水平数据（如体重范围、能量浓度和采食量）计算得来的。与NRC（1998）相比，公猪的每日采食量和营养需要量都提高了，这反映出现代瘦肉型种公猪需要更多的能量和氨基酸，以最大程度地发挥其繁殖性能。

表18-7 种公猪的营养需要量（饲喂基础）[1]

养分	% 或数量（kg）	数量（d）	数量（d）
体重（kg）	—	135 ~ 185	185 ~ 300
饲粮代谢能（kcal/kg）	3 300	3 300	3 300
代谢能摄入量（kcal/d）	—	8 085	9 075
采食量估测值（kg/d）	—	2.45	2.75
总氨基酸			
精氨酸	0.0%	0.0 g	0.0 g
组氨酸	0.17%	4.2 g	4.7 g
异亮氨酸	0.31%	7.7 g	8.6 g
亮氨酸	0.46%	11.2 g	12.6 g
赖氨酸	0.54%	13.2 g	14.8 g
蛋氨酸	0.14%	3.5 g	4.0 g
蛋氨酸 + 胱氨酸	0.38%	9.2 g	10.4 g
苯丙氨酸	0.30%	7.3 g	8.1 g
苯丙氨酸 + 酪氨酸	0.51%	12.5 g	14.1 g
苏氨酸	0.45%	11.0 g	12.3 g
色氨酸	0.11%	2.6 g	3.0 g
缬氨酸	0.36%	8.8 g	9.9 g

（续）

养分	% 或数量（kg）	数量（d）	数量（d）
矿物质			
钙	0.67%	16.5 g	18.5 g
总磷	0.54%	13.2 g	14.8 g
有效磷	0.31%	7.7 g	8.6 g
钠	0.13%	3.3 g	3.7 g
氯	0.11%	2.6 g	3.0 g
镁	0.04%	0.9 g	1.0 g
钾	0.18%	4.4 g	4.9 g
铜	4.49mg	12.3 mg	11.0 mg
碘	0.12mg	0.3 mg	0.3 mg
铁	71.8mg	176.0 mg	197.6 mg
锰	18.0mg	44.0 mg	49.4 mg
硒	0.12mg	0.3 mg	0.4 mg
锌	44.9mg	110.0 mg	123.5 mg
维生素			
维生素 A	3 591IU	8 800 IU	9 878 IU
维生素 D	179IU	440 IU	494 IU
维生素 E	39.6IU	97 IU	109 IU
维生素 K	0.44mg	1.1 mg	1.2 mg
生物素	0.16mg	0.4 mg	0.5 mg
胆碱	0.11%	2.75 g	3.09 g
叶酸	1.18mg	2.9 mg	3.2 mg
烟酸（可利用）	9.0mg	22 mg	25 mg
泛酸	10.6mg	26 mg	30 mg
核黄素	3.36mg	8.25 mg	9.26 mg
硫胺素	0.9mg	2.2 mg	2.5 mg

（续）

养分	% 或数量（kg）	数量（d）	数量（d）
维生素 B_6	0.9mg	2.2 mg	2.5 mg
维生素 B_{12}	13.5μg	33μg	37μg

[1]在NRC（1998）的基础上，根据《美国国家猪营养指南》（2010）中现代基因型猪的体重和生长性能进行调整。

3 生长猪

◆ 3.1 仔猪

给断奶仔猪提供充足的营养是一项既重要而又非常具有挑战性的任务，与其他任何阶段相比，这可能是最困难的阶段。环境应激、采食量减少以及饲粮的组成和物理状态的突然变化，通常都会引起刚断奶的仔猪生长性能下降（Liebbrandt等，1975）。仔猪饲粮的营养水平及所选用的饲粮原料质量，都会显著影响该阶段的生长性能，而且这个效应会持续到生长育肥期（Whang等，2000）。仔猪对各种营养的需求受许多因素的影响，包括年龄、消化道发育程度、自由采食量低和基因型等。Cromwell等（1996）总结了58个仔猪试验，发现仔母猪比阉公猪的生长速度快4.7%，表明性别也可能会影响仔猪的营养需要。但Hill等（2007）发现，仔猪对赖氨酸的需要量无性别差异。实践生产上，通常都不考虑性别对保育猪营养需要的影响。

随着现代猪种的瘦肉生长潜力的不断增长，在生长发育的各个阶段对营养的需要量也在不断增加。研究已证明，随着瘦肉沉积率的增加，仔猪对赖氨酸（Stahly等，1994）、P（Fredrick和Stahly，2000）和B族维生素（Stahly等，1995）的需要量也相应增加。

3.1.1 能量需要量

刚断奶的仔猪通常不能采食充足的饲粮来满足其能量需求（Bark等，1986；Pluske，1993）。NRC（1987）得出的结论是，仔猪采食量随断奶时间的推移呈线性增加，但在断奶后的24h内，仔猪采食量极低，或根本没不采食。假设饲粮ME约为3 265 kcal /kg，NRC（1998）根据下列公式估算出体重小于20 kg的仔猪每天的能量需要量。

$$\text{DE摄入量（kcal/d）}=(251\times BW)-(0.99\times BW^2)-133$$

$$\text{ME摄入量}=0.96\times \text{DE摄入量}$$

环境温度和地面空间大小等环境条件会影响仔猪自由采食量。NRC（1998）的生长模型可以根据地面空间大小、实际温度与适宜温度的偏离情况，来计算出相应的DE摄入量。

当仔猪自由采食量偏低时，通常在饲粮中补充脂肪来增加仔猪能量摄入。但刚刚断奶的仔猪对普通脂肪的利用效率非常低（Pettigrew和Moser，1991），而对含有很高的中短链脂肪酸或长链不饱和脂肪酸的脂肪的利用率较高（Cera等，1988；de Rodas和Maxwell，1990；Partridge和Gill，1993）。因此，仔猪的能量来源应该使用乳糖、葡萄糖或蔗糖等容易被利用的碳水化合物来提供（Maxwell和Carter，2001）。第一阶段的断奶仔猪饲粮应含15%～25%的乳糖，到第二阶段降至10%～15%（Maxwell和Carter，2001）。乳清粉、乳清透析物、脱脂奶粉和结晶乳糖等原料都是仔猪乳糖的重要来源。在断奶2～3个星期之后，饲粮中不需要再添加乳糖，能量只由植物性饲料和优质动植物油脂来提供。

3.1.2 **氨基酸需要量**

本章列出的仔猪标准回肠可消化氨基酸（SID）需要量是根据NRC（1998）的模型估算得到的。NRC（1998）的推荐值是根据生长遗传潜力低且健康状况差的猪品种试验得出的。现代养猪生产中通常使用生长遗传潜力高且健康状况好的猪品种，由于瘦肉率的提高和免疫系统刺激的降低，它们需要更高水平的氨基酸。对于健康的早期断奶仔猪，将赖氨酸总量在NRC（1998）的基础上提高15%～20%，可以提高生长速度和饲料转化效率（Owen等，1995；Chung等，1996；Williams等，1997）。Kendall等（2008）对5个仔猪试验进行了总结，发现11～27 kg的仔猪对标准回肠可消化赖氨酸的需要量为1.30%（3.80 g SID Lys/Mcal ME）。根据这些研究结果，考虑到现代猪种的生长速度更快、健康状况更好，表18-8给出的仔猪氨基酸需要量比NRC（1998）推荐值要高15%。

表18-8 仔猪能量和氨基酸的需要量[1]

项目	体重（kg）		
	3 ~ 5	5 ~ 10	10 ~ 20
采食量估测值（g/d）	170	750	1 000
饲粮代谢能（kcal/kg）	3 500	3 400	3 350
代谢能摄入量估算值（kcal/d）	600	2 550	3 350
预期体增重（g/d）	145	300	570
总赖氨酸（%）	1.70	1.54	1.32
标准回肠可消化氨基酸（%）[2]			
精氨酸	0.65	0.58	0.49
组氨酸	0.49	0.44	0.37
异亮氨酸	0.85	0.75	0.64
亮氨酸	1.54	1.37	1.16
赖氨酸	1.54	1.37	1.16
蛋氨酸	0.43	0.38	0.32
蛋氨酸 + 胱氨酸	0.89	0.79	0.67
苯丙氨酸	0.92	0.82	0.70
苯丙氨酸 + 酪氨酸	1.45	1.29	1.09
苏氨酸	0.95	0.85	0.72
色氨酸	0.26	0.23	0.20
缬氨酸	1.00	0.89	0.75

[1]根据NRC（1998）进行调整，即在NRC（1998）的基础上增加15%以满足现代基因型猪性能增加的需求。

[2]SID=标准回肠可消化氨基酸。

对于大多数实用饲粮，赖氨酸被公认为第一限制性氨基酸，因此其他氨基酸的需要量是根据理想氨基酸模式来确定的。《美国国家猪营养指南》（2010）提出了仔猪标准回肠可消化氨基酸比例：假设赖氨酸为100，精氨酸42，组氨酸32，异亮氨酸55，亮氨酸100，蛋氨酸28，蛋氨酸+胱氨酸58，苯丙氨酸60，苯丙氨酸+酪氨酸94，苏氨酸62，色氨酸17，缬氨酸65。含硫氨基酸（SAA）——蛋氨酸和半胱氨酸，通常认为是仔猪饲粮的第二或第三限制性氨基酸（Yi等，2006）。

在典型仔猪饲粮中，色氨酸通常是第二或第三限制性氨基酸。Guzik等（2002）提出，5.2～7.3 kg、6.3～10.2 kg和10.3～15.7 kg的仔猪对真可消化色氨酸的需要量分别为0.21%、0.20%和0.18%，这与NRC（1998）的推荐量类似。然而，仔猪在免疫应激状态下对色氨酸的需要量可能会大幅增加。Melchior等（2004）也发现，当仔猪患有肺炎时，色氨酸的分解代谢增加。随后，Le Floc'h等（2009）研究表明，当猪处于卫生条件差的环境时，色氨酸用于生长和其他代谢功能的量就会减少。Trevisi等（2009）表明，在大肠杆菌处理后的仔猪饲粮中补充色氨酸，能增加采食量，并且能保持适当的生长速度。养猪生产者和营养学家必须注意的是，当猪处于免疫应激条件下，必须调整相应的饲粮配方。表18-8列出的氨基酸需要量是针对健康的仔猪。

支链氨基酸——异亮氨酸和缬氨酸，往往被认为是生长猪第二限制性氨基氨（Figueroa等，2003）。Kerr等（2004）提出的回肠表观可消化异亮氨酸与赖氨酸之比例与NRC（1998）的完全相同，为55%，与Wiltafsky等（2009a）提出的54%也非常相似。同时，Wiltafsky等（2009b）还发现，以最佳生长性能为评价指标，8～25 kg猪的标准回肠可消化缬氨酸：赖氨酸为65%～67%，略低于NRC（1998）确定的比例。

3.1.3 矿物质和维生素的需要量

NRC（1998）推荐了仔猪对矿物质和维生素的需要量（表18-9）。此后，有关现代仔猪矿物质和维生素的需要量的研究非常少。Rincher等（2004，2005）发现，增加饲粮中铁、锌和铜的含量，尽管改善了仔猪体内矿物元素的营养状态，但并没有改善生长性能。近些年，针对仔猪矿物元素营养的研究多数集中于高锌（氧化锌形式的锌2 000～3 000 ppm）和高铜（硫酸铜形式的铜250 ppm）。通常在仔猪饲粮中使用高锌或高铜，以减少腹泻、提高生长性能（Carlson等，1999）。Hill等（2000）进行的特定区域试验表明，由添加3000 ppm锌或250 ppm铜引起的采食量增加，并不能完全解释生长性能的提高，还可以通过它们的抗菌作用来促进仔猪的生长；同时，高锌与高铜的促生长作用不具有可加性。

表18-9 保育猪矿物质和维生素需要（饲喂基础）[1]

项目	体重（kg）		
	3～5	5～10	10～20
矿物质（%）或数量（kg）			
钙（%）	0.90	0.80	0.70
总磷（%）	0.70	0.65	0.60
有效磷（%）	0.55	0.40	0.32
钠（%）	0.25	0.20	0.15
氯（%）	0.25	0.20	0.15
镁（%）	0.04	0.04	0.04
钾（%）	0.30	0.28	0.26
铜（mg）	6.00	6.00	5.00
碘（mg）	0.14	0.14	0.14
铁（mg）	100	100	80
锰（mg）	4.00	4.00	3.00
硒（mg）	0.30	0.30	0.25
锌（mg）	100	100	80
每千克饲料中维生素的含量			
维生素 A（IU）	2 200	2 200	1 750
维生素 D（IU）	220	220	200
维生素 E（IU）	16	16	11
维生素 K（mg）	0.50	0.50	0.50
生物素（mg）	0.08	0.05	0.05
胆碱（g）	0.60	0.50	0.40
叶酸（mg）	0.30	0.30	0.30
烟酸（mg）（平均）	20.00	15.00	12.50
泛酸（mg）	12.00	10.00	9.00
核黄素（mg）	4.00	3.50	3.00
硫胺素（mg）	1.50	1.00	1.00
维生素 B_6（mg）	2.00	1.50	1.50
维生素 B_{12}（μg）	20.00	17.50	15.00

[1]NRC（1998）。

肠道细菌能合成生物素和其他B族维生素，因此可以推测，使用高锌或高铜作为抗菌剂可能会无意间增加饲粮中生物素和其他B族维生素的添加量。针对这个假设的研究结果，各报道不一。Partridge和McDonald（1990）提出，生物素添加水平从55 μg/kg增加到500 μg/kg时，改善了生长猪的生长性能。Brooks等（1984）也发现，高铜饲粮中添加生物素，可以提高仔猪生长性能。但Wilt和Carlson（2009）在高锌饲粮中添加生物素，并没有发现类似的结果。Mahan等（2007）研究也表明，补充NRC推荐水平的B族维生素，就可以满足仔猪早期生长的需求，因此再添加更高水平的B族维生素也不能提高生长性能。然而，Stahly等（2007）报道，当猪处于瘦肉组织快速增长的时期，至少有一种B族维生素的需要量高于NRC（1998）的推荐量，这可能是由机体新陈代谢加快而不是由饲粮能量摄入量或体内能量沉积率的增加所导致的。由于科学研究目前还没有得出完全一致的结论，因此在实际生产中，根据NRC（1998）推荐量再加上一定的安全系数，添加维生素和矿物元素完全能够满足仔猪生长性能的需要。

◆ 3.2 生长育肥猪

仔猪阶段结束后，猪体重应达到20 kg。从20 kg到上市，猪对饲粮营养浓度的需求不断降低，而平均日采食量不断增加。既要满足猪各个阶段的营养需要量但又不过量是发展可持续养猪生产的关键。

3.2.1 瘦肉生长的营养需要量

营养需要量的最准确表示方式是以每天需要量来描述，但用营养素在饲粮中所占的比例来表示更为方便。因此，对于一个特定的饲粮，准确估测每天的摄入量非常重要。自由采食量与饲粮能量浓度关系密切。猪要采食足够量的饲粮来满足其能量需要。含有较高动植物油脂的高能饲粮，会降低猪的自由采食量；而以简单碳水化合物作为主要能量来源的低能饲粮，通常会增加自由采食量。因此，可以利用饲粮能量浓度与采食量之间的关系，来补偿热应激等环境因素引起的食欲下降。例如，在夏季最炎热的时期，可以通过提高生长育肥猪饲粮的能量和营养浓度来弥补由热应激引起的采食量降低。表18-10、表18-11和表18-12列出的营养需要量是基于NRC（1998）提出的，但又根据《美国国家猪营养指南》（2010）估测的日采食量偏低而瘦肉生长率偏高的情况进行了适当的调整。

表18-10 阉猪在生长育肥期对能量和氨基酸的需要量（饲喂基础）[1]

项目	体重（kg）			
	20 ~ 50	50 ~ 80	80 ~ 110	110 ~ 140
采食量估测值（kg/d）[2]	1.45	2.20	2.61	2.93
饲粮代谢能（kcal/kg）	3 350	3 350	3 350	3 350
代谢能摄入量（kcal/d）	4 857	7 370	8 744	9 815
预期增重（g/d）	703	993	1 007	970
总赖氨酸（%）	1.07	0.94	0.77	0.61
标准回肠可消化氨基酸（%）[3]				
精氨酸	0.38	0.30	0.23	0.18
组氨酸	0.30	0.26	0.21	0.17
异亮氨酸	0.52	0.45	0.37	0.29
亮氨酸	0.94	0.82	0.67	0.52
赖氨酸	0.94	0.82	0.67	0.52
蛋氨酸	0.27	0.24	0.19	0.16
蛋氨酸＋半胱氨酸	0.54	0.49	0.40	0.32
苯丙氨酸	0.56	0.49	0.40	0.31
苯丙氨酸＋酪氨酸	0.88	0.77	0.63	0.49
苏氨酸	0.59	0.52	0.44	0.35
色氨酸	0.15	0.13	0.11	0.08
缬氨酸	0.61	0.53	0.44	0.34

[1]基于NRC（1998）的生长模型，假设日增重为350 g。

[2]根据《美国国家猪营养指南》（2010）进行调整。

[3]SID=标准回肠可消化氨基酸；氨基酸比率来自美国国家猪营养指南（2010）。

表18-11 生长育肥猪能量和氨基酸的需要量（饲喂基础）[1]

项目	体重（kg）			
	20 ~ 50	50 ~ 80	80 ~ 110	110 ~ 140
采食量估测值（kg/d）[2]	1.45	2.00	2.40	2.68

（续）

项目	体重（kg）			
	20 ~ 50	50 ~ 80	80 ~ 110	110 ~ 140
饲粮代谢能（kcal/kg）	3 350	3 350	3 350	3 350
代谢能摄入量（kcal/d）	4 857	6 700	8 040	8 978
预期体增重（g/d）	703	867	941	895
总赖氨酸（%）	1.07	0.92	0.83	0.66
标准回肠可消化氨基酸（%）[3]				
精氨酸	0.38	0.29	0.24	0.19
组氨酸	0.30	0.26	0.23	0.18
异亮氨酸	0.52	0.44	0.40	0.31
亮氨酸	0.94	0.81	0.72	0.57
赖氨酸	0.94	0.81	0.72	0.57
蛋氨酸	0.27	0.23	0.21	0.17
蛋氨酸 + 胱氨酸	0.54	0.49	0.43	0.35
苯丙氨酸	0.56	0.49	0.43	0.34
苯丙氨酸 + 酪氨酸	0.88	0.76	0.68	0.54
苏氨酸	0.59	0.52	0.47	0.38
色氨酸	0.15	0.13	0.12	0.09
缬氨酸	0.61	0.53	0.47	0.37

[1]基于NRC（1998）生长模型，假设无脂瘦肉日增重为350 g。

[2]资料来源于《美国国家猪营养指南》（2010）。

[3]SID=标准回肠可消化氨基酸；氨基酸比率来自《美国国家猪营养指南》（2010）。

表18-12 阉猪和母猪在生长育肥阶段对维生素和矿物质的需要量（饲喂基础）[1]

项目	体重（kg）			
	20 ~ 50	50 ~ 80	80 ~ 110	110 ~ 140
矿物质（%）或数量（kg）				
钙（%）	0.65	0.55	0.51	0.46

（续）

项目	体重（kg）			
	20 ~ 50	50 ~ 80	80 ~ 110	110 ~ 140
总磷（%）	0.56	0.49	0.45	0.42
有效磷（%）	0.25	0.20	0.18	0.15
钠（%）	0.12	0.11	0.11	0.11
氯（%）	0.11	0.09	0.09	0.08
镁（%）	0.04	0.04	0.04	0.04
钾（%）	0.24	0.21	0.19	0.18
铜（mg）	4.55	3.80	3.34	3.04
碘（mg）	0.15	0.15	0.15	0.15
铁（mg）	68.6	53.7	45.1	39.4
锰（mg）	2.53	2.21	2.09	2.00
硒（mg）	0.20	0.17	0.15	0.14
锌（mg）	69.32	58.46	52.40	48.30
维生素（数量/kg）				
维生素 A（IU）	1 584	1 442	1 375	1 334
维生素 D（IU）	182	167	159	153
维生素 E（IU）	12.1	12.1	12.1	12.4
维生素 K（mg）	0.55	0.55	0.55	0.55
生物素（mg）	0.06	0.06	0.06	0.06
胆碱（g）	0.35	0.33	0.32	0.32
叶酸（mg）	0.33	0.33	0.33	0.33
烟酸（mg）（可利用）	10.28	8.46	7.52	6.93
泛酸（mg）	8.58	7.94	7.65	7.48
核黄素（mg）	2.70	2.35	2.15	2.02
硫胺素（mg）	1.10	1.10	1.10	1.10

（续）

项目	体重（kg）			
	20 ~ 50	50 ~ 80	80 ~ 110	110 ~ 140
维生素 B_6（mg）	1.21	1.11	1.06	1.06
维生素 B_{12}（μg）	10.24	6.8	5.08	4.04

[1]基于NRC（1998）生长模型，假设无脂瘦肉日增重为350 g。

3.2.2 饲养方案的应用

生长育肥猪的营养需要量根据体重不同分为四个阶段（表18-10、表18-11和表18-12），但这并不意味着更细的阶段划分就不好。阶段划分的数量要权衡两个方面：一是要精准满足猪的营养需求，二是要考虑饲料加工问题及饲粮配方的频繁改变对猪本身的应激问题。对于大型养殖业来说，所配制的饲粮至少要装满一货车，才会改变饲粮配方。因此，可持续养猪生产的关键是在提高生产效率的前提下尽可能节约饲料成本。

3.2.3 使用合适的饲料原料

用于配制生长育肥猪饲粮的饲料原料非常多。在早期生长阶段，能量摄入量常常是猪生产性能的限制性因素，所以可以通过提高饲粮能量浓度来获得最大的生长速度。因此，饲粮配制时，应选择高能饲料原料。通常，这个生长阶段的饲粮主要由谷物（如玉米、小麦、大麦和高粱）、蛋白原料（如豆粕、菜籽粕和葵花籽粕）及补充油脂类组成。为了提高油脂类的稳定性，应添加一些抗氧化剂，以防止它们发生酸败。在育肥后期，由于能量的摄入量不再是限制猪生长的因素，因此使用麦麸等原料配制低能饲粮可能更合适。对于任何一个养猪场来说，选择哪些饲料原料较为合适，取决于猪瘦肉生长的遗传潜力、当地可利用的饲料原料及养殖场面临的经济环境。一般情况下，多数养殖场会利用一些加工副产物来喂猪。这些副产物虽然被生产者视为废弃物，但用作饲料原料，可以提高生长育肥猪的养殖效益。因此，合理、有效地使用这些加工副产物，对促进现代养猪业的可持续发展具有重要意义。本书的其他章节已对加工副产物的应用进行了深入讨论。

3.2.4 影响生长猪饲养方案的环境因素

关于畜禽粪便作为肥料对外界环境的影响，早期研究多数集中于粪中氮含量、氮在土壤中的沉积及其对地表水质量的影响。因为在淡水养殖环境中，氮通常是第一限制性营养素。后来，人们又将注意力转移到磷，因为在咸水养殖环境中，磷通常被认为是第一限制性营养素。由于粪便的N：P比例一般都比大多数谷物生长所需的N：P比例低得多，如果单纯用粪便来满足氮的需要，会导致磷的过量。这样既造成了磷的浪费，又会污染环境。在低蛋白饲粮中使用合成氨基酸，会减少氮的排泄；而广泛使用外源性植酸酶，会提高磷的利用率。目前，人们开始关注饲粮高铜和高锌等高浓度微量元素的使用。

一般情况下，断奶仔猪饲粮中使用高铜和高锌可以促进生长性能。当以硫酸铜的形式在饲粮中添加250 ppm的铜具有促生长作用（Cromwell等，1978；Stahly，1980；Ribeiro de Lima等，1981）。这种高铜含量远远超过猪的营养需要（5 ~ 6 ppm）（NRC，1998）。研究发现，高铜可以减少猪粪的不良臭味（Armstrong等，2000），可能是因为高铜具有抗菌作用。锌通常是以氧化锌的形式添加，当饲粮中添加量达2 000 ~ 3 000 ppm时可提高仔猪的生长性能（Hill等，2000；Mavromichalis等，2000）。然而，同时添加高锌和高铜，并没有出现叠加效应（Smith等，1997；Hill等，2000）。

添加“预防性”剂量的无机铜和无机锌（通常分别为硫酸铜和氧化锌）绝大部分既没有被机体消化，也没有沉积到猪体内。这个观点过去一直被人们广泛接受，因为它们的抗菌效果与其消化率无关（Mavromichalis等，2000）。然而，这种做法在今后是不可能再被接受的，因为土壤中铜和锌的含量终究会成为监管对象，就像很多国家或地区已经开始土壤氮、磷的监管一样。幸运的是，猪饲粮中较低浓度的有机铜可能具有替代高水平硫酸铜的潜力。Veum等（2004）报道，与使用硫酸铜形式的250 ppm无机铜相比，饲喂蛋白盐形式的50 ~ 100 ppm铜可显著提高仔猪的生长性能和饲料转化效率，同时使铜的排泄量减少61% ~ 77%（Veum等，2004）。此外，使用有机铜还具有不向粪便中提供硫酸盐的其他作用，即降低粪便中臭味气体硫化氢的产生量（Armstrong等，2000）。随后，Armstrong等（2004）发现，以柠檬酸铜形式添加125 ppm铜具有类似于硫酸铜形式添加250 ppm铜的促进断奶仔猪生长作用。

使用高锌替代物的效果不是很理想。Buff等（2005）发现，猪以多糖锌螯合物形式饲喂300～450 ppm锌的效果，与以氧化锌形式饲喂2 000 ppm锌的效果类似；但在某些阶段，以氧化锌形式饲喂2 000 ppm组的体增重、采食量和饲料转化效率与负对照组也没有差异。Case和Carlson（2002）比较了3 000 ppm氧化锌与500 ppm多糖锌的使用效果，发现有2个试验在体增重、采食量和饲料转化效率方面没有差异，但在第3个试验，添加高水平的氧化锌的仔猪生长性能更好。后来，Carlson等（2004）也发现，与500 ppm多糖锌和800 ppm蛋白锌相比，添加2 000 ppm氧化锌对猪增重和采食量的改善作用更大。此外，Hollis等（2005）研究发现，与负对照组或2 000ppm氧化锌组相比，使用500 ppm的多糖锌、蛋白锌、锌复合物、锌螯合物和蛋氨酸锌对猪生长性能都没有改善作用。

因为有机铜具有促生长效果，而有机锌的促生长作用不稳定，而且以往报道高锌与高铜之间没有叠加效应，所以为了尽可能降低养猪生产对环境的负面影响，应该使用有机铜源作为断奶仔猪的促生长剂。这既可以让养猪生产者达到补充预防性剂量微量元素的促生长效果，同时又可以使养分排泄量降到最低。

减少微量元素排泄的另一种办法是在育肥猪饲粮中不补充微量元素添加剂。从理论上讲，接近上市体重的猪在屠宰前几周可以动用体内贮存的微量元素，在不耗尽体内贮存的情况不会抑制生长。Mavromichalis等（1999）发现，在屠宰前30日内，在育肥后期饲粮中不添加所有微量元素的预混料并只补充33%磷酸氢钙，对生长性能和胴体品质不产生负面影响。同样，Hernandez等（2008）表明，饲粮中的铜含量降低6倍和锌含量降低2.5倍，对生长育肥猪的生长性能和矿物质营养状态没有负面影响。另外，Shelton等（2005）认为，使用植酸酶后，没有必要在保育猪和生长育肥猪的饲粮继续添加微量元素添加剂。因为添加植酸酶可以提高养分利用率，进而减少微量元素的排泄。

限饲是另一种减小总养分排泄并提高养分利用率的方法。当养分摄入受到限制时，猪对饲粮中这些限制性营养物质的利用率更高。在限饲阶段提高的饲料利用率会一直持续到接下来的非限制阶段，使限饲动物在生产后期的生长速度比一直自由采食的动物更快。最简单的限饲方法是通过降低饲粮中赖氨酸等某些关键限制性氨基酸的浓度来实现。在猪生长的早期阶段，将饲粮可消化赖氨酸的含量从11.0 g/kg降到5.0 g/kg；随后，在育肥阶段饲喂营养充足的饲粮，可使猪达到上市体重时的胴体性状与用传统方法饲喂的猪没有差异（Fabian等，2002；Fabian等，2004）。与不限饲的猪相比，赖氨酸限饲使猪上市时间延长了约1周，但在整个生长育肥期间每头猪减少了753 g或21%的总氮排泄量（Fabian等，2004）。Collins等（2007）主要根据饲料成本的节约来计算，对10 ~ 14周龄的猪进行蛋白质限饲，发现每头猪的净利润可增加约4澳元。然而，由营养限饲所产生的确切经济效益将取决于当前当地的饲料成本、设备成本、当地生猪生产的消费需求以及屠宰场的定价。因此，养猪生产者和营养学家还需要进一步研究，以充分认识实行营养限饲的经济意义。

4 结 语

本章给出了种猪和生长猪的饲粮营养需要量。对于任何一个特定阶段与大小的猪，所列出的每种营养需要量都是个估算值。用户必须认识到，任何猪对各种营养物质的真实需要量都存在固有的变化，因为猪的遗传因素、健康状况、饲养环境以及使用的饲料原料都可能存在差异。因此，本章列出的需要量估算值只能为实际的养猪生产提供参考，在生产中需要根据养猪生产的实际情况对这些参数进行适当的调整，使猪生产性能得到最大程度的发挥。

作者：Lee J. Johnston,
Mark H. Whitney,
Samuel K. Baidoo和
Joshua A. Jendza
译者：李振田和黄金秀

参考文献

Armstrong, T. A., D. R. Cook, M. M. Ward, C. M Williams, and J. W. Spears. 2004. Effect of dietary copper source (cupric citrateand cupric sulfate) and concentration on growth performance and fecal copper excretion in weanling pigs. J. Anim. Sci.82:1234–1240.

Armstrong, T. A., C. M. Williams, J. W. Spears, and S. S. Schiffman. 2000. High dietary copper improves odor characteristics ofswine waste. J. Anim. Sci. 78:859–864.

Bark, L. J., T. D. Crenshaw, and V. D. Leibbrandt. 1986. The effect of meal intervals and weaning on feed intake of early weaned pigs. J. Anim. Sci. 62:1233.

Beeson, W. M., E. W. Crampton, T. J. Cunha, N. R. Ellis, and R. W. Leucke. 1953. *Nutrient Requirements of Swine*. National Research Council.

Beltranena, E., F. X. Aherne, G. R. Foxcroft, and R. N. Kirkwood. 1991. Effects of pre-and postpubertal feeding on production traits at first and second estrus in gilts. J. Anim. Sci. 69:886–893.

Bergeron, R., J. Bolduc, Y. Ramonet, M. C. Meunier-Salaun, and S. Robert. 2000. Feeding motivation and stereotypies in pregnant sows fed increasing levels of fiber and/or food. Appl. Anim. Behav. Sci. 70:27–40.

Brooks, P. H., D. T. Morgen, and K. E. Hastings. 1984. The effect of dietary biotin on the response of growing pigs to copper sulphate used as a growth promoter. Proc. 8th Int. Pig Vet. Soc. Congr. Ghent, Belgium. IPVS, Ghent, Belgium. Brouns, F., S. A. Edwards, and P. R. English. 1994. Effect of dietary fiber and feeding system on activity and oral behavior of group housed gilts. Appl. Anim. Behav. Sci. 39:215–223.

Buff, C. E., D.W. Bollinger, M. R. Ellersieck,W. A. Brommelsiek, and T. L. Veum. 2005. Comparison of growth performance and zinc absorption, retention, and excretion in weanling pigs fed diets supplemented with zinc-polysaccharide or zinc oxide. J. Anim. Sci. 83:2380–2386.

Carlson, M. S., C. A. Boren, C. Wu, C. E. Huntington, D. W. Bollinger, and T. L. Veum. 2004. Evaluation of various inclusion rates of organic zinc either as polysaccharide or proteinate complex on the growth performance, plasma, and excretion of nursery pigs. J. Anim. Sci. 82:1359–1366.

Carlson, M. S., G. M. Hill, and J. E. Link. 1999. Early and traditionally weaned nursery pigs' benefit from phase-feeding pharmacological concentrations of Zn oxide: Impact on metallothionein and mineral concentrations. J. Anim. Sci. 77:1199–1207.

Case, C. L. and M. S. Carlson. 2002. Effect of feeding organic and inorganic sources of additional zinc on growth performance and zinc balance in nursery pigs. J. Anim. Sci. 80:1917–1927.

Cera, K. R., D. C. Mahan, and G. A. Reinhart. 1988. Weekly digestibilities of diets supplemented with corn oil, lard, or tallow by weanling swine. J. Anim. Sci. 66:1430–1437.

Challinor, C. M., G. Dams, B. Edwards, and W. H. Close. 1996. The effect of body condition of gilts at first mating on long-term sow productivity. Anim. Sci. 62:660. (Abstr.)

Chung, J., B. Z. De Rodas, C. V. Maxwell, and M. E. Davis. 1996. The effect of increasing whey protein concentrate as a lysine source on performance of segregated early-weaned pigs. J. Anim. Sci. 74(Suppl. 1):195. (Abstr.)

Clowes, E. J., F. X. Aherne, G. R. Foxcroft, and V. E. Baracos. 2003. Selective protein loss in lactating sows is associated with reduced litter growth and ovarian function. J. Anim. Sci. 81:753–764.

Collins, C. L., D. J. Henman, and F. R. Dunshea. 2007. Reduced protein intake during the weaner period has variable effects on subsequent growth and carcass composition of pigs. Aust. J. Exper. Agric. 47:1333–1340.

Cromwell, G. L., R. D. Coffey, D. K. Aaron, M. D. Lindemann, J. L. Pierce, H. J. Moneque, V. M. Rupard, et al. 1996. Differences in growth rate of weanling barrows and gilts. J. Anim. Sci. 74(Suppl. 1):186. (Abstr.)

Cromwell, G. L., D. D. Hall, A. J. Clawson, G. E. Combs, D. A. Knabe, C. V. Maxwell, P. R. Noland, et al. 1989. Effects of additional feed during late gestation on reproductive performance of sows: A cooperative study. J. Anim. Sci. 67:3–14.

Cromwell, G. L., V. W. Hays, and T. L. Clark. 1978. Effect of copper sulfate, copper sulfide and sodium sulfide on performance and copper stores of pigs. J. Anim. Sci. 46:692–698.

Danielsen, V., and E. M. Vestergaard. 2001. Dietary fiber for pregnant sows: Effect on performance and behavior. Anim. Feed Sci. Technol. 90:71–80.

Davies, M. J., and R. J. Norman. 2002. Programming and reproductive functioning. Trends Endocrinol. Metab. 13:386–392.

de Rodas, B. Z., and C. V. Maxwell. 1990. The effect of fat source and medium-chain triglyceride level on performance of the early-weaned pig. Okla. Agric. Exp. Sta. Res. Rep. MP-129:278.

Dourmad, J. Y., M. Etienne, A. Prunier, and J. Noblet. 1994. The effect of energy and protein intake of sows on their longevity: A review. Livest. Prod. Sci. 40:87–97.

Dwyer, C. M., N. C. Strickland, and J. M. Fletcher. 1994. The influence of maternal nutrition on muscle fiber number development in the porcine fetus and on subsequent postnatal growth. J. Anim. Sci. 72:911–917.

Edwards, S. A. 1998. Nutrition of the rearing gilt and sow. Pages 361–382 in Progress in Pig Science. J. Wiseman, M. A. Varley, and J. P. Chadwick, eds. Nottingham University Press. Nottingham, UK. Fabian, J., L. I. Chiba, D. L. Kuhlers, L. T. Frobish, K. Nadarajah, C. R. Kerth, W. H. McElhenney, and A. J. Lewis. 2002. Degree of amino acid restrictions during the grower phase and compensatory growth in pigs selected for lean growth efficiency. J. Anim. Sci. 80:2610–2618.

Fabian, J., L. I. Chiba, L. T. Frobish, W. H. McElhenney, D. L. Kuhlers, and K. Nadarajah. 2004. Compensatory growth and nitrogen balance in grower–finisher pigs. J. Anim. Sci. 82:2579–2587.

Figueroa, J. L., A. J. Lewis, P. S. Miller, R. L. Fischer, and R. M. Diedrichsen. 2003. Growth, carcass traits, and plasma amino acid concentrations of gilts fed low–protein diets supplemented with amino acids including histidine, isoleucine, and valine. J. Anim. Sci. 81:1529–1537.

Foxcroft, G. R., W. T. Dixon, S. Novak, C. T. Putman, S. C. Town, and M. D. A. Vinsky. 2006. The biological basis for prenatal programming of postnatal performance in pigs. J. Anim. Sci. 84 (Suppl.):E105–E112.

Fredrick, B.R. and T. S. Stahly. 2000. Dietary available phosphorus needs of high lean pigs. J. Anim. Sci. 78(Suppl. 1):59. (Abstr.)

Gondret, F., L. Lefaucheur, H. Juin, I. Louveau, and B. Lebret. 2006. Low birth weight is associated with enlarged muscle fiber area and impaired meat tenderness of the longissimus muscle in pigs. J. Anim. Sci. 84:93–103.

Guzik, A. C., L. L. Southern, T. D. Bidner, and B. J. Kerr. 2002. The tryptophan requirement of nursery pigs. J. Anim. Sci. 80:2646–2655.

Hern'andez, A., J. R. Pluske, D. N. D'Souza, and B. P. Mullan. 2008. Levels of copper and zinc in diets for growing and finishing pigs can be reduced without detrimental effects on production and mineral status. Animal 2:1763–1771.

Hill, G. M., S. K. Baidoo, G. L. Cromwell, D. C. Mahan, J. L. Nelssen, and H. H. Stein. 2007. Evaluation of sex and lysine during the nursery period. J. Anim. Sci. 85:1453–1458.

Hill, G. M., G. L. Cromwell, T. D. Crenshaw, C. R. Dove, R. C. Ewan, D. A. Knabe, A. J. Lewis, et al. 2000. Growth promotion effects and plasma changes from feeding high dietary concentrations of zinc and copper to weanling pigs (regional study). J. Anim. Sci. 78:1010–1016.

Hollis, G. R., S. D. Carter, T. R. Cline, T. D. Crenshaw, G. L. Cromwell, G. M Hill, S. W. Kim, et al. 2005. Effects of replacing pharmacological levels of dietary zinc oxide with lower dietary levels of various organic zinc sources for weanling pigs. J. Anim. Sci. 83:2123–2129.

Johnston, L. J. 1993. Maximizing feed intake of lactating sows. Compendium Cont. Ed. Pract. Vet. 15:133–141.

Johnston, L. J., C. Bennett, R. J. Smits, and K. Shaw. 2007. Identifying the relationship of gilt rearing characteristics to lifetime sow productivity. Page 39 in *Manipulating Pig Production XI*. J. E. Paterson and J. A. Barker, eds. Australian Pig Science Association, South Perth, WA, Australia.

Johnston, L., and J. Holt. 2006. Improving pig welfare—The role of dietary fiber. Pages 171–182 in Proc. 67th Minnesota Nutr. Conf., Univ. of Minnesota, St. Paul, MN.

Jorgensen, B., and M. T. Sorensen. 1998. Different rearing intensities of gilts: II. Effects on subsequent leg weakness and longevity. Livest. Prod. Sci. 54:167–171.

Ju, J. C., S. P. Cheng, and H. T. Yen. 1985. Effects of amino acid additions in diets on semen characteristics of boars. J. Chinese Soc. Anim. Prod. 14:27–35.

Kemp, B., M. W. A. Verstegen, L. A. den Hartog, and H. J. G. Grooten. 1989a. The effect of environmental temperature on metabolic rate, and partitioning of energy intake in breeding boars. Livest. Prod. Sci. 23:329–340.

Kemp, B., L. A. den Hartog, and H. J. G. Grooten. 1989b. The effect of feeding level on semen quantity and quality of breeding boars. Anim. Reprod. Sci. 20:245.

Kendall, D. C., A. M. Gaines, G. L. Allee, and J. L. Usry. 2008. Commercial validation of the true ileal digestible lysine requirement for eleven- to twenty-seven-kilogram pigs. J. Anim. Sci. 86:324–332.

Kerr, B. J., M. T. Kidd, J. A. Cuaron, K. L. Bryant, T. M. Parr, C. V. Maxwell, and J. M. Campbell. 2004. Isoleucine requirements and ratios in starting (7 to 11 kg) pigs. J. Anim. Sci. 82:2333–2342.

Kim, S. W., W. L. Hurley, G. Wu, and F. Ji. 2009. Ideal amino acid balance for sows during gestation and lactation. J. Anim. Sci. 87(. Suppl.):E123–E132.

King, R. H. 1987. Nutritional anestrus in young sows. Pig News and Inform. 8:15–22.

Koketsu, Y., G. D. Dial, J. E. Pettigrew, and V. L. King. 1996. Feed intake pattern during lactation and subsequent reproductive performance of sows. J. Anim. Sci. 74:2875–2884.

Lauridsen, C., and V. Danielsen. 2004. Lactational dietary fat levels and sources influence milk composition and performance of sows and their progeny. Livest. Prod. Sci. 91:95–105.

Lawrence, A. B., and E. M. C. Terlouw. 1993. A review of behavioral factors involved in the development and continued performance of stereotypic behaviors in pigs. J. Anim. Sci. 71:2815–2825.

Le Floc'h, N., L. LeBellego, J. J. Matte, D. Melchior, and B. Seve. 2009. The effect of sanitary status degradation and dietary tryptophan content on growth rate and tryptophan metabolism in weaning pigs. J. Anim. Sci. 87:1686–1694.

Leibbrandt, V. D., R. C. Ewan, V. C. Speer, and D. R. Zimmerman. 1975. Effect of age and calorie: protein ratio on performance and body composition of baby pigs. J. Anim. Sci. 40:1070–

1076.

Leibbrandt, V. D., L. J. Johnston, G. C. Shurson, J. D. Crenshaw, G. W. Libal, and R. D. Arthur. 2001. Effect of nipple drinker water flow rate and season on performance of lactating swine. J. Anim. Sci. 79:2770–2775.

Lopez-Serrano, M., N. Reinsch, H. Looft, and E. Kalm. 2000. Genetic correlations of growth, back*f*at thickness and exterior with stayability in Large White and Landrace sows. Livest. Prod. Sci. 64:121–131.

Louis, G. F., A. J. Lewis, W. L. Weldon, P. S. Miller, R. J. Kittok, and W. W. Stroup. 1994a. Calcium levels for boars and gilts. J. Anim. Sci. 72:2038–2050.

Louis, G. F., A. J. Lewis, W. L. Weldon, P. S. Miller, R. J. Kittok, and W. W. Stroup. 1994b. The effect of protein intake on boar libido, semen characteristics and plasma hormone concentrations. J. Anim. Sci. 72:2051–2060.

Maxwell, C. V., Jr., and S. D. Carter. 2001. Feeding the weaned pig. Pages 691–715 in *Swine Nutrition*. 2nd Ed. A. J. Lewis and L. L. Southern, eds. CRC Press, Boca Raton, FL.

Mahan, D. C., S. D. Carter, T. R. Cline, G. M. Hill, S. W. Kim, P. S. Miller, J. L. Nelssen, et al. 2007. Evaluating the effects of supplemental B vitamins in practical swine diets during the starter and grower–finisher periods—A regional study. J. Anim. Sci. 85: 2190–2197.

Mavromichalis, I., J.D. Hancock, I. H. Kim, B.W. Senne, D. H. Kropf, G. A.Kennedy, R. H. Hines, and K. C. Behnke. 1999. Effects of omitting vitamin and trace mineral premixes and (or) reducing inorganic phosphorus additions on growth performance, carcass characteristics, and muscle quality in finishing pigs. J. Anim. Sci. 77:2700–2708.

Mavromichalis, I., C. M. Peter, T. M. Parr, D. Gnessunker, and D. H. Baker. 2000. Growth-promoting efficacy in young pigs of two sources of zinc oxide having either a high or a low bioavailability of zinc. J. Anim. Sci. 78:2896–2902.

Melchior, D., B. S`eve, and N. Le Floc'h. 2004. Chronic lung inflammation affects plasma amino acid concentrations in pigs. J. Anim. Sci. 82:1091–1099.

Meunier-Salaun, M. C., S. A. Edwards, and S. Robert. 2001. Effect of dietary fiber on the behavior and health of the restricted fed sow. Anim. Feed Sci. Technol. 90:53–69.

Milligan, B. N., D. Fraser, and D. L. Kramer. 2002. Within-litter birth weight variation in the domestic pig and its relation to pre-weaning survival, weight gain, and variation in weaning weights. Livest. Prod. Sci. 76:181–191.

Nimmo, R. D., E. R. Peo, Jr., B. D. Moser, and A. J. Lewis. 1981. Effect of level of dietary calcium-phosphorus during growth and gestation on performance, blood and bone parameters of swine. J. Anim. Sci. 52:1330–1342.

Noblet, J., J.Y. Dourmad, and M. Etienne. 1990. Energy utilization in pregnant and lactating sows: Modeling of energy requirements. J. Anim. Sci. 68:562–572.

NRC. 1987. *Predicting Feed Intake of Food-Producing Animals*. National Academies Press, Washington, DC. NRC. 1998. Nutrient Requirements of Swine. 10th rev. ed. National Academies Press, Washington, DC.

NSNG. 2010. National Swine Nutrition Guide. U.S. Pork Center of Excellence. Ames, IA.

Odberg, F. O. 1978. Abnormal behaviors: Stereotypies. Pages 475–480 in 1st World Congress on Ethology Applied to Zootechnics.Madrid, Spain. Orth, M. 2007. Optimizing skeletal health: The impact of nutrition. Pages 139–143 in Proc. Minnesota Nutr. Conf., Univ. of Minnesota, St. Paul, MN.

Owen, K. Q., J. L. Nelssen, R. D. Goodband, M. D. Tokach, B. T. Richert, K. G. Friesen, J. W. Smith, et al. 1995. Dietary lysine requirements of segregated early-weaned pigs. J. Anim. Sci. 73(Suppl. 1):68. (Abstr.)

Partridge, I. G., and B. P. Gill. 1993. New approaches with pig weaner diets. Page 221 in Recent Advances in Animal Nutrition. Nottingham University Press, Loughborough, UK.

Partridge, I. G., and M. S. McDonald. 1990.Anote on the response of growing pigs to supplemental biotin. Anim. Prod. 50:195–197.

Pettigrew, J. E. 1981. Supplemental dietary fat for peripartal sows: A review. J. Anim. Sci. 53:107–117.

Pettigrew, J. E., Jr., and R. L. Moser. 1991. Fat in swine nutrition. Page 133 in Swine Nutrition. Miller, E. R., D. E. Ullrey, and A. J. Lewis, eds. Butterworth-Heinemann, Stoneham, MA.

Pettigrew, J. E., and H. Yang. 1997. Protein nutrition of gestating sows. J. Anim. Sci. 75:2723–2730.

Pluske, J. R. 1993. Psychological and nutritional stress in pigs at weaning: Production parameters, the stress response, and histology and biochemistry of the small intestine. Ph.D. thesis. University of Western Australia, Perth, Australia.

Ribeiro de Lima, F., T. S. Stahly, and G. L. Cromwell. 1981. Effects of copper, with and without ferrous sulfide, and antibiotics on the performance of pigs. J. Anim. Sci. 52:241–247.

Rincker, M. J., G. M. Hill, J. E. Link, A. M. Meyer, and J. E. Rowntree. 2005. Effects of dietary zinc and iron supplementation on mineral excretion, body composition, and mineral status of nursery pigs. J. Anim. Sci. 83:2762–2774.

Rincker, M. J., G. M. Hill, J. E. Link, and J. E. Rowntree. 2004. Effects of dietary iron supplementation on growth performance, hematological status, and whole-body mineral concentrations of nursery pigs. J. Anim. Sci. 82:3189–3197.

Rozeboom, D. W. 2006. Nutritional aspects of sow longevity. Pork Information Gateway Factsheet. Accessed July, 1, 2001. http://umn.porkgateway.org/web/guest/home.

Rozeboom, D.W., R. D. Goodband, K. J. Stalder, and NCERA-89 Committee on Swine

Management. 2009. Effects of decreasing or increasing sow feed intake prior to farrowing on lactation performance. J. Anim. Sci. 87(E-Suppl. 3):96.

Rozeboom, D. W., J. E. Pettigrew, R. L. Moser, S. G. Cornelius, and S. M. El Kandelgy. 1996. Influence of gilt age and body composition at first breeding on sow reproductive performance and longevity. J. Anim. Sci. 74:138–150.

Shelton, J. L., F. M. LeMieux, L. L. Southern, and T. D. Bidner. 2005. Effect of microbial phytase addition with or without the trace mineral premix in nursery, growing, and finishing pig diets. J. Anim. Sci. 83:376–385.

Sinclair, A. G., V. C. Bland, and S. A. Edwards. 2001. The influence of gestation feeding strategy on body composition of gilts at farrowing and response to dietary protein in a modified lactation. J. Anim. Sci. 79:2397–2405.

Smith, J. W., M. D. Tokach, R. D. Goodband, J. L. Nelssen, and B. T. Richert. 1997. Effects of the interrelationship between zinc oxide and copper sulfate on growth performance of early-weaned pigs. J. Anim. Sci. 75:1861–1866.

Stahly, T. S., G. L. Cromwell, and H. J. Monegue. 1980. Effects of the dietary inclusion of copper and(or) antibiotics on the performance of weanling pigs. J. Anim. Sci. 51:1347–1351.

Stahly, T. S., N. H. Williams, T. R. Lutz, R. C. Ewan, and S. G. Swenson. 2007. Dietary B vitamin needs of strains of pigs with high and moderate lean growth. J. Anim. Sci. 85:188–195.

Stahly, T. S., N. H. Williams, and S. Swenson. 1994. Impact of genotype and dietary amino acid regimen on growth of pigs from 8 to 25 kg. J. Anim. Sci. 72(Suppl. 1):165. (Abstr.)

Stahly, T. S., N. H. Williams, S. G. Swenson, and R. C. Ewan. 1995. Dietary B vitamin needs of high and moderate lean growth pigs fed from 9 to 28 kg body weight. J. Anim. Sci. 73(Suppl. 1):193. (Abstr.)

Stalder, K. J., A. M. Saxton, G. E. Conatser, and T. V. Serenius. 2005. Effect of growth and compositional traits on first parity and lifetime reproductive performance in U.S. Landrace sows. Livest. Prod. Sci. 97:151–159.

Stevermer, E. J., M. F. Kovacs, R. C. Hoekstra, and H. L. Self. 1961. Effect of feed intake on semen characteristics and reproductive performance of mature boars. J. Anim. Sci. 20:858–865.

Trevisi, P., D. Melchior, M. Mazzoni, L. Casini, S. De Filipi, L. Minieri, G. Lalatta-Costerbosa, and P. Bosi. 2009. A tryptophanenriched diet improves feed intake and growth performance of susceptible weanling pigs orally challenged with *Escherichia coli* K88. J. Anim. Sci. 87:148–156.

Tritton, S. M., R. H. King, R. G. Campbell, A. C. Edwards, and P. E. Hughes. 1996. The effects of dietary protein and energy levels of diets offered during lactation on lactational and subsequent reproductive performance of first-litter sows. Anim. Sci. 62:573–579.

Tummaruk, P., N. Lundeheim, S. Einarsson, and A. M. Dalin. 2001. Effect of birth litter size, birth parity number, growth rate, back*f*at thickness and age at first mating of gilts on their reproductive performance as sows. Anim. Reprod. Sci. 66:225–237.

Veum, T. L., M. S. Carlson, C. W. Wu, D. W. Bollinger, and M. R. Ellersieck. 2004. Copper proteinate in weanling pig diets for enhancing growth performance and reducing fecal copper excretion compared with copper sulfate. J. Anim. Sci. 82:1062–1070.

Weldon, W. C., A. J. Lewis, G. F. Louis, J. L. Kovar, M. A. Giesemann, and P. S. Miller. 1994. Postpartum hypophagia in primiparous sows: I. Effects of gestation feeding level on feed intake, feeding behavior, and plasma metabolite concentration during lactation. J. Anim. Sci. 72:387–394.

Westendorf, P., and L. Richter. 1977. Nutrition of the boar. Ubersicht fur Tierernahrung. 5:161–184.

Whang, K. Y., F. K. McKeith, S. W. Kim, and R. A. Easter. 2000. Effect of starter feeding program on growth performance and gains of body components from weaning to market weight in swine. J. Anim. Sci. 78:2885–2895.

Whitney, M. H., and S. K. Baidoo. 2010. Breeding boar nutrient recommendations and feeding management, in *National Swine Nutrition Guide*. U.S. Pork Center of Excellence, Ames, IA.

Williams, N. H., J. Patterson, and G. Foxcroft. 2005. Non-negotiables in gilt development. Adv. In Pork Prod. 16:281–289.

Williams, N. H., T. S. Stahly, and D. R. Zimmerman. 1997. Effect of chronic immune system activation on the rate, efficiency, and composition of growth and lysine needs of pigs fed from 6 to 27 kg. J. Anim. Sci. 75:2463–2471.

Williams, B., D. Waddington, D. H. Murray, and C. Farquharson. 2004. Bone strength during growth: Influence of growth rate on cortical porosity and mineralization. Calcif. Tissue Int. 74:236–245.

Wiltafsky, M. K., J. Bartelt, C. Relandeau, and F. X. Roth. 2009a. Estimation of the optimum ratio of standardized ileal digestible isoleucine to lysine for eight- to twenty-five-kilogram pigs in diets containing spray-dried blood cells or corn gluten feed as a protein source. J. Anim. Sci. 87:2554–2564.

Wiltafsky, M. K., B. Schmidtlein, and F. X. Roth. 2009b. Estimates of the optimum dietary ratio of standardized ileal digestible valine to lysine for eight to twenty-five kilograms of body weight pigs. J. Anim. Sci. 87:2544–2553.

Wilt, H. D., and M. S. Carlson. 2009. Effect of supplementing zinc oxide and biotin with or without carbadox on nursery pig performance. J. Anim. Sci. 87:3253–3258.

Wu, G., F. W. Bazer, R. C. Burghardt, G. A. Johnson, S. W. Kim, X. L. Li, M. C.

Satterfield, and T. E. Spencer. 2010. Impacts of amino acid nutrition on pregnancy outcome in pigs: mechanisms and implications for swine production. J. Anim. Sci. 88:E195–E204.

Wu, G., F. W. Bazer, J. M. Wallace, and T. E. Spencer. 2006. Board–Invited Review: Intrauterine growth retardation: Implications for the animal sciences. J. Anim. Sci. 84:2316–2337.

Wu, G., and S. W. Kim. 2007. Functional amino acids in animal production. Encyclopedia Anim. Sci. doi:10.1081/E–EAS–120043422.

Yang, H., G. R. Foxcroft, J. E. Pettigrew, L. J. Johnston, G. C. Shurson, A. N. Costas, and L. J. Zak. 2000a. Impact of dietary lysine intake during lactation on follicular development and oocyte maturation after weaning in primiparous sows. J. Anim. Sci. 78:993–1000.

Yang, H., J. E. Pettigrew, L. J. Johnston, G. C. Shurson, and R. D. Walker. 2000b. Lactational and subsequent reproductive responses of lactating sows to dietary lysine (protein) concentrations. J. Anim. Sci. 78:348–357.

Yang, H., J. E. Pettigrew, L. J. Johnston, G. C. Shurson, J. E. Wheaton, M. E. White, Y. Koketsu, et al. 2000c. Effects of dietary lysine intake during lactation on blood metabolites, hormones, and reproductive performance in primiparous sows. J. Anim. Sci. 78:1001–1009.

Yi, G. F., A. M. Gaines, B.W. Ratliff, P. Srichana, G. L. Allee, K. R. Perryman, and C. D. Knight. 2006. Estimation of the true ileal digestible lysine and sulfur amino acid requirement and comparison of the bioefficacy of 2–hydroxy–4–(methylthio)butanoic acid and DL–methionine in eleven– to twenty–six–kilogram nursery pigs. J. Anim. Sci. 84:1709–1721.

Zak, L. J., J. R. Cosgrove, F. X. Aherne, and G. R. Foxcroft. 1997a. Pattern of feed intake and associated metabolic and endocrine changes differentially affect postweaning fertility in primiparous lactating sows. J. Anim. Sci. 75:208–216.

Zak, L. J., X. Xu, R. T. Hardin, and G. R. Foxcroft. 1997b. Impact of different patterns of feed intake during lactation in the primiparous sow on follicular development and oocyte maturation. J. Reprod. Fertil. 110:99–106.